D1156239

Natural Gas Processing

Natural Gas Processing
Technology and Engineering Design

Alireza Bahadori, Ph.D.
School of Environment, Science and Engineering,
Southern Cross University, Lismore, NSW, Australia

AMSTERDAM • BOSTON • HEIDELBERG • LONDON
NEW YORK • OXFORD • PARIS • SAN DIEGO
SAN FRANCISCO • SINGAPORE • SYDNEY • TOKYO

Gulf Professional Publishing is an imprint of Elsevier

Gulf Professional Publishing is an imprint of Elsevier
225 Wyman Street, Waltham, MA 02451, USA
The Boulevard, Langford Lane, Kidlington, Oxford, OX5 1GB, UK

Copyright © 2014 Elsevier Inc. All rights reserved.

No part of this publication may be reproduced, stored in a retrieval system or transmitted in any form or by
any means electronic, mechanical, photocopying, recording or otherwise without the prior written permission of
the publisher Permissions may be sought directly from Elsevier's Science & Technology Rights Department in
Oxford, UK: phone (+44) (0) 1865 843830; fax (+44) (0) 1865 853333; email: permissions@elsevier.com.
Alternatively you can submit your request online by visiting the Elsevier web site at http://elsevier.com/locate/
permissions, and selecting Obtaining permission to use Elsevier material.

Notice
No responsibility is assumed by the publisher for any injury and/or damage to persons or property as a matter
of products liability, negligence or otherwise, or from any use or operation of any methods, products, instructions
or ideas contained in the material herein. Because of rapid advances in the medical sciences, in particular,
independent verification of diagnoses and drug dosages should be made.

Library of Congress Cataloging-in-Publication Data
Application Submitted

British Library Cataloguing in Publication Data
A catalogue record for this book is available from the British Library

For information on all Gulf Professional publications
visit our web site at store.elsevier.com

Printed and bound in USA
14 15 16 17 18   10 9 8 7 6 5 4 3 2 1

ISBN: 978-0-08-099971-5

Working together
to grow libraries in
developing countries

ELSEVIER   Book Aid International

www.elsevier.com • www.bookaid.org

Dedicated to the loving memory of my parents, grandparents and to all who contributed so much to my work over the years

Contents

About the Author

Alireza Bahadori, Ph.D. is a research staff member in the School of Environment, Science & Engineering at Southern Cross University, Lismore, New South Wales, Australia. He received his Ph.D. from Curtin University, Western Australia. For the better part of 20 years, Dr. Bahadori has held various process engineering positions and involved in many large-scale projects at the National Iranian Oil Company, Petroleum Development Oman, and Clough AMEC Pty Ltd.

He is the author of over 250 articles and 12 books. Dr. Bahadori is the recipient of the highly competitive and prestigious Australian Government's Endeavour International Postgraduate Research award as part of his research in the oil and gas area. He also received the Top-Up award from the State Government of Western Australia through Western Australia Energy Research Alliance in 2009. Dr. Bahadori serves on many editorial boards for a number of journals. He was honored by Elsevier as an outstanding author for the *Journal of Natural Gas Science and Engineering* in 2009.

Preface

The demand for primary energy is ever growing. As the world struggles to find new sources of energy it is clear that the fossil fuels will continue to play a dominant role in the foreseeable future.

Many environmentalists view natural gas as a natural bridge fuel between the dominant fossil fuels of today and the renewable fuels of tomorrow. Within the hydrocarbon family the fastest growing hydrocarbon is natural gas. Most estimates put the average rate of growth of 1.5–2.0%.

For a given amount of heat energy, burning natural gas produces about half as much carbon dioxide, the main cause of global warming, as burning coal. One of the primary consumption of natural gas is as a source for electrical generation, and it is increasingly becoming popular because it burns cleaner than oil and coal and produces less greenhouse gases. This ability of natural gas raises the possibility that it could emerge as a critical transition fuel that could help to battle global warming.

The discovery of unconventional gas and, in particular, "Shale Gas" is perceived by many to be a game changer.

Unconventional gas refers to natural gas resources trapped in coals, shales, and tight sands. These resources differ markedly from conventional gas reservoirs, in that they are diffuse, continuous accumulations of natural gas, covering very large geographical areas. There are huge untapped unconventional reserves in many countries.

Developing unconventional gas resources requires a different approach from exploring for and developing conventional gas reservoirs. Exploration is focused on identifying productive fairways and developments that typically involve a relatively high number of wells, spread over a large development region. New technologies such as horizontal drilling, fracture stimulation, and dewatering have enabled the industry to develop these resources on a commercial scale. Some developments will use unconventional gas as a feedstock for liquefied natural gas (LNG). This new technology and commercial approach is reshaping gas markets throughout the world.

In nature, natural gas is much more in abundance than oil. Most oil economists put the natural gas reserves at least 50% higher than oil reserves at the current consumption rates.

At one point in the past, natural gas used to be a regionally based fuel, frequently flared off in oil fields because it was of little use, but now with the creation of pipelines and LNG, it is now fast becoming a major international commodity.

In the case of natural gas, the drilling of gas wells has some carbon footprint, as does the shipping of the gas by pipeline or in the form of LNG. A gas pipeline, for example, requires compressors, typically fueled by gas, to push the gas through the line. The production of gas from shale, in addition to requiring a relatively large number of wells, requires energy for the fracturing of the underground shale using high-pressure water.

But, unlike gasoline and diesel fuel, the production of natural gas does not require an energy-consuming refinery. Natural gas offers significant environmental, energy security, and economic benefits. It produces lower tailpipe emissions and greenhouse gases than diesel or gasoline (mainly because methane is less carbon-rich than petroleum). Also unlike gasoline, natural gas is nontoxic, noncorrosive, and noncarcinogenic and presents no threat to soil, surface water, or groundwater.

The comparison of a fuel such as natural gas with renewable energy sources also requires full life cycle analysis. For example, the production of corn ethanol requires energy for the operation of agricultural machinery; fertilizers, perhaps produced from natural gas; and the energy required for the extraction of alcohol from fermented corn. Solar cells require energy for their manufacture. And so on.

Then there are less tangible considerations, such as the relative impact on surface land of, say, a wind farm compared with a gas field.

So, figuring out the relative environmental impacts of different fuels can become a complex and sometimes uncertain exercise involving many different factors. And as well as encompassing full life cycle environmental impacts, those factors need to include cost comparisons between different ways of minimizing undesirable emissions—it could turn out, for example, to be more cost effective to remove pollutants from the exhaust from a cheaply produced fuel than to use an expensive fuel that does not require so much pollutant handling.

But there does seem to be a widely held view that natural gas, with its relatively benign exhaust products and ready availability, will play an important role in mankind's future energy mix, at least in the midterm.

As an abundant energy resource, an affordable energy choice, a safe and reliable fuel and the cleanest burning hydrocarbon, natural gas is a foundational element in the future energy supply mix. Many countries advocate for a diverse energy supply mix and the use of the right fuel in the right place at the right time and natural gas has a very important role to play in this equation.

Determining the correct size of equipment and facilities in natural gas processing is key to achieving perfect engineering design and saving on initial and operating costs. Size of natural gas processing facilities is particularly critical for optimal energy efficiency. When equipment is oversized, initial costs are higher, efficiency is reduced, energy costs increase, and operational costs must be compromised.

In view of the above it is an essential need to write a new book related to natural gas processing.

These design guidelines in this book are general and not for specific design cases. They were designed for engineers to do preliminary designs and process specification sheets. Of course, the final design must always be guaranteed for the service selected by the manufacturing vendor, but these guidelines will greatly reduce the amount of up-front engineering hours that are required to develop the final design. The guidelines in this book are a training tool for young engineers or a resource for engineers with experience.

The materials used in this book are compiled from various sources including high-quality reports, articles, catalogs, and other contributions in recent years, standards and recommendations published by several institutions.

Last but not least, I would like to extend my special thanks to Elsevier editorial team especially Mrs Katie Hammon and Ms Kattie Washington for their advice and editorial assistance during production of this book.

Alireza Bahadori

January 8, 2014

Overview of Natural Gas Resources

Natural gas is a vital component of the world's supply of energy. It is one of the cleanest, safest, and most useful of all energy sources.

As the world moves toward a lower carbon economy, gas is becoming a fuel of choice, particularly for power generation, in many regions. Gas is an attractive choice for emerging economies aiming to meet rapid growth in demand in fast-growing cities as urbanization increases.

The International Energy Agency (IEA) (2012) forecast that gas consumption is set to increase significantly, reflecting its greater use in power generation. Gas-fired electrical generation is typically characterized by lower capital expenditures, shorter construction times, greater flexibility in meeting peak demand, lower carbon emissions, and higher thermal efficiencies relative to other substitute fossil fuels.

Gas-fired generation can also serve to complement renewable energy sources and help to overcome intermittency problems associated with renewable energy sources, such as solar and wind. Although substantial growth in gas demand is projected to come from electrical generation, it will depend on the price of gas relative to substitute fuels, as well as domestic policy settings regarding nuclear energy and carbon pricing, and other carbon-limiting regulations or measures. Factors such as commitments to energy security, climate change, and local pollution issues will have substantial bearing on the setting and adaptation of policy.

Globally, natural gas has a proved reserves life index of 64 years. The IEA (2012) estimates that there are nearly 404 trillion cubic meters (tcm) (14,285 trillion cubic feet (tcf)) of remaining recoverable resources (including all resource categories) of conventional gas worldwide, a value that is equivalent to almost 130 years of production at 2011 rates. Russia, Iran, and Qatar together hold around half of the world's proved gas reserves.

The share of unconventional gas in total global gas production is projected to rise from 13% in 2009 to 22% in 2035. However, these projections are subject to a great deal of uncertainty, particularly in regions where unconventional gas production is yet to occur or is in its infancy. Environmental concerns and policy constraints also have the potential to limit unconventional gas output, particularly in Europe. The future of unconventional gas production and the extent to which it is developed over the coming decades is heavily dependent on government and industry response to environmental challenges, public acceptance, regulatory and fiscal regimes, and widespread access to expertise, technology, and water. Given that unconventional resources are more widely dispersed than conventional resources, patterns of future gas production and trade may change. This change is because all major consuming regions have estimated recoverable gas resources that are much larger than those estimated only 5 years ago.

Shale gas projects have recently contributed significantly to increased production in the United States. There is an expectation that rapid exploitation of shale gas developments is also likely to occur in other regions of the world. China is the only country with estimated shale gas resources greater than

Natural Gas Processing. http://dx.doi.org/10.1016/B978-0-08-099971-5.00001-5
Copyright © 2014 Elsevier Inc. All rights reserved.

the United States. The IEA has stated that Chinese shale reserves are the world's largest, estimated to be around 36.10 tcm (1275 tcf), although exploitation activities remain in their infancy due to challenges not present in the United States.

1.1 The formation of natural gas

Natural gas develops naturally over millions of years from the carbon and hydrogen molecules of ancient organic matter trapped within geological formations. Natural gas consists primarily of methane, but also ethane, propane, butane, pentanes, and heavier hydrocarbons.

Natural gas is a fossil fuel. Like oil and coal, this means that it is, essentially, the remains of plants and animals and microorganisms that lived millions and millions of years ago.

There are many different theories as to the origins of fossil fuels. The most widely accepted theory says that fossil fuels are formed when organic matter (such as the remains of a plant or animal) is compressed under the earth, at very high pressure for a very long time. This type of methane is referred to as thermogenic methane. Similar to the formation of oil, thermogenic methane is formed from organic particles that are covered in mud and other sediment. Over time, more and more sediment and mud and other debris are piled on top of the organic matter.

This sediment and debris put a great deal of pressure on the organic matter, compressing it. This compression, combined with high temperatures found deep underneath the earth (deeper and deeper under the earth's crust, the temperature gets higher and higher), breaks down the carbon bonds in the organic matter.

At low temperatures (shallower deposits), more oil is produced relative to natural gas. At higher temperatures, however, more natural gas is created, as opposed to oil. That is why natural gas is usually associated with oil in deposits that are 1609–3219 m (1–2 mi) below the earth's crust. Deeper deposits, very far underground, usually contain primarily natural gas, and in many cases, pure methane.

Natural gas can also be formed through the transformation of organic matter by tiny microorganisms. This type of methane is referred to as biogenic methane. Methanogens, tiny methane-producing microorganisms, chemically break down organic matter to produce methane. These microorganisms are commonly found in areas near the surface of the earth that are void of oxygen. These microorganisms also live in the intestines of most animals, including humans.

Formation of methane in this manner usually takes place close to the surface of the earth, and the methane produced is usually lost into the atmosphere. In certain circumstances, however, this methane can be trapped underground, recoverable as natural gas. An example of biogenic methane is landfill gas. Waste-containing landfills produce a relatively large amount of natural gas from the decomposition of the waste materials that they contain. New technologies are allowing this gas to be harvested and used to add to the supply of natural gas.

A third way in which methane (and natural gas) may be formed is through abiogenic processes. Extremely deep under the earth's crust, there exist hydrogen-rich gases and carbon molecules. As these gases gradually rise toward the surface of the earth, they may interact with minerals that also exist underground, in the absence of oxygen.

This interaction may result in a reaction, forming elements and compounds that are found in the atmosphere (including nitrogen, oxygen, carbon dioxide, argon, and water). If these gases are under

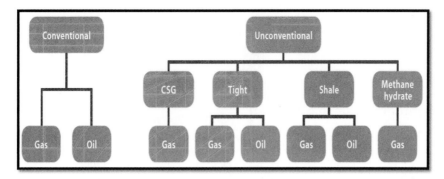

FIGURE 1.1

The range of conventional and unconventional hydrocarbons.

very high pressure as they move toward the surface of the earth, they are likely to form methane deposits, similar to thermogenic methane.

Natural gas is found overwhelmingly in sedimentary basins, in many geological settings and within various rock types. It is important to note that it is largely the rock type and the trapping mechanism that define whether a gas is regarded as "conventional" or "unconventional" (Figure 1.1), and not the composition of the gas. All natural gas is composed predominantly of methane (Chapter four), with variable but usually only minor quantities of other hydrocarbons.

1.2 Conventional natural gas resources

Natural gas that is economical to extract and easily accessible is considered "conventional." Conventional gas is a gas that is trapped in structures in the rock that are caused by folding and/or faulting of sedimentary layers. Exploration for conventional gas has been almost the sole focus of the oil and gas industry since it began around 100 years ago. Conventional gas is typically "free gas" trapped in multiple, relatively small, porous zones in various naturally occurring rock formations such as carbonates, sandstones, and siltstones.

Natural gas from conventional deposits is found in sandstone or limestone formations. These formations are very porous. By drilling a vertical gas well, the gas reservoir is accessed and gas flows freely to the surface. Natural gas from conventional deposits is often found along with oil. Gas streams produced from oil and gas reservoirs contain natural gas, liquids, and other substances. These streams are processed to separate the natural gas from the liquids and to remove contaminants.

Conventional gas is largely extracted through the drilling of a vertical well from surface into the gas accumulation in porous, permeable gas reservoirs. The gas is under pressure in the reservoir, and this pressure is released through the drilling of the well and the gas flows through the pore spaces in the rock, then into the well bore to surface. Figure 1.2 shows a typical conventional nonassociated gas reservoir.

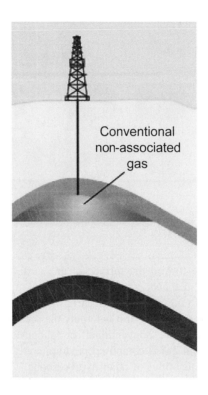

FIGURE 1.2

A typical conventional nonassociated gas reservoir.

There are several factors that need to be present for conventional gas accumulations, including the following.

- *Source*: an organic rock that is composed of either marine or terrestrial organic debris that has been compacted by layers of overlying rocks over long periods and subsequently "cooked" by the increased pressures and temperatures at depth to produce hydrocarbons.
- *Migration*: these hydrocarbons are released into pores or spaces of the rock and when these spaces connect, the rock is termed permeable. Permeable rocks allow the migration of hydrocarbons to travel upward toward lesser pressure until they reach a "trap."
- *Trap*: is required to accumulate the hydrocarbons into a specific area so that a well can be drilled into the area and extract the hydrocarbons. A trap or seal is commonly a nonporous or impermeable layer of rock that will not allow the penetration of any gas or fluid. It is also commonly folded to form an umbrella shape or faulted to juxtapose rocks that will restrict any gas or fluid flow.
- *Reservoir*: is the rock of high porosity and permeability that holds the hydrocarbons below the trap.

1.3 Gas reservoir fluids

Gas reservoir fluids fall into three broad categories: (1) aqueous solutions with dissolved salts, (2) liquid hydrocarbons, and (3) gases (hydrocarbon and nonhydrocarbon). In all cases, their compositions depend upon their source, history, and present thermodynamic conditions. Their distribution within a given reservoir depends upon the thermodynamic conditions of the reservoir as well as the petrophysical properties of the rocks and the physical and chemical properties of the fluids themselves.

Gas reservoirs are usually classified into the following three main types:

- Dry gas
- Wet gas
- Gas condensate

Gas reservoir fluids are discussed in Chapter two in more detail.

1.4 Unconventional natural gas resources

Most of the growth in supply from today's recoverable gas resources is found in unconventional formations. Unconventional gas reservoirs include tight gas, coal bed methane (coal seam gas), gas hydrates, and shale gas. The technological breakthroughs in horizontal drilling and fracturing hat have made shale and other unconventional gas supplies commercially viable.

Unconventional natural gas deposits are difficult to characterize overall, but in general they are often lower in resource concentration, more dispersed over large areas, and require well stimulation or some other extraction or conversion technology. Extremely large natural gas in-place volumes are represented by these resources, and only a fraction of their ultimate potential has been produced so far. Figure 1.3 is a schematic cross-section showing the general setting of basin-centered/low-permeability regional gas accumulations.

Unconventional gas is gas that is trapped in impermeable rock that cannot migrate to a trap and form a conventional gas deposit. Unconventional gas may be trapped in the source rock from which it is generated, or it has migrated to a formation of rock that has since become impermeable. Unconventional gas commonly requires hydraulic fracturing (HF) to allow the gas to flow into the well and be recovered. This process has been performed on many wells.

Gas that is trapped in formations that are less porous and permeable are unable to flow readily and require stimulation to enable the pores to connect and the gas to flow to a well bore. These types of accumulations are often trapped in the one rock unit.

To access this unconventional gas, the well is first drilled vertically to reach the required depth and then horizontally through the target unit, exposing as much of the gas-bearing rock to the well bore as possible. Horizontal wells may extend for hundreds of meters and can also be oriented so that they intersect natural fracture systems, resulting in increased flow rates. Figure 1.4 shows hydraulic fracture perforations in the horizontal well bore.

More than one horizontal section can be drilled from the one vertical well, increasing exposure to the target layer. This often occurs as the petroleum field is moving toward production and provides the additional benefit of accessing large target areas while minimizing the surface footprint of the operation.

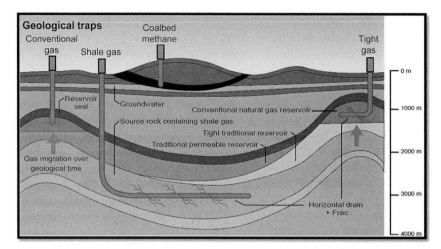

FIGURE 1.3

Schematic cross-section showing the general setting of basin-centered/low-permeability regional gas accumulations.

Taken from "Gas Fact Sheet—Gas Resource Types." Adopted from Department of Petroleum and Mines WA, Australia.

It is at this point that hydraulic fracture stimulation may then be used to microfracture the rock around the well bore and connect the pore spaces in the rock and further enable the flow of hydrocarbons into the well bore and then to surface.

Unconventional gas includes shale gas, tight gas, coal seam gas (coal bed methane), and methane hydrates, all of which are composed predominantly of methane. They are found in a variety of

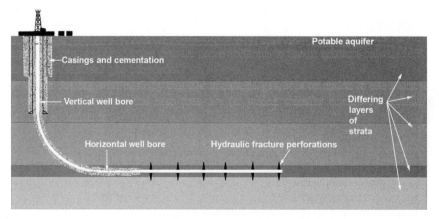

FIGURE 1.4

Hydraulic fracture perforations in horizontal well bore.

Adapted from Department of Petroleum and Mines WA, Australia.

geological settings. Methane hydrate occurs in vast quantities under the deep continental shelves in various parts of the world and in onshore or nearshore locations at high polar latitudes. It presents several unique technical challenges and is not currently being exploited. Methane hydrates may be an important energy source in the long term.

1.4.1 Shale gas

Shale gas is trapped in its source rock; therefore, the rock is also the reservoir. Shales are organically rich sedimentary rocks, very fine grained, and composed of many tiny layers (laminated). They typically have low permeability because of their laminated nature.

Gas within shale can be stored in three ways:

1. Adsorbed onto insoluble organic matter, kerogen, that forms a molecular or atomic film
2. Absorbed in the pore spaces
3. Confined in the fractures in the rock

Although shale and tight gas are both natural gas resources typically located two or more kilometers underground, they differ in the type of rock that traps the gas. Both require sophisticated extraction technology such as HF for gas production on a commercial basis. Hydraulic fracturing involves pumping fluid into gas-bearing geological formations to create tiny pathways that make the formations more permeable.

Shale gas is one of the most rapidly growing forms of natural gas. It, along with other nonconventional forms of natural gas, such as tight gas and coalbed methane, will make a major contribution to future gas production.

Shale gas is defined as natural gas from shale formations. The shale acts as both the source and the reservoir for the natural gas. Older shale gas wells were vertical, whereas more recent wells are primarily horizontal and need artificial stimulation, such as HF, to produce. Only shale formations with certain characteristics will produce gas. In large measure, this is attributable to significant advances in the use of horizontal drilling and well stimulation technologies and refinement in the cost-effectiveness of these technologies. Hydraulic fracturing is the most significant of these technologies. Figure 1.5 illustrates shale gas production techniques.

Shale has such low matrix permeability that it releases gas very slowly, and this is why shale is the last major source of natural gas to be developed. However, the upside is that shales can store an enormous amount of natural gas. Shale is a fissile, very fine grained sedimentary rock comprising clay minerals and very fine grained sand (quartz, feldspar, or carbonate), and it may contain organic material (kerogen as a hydrocarbon source).

Shale has been regarded as an impermeable seal (cap rock) for more porous and permeable sandstone and carbonate hydrocarbon reservoirs. However, in a shale gas play, it forms both the source rock and a low-permeability reservoir. Shale gas plays are not dependent on structural closure; hence, they can extend over large areas. The challenge is to find sweet spots that will produce commercially.

In gas shales, the gas is generated in place and the shale is both the source rock and the reservoir. The gas can be stored as free gas within pore spaces in both the inorganic sediment component and the organic carbon component of the rock, as free gas in fractures, and as gas adsorbed to the surface of organic components (kerogen).

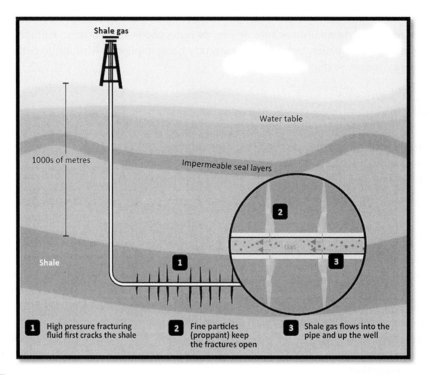

Shale gas

Water table

1000s of metres

Impermeable seal layers

Shale

Gas

1 High pressure fracturing fluid first cracks the shale

2 Fine particles (proppant) keep the fractures open

3 Shale gas flows into the pipe and up the well

FIGURE 1.5

Shale gas production techniques.

Adapted from International Energy Agency.

In situ generation and storage of hydrocarbons results in volume and pressure changes, and some overpressure is therefore characteristic of gas shales. Shale gas is produced from continuous gas accumulations that are regionally extensive, lack an obvious seal and trap, and have no defined gas–water contact.

Because shale matrix permeabilities are very low, operators generally seek to maximize the shale surface area exposed to production. This maximization is achieved by "gas farming" whereby multiple horizontal wells are drilled perpendicular to the direction of maximum horizontal stress and stimulated with multiple hydraulic fracture stages to access the largest volume of reservoir and to intersect the maximum number of (typically) subvertical fractures.

Microseismic monitoring can be used to identify fracture points in the reservoir during fraccing, to optimally orient follow up drilling, again as per geothermal reservoir stimulation. Natural fractures are beneficial, but they usually do not provide permeability pathways sufficient to support commercial production. Larger scale faults are generally identified using three-dimensional seismic and avoided because these faults complicate horizontal drilling (if the target shale bed is offset), inhibit HF, and can serve as water conduits.

The gas in natural (or induced) fractures, or gas that has migrated into thin sandstone interbeds, is produced first. After the initial flush, gas production declines exponentially. Production rates typically flatten out after 3–4 years, as the adsorbed gas is slowly produced and can continue at relatively low rates for decades. Ultimate recoveries are much lower than for conventional gas fields, but completion and production technology advances are increasing recovery factors.

1.4.2 Tight gas

Tight gas is gas that is held in low-permeability and low-porosity sandstones and limestones. The lack of permeability does not allow the gas to migrate out of the rock. A massive HF is necessary to produce the well at economic rates. Tight gas reservoirs are generally defined as having less than 0.1 millidarcy (mD) matrix permeability and less than 10% matrix porosity.

Although shales have low permeability and low effective porosity, shale gas is usually considered separate from tight gas, which is contained most commonly in sandstone, but sometimes in limestone. Tight gas is considered an unconventional source of natural gas.

In rock with permeabilities as little as 1 nanodarcy, reservoir simulation may be economically productive with optimized spacing and completion of staged fractures to maximize yield with respect to cost.

Tight gas is the term commonly used to refer to low-permeability reservoirs that produce mainly dry natural gas. Many of the low-permeability reservoirs developed in the past are sandstone, but significant quantities of gas also are produced from low-permeability carbonates, shales, and coal seams.

A tight gas reservoir is a low-porosity, low-permeability formation that must be fracture treated to flow at economic gas rates and to recover economic volumes of gas. Without a fracture treatment, gas from tight gas reservoirs will produce at low flow rates under radial flow conditions. After a successful fracture treatment, the gas flow mechanism in the reservoir will change from radial flow to linear flow, as shown in Figure 1.6.

As the propped fracture length increases, the well will produce more gas at higher flow rates provided that adequate fracture conductivity is also created.

Unconventional tight gas reservoirs have a distinct feature that drives the optimal use of HF: very low matrix permeability, which means that very long hydraulic fractures are desirable. Tight gas reservoirs are found at all depths in many geological basins around the world. The key to producing gas from a tight gas reservoir is to create a long, highly conductive flow path (a hydraulic fracture) to stimulate flow from the reservoir to the well bore. To maintain conductivity in the fracture, sufficient quantifies of propping agent need to be pumped into the fracture. Viscous fluids are used to carry high proppant concentrations deep into the fracture.

Tight gas (and tight oil) is not dissimilar to conventional gas, in terms of geological setting, except that the reservoir sand has a low permeability, meaning that it is more difficult to extract the gas than is the case for conventional high-permeability sands. Tight gas has been exploited for some decades and is fairly well understood. It also has several similarities with shale gas in terms of production processes such as the use of HF.

1.4.3 Coal bed methane (Coal seam gas)

Coal bed methane (CBM) or coal seam gas (CSG) is present within coal seams that is the source rock. The methane is stored in the matrix of the coal and the fracture spaces of the rock (cleats) and is held there by water pressure.

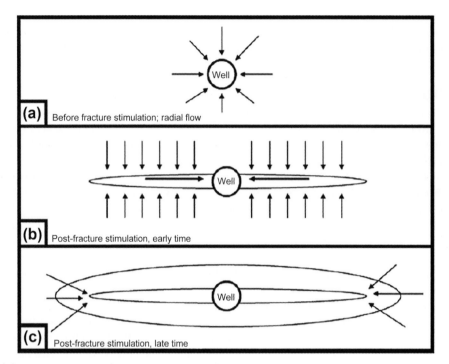

FIGURE 1.6

Gas flow mechanism before and after fracturing.

© Elsevier, Wang JY, Holditch SA, McVay DA. Effect of gel damage on fracture fluid cleanup and long-term recovery in tight gas reservoirs. J Nat Gas Sci Eng 2012;9:108–118.

Unlike shale and tight gas, methane desorbs from the micropores of the coal matrix when the hydrostatic pressure is reduced, such as the drilling of a well, and it flows through the cleats to a well bore.

Coal seams are usually quite close to the surface and may be synergous with surface aquifers, whereas shale and tight gas are commonly found at depths below 1 km. Water must be first pumped out to reduce pressure, and then as water production declines, gas production increases.

During coalification, large quantities of methane are generated. This gas is adsorbed onto the coal surface in cleats and pores and is held in place by reservoir and water pressure. Coal rank, reservoir pressure (related in part to depth), and temperature are important factors controlling the amount of methane held in a coal seam.

Gas content is a major economic consideration in assessing CBM potential. Estimates of coal gas contents without these measurements are difficult, and a range of indirect indicators has been used, such as mud gas measurements while drilling and water chemistry.

CBM forms by either biological or thermal processes. During the earliest stage of coalification (the process that turns plants into coal), biogenic methane is generated as a by-product of microbial action. Biogenic methane is generally found in near-surface low-rank coals such as lignite.

Thermogenic methane is generally found in deeper higher rank coals. When temperatures exceed about 50 °C due to burial, thermogenic processes begin to generate additional methane, carbon dioxide, nitrogen, and water. The maximum generation of methane in bituminous coals occurs at around 150 °C.

The methane produced is adsorbed onto micropore surfaces and stored in cleats, fractures, and other openings in the coals. It can occur also in groundwater within the coal beds. CBM is held in place by water pressure and does not require a sealed trap as do conventional gas accumulations. The coal acts as a source and reservoir for the methane gas, whereas the water is the seal.

CBM is produced by drilling a well into a coal seam, HF the coal seam then releasing the gas by reducing the water pressure by pumping away the water. Hydraulic fracturing of the coal seam fracture for distances of up to 400 m from the well is done by pumping large volumes of water and sand at high pressure down the well into the coal seam that causes it to fracture (See Figures 1.7 and 1.8). The sand carried in the water is deposited in the fractures to prevent them closing when pumping pressure

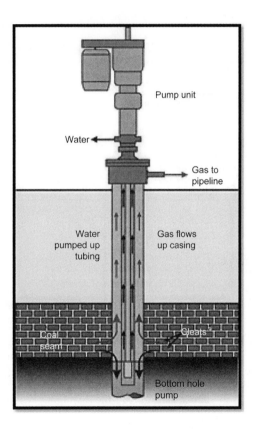

FIGURE 1.7

A schematic for production of coal bed methane (CBM) (*Adapted from instinct*) There are small natural fractures called "cleats" often filled with water within the coal seam. Natural coal seam pressure keeps methane "absorbed" or attached to the coal seam.

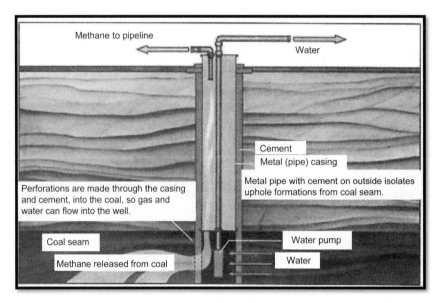

Methane to pipeline

Water

Cement

Metal (pipe) casing

Metal pipe with cement on outside isolates uphole formations from coal seam.

Perforations are made through the casing and cement, into the coal, so gas and water can flow into the well.

Coal seam

Methane released from coal

Water pump

Water

FIGURE 1.8

Coal bed methane production with a down-hole pump.

Adapted from Butte.

ceases. The gas then moves through the sand-filled fractures to the well. In Figures 1.7 and 1.8 the water pumps are installed in outside and inside of the well, respectively.

A commercial operation needs the right combination of coal thickness, gas content, permeability, drilling costs (number of wells, seam depth, and coal type), the amount of dewatering required to allow gas flow, and proximity to infrastructure.

Water pump reduces water pressure in the coal seam, allowing the methane molecules to detach from the coal and flow into the well bore and up to the surface facilities.

The methane gas is physically attached to the coal seam by pressure, usually water. The water must be removed (depressurization) so the gas can escape. If the water in the coal is saline (salty), it is "deep well disposed" deep into the earth so there can be no contamination. If the water is fresh, the company has to apply to relevant environmental department for a "diversion permit." The decision has to be made as to what will happen to this drinking water.

Problems could occur if fresh water is removed from aquifers, leaving water wells dry, or water tables could lose pressure if the water seeps down to fill the void caused by the depressurization.

Currently, there are dry coal seams that are the most sought after because there is less cost to production.

The coal seam is fractured to encourage the gas to flow toward the pipes. There may be a compressor at each hole to suck the methane up because it is such a low-pressure gas, and it may not come up on its own. It is then piped to a larger compressor station for distribution. Compressors are noisy unless they have noise suppression on them.

Because methane is such a low-pressure gas, many wells are usually necessary quite close together to get the most gas out of the coal. The cumulative impact of this many wells coupled with the already present conventional oil and gas sites is very concerning. Also of concern is the number of pipelines, roads, and power lines needed in these fields. Potential growth on these lands would be severely restricted. Flaring and venting of methane will also be necessary. Some unplanned venting of methane has been known to occur where there are fractures in the formations above the coal seams.

1.4.4 **Methane hydrate**

Methane hydrate is a naturally occurring frozen compound formed when water and methane combine at moderate-pressure and relatively low-temperature conditions. Methane hydrates represent a highly concentrated form of methane, with a cubic meter of idealized methane hydrate containing 0.8 m^3 of water and more than 160 m^3 of methane at standard temperature–pressure conditions. Ethane, propane, and carbon dioxide, and similar gases can also form gas hydrates, and individual molecules of these gases are often incorporated into gas hydrates that contain predominantly methane.

Both on a global volumetric basis and in terms of areal distribution, methane hydrates are the most important type of natural gas hydrate. For decades, gas hydrates have been discussed as a potential resource, particularly for countries with limited access to conventional hydrocarbons or a strategic interest in establishing alternative, unconventional gas reserves.

Methane has never been produced from gas hydrates at a commercial scale and, barring major changes in the economics of natural gas supply and demand, commercial production at a large scale is considered unlikely to commence within the next decade. Despite the overall uncertainty still associated with gas hydrates as a potential resource, gas hydrates remain a potentially large methane resource and must necessarily be included in any consideration of the natural gas supply beyond a few decades from now.

It is widely agreed that existing technology can be used to produce gas hydrates. The production methods being evaluated now have changed little since the early 1980s, with much discussion of the technical merit and economic feasibility of thermal stimulation, depressurization, and chemical inhibition for the production of gas from hydrates. This section reviews production methods and discusses some production scenarios.

- **Methods**

Here, we consider each of the primary production methods.

1. Thermal stimulation refers to warming the formation through the injection of heated fluid or potentially direct heating of the formation, as shown schematically in Figure 1.9(a). Thermal stimulation is energy intensive and will lead to relatively slow, conduction-limited dissociation of gas hydrates unless warmer pore fluids become mobilized and increase the volume of the formation exposed to higher temperatures. The endothermic nature of gas hydrate dissociation also presents a challenge to thermal stimulation; the cooling associated with dissociation (and, in some cases, gas expansion) will partially offset artificial warming of the formation, meaning that more heat must be introduced to drive continued dissociation and prevent formation of new gas hydrate. In terrestrial settings, thermal stimulation must be carefully controlled to minimize permafrost thawing that might lead to unintended environmental consequences and alter the permeability seal for the underlying gas hydrate deposits.

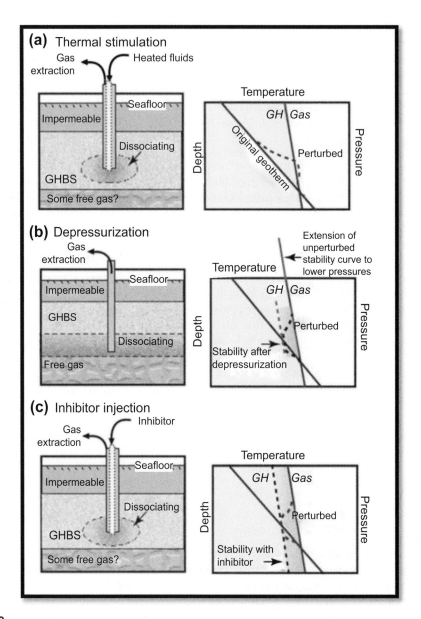

FIGURE 1.9

(a), (b), (c). Possible methods for producing gas from a marine gas hydrate deposit.

2. Depressurization, shown schematically in Figure 1.9(b), has emerged as the preferred and more economical means of producing gas from methane hydrates during most of a well's life. Depressurization does not require large energy expenditure and can be used to drive dissociation of a significant volume of gas hydrate relatively rapidly.
3. Chemical inhibition exploits the fact that gas hydrate stability is inhibited in the presence of certain organic (e.g., glycol) or ionic (seawater or brine) compounds (Figure 1.9(c)). Seawater or other inhibitors might be needed during some stages of production of gas from methane hydrate deposits, but they would not be the primary means of dissociating gas hydrate nor used for an extended period or on a large scale.

Figure 1.9(a–c) show possible methods for producing gas from a marine gas hydrate deposit, which is characterized by methane hydrate-bearing sediments (granular media labeled gas hydrate-bearing sediments (GHBS)) overlying sediments with free gas (channelled sediments). The impacts of the production methods on reservoir conditions are portrayed on the right. Sediments in which gas hydrate is dissociating are denoted in figures.

In real settings, the same well would not always be used to perturb the gas hydrate stability field and to extract the gas released by hydrate dissociation.

(a) Thermal stimulation introduces heat or warm fluids into the gas hydrate stability zone and dissociates gas hydrate. As shown on the right, the original geothermal will warm and dissociate gas hydrate in part of the reservoir.
(b) Depressurization lowers the pressure in the GHBS. The diagram at the right shows that the part of the stability curve (solid segment) that originally extended to pressures lower (shallower depths) than those at the seafloor applies to the gas hydrate reservoir after depressurization (dashed part). The pressure perturbation is shown schematically as the dashed curve.
(c) Inhibitors such as seawater generally shift the gas hydrate stability boundary toward lower temperatures, as shown on the right with the difference between the red stability boundary and the dashed curve for stability with inhibitor. Injection of an inhibitor will dissociate gas hydrate in the vicinity of the well and result in mixed stability conditions (parts of the reservoir with and without the inhibitor), as shown by the red dashed "perturbed" stability boundary.

1.5 **Hydraulic fracturing**

Hydraulic fracturing (also known as fraccing) is a process undertaken on a well after it has been drilled to depth. It involves pumping fluids (mostly water) under controlled pressure into the well bore to fracture the rock formations that hold gas accumulation and allow the gas to migrate out of the rock and be extracted through the well.

Wells that will be subjected to the fracturing process are cased and lined, with cementation of joins, throughout the well to prevent any interaction with geological formations along the well bore. Where the well bore intersects aquifers, at least three layers of casing and cement must be installed to prevent any interaction. The casing is pressure tested at intensities that are higher than any operating pressure to ensure that gas and HF fluids are contained inside the well.

Hydraulic fracture stimulation is only performed at the interval of gas-bearing rock. The casing in this specific zone is perforated, allowing fluids and proppants to interact with this formation of rock.

Fraccing fluids contain proppants that are forced into the natural fractures or fissures of the rock under pressure. They are typically sand or ceramic beads that hold the fractures open after the pressure is removed, allowing the hydrocarbons to flow into the well at a significantly increased rate.

Enterprises who apply to conduct fracture stimulation must demonstrate a comprehensive understanding of the geology and the stresses present in the subsurface rock. A range of physical tests of sample core as well as computer modeling of propagations are conducted to understand the characteristics of the rock before it is fractured.

Fraccing of rock normally extends to only 2–6 mm into the rock from the well bore along paths of least resistance. Shales are laminated rocks; therefore, they tend to fracture along the horizontal planes of the laminate.

Seismic monitoring of the fracture propagation during the stimulation process can determine the extent of the fractures. Fractures are undertaken in stages to control their length and ensure there is no interaction with aquifers.

1.5.1 Chemicals used in HF

Together with the proppant, chemicals may be added to the fluid for a variety of reasons:

- Carry the proppant
- Reduce the friction between the fluid and the pipe or casing of the well
- Stop growth of bacteria in the well and underground intervals
- Clean the well and increase permeability
- Prevent scaling; and
- Remove oxygen and prevent corrosion of the casing.

Any chemicals used in a stimulation fluid are diluted to a high extent, with 99.5% of water and sand in most fluids.

All chemicals used in hydraulic fracture stimulations in the Territory need to be fully disclosed, and these disclosure statements will be made publicly available. No benzene, toluene, ethylbenzene, and xylenes (commonly, BTEX) chemicals are permitted for use in HF in many countries.

Stimulation fluids are typically made up of commonly used compounds, including the following: acids, sodium chloride, polyacrylamide, ethylene glycol, borate salts, sodium/potassium carbonate, glutaraldehyde, guar gum, citric acid, and isopropanol.

After the pressure is reduced, most of the fraccing fluids flow back to the well head where they are treated and reused for subsequent fraccing or disposed of in accordance with approved environmental management plans. Unrecovered treatment fluids may be trapped in the fractured formation because of pore storage and stranding behind healed fractures.

1.5.2 Regulation of HF

Hydraulic fracturing in the horizontal plane is a new technology to the petroleum industry. It has been extensively used in many countries for decades. It is therefore prudent that the local governments control these practices and document each step in the process to ensure the protection of the environment and the community.

An application to conduct HF must provide detail on the five main areas of consideration, which is approved by the local environmental departments before HF may take place.

- **Water management**

Interaction with subsurface water is to be minimized; therefore, a significant amount of monitoring is required. All known aquifers are to be tested before, during, and after HF. Water levels, total dissolved sediments, and general water quality must be monitored. Water is monitored in water bores adjacent to the well, and results are to be made publicly available.

The use of water in drilling operations must be described in the application to drill or fracture. A complete description from source to transport and containment at site must be provided. Water quality should also be provided.

The disposal of water from drilling and fracturing operations must also be documented, detailing whether settlement ponds will be used, haulage, and final disposal methods and locations.

- **Use of chemicals in HF**

A complete list of chemicals that will or may be used in the fracturing process must be provided and made public before consent by the government of a HF event.

The type of chemical and approximate quantity must be stipulated, and material data safety sheets must be provided.

- **Well integrity**

Drilling and fracturing methods must use "best industry standards" that are amended from time to time to increase safety to people and the environment.

Specifically, for HF, wells must have installed surface casing to 1000 m, and the surface and conductor casings must be cemented to surface. Air drilling techniques or water-based drilling fluids must be used through aquifers. Each cementation job undertaken on casing must be tested and reported in daily drilling reports that are provided to the government.

Thorough physical and computer-generated modeling on the stress regime of the formation(s) that are being considered for fracturing must also be provided in the application.

- **Reporting**

Thorough reporting of everything that happens at a well site must be reported in a timely manner. Daily drilling reports must include all "off well" activities and describe every process undertaken as part of hydraulic fracture stimulations. Daily geological reports must provide information on testing, data obtained, research and results of fraccing, and final fracture length and orientation.

1.5.3 The HF water cycle

It is important to look at potential impacts of HF at each stage of the HF water cycle.

The steps mentioned below have been reported by the U.S. Environmental Protection Agency (EPA).

Stage 1: Water acquisition
- Large volumes of water are withdrawn from groundwater and surface water resources to be used in the HF process.

- Potential Impacts on Drinking Water Resources
  - Change in the quantity of water available for drinking
  - Change in drinking water quality
    Recently, some companies have begun recycling wastewater from previous HF activities, rather than acquiring water from ground or surface resources.

Stage 2: Chemical mixing

- Once delivered to the well site, the acquired water is combined with chemical additives and proppant to make the HF fluid.
- Potential Impacts on Drinking Water Resources
  - Release to surface water and groundwater through on-site spills, leaks, or both
    Groundwater is the supply of fresh water found beneath the Earth's surface, usually in aquifers, that supplies wells and springs. It provides a major source of drinking water.

Stage 3: Well injection

- Pressurized HF fluid is injected into the well, creating cracks in the geological formation that allow oil or gas to escape through the well to be collected at the surface.
- Potential Impacts on Drinking Water Resources
  - Release of HF fluids to groundwater due to inadequate well construction or operation
  - Movement of HF fluids from the target formation to drinking water aquifers through local artificial or natural features (e.g., abandoned wells and existing faults)
  - Movement into drinking water aquifers of natural substances found underground, such as metals or radioactive materials, that are mobilized during HF activities.
- Surface water resources include any water naturally open to the atmosphere, such as rivers, lakes, reservoirs, ponds, streams, impoundments, seas, and estuaries. Surface water provides a major source of drinking water.

Stage 4: Flow back and produced water (HF wastewaters)

- When pressure in the well is released, HF fluid, formation water, and natural gas begin to flow back up the well. This combination of fluids, containing HF chemical additives and naturally occurring substances, must be stored on-site—typically in tanks or pits—before treatment, recycling, or disposal.
- Potential Impacts on Drinking Water Resources
  - Release to surface water or groundwater through spills or leakage from on-site storage

Stage 5: Wastewater treatment and waste disposal

- Wastewater is dealt with in one of several ways, including but not limited to disposal by underground injection, treatment followed by disposal to surface water bodies, or recycling (with or without treatment) for use in future HF operations.
- Potential Impacts on Drinking Water Resources
  - Contaminants reaching drinking water due to surface water discharge and inadequate treatment of wastewater
  - By-products formed at drinking water treatment facilities by reaction of HF contaminants with disinfectants

Figure 1.10 shows the HF water cycle.

After the well is constructed, the targeted formation (shale, coal bed, or tight sands) is hydraulically fractured to stimulate natural gas production. The HF process requires large volumes of water that must be withdrawn from the source and transported to the well site (Figure 1.10).

Once on site, the water is mixed with chemicals and a propping agent (called a proppant). Proppants are solid materials that are used to keep the fractures open after pressure is reduced in the well. The most common proppant is sand, although resin-coated sand, bauxite, and ceramics have also been used.

Most, if not all, water-based fracturing techniques use proppants. There are, however, some fracturing techniques that do not use proppants. For example, nitrogen gas is commonly used to fracture coal beds and does not require the use of proppants.

After the production casing has been perforated by explosive charges introduced into the well, the rock formation is fractured when HF fluid is pumped down the well under high pressure. The fluid is also used to carry proppant into the targeted formation and enhance the fractures. As the injection pressure is reduced, recoverable fluid is returned to the surface, leaving the proppant behind to keep the fractures open.

The fluid that returns to the surface can be referred to as either "flow back" or "produced water," and it may contain both HF fluid and natural formation water. Flow back can be considered a subset of produced water. However, flow back is usually the fluid returned to the surface after HF has occurred

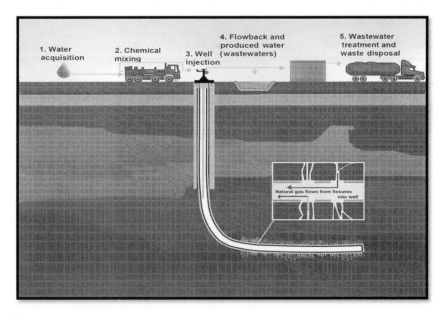

FIGURE 1.10

The hydraulic fracturing water cycle.

Courtesy of EPA.

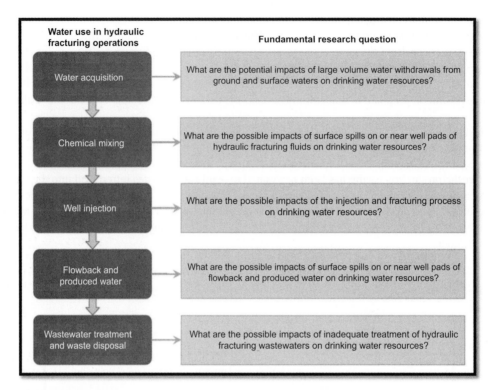

FIGURE 1.11

Fundamental research question for water use in HF operations.

but before the well is placed into production, whereas produced water is the fluid returned to the surface after the well has been placed into production. In this study plan, flow back and produced water are collectively referred to as "hydraulic fracturing wastewaters." These wastewaters are typically stored on-site in tanks or pits before being transported for treatment, disposal, land application, or discharge. In some cases, flow back and produced waters are treated to enable the recycling of these fluids for use in HF. Figure 1.11 illustrates fundamental research question for water use in HF operations.

Further reading

Baker GL, Slater SM. Coal seam gasdan increasingly significant source of natural gas in eastern Australia. APPEA J 2009;46(1):79–100.

Birchwood RA, Noeth S, Tjengdrawira MA, Kisra SM, Elisabeth FL, Sayers CM, et al. Modeling the mechanical and phase change stability of wellbores drilled in gas hydrates by the Joint Industry Participation Program (JIP) Gas Hydrates Project Phase II. In: SPE Ann. Tech. Conf., November 11–14, 2007; 2007. SPE 110796.

Boswell R, Collett TS. Current perspectives on gas hydrate resources. Energy Environ Sci 2011;4:1206–15.

Boyer C, Clark B, Jochen V, Lewis R, Miller CK. Shale gas: a global resource. Oilfield Rev 2011;23(3):28–39.

Bramoulle Y, Morin P, Capelle J. Differing market quality specs challenge LNG producers. Oil Gas J October 2004;11. and LNG Quality and Market flexibility challenges and solutions.

Buffett B, Archer D. Global inventory of methane clathrate: sensitivity to changes in environmental conditions. Earth Planet Sci Lett 2004;227:185–99.

Carter KM, Harper JA, Schmid KW, Kostelnik J. Unconventional natural gas resources in Pennsylvania: the backstory of the modern Marcellus shale play. Environ Geosci 2011;18(4):217–57.

CER Corp., Holditch SA, Assocs Inc. Staged field experiment no. 3: application of advanced technologies in tight gas sand stones–travis peak and cotton valley formations; February 1991. Waskom Field, Harrison City. Tex Gas Res Inst Report, GRI-91/0048 .

Cook A, Goldberg D, Kleinberg R. Fracture-controlled gas hydrate systems in the Gulf of Mexico. Mar Pet Geol 2008;25(9):932–41.

Elliot TR, Celia MA. Potential restrictions for CO2 sequestration sites due to shale and tight gas production. Environ Sci Technol 2012;46(7):4223–7.

Fisher WL, Brown Jr LF. Clastic depositional systemsda genetic approach to facies analysis. Austin: Bureau Economic Geology. U. of Texas; 1972.

Friedel T, Mtchedlishvili G, Behr A, Voigt H, Häfner F. Comparative analysis of damage mechanisms in fractured gas wells. In: Paper SPE 107662 presented at the European formation damage Conference. The Netherlands: Scheveningen; 2006. May 30–June 1, 2007.

Holditch SA. Factors affecting water blocking and gas flow from hydraulically fractured gas wells. SPEJ 1979; 7561:1515–24.

Holditch Stephen A. Tight gas sands. J Pet Technol; 2006:86–94.

Howard WE, Hunt ER. Travis peak: an integrated approach to formation evaluation. In: Paper SPE 15208 prepared for presentation at the SPE unconventional gas technology Symposium. Louisville: Kentucky; 1986. pp. 18–21.

Kuuskraa Vello A, Hoak Tom E, Kuuskraa Jason A. Hansen John. Tight sands gain as U.S. gas source. Oil Gas J 1996;94(12):102–7.

Leather DTB, Bahadori A, Nwaoha C, Wood D. A review of Australia's natural gas resources and their exploitation. J Nat Gas Sci Eng 2013;10:68–88.

Lolon EP, McVay D, Schubarth SK. Effect of fracture conductivity on effective fracture length. In: SPE 84311 presented at 2003 SPE Annual technical Conference and Exhibition in Denver, CO; 2003. October 5–8.

Luo D, Dai Y. Economic evaluation of coalbed methane production in China. Energy Policy 2009;37(10): 3883–9.

Maggio G, Cacciola G. When will oil, natural gas, and coal peak? Fuel 2012;98:111–23.

Makogon YF, Holditch SA, Makogon TY. Natural gas-hydrates–a potential energy source for the 21st Century. J Pet Sci Eng 2007;56(1–3):14–31.

Makogon YF. Natural gas hydrates–a promising source of energy. J Nat Gas Sci Eng 2010;2(1):49–59.

Mohr SH, Evans GM. Long term forecasting of natural gas production. Energy Policy 2011;39(9):5550–60.

Ren T, Daniëls B, Patel MK, Blok K. Petrochemicals from oil, natural gas, coal, and biomass: production costs in 2030–2050 Resources. Conserv Recycl 2009;53(12):653–63.

Rogner H-H. An assessment of world hydrocarbon resources. Laxenburg, Austria: IIASA, WP-96–26; May 1996.

Sandrea R. Evaluating production potential of mature US oil, gas shale plays. Oil Gas J 2012;110(12):58–67.

Sandu S, Copeland A. Natural gas. Aust Commod 2008;15(4):700–4.

Seljom P, Rosenberg E. A study of oil and natural gas resources and production. Int J Energy Sect Manag 2011; 5(1):101–24.

Ternes ME. Regulatory programs governing shale gas development. Chem Eng Prog 2012;108(8):60–4.

Wang JY, Holditch SA, McVay Duane A. Effect of gel damage on fracture fluid cleanup and long-term recovery in tight gas reservoirs. J Nat Gas Sci Eng 2012;9:108–18.

Wang X, Economides MJ. Natural gas hydrates as an energy source–revisited 2012, society of petroleum engineers–international petroleum technology conference 2012. IPTC 2012;1:176–86.

Yang X, Qin M. Natural gas hydrate as potential energy resources in the future. Adv Mater Res 2012;462:221–4.

Yost C. US gas market well-supplied: LNG or shale gas? Oil Gas J 2010;108(10):46–50.

Natural Gas Properties

Natural gas properties such as compressibility, density, viscosity, and others are important properties in the calculations of gas flow through reservoir rocks, material balance calculations, and the design of pipelines and production facilities.

The behavior of natural gas, whether pure methane or a mixture of volatile hydrocarbons and the nonhydrocarbons nitrogen, carbon dioxide, and hydrogen sulfide, must be understood by the engineer who is designing the operating equipment for its production, processing, and transportation. The constituents of natural gas are most likely to be found in the gaseous state but can occur as liquids and solids.

Gases and liquids are composed of molecules and may be treated on a molecular basis. The nature of the molecules and the forces existing between them control the properties of the fluid.

The kinetic theory of gases treats a gas as a group of molecules, each moving on its own independent path, entirely uncontrolled by forces from the other molecules, although its path may be abruptly altered in both speed and direction whenever it collides with another molecule or strikes the boundary of the containing vessel. In its simplest state, a gas may be considered as being composed of particles that have no volume and between which there are no forces.

Compressibility, density, and viscosity of natural gases must be known in most petroleum engineering calculations. Some of these calculations are gas metering, gas compression, design of processing units, and design of pipeline and surface facilities. Properties of natural gases are also important in the calculation of gas flow rate through reservoir rock, material balance calculations, and evaluation of gas reserves. Usually, the gas properties are measured in the laboratory. Occasionally, experimental data become unavailable and are estimated from equations of state (EOS) or empirical correlations.

The natural gas phase behavior is a plot of pressure versus temperature that determines whether the natural gas stream at a given pressure and temperature consists of a single gas phase or two phases (i.e., gas and liquid). The phase behavior for natural gas with a given composition is typically displayed on a phase diagram, an example of which is shown in Figure 2.1.

2.1 Fluid distribution in reservoir

The distribution of a particular set of reservoir fluids depends not only on the characteristics of the rock–fluid system now but also on the history of the fluids, and ultimately their source. A list of factors affecting fluid distribution would be manifold. However, the most important are as follows:

Depth: The difference in the density of the fluids results in their separation over time due to gravity (differential buoyancy).

Fluid composition: The composition of the reservoir fluid has an extremely important control on its pressure–volume–temperature properties, which define the relative volumes of each fluid in a reservoir. It also affects distribution through the wettability of the reservoir rocks.

Natural Gas Processing. http://dx.doi.org/10.1016/B978-0-08-099971-5.00002-7
Copyright © 2014 Elsevier Inc. All rights reserved.

FIGURE 2.1

Pressure–temperature diagram for a typical natural gas mixture.

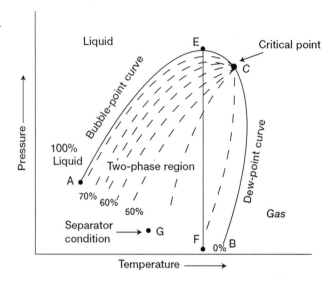

Reservoir temperature: This exerts a major control on the relative volumes of each fluid in a reservoir.

Fluid pressure: This exerts a major control on the relative volumes of each fluid in a reservoir.

Fluid migration: Different fluids migrate in different ways depending on their density, viscosity, and the wettability of the rock. The mode of migration helps to define the distribution of the fluids in the reservoir.

Trap type: Clearly, the effectiveness of the hydrocarbon trap also has a control on fluid distribution (e.g., cap rocks may be permeable to gas but not to oil).

Rock structure: The microstructure of the rock can preferentially accept some fluids and not others through the operation of wettability contrasts and capillary pressure. In addition, the common heterogeneity of rock properties results in preferential fluid distributions throughout the reservoir in all three spatial dimensions. The fundamental forces that drive, stabilize, or limit fluid movement are:

- gravity (e.g., causing separation of gas, oil, and water in the reservoir column).
- capillary (e.g., responsible for the retention of water in microporosity).
- molecular diffusion (e.g., small-scale flow acting to homogenize fluid compositions within a given phase).
- thermal convection (convective movement of all mobile fluids, especially gases).
- fluid pressure gradients (the major force operating during primary production).

Although each of these forces and factors varies from reservoir to reservoir and between lithologies within a reservoir, certain forces are of seminal importance. For example, it is gravity that ensures that when all three basic fluids types are present in an uncompartmentalized reservoir, the order of fluids with increasing depth is gas → oil → water.

2.2 **Phase behavior of hydrocarbon systems**

Figure 2.2 shows the pressure–versus–volume per mole weight (specific volume) characteristics of a typical pure hydrocarbon (e.g., propane). Imagine in the following discussion that all changes occur isothermally (with no heat flowing either into or out of the fluid) and at the same temperature. Initially, the component is in the liquid phase at 1000 psia and has a volume of about 2 ft^3/lb-mol (point A).

Expansion of the system (A → B) results in large drops in pressure with small increases in specific volume due to the small compressibility of liquids (liquid hydrocarbons and liquid formation waters have small compressibility values that are almost independent of pressure for the range of pressures encountered in hydrocarbon reservoirs).

On further expansion, a pressure will be attained where the first tiny bubble of gas appears (point B). This is the *bubble point* or *saturation pressure* for a given temperature. Further expansion (B → C) now occurs at constant pressure, with more and more of the liquid turning into the gas phase until no more fluid remains. The constant pressure at which this occurs is called the *vapor pressure* of the fluid at a given temperature.

Point C represents the situation where the last tiny drop of liquid turns into gas and is called the *dew point*. Further expansion now takes place in the vapor phase (C → D). The pistons in Figure 2.2

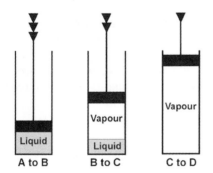

FIGURE 2.2

Pressure–volume phase behavior of a pure component.

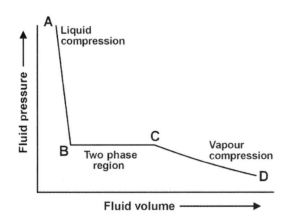

demonstrate the changes in fluid phase schematically. It is worth noting that the process A → B → C → D described earlier during expansion (reducing the pressure on the piston) is perfectly reversible. If a system is in state D, then application of pressure to the fluid by applying pressure to the pistons will result in changes following the curve D → C → B → A.

We can examine the curve in Figure 2.2 for a range of fluid temperatures. If this is done, the pressure–volume relationships obtained can be plotted on a pressure–volume diagram with the bubble point and dew point locus also included (Figure 2.3). Note that the bubble point and dew point curves join at a point (shown by a dot in Figure 2.3). This is the *critical point*. The region under the bubble point/dew point envelope is the region where the vapor phase and liquid phase can coexist and hence have an interface (the surface of a liquid drop or of a vapor bubble). The region above this envelope represents the region where the vapor phase and liquid phase do not coexist.

Thus, at any given constant low fluid pressure, reduction of fluid volume will involve the vapor condensing to a liquid via the two-phase region, where both liquid and vapor coexist. But at a given constant high fluid pressure (higher than the critical point), a reduction of fluid volume will involve the vapor phase turning into a liquid phase without any fluid interface being generated (i.e., the vapor becomes denser and denser until it can be considered a light liquid). Thus, the critical point can also be viewed as the point at which the properties of the liquid and the gas become indistinguishable (i.e., the gas is so dense that it looks like a low-density liquid, and vice versa).

Suppose that we find the bubble points and dew points for a range of different temperatures and plot the data on a graph of pressure versus temperature. Figure 2.4 shows such a plot. Note that the dew point and bubble points are always the same for a pure component, so they plot as a single line until the peak of Figure 2.3 is reached, which is the *critical point*.

The behavior of a hydrocarbon fluid made up of many different hydrocarbon components shows slightly different behavior (Figure 2.5). The initial expansion of the liquid is similar to that for the single-component case.

FIGURE 2.3

Pressure–volume phase behavior of pure component (Pc).

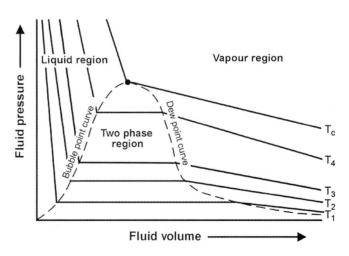

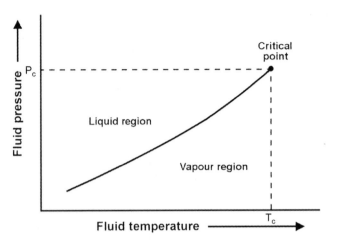

FIGURE 2.4

Pressure–temperature phase behavior of pure component (Pc).

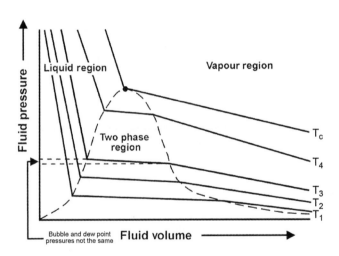

FIGURE 2.5

Pressure–volume phase behavior of multicomponent mixture. Pc, pure component.

Once the bubble point is reached, further expansion does not occur at constant pressure but is accompanied by a decrease in pressure (vapor pressure) due to changes in the relative fractional amounts of liquid to gas for each hydrocarbon in the vaporizing mixture. In this case, the bubble points and dew points differ, and the resulting pressure–temperature plot is no longer a straight line but rather a *phase envelope* composed of the bubble point and dew point curves, which now meet at the critical point (Figure 2.6). There are two other points on this diagram that are also of interest: the *cricondenbar*, which defines the pressure above which the two phases cannot exist together whatever the temperature, and the *cricondentherm*, which defines the temperature above which the two phases cannot exist together whatever the pressure. A fluid that exists above the bubble point curve is

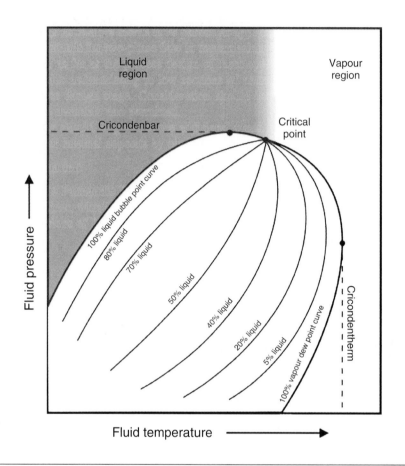

FIGURE 2.6

Pressure–temperature phase diagram for multicomponent system. Pc, pure component.

classified as *undersaturated* as it contains no free gas, while a fluid at the bubble point curve or below it is classified as *saturated* and contains free gas.

2.3 Pressure–volume–temperature properties of hydrocarbon fluids
2.3.1 Cronquist classification

Hydrocarbon reservoirs are usually classified into the following five main types:

- Dry gas
- Wet gas
- Gas condensate
- Volatile oil
- Black oil

Table 2.1 Typical Mol% Compositions of Fluids Produced from Cronquist Reservoir Types

Component or Property	Dry Gas	Wet Gas	Gas Condensate	Volatile Oil	Black Oil
CO_2	0.10	1.41	2.37	1.82	0.02
N_2	2.07	0.25	0.31	0.24	0.34
C1	86.12	92.46	73.19	57.6	34.62
C2	5.91	3.18	7.80	7.35	4.11
C3	3.58	1.01	3.55	4.21	1.01
iC4	1.72	0.28	0.71	0.74	0.76
nC4	-	0.24	1.45	2.07	0.49
iC5	0.5	0.13	0.64	0.53	0.43
nC5	-	0.08	0.68	0.95	0.21
C6s	-	0.14	1.09	1.92	1.16
C7+	-	0.82	8.21	22.57	56.4
GOR (SCF/STB)	Infinity	69,000	5965	1465	320
OGR (STB/MMSCF)	0	15	165	680	3125
API specific gravity, °API	-	65	48.5	36.7	23.6
C7+ specific gravity,	-	0.75	0.816	0.864	0.92

Note: Fundamental specific gravity γ_0 is equal to the density of the fluid divided by the density of pure water, and that for C7+ is for the bulked C7+ fraction. The API specific gravity γAPI is defined as; $\gamma API = (141.5/\gamma_0) - 131.5$.

Each of these reservoirs can be understood in terms of its phase envelope. The typical components of production from each of these reservoirs are shown in Table 2.1, and a schematic diagram of their pressure–temperature phase envelopes is shown in Figure 2.7.

2.3.2 Dry gas reservoirs

A typical dry gas fluid's phase behavior is shown in Figure 2.8. The reservoir temperature is well above the cricondentherm. During production, the fluids are reduced in temperature and pressure.

The temperature–pressure path followed during production does not penetrate the phase envelope, resulting in the production of gas at the surface with no associated liquid phase.

Clearly, it would be possible to produce some liquids if the pressure is maintained at a higher level. In practice, the stock tank pressures are usually high enough for some liquids to be produced (Figure 2.8).

2.3.3 Wet gas reservoirs

A typical wet gas reservoir fluid's phase behavior is shown in Figure 2.9. The reservoir temperature is just above the cricondentherm. During production, the fluids are reduced in temperature and pressure. The temperature–pressure path followed during production just penetrates the phase envelope, resulting in the production of gas at the surface with a small associated liquid phase.

FIGURE 2.7

Pressure–temperature for different types of hydrocarbon reservoirs.

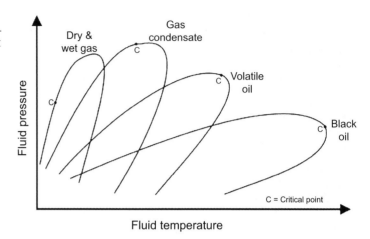

2.3.4 Gas condensate reservoirs

A typical gas condensate fluid's phase behavior is shown in Figure 2.10. The reservoir temperature is such that it falls between the temperature of the critical point and the cricondentherm. The production path then has a complex history. Initially, the fluids are in an indeterminate vapor phase, and the vapor expands as the pressure and temperature drop. This occurs until the dew point line is reached, when increasing amounts of liquids are condensed from the vapor phase. If the pressures and temperatures

FIGURE 2.8

A typical dry gas reservoir. PT, pressure–temperature.

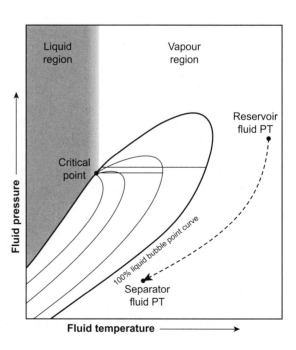

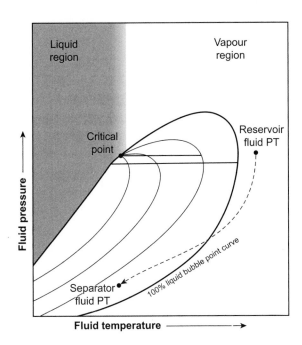

FIGURE 2.9

A typical wet gas reservoir fluid's phase behavior.
PT, pressure–temperature.

Liquid region

Vapour region

Critical point

Reservoir fluid PT

100% liquid bubble point curve

Separator fluid PT

Fluid pressure

Fluid temperature

reduce further, the condensed liquid may reevaporate, although sufficiently low pressures and temperatures may not be available for this to happen.

2.4 Gas compressibility factor

The kinetic theory of gases postulates that gases are composed of a very large number of molecules. For an ideal gas, the volume of these molecules is insignificant compared with the total volume occupied by the gas. It is also assumed that these molecules have no attractive or repulsive forces between them and that all collisions of molecules are perfectly elastic.

An ideal gas is defined as one for which both the volume of molecules and forces between the molecules are so small that they have no effect on the behavior of the gas. The ideal gas equation is:

$$PV = nRT \tag{2.1}$$

where:

P = absolute pressure, psia
V = volume, ft^3
T = absolute temperature, °R
n = number of moles of gas, lb-mol
R = the universal gas constant, which, for the above units, has the value 10.730 psia ft^3/lb-mol °R.

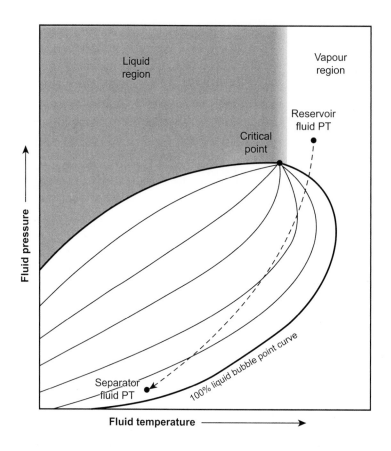

FIGURE 2.10

A typical gas condensate fluid's phase behavior. PT, pressure–temperature.

The number of pound-moles of gas (i.e., n) is defined as the weight of the gas m divided by the molecular weight MW, or:

$$n = m/\text{MW} \qquad (2.2)$$

Combining Eqn (2.1) and (2.2) gives:

$$PV = (m/\text{MW})RT \qquad (2.3)$$

where:

    m = weight of gas, lb
    MW = molecular weight, lb/lb-mol

Equation (2.3) can be rearranged to estimate the gas density at any pressure and temperature:

$$\rho_g = \frac{m}{V} = \frac{P\text{MW}}{RT} \qquad (2.4)$$

where

ρ_g = density of the gas, lb/ft^3

It should be pointed out that lb refers to pounds mass in any of the subsequent discussions of density in this text.

The ideal gas equation of state is a convenient and often a satisfactory tool when dealing with gases at pressures that do not exceed 1 atm. The errors associated with this equation are about 2–3% in this pressure range. However, the errors can escalate to hundreds of per cents at higher pressures. The Z-factor provides the ratio of the real gas volume to that of the ideal gas.

The Z-factor is a dimensionless parameter, independent of the quantity of gas, determined by the characteristics of the gas, the temperature, and the pressure. Knowing Z, calculations of pressure–volume–temperature relationships are as easy at high pressure as at low pressure.

Reduced temperature and reduced pressure have the same reduced volume. The reduced property is the property divided by the value of the property at the critical point. Thus, according to the theorem, different fluids that have the same reduced temperature and reduced pressure have the same Z-factor.

The theorem of corresponding states that fluids that have the same value of reduced temperature and reduced pressure have the same reduced volume. The reduced property is the property divided by the value of the property at the critical point. Thus, according to the theorem, different fluids that have the same reduced temperature and reduced pressure have the same Z-factor.

$$T_r = \frac{T}{T_c} \tag{2.5}$$

$$P_r = \frac{P}{P_c} \tag{2.6}$$

where P_r and T_r are reduced pressure and reduced temperature, respectively, and P_c and T_c are critical pressure and critical temperature of the gas, respectively. The values of critical pressure and critical temperature can be estimated from the following equations if the composition of the gas and the critical properties of the individual components are known [1]:

$$P_c = \sum_i^n P_{ci} y_i \ \text{ and } \ T_c = \sum_i^n T_{ci} y_i \tag{2.7}$$

where P_{ci} and T_{ci} are the critical pressure and critical temperature of component i, respectively; and y_i is the mole fraction of component i.

In cases where the composition of a natural gas is not available, the pseudo-critical properties, that is, "p_{pc}" and "T_{pc}," can be predicted solely from the specific gravity of the gas [2]. presented a graphical method for a convenient approximation of the pseudo-critical pressure and pseudo-critical temperature of gases when only the specific gravity of the gas is available. Standing (1977) [3] expressed this graphical correlation in the following mathematical forms:

Case 1: Natural Gas Systems

$$T_{pc} = 168 + 325\gamma_g - 12.5\gamma_g^2 \tag{2.8}$$

$$p_{pc} = 677 + 15.0\gamma_g - 37.5\gamma_g^2 \tag{2.9}$$

Case 2: Gas Condensate Systems

$$T_{pc} = 187 + 330\gamma_g - 71.5\gamma_g^2 \tag{2.10}$$

$$p_{pc} = 706 - 51.7\gamma_g - 11.1\gamma_g^2 \tag{2.11}$$

where:

T_{pc} = pseudo-critical temperature, °R
p_{pc} = pseudo-critical pressure, psia
γ_g = specific gravity of the gas mixture

The values of critical pressure and critical temperature can be estimated from its specific gravity if the composition of the gas and the critical properties of the individual components are not known.

For gas mixtures, the reduced conditions can be determined using pseudo-critical values instead of the true critical values [1]:

Pseudo-reduced temperature:

$$T_r = T \Big/ \sum (y_i T_{ci}) = T/T_{pc} \tag{2.12}$$

Pseudo-reduced pressure:

$$P_r = P \Big/ \sum_i (y_i P_{ci}) = P/P_{pc} \tag{2.13}$$

where:

T_{ci} = critical temperature for component "*i*"
T_{pc} = pseudo-critical temperature
T_r = reduced temperature
P_{ci} = critical pressure for component "*i*"
P_{pc} = pseudo-pressure
P_r = reduced pressure
y_i = mole fraction of component "*i*"

Any units of temperature or pressure are acceptable provided that the same absolute units are used for T as for T_c (T_{pc}) and for P as for P_c (P_{pc}). The "average molecular weight" for a gas mixture is

$$MW_{avg} = \sum_i y_i MW_i \tag{2.14}$$

where:

MW_{avg} = average molecular weight
MW_i = molecular weight of component *i*
y_i = mole fraction of component *i*

Figure 2.11 shows Z-factors for typical sweet natural gases. Using Z-factors from Figure 2.11 should yield mixture volumes (densities) within 2–3% of the true values for reduced temperatures from slightly greater than 1.0 to the limits of the chart for both temperature and pressure. The chart has been

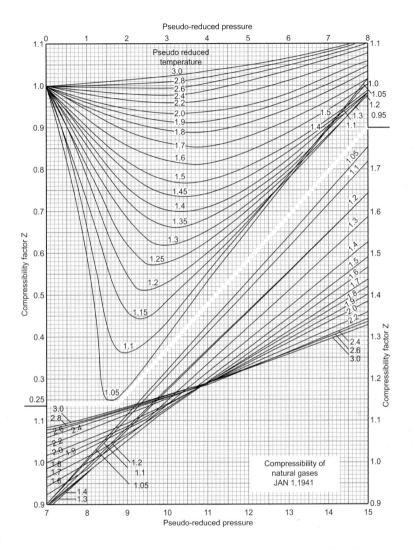

FIGURE 2.11

Compressibility factors for natural gas.

Courtesy of [1].

prepared from data for binary mixtures of methane with ethane, propane and butane and data for natural gas mixtures. All mixtures have average molecular masses less than 40, and all gases contain less than 10% nitrogen and less than 2% combined hydrogen sulfide and carbon dioxide.

Figure 2.11 does not apply for gases or vapors with more than 2% H_2S and/or CO_2 or more than 20% nitrogen. Use other methods for vapors that have compositions atypical of natural gases mixtures or for mixtures containing significant amounts of water and/or acid gases, and for all mixtures as saturated fluids, other methods should be used.

Particularly for natural hydrocarbon gases, the charts due to Standing and Katz (Figure 2.11, Ref. [4]) and Katz (Figure 2.12, Ref. [5]) are standards in the oil and gas (O&G) industry. The Standing and Katz chart and the Katz chart were developed from experimental data for natural hydrocarbon gases. Because these gases are not pure gases but a mixture of a large number of hydrocarbon gases, the pseudo H_2S reduced pressure (P_{pr}) and the pseudo H_2S reduced temperature (T_{pr}) were used. Different investigators have successfully reproduced those charts, at least under some limited conditions, by means of numerical correlations based on various EOS, and these have been widely used for numerical simulation in various applications. One of these correlations is due to Dranchuk and Abou-Kassem (DAK).

Some correlations to evaluate the Z-factor for natural gases were provided by Dranchuk and Abou-Kassem [6] and Nishiumi and Saito [7]. Takacs [8] compared eight correlations amenable for computer calculations, and the DAK correlation was found to most reliably fit the data. The DAK

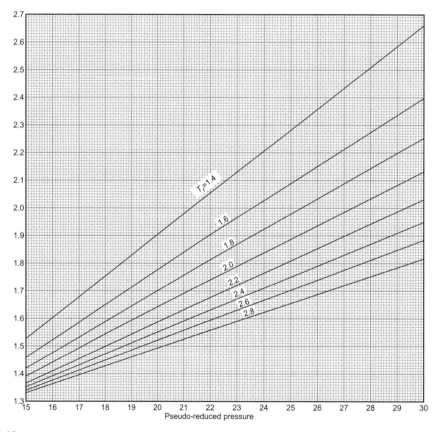

FIGURE 2.12

Compressibility factors for natural gases at pressures of 10,000–20,000 psia.

Ref. [1], © Elsevier.

correlation is based on a Han–Starling form of the Benedict–Webb–Rubin equation of state and is currently considered the standard for calculating gas compressibility and density. Dranchuk and Abu-Kassem [6] fitted an equation of state to the data of Standing and Katz [4] in order to estimate the gas deviation factor in computer routines. The DAK correlation is as follows:

$$Z = 1 + c_1(T_{pr})\rho_r + c_2(T_{pr})\rho_r^2 - c_3(T_{pr})\rho_r^5 + c_4(\rho_r, T_{pr}) \qquad (2.15)$$

where:

T_{pr} = pseudo-reduced temperature
P_{pr} = pseudo-reduced pressure

Other terms are calculated using Eqns (2.16)–(2.20):

$$\rho_r = 0.27P_{pr}/(zT_{pr}) \qquad (2.16)$$

$$c_1(T_{pr}) = A_1 + A_2/T_{pr} + A_3/T_{pr}^3 + A_4/T_{pr}^4 + A_5/T_{pr}^5 \qquad (2.17)$$

$$c_2(T_{pr}) = A_6 + A_7/T_{pr} + A_8/T_{pr}^2 \qquad (2.18)$$

$$c_3(T_{pr}) = A_9\left(A_7/T_{pr} + A_8/T_{pr}^2\right) \qquad (2.19)$$

$$c_4(T_{pr}) = A_{10}\left(1 + A_{11}\rho_r^2\right)\left(\rho_r^2/T_{pr}^3\right)\exp\left(-A_{11}\rho_r^2\right) \qquad (2.20)$$

where A_1–A_{11} are as follows:

$$\begin{aligned}
&A_1 = 0.3265 && A_2 = -1.0700 && A_3 = -0.5339 \\
&A_4 = 0.01569 && A_5 = -0.05165 && A_6 = 0.5475 \\
&A_7 = -0.7361 && A_8 = 0.1844 && A_9 = 0.1056 \\
&A_{10} = 0.6134 && A_{11} = 0.7210
\end{aligned}$$

Because the Z (compressibility factor) is on both sides of the equation, a trial-and-error solution is necessary to solve the above equation of state. Any one of the iteration techniques can be used with the trial-and-error procedure. One that is frequently used is the secant method, which has the following iteration formula:

$$X_{n+1} = X_n - f_n[(X_n - X_{n+1})/(f_n - f_{n-1})] \qquad (2.21)$$

To apply the secant method to the foregoing procedure, Eqn (2.20) is rearranged to the form

$$F(Z) = Z - \left(1 + c_1(T_{pr})\rho_r + c_2(T_{pr})\rho_r^2 - c_3(T_{pr})\rho_r^5 + c_4(\rho_r, T_{pr})\right) = 0 \qquad (2.22)$$

where:

ρ_r = reduced density
A_i = coefficients of DAK correlation
P_{pr} = pseudo-reduced pressure
P_r = reduced pressure
T_{pr} = pseudo-reduced temperature
T_r = reduced temperature

The left-hand side of Eqn (2.10) becomes the function f, and the Z (compressibility factor) becomes x. The iteration procedure is initiated by choosing two values of the Z and calculating the corresponding values of the function f. The secant method provides the new guess for Z, and the calculation is repeated until the function f is zero or within a specified tolerance (that is, $\pm 10^{-4}$).

2.4.1 Effect of nonhydrocarbon content

There are two methods that were developed to adjust the pseudo-critical properties of the gases to account for the presence of the nonhydrocarbon components. These two methods are the:

- Wichert–Aziz correction method.
- Carr–Kobayashi–Burrows correction method.

Natural gases containing H_2S and/or CO_2 exhibit different Z-factor behavior than do sweet gases. Wichert and Aziz present a calculation procedure to account for these differences. Their method uses the standard gas Z-factor chart and provides accurate sour gas Z-factors that contain as much as 85% total acid gas.

Wichert and Aziz define a "critical temperature adjustment factor," ε, that is a function of the concentrations of CO_2 and H_2S in the sour gas. This correction factor adjusts the pseudo-critical temperature and pressure of the sour gas according to the equations:

$$T'_{pc} = T_{pc} - \varepsilon \tag{2.23}$$

$$p'_{pc} = \frac{p_{pc} T'_{pc}}{T_{pc} + B(1 - B)\varepsilon} \tag{2.24}$$

$$\varepsilon = 120\left(A^{0.9} - A^{1.6}\right) + 15\left(B^{0.5} - B^{4.0}\right) \tag{2.25}$$

$$A = y_{CO_2} + y_{H_2S} \tag{2.26}$$

$$B = y_{H_2S} \tag{2.27}$$

where:

T_{pc} = pseudo-critical temperature, °R
p_{pc} = pseudo-critical pressure, psia
T'_{pc} = corrected pseudo-critical temperature, °R
p'_{pc} = corrected pseudo-critical pressure, psia
ε = pseudo-critical temperature adjustment factor

2.4.2 The Carr–Kobayashi–Burrows correction method

Carr, Kobayashi, and Burrows (1954) [9] proposed a simplified procedure to adjust the pseudo-critical properties of natural gases when nonhydrocarbon components are present. The method can be used when the composition of the natural gas is not available.

$$T'_{pc} = T_{pc} - 80y_{CO_2} + 130y_{H_2S} - 250y_{N_2} \tag{2.28}$$

$$p'_{pc} = p_{pc} + 440y_{CO_2} + 600y_{H_2S} - 170y_{N_2} \tag{2.29}$$

where:

T'_{pc} = the adjusted pseudo-critical temperature, °R
T_{pc} = the unadjusted pseudo-critical temperature, °R
y_{CO_2} = mole fraction of CO_2
y_{H_2S} = mole fraction of H_2S in the gas mixture
y_{N_2} = mole fraction of nitrogen
p'_{pc} = adjusted pseudo-critical pressure, psia
p_{pc} = unadjusted pseudo-critical pressure, psia

2.5 Equation of state

All fluids follow physical laws that define their state under given physical conditions. These laws are mathematically represented as equations, which are consequently known as EOS.

Several forms of the EOS have been presented to the O&G industry to calculate hydrocarbon reservoir fluid properties. These EOS can be written in general form as:

$$p = \frac{RT}{v - b} - \frac{a}{v^2 + uv - w^2} \tag{2.30}$$

In Eqn (2.30), p is the pressure, R is the gas constant, T is the absolute temperature, v is the molar volume, and u and w are constants. The parameters a and b are constants characterizing the molecular properties of the individual component. a is a measure of the intermolecular attraction, and b represents the molecular size. In two-parameter EOS, u and w are related to b, whereas in a three-parameter EOS, u and w are related to a third parameter, c. The two-parameter EOS are the most popular ones, where a and b are expressed as a function of critical properties as:

$$a = \frac{\Omega_a R^2 T_c^2}{P_c} \tag{2.31}$$

$$b = \frac{\Omega_b R T_c}{P_c} \tag{2.32}$$

where Ω_a and Ω_b are constants having different values for each of the EOS.

2.5.1 Two-parameter EOS

Van der Waals presented the first form of the two-parameter EOS. The two-parameter EOS assumes that the parameter a to be temperature independent. In this study, two of the mostly used two-parameter EOS are considered. The first is the one presented by Soave-Redlich-Kwong [10] (SRK-EOS), and the other is presented by Peng-Robinson [11] (PR-EOS). The SRK-EOS has the following form

$$p = \frac{RT}{v - b} - \frac{a\alpha}{v(v + b)} \tag{2.33}$$

In SRK-EOS, Ω_a and Ω_b are constants and equal to 0.42748, and 0.08664 respectively. Zudkevitch and Joffe, and Joffe showed that if Ω_a and Ω_b were treated as temperature dependent, SRK-EOS results would improve. Yarborough [12] showed the variation of Ω_a and Ω_b with reduced temperature (T_r) and correlated them with the accentric factor, ω. The dimensionless factor α is calculated from the following equation.

$$\alpha = \left[1 + m\left(1 - T_r^{0.5}\right)\right]^2 \tag{2.34}$$

where T_r is the reduced temperature (T/T_c) and m is defined as:

$$m = 0.48 + 1.574\omega - 0.176\omega^2 \tag{2.35}$$

To improve pure component vapor pressure prediction, Soave [10] and Graboski and Daubert [13] have proposed several modifications of m. SRK-EOS can be derived from the general form of the EOS by setting $u = b$ and $w = 0$. The gas compressibility factor (Z) is obtained from solving the following equation:

$$Z^3 - Z^2 + \left(A - B - B^2\right)Z - AB = 0 \tag{2.36}$$

where A and B are calculated from the following equations.

$$A = \frac{ap}{R^2T^2} \tag{2.37}$$

$$B = \frac{bp}{RT} \tag{2.38}$$

PR-EOS is obtained from the general form of the EOS by substituting u and w by $2b$ and b, respectively. PR-EOS has the following form:

$$p = \frac{RT}{v - b} - \frac{a\alpha}{[v(v + b) + b(v - b)]} \tag{2.39}$$

In PR-EOS, Ω_a and Ω_b are constants and equal 0.457235 and 0.077796, respectively. Peng and Robinson used the same Eqns 2.31 and 2.32 to calculate a and b but a different correlation to calculate m.

$$m = 0.37464 + 1.5422\omega - 0.26992\omega^2 \tag{2.40}$$

This correlation was later modified to improve prediction for heavier components [14].

$$m = 0.3796 + 1.485\omega - 0.1644\omega^2 + 0.01667\omega^3 \tag{2.41}$$

PR-EOS in term of gas compressibility factor (Z) is:

$$Z^3 - (1 - B)Z^2 + \left(A - 2B - 3B^2\right)Z - \left(AB - B^2 - B^3\right) = 0 \tag{2.42}$$

2.5.2 Volume shift

Comparison of SRK-EOS prediction and experimental liquid molar volume of pure components shows systematic deviation. For this reason, Peneloux et al. (1982) [15] introduced a volume shift to improve liquid density prediction of SRK-EOS. The volume shift has minimal effect on vapor density at low and moderate pressure, but it is advisable to adjust the gas volume by a third parameter to maintain consistency, particularly near the critical point. The corrected molar volume V^{corr} is;

$$V^{corr} = V - \sum x_i c_i \tag{2.43}$$

where x_i is the mole fraction of the component and c_i is a pure component correction factor. Similarly, Peneloux, Jhareni and Youngren (1982) [15] applied the volume shift concept to PR-EOS.

2.5.3 Three-parameter EOS

The two-parameter EOS (s) predicts the same critical compressibility factor, Z_c, for all substances. SRK-EOS predicts Z_c of 0.333 and PR-EOS predicts Z_c of 0.307 for all pure substances, whereas Z_c varies within a range of 0.2–0.3 for all hydrocarbons. Several forms of the three-parameter EOS are known in the O&G industry. In this study, the equation of state introduced at by Patel–Teja [16] is considered as it is widely used. The Patel–Teja equation of state (PT-EOS) is expressed as:

$$p - \frac{RT}{v-b} - \frac{a_c \alpha}{[v(v+b) + c(v-b)]} \tag{2.44}$$

In this equation, Patel–Teja modified the attraction term of PR-EOS by including a more flexible third parameter (c), as a function of accentric factor.

$$c = \Omega_c \frac{RT_c}{P_c} \tag{2.45}$$

$$\Omega_c = 1 - 3\zeta \tag{2.46}$$

$$\zeta = 0.329032 - 0.076799\omega - 0.0211947\omega^2 \tag{2.47}$$

PT-EOS can be reduced to PR-EOS or SRK-EOS by substituting the value of 0.307 or 0.333 for ζ, respectively. The general form of the EOS can be reduced to PT-EOS by substituting $u = b + c$ and $w^2 = cb$. Valderrama and Cristernas (1986) [17] modified PT-EOS using critical compressibility, Z_c, to correlate its parameters.

$$\Omega_a = 0.66121 - 0.76105 Z_c \tag{2.48}$$

$$\Omega_b = 0.02207 + 0.2086 Z_c \tag{2.49}$$

$$\Omega_c = 0.57765 - 1.870807 Z_c \tag{2.50}$$

Patel and Teja used the correlation to calculate m.

$$m = 0.46283 + 3.58230\omega Z_c - 0.819417(\omega Z_c)^2 \tag{2.51}$$

The gas compressibility factor (Z) is calculated from PT-EOS as:

$$Z^3 - (C-1)Z^2 + (A - 2BC - C - B^2)Z - (BC - B^2 C - AB) = 0 \tag{2.52}$$

2.5.4 Mixing rules

When EOS(s) are applied to multicomponent hydrocarbon mixtures, certain mixing rules are applied to determine their parameters. The parameters of the EOS (a, b, c) are considered to represent the attractive and repulsive forces between molecules of different substances forming the hydrocarbon mixture. These mixing rules are given by:

$$a = \sum \sum x_i x_j \left(a_i a_j \right)^{0.5} \left(1 - K_{i,j} \right) \tag{2.53}$$

$$b = \sum x_i b_i \tag{2.54}$$

$$c = \sum x_i c_i \tag{2.55}$$

The mixing rules, known as Van der Waals mixing rules, treat all components similarly, hence referred to as the random mixing rules. The binary interaction numbers $(K_{i,j})$ account for the attraction forces between pair of nonsimilar molecules. They are dependent on the difference in molecular size of the components in a binary system.

2.5.5 Binary interaction number (BIN)

Correlations to estimate the BIN for specific EOS such as SRK-EOS and PR-EOS as well as general ones have been suggested [18]. The inclusion of BIN in the EOS mixing rules provides more flexibility and in most cases reliability in EOS results. Making the BIN to be temperature dependent, pressure dependent, and composition dependent could enhance this flexibility in EOS calculations. However, it is well known that making BIN composition dependent causes additional complexity in EOS calculation, especially in compositional reservoir simulation. Groboski and Daubert [13] and Soave [10] suggest that no BINs are required for hydrocarbon mixture. On the other hand, a theoretical background was presented for the meaning and importance of the BIN [19,20]. Elliot and Daubert [21] presented a BIN for nonhydrocarbons such as N_2, CO_2, and H_2S to be used with SRK-EOS.

For N_2:

$$K_{i,j} \doteq 0.107089 + 2.9776 \, K_{i,j}^{\sim} \tag{2.56}$$

For CO_2:

$$K_{i,j} = 0.08058 - 0.77215 \, K_{i,j}^{\sim} - 1.8407 (K_{i,j}^{\sim})^2 \tag{2.57}$$

For H_2S:

$$K_{i,j} = 0.07654 + 0.01792 \, K_{i,j}^{\sim} \tag{2.58}$$

where:

$$K_{i,j}^{\sim} = \left[- \left(\varepsilon_i - \varepsilon_j \right)^2 \right] \Big/ \left(2 \varepsilon_i \varepsilon_j \right) \tag{2.59}$$

and

$$\varepsilon_i = [a_i \, \text{Ln}(2)]^{0.5} / b_i \tag{2.60}$$

Vatrotsis et al. (1986) [22] proposed a generalized correlation for evaluating the BIN for PR-EOS as a function of pressure, temperature, and accentric factors of the hydrocarbon. The generalized correlation is expressed as:

$$K_{i,j} = \delta_2 T_{rj}^2 + \delta_1 T_{rj} + \delta_0 \qquad (2.61)$$

where i refers to principal components N_2, CO_2, or CH_4 and j refers to the other hydrocarbon component of the binary. The coefficients δ_0, δ_1, and δ_2 are determined for each set of binaries from the following equations:

For N_2–HC:

$$\delta_0 = 0.1751787 - 0.7043 \log\left(\omega_j\right) - 0.0862066\left[\log\left(\omega_j\right)\right]^2 \qquad (2.62)$$

$$\delta_1 = -0.584474 + 1.328 \log\left(\omega_j\right) + 2.035757\left[\log\left(\omega_j\right)\right]^2 \qquad (2.63)$$

$$\delta_2 = 2.257079 + 7.869765 \log\left(\omega_j\right) + 13.50466\left[\log\left(\omega_j\right)\right]^2 + 8.3864\left[\log\left(\omega_j\right)\right]^3 \qquad (2.64)$$

The effect of the pressure on N_2–HC BIN is considered in the following equation:

$$K_{i,j} = K_{i,j}\left(1.04 - 4.2 \times 10^{-5}P\right) \qquad (2.65)$$

For CO_2–HC:

$$\delta_0 = 0.4025636 + 0.1748927 \log\left(\omega_j\right) \qquad (2.66)$$

$$\delta_1 = -0.941812 - 0.6009864 \log\left(\omega_j\right) \qquad (2.67)$$

$$\delta_2 = 0.741843368 + 0.441775 \log\left(\omega_j\right) \qquad (2.68)$$

The effect of the pressure on CO_2–HC is considered in the following equation:

$$K_{i,j} = K_{i,j}\left(1.044269 - 4.375 \times 10^{-5}P\right) \qquad (2.69)$$

2.6 Gas specific gravity

The specific gravity is defined as the ratio of the gas density to that of the air. Both densities are measured or expressed at the same pressure and temperature. Commonly, the standard pressure (p_{sc}) and standard temperature (T_{sc}) are used in defining the gas specific gravity:

Assuming that the behavior of both the gas mixture and the air is described by the ideal gas equation, the specific gravity can then be expressed as:

$$\gamma_g = \frac{\rho_g}{\rho_{air}} \qquad (2.70)$$

$$\gamma_g = \frac{\frac{p_{sc}MW_a}{RT_{sc}}}{\frac{p_{sc}MW_{air}}{RT_{sc}}} \qquad (2.71)$$

$$\gamma_g = \frac{MW_a}{MW_{air}} = \frac{MW_a}{28.96} \tag{2.72}$$

where :

γ_g = gas specific gravity
ρ_{air} = density of the air
MW_{air} = apparent molecular weight of the air is 28.96
MW_a = apparent molecular weight of the gas
p_{sc} = standard pressure, psia
T_{sc} = standard temperature, °R

2.7 Gas density

The density of an ideal gas mixture is calculated by simply replacing the molecular weight of the pure component with the average molecular weight of the gas mixture to give:

$$\rho_g = \frac{pMW_a}{RT} \tag{2.73}$$

where:

ρ_g = density of the gas mixture, lb/ft^3
MW_a = average molecular weight
p = absolute pressure, psia
T = absolute temperature
R = universal gas constant

2.8 Specific volume

The specific volume is defined as the volume occupied by a unit mass of the gas. For an ideal gas, this property can be calculated

$$v = \frac{V}{m} = \frac{RT}{pMW_a} = \frac{1}{\rho_g} \tag{2.74}$$

where:

v = specific volume, ft^3/lb
ρ_g = gas density, lb/ft^3

2.9 Isothermal compressibility of gases

The isothermal gas compressibility (c_g), which is given the symbol c_g, is a useful concept is used extensively in determining the compressible properties of the reservoir. The isothermal compressibility is also called the bulk modulus of elasticity. Gas usually is the most compressible medium in the reservoir. However, care should be taken so that it is not confused with the gas deviation factor, Z, which is sometimes called the super compressibility factor:

$$c_g = -\frac{1}{V_g}\left(\frac{\partial V_g}{\partial P}\right)_T \tag{2.75}$$

where V and P are volume and pressure, respectively, and T is the absolute temperature. For ideal gas, we can define the compressibility as:

$$c_g = \frac{1}{P} \tag{2.76}$$

Whereas, for nonideal gas, compressibility is defined as:

$$c_g = \frac{1}{P} - \frac{1}{Z}\left(\frac{\partial Z}{\partial P}\right)_T \tag{2.77}$$

2.10 Gas formation volume factor

The formation volume factor for gas is defined as the ratio of volume of 1 mol of gas at a given pressure and temperature to the volume of 1 mol of gas at standard conditions (P_s and T_s). The gas formation volume factor is used to relate the volume of gas, as measured at reservoir conditions, to the volume of the gas as measured at standard conditions (i.e., 60 °F and 14.7 psia). This gas property is then defined as the actual volume occupied by a certain amount of gas at a specified pressure and temperature, divided by the volume occupied by the same amount of gas at standard conditions. In an equation form, the relationship is expressed as

$$B_g = \frac{V_{p,T}}{V_{sc}} \tag{2.78}$$

where

B_g = gas formation volume factor, ft^3/scf
$V_{p,T}$ = volume of gas at pressure p and temperature T, ft^3
V_{sc} = volume of gas at standard conditions, scf

Applying the real gas EOS and substituting for the volume V gives:

$$B_g = \frac{\frac{znRT}{p}}{\frac{z_{sc}nRT_{sc}}{p_{sc}}} = \frac{p_{sc}}{T_{sc}}\frac{zT}{p} \tag{2.79}$$

where

z_{sc} = Z-factor at standard conditions
P_{sc}, T_{sc} = standard pressure and temperature

Assuming that the standard conditions are represented by $P_{sc} = 14.7$ psia and $T_{sc} = 520\ °R$, the above expression can be reduced to the following relationship:

$$B_g = 0.02827\frac{zT}{p} \tag{2.80}$$

where:

B_g = gas formation volume factor, ft^3/scf
Z = gas compressibility factor
T = temperature, °R

Gas formation volume factor can be expressed in terms of the gas density:

$$B_g = 0.02827\frac{MW_a}{R\rho_g} = 0.002635\frac{MW_a}{\rho_g};\ \ ft^3/scf \tag{2.81}$$

where

ρ_g = gas density, lb/ft^3
MW_a = apparent molecular weight of gas

In other field units, the gas formation volume factor can be expressed in bbl/scf to give:

$$B_g = 0.005035\frac{zT}{p} \tag{2.82}$$

Similarly, it can be expressed in terms of the gas density ρ_g by:

$$B_g = 0.00504\frac{MW_a}{R\rho_g} = 0.000469\frac{MW_a}{\rho_g};\ \ bbl/scf \tag{2.83}$$

The reciprocal of the gas formation volume factor is called the *gas expansion factor* and is designated by the symbol E_g or:

$$E_g = 35.37\frac{p}{zT};\ \ scf/ft^3 \tag{2.84}$$

Or in terms of the gas density ρ_g:

$$E_g = 35.37\frac{R\rho_g}{MW_a} = 379.52\frac{\rho_g}{MW_a};\ \ scf/ft^3 \tag{2.85}$$

In other units:

$$E_g = 198.6\frac{p}{zT};\ \ scf/bbl \tag{2.86}$$

or:

$$E_g = 198.6\frac{R\rho_g}{MW_a} = 2131.0\frac{\rho_g}{MW_a};\ \ scf/bbl \tag{2.87}$$

2.11 Standard volume

In many natural gas engineering calculations, it is convenient to measure the volume occupied by 1 lb-mol of gas at a reference pressure and temperature. These reference conditions are usually 14.7 psia and 60 °F and are commonly referred to as *standard conditions*. The standard volume is then defined as the volume of gas occupied by 1 lb-mol of gas at standard conditions. Applying the above conditions, the standard volume gives:

$$V_{sc} = \frac{(1)RT_{sc}}{p_{sc}} = \frac{(1)(10.73)(520)}{14.7} \tag{2.88}$$

or

$$V_{sc} = 379.4 \ \text{scf/lb-mol} \tag{2.89}$$

where:

V_{sc} = standard volume, scf/lb-mol
scf = standard cubic feet
T_{sc} = standard temperature, °R
p_{sc} = Standard pressure, psia

2.12 Acentric factor

The acentric factor, ω, is often a third parameter in corresponding states correlations. The acentric factor is a function of P_{vp}, P_c, and T_c. It is arbitrarily defined as

$$\omega = -\log\left(P_{vp}/P_c\right)_{T_r=0.7} - 1.0 \tag{2.90}$$

where:

P_{vp} = vapor pressure at a reduced temperature of 0.7
P_c = critical pressure, kPa (abs)
T_c = critical temperature, K

This definition requires knowledge of the critical (pseudo-critical) temperature, vapor pressure, and critical (pseudo-critical) pressure.

For a hydrocarbon mixture of known composition that contains similar components, a reasonably good estimate for the acentric factor is the molar average of the individual pure component acentric factors:

$$\omega = \sum_i y_i\omega_i \tag{2.91}$$

If the vapor pressure is not known, ω can be estimated for pure hydrocarbons or for fractions with boiling point ranges of 10 °C or less using:

$$\omega = \frac{3}{7}\left[\frac{\log P_c - \log 14.7}{(T_c/T_b) - 1}\right] - 1.0 \tag{2.92}$$

P_c = critical pressure, kPa (abs)
T_c = critical temperature, K
T_b = boiling point temperature, K

2.13 Viscosity

Just as the compressibility of natural gas is much higher than that of oil, water, or rock, the viscosity of natural gas is usually several orders of magnitude lower than oil or water. This makes gas much more mobile in the reservoir than either oil or water. Reliable correlation charts are available to estimate gas viscosity, and the viscosity of gas mixtures at 1 atm and reservoir temperature can be determined from the gas mixture composition

$$\mu_{ga} = \frac{\sum_{i=1}^{N} y_i \mu_i \sqrt{MW_{gi}}}{\sum_{i=1}^{N} y_i \sqrt{MW_{gi}}} \tag{2.93}$$

where μ_{ga} is the viscosity of the gas mixture at the desired temperature and atmospheric pressure, y_i is the mole fraction of the ith component, μ_i is the viscosity of the ith component of the gas mixture at the desired temperature and atmospheric pressure, MW_{gi} is the molecular weight of the ith component of the gas mixture, and N is the number of components in the gas mixture.

Knowledge of the viscosity of hydrocarbon fluids is essential for a study of the dynamic or flow behavior of these fluids through pipes, porous media, or, more generally, wherever transport of momentum occurs in fluid motion. The unit of viscosity is g/cm·s, or the poise. The kinematic viscosity is the ratio of the absolute viscosity to the density:

$$\frac{\mu}{\rho} = \frac{centipoise}{\dfrac{g}{cm^3}} = centistokes \tag{2.94}$$

Figure 2.13 shows the viscosity of paraffin hydrocarbon gases at 1 atm pressure. Figure 2.14 illustrates gas viscosity ratio with reduced temperature and pressure.

To correct for pressure, adjust the gas viscosity from Figure 2.13 from atmospheric pressure using Figure 2.14.

Figure 2.13 provides a rapid and reliable method for obtaining the viscosities of mixtures of hydrocarbon gases at 1.0 atm of pressure, given knowledge of the gravity and temperature of the gas. Insert plots show corrections to the value of hydrocarbon viscosity, which may be applied to take into account the effect of the pressure of low concentrations of hydrogen sulfide, nitrogen, or carbon dioxide. The effect of each of the nonhydrocarbon gases is to increase the viscosity of the gas mixture.

In most instances, the petroleum engineer is concerned with the viscosity of gases at pressures higher than 1 atm, so that we must now turn to a method for calculating the viscosity at pressures other than 1 atm.

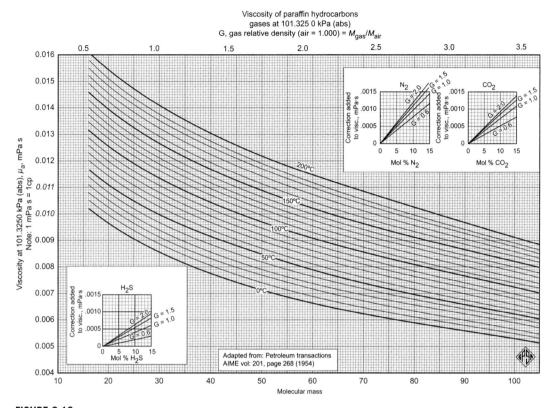

FIGURE 2.13

Viscosity of paraffin hydrocarbon gases at 1 atm pressure.

Courtesy of Ref. [1].

The theorem of corresponding states has been used to develop such a correlation; it is given as Figure 2.14. The pseudo-critical temperature and pressure are required to obtain reduced temperature and pressure for entry into the graph.

2.13.1 The Lee–Gonzalez–Eakin method

Lee, Gonalez, and Eakin [23] presented a semiempirical relationship for calculating the viscosity of natural gases. The authors expressed the gas viscosity in terms of the reservoir temperature, gas density, and the molecular weight of the gas. The proposed equation is given by:

$$\mu_g = 10^{-4} K \exp\left[X \left(\frac{\rho_g}{62.4} \right)^Y \right] \tag{2.95}$$

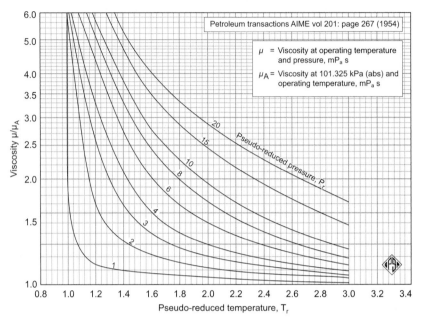

FIGURE 2.14

Correlation of gas viscosity ratio with reduced temperature and pressure.

Ref. [1].

where:

$$K = \frac{(9.4 + 0.02MW_a)T^{1.5}}{209 + 19MW_a + T} \tag{2.96}$$

$$X = 3.5 + \frac{986}{T} + 0.01MW_a \tag{2.97}$$

$$Y = 2.4 - 0.2X \tag{2.98}$$

ρ_g = gas density at reservoir pressure and temperature, lb/ft^3
T = reservoir temperature, °R
MW_a = apparent molecular weight of the gas mixture

The proposed correlation can predict viscosity values with a standard deviation of 2.7% and a maximum deviation of 8.99%. The correlation is less accurate for gases with higher specific gravities. The author points out that the method cannot be used for sour gases.

2.14 Thermal conductivity

Thermal conductivity for natural gas mixtures at elevated pressure can be calculated from an atmospheric value and a pressure correction. The pressure correction of Lenoir et al. [24] shown in Figure 2.15 applies to these low-pressure data in Figure 2.16.

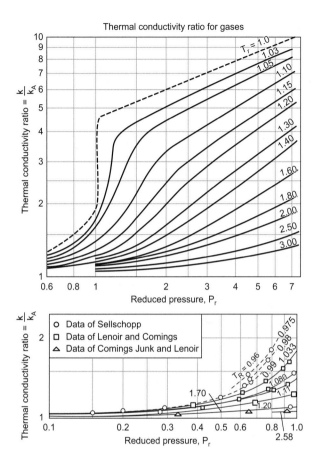

Thermal conductivity ratio for gases

FIGURE 2.15

Thermal conductivity ratio for gases.

Ref. [1].

EXAMPLE:

Find the thermal conductivity of a 25 molecular mass natural gas at 4826 kPa (abs) and 140 °C. $T_c = 244$ K, $P_c = 4550$ kPa (abs)

Solution steps

From Figure 2.16, at 140 °C:

$$kA = 0.0445 \text{ W/(m °C)}$$

$$T_r = (140 + 273)/244 = 1.69$$

$$P_r = 4826/4550 = 1.06$$

From Figure 2.15:

$$k/kA = 1.13$$

$$k = (1.13)(0.0445) = 0.0503 \text{ W/(m °C)}$$

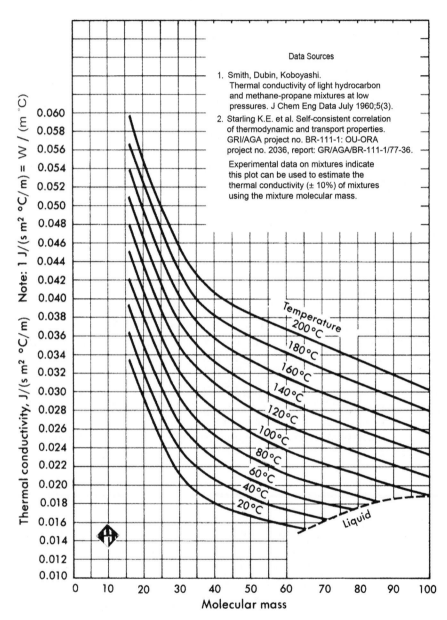

FIGURE 2.16

Thermal conductivity of natural and hydrocarbon gases at one atmosphere (101.325 kPa (abs)).

Ref. [1].

2.15 **Gross heating value of natural gases**

A complete compositional analysis of the mixture permits calculation of the gross heating value, specific gravity, and Z-factor of a natural gas mixture. Some definitions are in order before starting this discussion. It is necessary to understand the concepts: real gas, ideal gas, gross heating value (higher heating value, higher calorific value), net heating value (lower heating value, lower calorific value), and hypothetical state.

Gross heating value, by definition, is the total energy transferred as heat in an ideal combustion reaction at a standard temperature and pressure in which all water formed appears as liquid. The gross heating is an ideal gas property in a hypothetical state (the water cannot all condense to liquid because some of the water would saturate the CO_2 in the products):

$$Hv^{id} = \sum_i y_i Hv_i^{id} \qquad (2.99)$$

where:

Hv^{id} = gross heating value per unit volume of ideal gas, MJ/m^3
y_i = mole fraction in gas phase for component i

Calculation of the ideal energy flow requires multiplication of the gross heating value by the ideal gas volumetric flow rate of gas for the time period. To use a real gas flow to calculate the ideal energy flow requires converting the real gas flow rate to the ideal gas flow rate by dividing by the Z-factor. Often the heating value Hv^{id} appears in tables divided by the Z-factor in preparation for multiplying by the real gas flow rate.

Thus, Hv^{id}/Z is the ideal gross heating value per unit volume of real gas. Note that the Z-factor must be determined for the natural gas mixture and then divided into the gross heating value of the mixture. It is not correct to divide each pure component gross heating value by the pure component Z-factor and then take the molar average.

2.15.1 **Net heating value**

Net heating value, by definition, is the total energy transferred as heat in an ideal combustion reaction at a standard temperature and pressure in which all water formed appears as vapor. The net heating is an ideal gas property in a hypothetical state (the water cannot all remain vapor because, after the water saturates the CO_2 in the products, the rest would condense). It is a common misconception that the net heating value applies to industrial operations such as fired heaters and boilers. While the flue gases from these operations do not condense, the net heating value does not apply directly because the gases are not at 15 °C.

Where the gases to cool to 15 °C, some of the water would condense while the remainder would saturate the gases. It is possible to use either the gross or net heating value in such situations taking care to use the hypothetical state properly.

2.15.2 Cost of gas

This comes from a simple accounting equation:

$$c = \dot{V}^{id} Hv^{id} p^{id} \Delta t = \dot{V} Hv p \Delta t \tag{2.100}$$

where:

c = cost of gas
Hv^{id} = gross heating value per unit volume of ideal gas, MJ/m^3
Hv = gross heating value per unit volume of real gas, MJ/m^3
p^{id} = price of ideal gas
p = price of real gas
$\dot{V}^{id}$ = volumetric flow rate, ideal gas
$\dot{V}$ = volumetric flow rate, real gas
Δt = accounting period

This equation shows that for custody transfer it is not necessary to use real gas heating values (real gas heating values could be calculated from the ideal gas values using well known but computationally intense methods). If the price is set by contract, it is possible to set either a real gas price or an ideal gas price. For the former, it is necessary to use real gas values; for the latter, the ideal gas values are correct. Because it is much simpler to obtain the ideal gas values, it is preferable to establish contracts based on the ideal gas values. The ideal gas flow rate comes from the real gas flow rate

$$\dot{V}^{id} = \dot{V}/Z \tag{2.101}$$

where:

$\dot{V}^{id}$ = volumetric flow rate, ideal gas
$\dot{V}$ = volumetric flow rate, real gas
Z = Z-factor = PV/nRT

The Z-factor is the ratio of the real gas volume to the ideal gas volume. This paragraph represents the justification for using the Gas Processors Association (GPA) property tables for custody transfer. The GPA tables present ideal gas properties that are more accurate to use in mixture calculations and easier to present in tables than real gas properties.

Z-factor: When dealing with custody transfer of natural gas, it is common to use a simple equation of state for calculations because the base pressure (at which transfer occurs by definition) is near atmospheric. The equation is

$$Z = 1 + BP/RT \tag{2.102}$$

where:

$$Z = Z - \text{factor} = PV/nRT \tag{2.103}$$

B = second virial coefficient for a gas mixture, 1/(kPa (abs))
P = pressure, kPa (abs)

R = gas constant, 8.3145 J/mol K for all gases
T = absolute temperature, K
V = volume, m^3

It is common to assume that

$$- B/RT = \left[\sum_{i=1}^{N} x_i b_i \right]^2 \tag{2.104}$$

where:

b_i = summation factor for component i
x_i = mole fraction in liquid phase for component i

and

$$b_i = \sqrt{-B/RT} \tag{2.105}$$

This Z-factor is sufficiently accurate at low pressure for calculations involving natural gas mixtures.

Specific gravity (also termed relative density or gas gravity): By definition, this is the ratio of gas density (at the temperature and pressure of the gas) to the density of dry air (at the air temperature and pressure).

$$G = \rho/\rho_a = (PT_aZ_a/P_aTZ)(MW/MW_a) \tag{2.106}$$

where:

a = air
G = relative density (gas density)
P = pressure, kPa (abs)
ρ = density, kg/m^3
ρ_a = air density, kg/m^3

The ideal gas relative density is the ratio of the molecular mass of the gas to the molecular mass of dry air.

$$G^{id} = MW/MW_a \tag{2.107}$$

For a mixture:

$$G^{id} = \sum_i y_i G_i^{id} \tag{2.108}$$

The relative density G is measured and is generally used to calculate the molecular mass ratio G^{id} when the gas composition is not available.

$$G^{id} = MW/MW_a = GP_aTZ/PT_aZ_a \tag{2.109}$$

The temperatures and pressures used must correspond to actual measurement conditions or serious errors in G^{id} can occur.

2.15.3 Corrections for water content

When the gas contains water but the component analysis is on a dry basis, the component analysis must be adjusted to reflect the presence of water. The mole fraction of water in the mixture is estimated from the definition of relative humidity (on a 1-mol basis):

$$y_w = h^g P_w^\sigma / P = n_w / (1 + n_w) \tag{2.110}$$

where:

> n = number of moles
> n_w = number of water moles
> y_w = mole fraction of water
> P_w = partial pressure of water in gas phase, kPa
> P = total pressure, kPa
> h_g = relative humidity

where h_g is the relative humidity of the gas ($h_g = 1$ for saturated gas).

$$n_w = y_w / (1 + y_w) \tag{2.111}$$

Then, the corrected mole fractions of the gas become:

$$y_i(\text{cor}) = y_i \left[\frac{1}{1 + n_w} \right] = y_i \left[\frac{1}{1 + y_w / (1 - y_w)} \right] = (1 - y_w) y_i \tag{2.112}$$

and the gross heating value becomes:

$$Hv^{id} = (1 - y_w) \sum_i y_i Hv_i^{id} \tag{2.113}$$

If the compositional analysis determines water as a component, and the summation contains the water term, the heating value becomes:

$$Hv^{id} = \sum_i y_i Hv_i^{id} - y_w Hv_w^{id} \tag{2.114}$$

We remove the effect of water because, although water has a heating value (the ideal enthalpy of condensation), we assume that the water carried by wet gas (spectator water) does not condense while we assume that the water formed by the reaction does condense.

Accounting for water using Eqns (2.110)–(2.114) is sufficient for custody transfer purposes because the contracting parties can agree to accept the hypothetical state for the gross heating value. When trying to model actual situations, the question becomes more complex. It is obvious that all the water formed in a reaction cannot condense because in a situation in which both air and gas are dry, some of the reaction water must saturate the product gases while the remainder condenses.

References

[1] Gas processors and suppliers association engineering (GPSA) data book. 12th ed. 2004. Tulsa (OK, USA).
[2] Brown GG, Katz DL, Oberfell GG, Alden RC. Natural gasoline and the volatile hydrocarbons. Tulsa (OK): Natural Gas Assoc. of America; 1948.

[3] Standing MB. Volumetric and phase behavior of oil field hydrocarbon systems. Dallas (TX): Society of Petroleum Engineers of American Institute of Mining and Metallurgical Engineers; 1977. 127 p.

[4] Standing MB, Katz DL. Density of natural gases. Trans AIME 1942;146:140.

[5] Katz DL, Cornell D, Kobayashi R, Poettmann FH, Vary JA, Elenbaas JR, et al. Handbook of natural gas engineering. New York: McGraw-Hill; 1959. p. 107, Chap. 4.

[6] Dranchuk P, Abou-Kassem J. Calculation of z-factors for natural gases using equations-of-state. JCPT; July–September, 1975:34–6.

[7] Nishiumi H, Saito S. An improved generalized BWR equation of state applicable to low reduced temperatures. J Chen Engg Japan 1975;8(6):356–60.

[8] Takacs G. Comparisons made for computer z-factor calculations. Oil Gas J; December 20, 1976:64–6.

[9] Carr NL, Kobayashi R, Burrows DB. Viscosity of hydrocarbon gases under pressure. J Pet Technol 1954; 6(10):47–55.

[10] Soave G. Equilibrium constants from a modified Redlich-Kwong equation of state. Chem Eng Sci 1972;27: 1197–203.

[11] Peng D, Robinson DA. New two constant equation of state. Ind Eng Chem Fund 1976;15(1):59–64.

[12] Yarborough L. Application of a generalized equation of state to petroleum reservoir fluid. In: Choa KC, Robinson RK, editors. Equation of state in engineering, advances in chemistry series, vol. 182. American Chemical Soc; 1979. pp. 384–5.

[13] Graboski MS, Daubert TE. A modified Soave equation of state for phase equilibrium calculations 1. Hydrocarbon system. Ind Eng Chem Process Des Dev 1978;17:443–8.

[14] Robinson DB, Peng DY. The characterization of the heptanes and heavier fractions for the GPA Peng-Robinson programs. Gas processors association 1978 (Booklet only sold by the Gas Processors Association, GPA), RR-28 Research Report)

[15] Peneloux A, Rausy E, Freze R. A consistent correction for Redlich-Kwong-Soave volumes. Fluid Phase Equil. 1982;8:7–23.

[16] Patel N, Teja A. A new equation of state for fluids and fluid mixtures. Chem Eng Sci 1982;37(3):463–73.

[17] Valderrama J, Cisternas LA. A cubic equation of state for polar and other complex mixtures. Fluid Phase Equilibria 1986;29:431–8.

[18] Whitson CH, Brule MR. Phase behaviorIn SPE Monograph Series, vol. 20. Richardson (TX): Society of Petroleum Engineers Inc; 2000.

[19] Slot-Petersen C. A systematic and consistent approach to determine binary interaction coefficients for the Peng-Robinson equation of state. In: Paper SPE 16941, presented at the 62nd Annual Technical Conference of the SPE, held in Dallas, TX; September, 1987. pp. 27–30.

[20] Vidal J, Daubert T. Equations of state-reworking the old forms. Chem Eng Sci 1978;33:787–91.

[21] Elliot J, Daubert T. Revised procedure for phase equilibrium calculations with Soave equation of state. Ind Eng Chem Process Des Dev 1985;23:743–8.

[22] Varotsis N, Stewart G, Todd AC, Clancy M. Phase behavior of systems comprising north sea reservoir fluids and injection gases. J.P.T. 1986;38(11):121–1233.

[23] Lee AL, Gonzalez MH, Eakin BE. The viscosity of natural gases. J Pet Technol; August, 1966:997–1000.

[24] Lenoir JM, Junk WA, Comings EW. Chemical Engineering Progress 1953;49:539.

Further reading

Ahmed T. Hydrocarbon phase behavior. Gulf Publishing Co; 1989.

AIChE applications software survey for personal computers. New York (NY, USA): American Institute of Chemical Engineers; 1984.

API Research Project 44, Selected values of properties of hydrocarbons and related compounds, Thermodynamic Research Centre, Texas A&M University.

Benedict M, Webb GB, Rubin LC. An empirical equation for thermodynamic properties of light hydrocarbons and their mixtures. Chem Eng Prog 1951;47(8):419.

Bjorlykke OP, Firoozabadi A. Measurement and computation of near-critical phase behavior of a C1/nC24 binary mixture. SPE Res Eng; 1992:271.

Buxton TS, Campbell JM. Compressibility factors for lean natural gas-carbon dioxide mixtures at high pressures. SPEJ; March, 1967:80–6.

Carr NI, Kobayashi R, Burrows D. Viscosity of hydrocarbon gases under pressure. Trans AIME 1959;201:264.

Cengel YA, Boles MA. Thermodynamics: an engineering approach. 3rd ed. NY (USA): McGraw Hill; 2007.

Danesh A. PVT and phase behaviour of petroleum fluids. Nederlands: Elsevier Science B.V; 1998. pp. 10 and 80, Chap. 1 and 2.

George BA, editor. Proc. 61st Ann. Conv. GPA. Dallas: Texas; March 15–17, 1982. p. 171.

Hall KR, Yarborough L. A new equation of state for z-factor calculations. Oil Gas J June 18, 1973;71(25):82.

Hall KR, Yarborough L. How to solve equation of state for z-factors. Oil Gas J February 18, 1974;72(7):86.

Kamyab M, Sampaio Jr JHB, Qanbari F, Eustes AW. Using artificial neural networks to estimate the z-factor for natural, hydrocarbon gases. J Pet Sci Eng 2010;73:248–57.

Kamyab M, Sampaio Jr JHB, Qanbari F, Eustes AW. Using artificial neural networks to estimate the z-factor for natural hydrocarbon gases. J Pet Sci Eng 2010;73:248–57.

Kesler MG, Lee BI. Improve prediction of enthalpy of fractions. In: Hydrocarbon processing; March, 1976. pp. 153–8.

Londono Galindo FE. New correlations for hydrocarbon gas viscosity and gas density. Thesis from Texas A&M University; 2001. M.Sc.

McLeod WR. Application of molecular refraction to the principle of corresponding states. Ph. D. Thesis. University of Oklahoma; 1968.

Pedersen KS, Aa Fredensland, Thomassen P. Adv Thermodyn 1989;1:137.

Pedersen KS, Aa Fredenslund, Thomassen P. Properties of oils and natural gases. Houston: Gulf Publishing Co; 1989.

Physical Property Data Service, Institution of chemical engineers. Rugby (Warwickshire, U.K.): Railway Terrace. p. 165–171.

Pitzer KS, Lippman DZ, Curl Jr RF, Huggins CM, Petersen DE. The volumetric and thermodynamic properties of fluids. II. Compressibility factor, vapor pressure, and entropy of vaporization. J Am Chem Soc 1955;77:3433.

Pitzer KS. The volumetric and thermodynamics properties of fluids. J Am Chem Soc July, 1955;77(13):3427–33.

Starling KE. Fluid thermodynamic properties for light petroleum systems. Houston: Gulf Publ. Co.; 1973.

Sutton RP. Compressibility factors for high-molecular weight reservoir gases. In: Paper SPE 14265 presented at the annual tech. conf. & exhibit; September, 1985. Las Vegas (Nevada).

TRAPP by Ely JF. Hanley HJM. U.S. National Bureau of standards, National engineering laboratory. Boulder, Colorado 80303: Thermophysical properties Division; 1983.

Voros NG, Tassios DP. Vapor-liquid equilibria in nonpolar/weakly polar systems with different types of mixing rules. J Fluid Phase Equilibria 1985;91:1–29.

Whitson CH, Anderson TF, Soreide I. C7þ characterization of related equilibrium fluids using gamma distribution. In: Chorn LG, Mansoori GA, editors. C7þ Characterization. Taylor & Francis; 1989. pp. 35–56.

Wichert E, Aziz K. Calculate Z's for sour gases. Hyd Proc May, 1972;51:199.

Wichert E. Compressibility of sour natural gases, Ms. Thesis. Alberta: University of Calgary; 1970.

Zudkevitch D, Joffe E. Correlation and prediction of vapor-liquid equilibria with the Redlich-Kwong equation of state. AIChE 1970;16(1):112.

Single-phase and Multiphase Flow in Natural Gas Production Systems

This chapter covers process piping design and pipeline, in addition to the most popular pressure drop equations and fluid velocity. The subject of this chapter is mathematical relationships, based on which pipe size is calculated. The relationships presented cover Newtonian fluids and include the most useful process piping application.

This chapter covers single-phase liquid flow, single-phase gas flow, and specific cases requiring special treatment, and a brief discussion of two-phase, two-component flow calculations in short process pipes.

The flow of liquid, gases, and vapor, two-phase flow, and many other fluid systems have received sufficient study to allow definite evaluation of conditions for a variety of process situations for Newtonian fluids, discussed subsequently. For non-Newtonian fluids considerable data are available. However, their correlation is not as broad in application owing to the significant influence of physical and rheologic properties. Primary emphasis is given only to flow through circular pipe and tubes, because this is the usual means of movement for gases and liquids in process plants.

The basis for fluid flow follows Darcy and Fanning concepts. The exact transmission from laminar or viscous flow to turbulent condition is variously identified, or between a Reynolds number of 2000–3000.

The correlations included in this chapter are believed to fit average plant design with good engineering accuracy. However, other bases and correlations used by the designer should be mutually agreed upon.

As a matter of good practice with the exercise of proper judgment, the designer should familiarize himself with the background of the methods presented, to better select the conditions associated with a specific problem.

Most published correlations for two-phase pressure drop are empirical, and thus are limited by the range of data from which they were derived. A mathematical model to predict the flow regime and a procedure for calculating pressure drop in process pipelines are presented here.

Design conditions may be:

- Flow rate and allowable pressure drop are established to determine the pipe diameter for a fixed length.
- Flow rate and known length determine pressure drop and line size within the range of good engineering practice.

Usually either of these conditions requires a trial approach based on assumed pipe sizes to meet the stated conditions. Some design problems may require the determination of maximum flow for a fixed line size and length; however, this becomes the reverse of the conditions above.

Optimum economic line size is seldom realized in the average process plant. Unknown factors such as future flow rate allowance, actual pressure drops through certain process equipment, and so forth, can easily overbalance any design predicated on selecting the optimum. Certain guides as to the order of magnitude of costs and sizes can be established either by one of several correlations or by conventional cost-estimating methods. The latter is usually more realistic for a given set of conditions, because generalized equations often do not fit a plant system. Optimum criteria for pipe size should be subject to mutual agreement between the company and the designer.

Unless otherwise stated, equations presented here are used only to calculate pressure drop resulting from friction. Therefore pressure, loss or gain due to elevation must be taken into consideration where appropriate.

3.1 Basic fluid flow theory

The pressure drop equation, which is derived from the general energy equation, is given in its simplest form in Eqn (3.1a and b).

Equation (3.1a and b) shows that pressure drop is due to friction, elevation, and acceleration.

$$\Delta P_{\text{TOT}} = \Delta P_{\text{ELEV}} + \Delta P_{\text{FRIC}} + \Delta P_{\text{ACCEL}} \tag{3.1a}$$

3.1.1 Steady-state flow for single-phase fluids

A fluid passing from point 1 to point 2, in Figure 3.1, must be subject to the general principle of the conservation of energy.

This states that at steady state, the enthalpy of the fluid at point 1, plus any work performed on or by the fluid, plus any heat taken from or added to the fluid, must equal the enthalpy of the fluid at point 2.

$$\text{Energy input to system } = \text{ Energy output from system} \tag{3.1b}$$

$$Q + w_s + U_1 + p_1 V_1 + \frac{mv_1^2}{2g_c} + m Z_1 \frac{g}{g_c} = U_2 + p_2 V_2 + \frac{mv_2^2}{2g_c} + m Z_2 \frac{g}{g_c} \tag{3.2}$$

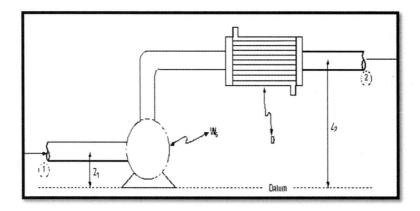

FIGURE 3.1

Conservation of energy–fluid flow in a simple system.

where:

$1/g_c$ = the proportionality/dimension constant
g = the gravitational constant
m = the fluid mass flow rate
p = the fluid pressure
Q = the heat added to, or taken away from, the fluid (e.g., by a heat exchanger or by transfer to or from the surroundings)
U = the fluid internal energy
v = the fluid velocity
V = the fluid specific volume
w_s = the shaft work done on or by the fluid (e.g., by a pump or in a turbine)
Z = the elevation of the fluid, referenced to one (arbitrary) datum level.

Note: Subscript "s" is used to differentiate work exchanged by the fluid through a shaft from that work done by the fluid itself on the system in entering and leaving the control volume under study.

Changing the basis of Eqn (3.2) to that of unit mass, and looking at differential changes rather than macro-system changes, results in Eqn (3.3):

$$du + d\left(\frac{p}{\rho}\right) + \frac{vdv}{g_c} + \frac{g}{g_c}dZ + dq + dw_s = 0 \tag{3.3}$$

where:

u = specific internal energy of the fluid
q = specific heat exchanged
w_s = specific shaft work
ρ = fluid density

The volume term, V, has been replaced by the density reciprocal, $1/\rho$ in Eqn (3.3). The only problematical term in this equation is the specific internal energy differential change, du. Internal energy refers to the energy of the molecules that make up the fluid, and is in no way connected to the energy the fluid has by virtue of its position (elevation) or movement (kinetics) as a whole.

Internal energy is believed to result from the three motion components of all molecules: the kinetic energy of rotation, translation, and internal vibration. These quantities are impossible to measure as an absolute quantity, and therefore, only the changes in internal energy are used in any thermodynamic analysis of flowing systems. However, the internal energy term in Eqn (3.3) is usually converted to its mechanical form before a practical version of this energy balance relationship can be derived using other basic thermodynamic relationships.

The specific enthalpy change, dh, in a system is defined as:

$$dh = du + d(pV) \tag{3.4}$$

Eqn (3.4) can also be written as:

$$dh = du + d\left(\frac{p}{\rho}\right) = du + \left(\frac{1}{\rho}\right)dp + pd\left(\frac{1}{\rho}\right) \tag{3.5}$$

For a reversible system:

$$du = d\,q_{rev} - dw_{rev} \tag{3.6}$$

From the Second Law of Thermodynamics, for reversible work w_{rev}, and heat q_{rev}:

$$dw_{rev} = PdV = Pd\left(\frac{1}{\rho}\right) \tag{3.7}$$

and

$$dq_{rev} = -Tds \tag{3.8}$$

where:

ds = the differential specific entropy change in a system
T = the absolute temperature value at which q is exchanged.

The resultant expression may then be directly substituted into Eqn (3.3), to create Eqn (3.8):

$$Tds + \frac{dp}{\rho} - d\left(\frac{p}{\rho}\right) + d\left(\frac{p}{\rho}\right) + \left(\frac{vdv}{g_c}\right) + \frac{g}{g_c}dZ + dq + dw_s = 0 \tag{3.9}$$

Eqn (3.8) is valid for a truly reversible process. In reality, because of irreversibilities such as friction, there will always be losses inherent in any process. Eqn (3.8) for irreversible processes then becomes:

$$dq + dL_w = -Tds \tag{3.10}$$

where:

L_w = lost friction work

If the assumption is made that no shaft work is performed by or on the fluid ($w_s = 0$), Expression (3.11) becomes:

$$\frac{dp}{\rho} + \frac{vdv}{g_c} + \frac{g}{g_c}dZ + dL_w = 0 \tag{3.11}$$

A pipe inclined at an angle θ to the horizontal, as shown in Figure 3.2, will effect a change to the vertical dZ term in Eqn (3.11). Because $dZ = dL \sin \theta$, multiplying Eqn (3.11) by ρ/dL gives:

$$\frac{dp}{dL} + \frac{\rho vdv}{g_c dL} + \frac{g}{g_c}\rho \sin \theta + \rho\frac{dL_w}{dL} = 0 \tag{3.12}$$

Eqn (3.12) can then be rearranged and integrated over a pipe length ΔL to give the familiar steady-state pressure drop Eqn (3.13). The resulting equation is known as Bernoulli's equation for the case of integrating for fluids of constant density, or as Euler's equation when used for an ideal fluid, where the loss term $\Delta(\rho L_w)$ is set to zero.

$$-\Delta p = \frac{g}{g_c}\Delta L\rho \sin \theta + \Delta(\rho L_w) + \frac{\Delta(\rho v^2)}{2g_c} \tag{3.13}$$

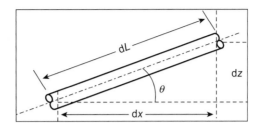

FIGURE 3.2

Schematic of piping inclination.

When computing the total pressure drop, Dp, the lost work as a result of friction must be expressed in more common terms. The loss component from Eqn (3.12), $\rho\frac{dL_w}{dL}$, can be more explicitly defined by performing a simple force balance on a section of circular pipe between the wall shear stress (which accounts for the frictional losses) and pressure forces (Figure 3.3).

$$p_1 - \left(p_1 - \frac{dp}{dL}dL\right)\frac{\pi g_c d_i^2}{4} = \tau_w(\pi d_i)dL \tag{3.14}$$

where:

$\tau_w =$ shear stress, or shear resistance to flow
$d_i =$ internal pipe diameter

Eqn (3.14) can be written as:

$$\left(\frac{dp}{dL}\right)_{\text{FRICTION}} = \frac{4g_c\tau_w}{d_i} = \rho\left(\frac{dL_w}{dL}\right) \tag{3.15}$$

Substituting for dL_w into Eqn (3.12) gives:

$$\frac{dp}{dL} + \frac{\rho v dv}{g_c dL} + \frac{g}{g_c}\rho\sin\theta + \frac{4g_c\tau_w}{d_i} = 0 \tag{3.16}$$

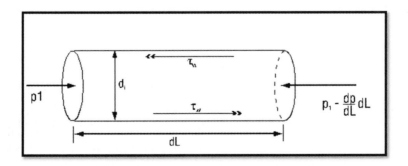

FIGURE 3.3

Section of circular pipe between wall shear stress and pressure forces.

For both laminar (streamline) and turbulent flow, a dimensionless friction factor, f, is defined as the ratio of shear stress to the fluid kinetic energy per unit volume:

$$f_{\text{FANNING}} = \frac{\tau_w}{\rho v^2/2} = \frac{2\tau_w}{\rho v^2} \tag{3.17}$$

Eqn (3.17) represents one of the original friction factor formulations as defined by Fanning. The Darcy-Weisbach (or Moody) friction factor, f_d, is four times larger than the Fanning factor:

$$f_d = \frac{\tau_w}{\rho v^2/8} = \frac{8\tau_w}{\rho v^2} \tag{3.18}$$

Substituting Eqn (3.18) into the pressure loss component of Eqn (3.16), gives:

$$\left(\frac{dp}{dL}\right)_{\text{FRICTION}} = \tau_w \frac{\pi d_i}{g_c A_i} = \left(\frac{f_d \rho v^2}{8g_c}\right)\left(\frac{\pi d_i}{\pi d_i^2/4}\right) = \frac{f_d \rho v^2}{2g_c d_i} \tag{3.19}$$

The general pressure drop expression for a single-phase fluid can then be rewritten as:

$$-\frac{dp}{dL} = \frac{g}{g_c}\rho \sin \theta + \frac{f_d \rho v^2}{2g_c d_i} + \frac{\rho v dv}{g_c dL} \tag{3.20}$$

3.2 Process pipe sizing for plants located onshore single phase

The optimum pipe size should be based on minimizing the sum of the costs of energy and piping. However, velocity limitations causing erosion or aggravating corrosion must be taken into consideration. Sometimes, the line size must satisfy process requirements such as the pump suction line.

Although pipe sizing is mainly concerned with pressure drop, for preliminary design purposes when pressure loss is not a concern, process piping may be sized on the basis of allowable velocity. When there is an abrupt change in the direction of flow (as in elbow or tees), local pressure on the surface perpendicular to the direction of flow increases dramatically. This increase is a function of fluid velocity, density, and initial pressure. Because velocity is inversely proportional to the square of the diameter, high-velocity fluids require special attention with respect to the size selection.

3.2.1 Fluid flow

In vapor systems, rule of thumb or approximate sizing methods can lead to critical flow and subsequent vibration and whistling. With two-phase systems, improper sizing can lead to slug flow, with its well-known vibration and pressure pulsations.

With both vapor and two-phase systems, approximate calculations often neglect the importance of momentum on total pressure drop. The result is that pressure drop available for controllability is reduced; rigorous calculations to determine pressure drop involving trial and error should be performed by computers. The problem is further complicated when a diameter will produce a specified pressure drop or outlet velocity for a given flow. In this situation, additional trial and error is required to determine the proper diameter.

The design problem as described above is correctly defined as line sizing. The opposite problem, of calculating velocity and pressure loss for a given diameter, is frequently encountered during hydraulic or spool checks. In general, evaluation of the total system equivalent length must be made based on fittings, valves, and straight line in the system. In addition, fitting and valve losses are not constant, but are functions of diameter. Preliminary line sizes often must be selected before accurate knowledge of the system equivalent length is available; spool check calculations are required before final specifications for prime movers can be written for the final diameter, and chosen.

3.2.2 Reynolds number

The relationship between pipe diameter, fluid density, fluid viscosity, and velocity of flow according to the Reynolds number is as follows:

$$Re = \frac{dV\rho}{\mu}$$

(3.21)

where:

$\rho =$ fluid density, kg/m^3
$Re =$ Reynolds number
$V =$ fluid velocity, m/s
$d =$ internal diameter of pipe, m
$\mu =$ fluid viscosity, Pa s

3.2.3 Friction factor

Flow is always accompanied by friction, which results in a loss of energy available for work. A general equation for press pressure drop resulting from friction is the Darcy–Weisbach equation (often referred to simply as the Darcy equation). This equation can be rationally derived by dimensional analysis, with the exception of the friction factor, f_m, which must be determined experimentally. Expressed in meters of fluid, this equation is:

$$h_L = \frac{f_m L V^2}{2gD}$$

(3.22)

where:

$D =$ internal diameter of pipe, m
$g =$ acceleration due to gravity, 9.8067 m/s^2
$h_L =$ loss of static pressure head owing to fluid flow, m
$L =$ length of line, m
$f_m =$ Moody friction factor ($f_m = 4.0 \, f_f$)
$L =$ length of line, m
$V =$ fluid velocity, m/s

Converting to kPa, the equation becomes:

$$\Delta P_f = \frac{0.5\rho f_m L V^2}{d}$$

(3.23)

where:

ΔP_f = frictional component of pressure drop, kPa
d = internal diameter of pipe, m
f_m = Moody friction factor ($f_m = 4.0\, f_f$)
L = length of line, m
V = fluid velocity, m/s
ρ = fluid density, kg/m^3

The Moody friction factor, f_m, is used in the above equations. Some equations are shown in terms of the Fanning friction factor, f_f, which is one-fourth of f_m ($f_m = 4.0\, f_f$).

The Darcy–Weisbach equation is valid for both laminar and turbulent flow of any liquid, and may also be used for gases with certain restrictions. When using this equation, changes in elevation, velocity, or density must be accounted for by applying Bernoulli's theorem. The Darcy–Weisbach equation must be applied to line segments sufficiently short such that fluid density is essentially constant over that segment. The overall pressure drop is the sum of the ΔP_f values calculated for the individual segments. For gas applications, the segmental length may be relatively short compared with liquid applications, because many gas applications involve compressible gases in which gas densities vary with pressure.

When the fluid flow is laminar (Re < 2000), the friction factor has a direct relationship on the Reynolds number, such that:

$$f_m = 64/\text{Re} \text{ or } f_f = 16/\text{Re}.$$

Pipe roughness has no effect on the friction factor in laminar flow. Substitution of the formula for Reynolds number yields the following:

$$f_m = \frac{64\mu}{DV\rho} \tag{3.24}$$

where:

f_m = Moody friction factor ($f_m = 4.0\, f_f$)
D = internal diameter of pipe, m
μ = fluid viscosity, Pa s
V = fluid velocity, m/s
ρ = fluid density, kg/m^3

This resulting in the following formula for pressure loss, in kPa:

$$\Delta P_f = \frac{32,000\mu LV}{d^2} \tag{3.25}$$

where:

ΔP_f = frictional component of pressure drop, kPa
μ = fluid viscosity, Pa s
L = length of line, m
V = fluid velocity, m/s

Eqn (3.25) is commonly known as Poiseuille's law for laminar flow.

When the flow is turbulent, the friction factor depends on the Reynolds number and the relative roughness of the pipe, ε/D, which is the roughness of the pipe, ε, over the pipe diameter, D. Figure 3.4 incorporates the relative roughness of the pipe into the determination of the friction factor.

These figures are based on the iterative solution of the following equation, developed by Colebrook:

$$\frac{1}{\sqrt{f_m}} = -2\log_{10}\left(\frac{\varepsilon}{3.7d} + \frac{2.51}{Re\sqrt{f_m}}\right) \tag{3.26}$$

The basis of the Moody friction factor chart is the Colebrook equation, where:

$f_m = $ Moody friction factor $(f_m = 4.0\, f_f)$
$Re = $ Reynolds number
$d = $ internal diameter of pipe, m
$\varepsilon = $ absolute roughness, m

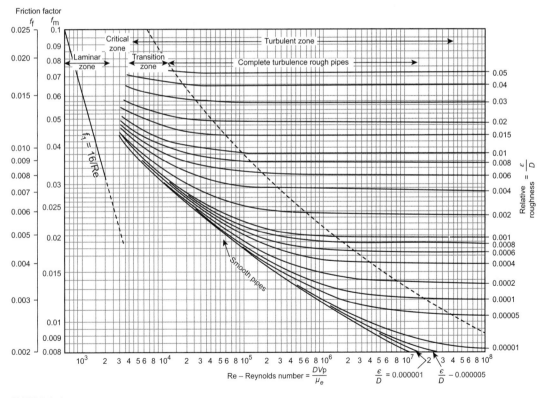

FIGURE 3.4

Moody friction factor chart.

Courtesy of Ref. [1].

3.2.4 Fluid flow calculations

This chapter discusses calculation of pressure loss for a single-phase (liquid–gas–vapor) fluid at isothermal conditions when flow rate and system characteristics are given, presented through the application of Darcy–Weisbach (often referred to as simply Darcy) and Fanning principles.

For compressible gas and vapor lines, where the pressure losses are small relative to line pressure, reasonable accuracy can often be predicted provided the following conditions are met: Average gas density of flow in use, i.e.,

$$\rho = \frac{(\rho_1 + \rho_2)}{2} \tag{3.27}$$

and pressure drop is $\leq 40\%$ of upstream

$$\text{i.e.,} \quad (P_1 - P_2) \leq 0.4P_1 \tag{3.28}$$

where:

$P_1 =$ pressure upstream
$P_2 =$ pressure downstream

This is because energy loss resulting from acceleration and density variations can be neglected up to this limit. In cases in which the pressure loss is $<10\%$ of the upstream pressure, an average value of density is not required, and either the downstream or upstream density can be used.

3.2.5 Single-phase liquid flow

Flow is considered to be laminar at a Reynolds number of ≤ 2000; therefore, before using the formula for pressure drop, the Reynolds number should be determined for flow.

To calculate pressure drop in liquid lines, the Darcy–Weisbach method can be used. The calculation is simplified for liquid flows because the density can reasonably be assumed to be a constant. As a result, the Darcy–Weisbach calculation can be applied to a long run of pipe, rather than segmentally, as dictated by the variable density in gas flow. In addition, several graphic aids are available to calculate pressure drop.

Elevation pressure drops must be calculated separately using Eqn (3.29). These elevation pressure gains or losses are added algebraically to the frictional pressure drops.

$$\Delta P_e = (0.00981)\rho_L Z_e \tag{3.29}$$

where:

$Z_e =$ pipeline vertical elevation rise, m
$\rho_L =$ liquid density, kg/m^3
$\Delta P_e =$ elevation component of pressure drop, kPa

3.2.6 Special conditions

- **Water flow**

The pressure loss for water flow is calculated by Hazen–Williams' formula. The Hazen-Williams' relationship is one of the most accurate formulas for calculating pressure loss in a water line. For the design of new water pipelines, constant C is taken to be 100. The Hazen–Williams formula is as follows:

$$q = 3.765(10)^{-6}d^{2.63}C\left(\frac{P_1 - P_2}{L}\right)^{0.54} \tag{3.30}$$

where:

    $C = 140$ for new steel pipe
    $C = 130$ for new cast iron pipe
    $C = 100$ is often used for design purposes to account for pipe fouling, etc.
    $q =$ flow rate, m³/h
    $d =$ diameter, m
    $L =$ length of line, m
    $P_1 =$ inlet pressure, kPa
    $P_2 =$ outlet pressure, kPa

- **Pump suction lines**

Generally, the pressure drops in pump suction lines should be held below 4.5 kPa/100 m in the case of liquid at the boiling point, and below 7.9 kPa/100 m in the case of liquid below the boiling point.

The maximum velocity of boiling-point liquids is be 1.2 m/s, and for subcooled liquids, 2.4 m/s. For corrosive liquids, these values may be reduced by 50%. Allowable pressure drops can be determined by the following formula:

$$\Delta P = 9.835\,S\,[H - (\text{NPSHR} + \alpha)] + (P_1 - P_v) \tag{3.31}$$

where:

    $\Delta P =$ friction loss in piping to pump inlet, kPa
    $S =$ relative density (water $= 1$)
    $H =$ height from datum to pump center, m (the term "datum" refers to the bottom tangent line in the case of vertical vessels, and to the bottom level in the case of horizontal vessels)
    NPSHR $=$ net positive suction head required, m
    $\alpha = 0.305$ m (1 ft) for liquid at the boiling point and 0.2134 m for liquid below the boiling point
    $P_1 =$ pressure working on suction liquid surface, kPa
    $P_v =$ vapor pressure of liquid at suction temperature, kPa

When permanent strainers are to be provided, a minimum pressure drop of 3.45 kPa (0.5 psi) should be added in the case of dirty service. No addition is required in the case of clean service.

The equivalent length to be used for pressure drop calculations is assumed to be 46 m (150 ft).

A suction liquid line to a centrifugal pump should be short and simple. Velocities are usually between 0.3 and 2.13 m/s. Higher velocities and unit losses can be allowed within this range when subcooled liquid is flowing, compared with when the liquid is saturated.

Note that the longer payout times favor larger pipe diameters. A pipe smaller than the pump discharge nozzle size is not used.

- **Cooling water**

Cooling water discharge headers are usually sized with unit pressure losses in decimals of 7 kPa (1 psi). An economical comparison is justified with large diameter piping, in which most of the pump pressure is used for pipe and equipment resistance. Of course, piping costs increase with diameter, whereas utility costs decrease. Between alternate designs, the best size can be determined by adding the total cost of utilities over the period of capital payout to the capital cost of each installation. The lowest overall figure will give the most economical solution.

- **Limitations owing to erosion preventive measures**

Velocity of the fluid has an important role in erosion–corrosion. Velocity often strongly influences the mechanism of the corrosion reactions. Mechanical wear is effected at high values, and particularly when the solution contains solids in suspension.

- **Amine solution**

The following limitations should be considered for amine solutions.

For carbon steel pipe:

Liquid, 3 m/s
Vapor–liquid, 30 m/s

For stainless-steel pipe:

Liquid, 9 m/s
Vapor–liquid, 36 m/s

- **Ammonium bisulfate (NH_3–H_2S–H_2O) solution**

Aqueous solutions of ammonium bisulfate produced in the effluent line of hydrocracking and hydrotreating processes often cause rapid erosion–corrosion of carbon steel pipes, especially for nozzles, bend, tees, and reducer and air cooler tube inlet parts after water injection points. Care must be taken not to exceed the highest fluid velocity in pipe tubes.

3.2.7 Gravity flow

- **Side cut draw-off**

In cases in which no controller is provided for the liquid level in the liquid draw-off tray, the flow velocity in the first 3 m of the vertical line is <0.762 m/s. This value is intended for vapor liquid separation, based on the particle diameter 200 μm (1000 μm = 1 mm) when the operating pressure is high or the difference between the vapor and liquid densities is small.

The line size should also be checked so that the control valve size does not become larger than the line size.

Line sizes in cases where the liquid enters a control valve at the boiling point should be determined on the following basis.

With consideration given to the static head and length of the line from the liquid level to the control valve, the line size should be determined so that no vaporization occurs at the inlet of control valve. In this case, the following should be satisfied:

$$0.1412\rho H > \Delta P(\text{flowmeter}) + \Delta P(\text{line friction}) \tag{3.32}$$

or

$$2.26Sh > \Delta P(\text{flowmeter}) + \Delta P(\text{line friction}) \tag{3.33}$$

where:

ρ = relative density, kg/m^3
H = static head, m
h = static head, mm
S = relative density
ΔP (flowmeter) = kPa pressure drop in flowmeter
ΔP (line friction) = kPa pressure drop due to friction

- **Vacuum tower overhead line**

1. The line cost and steam and cooling water consumptions should be calculated and the line size should be decided so that the annual cost will be minimal.
2. A precondenser should be provided in wet cases in which pressure loss in the precondenser is 4 mm Hg.

- **Steam condensate lines**
- **Line from heat exchanger to steam trap or control valve**

The pressure drop in this line should be smaller than 11.3 kPa/100 m (0.1 kg/cm^2/100 m) and should be checked that no condensate vaporizes there.

- **Line from steam trap or control valve to following vessel**

1. Steam condensate return lines must be sized to avoid excessive pressure loss. Part of the hot condensate flashes into steam when it is discharged into the condensate return system.
2. In this case, flow velocity V must be limited to 1524 m/min to prevent erosion.

3.2.8 Single-phase gas flow

In general, when considering compressible flow, as pressure decreases along the line, so does the density (assuming isothermal flow). Variation in density implies variation in the Reynolds number on which the friction factor depends. Rigorous calculation of pressure loss for a long pipeline involves dividing it into segments, performing the calculation for each segment (considering variable parameters) and integrating it over the entire length. For process piping, however, because pipe lengths are

Table 3.1 Recommendations for Single-Phase Pressure Drop Correlations

Correlation	Recommendations
Gases	
Panhandle B	Good for long and/or large diameter pipes
Weymouth	Good for short and/or small diameter pipes
Moody	Applicable for all diameters and lengths. Especially good for high-velocity lines because acceleration losses are considered
American Gas Association	Recommended by the American Gas Association
Liquids	
Hazen–Williams	Applicable for low-viscosity fluids such as water and gasoline. Should not be used for high-viscosity fluids
Moody	Applicable for fluids over a wide range of conditions

generally short, rigorous calculation would not be necessary and the equations outlined subsequently are considered adequate.

As mentioned earlier, to estimate pressure drop in a short run of gas piping, such as within a plant or battery limit, a simplified formula for compressible fluids is accurate for fully turbulent flow, assuming the pressure drop through the line is not a significant fraction of the total pressure (i.e., no more than 10%).

Table 3.1 lists recommendations for some single-phase pressure drop correlations. The recommendations are provided to help the user choose the most appropriate correlation for a given application.

3.2.9 Transmission line gas flow

Isothermal Flow: The steady-state, isothermal flow behavior of gas in pipelines is defined by a general energy equation of the form:

$$Q = 0.018 \left(\frac{T_s}{P_s}\right) E \sqrt{\frac{1}{f_f} \left[\frac{P_1^2 - P_2^2}{\gamma L_m T_{avg} Z_{avg}}\right]^{0.5}} d^{2.5} \qquad (3.34)$$

where:

Q = flow rate of gas, m³/day at base conditions
T_s = base absolute temperature, K: $T_s = 288.9$ K
P_s = base absolute pressure, kPa (abs)
$P_s = 101.56$ kPa (abs)
f_f = Fanning friction factor
γ = relative density of flowing gas (air = 1.0)
L_m = length of line, m
P_1 = inlet pressure, kPa (abs): P_2 = outlet pressure, kPa (abs)
d = internal diameter of pipe, m

T_{avg} = average temperature, K [T_{avg} = 1/2 (T_{in} + T_{out})]
Z_{avg} = average compressibility factor
E = pipeline efficiency factor (fraction)

This equation is completely general for steady-state flow, and adequately accounts for variations in compressibility factor, kinetic energy, pressure, and temperature for any typical line section. However, the equation as derived involves an unspecified value of the transmission factor, $\sqrt{1/f_f}$. Correct representation of this friction factor is necessary for the validity of the equation.

The friction factor is fundamentally related to the energy lost as a result of friction. In deriving the general energy equation, all irreversibilities and non-idealities, "A Practical Way to Calculate Gas Flow in Pipeline," which are covered by the real gas law, are collected into the friction loss term.

Empirical methods historically and currently used to calculate or predict the flow of gas in a pipeline are the result of various correlations of the transmission factor substituted into the general energy equation.

Examination of the relationships presented by various authors shows that their forms differ primarily in the inherent or specified representation of the transmission factor that defines the energy lost in resistance to flow for various pipe sizes, roughnesses, flow conditions, and gases.

To obtain Eqn (3.14), which is convenient for general calculations, a number of simplifying assumptions have been made. For other than pipeline sections with a very high pressure gradient, the change in the kinetic energy of the gas is not significant, and is assumed to be equal to zero.

It is also assumed that the gas temperature is constant at an average value for the section considered; the compressibility factor is constant at the value characterized by the average gas temperature and pressure; and in the term giving the effect of elevation change, the pressure is constant at the average value. In the range of conditions to which pipeline flow equations are ordinarily applied, averages are usually sufficiently accurate.

The average pressure in the line can be computed by:

$$P_{avg} = \frac{2}{3}\left(P_1 + P_2 - \frac{P_1 P_2}{P_1 + P_2}\right) \tag{3.35}$$

where:

P_{avg} = average pressure, kPa (abs)
P_1 = inlet pressure, kPa (abs)
P_2 = outlet pressure, kPa (abs)

In the absence of field data indicating otherwise, an efficiency factor, E, of 1.0 is usually assumed.

3.2.10 The American gas association equations

The American Gas Association (AGA) equations were developed to approximate partially and fully turbulent flow using two different transmission factors. The fully turbulent flow equation accounts for the relative pipe roughness, ε/D, based on the rough-pipe law. This equation uses the following transmission factor:

$$\sqrt{1/f_f} = 4\log_{10}\left(\frac{3.7d}{\varepsilon}\right) \tag{3.36}$$

where:

f_f = Fanning friction factor
d = diameter, m
ε = absolute roughness, m

When the transmission factor for fully turbulent flow is substituted into the general energy equation (Eqn (3.34)), the AGA equation for fully turbulent flow becomes:

$$Q = 0.018 \left(\frac{T_s}{P_s}\right) E \left[4 \log_{10}\left(\frac{3.7d}{\varepsilon}\right)\right]\left[\frac{P_1^2 - P_2^2}{\gamma L_m T_{avg} Z_{avg}}\right]^{0.5} d^{2.5} \qquad (3.37)$$

where:

Q = flow rate of gas, m³/day at base conditions
T_s = base absolute temperature, K: T_s = 288.9 K
P_s = base absolute pressure, kPa (abs): P_s = 101.56 kPa (abs)
γ = relative density of flowing gas (air = 1.0)
L_m = length of line, m
P_1 = inlet pressure, kPa (abs)
P_2 = outlet pressure, kPa (abs)
d = internal diameter of pipe, m
T_{avg} = average temperature, K [T_{avg} = 1/2 (T_{in} + T_{out})]
Z_{avg} = average compressibility factor
E = pipeline efficiency factor (fraction)

The partially turbulent flow equation is based on the smooth-pipe law and is modified to account for drag-inducing elements. The transmission factor for this equation is:

$$\sqrt{1/f_f} = 4 \log_{10}\frac{Re}{\sqrt{1/f_f}} - 0.6 \qquad (3.38)$$

Substituting $\sqrt{1/f_f}$ from Eqn (3.38) into Eqn (3.34) does not provide an equation that can be solved directly. For partially turbulent flow, a frictional drag factor must also be applied to account for the effects of pipe bends and irregularities. These calculations are beyond the scope of this book, and the AGA *Steady Flow in Gas Pipelines* should be consulted for a detailed treatment of partially turbulent flow.

3.2.11 The Weymouth equation

The Weymouth equation, published in 1912, evaluates the coefficient of friction as a function of the diameter.

$$f_f = \frac{0.0235}{d^{1/3}} \qquad (3.39)$$

$$\sqrt{1/f_f} = 6.523d^{1/6} \qquad (3.40)$$

When the friction factor, f_f, is substituted into the general energy equation, the Weymouth equation becomes:

$$Q = (0.1182)\left(\frac{T_s}{P_s}\right)d^{2.667}\left[\frac{P_1^2 - P_2^2}{L\gamma T}\right]^{1/2} \tag{3.41}$$

where:

Q = flow rate of gas, m^3/day at base conditions
T_s = base absolute temperature, K: T_s = 288.9 K
P_s = base absolute pressure, kPa (abs): P_s = 101.56 kPa (abs)
γ = relative density of flowing gas (air = 1.0)
L = length of line, m
P_1 = inlet pressure, kPa (abs)
P_2 = outlet pressure, kPa (abs)
d = internal diameter of pipe, m
T = absolute temperature of flowing gas, K

The Weymouth formula for short pipelines and gathering systems agrees more closely with metered rates than those calculated by most other formulas. However, the degree of error increases with pressure. If the Q calculated from the Weymouth formula is multiplied by $\sqrt{1/Z}$, where Z is the compressibility factor of the gas, the corrected Q will closely approximate the metered flow.

The equation cannot be generally applied to any variety of diameters and roughness, and in the flow region of partially developed turbulence, it is not valid. The Weymouth equation may be used to approximate fully turbulent flow by applying correction factors determined from the system to which it is to be applied.

In gas transmission lines, changes in elevation may seem to have a negligible contribution to the overall pressure drop, but it turns out that, particularly in high pressure lines, this contribution could be appreciable.

When corrosion inhibitor is being injected into gas transmission line, particular attention must be paid to gas velocity. High gas velocities tend to decrease the effectiveness of corrosion inhibitors. At the design stage, it would be helpful to consult the inhibitor manufacturer for limiting velocities.

In long gas transmission lines when excessive pressure drop is encountered, the final temperature might even drop below ambient. This phenomenon, called the Joule-Thomson effect in pipelines, should be watched for, particularly when the gas contains water vapor and hydrate formation is suspected.

Because of operating problems, normally a transmission line is not designed to handle two-phase (gas–liquid).

Exceptions are flow and gas lines between oil and gas wells and a separation unit or system. Sometimes, rich gas gathering networks also exhibit two-phase behavior.

To obtain a reasonable evaluation, one must resort to computer application. In recent years, several computer methods have been developed for predicting the behavior of two-phase flow in pipelines. However, because of the complex nature of two-phase phenomena, interphase changes along the line, the effect of elevation changes, and so forth, general agreement on the best methods available does not exist. Each method has its relative merits in its particular applications. Although the intention is not to present calculation methods for two-phase flow in transmission pipelines in this chapter, the following points are noteworthy as far as the process design of these systems is concerned.

The effect of liquids accumulating in low sections of natural gas pipelines should be taken into account in the process design stage. Consideration should be given to the incorporation of liquid knockout traps where desirable and permitted.

Particular attention should be given to the additional pressure required if it is intended to remove accumulated liquids or internal deposits by pigs. The factor can be critical in hill country.

• **Panhandle A equation**

In the early 1940s, the Panhandle Eastern Pipe Line Company developed a formula to calculate gas flow in transmission lines, which has become known as the Panhandle A equation. This equation uses the following expressions of Reynolds number and transmission factor:

$$\text{Re} = 1734.55 \frac{Q\gamma}{d} \tag{3.42}$$

$$\sqrt{1/f_f} = 11.85 \left(\frac{Q\gamma}{d}\right)^{0.07305} = 6.872(\text{Re})^{0.07305} \tag{3.43}$$

where:

d = internal diameter of pipe, m
f_f = Fanning friction factor
Q = flow rate of gas, m^3/day at base conditions
Re = Reynolds number
γ = relative density of flowing gas (air = 1.0)

The transmission factor assumes a Reynolds number value from 5 to 11 million based on actual metered experience.

Substituting Eqn (3.23) for $\sqrt{1/f_f}$ in the general energy equation, the Panhandle A equation becomes:

$$Q = 0.191 \left(\frac{T_s}{P_s}\right)^{1.0788} E \left[\frac{P_1^2 - P_2^2}{\gamma^{0.853} L_m T_{avg} Z_{avg}}\right]^{0.5392} d^{2.6182} \tag{3.44}$$

where:

Q = flow rate of gas, m^3/day at base conditions
T_s = base absolute temperature, K: T_s = 288.9 K
P_s = base absolute pressure, kPa (abs): P_s = 101.56 kPa (abs)
γ = relative density of flowing gas (air = 1.0)
L_m = length of line, m
P_1 = inlet pressure, kPa (abs)
P_2 = outlet pressure, kPa (abs)
d = internal diameter of pipe, m
T = absolute temperature of flowing gas, K

T_{avg} = average temperature, K [T_{avg} = 1/2 (T_{in} + T_{out})]
Z_{avg} = average compressibility factor
E = pipeline efficiency factor (fraction)

This equation was intended to reflect the flow of gas through smooth pipes. When adjusted with an efficiency factor, E, of about 0.90, the equation is a reasonable approximation of the partially turbulent flow equation. The equation becomes less accurate as flow rate increases. Many users of the Panhandle A equation assume an efficiency factor of 0.92.

- **Panhandle B equation**

A new or revised Panhandle equation was published in 1956. This revised equation is known as the Panhandle B equation and depends only slightly on the Reynolds number. Therefore, it more nearly approximates fully turbulent flow behavior. The transmission factor used here is:

$$\sqrt{1/f_f} = 19.08\left(\frac{Q\gamma}{d}\right)^{0.01961} = 16.49(\text{Re})^{0.01961} \tag{3.45}$$

Substituting 3.25 for $\sqrt{1/f_f}$ in the general energy equation, the Panhandle B equation becomes:

$$Q = 0.339\left(\frac{T_s}{P_s}\right)^{1.02} E\left[\frac{P_1^2 - P_2^2}{\gamma^{0.961} L_m T_{avg} Z_{avg}}\right]^{0.51} d^{2.53} \tag{3.46}$$

where:

Q = flow rate of gas, m³/day at base conditions
T_s = base absolute temperature, K: T_s = 288.9 K
P_s = base absolute pressure, kPa (abs): P_s = 101.56 kPa (abs)
γ = relative density of flowing gas (air = 1.0)
L_m = length of line, m
P_1 = inlet pressure, kPa (abs)
P_2 = outlet pressure, kPa (abs)
d = internal diameter of pipe, m
T = absolute temperature of flowing gas, K
T_{avg} = average temperature, K [T_{avg} = 1/2 (T_{in} + T_{out})]
Z_{avg} = average compressibility factor
E = pipeline efficiency factor (fraction)

The equation can be adjusted through the use of an efficiency term that makes it applicable across a relatively limited range of Reynolds numbers. Other than this, however, there are no means for adjusting the equation to correct it for variations in pipe surface. Adjusted to an average flowing Reynolds number, the equation will predict low flow rates at low Reynolds numbers and high flow rates at high Reynolds numbers, compared with a fully turbulent flow equation.

Efficiencies based on the Panhandle B equation decrease with increasing flow rate for fully turbulent flow. The efficiency factor, E, used in the Panhandle B equation generally varies between about 0.88 and 0.94.

Successful application of these transmission line flow equations in the past largely involved compensation for discrepancies through the use of adjustment factors, usually termed efficiencies. These efficiencies are frequently found in practice by determining the constant required to cause predicted gas equation behavior to agree with flow data. As a result, the values of these factors are specific to particular gas flow equations and field conditions, and under many circumstances, vary with flow rate in a fashion that obscures the real nature of flow behavior in the pipe.

Reynolds number–dependent equations such as the Panhandle equations use a friction factor expression that yields an approximation of partially turbulent flow behavior in the case of the Panhandle A equation and an approximation of fully turbulent behavior in the case of the Panhandle B equation.

These equations suffer from the substitution of a fixed gas viscosity value into the Reynolds number expression, which in turn, substituted into the flow equation, results in an expression with a preconditioned bias.

Regardless of the merits of various gas flow equations, past practices may dictate the use of a particular equation to maintain continuity of comparative capacities through application of a consistent operating policy.

Reference should be made to *Steady Flow in Gas Pipelines*, published by the AGA, for a complete analysis of steady flow in gas pipelines.

- **Low-pressure gas flow**

Gas gathering often involves operating pressures below 690 kPa. Some systems flow under vacuum conditions. For these low-pressure conditions, equations have been developed that give a better fit than the Weymouth or Panhandle equations. Two such formulas are as follows.

The Oliphant formula for gas flow between vacuum and 690 kPa is:

$$Q = 0.051 \left(d^{2.5} + \frac{d^3}{151.19} \right) \left(\frac{99.3}{P_s} \right) \left(\frac{T_s}{288.9} \right) \left[\left(\frac{0.6}{\gamma} \right) \left(\frac{288.9}{T} \right) \left(\frac{P_1^2 - P_2^2}{L} \right) \right]^{0.5} \tag{3.47}$$

The Spitzglass formula for gas flow below 7 kPa (ga) at 15 °C is:

$$Q = 0.821 \left[\frac{(P_1 - P_2)d^5}{\gamma L \left(1 + \frac{91.44}{d} + 0.00118d \right)} \right]^{1/2} \tag{3.48}$$

where:

Q = flow rate of gas, m³/day at base conditions
T_s = base absolute temperature, K: $T_s = 288.9$ K
P_s = base absolute pressure, kPa (abs): $P_s = 101.56$ kPa (abs)
γ = relative density of flowing gas (air = 1.0)
L_m = length of line, m
P_1 = inlet pressure, kPa (abs)
P_2 = outlet pressure, kPa (abs)
d = internal diameter of pipe, m
T = absolute temperature of flowing gas, K

Table 3.2 Allowable maximum flow velocities under sound pressure level of background noise	
Normal Background Sound Pressure, dBA	**Maximum Fluid Velocity to Prevent Noise, m/s (Obviously these Velocity Limitations Refer to Compressible Flow)**
60	30
80	41
90	52

Single-phase pressure drop correlations tend to be based more on theory than experiments (with the exception of the AGA method for single-phase gas and the Hazen–Williams expression for single-phase liquid), and therefore, less deviation can be expected for most simulations between the methods. The default methods invoked will produce reasonable results in most cases and are therefore recommended for all such situations.

In the case of high-velocity gas flow, the Panhandle B and Weymouth methods both ignore the accelerational component of the pressure drop, and results should be inspected carefully and compared with those produced by using the Moody equation. If the flow is reported to be at or near critical, these correlations cannot be relied upon to be consistent in this region and the user should switch the fluid model to compositional and use one of the special high-velocity correlations available (such as Beggs and Brill high velocity), possibly in conjunction with the flare algorithm, which can model physical occurrences such as critical pressure discontinuities.

3.2.12 Flow-induced noise

The allowable maximum flow velocities in cases in which the maximum sound pressure levels of the piping noises must be kept at 8–10 dB (A) under the sound pressure level of the background noise, are as follows in Table 3.2.

3.3 Process pipe sizing for plants located offshore

This section recommends minimum requirements and guidelines for the sizing of a new piping system on production platforms located offshore. The maximum design pressure within the scope of this section is 69,000 kPa gauge (10,000 psig) and the temperature range is −29 °C (−20 °F) to 343 °C (650 °F). For applications outside these pressures and temperatures, special consideration should be given to material properties (ductility, carbon migration, etc.).

The recommended practices presented are based on years of experience in developing oil and gas losses. Practically all of the offshore experience has been in hydrocarbon service free of hydrogen sulfide. However, recommendations based on extensive experience onshore are included for some aspects of hydrocarbon service containing hydrogen sulfide.

In determining the transition between risers and platform piping to which these practices apply, the first incoming and last outgoing valve that block pipeline flow are the limit of this document's application.

3.3.1 Sizing criteria general

In determining the diameter of pipe to be used in platform piping systems, both the flow velocity and pressure drop should be considered. The following sections present equations for calculating pipe diameters for liquid lines, single-phase gas lines, and gas/liquid two-phase lines, respectively.

When determining line sizes, the maximum flow rate expected during the life of the facility should be considered, rather than the initial flow rate. It is also usually advisable to add a surge factor of 20–50% to the anticipated normal flow rate, unless surge expectations have been more precisely determined by pulse pressure measurements in similar systems or by a specific fluid hammer calculation.

Determination of pressure loss in a line should include the effects of valves and fittings.

Calculated line sizes may need to be adjusted in accordance with good engineering judgment.

3.3.2 Sizing criteria for liquid lines

Single-phase liquid lines should be sized primarily on the basis of flow velocity. For lines transporting liquids in single phase from one pressure vessel to another by pressure differential, the flow velocity should not exceed 4.6 m/s at maximum flow rates, to minimize flashing ahead of the control valve. If practical, flow velocity should not be <0.91 m/s, to minimize deposition of sand and other solids. At these flow velocities, the overall pressure drop in the piping will usually be small. Most of the pressure drop in liquid lines between two pressure vessels will occur in the liquid dump valve and/ or choke.

3.3.3 Pump piping

Reciprocating, rotary, and centrifugal pump suction piping systems should be designed so that the available net positive suction head (NPSH) at the pump inlet flange exceeds the pump required NPSH. In addition, provisions should be made in reciprocating pump suction piping to minimize pulsations. Satisfactory pump operation requires that essentially no vapor be flashed from the liquid as it enters the pump casing or cylinder.

In a centrifugal or rotary pump, the liquid pressure at the suction flange must be high enough to overcome the pressure loss between the flange and the entrance to the impeller vane (or rotor) and maintain the pressure on the liquid above its vapor pressure. Otherwise, cavitation will occur.

In a reciprocating unit, the pressure at the suction flange must meet the same requirement; but the pump required NPSH is typically higher than for a centrifugal pump because of a pressure drop across the valves and a pressure drop caused by pulsation in the flow. Similarly, the available NPSH supplied to the pump suction must account for the acceleration in the suction piping caused by the pulsating flow, as well as the friction, velocity, and static head.

The necessary available pressure differential over the pumped fluid vapor pressure may be defined as net positive suction head available ($NPSH_a$), which is the total head in meter absolute determined at the suction nozzle, less the vapor pressure of the liquid in meter absolute. Available NPSH should always be equal or exceed the pump's required NPSH. Available NPSH for most pump applications may be calculated using Eqn (3.49):

$$NPSH_a = h_p - h_{vpa} + h_{st} - h_f - h_{vh} - h_a \tag{3.49}$$

where:

h_p = absolute pressure head due to pressure, atmospheric or otherwise, on the surface of liquid going to suction, m liquid

h_{vpa} = absolute vapor pressure of the liquid at suction temperature, m liquid

h_{st} = static head, positive or negative, due to liquid level above or below datum line (center line of pump), m liquid

h_f = friction head, or head loss due to flowing friction in the suction piping, including entrance and exit losses, m liquid

h_{vh} = velocity head, m liquid

h_a = acceleration head, m liquid

V_L = velocity of liquid in piping, m/s

g = gravitational constant (usually 9.81 m/s^2)

For a centrifugal or rotary pump, the acceleration head, h_a, is zero. For reciprocating pumps, the acceleration head is critical and may be determined by the following equation from the Hydraulics Institute:

$$h_a = \frac{LV_L R_P C}{kg} \tag{3.50}$$

where:

h_a = acceleration head, m liquid

L = length of suction line, m (actual length, not equivalent length)

V_L = average liquid velocity in suction line, m/s

R_p = pump speed, rpm

C = empirical constant for the type of pump:

= 0.200 for simplex double acting

= 0.200 for duplex single acting

= 0.115 for duplex double acting

= 0.066 for triplex single or double acting

= 0.040 for quintuplex single or double acting

= 0.028 for septuplex single or double acting

Note: The constant C will vary from these values for unusual ratios of connecting rod length to crank radius.

k is a factor representing the reciprocal of the friction of the theoretical acceleration head that must be provided to avoid a noticeable disturbance in the suction piping:

= 1.4 for liquid with almost no compressibility (de-aerated water)

= 1.5 for amine, glycol, water

= 2.0 for most hydrocarbons

= 2.5 for relatively compressible liquid (hot oil or ethane)

g = gravitational constant (usually 9.81 m/s^2)

There is no universal acceptance for Eqn (3.50). However, Eqn (3.50) is believed to be a conservative basis that will ensure adequate provision for the acceleration head.

When more than one reciprocating pump is operated simultaneously on a common feed line, at times, all crankshafts are in phase and, to the feed system, the multiple pumps act as one pump of that type with a capacity equal to that of all pumps combined. In this case, the maximum instantaneous velocity in the feed line would be equal to that created by one pump with a capacity equal to that of all pumps combined.

If the acceleration head is determined to be excessive, the following should be evaluated:

1. Shorten suction line. Acceleration head is directly proportional to line length, L.
2. Use larger suction pipe to reduce velocity. This is helpful because velocity varies inversely with the square of the pipe's inside diameter. The acceleration head is directly proportional to fluid velocity, V_L.
3. Reduce required pump speed by using a larger size piston or plunger, if permitted by pump rating. Speed required is inversely proportional to the square of piston diameter. Acceleration head is directly proportional to pump speed, R_p.
4. Consider a pump with a larger number of plungers. For example: $C = 0.04$ for a quintuplex pump. This is about 40% less than $C = 0.066$ for a triplex pump. The acceleration head is directly proportional to C.
5. Consider using a pulsation dampener if these remedies are unacceptable. The results obtainable using a dampener in the suction system depend on the size, type, location, and charging pressure used. A good, properly located dampener, if kept properly charged, may reduce L, the length of pipe used in the acceleration head equation, to a value of 5–15 nominal pipe diameter. Dampener should be located as close as possible to the pump suction.
6. Use a centrifugal booster pump to charge the suction of the reciprocating pump.

The following requirements are recommended for designing suction piping:

1. Suction piping should be one or two pipe sizes larger than the pump inlet connection.
2. Suction lines should be short, with a minimum number of elbows and fittings.
3. Eccentric reducers should be used near the pump, with the flat side up to keep the top of line level. This eliminates the possibility of gas pockets forming in the suction piping. If the potential for accumulation of debris is a concern, means for removal is recommended.
4. For reciprocating pumps, provide a suitable pulsation dampener (or make provisions for adding a dampener at a later date) as close as possible to the pump cylinder.
5. In multi-pump installations, size the common feed line so the velocity will be as close as possible to the velocity in the laterals going to the individual pumps. This will avoid velocity changes and thereby minimize acceleration head effects.

Reciprocating, centrifugal, and rotary pump discharge piping should be sized on an economical basis. In addition, reciprocating pump discharge piping should be sized to minimize pulsations. Pulsations in reciprocating pump discharge piping are also related to the acceleration head, but are more complex than suction piping pulsations. The following guidelines may be useful in designing discharge piping:

1. Discharge piping should be as short and direct as possible.
2. Discharge piping should be one or two pipe sizes larger than the pump discharge connection.

Table 3.3 - Typical Flow Velocities

	Suction Velocity, m/s (ft/s)	Discharge Velocity, m/s (ft/s)
Reciprocating Pumps		
Speeds up to 250 rpm	0.61 (2 ft/s)	1.83 (6 ft/s)
Speeds 251–330 rpm	0.46 (1.5 ft/s)	1.37 (4.5 ft/s)
Speeds above 330 rpm	0.305 (1 ft/s)	0.91 (3 ft/s)
Centrifugal pumps	0.61–0.91 (2–3 ft/s)	1.83–2.74 (6–9 ft/s)

3. Velocity in discharge piping should not exceed three times the velocity in the suction piping. This velocity will normally result in an economical line size for all pumps, and will minimize pulsations in reciprocating pumps.

4. For reciprocating pumps, include a suitable pulsation dampener (or make provisions for adding a dampener at a later date) as close as possible to the pump cylinder.

The Table 3.3 may be used to determine preliminary suction and discharge line sizes.

3.3.4 Sizing criteria for single-phase gas lines

• **Process lines**

When pressure drop is a consideration (lines connecting two components operating at essentially the same pressure, etc.), single-phase gas lines should be sized on the basis of acceptable pressure loss.

The pressure drops listed in Table 3.4 have been found by experience to be an acceptable balance for short lines, when capital costs (pipe or compression) and operating costs are considered. When velocities in gas lines exceed 18.3 m/s (60 ft/s), noise may be a problem.

Table 3.2 may be used to determine pressure loss if the total pressure drop is <10% of inlet pressure. If the total pressure drop is >10%, an equation such as Weymouth should be used.

• **Compressor lines**

Reciprocating and centrifugal compressor piping should be sized to minimized pulsation, vibration, and noise. The selection of allowable velocities requires an engineering study for each specific application.

Table 3.4 - Acceptable Pressure Drops for Single-phase Gas Process Lines

Operating Pressure kPa (ga)	Acceptable Pressure Drop kPa/100 m (psi/100 ft)
100–690	1.13–4.30 (0.05–0.19)
696–3447	4.52–11.082 (0.2–0.49)
3454–13,790	11.3–27.14 (0.5–1.2)

3.3.5 Sizing criteria for gas/liquid two-phase lines

Erosional velocity, flow lines, production manifolds, process headers, and other lines transporting gas and liquid in two-phase flow should be sized primarily on the basis of flow velocity. Flow velocity should be kept at least below fluid erosional velocity. If solid (sand) production is anticipated, fluid velocity should be reduced accordingly.

The velocity above which erosion may occur can be determined by the following empirical equation:

$$V_e = \frac{1.22C}{\sqrt{\rho_m}} \tag{3.51}$$

where:

V_e = fluid erosional velocity, m/s
C = empirical constant:
   = 125 for noncontinuous service
   = 100 for continuous service
ρ_m = gas/liquid mixture density at operating pressure and temperature, kg/m^3

3.3.6 Minimum velocity

If possible, the minimum velocity in two-phase lines should be about 3 m/s, to minimize slugging of separation equipment. This is particularly important in long lines with elevation changes.

3.4 Transmission pipelines

The transmission line as related to the requirements of this section is a pipeline transporting gas or liquid and also two-phase flow from oil fields to the ship loading points or production units such as refineries and natural gas plants. This section presents different methods for economical calculations of pressure loss and required diameter for transmission of a specific quantity of crude oil, products, and natural gas to the terminal under consideration.

3.4.1 Sizing criteria

Although pressure loss is a primary criterion in determining line size, the following points should be taken into consideration when designing a pipeline.

Design consideration should be given to flow velocity within a range that will minimize corrosion. The lower limit of the flow velocity range should be a velocity that will keep impurities suspended in the commodity, thereby minimizing the accumulation of corrosion matter within the pipeline.

The upper limit of the velocity range should be such that erosion-corrosion cavitation or impingement attack will be minimal.

Intermittent flow conditions should be avoided when possible. If operating criteria dictate the need for intermittent flow, design consideration should be given to obtaining an operating velocity that will pick up and sweep away water or sediment that accumulates in lower places in the line during periods of no flow.

If water, sediment, or other corrosive contaminates are expected to accumulate in the pipeline, the design should include a loading and receiving pig trap. Operating procedures should be developed and implemented for adequate cleaning (see also NACE MR 0175–2002).

Crude oils generally tend to carry free water along the pipeline; this water is a potential source of corrosion. If the velocity of crude oil is too low, the water stratifies to the bottom of pipe and corrosion may occur. It is advisable to maintain a certain minimum velocity, to keep the water from stratification.

3.4.2 Natural gas liquids pipelines

Natural gas liquids (NGL) are generally in natural gas processing plants (dew point depression process). Their components typically range from C_2 to C_9 and the relative density is typically about 0.55.

Pressure drop calculations should be made using an appropriate method as described for liquid handling.

In transmission of NGL by pipe, pressure loss should not cause vaporization and consequently create two-phase flow. Therefore, when calculating pressure loss, the actual temperature and pressure of the line should be regarded.

Because of its higher vapor pressure, a two-phase flow condition must be avoided by maintaining adequately high minimum pressure along the pipelines.

3.4.3 Two-phase fluid flow in pipes

Up to this point in our discussion of fluid flow, assumptions were made for single-phase fluids. In topics covering laminar and turbulent flow, fluid flow theory can take us only as far as single-phase, laminar flow. Once the laminar to turbulent flow transition is made for single-phase fluids, modeling the flow becomes complex. Fluid flow theory then enters a gray area in which empirical correlations and experimental observation account for a large portion of key areas in the prediction of pressure drop. All salient equations describing two-phase flow build on existing single-phase equation forms. However, they are far more empirical than their single-phase equivalents owing to the complex nature of multiphase flow.

In single-phase laminar flow, fluid properties such as density and viscosity must be known before any pressure drop formulation is applied. This concept is relatively simple for a single-phase fluid; simple temperature- and pressure-dependent correlations and mixing rules can be applied with confidence for all but the heaviest and nonideal of fluids.

However, the definition and meaning of two-phase fluid physical properties such as density or viscosity require further discussion.

Clearly, some type of viscosity and density must be used in the overall pressure drop expression, and because the only physical data available to the researcher and engineer are viscosities of both phases in isolation, a suitable mixing rule for the two phases must be formulated. Before this can be accomplished, there must be knowledge of the relative amounts of both phases existing at the point of interest in the pipe. The term "holdup" is commonly used to describe this quantity.

3.4.4 Liquid holdup and superficial velocities

Two quantities are used extensively in two-phase flow analyses: the superficial velocity of each phase, and the liquid or gas holdup. The superficial velocity is defined as the velocity at which one phase would travel if it solely occupied the whole pipe, and can be calculated from volumetric flow rates:

$$v_{sL} = \frac{q_L}{A} \quad \text{(for liquids)} \tag{3.52}$$

$$v_{sG} = \frac{q_G}{A} \quad \text{(for gases)} \tag{3.53}$$

where:

q = the volumetric flow rate of the phase
A = the cross-sectional area of the pipe
v_s = the superficial velocity of the phase

Subscripts G and L refer to the gas (vapor) and liquid phase, respectively. The fluid mixture velocity is defined as the velocity of the total mixture, and can be shown to be equal to the sum of the component superficial velocities:

$$v_M = \frac{q_L + q_G}{A} \tag{3.54}$$

$$v_M = v_{sL} + v_{sG} \tag{3.55}$$

where subscript M refers to the mixture.

3.4.5 Liquid and gas holdup: slip and no-slip

Liquid holdup, H_L, is defined as the fraction of the pipe's cross-sectional area occupied by liquid:

$$H_L = \frac{A_L}{A} \tag{3.56}$$

For the complementary gas holdup:

$$H_G = \frac{A_G}{A} \tag{3.57}$$

Figure 3.5 shows liquid and vapor holdup. For a given steady-state snapshot of a two-phase fluid flowing in a pipe, there can be two feasible liquid holdup scenarios: slip liquid holdup (in which the gas and liquid are traveling at different velocities) and no-slip liquid holdup (in which both phases are actually traveling at the same speed). These concepts are important in the study of two-phase flow systems, and will be used fully when describing two-phase pressure drop correlations. In either scenario, the liquid holdup varies from a value of 1 for all liquid flow to 0 for all gas flow. The term "positive slip" is used to describe situations in which the gas is traveling faster than the liquid, and "negative slip" for the reverse situation.

The slip velocity is defined as the difference between the actual gas and liquid velocities:

$$v_S = v_G - v_L \tag{3.58}$$

Now,

$$v_G = \frac{q_G}{A_G} \tag{3.59}$$

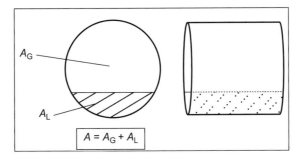

FIGURE 3.5

Liquid and vapor holdup.

$$v_L = \frac{q_L}{A_L} \tag{3.60}$$

Then, from Eqns (3.23) and (3.25–3.26), it is simple to show that for cases of no-slip here:

$$v_G = v_L \tag{3.61}$$

$$\lambda_L = H_{L,NO-SLIP} = \frac{v_{SL}}{v_M} \tag{3.62}$$

where:

$$\lambda_L = No - slip \ liquid \ hold \ up \tag{3.63}$$

Therefore, the no-slip liquid holdup can be shown to be a function only of the superficial velocities of each phase, and is thus simple to calculate if the volumetric flow rate of each phase is known.

In most two-phase flow situations, however, the no-slip assumption is false and should not be used for reasonable pressure drop predictions. This is typically because of the flow pattern and/or the topology of the piping system (which can encourage slip conditions through inclined and vertical pipe sections).

In these cases, liquid holdups must be calculated from empirical correlations.

Once the liquid holdup value has been determined for the segment of piping under study, its first use is in calculating actual gas and liquid velocities from their corresponding superficial values:

$$v_L = \frac{q_L}{AH_L} \tag{3.64}$$

3.5 Two-phase mixture properties

A two-phase fluid mixture density ρ_m, can be defined using the holdup as the weighting factor between both phases. Although some researchers use different formulations, Eqn (3.31) represents one of the most common relationships for density used to calculate the elevation term in the pressure drop equation (in this equation, H_L refers to either slip or no-slip liquid holdup):

$$\rho_M = \rho_L H_L + (1 - H_L)\rho_G \tag{3.65}$$

The viscosity of a fluid is used in all pressure drop correlations, at least in determining the friction factor, f_d, via the Reynolds number, Re. It is also typically used within correlating parameters in more empirically based methods. This can be seen in the two-phase pressure drop correlations.

The types of equation formulated for two-phase viscosities are more varied than those developed for density, because the concept of a mixture viscosity is more difficult to comprehend. A typical formulation for mixture viscosity is shown in Eqn (3.32):

$$\mu_M = \mu_L^{H_L} \mu_G^{H_G} \tag{3.66}$$

In dealing with two-phase flow, surface tension is another physical property that affects pressure drop. It is typically used as a correlating property in determining slip liquid holdup as well the different flow regimes.

In systems involving a wet hydrocarbon mixture, there will often be enough water present to form a separate, second liquid phase. Most of the current, accredited industry-standard methods are available in the literature. In each method, only one liquid phase is considered to be present. Therefore, when two liquid phases actually exist, they are bulked together to form one liquid phase.

The physical transport properties required by the pressure drop method for the bulk liquid are then calculated by weighing the relative amounts of each phase present. Liquid viscosity, density, and surface tension are all calculated in this fashion. In Eqn (3.33), any of these properties can substituted for x:

$$x_L = x_{HC} VOLFR_{HC} + x_{WAT} VOLFR_{WAT} \tag{3.67}$$

where:

x = liquid viscosity, density, or surface tension
VOLFR = volume fraction

Subscripts HC and WAT refer to the hydrocarbon and aqueous phase, respectively. Other viscosity mixing rules, e.g., the American Petroleum Industry (API) methods and the Woelflin procedure (for cases involving water-in-oil emulsions) are also available for various engineering calculations.

3.6 Two-phase flow pressure drop

The formula for calculating the pressure drop for two-phase fluids in pipes is analogous to the equivalent single-phase flow version shown in Eqn (3.19), except that the friction factor and physical properties are replaced by their two-phase equivalents:

$$-\frac{dP}{dL} = \frac{g}{g_c} \rho_M \sin\theta + \frac{f_M \rho_M v_M^2}{2g_c d_i} + \frac{\rho_M v_M dv_M}{g_c d_L} \tag{3.68}$$

The pressure loss term in Eqn (3.34) includes a two-phase friction factor, f_M. In the past, researchers have spent much effort in developing predictive correlations for the two-phase friction factor, as well as the slip liquid holdup term.

Definitions of the mixture density (ρ_m) and mixture friction factor (f_m) terms are specific to the correlation to which they are employed.

3.6.1 **Flow regimes**

When two fluids with different physical properties flow together in the same pipe, there will be a wide range of possible flow regimes. A flow regime (or flow pattern) is essentially a description of the flow structure, or distribution of one fluid phase relative to the other. For example, for upward flow of air and water in an inclined pipe, the dominant flow regime is generally described as either mist flow or slug flow.

Different inclinations of pipe, together with the direction of flow, have a major effect on the actual flow regime. Several correlations have been developed specifically for one type of topology and flow direction. The types of flow regime encountered include, but are not limited to, those shown in Figure 3.6. Depending on the researcher, categories of flow regime may be labeled differently.

A flow regime map is generally included as a front end to most modern two-phase pressure drop calculation methods. The Taitel–Dukler–Barnea map, for example, is a graphic representation of flow regime correlated against superficial gas and superficial liquid velocities. Once the respective superficial velocities are known, the flow regime can be easily read off the graph. Depending on which flow regime is suggested for those superficial velocities, a regime-specific correlation for liquid holdup is then invoked. Different pressure drop correlations begin with a different flow map, because the link between the resultant pressure drop and flow regime has typically been made by regression during the experimental phase of the research.

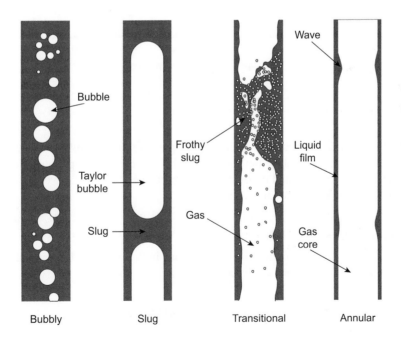

FIGURE 3.6

Flow regimes in two-phase flow.

From Ref. [2].

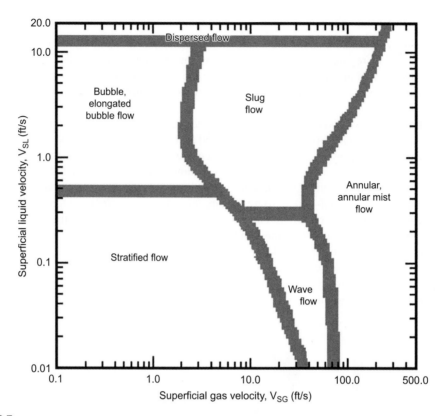

FIGURE 3.7

Sample of simplified two-phase flow regime map.

Note: The flow regime predicted by an individual pressure drop correlation may not reflect the actual regime encountered in reality, but the use of this regime in that particular correlation will produce the results as correlated and expected by the researchers.

For an example of a flow regime map, see Figure 3.7.

3.6.2 Two-phase flow correlation

Table 3.5 lists available two-phase empirical correlations in the literature for vertical, horizontal, or inclined flow. Available correlations are listed.

3.6.3 Recommendations on pressure drop correlations

• **Two-phase and compositional methods**

This section also contains a table that shows the merits for each of the major two-phase pressure drop correlations. Because of the heavy experimental basis of each of the two-phase methods, no single correlation can be recommended for all systems. Most piping systems contain topologies within which fluids flow in all directions and through a variety of valves and fittings. Therefore, if one were to select

Table 3.5 Two-Phase Pressure Drop Correlations

Name of Correlation	Flow Regime	Liquid Holdup (H_L)	Friction (dp/dL)	Elevation (dp/dL)	Acceleration (dp/dL)
Duns & Ross	Bubble and slug flow	$H_l = V_s - V_m + \dfrac{\sqrt{(V_m - V_s)^2 + 4V_s} + V_{sl}}{2V_s}$ V_s calculated from correlation	$\left(\dfrac{dp}{dL}\right)_f = \dfrac{f_m \rho_l V_{sl} V_m}{2g_c d144}$ f_m from correlation	$\left(\dfrac{dp}{dL}\right)_\varepsilon = \dfrac{\rho_{tp} V_{sl} g \sin\varphi}{g_c 144}$ $\rho_{tp} = \rho_l H_l + \rho_g H_g$	$\left(\dfrac{dp}{dL}\right)_{acc} = 0$ $\rho_{tp} = \rho_l H_l + \rho_g H_g$
	Mist low	$H_l = \lambda_l = \dfrac{V_{sl}}{V_{sl} + V_{sg}}$	$\left(\dfrac{dp}{dL}\right)_f = \dfrac{f \rho_g V_{sg}^2}{2g_c d144}$ $V_{sg}^o = \dfrac{V_{sg} d^2}{(d - \varepsilon)^2}$ f is calculated from the Moody diagram ε = roughness from pipe and liquid film	Same as bubble and slug flow	$\left(\dfrac{dp}{dL}\right)_{acc} = \dfrac{V_m V_{sg} \rho_n}{g_c P144}\left(\dfrac{dp}{dL}\right)_T$ $\rho_n = \lambda_l \rho_l + (1 - \lambda_l)\rho_g$
	Transition flow	$H_l = A_1(H_l)_{slug}$ $\quad + B_1(H_l)_{slug}$ A_1 f(dimensionless numbers), $0 \le A_1 \le 1$ $B_1 = 1 - A_1$	$\left(\dfrac{dp}{dL}\right)_f = A_1\left(\dfrac{dp}{dL}\right)_{f,slug}$ $\quad + B_1\left(\dfrac{dp}{dL}\right)_{f,mist}$	$\left(\dfrac{dp}{dL}\right)_\varepsilon = A_1\left(\dfrac{dp}{dL}\right)_{\varepsilon,slug}$	$\left(\dfrac{dp}{dL}\right)_{acc} = A_1\left(\dfrac{dp}{dL}\right)_{acc,slug}$ $\quad + B_1\left(\dfrac{dp}{dL}\right)_{acc,mist}$
Hagedorn and Brown	—	$H_l = f$(dimensionless numbers), calculated from correlation $H_l = f$(dimensionless numbers), calculated from correlations	$\left(\dfrac{dp}{dL}\right)_f = \dfrac{f \rho_t V_m^2}{2g_c d144}$ $\rho_t = \dfrac{\rho_n^2}{\rho_s}$ $\rho_s = \rho_l H_l + \rho_g H_g$ $\mu_s = \mu_l H_l + \mu_g H_g$ $\quad = \dfrac{1488 \rho_n V_m d}{N_{Re}}$	$\left(\dfrac{dp}{dL}\right)_\varepsilon = \dfrac{\rho_s g \sin\varphi}{g_c 144}$	$\left(\dfrac{dp}{dL}\right)_{acc} = \dfrac{V_m V_{sg} \rho_n}{g_c P144}\left(\dfrac{dp}{dL}\right)_T$

Continued

Table 3.5 Two-Phase Pressure Drop Correlations—cont'd

Name of Correlation	Flow Regime	Liquid Holdup (H_L)	Friction (dp/dL)	Elevation (dp/dL)	Acceleration (dp/dL)
Orkiszewski	Bubble flow	$H_l = 1 - \dfrac{1}{2}\left[1 + \dfrac{V_m}{V_s} - \sqrt{\left(1 + \dfrac{V_m}{V_s}\right)^2 - 4\dfrac{V_{sg}}{V_s}}\right]$ $V_s = 0.8$ ft/s	$\left(\dfrac{dp}{dL}\right)_f = \dfrac{f\rho_l\left(\frac{V_l}{H_l}\right)^2}{2g_c d\,144}$	Same as Hagedorn & Brown elevation	Same as Duns & Ross bubble flow acceleration
	Slug flow	$H_l = \dfrac{\rho_s - \rho_g}{\rho_l - \rho_g}$ $\rho_s = \rho_l(V_{sl} + V_b) + \rho_s V_{sg} + \rho_l\delta$ $V_b = f(N_{Re})$ $N_{Re} = f(V_b)$ d = distribution coefficient based on the continuous phase and mixture velocity. The liquid phase is oil continuous for water cuts <50% and continuous water For water cuts >50%	$\left(\dfrac{dp}{dL}\right)_f = \dfrac{f\rho_l V_m^2}{2g_c d\,144}$	Same as bubble flow	—
	Mist flow	Same as Duns and Ross		Same as bubble flow	$\left(\dfrac{dp}{dL}\right)_{acc} = \dfrac{V_m V_{sg}\rho_m}{g_c P\,144}\left(\dfrac{dp}{dL}\right)_T$
	Transition flow	Same as Duns and Ross transition flow holdup	Same as Duns and Ross transition flow friction	Same as bubble flow elevation	Same as Duns & Ross transition flow acceleration
Angel–Welchon–Ross	—	$H_l = \lambda_l$ (No-slip holdup)	$\left(\dfrac{dp}{dL}\right)_f = \dfrac{f_{tp}\rho_n V_m^2}{2g_c d\,144}$	Same as Orkiszewski or Hagedorn & Brown elevation	Same as Orkiszewski bubble flow or Hagedorn & Brown elevation
Aziz	Bubble flow	$H_l = 1 - \dfrac{V_{sg}}{V_{bf}}$ $V_{bf} = 1.2V_m + V_{bs}$	$\left(\dfrac{dp}{dL}\right)_f = \dfrac{f\rho_s V_m^2}{2g_c d\,144}$ $N_{Re} = \dfrac{1488\rho_l V_m d}{\mu_l}$ f is from the Moody diagram	$\left(\dfrac{dp}{dL}\right)_e = \dfrac{\rho_s g\sin\varphi}{g_c\,144}$ $\rho_{tp} = \rho_l H_l + \rho_g H_g$	Same as Duns & Ross

	Slug flow	$H_l = 1 - \dfrac{V_{sg}}{V_{bf}}$ $V_{bfsl} = 1.2V_m + V_{bsl}$ $V_{bsl} = C\sqrt{\dfrac{dg(\rho_l - \rho_g)}{\rho_l}}$	$\left(\dfrac{dp}{dL}\right)_f = \dfrac{f\rho_l V_m^2 H_l}{2g_c d144}$ $N_{Re} = \dfrac{1488\rho_l V_m d}{\mu_l}$	Same as bubble flow	Same as bubble flow
	Mist flow	Same as Duns and Ross	Same as Duns and Ross	Same as bubble flow	Same as Hagedorn & Brown
	Transition flow	$H_l = A_3(H_l)_{slug} + B_3(H_l)_{slug}$ $A_3 = f(\text{dimensionless numbers})$ $0 \le A_3 \le 1$ $B_3 = 1 - A_3$	$\left(\dfrac{dp}{dL}\right)_f = A_4\left(\dfrac{dp}{dL}\right)_{f,slug}$ $\quad + B_4\left(\dfrac{dp}{dL}\right)_{f,mist}$	Same as bubble flow	$\left(\dfrac{dp}{dL}\right)_{acc} = A_3\left(\dfrac{dp}{dL}\right)_{acc,\,slug}$ $\quad + B_3\left(\dfrac{dp}{dL}\right)_{acc,\,mist}$ Same as Hagedorn & Brown
Beggs and Brill	Segregated, intermittent, distributed flow	$H_l = \left(\dfrac{a\lambda_l^b}{N_{Fr}^c}\right)\Psi$ $\varphi = 0 \ \ \Psi = 0$ $\Psi = 1 + C[\sin(1.8\varphi) - 0.33\sin^3(1.8\varphi)]$ $C = (1 - \lambda_l)\ln(d\,\lambda_l^e N_{lv}^f N_{Fr}^g)$ $a\text{–}g = $ constants, f(flow pattern)	$\left(\dfrac{dp}{dL}\right)_f = \dfrac{f\rho_s V_m^2}{2g_c d144}$ $f_{tp} = e^s f_n$ f_n is from the Moody diagram for smooth pipe. $S = \dfrac{y}{-0.0523 + 3.1825y^2 - 0.01863y^4}$ if $1 < e^y < 1.2$ $S = \ln(2.2e^y - 1.2)$ $y = \ln\left(\dfrac{\lambda_l}{\mu_l^2}\right)$	Same as Aziz	Same as Hagedorn & Brown
	Transition flow	$H_l = A_4(H_l)_{seg} + B_4(H_l)_{int}$ $A_4 = f(\text{dimensionless numbers})$ $0 < A_4 < 1$ $B_3 = 1 - A_4$	$\left(\dfrac{dp}{dL}\right)_f = A_4\left(\dfrac{dp}{dL}\right)_{f,seg}$ $\quad + B_4\left(\dfrac{dp}{dL}\right)_{f,int}$	Same as segregated flow	Same as segregated flow

Continued

Table 3.5 Two-Phase Pressure Drop Correlations—cont'd

Name of Correlation	Flow Regime	Liquid Holdup (H_L)	Friction (dp/dL)	Elevation (dp/dL)	Acceleration (dp/dL)
Gray	—	$H_l = 1 - H_g$ $$H_g = \frac{1 - e^{A^1}}{R + 1}$$ $$A^1 = -2.314\left[N_v\left(1 + \frac{205}{N_d}\right)\right]^{B^1}$$ $B^1 = 0.0814$ $$\left[1 - 0.05554\ln\left(1 + \frac{0.730R}{R + 1}\right)\right]$$ $$R = \frac{V_{sl}}{V_{sg}} \quad N_v = \frac{\rho_m^2 V_{sm}^2}{g\sigma_l(\rho_l - \rho_g)}$$ $$N_d = \frac{g(\rho_l - \rho_g)D^2}{\sigma_l}$$ $$\sigma_l = \frac{\sigma_o q_o + 0.617\sigma_w q_w}{q_o + 0.617 q_w}$$ s = surface tension	$$\left(\frac{dp}{dL}\right)_f = \frac{f_{tp}\rho_s v_m^2}{2g_c d\,144}$$ $$N_{Re,l} = \frac{1488\rho_n V_m d}{\mu_{nl}}$$ f_n is from the Moody diagram	Same as Beggs & Brill	Same as Hagedorn & Brown
Flannigan		$$H_l = \frac{1}{1 + 0.3264 V_{sg}^{1.05}}$$			
Lockhart and Martinelli		$H_l = f(x)$ calculated from correlation $$x = \left[\frac{\left(\frac{dp}{dL}\right)_l^s}{\left(\frac{dp}{dL}\right)_g^s}\right]^{\frac{1}{2}}$$ $$\left(\frac{dp}{dL}\right)_l^s = \frac{f_l\rho_l v_{sl}^2}{2g_c d\,144}$$ $$N_{Re,l} = \frac{1488\rho_l V_{sl} d}{\mu_l}$$ $$\left(\frac{dp}{dL}\right)_g^s = \frac{f_l\rho_g v_{sg}^2}{2g_c d\,144}$$ $$N_{Re,g} = \frac{1488\rho_g V_{sg} d}{\mu_l}$$	$$\left(\frac{dp}{dL}\right)_f = \frac{\phi_g^2\left(\frac{dp}{dL}\right)_g^s + \phi_l^2\left(\frac{dp}{dL}\right)_l^s}{2}$$ $\phi_g = f(X_l, N_{Re,g})$ $\phi_l = f(X_l, N_{Re,l})$	Same as Aziz	Same as Duns & Ross bubble flow

Eaton	$H_l = f(\text{dimensionless numbers})$, calculated from correlation	$\left(\dfrac{dp}{dL}\right)_f = \dfrac{f_{tp}\rho_n V_m^2}{2g_c d\,144}$ f_{tp} calculated from correlation	Same as Aziz	$\left(\dfrac{dp}{dL}\right)_{acc} = \dfrac{W_l D V_l^2 + W_g D V_g^2}{2g_c q_m D P_L\,144}$
Dukler	H_l = calculated from correlation iteratively $H_l = f(H_l)$ $N_{Re,k} = \dfrac{1488\rho_k V_m d}{\mu_n}$ $\rho_k = \dfrac{\rho_l \lambda_l^2}{H_l} + \dfrac{\rho_g \lambda_g^2}{H_g}$	$\left(\dfrac{dp}{dL}\right)_f = \dfrac{f\rho_k V_m^2}{2g_c d\,144}$ $\dfrac{f}{f_n} = 1 + \dfrac{y}{1.281+(-0.478y)+0.444y^2-0.094y^3+0.00843y^4}$ $y = \ln(\lambda_l)$ $f_n = 0.0056 + 0.5 N_{Re,k}$	Same as Aziz	$\left(\dfrac{dp}{dL}\right)_{acc} = \dfrac{1}{2g_c D P_L\,144}$ $D\left(\dfrac{\rho_g V_{eg}^2}{H_g} + \dfrac{\rho_l V_{sl}^2}{H_l}\right)$
Mukherjee and Brill	Bubble, slug, mist flow: $f < 0$ $H_L = e^{H_2}$ $H_2 = H_1\left\{\dfrac{N_{gV}^{0.371771}}{N_{lV}^{0.393952}}\right\}$ $H_1 = -0.51664$ $\quad + 0.789805\sin\phi$ $\quad - 0.551627\sin^2\phi$ $\quad + 15.519214 N_l^2$	$\left(\dfrac{dp}{dL}\right)_f = \dfrac{f\rho_m V_m^2}{2g_c d\,144}$ f is from the Moody diagram $N_{Re} = \dfrac{1488\rho_n V_m d}{\mu_g}$	Same as Aziz	Same as Hagedorn & Brown
	Slug flow: $f < 0$	Same as bubble flow	Same as bubble flow	Same as bubble flow
	Mist flow: $f < 0$	Same as bubble flow	Same as bubble flow	—
	Stratified flow: $f < 0$ $H_L = e^{H_2}$ $H_2 = H_1\left\{\dfrac{N_{gV}^{0.079951}}{N_{lV}^{0.504887}}\right\}$	$\left(\dfrac{dp}{dL}\right)_f = \dfrac{f\rho_g V_g^2}{2g_c d_h\,144}$ $N_{Re} = \dfrac{1488\rho_g V_g d_h}{\mu_g}$ f is from the Moody diagram d_h = hydraulic diameter of gas phase	Same as bubble flow	Same as bubble flow

the appropriate correlation for each pipe run, should the answer be necessarily more accurate? The answer to this question is probably not. Most the two-phase correlations available today have been produced to address much larger piping systems, typically transporting oil, gas, and/or water, in which the flow rates are relatively large; and much work has been performed on studying the effects of flow regime and other large topologic changes on overall pressure drop.

When this analysis is brought down to the level of refinery- or chemical plant–sized piping runs and typical simulations, these large-scale phenomena begin to lose meaning in translation, and the choice of pressure drop correlation outside the arena of critical flow therefore tends to have correspondingly less impact on the results. With any two-phase flow system, the user is recommended to bracket the solution to the simulation by using two or more appropriate correlations. Bracketing means that the user may have more confidence in the true solution being between the values reported using the different correlations.

The following are general guidelines:

For gas-dominated two-phase pipe runs with subcritical flow:

- Use the Dukler–Eaton correlation, and for cases with very low liquid loadings (<10 bbl/MMSCF, or 0.056 m^3/1000 s m^3) bracket with the Beggs, Brill, and Moody correlation. Mukherjee-Brill is better for $0.1 < H_L < 0.35$.

For single-phase liquid and liquid-dominated fluid lines, such as crude oil and its products and water:

- Use the Beggs, Brill, and Moody correlation.

For dense-phase gas pipelines, such as CO_2 or NH_3:

- Use the Beggs, Brill, and Moody correlation.

For downward-flowing pipes contain two-phase fluids or steam:

- Use the Beggs and Brill No-Slip correlation.

For all steam piping, except downward flow pipes (as above):

- Use the Beggs, Brill, and Moody correlation.

For high-velocity and critical flow systems:

- Use the high-velocity modifications to the standard Beggs and Brill, and Beggs, Brill, and Moody correlations.

Table 3.6 shows some recommendations for two-phase pressure drop correlations. Table 3.7 summarizes experimental information for some two-phase pressure prop correlations.

3.6.4 Two-phase flow patterns

Figure 3.8 shows some two-phase flow patterns.

- **Bubble or froth flow**

This pattern, characterized by bubbles of gas moving along the upper part of the pipe at approximately the same velocity as the liquid, develops when bubbles of gas are dispersed throughout the liquid. It

Table 3.6 Recommendations for Two-phase Pressure Drop Correlations

Correlation	Correlation Recommendations
Horizontal Flow	
Lockhart–Martinelli	Widely used in the chemical industry. Applicable for annular and annular mist flow regimes if flow pattern is known a priori. Do not use for large pipes. Generally overpredicts pressure drop
Eaton	Do not use for diameters <2 in. Do not use for very high or low liquid holdup. Underpredicts holdup for $H_L < 0.1$. Works well for $0.1 < H_L < 0.35$
Dukler	Good for horizontal flow. Tends to underpredict pressure drop and holdup. Recommended by API for wet gas lines
Beggs and Brill	Use the no-slip option for low holdup. Underpredicts holdup. Most consistent and well-behaved correlation
Inclined Flow	
Mukherjee–Brill	Recommended for hilly terrain pipelines. New correlation based heavily on in situ flow pattern. Only available model that calculates flow patterns for all flow configurations and uses this information to determine modeling technique

Table 3.7 Experimental Information for Two-phase Pressure Drop Correlations

Correlation	Date	Basis	Pipe Size(s), in	Fluids
Horizontal Flow				
Lockhart–Martinelli	1949	Laboratory data	0.0586–1.1017	—
Eaton	1966	Laboratory and field data	2–4	—
Dukler	1969	Data and similarity Analyses	Wide range	Oil, gas, water
Beggs and Brill	1973	Laboratory data	1–1.5	Gas, water
Inclined Flow				
Mukherjee–Brill	1983	Laboratory data	1.5	Kerosene, lube oil, gas

occurs for liquid superficial velocities of about 1.5–4.6 m/s (5–15 ft/s) and gas superficial velocities of about 0.305–3.05 m/s (1–10 ft/s).

- **Plug flow**

Alternate plugs of liquid and gas move along the upper part of the pipe and liquid moves along the bottom of the pipe. Plug flow occurs for liquid velocities less than 0.61 m/s (2 ft/s), and gas velocities, less than about 0.91 m/s (3 ft/s).

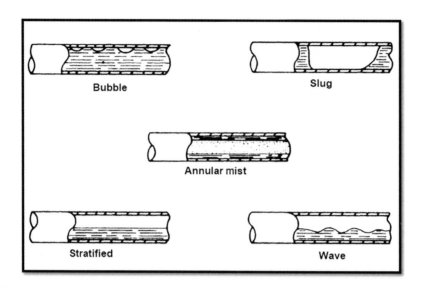

FIGURE 3.8

Sample two-phase flow in horizontal pipes (GPSA 2004).

- **Stratified flow**

The liquid phase flows along the bottom of the pipe, whereas the gas flows over a smooth liquid–gas interface. It occurs for liquid velocities less than 0.15 m/s (0.5 ft/s) and gas velocities of about 0.61–3.05 m/s (2–10 ft/s).

- **Wave flow**

Wave flow is similar to stratified flow, except that the gas is moving at a higher velocity and the gas–liquid interface is distributed by waves moving in the direction of flow. It occurs for liquid velocities less than 0.305 m/s and gas velocities from about 4.6 m/s.

- **Slug flow**

This pattern occurs when waves are picked up periodically by the more rapidly moving gas. These form frothy slugs that move along the pipeline at a much higher velocity than the average liquid velocity. This type of flow causes severe and in most cases dangerous vibration in equipment because of the impact of the high-velocity slugs against fittings.

- **Annular flow**

In annular flow, liquid forms around the inside wall of the pipe and gas flows at a high velocity through the central core. It occurs for gas velocities >6.1 m/s (20 ft/s).

- **Dispersed spray or mist flow**

Here, all of the liquid is entrained as fine droplets by the gas phase. Dispersed flow occurs for gas velocities >61 m/s (200 ft/s).

3.6.5 **Velocity limitations**

Depending on the flow regime, the liquid in a two-phase flow system can be accelerated to velocities approaching or exceeding the vapor velocity. In some cases, these velocities are higher than what would be desirable for process piping. Such high velocities lead to a phenomenon known as erosion–corrosion, in which the corrosion rate of material is accelerated by an erosive material or force (in this case, the high-velocity liquid).

Corrosional velocity limitations may be determined experimentally. The limitation for corrosional velocity is based on the inhibitor film resistance and experiments. It is normally less than erosional velocity and is basis for design velocity in pipelines.

3.6.6 **Maintain the proper regime**

In addition to keeping the velocity-density product within the acceptable range, one must also maintain the proper flow regime.

Most important, slug flow must be avoided. Slug flow unit loss in process piping is generally not calculated, because it causes various mechanical and process problems.

First, water hammer may occur as the slug of liquid impinges on pipe and equipment walls at every change of flow direction. This could result in equipment damage owing to erosion–corrosion. Second, if slug flow enters a distillation column, the alternating composition and density of the gas and liquid slugs cause cycling of composition and pressure gradients along the length of the column. The cycling causes problems with product quality and process control.

Slugs can form in a variety of ways. They may be created as a result of wave formation of the liquid–gas interface in a stratified flow. As the liquid waves grow large enough to bridge the entire pipe diameter, the stratified flow pattern breaks down into slug flow. Slugs can also form as a result of terrain effects such as liquid collecting at a sag in a pipeline and blocking gas flow.

The pressure in this blocked gas rises until it forces the accumulated liquid downstream in the form of a slug. Changes in the inlet flow rate can cause slugs as well. When the flow rate increases, the liquid inventory in the pipeline decreases and the excess liquid forms a slug or a series of slugs. Pigging—the removal of water from the line to minimize corrosion—can cause very large slugs as the line's entire liquid inventory is swept ahead of the pig.

3.6.7 **Slug flow can be avoided in several ways**

- By reducing lines sizes to the minimum permitted by available pressure differentials.
- By designing parallel pipelines that will increase flow capacity without increasing overall friction loss.
- By using valves auxiliary pipe runs to regulate alternative flow rates and avoid slug flows.
- By using a low-point effluent drain or bypass, or other solutions.
- By arranging the pipe configurations to protect against slug flow.

3.6.8 **Design considerations**

The significance of two-phase flow theories and experiments to the process piping designer is threefold. First, it has been shown that if the liquid content of a vapor line increases, friction loss can be many times its original single-phase pressure loss. Second, it has been shown that for a given set of vapor liquid properties and physical properties, a characteristics flow pattern develops. Third, among the various flow patterns, unit losses can differ.

- The flow patterns should be checked for normal flow, maximum flow, and turn-down flow.
- Two-phase calculations normally are done by computer packages. They use different correlations for flow patterns and equilibrium of states equations.
- The package and the selected correlations should be confirmed by the company.

3.6.9 Two-phase flow piping design through rational and empirical steps

Limitations, generalizations, and simplifications have been introduced to provide practical methods of design. Assumptions connected with this section are that two-phase flow is isothermal, two-phase flow is turbulent in both liquid and vapor phases, the flow is steady (liquid and vapor move with the same velocity), and the pressure loss is not >10% of the absolute downstream pressure.

In long lines, vapor moves faster than liquid. Thus, there are varying densities along the pipe length. In vertical lines, the static head back pressure will not be the same as that calculated with average densities.

Process piping systems have flexibility in the distribution of pressure loss, and control valves can operate within a wide range of available pressure differentials.

Designer can capitalize on the characteristics to obtain the optimum piping and component size and layout.

In general, the criterion for selecting a suitable line size is that the pipe diameter must be sufficiently small to have the highest possible velocity, but large enough to stay within available pressure differentials and allowable pressure drops.

Normally, slug flow is undesirable in two-phase flow pipelines. Because flow velocity is one of the factors that influences the flow regime, due consideration should be given to this phenomenon at the design stage.

Pipe size over design must be watched for, in particular. Low velocities tend to increase the chances for slug flow.

Two-phase flow pipelines are unstable. When changes are made to pressures or to flow rates, the pipeline readjusts itself gradually and equilibrium may not be reestablished for several days. A two-phase pipeline is both a pipeline and an extremely long storage tank; changes in flow conditions cause the liquid to go into or come out of storage.

Liquid in two-phase pipes passing over hills tends to run back downhill and accumulate in valleys. Increasing the velocity in the uphill portion of the line reduces the liquid holdup in the valleys and lowers the pressure drops.

3.7 General aspects in design of piping systems in oil, gas, and petrochemical plants

This section covers minimum requirement(s) for general aspects to be considered when designing piping for oil, gas, and petrochemical plants in accordance with *American Society of Mechanical Engineers* (ASME) B 31.3, which includes but is not limited to the following:

1. Loading and unloading terminals
2. Crude oil and gas gathering central facilities

3. Process units
4. Package equipment, in accordance with ASME B 31.3
5. Pump house and compressor stations (booster stations)
6. Tank farms and oil/gas depots.

3.7.1 Design procedure

The design of piping is characterized by two successive phases:

- **Basic design**

The following documents are minimum requirements for piping design in this stage:

- Plot plan and/or equipment layout
- Piping and instruments diagrams
- Piping specifications relating to individual project
- Line identification list

- **Detail engineering design**

Layout for erection and detailed piping drawings for construction should be produced during this stage.
   The detail design of piping should include, but is not limited to the following:

1. Final (detailed) piping and instrument diagram (P&ID)
2. General plot plan
3. Unit plot plan or equipment layout
4. Above-ground piping layout
5. Underground piping and foundation layout
6. Piping plans (erection drawings)
7. Isometric drawings
8. Line identification list
9. Material Take-off list
10. Piping material specification
11. Pipe support schedule
12. Stress analysis calculation
13. Design model (optional)
14. Pressure testing P&ID
15. Tie-in diagram

- **Piping and instrumentation diagram**

The following items should be considered and shown in the P&ID:

1. Data and information about equipment
2. Line identification
3. Nozzle's position and size, for vessels and towers
4. Types of valve
5. Vents, drains, and relief systems for lines and equipment
6. Insulation and tracing on lines

7. Pipe class (wall thickness and material)
8. Control systems and loops (instrumentation)

The utility flow diagram is a type of P&ID that represents the utility systems within a plant and shows all equipment and piping with respect of utilities (water, air, steam, etc.).

- **General plot plan**

The spacing and arrangement of the unit should be designed in accordance with the requirements of available standards. The general plot plan should give the layout of the whole plant(s). It could be prepared to one of the following scales: 1:500, 1:1000, or 1:2000.
The following items should be shown in the plot plan:

1. Battery limits of the complex (area boundary)
2. Geographic and conventional or plant north
3. Elevation with regard to the nominal plant 0 elevation
4. Coordinates of main roads, process units, utility units, buildings, storage tanks, and main pipe rack
5. Location of flares and burn pit
6. Direction of the prevailing wind

The arrangement of unit areas, storage areas, buildings, and devices for shipment to be provided within the plant should be decided on the basis of the following factors:

1. Soil characteristics
2. Main road or rail access ways
3. Location of pipelines to and from the plant
4. Direction of the prevailing wind
5. Local laws and regulations that may affect the location of units and storage facilities
6. Natural elevation for location of units and equipment (storage tanks, waste water unit, oil/water separator, etc.)

The units should be separated by roads. Major roads should have a minimum width of 6 m and a maximum length of 400 m. Minor roads should have a minimum width of 4 m. (Minor roads should not be in an area classified as zone 0 or 1).
A plant may contain one or several process units. Where any unit processes flammable fluids and may be operated independently (i.e., one unit may be shut down with others still in operation), the minimum spacing between equipment on the two adjacent units should be at least 20 m.
For units processing flammable fluids, the central control building should be adjacent to a road. It should not be located in any area classified as zone 0 to 1.

- **Security fence**

1. All sites (plants or complex) should be within a security fence.
2. Any public building, such as an administration office, restaurant, or clinic, should be located outside the process area boundary.
3. Except for case (d), the minimum space between the security fence and the units' boundary should be 20 m, and between the security fence and equipment, 30 m.

4. In case of special units such as flammable material storage with vapor release and toxic materials, the minimum space should be at least 60 m from site boundaries adjacent to centers of population (domestic, work, or leisure).

Except where they are an integral part of a process unit, site utility units should be grouped together in an area classified as non-hazardous.

Fire water pumps and equipment should be sufficiently remote from the processing, storage, and loading area, where a major fire could occur.

Waste water treatment facilities should be located at the lowest points of the plant.

Loading/unloading areas for road transport should have adequate space to provide access for filling, parking, and maneuvering. A drive-through rack arrangement is preferred. Loading and unloading facilities should be downwind or crosswind from process units and sources of ignition, based on the direction of the prevailing wind.

- **Flare access requirements**

1. Access ways within the plant should be provided for maintenance and emergencies, and for firefighting from the road around the plant. The piping system should be laid out in such a way as to make the passage of mobile equipment possible.
2. Minimum widths of the access way should be as follows:
   a. Vehicular access ways within units: 4.0 m
   b. Pedestrian access ways and elevated walkway: 1.2 m
   c. Stairways and platforms: 0.8 m
   d. Footpaths in tankage areas: 0.6 m
   e. Maintenance access around equipment: 1 m
3. Minimum headroom clearance for access ways should be as follows:
   a. Over railways or main road: 6.8 m
   b. Over access roads for heavy trucks: 6 m
   c. For passage of trucks: 4 m
   d. For passage of personnel: 2.1 m
   e. Over forklift truck access: 2.7 m

3.7.2 Unit plot plan

The unit plot plans should be designed based on the general plot plan. The drawing should show the following items:

1. Conventional north
2. Coordinates of battery limits and roads
3. Symbols for equipment and coordinates of their center lines
4. Finished floor elevation
5. Equipment index list

Normally, the area of any unit should not exceed 20,000 m^2, and the length of each side should not exceed 200 m.

The piping layout should minimize piping runs on very high pressure and corrosive/toxic services such as acidic gases, and should consider economy, accessibility for operation, maintenance, construction, and safety.

3.7.3 Layout of equipment

1. Compressors
   a. Generally, compressors should be installed outdoors. In case a shelter is required, the ventilation of the room should be taken into consideration.
   b. Insofar as it is practical, all compressors should be positioned under one shelter. This arrangement makes work easier for operators and maintenance crews; in addition, one crane may serve all compressors if its deployment becomes necessary.
   c. Minimum spacing between gas compressor and open flames should be 30 m.
2. Pumps
   a. Pumps should generally be located in an open area, at or near grade level. Adequately ventilated shelters should be provided for large machines requiring in situ maintenance. The pumps should also be located under the pipe rack.
   b. All pumps should be accessible for operation and maintenance. Adequate space for lifting and handling facilities for maintenance should be provided.
   c. Pumps should be located and specified so that an acceptable NPSH can be obtained without undue elevation of suction vessels or columns. Pumps on flammable or toxic duties should not be located in pits to meet this requirement.
   d. In flammable fluid service, the horizontal distance between the related pump and adjacent heat source of 650 °C or more should be 30 m.
3. Fired heaters
   a. A heater or group of heaters should be located on the periphery of a plot and immediately adjacent to an unrestricted road. There should be adequate access for firefighting from all sides of a heater.
   b. The layout and design of heaters normally should be such that tube removal can be effected with mobile lifting equipment, for which there should be proper access.
4. Air cooled heat exchangers (fin fan)
   a. The location of air-cooled heat exchangers should be specifically considered with respect to any areas of special fire risk. Such consideration should include:
      – The effect of the exchanger on air movement and increased fire spread
      – The possibility of failure of the exchanger tubes releasing more combustible fluid to the fire
   b. Air-cooled heat exchangers may be located above pipe racks, where practicable and economical.
   c. The air cooler should not be located within 7.5 m horizontally from pumps on hydrocarbon service, and where practicable, they should be at least 20 m horizontally from fired heaters, to minimize the possibility of circulation of hot air.
5. Shell and tube heat exchangers
   a. The heat exchanger should be located so that when their tube bundles are withdrawn, they do not project into an emergency escape route or any road with unrestricted vehicle access. They should be arranged so that they can be readily dismantled for cleaning and maintenance.
   b. Heat exchangers should be located collectively, at one point if possible, and their tube bundle pulling area should be provided.

6. Cooling towers

The direction of the prevailing wind should be considered in selecting the location of cooling towers. The towers should be located to minimize nuisance, both within and outside the site, from the water blowout, evaporation, drift, and ice formation. The requirements of BS-4485:1988 should be met.

7. Air intakes and discharges
   a. Air intakes, including intakes to heating and ventilating systems; to air compressors for process, instrument, plant and breathing air; and to gas turbines should be located as far as is practicable away from areas where air contamination by dust or by flammable or toxic material can occur. They should not be located in any area classified as zone 0, 1 or 2 (except for gas turbine air intakes, which should be in accordance with the manufacturer's requirements), or located above or below an area classified as zone 0 or 1.

Note: Intakes and discharges should be separated to prevent cross-contamination by recirculation, taking into account natural wind effects. The distance between intakes and discharges should be not less than 6 m.

8. Storage tanks (liquids)
   a. Storage tanks in a tank farm should be laid out in a separate area (unit) and should be completely surrounded by a bund or dyke, as specified in NFPA 30:2003 (for minimum tank spacing) and inside pipe marketing code, part 4, section 7 (for a bund).
   b. For tanks with a diameter <48 m, individual bounded compounds are not required, but for each crude oil tank with a diameter of ≥48 m, a separate bounded compound should be provided. In no case should the number of tanks in any bounded compound exceed six; nor should the total capacity should exceed 60,000 m^3. Intermediate walls of lesser height than the main bunds may be provided to divide tankage into groups of a convenient size, to contain small spillage and act as firebreaks.
   c. Tanks should be laid out to provide access for firefighting. There should be no more than two rows of tanks between adjacent access roads.
   d. Pumps associated with tankage operation should not be located inside a bounded tank compound.
9. Pressurized liquefied petroleum gas (LPG) storage
   a. Any site boundary to third-party property should have such a distance that radiation at ground level, in the event of ignition of leakage from a single relief valve and/or from a fire in a spill-contaminated area, does not exceed 4.7 kW/m^2.

Ground-level radiation should be calculated using the method in API Recommended Practice 521.

10. Sour NGL storage

In sour NGL storage tanks, in addition to heat radiation mentioned earlier, safe distance with regard to an H_2S-contaminated area should be considered.

3.8 Isometric drawings

* All lines DN 50 (NPS two) and larger in the process and utility areas should have isometric (spool) drawings; utility and instrument piping DN 50 (NPS two) are exempted, unless otherwise specified.

- Isometric drawings should be prepared for the construction of each pipe (prefabrication or site fabrication), as per piping plan drawings.
- Drawings should be designed without using scale and should include a graphic part, dimensional tables, a list of materials, and plant north, design data, insulation, test, and so forth.

3.9 Line identification list

A line identification list (line list) should include, but not be limited to, the following information:

1. Start point and end of line (connected to equipment or other lines)
2. Medium service
3. Phase of flow (liquid, vapor, etc.)
4. Pressure and temperature (design and operating)
5. P&ID and reference drawing
6. Line number
7. Piping specification code (line class)
8. Type of insulation
9. Pipe size
10. Corrosion allowance
11. Heat treatment
12. Branch reinforcement
13. Special information, if required

3.10 Pipe supports

A pipe support schedule should be prepared with the following data:

1. Type of support
2. Reference drawing for fabrication and installation
3. Line number
4. Location of installation (unit, area, and coordinates)
5. Piping plan and civil drawing number

The location and identification of all pipe supports should be shown on the piping plan and isometric drawings.

3.11 Pressure testing diagram

This drawing should be prepared based on the final P&ID with the following considerations:

1. Position of valves (closed or open)
2. Isolation of equipment nozzles and limit of test section with the spectacle blind or similar facilities

3. Installation of vent and drain connections for test
4. Isolation or removal of all instruments
5. Test pressure and test medium
6. Test procedure

3.12 Tie-in diagram

1. In case of modification or expansion of existing plants, tie-in diagrams should be prepared to clarify piping connection points and their tie-ins between the existing plant and its expansion parts.
2. This diagram should be as detailed as a P&ID.
3. The tie-in diagram should show the location points and procedure of the tie-in.

3.13 Above-ground piping systems

This section sets forth minimum engineering requirements for the safe design of above-ground piping within the property limits of plants handling oil, gas, and petrochemical products.

3.13.1 Piping design

• **Design pressure**

Design pressure for the piping system should be determined in accordance with ASME B 31.3, with the following additions:

1. Where the pressure is limited by a relieving device, the design pressure should not be less than the pressure that will exist in the piping systems when the pressure-relieving device starts to relieve, or the set pressure of the pressure-relieving device, whichever is greater.

The maximum differences in pressure between the inside and outside of any piping component or between chambers of a combination unit (e.g., a jacketed pipe) should be considered, including the loss of external or internal pressure.

Piping subject to vacuum should be designed for a negative pressure of 100 kPa (1 bar) unless a vacuum breaker or similar device is provided, in which case a higher design pressure may be approved.

The value of the design pressure to be used should include the static head, where applicable, unless this is taken into account separately.

The design pressure of a piping system subject to internal pressure is defined as one of the following:

1. Design pressure of the equipment to which the piping is connected
2. Set pressure of the relief valve of the piping equipment system (if lower than [a])
3. A pressure not lower than the shutoff pressure or that resulting from the sum of the maximum suction pressure plus 1.2 times the design differential pressure, for discharge lines of pumps and/or compressors not protected by a relief valve

4. Maximum differences in pressure between inside and outside of any piping component or between chambers of a combination unit (e.g., a jacketed pipe) should be considered, including the loss of external or internal pressure

- **Vacuum piping**

The piping subject to vacuum should be designed for a negative pressure of 100 Kpa (15 psi) unless a vacuum breaker or similar device is provided.

- **Design temperature**

The design temperature should be determined in accordance with ASME Code B 31.3, with following additions:

1. The design temperature should include an adequate margin to cover uncertainty in temperature prediction.
2. The design maximum temperature should not be less than the actual metal temperature expected in service, and should be used to determine the appropriate design stress "S" for the selected material.

In case the exterior of components are thermally insulated, the lowest metal temperature should be taken as the minimum temperature of the contents of the pipe.

- **Operating temperature**

The operating temperature of piping should be determined as the temperature corresponding to that of the fluid under normal operating conditions.

In case of steam-traced piping, the operating temperature should be assumed to be equal to one of the following:

1. Temperature equal to 70% of the steam operating temperature if conventional tracing is employed without the use of heat-conductor cement, and when the steam temperature is higher than the operating temperature of the process fluid
2. The steam operating temperature, in case of tracing with the use of heat-conductor cement
3. The steam operating temperature, in case of jacketed piping

- **Piping components**

The selection of type and material of pipe should be in accordance with standards. For sour services, requirements of NACE.MR 01-75 should be considered.

- **Pipe wall thickness**

The required thickness of pipes should be determined in accordance with ASME B 31.3:2004.

The selection of standard wall thicknesses of pipes should be in accordance with ASME B 36.10M: 2004.

- **Pipe bending**

1. The bending radius should be given on the isometric drawing, but in principle, it should not be less than five times the nominal pipe diameter.

2. All bending of stainless-steel and nickel alloy pipe should be done cold. Where the size and schedule of pipe are such that cold bending becomes impracticable, a hot bending and subsequent solution heat treatment procedure should be prepared for the engineer's review and approval.
3. Stress relieving is not normally considered necessary for cold-bended stainless steel and high nickel alloys, but in the case of austenitic stainless steels, it may be necessary to obtain some reduction in the residual stress level after cold bending (e.g., in a case in which the bend is subject to chloride or polythionic acid attack).

Where stress relief is specified by the engineer, the bend should be stress relieved by heating rapidly to a temperature of 900–950 °C and holding for a period of 1 h/25-mm thickness for a minimum period of 1 h, followed by cooling in still air. Heating should be either by local electric heating blankets or by the use of a furnace. In the latter case, the furnace gas should have controlled sulfur content.

4. For ferritic steels, with the exception of the quenched and tempered grades, a normalizing heat treatment should be applied if the cold deformation is more than 15%, or when the hardness increase in Vickers or Brinnell is more than 100.
5. For quenched and tempered ferritic steels, an appropriate stress-relieving heat treatment should be applied, if the cold deformation is more than 15% or when the hardness increase in Vickers or Brinnell is more than 100. Stress relieving should be at least 10 °C below the tempering temperature.

- **Pull bends**

Pull bends with a center line bend radius not less than five times nominal pipe size (NPS) may be made in pipe sizes up to and including DN 40 (NPS 1.5).

- **Miter bends (addition to ASME B 31.3)**

Miter bends should not be used where the pressure exceeds 500 kPag (72.5 psig) or where the stress range reduction factor "f" in the case of thermal or pressure cycling, would be <1.

3.14 Valves

For economy and interchange ability, types of standard valves to be selected should be kept at a minimum.

All valves for steam services in piping designation PN100 (classes 600) and higher should be of the butt weld end type, except for instrument isolation valves and blow-down valves.

3.14.1 Low-temperature services

1. The following types of valve may be used for services down to –20 °C:
   a. Gate, globe, and check valves, with metal seats or with soft insert
   b. Ball valves and floating and trunnion mounted ball with soft and metal seats
   c. Butterfly valves, offset type, soft and metal seated
2. Gate and ball valves may have cavities where cryogenic liquid could be trapped and cause excessive pressure buildup during warmup of the valve body. For such valves, in services of $\leq$ –20 °C, a cavity relief should be provided by drilling a hole of 3–5 mm in the upstream side

of the wedge or ball outside the seat-facing area. These valves should be clearly marked to indicate the cavity relief side.

3.14.2 Plug valves lubricated

These valves should not be used for general purposes and should be used only when the product allows the use of a plug lubricant. Plug valves should not be used as a single item, because they require periodic maintenance by trained staff. Alternatively, in high-pressure gas systems, the use of pressure-balanced (non-lubricated) plug valves may be considered.

3.14.3 Valves of special types

Many special valves have been developed and proven suitable for process requirements and special services. Care should be taken to select the correct valves, with a view to the design, materials, fabrication, and testing.

1. Flush bottom valve

Drain valve on piping or equipment without a dead nozzle end, for viscous or solidifying products

2. Multi-port valves

A multi-port ball valve or plug valve with a sleeve can be selected to divert flows. However, the use of two normal valves is preferred.

3. Iris-type valve with flexible diaphragm sleeve

For pneumatic or gravity feed of solids and powders

4. Steam stop valves and hydrogen valves

Flexible wedge gate valves should be used in main steam and hydrogen lines, DN 150 (NPS 6) and above. The double disc parallel-seat type is an acceptable alternative.

3.14.4 Location of valves

1. Piping layout should ensure that valves are readily accessible to allow operation and maintenance at the site.
2. Valves for the emergency isolation of equipment, and valves that must be frequently operated or adjusted during operation, should be accessible from a grade, platform, or permanent ladder.

3.14.5 Elevated valves

1. Generally, chain-operated valves should not be used. When this is not practical, elevated valves with a centerline elevation >2.2 m should be chain operated.
2. Hand wheels and stems of valves should be kept out of operation aisles. When this is not practical, the elevation of valves should be 2 m from grade to the bottom of the hand wheel.
3. Valves above roads and in an overhead pipe rack should be avoided.

3.14.6 **Control valves**

1. Control valves should be located so as to facilitate their maintenance and manual operation from the operating floor. Except for process reasons, the valves should be located at grade level.
2. The control valve should normally be installed in horizontal piping with the valve stem vertically upward.
3. Sufficient clearance should be provided above the diaphragm and below the bottom of the body for easy maintenance.
4. Control valves should not be bolted up against stop valves on both sides.

3.14.7 **Check valves**

All check valves should preferably be located in a horizontal position. The valves may be mounted in the vertical position with upward flow.

3.14.8 **Safety and pressure relief system**

The location and arrangement of safety and pressure relief valves and relief systems should be in accordance with standards. Consideration should be given to the following:

1. Safety valves should be located as close as possible to the equipment they are to protect, and be accessible for check and maintenance purposes.
2. Discharge of relief/safety valves to the atmosphere should be arranged as follows:
   a. Minimum height from grade: 15 m
   b. Height higher by 3 m with respect to any equipment or service platform that is located within a radius of 15 m (7.5 m for steam)
   c. Minimum distance from open flames: 30 m (such as from furnace burners)
   d. Each pipe discharging to atmosphere should have a 6 mm (0.025-in)-diameter weep hole at the lowest point
3. Safety valve thermal relief should always be installed on lines that may be intercepted at ends, when the line internal fluid, at maximum ambient temperature, may reach conditions higher than the rating of the line itself.

When discharge of a safety valve is connected to a close circuit (blown down), the valve should be higher than the header, to create natural drainage.

4. The pressure drop between the vessel and the safety valve should not exceed 3% of the operating pressure of the vessels.
5. Installation of block valves or spades in any location where they would isolate a vessel or system from its pressure or vacuum relief device should be in accordance with ASME B 31.3:2004.
6. The discharge piping and overpressure protection system should be in accordance with ASME B 31.3: 2004.

3.14.9 **Block and bypass valves**

1. Unless otherwise required by process, block and bypass valves should be provided for control valve installation, as per standards.

Table 3.8 Types of Valves

Type	Size	Service
Ball valves	DN 15 and larger	General service
Gate valves	DN 15 and larger	General service
Butterfly valves, lined	DN 100 and larger	General (water) service, class 150#
Butterfly valves, lined	DN 100 and larger	150# and 300# corrosive service
Butterfly valves, soft or metal seated (high performance)	DN 80 and larger	General service and special applications, e.g., cryogenic, high-temperature
Diaphragm valves, lined	DN 15 to DN 300	150#, corrosive service
Ball valves, lined	DN 15 to DN 150	150# and 300# corrosive service
Plug valves (pressure balanced)	DN 15 and larger	High pressure gas systems (e.g., hydrogen)
Needle valves	DN 15 to DN 40	Accurate control
Globe valves	DN 15 to DN 200	General service
Diaphragm valves, lined	DN 15 to DN 300	Low pressure corrosive service
Butterfly valves, lined	DN 50 and larger	Moderate-pressure corrosive service
Choke valves	DN 50 to DN 200	For high-pressure difference and/or erosive service
Butterfly valves (high-performance)	DN 80 and larger	General service
Check valves	DN 15 to DN 40	Piston type, horizontal flow
	DN 15 to DN 40	Ball type, horizontal flow
	DN 50 and larger	Swing type
	DN 50 and larger	Dual plate type, spring energized

2. Block and bypass assemblies should have means of depressurizing and draining the associated valve and pipe work.
3. A valved drain connection should be provided upstream of each control valve between the control valve and block valve. Table 3.8 provides a summary of types of valves.

3.15 Flanges
3.15.1 Flange types
Flange types should be in accordance with the following considerations:

1. Flanges should normally be of the welding neck type.
2. Slip-on flanges should not be welded directly onto elbows or other fittings, and should be double-welded for all services.
3. PN68 (class 400) flanges should not be used.
4. Where flanges in accordance with other standards such as BS 3293 are required, because of adjacent equipment, a check calculation on the suitability of the flange design for hydrostatic test conditions should be made.

3.15.2 Limitation on flange facings

1. The facings of PN 20 (class 150), PN50 (class 300), PN100 (class 600), and PN 150 (class 900) flanges should be raised except where ring type joints are required, as given in (b).
2. Unless otherwise specified, the facings of flanges in PN250 (class 1500) and PN420 (class2500) should be of the ring joint type. When used on hydrogen service, class 900 flanges should also be of the ring joint type.
3. A steel flange should have a plain (flat) face at a joint with a cast iron or nonferrous flange with a plain (flat) face.
4. For flanges, finishing should be considered.

3.15.3 Blind flanges

1. Blind's size DN 300 (NPS 12) and larger should be supplied with jack screws. All heavy flanges (DN 300 [NPS 12] and larger) should be equipped with facilities for jack bolting.

3.15.4 Limitations on gaskets

The use of asbestos should be avoided. However, if its use is inevitable, the following should be considered:

1. To avoid galvanic corrosion, a graphite-compressed asbestos fiber (CAF) gasket should not be used in austenitic stainless-steel piping on corrosive aqueous duties.
2. A spiral-wound gasket for use with class 900 RF flanges should be provided with both inner and outer guide rings. (For class 600 and lower, an outer ring would suffice.)
3. Unless required with special properties, a CAF gasket should be specified as oil resistant, to be suitable for oil refinery and chemical plant duty.
4. Compressed asbestos fiber flat gaskets should be specified as 1.5 mm thick for flanges up to DN 600 (NPS 24).

3.15.5 Piping joints

Piping joints not covered by standards should be designed in accordance with ASME B 31.3.

• **Threaded joints**

Threaded joints may be used for normal fluid service and when:

1. The fluid handled is nonflammable and nontoxic, nonhazardous, and non-erosive, and the duty is noncyclic.
2. The design pressure does not exceed 1000 kPag (150 psig).
3. The design temperature is between $-29\,°C$ ($-20\,°F$) and $186\,°C$ ($366\,°F$); steam is not included in this category.
4. The connection is provided for a pressure test.

Where seal welding of threaded joints is used, the material should be weldable.

Threaded joints and fittings should not be used for:

1. General chemical service
2. Corrosive fluid
3. Steam service

To reduce the incidence of leakage, the use of threaded joints and unions where permitted should be minimized consistent with the needs of pipe work fabrication. Sufficient threaded joints or unions, where permitted, should be provided to facilitate dismantling of pipe work for all operational, maintenance, and inspection purposes, including requirements for shutdown and gas freeing.

With the exception of connections to instruments and instrument valve manifolds, threaded joints should not be used in stainless-steel, alloy steel, or aluminum piping systems.

No threaded joints or fittings should be used between a pressure vessel or main pipe DN 50 (NPS 2) or above and the first block valve isolating a piping system. This valve should be flanged or may be socket-welded with a flanged joint immediately downstream.

Layout of piping employing threaded joints should minimize stress on joints as far as possible.

- **Socket weld joints**

1. Socket welding connections should be used wherever possible, up to the limiting size of DN 40 (NPS 1.5), except for water service, which may be used up to DN 80 (NPS 3).
2. Where permitted, socket welding joints are preferred to threaded joints, except for nonhazardous service.
3. Socket welding rather than threaded fittings should be used for searching fluid service (e.g., for glycol service).

- **Flanged joints**

The use of flanged joints should be kept to a minimum, particularly on hazardous service. However, sufficient break flanges should be provided to allow removal and replacement of piping where:

1. Process duties may be fouling
2. Deterioration of piping or valves is anticipated in service owing to corrosion, erosion, and so forth

- **Expansion joints**

Guidance on the selection and application of expansion joints is contained in BS 6129, ASME B 31.3 and the Expansion Joint Manufacturers Association.

- **Vibration**

Attention should be given to any expected vibration (e.g., from associated machinery), when specifying the design requirements of the bellows.

- **External shrouds**

External shrouds should be provided for mechanical protection; the shrouds should be designed to minimize the ingress and prevent the retention of water within the convolutions.

- **Internal sleeves**

Internal sleeves are required when:

1. Flow induced vibration may occur
2. It is possible for solid debris to collect in the convolutions. This debris may result from insufficient flushing of lines or equipment after initial construction or from internals coming loose because of corrosion, vibration, or erosion.
3. The duty is fouling or corrosive or contains solids. In this case, purging behind the sleeve should be specified.

When an internal sleeve is provided, the bellows should be installed in the vertical position with the sleeve pointing downward, and the convolutions should be self-draining.

- **Spool pieces**

For bellows to be welded directly into a line, they should be purchased with spool pieces of the same material as that of the line, welded to the ends of the bellows.

- **Lubricants**

The use of molybdenum disulfide lubricants should be avoided on external tie bars, and so forth, if the bellows operate at a high temperature.

- **Support**

The support and guidance of piping systems containing bellows expansion joints should be fully assessed at the design stage, preferably also by the bellows expansion joint manufacturer, to ensure that the bellows expansion joint deflects in the manner in which it was designed, and that the system retains structural stability.

3.16 Instrument piping

Piping specification should apply up to and including the first block valve(s) from the process lines or equipment.

The limit and type of piping for instrument connections should be in accordance with ASME B 31.3.

Pressure points should be as short as possible. Long connections, if unavoidable, and connections to vibrating lines should be properly braced.

3.16.1 Instrument connections
3.16.2 Level and contents (level gages)

1. The connection for external float and displacer chambers should be in accordance with standards. Branch connection should be DN 50 (NPS 2).

2. Level gages should be connected with block valves to the equipment. Level gages in PN 150 (pressure rating, 900) and higher should have a double block valve.
3. A stand pipe should be applied if more than two pairs of level gage connection are required. The minimum diameter of the stand pipe should be DN 80 (NPS 3) and the equipment connection should be DN 50 (NPS 2).
4. If the required level gage is too large for a single gage, a multiple-level gage should be used. The connection of stages should have a minimum overlap of 25 mm.
5. Loads on equipment nozzles from the weight of a long stand pipe with level gages and/or thermal expansion forces should be considered.
6. There are two alternatives for connecting level gages: (the second case is the preferred design):
    a. with block valves between the level gage and the stand pipe
    b. with block valves between the stand pipe and the equipment
7. Connections for internal level instruments and flange-mounted differential pressure instruments should be sized to permit float removal.
8. Drainage from gage glasses, the external float, and displacer chambers should be led via a tundish to a suitable drain at grade.

A closed disposal system should be provided for drainage of vaporizing liquids that evolve H_2S, other toxic gas, or large volumes of flammable vapor.

3.16.3 Temperature measurement

1. A connection for thermowells should be in accordance with standards.
2. Branches for flanged thermowells should be DN 40 (NPS 1.5) minimum and should conform to the process duty line specifications.
3. Thermowells should be located at least 10 times the pipe's inside diameter downstream from the mixing point of two different temperature streams.

3.16.4 Pressure measurement

1. Connections for pressure instruments should be in accordance with standards
2. Isolation valves should be full bore when rodding is specified.
3. Instrument connections should be orientated horizontally in vertical lines and on top of the horizontal lines. Tappings at the side or 45° from horizontal are permitted only where necessary to prevent fouling from adjacent pipe work, structure, or equipment. The selected orientation should permit remote mounted instruments to be self-venting or self-draining. Connections for instruments on immiscible fluids should be horizontal.

3.16.5 Flow measurement

1. Orifice flange connections should be in accordance with standards.
2. For in-line flow measurement devices, the manufacturer's installation requirements should be followed.

3. Piping and primary valving at the orifice fitting or other flow measurement device should conform to the line specification.
4. Line tap connections should conform to the requirements for pressure tapping.
5. Orifice flanges should be installed in horizontal pipes, as far as possible. When it is impossible to find a sufficient meter run in a horizontal piping, orifices may be installed in a vertical piping with upward flow for liquids and downward flow for gas and steam.
6. The minimum meter run required for both upstream and downstream of the orifice flanges should be in accordance with standards.

3.16.6 Analyzers

1. Process line connections for analyzer sampling systems should be:
   a. A flanged connection to the line specification, to accommodate a probe
2. Fast loops up to and including the main sample filter or pressure-reducing valve should be to line specification.
3. Sample lines from the process connection or off-take from the fast loop should be under instrument impulse lines.
4. For an in-line analyzer, the manufacturer's installation requirements should be followed.

3.17 Sample systems
3.17.1 Manual sampling systems

1. The system should be designed to minimize the possibility of loss of containment, taking into particular account the process conditions and hazardous nature of the material to be sampled.
2. A representative sample should be obtained by placing the sample connection at a suitable point on the process line.
3. The effect on area classification should be considered in positioning sample points.
4. Sample piping should be as short as possible.
5. All operational parts of sampling systems should be easily accessible, including the primary isolation valve.
6. Sample points should either be approximately 1 m above ground or 1 m above an accessible level.

3.17.2 Sample connections

1. A liquid sample connection should not be located at the dead ends of piping.
2. Gas sample connections should be located at the top of a horizontally run piping, or at the side of vertical piping.
3. A minimum-size DN 20 (NPS 0.75) sample connection should be provided for nonhazardous duties on feed and product lines, and elsewhere specified by process requirement in P&ID diagrams.

4. Sample points with two valves should have one valve at the takeoff point of the process line with the same size as the standard drain valve, and another at the sampling point with a size maximum of DN 15 (NPS 0.5).
5. Unless otherwise specified, sample coolers should be provided for sample connections when the fluid is above 70 °C (160 °F). Requirements for the sample cooler should conform to ASME B 31.3.

3.18 Vents and drains

All piping systems should have adequate numbers of vent and drain connections to ensure effective venting and draining of the system.

The sizes of vent and drain valves should be selected as a function of the characteristics of the liquid to be drained, but they should not be smaller than those cited in the following:

1. For lines DN 40 (NPS 1.5) and smaller: DN 15 (NPS 0.5)
2. For lines DN 50 (NPS 2) and larger: DN 20 (NPS 0.75) (for nonhazardous service)
3. Small connections from hazardous service lines (e.g., nipples for vents, drains, pressure tappings, sample points, excluding connections from orifice flanges, carriers, and other instrumentation) should be DN 25 (NPS 1) minimum for strength

If vent and drain connections are not provided on the equipment, blanked branches should be provided on the attached piping, provided that effective venting and draining of the equipment through piping system are practicable. Minimum sizes of vent and drain connections are given in Table 3.9.

Operational vents, sample points, and drains that may discharge flammable fluids should be minimized and should be not <15 m (50 ft) minimum from possible sources of ignition (e.g., furnaces, hot lines, rotating machinery).

Drainage lines for hazardous or valuable process chemicals should be accessible for inspection and maintenance.

Drainage from such equipment should be permanently piped to a sump tank provided with means for emptying.

For other services, drainage from sample points, gage glasses, and level controllers should be led via tundishes into an adjacent drain. Sample points and, when necessary, drains should be protected against static-induced ignition of flammable materials.

Table 3.9 Minimum Sizes of Vent and Drain Connections

Equipment Capacity		Vent		Drain	
m^3	ft^3	DN	NPS	DN	NPS
≥1.5	≥50	25	1	25	1
≥6	≥200	25	1	40	1.5
>6–17	>200–600	25	1	50	2
>17–71	>600–2500	40	1.5	80	3
>71	>2500	50	2	80	3

When a significant release of H$_2$S, other toxic gas, or large volumes of flammable vapor could occur, vents and drains should be piped to a closed disposal system.

Separate arrangements should be made for drainage and the tracing of lines containing materials likely to solidify. Provision should be made for rodding and high-pressure water cleaning.

Vents and drains for pressure test purposes should be sized as follows:

- Vents and drains for custody metering systems should include double isolation with one lockable valve, and drip-type sight glass for leakage detection.
- In case vent connection is provided for pressure test only, the valve should be replaced with a plug.
- On process piping (conveying hazardous fluids), vents and drains should be designed as follows:
  - For piping DN 40 (NPS 1.5) and smaller: valve and cap (plug).
  - For piping DN 50 (NPS 2) and larger: valve with blind flange.

3.19 Blow-down

A Blow-down header should run with a minimum slope of 1:500 toward the separator, avoiding intermediate piping components restricting flow.

Branching of a discharge pipe to the blow-down header should always be at the top of a pipe with configuration of 45° in the direction of flow.

3.20 Utility piping
3.20.1 Utility services

- In process areas and where necessary for maintenance and cleaning, valved hose connections should be installed for compressed air, low-pressure steam, service water, and inert gas (when required).
- The diameter of service connections should be DN 20 (NPS 0.75) and the hose length should be 15 m.
- All branch piping from utility headers should be taken from the top of the header to prevent plugging.
- Hose connections should be located with respect to the operator: air on the left, steam at the middle, and water on the right.
- Emergency showers and eyewash fountains should be provided for process plants handling dangerous chemicals.

3.20.2 Cooling water

1. Cooling water lines should have block valves at the unit limit.
2. Restriction orifices should be installed in open cooling water outlet lines to maintain a slight overpressure, to avoid vapor locks at the channel side of coolers and condensers in elevated positions.

3. In freezing areas, a closed cooling water system should have a bypass with a globe valve upstream of the supply and downstream of the return block valve for each unit, for winterization purposes.
4. A cooling water system with a cooling tower should have block valves at the inlet to each cooling tower cell.

3.20.3 Cooling water pipe work to and from heat exchangers

1. Where seawater or aggressive water is used as a cooling medium, the water velocity in the exchanger tubes should be controlled, given the design limits to prevent scaling, erosion, corrosion, or the formation of deposits. A control or butterfly valve should be used in the outlet piping to achieve this purpose.
2. Where further means of controlling water velocity in the tubes is necessary, some means of measuring device should be used. If the orifice plate is used, it should be of type 316 stainless steel, Monel, or Incoloy 825.
3. When the water pipe work is associated with elevated condensers or coolers, irrespective of the type of water used, the control should be positioned on the outlet side of the equipment, to ensure that the equipment and pipe work always run full of water, to avoid vacuum conditions leading to boiling, impingement, and scale formation.

3.20.4 Fuel oils and fuel gas

1. For low-flashpoint fuel, a separate all-welded supply system should be provided with steam connections for purging.
2. Fuel gas pipe work should be provided with a condensate knockout drum with a means of disposing of the condensate to a closed system.
3. If necessary, fuel gas lines should be steam traced and insulated downstream of the condensate knockout drum.
4. Burner piping below fired heaters should not restrict access to or exit from the area, for reasons of safety and for the removal and cleaning of gas and oil burners.
5. Fuel oil piping loops to burners should be fully steam traced, with the steam supply kept separate from the atomizing steam. It is not permissible to insulate the fuel oil and atomizing steam lines in a common jacket in lieu of providing steam tracing. Pressure tappings to fuel oil pressure controllers should also be adequately traced.

3.20.5 Hydrogen service

1. Leakage should be minimized by using welded joints and excluding threaded connections. Vents and drains should be kept to a minimum and their valves should be blanked.
2. Purging connections permanently piped to a supply of nitrogen should be provided on all units on hydrogen service and on separate pieces of equipment that may have to be isolated during operation of the unit. Such connections should incorporate double-block valves with a valved bleed between.
3. Flange facing in hydrogen service should have a smooth finish to 100–150 AARH (see MSS-SP-6).

3.20.6 **Instrument air**

1. The main instrument air supply should be an independent, self-contained system.
2. Headers or manifolds should be fitted with isolation valves and removable caps, plugs, or flanges to allow for blow-down.
3. Main header isolating valves should be socket weld or flanged gate types. Isolating valves local to each individual instrument off-take may be threaded and of brass construction.
4. All low points on main and subheaders should be provided with an accessible drain valve DN 25 (NPS 1) or header size, whichever is less. Individual supply lines do not normally require drain valves.
5. Distribution off-takes should be taken from the top of horizontal headers.

3.20.7 **Process air**

1. Process air should be supplied by permanently installed compressors, which may also supply service air.
2. Process air header should be taken from upstream of an air dryer with separate piping from the instrument air system.
3. Service air may be supplied by means of portable compressors when a permanent system is specified in standards.
4. A breathing air system used to prevent vacuum in a nonpressurized vessel should be of a high-integrity type.

3.20.8 **Steam system**

1. The main steam distribution header should have a block valve at the main header off-take. The lineup should allow for spading.
2. The main steam distribution header entering process units should have a double-block and bleed valve at an easily accessible location at the unit battery limit.
3. Steam distribution header(s) should have a block valve with a spectacle spade at the off-take of the main steam distribution header. The off-take should be located at the top of the line. Instruments and recorder connections for flow, pressure, and temperature should be installed downstream of the block valves to the plant or unit.
4. Steam required for smothering, snuffing, tracing, and similar services should be supplied through separate distribution header(s).
5. Block valves should be of parallel slide type on the main steam distribution system serving the refinery or works, together with those at the battery limit of any other area (e.g., tank farm, administration area). Within any process unit where any section can be taken out of service for maintenance with normal operation continuing on the remaining sections, the section isolating valves should also be of the parallel slide type.
6. If it is necessary to discharge large quantities of steam, noise suppressors should be provided.
7. The draining facilities of a steam supply line should not discharge into sewer systems. They should run to a safe location such as collecting condensate pits, contaminated water rundown systems, gravel pits, gullies, and so forth, and should be combined as far as is practical.

Situations jeopardizing personnel and goods should be avoided. In cold areas, icing up of personnel access surfaces should be avoided.

8. Stagnant and reverse-flow conditions should be avoided in steam distribution systems.
9. For steam services, valves DN 150 (NPS 6) and larger with a pressure rating of PN100 (class 600) and higher should have a bypass valve for preheating and pressure balancing. The bypass size should be based on below Table 3.10:
10. Steam lines connected to process lines should be fitted with a block valve. A check valve should be installed upstream of the block valve, with a bleeder in between. The block valve and check valve should be close together and close to the process line.
11. In-line silencers should be fitted with a small drain line at the bottom of the silencer, to prevent the accumulation of condensate.
12. Vent facilities should be installed to permit warming up of the lines before commissioning.
13. All steam supply lines should have drain facilities at the low points and at the end to remove condensate (e.g., during commissioning).

- **Steam connections**

1. Branch connections for steam systems at 4500 kPa (gage) (650 psig) and above should be DN 25 (NPS 1) minimum and should be taken off the top of the main steam distribution system. A block valve between branch and steam mains should be provided.
2. Process steam connections to fired heaters should be provided with a check valve and block valve in series; the check valve should be located between the block valve and the fired heater.
3. Process steam connections to fractionating columns and similar process equipment should be provided with a check valve and a block valve in series, with the block valve located at the column.
4. Utility connection up to DN 50 (NPS 2) should not be connected permanently to the steam header.
5. Steam lines to groups of pumps should have individual block valves for independent shutoff.

Table 3.10 Bypass Size for Steam Systems

Nominal Size of Main Valve, in		Nominal Size of Bypass Valve for Warming Up of Pipe and Pressure Balancing of Lines with Limited Volumes, in		Nominal Size of Bypass Valve for Pressure Balancing, in	
DN	NPS	DN	NPS	DN	NPS
150	6	20	0.75	25	1
200	80	20	0.75	40	1.5
250	10	25	1	40	1.5
300	12	25	1	50	2
350	14	25	1	50	2
400	16	25	1	80	3
450	18	25	1	80	3
500	20	25	1	80	3
600	24	25	1	100	4

- **Steam-out connections**

1. Piping installed for steaming-out fractionating columns and similar process equipment should be provided with two valves (one gate and one check valve). The gate valve may be that required and a drain valve should be installed between them.
2. The steaming-out connection should be independent of the drain from the vessel and should be provided with a spade for positive isolation.

Steam-out connection sizes should be as listed in Table 3.11:

3. Process steam and steam-out connections should be provided with drainage arrangements.
4. Exhaust lines from steam machinery discharging to atmosphere should be fitted with an exhaust head suitably drained.
5. Exhaust steam lines should enter into the top of the exhaust steam collecting header.

- **Steam trapping**

1. Steam traps should not be installed in superheated main steam headers or superheated main steam distribution headers. For saturated steam service, steam traps should be fitted to drain pockets at low points of main steam headers and main steam distribution headers.
2. Sections of steam distribution headers, heating elements, coils, tracers, and so forth, each should have a steam trap.
3. Steam traps should be as near as possible to the condensate outlet of the unit to be drained, unless a cooling leg is required. Traps should be at all low points or at natural drainage points (e.g., in front of risers, expansion loops, changes of direction, valves, and regulators).
4. Steam traps should have a bypass arrangement if the system cannot accommodate replacement and/or repair time without causing a process problem.
5. Steam traps should be easy to maintain and replace. The connecting piping up to and including the first downstream block valve should be designed for the full steam pressure and temperature. Steam traps inside buildings should have a bypass and should not discharge into an open drain inside the building.
6. Open steam trap discharges should be located away from doors, windows, air intakes, ignition sources, stairs, and access ways.
7. All trap pipe work should be designed to provide flexibility to allow for thermal movement between the main, trap, and condensate return main.

Table 3.11 Steam-out Connection Sizes

Vessel Capacity		Connection	
m^3	ft^3	DN	NPS
≤28	≤1000	25	1
>28–57	>1000–2000	40	1.5
>57–1400	>2000–50,000	50	2
>1400	>50,000	80	3

8. Trap size should be based on the maximum quantity to be discharged at the minimum pressure difference between the inlet and outlet.
9. Traps should be fitted with a strainer on the inlet, unless they are an integral part of the trap, and should be of cast or forged steel, according to the duty.
10. Socket weld trap assemblies should be provided with flanges to allow for maintenance. The flanges should be so arranged that the upstream atmospheric blow-down will be effective while the trap assembly is removed for maintenance.
11. No steam trap should be connected to more than one steam line or to more than one section of the same steam line.
12. Open tail pipes should terminate 75 mm (3 in) above ground level and should not discharge onto stanchions or pipe supports, or directly into salt-glazed drains, and so forth, that might be adversely affected by the discharge. They should be directed in such a way as not to present a hazard, and in paved areas should be directed so that the condensate does not run across the paving.
13. Traps discharging to atmosphere should be mounted to be self-draining, to avoid frost damage.
14. Traps operating on different steam pressures may discharge into the same header, provided the condensate line is adequately sized to accommodate the flash steam.
15. Traps should be located adjacent to the equipment they serve, and should be accessible for maintenance and be firmly supported.
16. Multiple traps should be grouped together and installed in enclosures so that the operation of each trap can be checked and prevent frost damage to traps not in use. Tail pipe discharges should be arranged to allow maintenance on any one trap while all others are operating
17. Trapping systems not detailed on the pipe work drawings should be site run, ensuring that steam and condensate lines do not interfere with normal operation and maintenance, and, in particular, with access to valves and other equipment.
18. Drainage from large steam consumers such as heaters, condensers, and reboilers should use level-controlled collection pots.
19. Condensate pots should be sized as follows in Table 3.12:
20. Valves on condensate pots should be DN 25 (NPS 1) in the minimum nominal bore; they may be of the gate, parallel slide, or globe type. For high-pressure, superheated steam, large, or important steam mains, globe or parallel slide types should be used. Globe valves should be used where it may be necessary to control flow.
21. All globe valves should be capable of passing the full rated flow of condensate.

• **Steam tracing**

1. Steam tracing of piping should be installed generally in accordance with available standards.
2. Materials for steam tracers should comply with the appropriate line specification.

Table 3.12 Condensate pot sizing

Main Size	Pot Size
DN 100 (NPS 4) and below	Main size
DN 150 (NPS 6)	DN 100 (NPS 4)
DN 200 (NPS 8) and above	DN 150 (NPS 6)

3. Flattening or crimping of the tracer line should be avoided.

4. Fittings between steam supply pipes or condensate drain pipes and the copper tracer should be carbon steel adaptors socket welded to the carbon steel pipe and brazed to the copper tube. The fittings should be separately insulated from the traced line.

5. Fittings should be used only where necessary to join the longest possible length of tracer, and not only for ease of installation. Essential joints should be located at the pipe flanges. Loops should be provided adjacent to pipe flanges to allow for future use of compression fittings.

6. Piping DN 40 (NPS 1.5) and smaller may be grouped together with a single tracer. Piping DN 50 (NPS 2) and larger should be individually traced.

7. Piping on corrosive services and piping liable to blockage resulting from deposition of solids or the formation of solid polymers should have individual steam tracing irrespective of the pipe size, and should not be grouped together with other pipes.

8. External tracing should consist of a single steam line run at the bottom of the line to be traced, and the pipe and tracer insulated with the standard insulation for the next larger size pipe. Where heat requirements dictate, however, multiple tracers should be provided. Tracers for vertical lines may be coiled around the lines.

9. Expansion loops should be installed where necessary in tracers, and should coincide with flanged joints in traced lines. Loops coinciding with flanges should be such as to allow flanges to be sprung apart, and at spaded flanges, should allow the spades to be swung.

10. Expansion loops should be installed in the horizontal plane and pockets should be avoided.

11. Each steam distribution or supply point should be located above the highest point of the piping system being traced. Each condensate collection header should be located at an elevation low enough to permit gravity flow of condensate from all connected lines.

12. Each individual tracing line should be provided with a block valve located at the steam header or subheader. Valves should be of the steel socket welding type. Valves should be readily accessible from ground or platform level and positioned on the subheader for ease of maintenance. Each tracer or leg of parallel tracers should be provided with its own trap, except that groups of tracers that are self-draining may be drained to a level-controlled condensate pot or a collection header. Tracing on control valves and bypasses should allow control valve removal without interfering with the tracing of the bypass.

13. Tracers should be attached to lines by strapping or binding wires. Heat transfer cement may be used to improve the transmission of heat from the tracer to the traced line.

14. Where degradation of a product or metallurgical deterioration of the pipe may occur owing to local hot spots pipe or where an internal lining may be damaged, direct contact with an external tracing line should be prevented by a suitable insulating strip between the pipe and the tracing line.

15. For pipe work on which moisture may form as a result of low-temperature operating conditions or intermittent service that would allow the pipe work to cool down during idle periods, corrosion owing to a galvanic couple between the pipe and/or support clamps and the tracing line should be avoided by using an approved insulating strip.

16. Steam-heated pumps and other equipment should have their own individual steam supply independent of the line tracing. On pumps, valves, or other equipment requiring removal for maintenance, steam tracing and condensate connections should be flanged.

17. Each tracing circuit should be labeled clearly and permanently immediately upstream of the supply isolating valve and immediately before the steam trap. The label should be in stainless steel and should have the following information:
    a. Steam supply identification code
    b. Line designation
    c. Steam trap designation
18. A steam or condensate line should not be attached to, or supported from, any line other than the one it is tracing.

3.20.9 Jacketed piping

Jacketed piping is classified as either partly jacketed or fully jacketed.

3.20.9.1 General points common to all jacketed piping

1. The design of piping and flanges should consider differential expansion between the inner pipe and jacket during startup, shutdown, normal operation, or any abnormal conditions. The design should ensure no buckling of the inner pipe owing to external pressure or differential expansion. The jacket, connections, and inner pipe should all be of the same material of construction, to avoid problems resulting from thermal stresses or the welding of dissimilar metals.
2. Line/jacket sizes should be as follows in Table 3.13
3. Spacers should be used to ensure that the inner pipe is concentric with the jacket.
4. The radius of pulled bends should not normally be less than five times the jacket nominal inner diameter.
5. Forged elbows may also be used to fabricate bends in certain instances in which the radius of the bends coincides and when the tangent lengths allow assembly.
6. Where the main process line is sectionalized by block valves, the supply and trapping arrangement should be similarly arranged, to facilitate maintenance on isolated sections.

Table 3.13 Line/Jacket Sizes

Line Size		Jacket Size	
DN, mm	NPS, in	DN, mm	NPS, in
20	0.75	40	1.5
25	1	50	2
40	1.5	80	3
50	2	80	3
80	3	100	4
100	4	150	6
150	6	200	8
200	8	250	10
250	10	300	12

7. Transfer of steam from one jacket to another should be arranged as follows:
   A single jump-over connected into the jackets by radial branches should be used on vertical jacketed pipe. These should be positioned as low as possible on the upper jacket and as high as possible in the lower one. Where the jacketed pipe run is not vertical, one of the following alternatives may be used:
   a. Single jump-over from the lowest part at upstream to the highest part of downstream of steam
   b. Single jump-over connections branched into the highest part of the jackets where absence of condensate is ensured
   c. Double jump-over connections: one for steam at the highest point and one for condensate at the lowest point
8. Jump-over connections at main line flanges should include break flanges. Jump-over at other locations should be all welded
9. Main line flange bolt holes positioned on-center to allow bolt access may be necessary at tangentially branched jump-over
10. Heat transfer cement may be used on any unjacketed parts between the tracer and pipe to eliminate cold spots
11. Steam jacket supply valves and traps should be identified by labels as required for steam tracing circuits,

- **Partly jacketed piping**

1. Partial jacketing is suitable for lines carrying materials where there is no risk of blockage at cold spots and may be specified where contamination of the process fluid with water cannot be tolerated and all butt welds on the jacketed line are required to remain uncovered by the jacket.
2. Jackets should be swaged on to the inner pipe adjacent to flanges and at inner line butt welds where these are required to remain uncovered. Alternatively, butt-welded caps bored to suit the outside diameter of the inner line may be used. Where branch welds on the inner line remain uncovered, the main jacket should be locally swaged to the inner line, and the branch separately jacketed.
3. Main line flanges should be standard raised face slip-on flanges to ASME B 16.5 (inch dimensions) sized to suit the inner line.

- **Fully jacketed piping**

1. Full jacketing should be used for duties in which it is essential that cold spots are eliminated, and butt welds on the inner line may be covered by the jacket.
2. The jacket should be welded to the back of the main line flange. Main line flanges should be raised face slip-on weld flanges to ASME B 16.5 (inch dimensions) sized to suit the jacket, but with a bore to suit the inner line.
3. Reducing orifice-type flanges complete with a telltale hole drilled radially through the rim may be used where leak detection at flange welds is necessary.
4. The use of blinds or blanks bored out for this duty is not acceptable.
5. Split-forged tees should be used for the jackets of branch connections.

3.21 Piping adjacent to equipment
3.21.1 Isolation of equipment
In this section, isolation is subdivided into two categories:

1. Positive isolation, in which no leakage can be tolerated (e.g., for safety or contamination reasons)
2. General isolation, in which the requirements are less critical than for (a).

In certain circumstances, it may be necessary for operational reasons or additional security to provide a combination of (a) and (b).

3.21.2 Positive isolation

1. Positive isolation should be accomplished by one of the following:
  a. Removal of a flanged spool piece or valve and fitting of blind flanges to the open-ended pipes
  b. Line blind
  c. A spade in accordance with standards. The arrangements of spading points, together with venting, draining, and purging facilities, should enable a section of line containing a spade to be checked as free from pressure before spade insertion or removal
2. Positive isolation methods should be provided:
  a. To permit isolation of major items of equipment or group of items for testing, gas freeing, safety, and so forth. A group of items containing no block valves in the interconnecting pipe work may be considered one item of equipment
  b. To isolate a section of plant for overhaul
  c. To isolate utility services (e.g., fuel gas, fuel oil, atomizing, snuffing, or purge steam to individual fired heaters)
  d. To prevent contamination of utility supplies (e.g., steam, water, air, and nitrogen) where permanently connected to a process unit
  e. Steam and air connections for regeneration and steam/air decoking should be positively isolated from the steam and air systems, preferably by spool pieces or swing bends

3.21.3 General isolation

1. General isolation should be accomplished by one of the following:
  a. A bidirectional block valve
  b. A unidirectional block valve, where isolation in only one direction is required for all conditions and where internal relief of the valve cavity is required
  c. Double block valves with a bleed valve mounted on the pipe work between them in the following cases and in (d) below:
   - Drain connection in gas services when ice formation or freezing can occur
   - All steam piping entering or leaving process units
   - High-pressure steam supply lines 4137 kPag (600 psig) to turbines over 300 kW (400 HP)
   - All manifolds in the tank age area should have a double block and bleed valve where there is a possibility of product contamination

d. Double valves without a bleed valve mounted on the pipe work between them should be provided for sample connection, drains, and vents:
  - In hazardous fluid service
  - In PN 150 (class 900) piping or higher
  - In an unmanned installation where vibration is anticipated
  - Where freezing resulting from cooling on expansion may occur

The block valves should be far enough apart to allow safe access to the upstream valve with material discharging from the point of emission.

For fluids such as LPG, the valves should be separated by at least 1 m to reduce the risk of simultaneous obstruction of both valves by ice or hydrate formation.

2. Block valves should be provided:
  a. In all lines at recognized separation of process units or limits of operating areas
  b. At vessel branches, excluding:
   - Connections for overhead vapor lines, transfer lines, reboiler lines, and side stream vapor return lines
   - Pump suction and reflux line connections on vessels that have a block valve located within 10 m (30 ft) in a horizontal direction from the vessel branch. This exception should not apply when there is a particular process requirement to isolate a vessel inventory
   - Vents on tanks and vessels open to the atmosphere
   - Relief device connections and/or connections to other vessels or piping systems fitted with relief devices that protect the vessel in question
   - Overflow connections
  c. At suction and discharge of pumps and compressors, but not at suctions of air compressors taking suction from the atmosphere
  d. For isolating equipment (e.g., individual or groups of heat exchangers) requiring servicing during plant operation
  e. For isolation of instruments
  f. Where required to prevent flow at vents and at drains, sample points, and steam-out points, and to divert flow through alternative routes and bypasses
  g. At inlets, outlets, and drains of storage tanks
  h. To isolate utility services (i.e., fuel oil, fuel gas, atomizing, snuffing) and purge steam to individual fired heaters
  i .On supply lines to road and rail car filling stations handling toxic or flammable materials; the block valve should be located remote from the filling point for emergency shutoff.
  j. In all utility services at:
   - Each branch on the refinery or process plant main
   - The inlet at each individual user
   - The outlet of each cooling water user
  k. In steam and condensate systems at each branch on the header in each plant or unit, if the length of pipe between header and user is ≥5 m (16 ft).
  l. At vents placed in piping for operational purposes. These vents should be spaded or blanked in accordance with the line specification. Vents for testing purpose only should not be valved

   m. At drain connections in the low point of piping systems, where the arrangements should be as in (xii) above

   n. At suitable points in ring, main, or distribution systems to allow sectionalizing

   o. At the outlet from each non-condensing steam user, where the user discharges to a pressurized system

3. Process area limit block valves for relief and blow-down piping should be provided with a purpose-built locking device, to lock valves open. Gate valves fitted on this duty should be installed in the horizontal or inverted position so that the valves tend to fail in the open position.

4. Double-block valves with an intermediate valve to vent the space between the block valves should be provided:

   a. Where frequent isolation is required, and temporary, not positive, isolation is acceptable for operational safety (e.g., in segregating products).

   b. For temporary isolation between permanently connected utility supplies and process units where the utility requirement is in frequent use and there is a need to make the utility readily available for injection into the process (e.g., stripping steam where the system has to be made condensate-free right up to the point of injection). (Non-return valves will also be required in all utility services connected directly with process equipment).

   c. Where equipment may be taken out of service while the unit remains in operation (e.g., compressors) or where there is a requirement to isolate equipment, yet hold it readily available for use with the operating unit (e.g., hydrogen storage vessels).

3.21.4 Pump emergency isolation

1. Except for glandless-type pumps, an emergency isolation valve should be provided in the suction line between a vessel and a pump when any of the following apply:

   a. The suction inventory at normal operating level is $\geq 30 \text{ m}^3$ (1060 ft^3) and the pumping temperature is greater than the auto-ignition temperature.

   b. The liquid is toxic.

2. Where a pump is paired or spared, the common line should be fitted with the valve.

3. The valve should be located as close to the vessel as possible outside the vessel supports and not <3 m (10 ft) from the pump.

4. The valve should be remotely operable when it is <15 m (50 ft) horizontally from the pump, unless a firewall is installed between the valve and the pump.

5. For remotely operated valves, the valve control station should be located not <15 m (50 ft) from the pump and, if practicable, within sight of the valve. The control station should not be located in a special fire risk area, and should be accessible through a low–fire risk area.

3.21.5 Provision of strainers and filters

Provision of strainer should be in accordance with standards with following considerations:

1. Permanent strainers or filters should be fitted in the following instances if they are not already an integral part of the equipment:

   a. In a fuel oil supply to burners and for each set of gas pilots

   b. In hydraulic systems for remote control of valves

    c. At the inlet of steam turbines, jet ejectors, and trapping systems

    d. In loading installations, when necessary to maintain product quality

    e. In the suction lines of pumps, where the liquid may contain solids liable to damage pumps

    f. In compressor suction lines

    g. In any lubricating system (i.e., sealing oil, gland oil, and gear coupling lubricators).

    h. In any flushing oil system

2. Permanent strainers in compressor suction lines should be provided with isolation to permit easy removal. Maximum design pressure drops and flow direction should be indicated by a nameplate permanently attached to the strainer.

3. Temporary strainers should be fitted between the suction valve and the equipment, and should ensure that the debris is completely removed from the system when the strainer is cleaned, and be cleaned easily without disturbing the main pipe work. Pipe work should be designed to incorporate the strainer, and no pipe springing should be allowed for retroactive installation. A suitable spacer should be provided for use on removing the strainer.

4. If permanent strainers are fitted, the temporary ones should normally be located between the pump or compressor and its associated suction-side block valves. On main crude oil charge pumps, twin strainers arranged in parallel, and each with its own block valve should be installed upstream of the suction header.

5. Twin parallel filters should be provided on vital process pumps that do not have a standby and for which the shutdown for strainer cleaning could lead to a shutdown of the refinery or works, or of an important unit.

6. Pressure tappings should be provided for measuring pressure drop at all permanent and temporary strainers.

3.21.6 Piping to equipment

- **Pumps**

1. Pump suction piping should cause minimum flow turbulence at the pump nozzle. Suction piping should not have pockets where gas can accumulate. However, if this is unavoidable, venting facilities should be provided.

2. If the suction nozzle of a pump is smaller in size than the connecting piping and a reducer is required in a horizontal line, it should be eccentric, installed with the belly down (top flat). This may require an additional drain.

3. If the discharge line size differs from the pump discharge nozzle, a concentric reducer should be applied.

4. A block valve should be installed in the suction line of each pump upstream of the strainer. The discharge line should also have a block valve. A non-return valve should be installed upstream of the discharge block valve.

5. The discharge valve, suction strainer, and suction valve may be of the same size as the pump nozzles for economic reasons, and also to avoid comparatively heavy attachments, unless the pressure drop is too high

6. For spared pumps that have common suction and discharge lines, a DN 20 (NPS 0.75) bypass with throttling valve should be installed around the discharge non-return valve in the following cases:
  a. If discharge and suction line working temperatures are above 230 °C
  b. If process fluid can solidify at ambient temperature (e.g., water lines in frost areas)
  c. If discharge/suction line working temperature is below −100 °C
  d. If draining of the space upstream of the non-return valve is required

7. When the discharge and suction lines are working at or below ambient temperatures, a 3- to 5-mm hole in the closing member of the non-return valve may be considered instead of a bypass around the non-return valve. Valves with such a hole in the closing member require marking on the valve body and on the process engineering flow schemes and isometric drawings.

8. Permanent strainers should be installed in all pump suction lines. Y- or T-type strainers should be used for permanent installation in vertical suction lines; in services with a high content of impurities, Y-type is preferred. In horizontal suction lines, Y-type or bucket-type strainers may be used. For suction lines ≥DN 450 (NPS 18), bucket-type strainers should be used. Installation of the Y-type strainer for double suction pumps should not disturb even flow to the two suction nozzles of the pump. In a vertical suction line, the Y-type strainer should be installed pointing away from the pump. In a horizontal suction line, the Y-type strainer should be installed pointing downward.

9. Warming-through connections should be supplied for pumps as per standards.

10. For a multistage pump fitted with a pressure relief device for pump casing protection, the suction-side design pressure need not be greater than (but should not be less) the relief set pressure.

11. Pump vents should be connected to the vapor space of the suction vessel for operation under vacuum. This allows filling of the pump before startup. The vent line should have two valves: one at the pump and one at the vessel. Pump vent and drain nozzles should be fitted with valves; if not connected to a drain system, the valves should be fitted with plugs. Pumps handling butane or lighter process fluids should have a vent line to the flare system. The vent line should have a spectacle or spade blind.

12. To avoid a fire hazard, lubricating oil, control oil, and seal oil lines should not be routed in the vicinity of hot process and hot utility lines.

13. Cooling water lines to pumps and compressors should not be less than DN 20 (NPS 0.75). Lines DN 25 (NPS 1) or less should have the takeoff connection from the top of the water main line to prevent plugging during operation.

14. Pumps for vacuum service require a sealing liquid on the stuffing boxes and a vent line to the process system to prevent dry-running.

15. Reciprocating, positive displacement, and centrifugal pumps (if required) should be safeguarded against a blocked outlet with a pressure-relief device. This should not be an integrated part of the pump. The relief valve should be installed in a bypass between the discharge line upstream of the check valve and the suction vessel. Alternatively, the relief valve may be installed in a bypass between the discharge line upstream of the check valve and the suction line downstream of the block valve. However, it should be ensured that this will not create overpressure of the suction system.

16. Spools should be provided between the pump suction strainer/pump suction connection and pump discharge nozzle/non-return valve to facilitate easy removal of the pump for maintenance.

- **Compressors**

1. To prevent fatigue failure of the compressor piping, the effect of vibrations and pressure surge should be considered. Piping should have a minimum of overhung weight.
2. Suction line should be designed with special consideration for straight and minimum length. Interstage and discharge piping should be sufficiently flexible to allow expansion owing to the heat of compression.
3. Block valves should be in the suction and discharge lines, except for air compressors, which should have block valves in the discharge lines only. Discharge lines should have a check valve between block valve and discharge nozzle. In the case of reciprocating compressors, a check valve may not be required.
4. In each compressor suction line, a suction strainer should be installed downstream of the block valve of the compressor and as close as possible to the compressor suction nozzle. Screens and filters should be reinforced to prevent failure and subsequent entry into the compressor. Provision should be made to measure the pressure difference across the filter.
5. Reciprocating compressors should be safeguarded against a blocked outlet with a pressure-relieving device installed in a bypass between the discharge line upstream of the block valve and the suction vessel. Interstage sections should also be protected by relief valves.
6. The suction line between a knockout drum and the compressor should be as short as practicable, without pockets, and slope toward the knockout drum.
7. In the single-stage compressor, the pressure rating of the suction valve and piping between this valve and the suction nozzle should be equal to the rating of the discharge line. The pressure rating of the suction piping of a reciprocating compressor should have the same rating as the discharge of that stage, including valves and suction pulsation dampeners.
8. In case of multistage compression, the suction design pressure should be equal to the highest design pressure of the equipment from which it takes suction. If the design pressure turns out to be lower than the maximum shut-in pressure and/or the discharge pressure, the suction piping must be relief protected.
9. The two-design pressure system is not preferred for a less than three-stage station.
10. Suction lines should be connected to the top of the header, except for suction lines at least one pipe size smaller than the header, which may be connected concentrically at the side of the header.
11. Compressors in hydrocarbon or very toxic service should have purge facilities. The possibility of spading should be provided by spectacle blinds, removable spool pieces, or elbows.

- **Steam turbines**

1. The set pressure of the relief valve in the turbine exhaust system should not exceed either the turbine design pressure or the pressure of the exhaust piping. The calculation for the relief valve orifice should be based on the turbine inlet nozzle.
2. Warming-up facilities for the turbine should be provided.

3. Piping should be designed to permit steam to blow up to the inlet and outlet flanges of the turbine before startup.
4. Steam vents should be routed to a safe location and should not be combined with any lubricating oil, seal oil, or process vent.

- **Heat exchanger**

1. The nozzle positions of heat exchangers should allow an optimum piping layout.
2. Sufficient space should be kept between adjacent heat exchanger inlet and outlet (control) valve manifolds, as per standards.
3. Heat exchanger piping should not be supported on the shell and should not hamper the removal of the tube bundle and shell/channel covers. A removable pipe spool may be required.
4. When shell-and-tube exchangers can be blocked in by valves, causing trapped liquid, attention should be paid to:
   a. Preventing exposure of the low-pressure side piping to the maximum pressure of the high-pressure side, irrespective of whether it is caused by internal failure or otherwise
   b. A potential increase in pressure as the result of thermal expansion of the trapped liquid on the cold side or solar radiation
5. The equipment and the connected piping should be protected by thermal expansion relief valves, if pressures and/or pressure differences can increase beyond the design limits.
6. Steam heat exchangers should have a non-return valve in the steam inlet if the normal steam pressure is <110% of the process relief valve set pressure or, without a relief valve, 110% of the process design pressure.
7. Where contamination is critical in heat exchangers, a check valve should be installed in the inlet of low pressure side if the normal pressure of low pressure side is <110% of the design pressure of the high-pressure side.

- **Pressure vessel**

1. Piping to columns should drop or rise immediately after the nozzle and run parallel and close to the column. For ease of support, a number of lines can be routed together, parallel in one plane.
2. If a tall slender vessel (length over diameter (L/D) ≥ 10) is susceptible to aerodynamic oscillations, the piping platforms and ladders of the top of the vessel should be located so that the platform projected area against wind is kept at a minimum.
3. Pressure vessels that are grouped together should have platforms and interconnecting walkways at the same elevation. The number of stairways and ladders to the platforms should be sufficient to meet safety requirements.
4. If not controlled in another way, process steam lines to pressure vessels should have a regulating globe valve direct to the pressure vessel nozzle. A check valve to prevent the product from entering the steam line should be installed close to and upstream of the regulating valve with a valved low point drain between them. A gate valve upstream of the check valve should isolate the line from the main steam header.
5. The steaming-out pressure for columns should be 350 kPag (50.8 psig) for tall columns; a higher pressure may be considered if the design permits.
6. Pressure vessel drain valve should be located outside the skirt.

- **Fired heaters (furnaces)**

1. Burner utilities headers (fuel oil, fuel gas, or atomization steam) should be arranged with a vertical bundle along furnace walls.
2. The set of valves controlling the feeding of smothering steam should be located in a safe location at least 15 m from the furnace.
3. A throttling balance valve should be provided in the inlet to each coil of a set of parallel coils.
4. Outlet lines should be provided with the following:
   a. A check valve should be installed in the outlet from each heater with the check valve nearer to the furnace.
   b. Drain valves should be provided to drain each coil.

- **Piping layout design**

Generally, all process and utility piping should be installed above ground.

Inside plants (process units and utilities areas) piping should be routed on overhead pipe bridges (pipe rack).

Equipment that is a potential source of fire should not be located under a pipe rack.

If firefighting water lines are installed above ground, they should not run along pipe bridges or pipe racks.

Outside plants and in the interconnection areas (manifold, tank farms, flares, etc.), piping should preferably be installed on the ground on concrete sleepers in pipe racks.

Pipe trenches close to process equipment should be avoided. Where it is not practicable to run pipe over a rack, and trenches below paving level are unavoidable, a trench should be divided into sections about 10 m minimum length by a fire break.

- **Slope in piping**

Except for branch and equipment connections, all lines should run in a horizontal direction. Where lines need to be drained completely, the piping should be sloped and provided with drainage points.

The minimum slope of lines should be as follows:

1. Process lines on sleeper 1: 120
2. Process lines on pipe racks 1: 240
3. Service lines 1: 200 to 1: 240
4. Drain lines 1: 100

The slope of lines should be indicated on the P&ID.

- **Pipe racks**

Overhead racks may contain more than one level. For steel pipe racks, the height of levels should have one of the following elevations:

1. Main pipe racks: 4.60, 6.20, and 7.80 m.
2. Individual or secondary pipe rack: 3.80, 5.40, and 7.00 m.

In special cases, for large size pipes or concrete pipe racks, the distance between the various floors may be increased.

Except for special cases, the minimum width of the pipe rack should be 6 m. The width of the pipe rack should be designed to accommodate all pipes involved, plus 20% space for future expansion or modification. Where the pipe rack supports air coolers, the preferred width should be the width of the air coolers.

In multilevel pipe racks, pipes carrying corrosive fluids should be on the lower level, and utility lines should be at the upper floor. Large-size or heavy-weight pipes should be located at the lower level and on the extreme sides.

- **Pipes space**

The space between the axis of two adjacent pipes should be at least equal to the sum of one-half the outer diameter of flanges (with a higher rating) or one-half the outer diameter of each pipe plus 25 mm. This dimension (25 mm) should be increased to 50 mm for insulated pipes.

Sufficient space should be allowed between adjacent lines at points of change in direction to prevent damage of one line by another as a result of expansion or contraction.

At crossings, lines should have a minimum clearance of 25 mm, after allowing for insulation and deflection.

Hot lines in pipe racks should be grouped together and consideration should be given to the expansion loops.

- **Pipe branches**

For gases or vapors, the branches should be taken from the top of the main lines.

Where main trunk lines for steam, water, and other common systems run through a number of process units, take-offs to the users in any single unit should not be taken directly from the main trunk lines. Take-offs should be from a header or headers supplying each unit- and should be provided with a valve at their junction with the main trunk lines.

Where the main trunk lines run in offsite pipe racks, take-offs into each unit should be fitted with a valve at the plot limit of each unit.

Type of branches should be in accordance with piping specification conforming to standards.

Take-offs from a pipe rack to process areas should rise from the rack level, run low, and rise at the battery limit to the elevation of the piping within the process area.

- **Battery limits**

All process and utilities piping at the battery limits of one unit or groups of units linked should be equipped with block valves and a spectacle blind, as specified in standards.

Battery limit valves, spades, and blinds should preferably be located in the vertical riser.

3.22 Piping flexibility

3.22.1 Expansion and contraction

The piping system should be designed for thermal expansion or contraction in accordance with standards, taking the following into consideration as a minimum:

1. Depressurizing temperature
2. Drying-out temperature

3. Minimum/maximum working temperature
4. Defrosting temperature

3.22.2 Flexibility design

Consideration and calculation of stress analysis should be made in accordance with standards with the following requirements:

1. Startup, shutdown, steam-out, where applicable, and upset conditions, including short-term excursions to higher temperatures or pressure, as well as normal operating conditions, should be considered in flexibility analysis.
2. Sufficient flexibility should be provided in the piping to enable spade, line blinds, or bursting discs to be changed.
3. Vessels or tanks at which piping terminates should be considered inflexible for the initial piping analysis.
4. Only where it is impractical to increase flexibility sufficiently, to reduce the stress range or anchor loads to acceptable levels, should the use of bellows or expansion joints be considered.
5. Specific attention should be given in the design and flexibility analysis of piping connecting to machinery to ensure that piping loads transmitted to the machine are within the acceptable limit under all operating conditions. Hanger supports should be used on these lines wherever practicable. Where impractical, low-friction pads (e.g., polytetrafluoroethylene(PTFE)) may be used.
6. A flexibility analysis for pipe work connecting to machinery should include the mismatch to the allowed tolerance between pipe work and the machine to ensure that the calculated loads represent the worst loads that might be generated. A number of calculations may be necessary to determine the conditions of mismatch that will generate the maximum loads. Specific attention should be given to mismatch in the flexibility analysis of smaller lines that are unusually rigid.

3.23 Piping supports

1. Piping should not be supported off other pipes, particularly if either or both pipes are subject to thermal expansion or vibration. Nor should it be supported from vessels or other equipment, except where brackets have been specifically provided. Piping should not be placed in direct contact with concrete, nor should it be supported from removable flooring, flexible flooring, deck plating, or hand rails.
2. Stainless-steel piping should be protected at pipe supports against galvanic and crevice corrosion.
3. The line should be designed to be self-supporting when pressure relief valves are removed. In case of spades, line blinds, bursting discs, and others, temporary support may be used.
4. Pipes in a pipe rack or pipe track (sleeper way) should be grouped according to size, to permit longer spans for the larger pipes.
5. Local stresses in piping resulting from pipe support should be considered.

3.23.1 Design and selection

In general, the location and design of pipe supporting elements may be based on simple calculations and engineering judgment. However, when a more refined analysis is required and a piping analysis, which may include support stiffness, is made, the stresses, moments, and reactions determined should be used to design supporting elements.

Piping should be supported, anchored, or guided to prevent line deflection, vibration, or expansion/contraction, which could result in stresses in excess of those permitted by ASME B 31.3 in the piping or in-line connected to equipment.

In designing supporting elements, the following considerations should apply:

Each support assembly, including the spring supports, should be designed to sustain the hydrostatic test load.

Field welding of pipe supports to piping should be kept to a minimum. Structural provisions to connect pipe supports should be made as much as possible during the civil and mechanical engineering phase (e.g., concrete plinths, inserts, cleats and brackets to steel structure, clips to vessels). Welded guides and other welded support elements in hot dip–galvanized steel structures (e.g., pipe racks/bridges) should be attached to steel members before they are hot dip–galvanized.

Pipes should be supported in groups at a common support elevation of the supporting structure. Inserts should be poured into vertical and horizontal concrete beams, allowing supports and hangers to be bolted. Because civil design is often well ahead of the pipe support design, provision should be made to incorporate these inserts at standard locations.

To prevent galvanic corrosion, carbon steel clamps on pipes of other metallic materials should be separated from the pipe using synthetic rubber, glass fiber paper tape, or other insulating material between the clamp and the pipe.

Individual lines may be suspended by hanger supports only when no other methods of support are practical. Suspension of one line from another should be avoided.

Special attention should be given to locations with potentially high load concentrations, such as valves, strainers, in-line instruments, and equipment. The supports should be suitable for these high loads and should facilitate maintenance exchange of the heavy valves/equipment.

For piping containing gas or vapor, weight calculations need not include the weight of liquid if the designer has taken specific precautions against the entrance of liquid into the piping, and if the piping is not to be subjected to hydrostatic testing at the initial construction or subsequent inspections.

In addition to the weight effects of piping components, consideration should be given in the design of pipe supports to other load effects introduced by service, such as pressure and temperatures, vibration, deflection, shock, wind, earthquake, and displacement strain.

The layout and design of pipe supporting elements should be directed toward preventing the following:

1. Excessive thrusts and moments on connected equipment (such as pumps and turbines)
2. Excessive stresses in the supporting (or straining) elements
3. Resonance with imposed or fluid-induced vibrations
4. Excessive interference with thermal expansion and contraction in a piping system that is otherwise adequately flexible
5. Unintentional disengagement of piping from its supports

6. Excessive heat flow, exposing supporting elements to temperature extremes outside their design limits

Design of the elements for supporting or restraining piping systems or components should be based on all of the concurrently acting loads transmitted into the supporting elements.

In load calculations, where required, consideration should be given to the following:

1. Weights of pipe, valves, fittings, insulating materials, suspended hanger components, and normal fluid contents
2. Weights of hydrostatic test fluid or cleaning fluid if normal operating fluid contents are lighter
3. Additional loading that may occur during erection
4. Intentional use of restraints against normal thermal expansion
5. The effects of anchors and restraints to provide for the intended operation and protection of expansion joints
6. Reaction forces resulting from operation of safety or relief valves
7. Wind, snow, or ice loading on outdoor piping
8. Additional loadings due to seismic forces

3.23.2 Additional requirements for lines connected to equipment

Lines connected to columns and other vertical vessels should have a resting support as close as possible to the column or vessel nozzle, and be guided at regular intervals to safeguard the line against wind load and/or buckling.

The maximum guide distance should be 6 m for lines smaller than DN 200 (NPS 8) and 10 m for lines DN 200 (NPS 8) and larger.

Pipe supports on equipment should be bolted to cleats welded to the equipment. The cleats should be supplied by the equipment manufacturer. The executor should use standard cleats for the connection of pipe supports, ladders, and platforms.

To support piping systems connected to equipment, maximum use should be made of platforms, fire decks, and so forth.

To allow adequate clearance for the removal of covers, heads, channels, bundles, and shells, lines should not be supported on heat exchanger shells and heads.

Onshore reciprocating compressors and integral piping should be supported on a common slab.

Piping connected to rotating equipment should have adjustable supports to facilitate alignment, spading, and equipment exchange. The supports should allow for thermal expansion and vibration and should be modeled in the pipe stress analysis.

To prevent damage to lines and tank connections caused by settlement of the tank, the first pipe support should be located sufficiently far from the tank. The distances in Table 3.14 should be adhered to.

3.23.3 Material selection

Permanent supports and restraints should be of a material suitable for the service conditions. If steel is cold formed to a center line radius less than twice its thickness, it should be annealed or normalized after forming.

Table 3.14 Distances to Prevent Damage to Lines and Tank Connections Caused by Settlement of the Tank

Nominal Pipe Size, DN	Distance between Tank and First Support, m
≤100	5
150	6
200	7
250	8
300	9
350	10
400	10
450	10
≥500	12

Cast, ductile, and malleable iron may be used for rollers, roller bases, anchor bases, and other supporting elements subject chiefly to compressive loading. Cast iron is not recommended if the piping may be subject to impact-type loading resulting from pulsation or vibration. Ductile and malleable iron may be used for pipe and beam clamps, hanger flanges, clips, brackets, and swivel rings.

Steel of an unknown specification may be used for pipe-supporting elements that are not welded directly to pressure-containing piping components. (Compatible intermediate materials of known specification may be welded directly to such components.) Basic allowable stress in tension or compression should not exceed 82 MPa, and the support temperature should be within the range of −29 °C to 343 °C.

Attachments welded or bonded to the piping should be of a material compatible with the piping and service.

Materials commonly used in the design of pipe supporting elements should be selected from Table 3.15.

3.23.4 Anchors and guides

• **Anchors and guides for pipe work**

A supporting element used as an anchor should be designed to maintain an essentially fixed position.

To protect terminal equipment or other (weaker) portions of the system, restraints (such as anchors and guides) should be provided where necessary to control movement or direct expansion into those portions of the system that are designed to absorb them.

The design, arrangement, and location of restraints should ensure that expansion joint movements occur in the directions for which the joint is designed. In addition to the other thermal forces and moments, the effects of friction in other supports of the system should be considered in the design of such anchors and guides.

Where corrugated or slip-type expansion joints, or flexible metal hose assemblies are used, anchors and guides should be provided where necessary to direct the expansion into the joint or hose assembly.

Table 3.15 — Steel Materials Mainly Used for Pipe Supports: Basic Allowable Stresses in kPa (kSi) at Metal Temperature, °C (°F)

Material	Min Temp °C (°F)	Tensile Strength kPa (KSI)	Yield Strength kPa (KSI)	Min Temp to 37.8°C (100°F)	Min Temp to 93.3°C (200°F)	Min Temp to 148.9°C (300°F)	Min Temp to 204.4°C (400°F)	Min Temp to 260°C (500°F)	Min Temp to 315.6°C (600°F)	Min Temp to 371.1°C (700°F)
A-36	-28.89 (-20)	399,900 (58)	248,200 (36)	1227 (17.8)	1165 (16.9)	1165 (16.9)	1165 (16.9)	1165 (16.9)	1165 (16.9)	1165 (16.9)
A-53 Gr.B	-28.89 (-20)	413,700 (60)	206,900 (30)	1379 (20)	1379 (20)	1379 (20)	1379 (20)	1303 (18.9)	1193 (17.3)	1138 (16.5)
A-105	-28.89 (-20)	482,700 (70)	248,200 (36)	1607 (23.3)	1510 (21.9)	1469 (21.3)	1420 (20.6)	1338 (19.4)	1227 (17.8)	1193 (17.3)
A-106 Gr.A	-28.89 (-20)	331,000 (48)	206,900 (30)	1103 (16)	1103 (16)	1103 (16)	1103 (16)	1103 (16)	1020 (17.8)	993 (14.4)
A-106 Gr.B	-28.89 (-20)	413,700 (60)	241,300 (35)	1379 (20)	1379 (20)	1379 (20)	1379 (20)	1303 (18.9)	1193 (17.3)	1138 (16.5)
A-106 Gr.C	-28.89 (-20)	482,700 (70)	275,800 (40)	1607 (23.3)	1607 (23.3)	1607 (23.3)	1579 (22.9)	1489 (21.6)	1358 (19.7)	1324 (19.2)
A-283 Gr.B	-28.89 (-20)	344,800 (50)	186,200 (27)	1055 (15.3)	1007 (14.6)	965 (14)	917 (13.3)	862 (12.5)	814 (11.8)	765 (11.1)
A-307 Gr.B	-28.89 (-20)	413,700 (60)	—	945 (13.7)	945 (13.7)	945 (13.7)	945 (13.7)	945 (13.7)	—	—
A-515 Gr.60	-28.89 (-20)	413,700 (60)	220,600 (32)	1379 (20)	1345 (19.5)	1303 (18.9)	1262 (18.3)	1193 (17.3)	1089 (15.8)	1062 (15.4)
A-516 Gr.60	-28.89 (-20)	413,700 (60)	220,600 (32)	1379 (20)	1345 (19.5)	1303 (18.9)	1262 (18.3)	1193 (17.3)	1089 (15.8)	1062 (15.4)
A-387 Gr.11	-28.89 (-20)	413,700 (60)	241,300 (35)	1379 (20)	1379 (20)	1379 (20)	1358 (19.7)	1303 (18.9)	1262 (18.3)	1214 (17.6)
A-387 Gr.22	-28.89 (-20)	413,700 (60)	206,900 (30)	1379 (20)	1276 (18.5)	1241 (18)	1234 (17.9)	1234 (17.9)	1234 (17.9)	1234 (17.9)
A-387 Gr.5	-28.89 (-20)	413,700 (60)	206,900 (30)	1379 (20)	1218 (18.1)	1200 (17.4)	1186 (17.2)	1179 (17.1)	1158 (16.8)	1124 (16.3)
A-210 TP 304	-28.89 (-20)	517,100 (75)	206,900 (30)	1379 (20)	1379 (20)	1379 (20)	1289 (18.7)	1207 (17.5)	1131 (16.4)	1103 (16)
A-240 TP 347	-28.89 (-20)	517,100 (75)	206,900 (30)	1379 (20)	1379 (20)	1379 (20)	1379 (20)	1372 (19.9)	1331 (19.3)	1282 (18.6)
A-312 TP 304	-28.89 (-20)	517,100 (75)	206,900 (30)	1379 (20)	1379 (20)	1379 (20)	1289 (18.7)	1207 (17.5)	1131 (16.4)	1103 (16.0)

Such anchors should be designed to withstand the force specified by the manufacturer for the design conditions at which the joint or hose assembly is to be used. If this force is otherwise unknown, it should be taken as the sum of the product of the maximum internal area times the design pressure, plus the force required to deflect the joint or hose assembly.

Where expansion joints or flexible metal hose assemblies are subjected to a combination of longitudinal and transverse movements, both movements should be considered in the design and application of the joint or hose assembly.

Flexible metal hose assemblies should be supported in such a manner as to be free from any effects resulting from torsion and undue strain, as recommended by the manufacturer.

3.23.5 Bearing-type supports

Bearing-type supports should permit free movement of the piping, or the piping should be designed to include the imposed load and frictional resistance of these types of supports, and dimensions should provide for the expected movement of the supported piping.

To ensure unrestricted movement of sliding supports, bearing surfaces should be clean.

3.23.6 Structural attachments

External and internal attachments to piping should be designed so that they will not cause undue flattening of the pipe, excessive localized bending stresses, or harmful thermal gradients in the pipe wall. It is important that attachments be designed to minimize stress concentration, particularly in cyclic services.

- **Non-integral attachments**

Non-integral attachments, in which the reaction between the piping and the attachment is by contact, include clamps, slings, cradles, U-bolts, saddles, straps, clevises, and pickup supports. If the weight of a vertical pipe is supported by a clamp, to prevent slippage, it is recommended that the clamp be located below a flange, fitting, or support lugs welded to the pipe.

In addition, riser clamps to support vertical lines should be designed to support the total load on either arm in the event the load shifts owing to pipe and/or hanger movement.

- **Integral attachments**

Integral attachments include ears, shoes, lugs, dummy supports, rings, and skirts that are fabricated so that the attachment is an integral part of the piping component. Integral attachments should be used in conjunction with restraints or braces where multiaxial restraint in a single member is to be maintained.

Consideration should be given to the localized stresses induced into the piping component by the integral attachments.

The design of hanger lugs for attachment to piping for high-temperature service should be such as to provide for differential expansion between the pipe and the attached lug.

To prevent lines subjected to thermal expansion/contraction from moving off their supports, consideration should be given to the actual length of the cradle or pipe shoes to be used.

Pipe stanchions, pipe dummies and trunnions should have welded end plates.

Weld-on support attachments, such as cradles or pipe shoes, pipe stanchions, pipe dummies, trunnions, and lugs, should not be attached to tees, reducers, and elbows. When stress analysis permits, pipe stanchions, pipe dummies, and lugs may be attached to elbows.

Field welding to pipes for pipe-supporting purposes should be limited as far as possible. Field welding for pipe support purposes should not be performed on the following pipe materials:

- Materials requiring post-weld heat treatment
- Lined carbon steel (glass, PTFE, rubber, cement, etc.)
- Nonferrous materials

For pipes requiring post-weld heat treatment, attachments required for supporting purposes should be indicated on the piping isometric drawings, and welding should be executed at the pipe shop before post-weld heat treatment.

All welds of support elements and of supports to piping should be continuous. The fabricated and supplied supports should conform to the Bill of Material for Supports drawings, and standards should be able to withstand the allowable loads.

Welds should be proportioned so that the shear stresses do not exceed either 0.8 times the applicable S values for the pipe material shown in the allowable stress tables, or the allowable stress values determined in accordance with standards.

If materials for attachments should have allowable stress values different from the pipe, the lower allowable stress value of the two should be used.

3.23.7 Supports for insulated pipes and attachments

Insulated lines running in pipe trenches should be supported high enough to ensure the insulation will remain above the highest expected storm water levels.

Clamped cradles or pipe shoes should be used on the following insulated lines:

- Piping lined with glass, rubber, plastics, etc.
- Piping requiring post-weld heat treatment
- Piping requiring approval of the engineer
- Expensive materials such as titanium, Hastelloy, Monel, etc.
- Piping with corrosion-resistant coating (e.g., galvanized piping)

For all other insulated lines, welded cradles or pipe shoes should be used.

3.23.8 Insulated lines for hot service

- Piping DN 40 (NPS 1.5) and smaller should be supported directly from the insulation by the addition of a load-bearing metal sleeve or saddle outside the insulation.
- Piping DN 50 (NPS 2) and larger should be supported on pipe shoes or saddles that allow for the full insulation thickness.
- The lines should be set above the supporting structure by cradles or pipe shoes to provide adequate clearance for painting and insulation. The clearance between the insulation and the supporting structure should be at least 50 mm.

- To maintain common support levels, the back or underside of pipes (or underside of supports) should be on the same plane or level, irrespective of pipe size or insulation thickness.
- Supports for lines with heat tracing should be shaped so that they will not obstruct the tracing or impede dismantling of supports and tracing.
- To limit water ingress into insulation, welded, rather than clamped, cradles/pipe shoes should be used where the latter could pierce through the insulation.
- Cradles and pipe shoes of lines operating at temperatures above 400 °C should be isolated from the supporting structure by incombustible insulating blocks of sufficient load-bearing and insulation capabilities.
- Alternatively, clamped cradles or pipe shoes can be installed around the insulation. At the location of these supports, the insulation should have sufficient load-bearing capabilities.

3.23.9 Uninsulated lines

Uninsulated lines should rest directly on the supporting structure. However, cradles or pipe shoes should be considered if:

- The operating temperature of the line is below ambient, and therefore the line will often have surface condensation.
- The line is a permanently operating transport line, operating at ambient temperatures, without switch-over possibilities.
- The line requires a slope.

Note: This is only for small slope corrections. The height of cradles or pipe shoes measured from underside of pipe should be a maximum of 400 mm.

- The line may operate (even temporarily) at such a low temperature that this may cause embrittlement of the supporting member.
- They are needed to avoid unacceptable pipe corrosion in high-corrosion areas (e.g., owing to coating damage caused by movement and water collection on top of the supporting structure).

Note: The application of cradles or pipe shoes in these situations does not alleviate corrosion problems of the supporting members themselves. In very corrosive environments, saddles may be considered instead of cradles or pipe shoes.

If the span between the supports (e.g., where the economic span is dictated by a majority of bigger lines) is too large for a pipe, the size of that pipe may be increased to meet acceptable deflection and stresses, provided an economic evaluation justifies such an increase versus the cost of intermediate supports. Such a decision is subject to approval.

3.23.10 Spacing (span)

1. Where line self-draining is essential, the deflection of the pipe owing to self-weight at the point of maximum deflection should not exceed the vertical fall over the span resulting from the set slope.

Where supports have vertical adjustment, any bolt projection below the foot plate should not exceed two bolt diameters. The use of metric threads on pipe supporting elements is permitted (addition to ASME B 31.3).

3.23.11 **Fixtures**

Inextensible supports other than anchors and guides.

1. Pipe support clips for carbon steel lines should be made from round bars (i.e., U-bolts) rather than flat strips, to reduce corrosion (addition to ASME B 31.3).
2. Shoes should be provided at supports on insulated lines where the normal operating fluid temperature exceeds 100 °C (212 °F). Shoes should be sufficiently large to prevent disengagement from the supporting structure under abnormal temperature cycle or hydraulic shock conditions (addition to ASME B 31.3).

3.24 **Insulation**

1. Insulation should be calculated as a function of piping operating temperature and ambient conditions. The ambient reference temperature should be calculated as the average of the minimum values occurring in the coldest month of the year for hot insulation and the average of the maximum values occurring in the hottest month of the year for cold insulation.
2. The design for thickness and type of various insulations should be made in accordance with standards.
3. To reduce the possibility of condensation within the bellows, thermal insulation or screening may be used, but the bellows material should not be subjected to a continuously high temperature that would lead to an unacceptable fatigue life.

3.24.1 **Winterization**

Equipment and piping should be winterized if any of the following conditions applies in a stagnant system for the fluid being handled:

1. The lowest ambient temperature is below the pour point or freezing point
2. Undesirable phase separation, deposition of crystals, or hydrate formation will occur at any ambient temperature
3. In gas systems where condensation, hydrate, or ice formation can occur at any ambient temperature (or owing to cooling caused by expansion of the gas)
4. Viscosity at low ambient temperature is so high that an inadequate flow rate is obtained with the pressure available for starting circulation
5. Lines that are normally dry (e.g., flare lines or instrument air lines), but that may carry moisture during an operating upset

3.24.2 **Painting**

1. Painting should be applied to uninsulated piping to protect it from corrosion resulting from environmental agents.
2. Any piping component should be furnished completely painted by the manufacturer, as specified by purchase order.
3. Painting should not be applied to galvanized and stainless-steel pipes.

3.25 Piping connections to existing plant

1. Cut-ins and tie-ins should be designed so that:
   a. Fabrication and field work is minimized consistent with plant operating and shut-down limitation requirements.
   b. Hot work (field welds, etc.) should be minimized in restricted areas.
2. The method and location of cut-ins and tie-ins should be approved by the operating management at the design stage.

3.26 Underground piping systems

This part covers the minimum requirements for the design of underground piping in oil, gas, and petrochemical plants.

The design pressure and temperature of underground piping should be determined in accordance with ASME B 31.3.

3.26.1 Underground piping drawings

The following items should be shown in underground piping drawings:

1. All buried sewers
2. Underground process and water lines
3. Equipments and structures foundations
4. Cables trenches for electricity and instruments
5. Coordinates of roads, buildings, storage tanks, and pipe racks
6. Natural elevation

3.26.2 Underground piping layout design

The minimum clear space between underground piping and sewer lines is 300 mm, except for cooling water lines where heat transfer might occur, which should have 460 mm minimum clearance. The minimum cover should be 460 mm (or greater where frost or loading may govern below the high point of the finished surface). The minimum cover for water lines is 1 m.

Embedded lines should be arranged as horizontally and plainly as possible.

The external surface of embedded piping should be spaced at a minimum distance of 0.3 m from other installations.

Where underground piping crosses each other, about 150–300 mm differential elevation (higher or lower) should be provided at a crossing point during installation.

As a rule, underground piping should not be installed on two levels and on same route.

As a rule, branch-off from the underground piping should be taken out from the upper side of main header; however, water piping can be taken out from the edge side when possible.

Underground piping should be shaped so as not to permit air pockets or drain pockets, as much as possible.

Road crossings should be properly constructed concrete underpasses, allowing ample space for maintenance purposes and 20% allowance for the installation of future lines. These crossings should be specified on the drawings.

Where piping is embedded below the surface of subterranean water, a study of floating force to piping should be performed.

3.26.3 Valve installation

Block valves should be provided near the header of each branch.

All valves should be installed in valve pits according to standard drawings.

3.26.4 Water lines

The cooling lines should be arranged at the equipment side.

Fire hydrant lines should be arranged in the outer pipe rack route and near an access road.

3.26.5 Protection of underground piping

All buried steel piping should be primed, coated, and wrapped in accordance with standards.

Underground water lines outside process unit boundaries should also be cathodically protected.

References

[1] Gas processors and suppliers association (GPSA) engineering data-book. 12th ed. 2004. Tulsa (OK, USA).
[2] Lowe DC, Rezkallah KS. Flow regime identification in microgravity two-phase flows, using void fraction signals. Int J Multiphase Flow 1999;25:433–57.

Further reading

Andreussi P, Bendiksen K. An investigation of void fraction in liquid slugs for horizontal and inclined gas-liquid flow. Int J Multiph Flow 1989;15(2):937–46.

Andritsos N, Hanratty TJ. Influence of interfacial waves in stratified gas-liquid flow. AIChE J 1987;33(3): 444–54.

Angel RR, Welchon JK. Low ratio gas-lift correlation for casing-tubing annuli and large-diameter tubing. Drill Prod Prac API; 1964:100.

Ansari AM. A comprehensive mechanistic model for upward two-phase flow [M.S. thesis]. University of Tulsa; 1988.

API (American Petroleum Institute) API Publication 2564. 3rd ed. December 2001.

API (American Petroleum Institute) RP-521 Guide for pressure-relieving and depressuring systems.

API Recommended Practice 520. Recommended practice for the design and installation of pressure-relieving systems in refineries: Part I-sizing and selection. 6th ed 1993.

API recommended practice 520. Recommended practice for the design and installation of pressure-relieving systems in refineries: Part II–design. 6th ed 1993.

API standard 526. Flanged steel safety–relief valves. 3rd ed 1989.

API technical data book for petroleum refining; 1980. Procedure2B2.1.

ASME (American Society of Mechanical Engineers) B-31.1:2004 Power piping, B-31.3:2004 Process piping, B-36.10M:2004 Welded and seamless wrought steel pipe, B-16.5:2003 Pipe flanges and flanged fittings.

Aziz K, Govier GW, Fogarasi M. Pressure drop in wells producing oil and gas. J Can Pet Technol; July–September 1972:38–48.

Baker A, Nielsen K, Gabb A. Pressure loss, liquid holdup calculations developed. Oil Gas J; March 14, 1988:55–9.

Baker O, et al. Gas-liquid flow in pipelines. II. design manual. AGA-API Project NX; October 1970:28.

Barnea D, Taitel Y, Shoham O. Flow pattern transition for downward inclined two-phase flow horizontal to vertical. Chem Eng Sci 1982;37(5):735–40.

Barnea D, Taitel Y, Shoham O. Flow pattern transition for vertical downward two-phase flow. Chem Eng Sci 1982; 37(5):741–4.

Barnea D. A unified model for predicting flow-pattern transition for the whole range of pipe inclinations. Int J Multiph Flow 1987;13:1–12.

Barua S, Sharma Y, Brosius MG. Two-phase flow model aids flare network design. Oil Gas J; January 27, 1992: 90–4.

Beggs HD. An experimental study of two-phase flow in inclined pipes [Ph.D. dissertation]. The University of Tulsa; 1972.

Bendiksen KH, Malnes D, Moe R, Nuland S. The dynamic two-fluid model OLGA: theory and application. SPE Prod Eng; May 1991.

Browne EJP. Practical aspects of predicting errors in two-phase pressure loss calculations. SPE 5000, presented at the 49th annual SPE fall Meeting. Houston: Texas; 1974.

BSI (British Standard Institution) BS-3274. Tubular heat exchangers for general purposes, BS-3293:1960 Specification for carbon steel pipe flanges (over 24 nominal size) for the petroleum industry, BS-4485:1988 Water cooling towers, BS-6129–1:1981 Selection and application of bellows expansion joints; 1960.

Caetano EF. Upward vertical two-phase flow through an annulus [Ph.D. dissertation]. University of Tulsa; 1986.

Chen NH. An explicit equation for friction factor in pipe. Ind Eng Chem Fund 1979;18(3):296.

Churchill SW. Comprehensive correlating equations for heat, mass and momentum transfer in fully-developed flow in smooth tubes. Ind Eng Chem Fundam 1977;16(1):109–15.

Colebrook CF. Turbulent flow in pipes with particular reference to the transition region between the smooth and rough pipe law. J Inst Civ Eng London; 1939.

Crane Ltd. Flow of fluids through valves, fittings and pipe. Technical Paper No. 410M; 1988.

Crocker S. Piping handbook. McGraw-Hill Publishing Co., Inc; 1945.

Cunliffe RS. Condensate flow in wet gas lines can be predicted. Oil Gas J; October 30, 1978:100–8.

Degance AE. Atherton RW. Horizontal–flow correlations. Chem Eng; July 13, 1970:95.

Dukler AE, et al. Gas-liquid flow in pipelines, I. research results. AGA-API Project NX; May 1969:28.

Duns Jr H, Ros NCJ. Vertical flow of gas and liquid mixtures in wells. In: Proc 6th World Pet Congr; 1963. p. 451.

Eaton BA. The prediction of flow patterns, liquid holdup and pressure losses occurring during continuous two-phase flow in horizontal pipelines [Ph.D. thesis]. The University of Texas; 1966.

Fernandes RC, Semait T, Dukler AE. Hydrodynamic model for gas-liquid slug flow in vertical tubes. AIChE J 1986;29:981–9.

Fisher HG, Forrest HS, Grossel SS, Huff JE, Muller AR, Noronha JA, et al. Emergency relief system design using DIERS technology. DIERS Project Manual (6th edn.). AIChE; 1992.

Flanigan O. Effect of uphill flow on pressure drop in design of two-phase gathering systems. Oil Gas J; March 10, 1958:56.

Fortunati F. Two-phase flow through wellhead chokes. Amsterdam, The Netherlands: SPE 3742, presented at SPE European spring meeting; May 1972.

Gray HE. Vertical flow correlation in gas wells. In: User manual for API 14B, subsurface controlled safety valve sizing computer Program, App. B; June 1974.

Greskovich EJ, Shrier AL. Slug frequency in horizontal gas-liquid slug flow. Ind Eng Chem Proc Dev 1972;11(2): 317–8.

Hagedorn AR, Brown KE. Experimental study of pressure gradients occurring during continuous two-phase flow in small-diameter vertical conduits. J Pet Technol; April 1965:475–84.

Hydraulic Institute Standard. Centrifugal, rotary and reciprocating pumps. 14th ed. January 1982.

Lasater JA. Bubble point pressure correlation. Trans AIME; 1958:379.

Lawson JD, Brill JPA. Statistical evaluation of methods used to predict pressure losses for multiphase flow in vertical oil well tubing. SPE 4267 to be presented at 48th Annual SPE Fall meeting. Las Vegas: Nevada; September 30–October 3, 1973.

Leung J. Size safety relief valves for flashing liquids. Chem Eng Prog; Febraury 1992:70–5.

Leung JC. A generalized correlation for one-component homogeneous equilibrium flashing choked flow. AIChE J; October 1986:1743–6.

Lockhart RW, Martinelli RC. Proposed correlation of data for isothermal two-phase, two component flow in pipes. Chem Eng Prog January 1949;45:39.

Mandhane JM, Gregory GA, Aziz K. Critical evaluation of holdup prediction methods for gas-liquid flow in horizontal pipes. SPE 5140, presented at SPE annual fall meeting. Houston: Texas; October 1974.

Manual of Petroleum Measurement. Chapter 14.3, orifice metering of natural gas and other related hydrocarbon fluids. 2nd ed. September 1985 (AGA Report #3)(GPA 8185–85)(ANSI/API 2530).

McDonald AE, Baker O. Multiphase flow in pipelines. Oil Gas J; June 15, June 22, June 29, and; July 6, 1964.

McLeod WR, Rhodes DF, Day JJ. Radiotracers in gas-liquid transportation problems–a field case. J Pet Technol; August 1971:939–47.

Moody LF. Friction factors for pipe flow. Trans ASME 1944;66:671.

Mukherjee HK. An experimental study of inclined two phase flow [Ph.D. dissertation]. University of Tulsa; 1979.

NACE (National Association of Corrosion Engineers) NACE MR-0175. Standard material requirements sulfide stress cracking resistant metallic; 2002. USA.

Nazario FN, Leung JC. Sizing pressure relief valves in flashing two-phase service: an alternative procedure. J Loss Prev Proc Ind 1992;5:263–9.

Nicklin DJ, Wilkes JO, Davidson JF. Two-phase flow in vertical tubes. Trans Inst Chem Eng 1962;40:61–8.

Oliemans RVA, Potts BF, Trope N. Modelling of annular dispersed two-phase flow in vertical pipes. Int J Multiph Flow 1986;12(5):711–32.

Orkiszewski J. Predicting two-phase pressure drops in vertical pipes. J Pet Technol; June 1967:829–38.

Palmer CM. Evaluation of inclined pipe two-phase liquid holdup correlations using experimental data [M.S. thesis]. The University of Tulsa; 1975.

Pauchon C, Dhulesia H, Lopez D, Fabre JTACITE. a comprehensive mechanistic model for two-phase flow. BHRG conference on multiphase production. Cannes; June 16–19, 1993.

Payne GA. Experimental evaluation of two-phase pressure loss correlations for inclined pipe [M.S. thesis]. The University of Tulsa; 1975.

Ramey Jr HJ. Wellbore heat transmission. J Pet Technol; April 1962:427–40.

Reid Prausnitz. Sherwood. The properties of gases and liquids. McGraw-Hill; 1977.

Simpson LL. Estimate two-phase flow in safety devices. Chem Eng; August 1991:98–102.

Smith JM, Van Ness HC. Introduction to chemical engineering thermodynamics. 3rd ed. McGraw-Hill; 1981.

Standing MB. A general pressure-volume-temperature correlation – for mixtures of California oils and greases. Drill Prod Prac API; 1947:275.

Steady flow in gas pipelines. In: American Gas Association. Chicago: IGT Technical Report 10; 1965.

Steady-state flow computation manual for natural gas transmission lines. New York: American Gas Association; 1964.

Sylvester ND. A mechanistic model for two-phase vertical slug flow in pipes. ASME J Energy Resour Technol 1987;109:206–13.

Taitel Y, Barnea D, Dukler AE. Modeling flow pattern transitions for steady upward gas-liquid flow in vertical tubes. AIChE J May 1980;26(3):345–54.

Taitel Y, Barnea D. A consistent approach for calculating pressure drops in inclined slug flows. Chem Eng Sci 1990;45(5):1199–206.

Taitel Y, Dukler AE. A model for predicting flow regime transitions in horizontal and near horizontal gas-liquid flow. AIChE J January 1976;22(1):47–55.

Taitel Y, Dukler AE. A model for predicting flow regime transitions in horizontal and near horizontal gas-liquid flow. AIChE J 1980;22(1):47–55.

Vazquez AME. Correlations for fluid physical property prediction [M.S. thesis]. Tulsa University.

Vohra IR, Marcano N, Brill JP. Comparison of liquid holdup correlations for gas-liquid flow in horizontal pipes. SPE 4690, presented at the SPE annual fall meeting. Las Vegas: Nevada; October 1973.

Vohra IR, Robinson JR, Brill JP. Evaluation of three new methods for predicting pressure losses in vertical oil well tubing. SPE 4689, presented at the 48th annual SPE fall meeting. Las Vegas: Nevada; 1973.

Weymouth TR. Trans Am Soc Mech Eng; 1912. 34.

Xiao JJ, Shoham O, Brill JPA. Comprehensive mechanistic model for two-phase flow in pipelines. New Orleans: 65th Annual SPE conference; September 23–26, 1990.

Gas–Liquid Separators

This chapter covers minimum requirements for the process design (including criteria for type selection) of gas (vapor)–liquid separators used in production in the oil and/or gas refineries and other gas processing and petrochemical plants. For the purpose of this chapter, separation techniques are defined as those operations that isolate specific immiscible ingredients of a mixture mechanically, i.e. without a chemical reaction or a mass transfer process taking place.

Gas–liquid separators can be generally divided into two main groups, high gas to liquid ratio (e.g., flare knock-out drums, scrubbers), and low gas to liquid ratio (e.g., oil/gas separators, flash tanks) separators. In this chapter, process aspects of three types of the most frequently used gas (vapor)–liquid separators are discussed more or less in detail.

4.1 Gravity settling

Liquid droplets will settle out of a gas phase if the gravitational force acting on the droplet is greater than the drag force of the gas flowing around the droplet (see Figure 4.1). These forces can be described mathematically using the terminal or finite-settling velocity calculation, Eqn (4.1).

$$V_t = \sqrt{\frac{2gM_p(\rho_1 - \rho_g)}{\rho_1\rho_g A_p C_D}} = \sqrt{\frac{4gD_p(\rho_1 - \rho_g)}{3\rho_g C_D}} \tag{4.1}$$

where:

A_p = particle or droplet cross sectional area, m^2
C_D = drag coefficient of particle, dimensionless
g = acceleration due to gravity, 9.81 m/s^2
M_p = mass of droplet or particle, kg
V_t = critical or terminal gas velocity necessary for particles of size D_p to drop or settle out of gas, m/s
ρ_g = gas phase density, kg/m^3
ρ_1 = liquid phase density, droplet or particle, kg/m^3

The drag coefficient has been found to be a function of the shape of the particle and the Reynolds number of the flowing gas. For the purpose of this equation, particle shape is considered to be a solid, rigid sphere. The Reynolds number is defined as:

$$Re = \frac{1000D_p V_t \rho_g}{\mu} \tag{4.2}$$

Natural Gas Processing. http://dx.doi.org/10.1016/B978-0-08-099971-5.00004-0
Copyright © 2014 Elsevier Inc. All rights reserved.

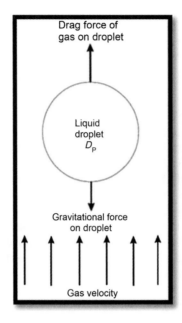

FIGURE 4.1

Forces on liquid droplet in gas stream.

where:

D_p = droplet diameter, m
Re = Reynolds number, dimensionless
V_t = critical or terminal gas velocity necessary for particles of size D_p to drop or settle out of gas, m/s
ρ_g = gas phase density, kg/m^3
μ = viscosity of continuous phase, mPa s

Figure 4.2 shows the relationship between drag coefficient and particle Reynolds number for spherical particles. In this form, a trial and error solution is required since both particle size (D_p) and terminal velocity (V_t) are involved. To avoid trial and error, values of the drag coefficient are presented in Figure 4.3 as a function of the product of drag coefficient (C_D) times the Reynolds number squared; this technique eliminates velocity from the expression.[1] The abscissa of Figure 4.3 is given by:

$$C_D(\mathrm{Re})^2 = \frac{(1.31)(10^7)\rho_g D_p^3 (\rho_1 - \rho_g)}{\mu^2} \tag{4.3}$$

As with other fluid flow phenomena, the gravity settling drag coefficient reaches a limiting value at high Reynolds numbers. As an alternative to using Eqn 7.3 and Figure 4.3, the following approach is commonly used. The curve shown in Figure 4.2 can be simplified into three sections from which

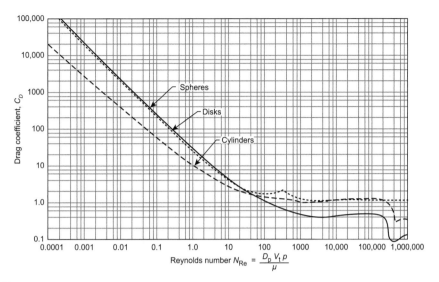

FIGURE 4.2

Drag coefficient and Reynolds number for spherical particles.

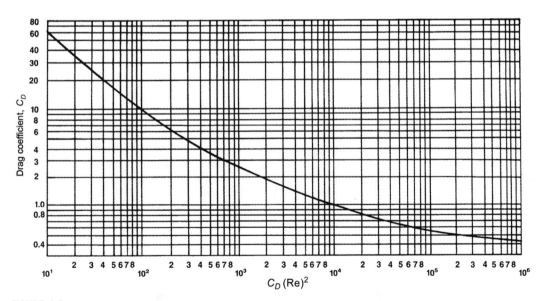

FIGURE 4.3

Drag coefficient of rigid spheres.

curve-fit approximations of the C_D versus Re curve can be derived. When these expressions for C_D versus Re are substituted into Eqn (4.1), three settling laws are obtained as described below [1].

4.1.1 Stokes law

At low Reynolds numbers (less than 2), a linear relationship exists between the drag coefficient and the Reynolds number (corresponding to laminar flow). Stokes law applies in this case and Eqn (4.1) can be expressed as [1]

$$V_t = \frac{1000gD_p^2(\rho_1 - \rho_g)}{18\mu} \tag{4.4}$$

where:

g = acceleration due to gravity, 9.81 m/s^2
V_t = critical or terminal gas velocity necessary for particles of size D_p to drop or settle out of gas, m/s
ρ_g = gas phase density, kg/m^3
ρ_1 = liquid phase density, droplet or particle, kg/m^3
μ = viscosity of continuous phase, mPa s

The droplet diameter corresponding to a Reynolds number of 2 can be found using a value of 0.033 for K_{CR} in Eqn (4.5) [1].

$$D_p = K_{CR} \left[\frac{\mu^2}{g\rho_g(\rho_1 - \rho_g)} \right]^{0.33} \tag{4.5}$$

where:

D_p = droplet diameter, m
g = acceleration due to gravity, 9.81 m/s^2
ρ_g = gas phase density, kg/m^3
ρ_1 = liquid phase density, droplet or particle, kg/m^3
μ = viscosity of continuous phase, mPa s
K_{CR} = proportionality constant from Figure 4.3

By inspection of the particle Reynolds number equation (Eqn (4.2)), it can be seen that Stokes law is typically applicable for small droplet sizes and/or relatively high viscosity liquid phases [1].

4.1.2 Intermediate law

For Reynolds numbers between 2 and 500, the intermediate law applies, and the terminal settling law can be expressed as:

$$V_t = \frac{2.94g^{0.71}D_p^{1.14}(\rho_1 - \rho_g)^{0.71}}{\rho_g^{0.29}\mu^{0.43}} \tag{4.6}$$

where:

D_p = droplet diameter, m
g = acceleration due to gravity, 9.81 m/s^2
V_t = critical or terminal gas velocity necessary for particles of size D_p to drop or settle out of gas, m/s
ρ_g = gas phase density, kg/m^3
ρ_l = liquid phase density, droplet or particle, kg/m^3
μ = viscosity of continuous phase, mPa s

The droplet diameter corresponding to a Reynolds number of 500 can be found using a value of 0.435 for K_{CR} in Eqn (4.5). The intermediate law is usually valid for many of the gas–liquid and liquid–liquid droplet settling applications encountered in the gas business.

4.1.3 Newton's law

Newton's law is applicable for a Reynolds number range of approximately 500–200,000, and finds applicability mainly for separation of large droplets or particles from a gas phase, e.g., flare knockout drum sizing. The limiting drag coefficient is approximately 0.44 at Reynolds numbers above about 500. Substituting $C_D = 0.44$ in Eqn (4.1) produces the Newton's law equation expressed as:

$$V_t = 1.74\sqrt{\frac{gD_p(\rho_l - \rho_g)}{\rho_g}} \tag{4.7}$$

where:

D_p = droplet diameter, m
g = acceleration due to gravity, 9.81 m/s^2
V_t = critical or terminal gas velocity necessary for particles of size D_p to drop or settle out of gas, m/s
ρ_g = gas phase density, kg/m^3
ρ_l = liquid phase density, droplet or particle, kg/m^3

An upper limit to Newton's law is where the droplet size is so large that it requires a terminal velocity of such magnitude that excessive turbulence is created. For the Newton's law region, the upper limit to the Reynolds number is 200,000 and $K_{CR} = 23.64$.

4.2 Gas–liquid separators in oil and gas processing

Gas–liquid separator types often used in oil and gas processes which are discussed in this chapter are:

- Conventional gas/liquid separators.
- Cyclones.
- Oil/gas separators.
- Flare knock-out drums.
- Filter separators.

Figure 4.4 shows a summary of different types of mechanical separators.

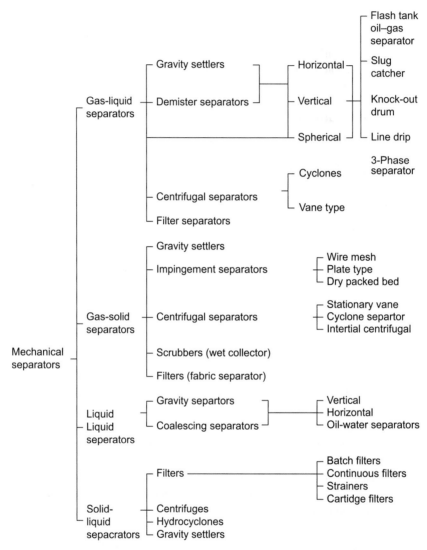

FIGURE 4.4

Types of mechanical separators.

4.2.1 Separation principles

Gas–liquid separation processes most frequently employed in oil, gas, and petrochemical (OGP) industries are based on either one or a combination of gravity settling, impingement, and centrifugation principles. Some types of filtration are seldom employed in this field. The principles of mechanical separation are briefly described in the following sections. Note that as a general rule, mechanical separation occurs only when the phases are immiscible and/or have different densities.

4.2.2 **Mechanical separation by momentum**

Fluid phases with different densities will have different momentum. If a two-phase stream changes direction sharply, greater momentum will not allow the particles of heavier phase to turn as rapidly as the lighter fluid, so separation occurs. Momentum is usually employed for bulk separation of the two phases in a stream.

4.2.3 **Mechanical separation by gravity**

Liquid droplets or solid particles will settle out of a gas phase if the gravitational force acting on the droplet or particle is greater than the drag force of the gas flowing around the droplet or particle. The same phenomenon happens for solid particles in liquid phase and immiscible spheres of a liquid immersed in another liquid. Rising of a light bubble of liquid or gas in a liquid phase also follows the same rules, i.e., results from the action of gravitational force.

4.2.4 **Mechanical separation by filtration**

Filtration is the separation of a fluid-solid mixture involving passage of most of the fluid through a porous barrier that retains most of the solid particulates contained in the mixture.

Filtration processes can be divided into three broad categories: cake filtration, depth filtration, and surface filtration.

4.2.5 **General notes on separator piping**

Piping to and from the separator must interfere as little as possible with the good working of the separator. The following constraints should be observed:

1. There should be no valves, pipe expansions, or contractions within 10 pipe diameters of the inlet nozzle.
2. There should be no bends within 10 pipe diameters of the inlet nozzle except the following:
   a. For knock-out drums and demisters, a bend in the feed pipe is permitted if this is in a vertical plane through the axis of the feed nozzle.
   b. For cyclones a bend in the feed pipe is permitted if this is in a horizontal plane and the curvature is in the same direction as the cyclone vortex.
3. If desired, a pipe reducer may be used in the vapor line leading from the separator, but it should be no nearer to the top of the vessel than twice the outlet pipe diameter.

If these conditions cannot be complied with, some loss of efficiency will result.

If a valve in the feed line near to the separator cannot be avoided, it should preferably be of the gate or ball type fully open in normal operation. High pressure drops that cause flashing and atomization should be avoided in the feed pipe. If a pressure-reducing valve in the feed pipe cannot be avoided, it should be located as far upstream of the vessel as practicable.

4.3 **Conventional gas–liquid separators**

Conventional gas–liquid separators are usually characterized as vertical, horizontal, or spherical. Horizontal separators can be single or double barrel and can be equipped with sumps or boots.

They may be further classified as two-phase (gas–liquid) or three-phase (gas–liquid–liquid). Horizontal separators can be single- or double-barrel and can be equipped with sumps or boots.

Regardless of shape, separation vessels usually contain four major sections plus the necessary controls. These sections are shown for horizontal and vertical vessels in Figure 4.5.

As shown in Figure 4.5, the inlet device (A) is used to reduce the momentum of the inlet flow stream, perform an initial bulk separation of the gas and liquid phases, and enhance gas flow distribution. There are a variety of inlet devices available and these are discussed in more detail in a later section.

FIGURE 4.5

Gas–liquid separators (GPSA 2004).

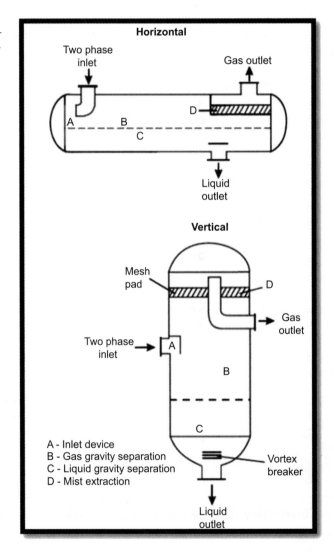

The gas gravity separation section (B) is designed to utilize the force of gravity to separate entrained liquid droplets from the gas phase, preconditioning the gas for final polishing by the mist extractor. It consists of a portion of the vessel through which the gas moves at a relatively low velocity with little turbulence. In some horizontal designs, straightening vanes are used to reduce turbulence. The vanes also act as droplet coalescers, which reduces the horizontal length required for droplet removal from the gas stream [1].

The liquid gravity separation section (C) acts as a receiver for all liquid removed from the gas in the inlet, gas gravity, and mist extraction sections. In two-phase separation applications, the liquid gravity separation section provides residence time for degassing the liquid. In three-phase separation applications, the liquid gravity section also provides residence time to allow for separation of water droplets from a lighter hydrocarbon liquid phase and vice versa. Depending on the inlet flow characteristics, the liquid section should have a certain amount of surge volume, or slug catching capacity, to smooth out the flow passed on to downstream equipment or processes [1].

Efficient degassing may require a horizontal separator while emulsion separation may also require higher temperature, use of electrostatic fields, and/or the addition of a demulsifier. Coalescing packs are sometimes used to promote hydrocarbon liquid—water separation, though they should not be used in applications that are prone to plugging, e.g., wax, sand.

The mist extraction section (D) utilizes a mist extractor that can consist of a knitted wire mesh pad, a series of vanes, or cyclone tubes. This section removes the very small droplets of liquid from the gas by impingement on a surface where they coalesce into larger droplets or liquid films, enabling separation from the gas phase. Quoted liquid carryover from the various types of mist extraction devices are usually in the range of 0.01–0.13 m^3/Mm^3 [1].

4.3.1 Vertical separators

Vertical separators (see Figures 4.6 and 4.7) are selected when the gas–liquid ratio is high. In cases where there is a frequent fluctuation in inlet liquid flow or where revaporization or remixing of fluids in the vessel should be prevented, vertical separators are preferred.

A vertical separators, shown in Figure 4.8, is usually selected when the gas–liquid ratio is high or total gas volumes are low. In a vertical separator, the fluids enter the vessel through an inlet device whose primary objectives are to achieve efficient bulk separation of liquid from the gas and to improve flow distribution of both phases through the separator. Liquid removed by the inlet device is directed to the bottom of the vessel.

The gas moves upward, usually passing through a mist extractor to remove any small entrained liquid droplets, and then the vapor phase flows out of the vessel. Liquid removed by the mist extractor is coalesced into larger droplets that then fall through the gas to the liquid reservoir in the bottom (GSPA 2004).

The ability to handle liquid slugs is typically obtained by increasing vessel height to accommodate additional surge volume. Level control is normally not highly critical, and liquid level can fluctuate several inches without affecting the separation performance or capacity of the vessel. Except for knockout drum applications, mist extractors are normally used to achieve a low liquid content in the separated gas in vessels of reasonable diameter (GSPA 2004).

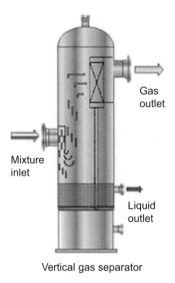

Mixture inlet

Gas outlet

Liquid outlet

Vertical gas separator

FIGURE 4.6

A vertical separator.

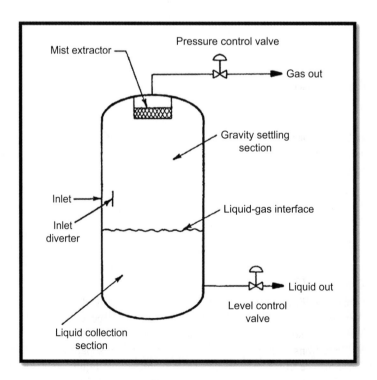

Mist extractor

Pressure control valve

Gas out

Gravity settling section

Inlet

Inlet diverter

Liquid–gas interface

Liquid out

Level control valve

Liquid collection section

FIGURE 4.7

A vertical separator schematic.

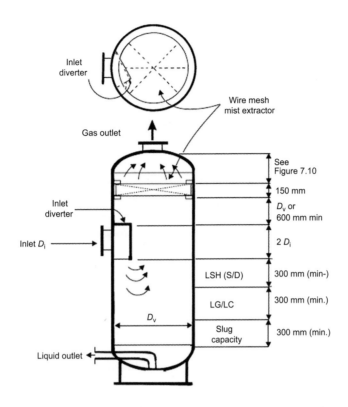

FIGURE 4.8

Vertical separator with wire mesh mist extractor (GPSA 2004).

Typical vertical separator length over diameter (L/D) ratios are normally in the two to four range. As an example of a vertical separator, consider a compressor suction scrubber. In this service, the vertical separator:

- Does not need significant liquid retention volume
- Has a properly designed liquid level control loop that responds quickly to any liquid that enters, thus avoiding tripping an alarm or shutdown
- Occupies a small amount of plot space

4.3.2 Horizontal separators

Horizontal separators (see Figure 4.9) are used where large volumes of total fluids and large amounts of dissolved gas are present with the liquid. They are also preferred where the vapor–liquid ratio is small or where three-phase separation is required.

In cases where limitations exist in vessel size, a double barrel separator can be employed.

Horizontal separators are most efficient when large volumes of liquid are involved. They are also generally preferred for three-phase separation applications. In a horizontal separator, shown in

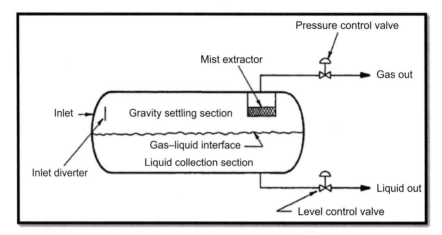

FIGURE 4.9

Horizontal separator schematic.

Figure 4.10, the liquid that has been separated from the gas moves along the bottom of the vessel to the liquid outlet.

The gas and liquid occupy their proportionate shares of the shell cross-section. Increased slug capacity is obtained through shortened retention time and increased liquid level [1].

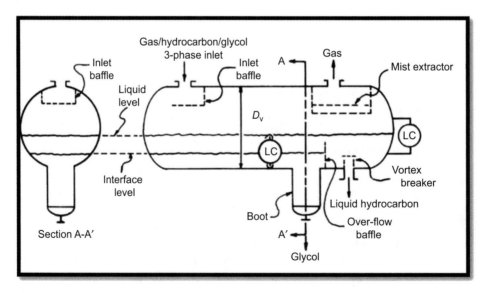

FIGURE 4.10

Horizontal Three-phase separator with wire mesh mist extractor.

Figure 4.10 also illustrates the separation of two liquid phases (glycol and hydrocarbon). The denser glycol settles to the bottom and is withdrawn through the boot. The glycol level is controlled by an interface level control instrument. Horizontal separators have certain advantages with respect to gravity separation performance in that the liquid droplets or gas bubbles are moving perpendicular to the bulk phase velocity, rather than directly against it as in vertical flow, which makes separation easier.

In a double-barrel separator, the liquids fall through connecting flow pipes into the external liquid reservoir below. Slightly smaller vessels may be possible with the double-barrel horizontal separator, where surge capacity establishes the size of the lower liquid collection chamber.

Typical L/D ratios for horizontal separators normally fall in the range of 2.5–5.

As an example of a horizontal separator, consider a rich amine flash tank. In this service: [1]

- There is relatively large liquid surge volume required for longer retention time. This allows more complete release of the dissolved gas and, if necessary, surge volume for the circulating system.
- There is more surface area per liquid volume to aid in more complete degassing.
- The horizontal configuration handles a foaming liquid better than vertical.
- The liquid level responds slowly to changes in liquid inventory, providing steady flow to downstream equipment.

4.3.3 Spherical separators

These separators are used for high-pressure service where compact size is desired and liquid volumes are small, (see Figures 4.11 and 4.12).

Mist extractor can be installed in this type of separators.

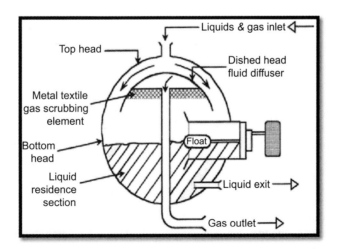

FIGURE 4.11

A spherical separator schematic.

FIGURE 4.12

A spherical separator diagram.

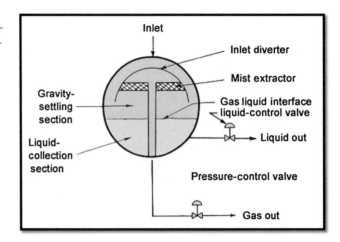

4.4 Design criteria of separators

Separators without mist extractors may be designed for gravity settling using Stokes law equation.

Typically the sizing is based on removal of 150 μm and larger diameter droplets. For practical purposes, to be on the safe side, use of the methods presented in the following sections is recommended.

4.4.1 Design flow rates

A separator may be incorporated in a process scheme for which there are different modes of operation. In this case, the separator design is based on the operation mode with the severest conditions. For the gas–liquid separation in a knock-out drum or demister, the severest condition is that with the highest value of Q, (m³/s), where this is defined by:

$$Q = Q_g \left(\frac{\rho_g}{\rho_1 - \rho_g} \right)^{1/2} \tag{4.8}$$

where:

Q = Volumetric load factor, (m³/s)
Q_g = Volumetric flow rate of gas, (m³/s)
Q_{max} = Maximum value of Q, in (m³/s)
ρ_g = gas phase density, kg/m³
ρ_1 = liquid phase density, droplet or particle, kg/m³

Having identified the severest mode from the highest value of Q, it is then necessary to add on a margin to give the value on which the separator design shall be based. This value, Q_{max} (maximum value of Q, in (m³/s)) should include margins for overdesign, safety, or surging. It represents the 'flooding' condition. The margin to be applied over the normal (process design) flow rate depends on the application. For oil processing, a margin of 15–25% is common; for oil production, margins up to 50% may be required.

4.4.2 **Nature of the feed**

The type of flow in feed pipe to the separator, transition from one flow regime to another, formation of droplets, foaming tendency of the liquid, which may lead to carryover of the liquid or carry-through of gas and other factors such as existence of sands, rust, wax, and other solids, and coking tendency of the liquid are points which should be taken into account when fixing design conditions.

4.4.3 **Efficiency and separator type**

The efficiency of a separator is defined here as the fraction (or percentage) of the liquid entering the vessel that is separated off.

Table 4.1 gives the preferred choice of separator according to feed description and the approximate efficiency expected:

Although a demister separator fitted with a wire mesh demister mat will often have efficiency better than 99%, it should be noted that a substantial fraction of droplets less than 20 μm in diameter will pass through a wire mesh demister.

4.4.4 **Liquid handling requirements**

In checking the capacity of the separator to handle liquid, the following points should be considered.

- **Degassing**

In order to prevent carry-through of the gas bubbles into the liquid outlet stream in vertical vessels, the liquid velocity should satisfy the following requirement:

$$\frac{Q_g}{\frac{\pi}{4}D_v^2} < \frac{2.18 \times 10^{-2}}{\gamma_1} \times \frac{\rho_1 - \rho_g}{\rho_1} \tag{4.9}$$

where:

D_v = internal diameter of a separator vessel, in m
Q_g = volumetric flow rate of gas, in m^3/s
γ_1 = liquid kinematic viscosity, in mm^2/s or cSt
ρ_g = gas phase density, kg/m^3
ρ_1 = liquid phase density, droplet or particle, kg/m^3

Table 4.1 Efficiency versus Type of Flow of Separators

Type of Flow	Efficiency Expected	
	<90%	>90%
Bubbly	Knock-out drum	—
Stratified, smooth	—	Knock-out drum
Stratified, wavy	Knock-out drum	Demister or cyclone
Intermittent	Knock-out drum	Demister or cyclone
Annular	—	Demister or cyclone

- **Control volume**

  For degassing (and also to counter foam), the liquid velocity should be limited. When a liquid level is required in the separator, a certain liquid volume is necessary for control purposes. The following recommendations are intended as a guide:

  - Hold-up time for control

    The minimum requirement between level alarm (low) LA(L) and level alarm (high) LA(H) shall be applied in the absence of other overriding process considerations is as follows:

    Automatic control

    > 4 min for product to storage
    > 5 min for feed to a furnace
    > 4 min for other applications

    Manual control

    > 20 min

  - Hold-up time for operator intervention

    When operator action is necessary to avoid upsets or a shutdown of plant operation, a realistic estimate shall be made of the time required.

    For inside plot separators a minimum hold-up of 5 min between pre-alarm level alarm (LA) and shutdown action (LZA) is recommended.

    For off-plot separators, the time the operator needs to reach the equipment should be added. No hold-up is required for switches that do not directly affect plant operation, e.g. automatic starting of standby pump. The pre-alarm may then be omitted.

- **Liquid slugs**

If the feed contains slugs of liquid, extra hold-up volume is required. In the absence of other information, the slug volume should be taken as 2–5 s of flow with the normal feed velocity and 100% liquid filling of the feed pipe.

The separator should be able to accommodate this slug volume between no level (NL) and LA(H) or between NL and LZA (HH), depending on whether or not it is required for the arrival of a slug to sound the high-level alarm. To increase the volume for accommodation of the slug the NL may be set closer to LA(L).

Note:

When slugs are expected in the feed, consideration should be given to strengthening the piping and feed inlet device, if one is to be fitted.

4.5 Gas–liquid separator sizing

4.5.1 Specifying separators

Separator designers need to know pressure, temperature, flow rates, and physical properties of the streams as well as the degree of separation required. It is also prudent to define if these conditions all occur at the same time or if there are only certain combinations that can exist at any time. If known, the type and amount of liquid should also be given, and whether it is mist, free liquid, or slugs.

For example, a compressor suction scrubber designed for 2–4.3 Mm^3/day gas at 2750–4100 kPa (ga) and 20–40 °C would require a unit sized for the worse conditions, i.e., 4.3 Mm^3/day at 2700 kPa (ga) and 40 °C. But if the real throughput of the compressor varies from 4.3 Mm^3/day at 4100 kPa (ga), 40 °C to 2 Mm^3/day at 2750 kPa (ga), 20 °C, then a smaller separator is acceptable because the high volume occurs only at the high pressure [1].

An improperly sized separator is one of the leading causes of process and equipment problems. Inlet separation problems upstream of absorption systems (e.g., amine and glycol) can lead to foaming problems, and upstream of adsorption systems (e.g., molecular sieve, activated alumina, and silica gel) can cause fouling, coking, and other damage to the bed [1].

Equipment such as compressors and turbo-expanders tolerate little or no liquid in the inlet gas steam, while pumps and control valves may have significant erosion and/or cavitation when vapors are present due to improper separation. In addition, direct-fired reboilers in amine and glycol service may experience tube failures due to hot spots caused by salt deposits caused by produced water carryover into the feed gas [1].

4.5.2 Design approach

There is as much art as there is science to properly design a separator. Three main factors should be considered in separator sizing: (1) vapor capacity, (2) liquid capacity, and (3) operability.

The vapor capacity will determine the cross-sectional area necessary for gravitational forces to remove the liquid from the vapor. The liquid capacity is typically set by determining the volume required to provide adequate residence time to "degas" the liquid or allow immiscible liquid phases to separate. Operability issues include the separator's ability to deal with solids if present, unsteady flow/liquid slugs, turndown, etc. Finally, the optimal design will usually result in an aspect ratio that satisfies these requirements in a vessel of reasonable cost. These factors often result in an iterative approach to the calculations [1].

4.5.3 Vapor handling

- **Separators without mist extractors**

Separators without mist extractors are not frequently utilized. The most common application of a vapor–liquid separator that does not use a mist extractor is a flare knockout drum.

Mist extractors are rarely used in flare knockout drums because of the potential for plugging and the serious implications this would have for pressure relief. Typically a horizontal vessel utilizes gravity as the sole mechanism for separating the liquid and gas phases. Gas and liquid enter through the inlet nozzle and are slowed to a velocity such that the liquid droplets can fall out of the gas phase. The dry gas passes into the outlet nozzle and the liquid is drained from the lower section of the vessel [1].

To design a separator without a mist extractor, the minimum size diameter droplet to be removed must be set. Typically this diameter is in the range of 300–2000 μm (1 μm or micron = 10-4 cm or 0.00003937 in).

The length of the vessel required can then be calculated by assuming that the time for the gas flow from inlet to outlet is the same as the time for the liquid droplet of size D_p to fall from the top of the vessel to the liquid surface. Eqn (4.10) then relates the length of the separator to its diameter as a function of this settling velocity (assuming no liquid retention):

$$L = \frac{4\left(10^3 \text{ mm/m}\right)^2 Q_A}{\pi V_t D_v} \tag{4.10}$$

where:

L = seam to seam length of vessel, mm
Q_A = actual gas flow rate, m³/s
D_v = inside diameter of vessel, mm
V_t = critical or terminal gas velocity necessary for particles of size D_p to drop or settle out of gas, m/s

If the separator is to be additionally used for liquid storage, this must also be considered in sizing the vessel [1]

Example: Calculate the size for a horizontal gravity separator (without mist extractor) that is required to handle 2 Mm³/day of 0.75 relative density gas (molecular weight = 21.72) at a pressure of 3500 kPa (ga) and a temperature of 40 °C. Compressibility is 0.9, viscosity is 0.012 mPa s, and liquid relative density is 0.50. It is desired to remove all entrainment greater than 150 μm in diameter. No liquid surge is required.

Gas density:

$$\rho_g = \frac{P(\text{MW})}{RTZ} = \frac{(3601)(21.72)}{(8.31)(313)(0.90)} = 33.4 \text{ kg/m}^3$$

Liquid density:

$$\rho_l = 0.5(1000) = 500 \text{ kg/m}^3$$

Mass flow,

$$M = \frac{(2)\left(10^6\right)(21.72)}{(23.7)(24)(3600)} = 21.2 \text{ kg/s}$$

Particle diameter

$$D_p = (150)\left(10^{-6}\right) = 0.000150 \text{ m}$$

$$C_D(\text{Re})^2 = \frac{(1.31)\left(10^7\right)\rho_g D_p^3 \left(\rho_l - \rho_g\right)}{\mu^2}$$

$$= \frac{(1.31)\left(10^7\right)(33.4)(0.000150)^3(500 - 33.4)}{(0.012)^2}$$

From Figure 4.2, Drag coefficient, $C_D = 1.40$

Terminal velocity,

$$V_t = \sqrt{\frac{4gD_p(\rho_1 - \rho_g)}{3\rho_g C_D}}$$

$$= \sqrt{\frac{4(9.81)(0.000150)(467)}{3(33.4)1.40}}$$

$$= \sqrt{0.0196} = 0.14 \text{ m/s}$$

Gas flow,

$$Q_A = \frac{M}{\rho_g} = \frac{21.2}{33.47} = 0.63 \text{ m}^3/\text{s}$$

Assume a diameter, $D_v = 1000$ mm.
   Vessel length,

$$L = \frac{4(10^3 \text{ mm/m})^2 Q_A}{\pi V_t D_v} = \frac{4(10^3 \text{ mm/m})^2(0.63)}{\pi(0.14)(1000)}$$

Other reasonable solutions are as follows:

Diameter, mm	Length, mm
1200	4800
1500	3800

1. What size vertical separator without mist extractor is required to meet the conditions

$$A = \frac{Q_A}{V_t} = \frac{0.63}{0.14} = 4.5 \text{ m}^2$$

$$D_v = 2400 \text{ mm minimum}$$

Recommended methods for basic (process) design of conventional gas–liquid separators are presented in the following sections. General references are available in literature. For symbols and abbreviations, see Section 4.4.

4.5.4 Conventional separators without mist extractors

This type of separator vessel utilizes gravity as the sole mechanism for separating the liquid and gas phases.

   To design a separator without a mist extractor, the minimum size diameter droplet to be removed must be set. Typically this diameter is in the range of 150–2000 μm. Although the calculation method for gravity settling is valid for this type, the methods given in the following sections can also yield a safe basic design.

4.5.5 Vertical knock-out drum without mist extractors

1. Diameter

The vessel diameter, D_v, can be estimated from:

$$\frac{Q_{max}}{\pi/4D_v^2} \leq 0.07 \tag{4.11}$$

where:

Q_{max} is the process design value of Q plus margin

2. Height

If h is the height of vessel required for liquid hold-up, then the total vessel height (tangent to tangent) is

$$H = h + d_n + X + Y \tag{4.12}$$

where:

d_n is inlet nozzle diameter, in m
X is 0.3 D_v with a minimum of 0.3 m
Y is 0.9 D_v with a minimum of 0.9 m

3. Nozzles

a. Feed nozzle
The feed nozzle shall be fitted with a half open pipe or a flow diverting box inlet device. The nozzle diameter, d_n, may be taken equal to that of the feed pipe but the product $\rho_m \times V_m$ shall not exceed 1500 kg/m s^2.

b. Gas outlet nozzle
The diameter of the gas outlet nozzle should normally be taken equal to that of the outlet pipe, but the product $\rho_g \times V_g^2$ shall not exceed 3750 kg/m s^2.

c. Liquid outlet nozzle
The diameter of the liquid outlet nozzle shall be chosen such that the velocity in it does not exceed 1 m/s, but should preferably be lower. The nozzle shall be equipped with a vortex breaker.

4.5.6 Horizontal knock-out drum without mist extractor

For a horizontal knock-out drum, the following design method can be applied:

1. Size

The cross-sectional area for gas flow, A, follows from:

$$\frac{Q_{max}}{A} \leq 0.1 \text{ m/s} \tag{4.13}$$

where A is taken above the LZA(HH) liquid level and Q_{max} is the process design value of Q plus a margin. Horizontal vessels are usually designed to be between about one-third and one-half full of liquid.

Note:

The liquid hold-up volume is determined by other considerations,. The design method involves trial and error. L_v and D_v are fixed, and a fractional filling chosen to satisfy liquid hold-up requirements. It is then necessary to check that Q_{max}/A does not exceed 0.1 m/s. A starting value for the ratio L_v/D_v of 3.0 is suggested; values of 2.5–6.0 for this ratio are normal; values of 6.0 or higher are recommended for high-pressure applications.

2. **Nozzles**
   a. Feed nozzle
      The nozzle diameter, d_n, may be taken equal to that of the feed pipe, but the product $\rho_m \times V_m^2$ shall not exceed 1000 kg/m s².
      where:
      ρ_m = Mixture density at inlet in (kg/m^3)
      V_m = Mixture velocity at inlet $= (Q_g + Q_l)/(\pi d_n^{2/4})$, in (m/s)
   b. Gas outlet nozzle
      For gas outlet nozzles, the same method can be applied as for a vertical vessel.
   c. Liquid outlet nozzle
      For liquid outlet nozzles, the same method can be applied as for a vertical vessel.

4.5.7 Conventional gas–liquid separators with wire mesh mist extractors

Wire mesh pads are frequently used as entrainment separators for the removal of very small liquid droplets and, therefore, a higher over-all percentage removal of liquid. Removal of droplets down to 10 μm or smaller may be possible with these pads.

In plants where fouling or hydrate formation is possible or expected, mesh pads are typically not used. In these services vane or centrifugal type separators are generally more appropriate.

Most installations will use a 150 mm thick pad with 145–190 kg/m³ bulk density. Minimum recommended pad thickness is 100 mm. Manufacturers should be contacted for specific designs.

If the mist eliminators need to be designed for an excessively large size, the vessels shall be designed for a horizontal type.

4.5.8 Vertical demister separators

Most vertical separators that employ mist extractors are sized using equations that are derived from gravity settling equations. The most common equation used is the critical velocity equation, shown here in Eqn (4.14):

$$V_c = K\sqrt{\frac{\rho_l - \rho_g}{\rho_g}} \tag{4.14}$$

V_c = Critical gas velocity necessary for particles to drop or settle, in (m/s)
K = K factors for sizing wire mesh demisters from Table 4.2

Table 4.2 Typical K Factors for Sizing Wire Mesh Demisters

Separator Type	K Factor, m/s
Horizontal (with vert. pad)	0.122–0.152
Spherical	0.061–0.107
Vertical or horizontal (with horiz. pad)	0.055–0.107
At atm. Press.	0.107
At 2100 kPa	0.101
At 4100 kPa	0.091
At 6200 kPa	0.082
At 10300 kPa	0.064
Wet steam	0.076
Most vapors under vacuum	0.061
Salt and caustic evaporators	0.046

Notes.
1. $K = 0.107$ at 700 kPa–Subtract 0.003 for every 700 kPa above 700 kPa.
2. For glycol and amine solutions, multiply K by 0.6–0.8.
3. Typically use one-half of the above K values for approximate sizing of vertical separators without wire mesh demisters.
4. For compressor suction scrubbers and expander inlet separators, multiply K by 0.7–0.8.

ρ_g = gas phase density, kg/m^3
ρ_l = liquid phase density, droplet or particle, kg/m^3

Some typical values of the separator sizing factor, K, are given in Table 4.2. Note that this equation actually gives the size of the separation element (mist extractor), and does not size the actual separator containment vessel. That means the vessel may be selected larger than the element, (e.g. for surge requirements).

The following is a rough, but safe sizing method for vertical demister separators:

• **Diameter**
   The vessel diameter, D_v (in meters) follows from:

$$\frac{Q_{max}}{\pi D_v^2/4} \leq 0.105 \text{ m/s} \tag{4.15}$$

where:

Q_{max} = Maximum value of Q, in (m^3/s)
D_v = Internal diameter of a separator vessel, in (m)
When a standard support ring for the demister mat is designed, then the width of the ring is considered to be negligible and D_v calculated from Eqn (4.15) will be the vessel internal diameter.

Notes:
- For viscous liquids, the maximum capacity of a horizontal demister mat is less. For $\mu_g = 100$ cP, reduce the value 0.105 m/s by 10%. For $\mu_g = 1000$ cP, apply 20% reduction.
- Maximum capacity of the mat decreases as the rate of liquid fed to it increases. The above values apply to a lightly loaded mat, as encountered in most separators.

- **Height**

  Let h be the height of vessel required for liquid hold-up. Then the total vessel height (tangent to tangent) is

  $$H = h + d_n + t + X + Y + 0.15D_v \tag{4.16}$$

where:

  H = Height, (tangent to tangent) of vessel, in (m)
  h = Height of vessel required for hold-up, in (m)
  d_n = Nozzle diameter, in (m)
  D_v = Internal diameter of a separator vessel, in (m)
  t = Thickness of demister pad, in (m) (usually 0.1 m)
  $X = 0.3 D_v$ with a minimum of 0.3 m (for a vessel equipped with half-open pipe inlet device)
  $Y = 0.45 D_v$ with a minimum of 0.9 m (for a vessel equipped with half-open pipe inlet device)

- **Nozzles**
  - Feed nozzle

    When the vessel diameter is less than 0.8 m, the feed nozzle should be fitted with a half-open pipe inlet device.

    For vessel diameters of 0.8 m and greater, a vane-type inlet device is recommended. The diameter of the nozzle d_n, may be taken equal to that of the feed pipe, but the following two criteria shall also be satisfied:

    $$\rho_m V_m^2 \leq 6000 \text{ kg}/(\text{m s}^2) \tag{4.17}$$

    $$\rho_g V_{g,in}^2 \leq 3750 \text{ kg}/(\text{m s}^2) \tag{4.18}$$

where:

    $$\rho_m = (M_g + M_l)/(Q_g + Q_l) \tag{4.19}$$

is mean density of mixture in the feed pipe, and velocity of mixture in the inlet nozzle is

    $$V_m = \frac{Q_g + Q_l}{\pi \frac{d_n^2}{4}} \tag{4.20}$$

  - Gas outlet nozzle

    For gas outlet nozzle, the same method may be applied as for a vertical knock-out drum (without mist extractor).
  - Liquid outlet nozzle

- **Pressure drop**

  In addition to the pressure drop, it is sufficient for most purposes to assume that the extra pressure drop over the demister mat is equivalent to 10 mm of liquid. The vane type inlet does not cause any significant pressure drop.

- **Demister mat specifications**

The mat shall be made of knitted wire formed to give the correct shape, and not cut so as to leave raw edges and loose pieces of wire that could become detached.

The wire mesh shall have a free volume >97%, a wire surface area >350 m^2/m^3, and a wire thickness ≥0.23 mm and ≤0.28 mm.

The thickness of 0.1 m for the mat is recommended. The wire mat shall be placed between two grids and shall be fastened in such a way that it cannot be compressed when being mounted.

4.5.9 Horizontal demister separators

Horizontal separators with mist extractors are sized using the equation presented in a previous section except that a factor is added for the length, L_v, of the gas flow path.

Separators can be any length, but the ratio of seam-to-seam length to the diameter of the vessel, L_v/D_v, is usually in the range of 2:1–4:1:

$$V_c = K \sqrt{\frac{\rho_1 - \rho_g}{\rho_g}} \left(\frac{L_v}{10}\right)^{0.56} \tag{4.21}$$

where:

V_c = critical gas velocity necessary for particles to drop or settle, in m/s
K = K factors for sizing wire mesh demisters from Table 4.2
L_v = length of a horizontal separator vessel, (tangent to tangent), in m
ρ_g = gas phase density, kg/m^3
ρ_1 = liquid phase density, droplet or particle, kg/m^3

K values given in Table 4.2 can be used for horizontal demister separators. Note that the preferred orientation of the mesh pad in horizontal separators is in the horizontal plane, and it is reported to be less efficient when installed in vertical position. But both designs are actually used for specified applications. The method presented in the following sections can be followed as a quick and safe method for basic design of horizontal demister separators.

1. **Size**
   a. Horizontal mat

      For sizing a horizontal separator with a horizontal mat, the equation and method presented above in this section is recommended.
   b. Vertical mat

      For a horizontal vessel fitted with a vertical mat, Eqn (4.22) can be applied:

$$\frac{Q_{max}}{A} \leq 0.15 \text{ m/s} \tag{4.22}$$

where:

    A is taken above the LZA (HH) liquid level

    Q_{max} is the process design value of Q plus a margin (see 7.2.1)

    Horizontal vessels are usually designed to be between about one-third and one-half full of liquid

Figure 4.13 shows a horizontal demister vessel with vertical mat and vane type inlet device. The following notes should be considered in Figure 4.13:

    The design method involves trial and error:

- L_v and D_v fixed, and a fractional filling chosen to satisfy liquid hold-up requirements. It will then be necessary to check that Q_{max}/A does not exceed 0.15 m/s. A starting value for the ratio L_v/D_v of 3.0 is suggested; values of 2.5–6.0 for this ratio are normal.

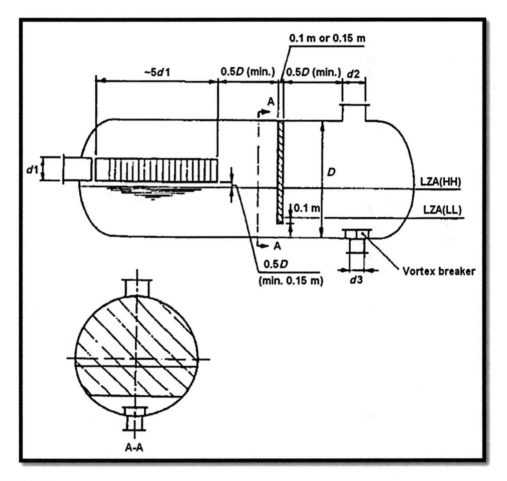

FIGURE 4.13

Horizontal demister vessel with vertical mat and vane type inlet device.

- The vertical demister mat shall extend from the top of the vessel to 0.1 m below the LZA (LL) level. The area between the mat and the bottom of the vessel shall be left substantially open, to allow free passage of liquid.

2. **Nozzles**

   a. Feed nozzle

   The feed nozzle shall be fitted with a vane-type or another type of inlet device. The diameter of the nozzle, dn1, may be taken equal to that of the feed pipe, but the following two criteria shall also be satisfied:

$$\rho_m V^2 \leq 6000 \text{ kg/m s}^2 \tag{4.23}$$

$$\rho_m V_{g,in}^2 \leq 3750 \text{ kg/m s}^2 \tag{4.24}$$

The length of the vane-type inlet nozzle should be taken equal to approximately five times the feed nozzle diameter.

   b. Gas outlet nozzle

   For the gas outlet nozzle, the same method applies as for a vertical knock-out drum,

   c. Liquid outlet nozzle

   For the liquid outlet nozzle, the same method applies as for a vertical knock-out drum.

3. **Pressure drop**

   In addition to the pressure drop, it should be sufficient for most purposes to assume that the extra pressure drop over the demister is equivalent to 10 mm of liquid. There will be no additional pressure loss over the vane-type inlet device.

4. **Demister mat specifications**

   it is recommended that for a vertical mat of height greater than 1.5 m (from NL to top), the thickness should be increased to 0.15 m

5. **Boots**

Boots, if necessary, shall be sized for a minimum residence time of 5 min as a guideline, and their diameters shall be the same as the commercial pipe sizes as far as possible.

The height-diameter ratio shall be 2:1–5:1, and it shall be determined with consideration given to operability and the minimum sizes of level instruments and equipment.

Boot diameters shall be 300 mm minimum because good operability cannot be provided if the boots are smaller than 300 mm.

Maximum boot diameters shall be one-third of the drum inside diameter.

4.5.10 Separators with vane-type mist extractors

Vanes differ from wire mesh in that they do not drain the separated liquid back through the rising gas stream. Rather, the liquid can be routed into a downcomer, which carries the fluid directly to the liquid reservoir. A vertical separator with a typical vane mist extractor is shown in Figure 4.14.

The vanes remove fluid from the gas stream by directing the flow through a torturous path. A cross-section of a typical vane unit is shown in Figure 4.15.

Vane-type separators generally are considered to achieve the same separation performance as wire mesh, with the added advantage that they do not readily plug and can often be housed in smaller

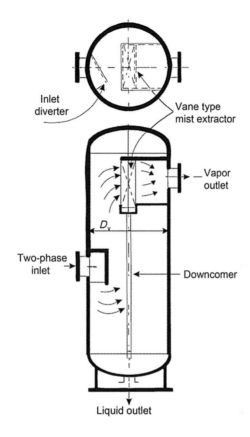

FIGURE 4.14

Example of vertical separator with vane-type mist extractor (GPSA 2004).

vessels. Vane-type separator designs are proprietary and are not easily designed with standard equations. Manufacturers of vane-type separators should be consulted for detailed designs of their specific equipment.

4.6 Specification sheet

The following points should be considered when preparing a specification sheet (or process data sheet) for a wire mesh gas–liquid separator.

1. Application service:
   a. Source of entrainment
   b. Operating conditions (normal, minimum, maximum):
      – Temperature;
      – Pressure;
      – Vapor phase:

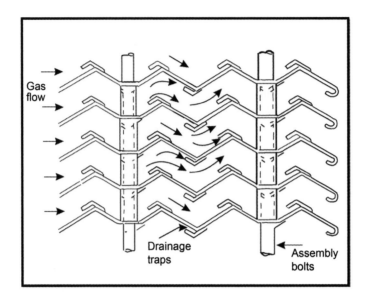

FIGURE 4.15

Cross section of example vane element mist extractor showing corrugated plates with liquid drainage traps (GPSA 2004).

Flow rate.
Velocity.
Density (at operating conditions).
Molecular mass.
Composition or nature of phase.
– Liquid entrainment phase:
Quantity (if known).
Density.
Viscosity.
Surface tension.
Composition or nature of entrainment.
Droplet sizes or distribution (if known).
Solids content (composition and quantity):
Dissolved.
Suspended.
c. Performance
– Allowable total separator pressure drop.
– Allowable mesh pressure drop.
– Allowable entrainment.
– Mesh thickness recommended.

2. Construction and installation:
   a. Vessel
   – Diameter (ID), and length.
   – Position (horizontal, vertical, inclined).
   – Shape (circular, square, etc.).
   – Type (evaporator, still, drum, etc.).
3. Special conditions

4.7 Mist eliminator type and installation point

1. Type

Depending on the quality of the crude, either wire mesh or vane-type eliminators could be used. However, the use of the vane type is usually preferred due to nonplugging aspect. This becomes a must if waxy crudes are encountered.

2. Installation point

After entering the separator, the associated gas travels in a longitudinal plane along the top half of the separator. Therefore it is recommended that the gas outlet mist eliminator be placed at the farthest point away from the inlet diverter. This allows the gas maximum residence time in the vessel to ensure that the majority of the liquid droplets entrained in the gas are released prior to the gas outlet from the vessel.

4.7.1 Inlet diverter

An inlet diverter should always be included as this will break up the bulk of the inlet stream into smaller particles.

There are various types available all of which are used by manufacturers. These types are listed:

1. The dished end type inlet diverter directs the inlet fluid back into the vessel head. This is used in cases where the inlet nozzle is in the head of the vessel.
2. The fluid is directed back against the vessel head by a 90° elbowed pipe. This type is used for gases.
3. Presently the most common of all types is either a vane, angle, or pipe type inlet diverter in a box type arrangement.

4.7.2 Coalescing packs

Coalescing packs (or pads) have special designs and are used by some manufacturers (in three-phase separators) in order to assist the separation of liquid phases. These are usually installed in the oil layer; however, in certain instances where produced water cleanup is important, the pack can be installed across both layers.

4.7.3 Operating levels

As a general rule, maximum operating level in a horizontal crude oil separator should not exceed 50%, and minimum level should not be less than 200 mm.

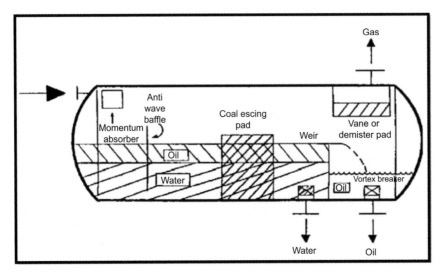

FIGURE 4.16

A typical horizontal oil–gas–water separator.

In certain circumstances, however, the maximum level can be raised to 60%; i.e. if the gas/oil ratio is particularly low or if in conjunction with a relatively low gas rate, the required liquid removal efficiency from the gas can be relaxed.

The minimum liquid level of 200 mm from the bottom inside of the shell cannot practically be reduced, as it would be very difficult to control, in most cases this would be the low level shutdown position.

Alarm/shutdown positions are usually specified by engineering judgment, but the following general rules for high and low levels based on previous operating experience are recommended:

1. The high liquid level shutdown positions should not be greater than 70% in any circumstances, and in addition it should not be more than 100 mm lower than the inlet nozzle.
2. The low liquid level shutdown position cannot be less than 200 mm above the bottom inside wall of the separator shell due to practical control installation problems.

4.7.4 Vortex breakers

Vortex breakers should always be installed on liquid outlet nozzle (two-phase), or water and oil nozzle in the case of three-phase separators.

4.7.5 Three-phase vessel weir plates

The weir plate is a device that separates oil and water into two compartments.

Weir plates can be either fixed or adjustable. Fixed weir plates should be used in cases where the water content is constant. Adjustable weir plates are required when the water content is expected to increase.

Generally, the weir plate should always be 150 mm (minimum) above the oil/water interface. It can vary in height from the bottom inside shell wall to the top of the plate from 300 mm to the midpoint of the vessel.

4.7.6 Antiwave baffles

In large volume three-phase separators, it is sometimes necessary to install one or more antiwave baffle to eliminate disturbances of oil/water interface. This is a partial cross sectional area plate with punched holes that act as a wave breaker while still letting liquid pass through (see Figure 4.16).

4.8 Centrifugal gas–liquid separators

Centrifugal separators utilize centrifugal action for the separation of materials of different densities and phases. They are built in stationary and rotary types. Various modifications of stationary units are used more than any other kind. Centrifugal separators are generally divided into three types:

1. Stationary vane separators.
2. Cyclone separators.
3. Inertial centrifugal separators.

The efficiency of each of three types can be estimated using Table 4.3. Because of wide usage of cyclone separators for separation of liquid entrainment from gas streams, this type is discussed in the following sections.

4.8.1 Cyclone separators

The cyclone type unit is well recognized and accepted in a wide variety of applications from steam condensate to dusts from kilns. In this unit, the carrier gas and suspended particles enter tangentially or volutely into a cylindrical or conical body section of the unit, then spiral downward, forcing the heavier suspended matter against the walls.

Solids tend to slide down the wall while liquid particles wet the wall, form a running film, and are removed at the bottom.

• **Design criteria**
  The design incorporates two special internals, the bottom plate with vortex spoilers, and the cylindrical baffle at the top of the cyclone.

Table 4.3 The Efficiency of Each Centrifugal Separators

Type	Efficiency Range
High velocity stationary vanes	99% or higher of entering liquid. Residual entrainment 1 mg/kg (ppm) or less.
Cyclone separator	70–85% for 10 μm, 99% for 40 μm, and larger. For high entrainment, efficiency increases with concentration
Rotary	98% for agglomerating particles

• Diameter

The diameter of the cyclone should be related to the size of the inlet, as follows:
Type I:

$$a = 7 \times 10^{-5} \left[\frac{Q_g (\rho_l - \rho_g)}{\mu_g} \right]^{1/3} \tag{4.25}$$

$$a = 0.13 \rho_g^{1/4} Q_g^{1/2} \tag{4.26}$$

The value "a" should be taken as the larger of the values calculated from Eqns (4.25 and 4.26). The diameter is then given by:

$$D_v = 2.8a \tag{4.27}$$

Type II

$$d_1 = 7 \times 10^{-5} \left[\frac{Q_g (\rho_l - \rho_g)}{\mu_g} \right]^{1/3} \tag{4.28}$$

$$d_1 = 0.15 \rho_g^{1/4} Q_g^{1/2} \tag{4.29}$$

d_1 should be taken as the larger of the values calculated from Eqns (4.28 and 4.29). The diameter is then given by:

$$D_v = 3.5 d_1 \tag{4.30}$$

where:

a = length of side, square cyclone inlet, (type I), in m
d_1 = internal diameter of feed inlet, in m
D_v = internal diameter of a separator vessel, in m
Q_g = volumetric flow rate of gas, in m^3/s
μ_g = dynamic viscosity of gas, in cP, mPa s
ρ_g = gas phase density, kg/m^3
ρ_l = liquid phase density, droplet or particle, kg/m^3

The superficial liquid velocity in the inlet nozzle should not exceed 1 m/s.

4.8.1.1 Height
The following applies to both Types I and II:

$$\text{When } \frac{M_l}{M_g} \geq 1 \quad \text{take } \frac{H_c}{D_v} = 0.5 \text{ to } 0.7 \tag{4.31}$$

$$\text{When } \frac{M_l}{M_g} = 1 \text{ to } 0.1 \quad \text{take } \frac{H_c}{D_v} = 0.7 \text{ to } 1.2 \tag{4.32}$$

$$\text{When } \frac{M_l}{M_g} = 0.1 \text{ to } 0.01, \quad \text{take} \frac{H_c}{D_v} = 1.2 \text{ to } 1.7 \tag{4.33}$$

$$\text{When } \frac{M_l}{M_g} = 0.01 \text{ to } 0.001, \quad \text{take } \frac{H_c}{D_v} = 1.7 \text{ to } 2.2 \tag{4.34}$$

where:

D_v = internal diameter of a separator vessel, in m
H_c = height of cyclone (from bottom plate to outlet pipe), in m
M_l = mass flow rate of liquid, in kg/s
M_g = mass flow rate of gas, in kg/s
M_l and M_g denote the mass flow rates of liquid and gas, respectively, in kg/s

The shape and size of the space below the bottom plate have no effect on the working of the cyclone and shall be determined in connection with liquid hold-up requirements. It is important that any liquid level shall always be kept below the bottom plate.

4.8.1.2 Nozzles

The liquid drain shall be sized so that the velocity in it does not exceed 1 m/s but should preferably be lower. A vortex breaker shall be used.

4.8.1.3 Pressure drop

Pressure differentials between inlet, vapor outlet, and liquid drain are as follows:

$$P_{in} - P_{out} = X \rho_g V_{g,in}^2 \tag{4.35}$$

$$P_{in} - P_{drain} = Y \rho_g V_{g,in}^2 \tag{4.36}$$

where:

ρ_g is in kg/m^3, V_g is in m/s, and P is in kPa
For Type I: $X = 0.015$ and $Y = 0.005$
For Type II: $X = 0.10$ and $Y = 0.003$

4.8.1.4 Liquid drain sealing

For satisfactory operation of any cyclone, it is essential that the liquid drain will be sealed; i.e. there shall be no flow of vapors in either direction in the drain.

Liquid from the cyclone will sometimes be drained via a dip leg to a receiving vessel. In such cases it is essential that a pressure balance shall be calculated to ensure that the liquid level always remains below the bottom plate. If the level rises above this plate, severe entrainment will result.

4.8.1.5 Foaming service

For foaming systems, the width of split between the bottom plate and the wall, S, should be increased such that

$$\frac{S}{D_v} = 0.05 \tag{4.37}$$

A minimum width of split of 20 mm is recommended in foaming service.

4.8.2 Multi-cyclone separators

This type of centrifugal separator is a high-efficiency one, which is claimed by manufacturers to be capable of removing almost all liquid droplets of 5 μm diameter and larger. This type can also be used in gas–solid separation services.

4.8.3 Specification sheet

The following points should be considered when preparing a spec. sheet (or process data sheet) for a cyclone separator in liquid entrainment separation service.

1. Application: (service application of unit should be described if possible).
2. Fluid stream.
3. Fluid composition (vol%).
4. Entrained particles:
   a. Size range (micrometers or mesh).
   b. Size percentage distribution.
   c. True relative density (specific gravity) of particle, referred to water $= 1.0$.
   d. Source of entrainment, (boiling liquid, etc.).
   e. Composition.
5. Operating conditions, (minimum, maximum, and normal):
   a. Gas flow rate.
   b. Entrained flow rate.
   c. Temperature, (°C).
   d. Pressure, (kPa).
   e. Moisture content.
   f. Dew point, (°C).
6. Installation altitude:
   a. Normal barometer, (mm Hg).
7. Nature of entrained liquid:
   a. Description, (oily, corrosive, etc.).
   b. Surface tension at operating conditions.
   c. Viscosity at operating conditions.
8. Insulation required and reason.
9. Construction features:
   a. Storage required for collected liquid, (h).
   b. Preliminary size of the inlet connection: (mm).
10. Special conditions (if any).

4.9 Flare knock-out drums

Flare knock-out (KO) drums are one type of gas–liquid separators that are used specifically for separation of liquids carried with gas streams flowing to the flares in OGP plants.

The main difference between flare KO drums and other conventional gas/liquid separators lies in the size of the droplets to be separated; i.e. separation of 300, 600 μm droplets fulfills the requirements

Table 4.4 Values of C^* Used in Equations 7.13 and 7.14

Emulsion Characteristic	Droplet Diameter, μm	Constant, C^*
Free liquids	200	1100
Loose emulsion	150	619
Moderate emulsion	100	275
Tight emulsion	60	99

of flare gas disengagement. Therefore usage of mist eliminating device is not usually necessary in flare KO drums, except for cases where the results of calculations lead to an abnormally large drum size. In such cases, application of vane-type or multicyclone separators may help avoid employing an extremely large drum.

4.9.1 Design criteria

For detail sizing procedure, please refer to chapter related to Process Design of Flare and Blowdown Systems.

- **Three-phase and liquid–liquid separation**

The gas handling requirements for three-phase separation are dealt with in a similar manner as discussed for two-phase separation. Traditionally, sizing for liquid–liquid separation has involved specification of liquid residence times.

Table 4.4 provides suggested residence times for various liquid–liquid separation applications. These figures generally assume equal residence times for both the light and heavy liquid phases.

While the residence time approach for liquid–liquid separation equipment design has been widely used in industry for years, it does have some limitations.

- The typical approach of assuming equal residence times for both liquid phases may not be optimum. For example, it is generally much easier to separate oil droplets from water than vice versa. Settling theory (Eqn (4.1)) explains this as being due to the lower viscosity of water compared to oil.
- Residence times do not take into account vessel geometry; i.e. 3 min residence time in the bottom of a tall, small diameter vertical vessel will not achieve the same separation performance as 3 min in a horizontal separator, again according to droplet settling theory.
- The residence time method does not provide any direct indication as to the quality of the separated liquids, e.g. amount of water in the hydrocarbon or the amount of hydrocarbon in the water. Droplet settling theory cannot do this either in most cases, but there is some empirical data available that allow for approximate predictions in specific applications.

Removal of very small droplets may require the use of specialized internals or the application of electrostatic fields to promote coalescence.

Liquid–liquid separation may be divided into two broad categories of operation. The first is defined as "gravity separation," where the two immiscible liquid phases separate within the vessel by the differences in density of the liquids. Sufficient retention time must be provided in the separator to allow for the gravity separation to take place. The second category is defined as "coalescing separation." This is where small particles of one liquid phase must be separated or removed from a large quantity of another liquid phase. Different types of internal construction of separators much be provided for each type of liquid–liquid separators. The following principles of design for liquid–liquid separation apply equally for horizontal or vertical separators. Horizontal vessels have some advantage over verticals for liquid–liquid separation, due to the larger interface area available in the horizontal style, and the shorter distance particles must travel to coalesce.

There are two factors that may prevent two liquid phases from separating due to differences in specific gravity:

- If droplet particles are so small that they may be suspended by Brownian movement. This is defined as a random motion that is greater than directed movement due to gravity for particles less than 0.1 μm in diameter.
- The droplets may carry electric charges due to dissolved ions. These charges can cause the droplets to repel each other rather than coalesce into larger particles and settle by gravity.

Effects due to Brownian movement are usually small and proper chemical treatment will usually neutralize any electric charges. Then settling becomes a function of gravity and viscosity in accordance with Stokes law. The settling velocity of spheres through a fluid is directly proportional to the difference in densities of the sphere and the fluid, and inversely proportional to the viscosity of the fluid and the square of the diameter of the sphere (droplet), as noted in Eqn (4.3). The liquid–liquid separation capacity of separators may be determined from Eqns (4.38) and (4.39), which were derived from Eqn (4.3). Values of C^* are found in Table 4.4.

Vertical vessels:

$$W_{cl} = C^* \left(\frac{S_{hl} - S_{ll}}{\mu} \right) (0.785) D_v^2 \tag{4.38}$$

where:

W_{cl} = flow rate of light condensate liquid, m³/day
C^* = empirical constant for liquid–liquid separators, (m³ mPa s)/(m² day)
S_{hl} = relative density of heavy liquid, water = 1.0
S_{ll} = relative density of light liquid, water = 1.0
D_v = inside diameter of vessel, mm
μ = viscosity of continuous phase, mPa s

Horizontal vessel:

$$W_{cl} = C^* \left(\frac{S_{hl} - S_{ll}}{\mu} \right) L_l H_l \tag{4.39}$$

where:

W_{cl} = flow rate of light condensate liquid, m³/day
C^* = empirical constant for liquid–liquid separators, (m³ mPa s)/(m² day)
H_l = width of liquid interface area, m
L_l = length of liquid interface, mm

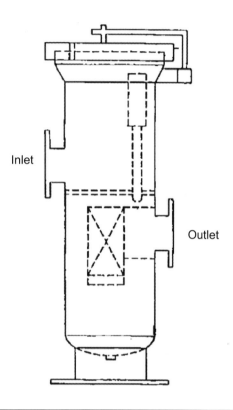

Inlet

Outlet

FIGURE 4.18

Sketch of vertical gas filter separator.

- **Filter elements**

The most common and efficient filter element (agglomerator) is composed of a fiber glass tubular filter pack that is capable of holding the liquid particles through submicron sizes.

- **Mist elimination**

Mist elimination from a gas stream is performed in the liquid compartment of the filter using a wire mesh or vane-type mist extractor installed in the gas outlet. Rules presented in the previous sections for demister gas–liquid separators govern here.

- **Liquid removal**

For heavy liquid loads, or where free liquids are contained in the inlet stream, a horizontal filter separator with a liquid sump, which collects and dumps the inlet free-liquids, is often preferred. In such cases, another sump drum for the separated liquid in the demister compartment should also be prepared.

When the liquid quantity is guaranteed to be small (less than 30 l/1000 cm^3), there is an economic saving to be made by having a lower barrel in the demister or vane compartment only. If the liquid quantity is large, then liquid from the first compartment must be dumped also.

S_{hl} = relative density of heavy liquid, water = 1.0
S_{ll} = relative density of light liquid, water = 1.0
μ = viscosity of continuous phase, mPa s

Since the droplet size of one liquid phase dispersed in another is usually unknown, it is simpler to size liquid–liquid separation based on retention time of the liquid within the separator vessel. For gravity separation of two liquid phases, a large retention or quiet settling section is required in the vessel. Good separation requires sufficient time to obtain an equilibrium condition between the two liquid phases at the temperature and pressure of separation. The liquid capacity of a separator or the settling volume required can be determined using the retention time given in Table 4.4. The following example shows how to size a liquid–liquid separator.

Example: Determine the size of a vertical separator to handle 100 m^3/day of 0.76 relative density condensate and 10 m^3/day of produced water. Assume the water particle size is 200 μm. Other operating conditions are as follows:

Operating temperature = 25 °C
Operating pressure = 6900 kPa (ga)
Water relative density = 1.01
Condensate viscosity = 0.55 mPa s @ 25 °C
Condensate relative density = 0.76
From Eqn (4.38):

$$W_{cl} = C^* \left(\frac{S_{hl} - S_{ll}}{\mu} \right) (0.785) D_v^2$$

From Table 4.4 for free liquids with water particle diameter = 200 μm, $C^* = 1880$.

$$100 \text{ m}^3/\text{day} = 1880 \frac{1.01 - 0.76}{(0.55)} (0.785) D_v^2 (10^{-6})$$

$$(D_v^2) = (1.43)(10^5)$$

$$D_v = 390 \text{ mm}$$

Using the alternate method of design based on retention time would give:

$$U = \frac{W(t)}{1440} \tag{4.40}$$

U = volume of settling section, m^3
W = total liquid flow rate, m^3/day
t = retention time, min

From Table 4.5, use 3 min retention time.

$$U = \frac{(110)(3)}{1440} = 0.23 \text{ m}^3$$

Table 4.5 Typical Retention Times for Liquid–Liquid Separation

Type of Separation	Retention Time, min
Hydrocarbon/Water Separators	
Above 35° API hydrocarbon	3–5
Below 35° API Hydrocarbon	
100 °F and above	5–10
80 °F	10–20
60 °F	20–30
Ethylene glycol/hydrocarbon Separators (cold separators)	20–60
Amine/hydrocarbons separators	20–30
Coalescer, Hydrocarbon/Water Separators	
100 °F and above	5–10
80 °F	10–20
60 °F	20–30
Caustic/propane	30–45
Caustic/heavy gasoline	30–90

A 390 mm diameter vessel will hold about 0.12 m³/1000 mm of height. The small volume held in the bottom head can be discounted in this size vessel. The shell height required for the retention volume required would be:

$$\text{Shell height} = \frac{0.23}{0.12} = 1.9 \text{ m} = 1900 \text{ mm}$$

Another parameter that should be checked when separating amine or glycol from liquid hydrocarbons is the interface area between the two liquid layers. This area should be sized so the glycol or amine flow across the interface does not exceed approximately 100 m³/day/m².

The above example indicates that a relatively small separator would be required for liquid–liquid separation. It should be remembered that the separator must also be designed for the vapor capacity to be handled. In most cases of high vapor–liquid loadings that are encountered in gas processing equipment design, the vapor capacity required will dictate a much larger vessel than would be required for the liquid load only. The properly designed vessel has to be able to handle both the vapor and liquid loads. Therefore, one or the other will control the size of the vessel used.

4.10 Gas–liquid filter separators

Gas–liquid filter separator (usually called gas filter separator) is used in separation of liquid and solid particles from a gas stream. A gas filter separator has a higher separation efficiency than the centrifugal

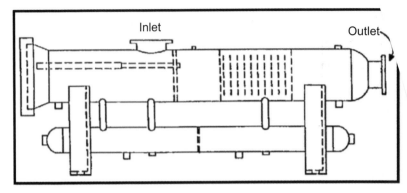

FIGURE 4.17

Sketch of horizontal gas filter separator.

separator, but it uses filter elements that must periodically be replaced. Gas filter separators can handle more than small quantities of liquid (greater than 30–170 l/1000 Standard m³).

This type of separator is usually made horizontal with a lower liquid barrel (see Figure 4.17), but a vertical type, especially when saving space is important, is offered by manufacturers (see Figure 4.18).

4.10.1 Applications

Typical applications of gas filter separators are as follows:

- Compressor stations to protect compressors from free liquid and prevent cylinder wear from solids.
- Metering and pressure reduction stations at city gates: to remove liquid hydrocarbons, water, sand, and pipe scale.
- Protection of desiccant beds and collection of dust carryover from beds.
- Gas storage systems: to prevent injection or withdrawal of solids, dust, and small amounts of liquids.
- Fuel lines to power plants and engines.
  Note: This type of separator shall not be used on a stream carrying wax or a congealing liquid, otherwise filter elements will plug.

4.10.2 Operation

- **Operating principles**

As shown in Figure 4.18, gas enters the inlet nozzle and passes through the filter section where solid particles are filtered from the gas stream and liquid particles are coalesced into larger droplets. These droplets pass through the tube and are entrained into the second section of the separator, where a final mist extraction element removes these coalesced droplets from the gas stream.

Vertical vessels should be used only for the removal of solid particles, with very small traces of liquid present.

4.10.3 Efficiency

The efficiency of a filter separator largely depends on the proper design of the filter pack, i.e. a minimum pressure drop while retaining an acceptable extraction efficiency. Regarding the filtration capability, various guarantees are available from filter separator manufacturers, among them, ability of a filter to remove 100% of 8 μm and larger liquid droplets or solid particles, 99.5% of 3 μm and larger, and 98% of droplets (or particles) larger than 1 μm is recommended to be specified.

However, note that guarantees for the performance of separators and filters are very difficult to verify in the field.

4.10.4 Design criteria

The design of filter separators is proprietary and a manufacturer should be consulted for specific size and recommendations. However, the following points may be useful in basic process design stage.

- **Vessel size**

The body size of a horizontal filter separator for a typical application can be estimated by using 0.4 for the value of K. This provides an approximate body diameter for a unit designed to remove water (other variables such as viscosity and surface tension enter into the actual size determination). Units designed for water will be smaller than units sized to remove light hydrocarbons.

In many cases the vessel size will be determined by the filtration section rather than the mist extraction section.

- **Filter section**

The filter cartridges coalesce the liquid mist into droplets which can be easily removed by the mist extractor. A design consideration commonly overlooked is the velocity out of these filter tubes into the mist extraction section. If the velocity is too high, the droplets will be sheared back into a fine mist that will pass through the extractor element. No published data can be cited since this information is proprietary with each filter separator manufacturer.

The approximate filter surface area for gas filters can be estimated from Figure 4.19. The figure is based on applications such as molecular sieve dehydrator outlet gas filters. For dirty gas service, the estimated area should be increased by a factor of two or three.

- **Pressure drop**

A pressure drop of 7–15 kPa is normal in a clean filter separator, but 7 kPa is often recommended to be specified in specification sheets. If excessive solid particles are present, it may be necessary to clean or replace the filters at regular intervals when a pressure drop in excess of 70 kPa is observed.

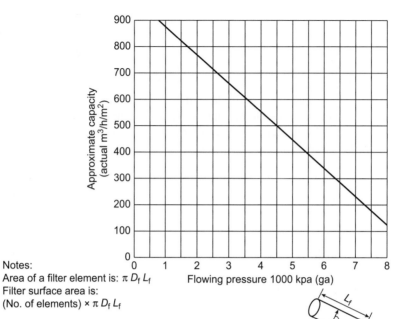

Notes:
Area of a filter element is: $\pi\, D_f\, L_f$
Filter surface area is:
(No. of elements) × $\pi\, D_f\, L_f$

Where $D_f\, L_f$ are the filter element outside diameter and length respectively.

FIGURE 4.19

Approximate gas filter capacity.

Although the filter coalescing elements are usually designed for a collapsing pressure of 250–350 kPa, as a rule, 170 kPa is recommended as a maximum for allowable differential pressure.

4.10.5 Specification sheet

In addition to the information necessary for a demister gas–liquid separator, which was mentioned in Section 7.3, the following data should also be written in the specification sheet of a gas filter separator:

1. Fluid filtered.
2. Molecular mass of gas and liquid density, in kg/m^3.
3. Gas flow rate, (range at expected pressures), in m^3/day.
4. Turn-down requirements.
5. Temperature range, in °C.
6. Normal liquid rate, in m^3/day.
7. Expected liquid slug size.
8. Expected solid content.
9. Other contaminants

4.11 **Process requirements of vessels, reactors, and separators**

Process design of gas (vapor)–liquid separators should be based on the discussions presented and should be amended with the following selection and design criteria.

1. Selection criteria

Various criteria and features play a role in separator performance and selection. Table 4.6 summarizes the relative performance of various types of separators.

2. Orientation

In general, a vertical vessel is preferred for gas–liquid separation for the following reasons:

- When the gas–liquid ratio is high;
- A smaller plan area is required (critical on offshore platforms);
- Easier solids removal;
- Liquid removal efficiency does not vary with liquid level;
- Vessel volume is generally smaller.

However, a horizontal vessel should be chosen if:

- Large volume of total fluid is available;
- Large amount of dissolved gas is available;
- Large liquid slugs have to be accommodated;
- There is restricted head room;
- A low downward liquid velocity is required (for degassing purposes, foam breakdown, or in case of a difficult liquid–liquid separation).

3. Components

Following the gas–liquid flow path through the separator, the following parameters should be identified:

- Feed inlet
  This comprises the upstream piping, inlet nozzle, and the inlet devices (if any).
  - The diameter of the inlet nozzle is a function of the feed flow rate and pressure.
  - Information on the nature of the feed (foaming tendency, feeds with solids, wax or coking tendency) are given below.

4.12 **Nature of the feed**

- **Foaming tendency**

For foaming to occur it is necessary for gas bubbles to be formed, and for the drainage of the liquid films surrounding the bubbles to be retarded. Drainage of the films is slower in highly viscous liquids, but the chief causes of foaming are surface properties, which are usually unpredictable. For this reason the foaming tendency is best judged on the basis of experience of similar cases. Laboratory tests may also give an indication of the foaminess of the system.

Table 4.6 Performance Comparison of Various Separators

	KO Drum		Wire Mesh Demister		Vane-Type Demister			Cyclone	Vertical Multi-Cyclone Separator	Filter Separator
					Vertical		Horizontal			
	Vertical	Horizontal	Vertical	Horizontal	In-line	Two-stage	Two-stage			
Gas handling max capacity (λ)	Low	Low	Moderate	Moderate	High	High	High	Very high	Very high	Low
Gas handling turndown (max/min)	Infinity	Infinity	4	4	3	3	3	2	2	Infinity
Liquid removed efficiency overall, %	80–90	80–90	>98	>98	>96	>96	>96	>96	>93	>99
Liquid removed with respect to fine mist	Very low	Very low	High	High	Low	Low	Low	Low	Low	Very high
Liquid removed flooding above λ_{max} (This will cause a sharp decrease in efficiency)	No	No	Yes	Yes	Yes, unless double-pocket vanes are used			No	Yes	Yes
Liquid handling capacity as slugs	High	Very high	High	Very high	Very Low	High	High	High	Low	Low
Liquid handling capacity as droplets (Q_{max})	High	High	High	High	Low	Moderate	Moderate	High	Low	Low
Fouling tolerance sand	High	High	Low	Low	Very low if double-pocket vanes, low if single-pocket vanes, moderate if no-pocket vanes.			High	High	High
Fouling tolerance sticky material	High	High	Very low	Very low				High	Low	Low
Pressure drop	Very low	Very low	Low	Low	Low	Low	Low	High	High	Very high

Examples of foaming systems are some crude oils, heavy residues, absorption and extraction solvents.

Foaming in the separator may lead to carryover of liquid (when foam reaches the gas–liquid separation internal and/or the gas outlet) or to carryunder of gas. It will also upset the level control system.

Note that foaming is more likely to be a problem at high liquid loads, when flow in the inlet pipe is in the froth or intermittent flow regimes.

Installation of internals to combat foam is normally not effective and may even be counterproductive.

Foaming in the vessel is minimized by decreasing the downward liquid velocity, for instance, by increasing the diameter of the separator vessel.

Sometimes an antifoam agent can be injected to suppress foaming.

- **Feeds with solids, wax, or coking tendency**

Sand, rust, scale, or other solids present in the feed will leave the separator together with the liquid. However, solids will also settle out in the separator and tend to accumulate. For this reason care should be taken with the location of instrument connections that could become plugged. Provision should be made for cleaning the separator during shutdowns, and if necessary during operation, by the installation of a water spray and drain.

When solids are present in the feed, consideration should be given to reducing the inlet velocity and adding an "erosion allowance" of 1–2 mm extra material thickness to the inlet device (if fitted).

Wax in the feed will be deposited on any surfaces where the velocities are low. Also, narrow openings will tend to become plugged.

Coke will also accumulate on surfaces where the velocities are low, and will tend to form on any surfaces that are not continuously wetted.

Knock-out drums are recommended for waxy feeds and cyclones for coking feeds. Multicyclone separators and filter separators can handle solids as long as they are not sticky.

Vane pack separators may be used provided the service is only slightly fouling and the vane structure is sufficiently open. Demister mats shall not be used because of the danger of plugging.

- The criterion for the nozzle sizing is that the momentum of the feed shall not exceed prescribed levels. The maximum allowable inlet momentum can be increased by applying inlet devices.
- The momentum criteria are given in the next section.
- The function of the inlet device is to initiate the gas–liquid separation and to distribute the gas flow evenly in the gas compartment of the vessel.
- Commonly used inlet devices are the half-open pipe and their specific proprietary inlet devices designed for introducing gas–liquid mixtures into a vessel or column.

4.12.1 Sizing of the feed and outlet nozzles

The sizing of the nozzles shall be based on the actual flow rates (i.e. excluding the appropriate design margin).

- **Feed inlet nozzle**

The internal nozzle diameter, d_1, may be taken equal to that of the feed pipe, but also a momentum criterion (dependent on the inlet device, if any) shall be satisfied.

   If no inlet device is used:

$$\rho_m V_{m,in}^2 < 1000 \text{ Pa} \tag{4.41}$$

where:

   ρ_m is mean density of the mixture in the feed pipe

$$= (M_g + M_l)/(Q_g + Q_l); \tag{4.42}$$

and

   $V_{m,in}$ is velocity of the mixture in the inlet nozzle

$$= (Q_g + Q_l)/(\pi d_1^2/4) \tag{4.43}$$

If a half-open pipe is used as inlet device:

$$\rho_m V_{m,in}^2 \le 1500 \text{ Pa} \tag{4.44}$$

If a specific inlet proprietary device is used, the respective $\rho_m V_{m,in}$ value should be specified.

- **Gas outlet nozzle**

The diameter of the gas outlet nozzle, d_2, should normally be taken equal to that of the outlet pipe, but the following criterion shall be satisfied:

$$\rho_g V_{g,out}^2 \le 3750 \text{ Pa} \tag{4.45}$$

In high vacuum units where a precondenser is used in overhead systems, this criterion may result in a high outlet velocity, leading to a pressure drop that is too high.

   In that case it is recommended to size the gas outlet nozzle such that the pressure drop requirements between column and downstream system are met.

- **Liquid outlet nozzle**

The diameter of the liquid outlet nozzle, d_3, shall be chosen such that the liquid velocity does not exceed 1 m/s. The minimum diameter is 0.05 m (2 in). The nozzle shall be equipped with a vortex breaker.

- Separator internals
  - In knock-out vessels, the diameter should be selected sufficiently large to keep the gas velocity low at which the major portion of the droplets could be settled by gravity.
  - In all other types of gas–liquid separators, internals should be considered. The required duty, wire mesh, vane-pack (either horizontal or vertical flow), multicyclones axial or reversed flow, filter candles, etc. should be studied.

- Gas and liquid outlets

After completion of the gas–liquid separation process, the two phases will leave the vessel via the gas and liquid outlet, respectively.

- Gas handling capacity

The separator shall be large enough to handle the gas flow rate under the most severe process conditions. The highest envisaged gas flow rate should be determined by including a margin for surging, uncertainties in basic data. This margin is typically between 15 and 50%, depending on the application. For the recommended margin see the next section.

4.12.2 Design margin *S* for separator

To determine the highest envisaged volumetric load factor for vessel design, the following design margins (surge factor) are recommended, in Table 4.7:

- Selection strategy

To facilitate the choice of a separator for a given application, the performance characteristics of various separators are summarized in Table 4.7.

Table 4.7 In Exploration and Production	**Design Margins (Surge Factor)**
(a) Offshore Service	
Separator handling natural-flowing production from	
Its own platform	1.2
Another platform or well jacket in shallow water	1.3
Another platform or well in deep water	1.4
Separator handling gas lifted production from	
Its own platform	1.4
Another platform or well jacket	1.5
(b) Onshore Service	
Separator handling natural-flowing production, or gas plant inlet separator in	
Flat or low rolling country	1.2
Hilly country	1.3
Separator handling gas lifted production in	
Flat or low rolling country	1.4
Hilly country	1.5
In Oil, Gas, and Chemicals Manufacturing	
The design margin ranges typically from	1.15 to 1.25

1. Gas handling capacity:
   a. max. capacity (gas load factor);
   b. turndown ratio.
2. Liquid removal efficiency:
   a. overall;
   b. with respect to fine mist;
   c. with respect to the possible flooding above the maximal load factor (which will affect the sharpness of the efficiency decline above the maximum capacity).
3. Liquid handling capacity:
   a. slugs;
   b. droplets.
4. Fouling tolerance:
   a. sand;
   b. sticky material.
5. Pressure drop:

The following selection strategy is suggested:

First define the mandatory requirements that the separator shall satisfy. With the aid of Table 4.7, a number of separators can then be ruled out.

4.12.3 Design criteria

Unless explicitly stated otherwise, both the maximum gas and liquid flow rates should contain a design margin or surge factor.

- **Vertical and horizontal separators**

Specific process application, characteristics, and recommended and non-recommended use of various vertical/horizontal separators in OGP production plants are given here for design consideration:

- Vertical knock-out drum
  Application:
  - Bulk separation of gas and liquid.
  Characteristics:
  - unlimited turndown;
  - high slug-handling capacity;
  - liquid removal efficiency typically 80–90% (ranging from low to high liquid load);
  - Liquid removal efficiency for mist is very poor;
  - very low pressure drop;
  - insensitive to fouling.
  Recommended use:
  - vessels where internals have to be kept to a minimum (e.g. flare knock-out drums);
  - fouling service, e.g. wax, sand, asphaltenes;
  - foaming service.
  Non-recommended use:
  - where efficient demisting of gas is required.

Typical process applications:
- – vent and flare stack knock-out drums;
- – production separator;
- – bulk separator (e.g. upstream of gas coolers);
- – flash vessel.

- Horizontal knock-out drum

Application:
- – Bulk separation of gas and liquid.

Characteristics:
- – can handle large liquid fractions;
- – unlimited turndown;
- – very high slug-handling capacity;
- – liquid removal efficiency typically 80–90% (ranging from low to high liquid load);
- – Liquid removal efficiency for mist is very poor;
- – insensitive to fouling;
- – very low pressure drop.

Recommended use:
- – vessels where internals have to be kept to a minimum and where there are height limitations;
- – slug catchers;
- – fouling service, e.g. wax, sand, asphaltenes;
- – for foaming or very viscous liquids.

Non-recommended use:
- – where efficient demisting of gas is required.

Typical process applications:
- – vent and flare stack knock-out drums;
- – production separator-low gas oil ratio (GOR);
- – bulk separator;
- – slug catcher.

- Vertical wire mesh demister

Application:
- – demisting of gas.

Characteristics:
- – high turndown ratio;
- – high slug-handling capacity;
- – liquid removal efficiency >98%;
- – sensitive to fouling;
- – low pressure drop.

Recommended use:
- – for demisting service with a moderate liquid load;
- – where slug-handling capacity may be required.

Non-recommended use:
- – fouling service (wax, asphaltenes, sand, hydrates);
- – for viscous liquids where degassing requirement determines vessel diameter;

    – for compressor suction scrubbers unless precautions are taken to prevent the possibility of loose wire cuttings entering the compressor or plugging of the demister mat, increasing suction pressure drop.

Typical process applications:

    – production/test separator;
    – moderate GOR;
    – non-fouling.
    – inlet/outlet scrubbers for glycol contactors;
    – inlet scrubbers for gas export pipelines;
    – for small-diameter and/or low pressure vessels, where extra costs of vane of SMS internals cannot be justified.

- Horizontal wire mesh demister

Application:

    – demisting of gas where a high liquid handling capacity is required.

Characteristics:

    – high turndown ratio;
    – very high slug-handling capacity;
    – liquid removal efficiency >98%;
    – sensitive to fouling;
    – low pressure drop.

Recommended use:

    – typically for demisting service with a high liquid load and low GOR;
    – applied where slug-handling capacity may be required;
    – for viscous liquids where liquid degassing requirement determines vessel diameter;
    – in situations where head room is restricted;
    – for foaming liquids.

- Vertical vane-type demister

Application:

    – demisting of gas.

Characteristics:

    – liquid removal efficiency >96%;
    – moderate turndown ratio;
    – suitable for slightly fouling service (if without double-pocket vanes);
    – robust design;
    – sensitive to liquid slugs (in-line separator cannot handle slugs).

Recommended use:

    – typically for demisting service;
    – in-line separator to be used only with relatively low flow parameter ($\varphi_{feed} < 0.01$);
    – two-stage separator to be used if $\varphi_{feed} \geq 0.01$;
    – attractive for slightly fouling service (if without double-pocket vanes);
    – may be used where demister mats may become plugged, i.e. waxy crudes.

Non-recommended use:

    – heavy fouling service (heavy wax, asphaltenes, sand, hydrates);
    – for viscous liquids where degassing requirement determines vessel diameter;

- the in-line vertical flow vane pack separator shall not be used where liquid slugging may occur or where $\varphi_{feed} \geq 0.01$;
- if pressure exceeds 100 bar (abs), due to the consequent sharp decline in liquid removal efficiency.

Typical process applications:
- compressor suction scrubbers, where vane packs are preferred to demister mats since their construction is more robust;
- demisting vessels with slightly fouling service.

- Horizontal vane-type demister

Application:
- demisting of gas where a high liquid-handling capacity is required.

Characteristics:
- liquid removal efficiency >96%;
- moderate turndown ratio;
- suitable for slightly fouling service (if without double-pocket vanes);
- high slug handling capacity;
- robust design.

Non-recommended use:
- heavy fouling service (heavy wax, asphaltenes, sand, hydrates);
- if pressure exceeds 100 bar (abs).

Typical process applications:
- production separator where GOR is low and the service is slightly fouling.

- Cyclone

Application:
- demisting of gas in fouling service.

Characteristics:
- liquid removal efficiency >96%;
- insensitive to fouling;
- limited turndown ratio;
- high pressure drop.

Recommended use:
- typically for use in a fouling (e.g. coke-formation) environment and where a high demisting efficiency is still required.

Non-recommended use:
- if high pressure drop cannot be tolerated.

Typical process application:
- in oil refineries: thermal gas–oil unit; visbreaker unit;
- in chemical plants: thermoplastic rubber plants.

- Vertical multicyclone separator

Application:
- demisting and de-dusting of gas in slightly fouling service and high pressure.

Characteristics:
- liquid removal efficiency >93%;
- suitable for slightly fouling service (e.g. low sand loading);

- high pressure drop;
- compact separator;
- sensitive to high liquid loading or slugs.

Recommended use:
- typically for use in a slightly fouling environment where the gas pressure is higher than 100 bar (abs) and a compact separator is required.

Non-recommended use:
- low gas pressure;
- heavy fouling service (high sand loading will cause erosion);
- high liquid loading;
- slug;
- when high liquid removal efficiency is required.

Typical process application:
- wellhead separators;
- primary scrubbers under slightly fouling service and when the liquid loading is low;
- compressor suction scrubbers if sand is present in the feed.

- Filter Separator

Application:
- after cleaning (liquid and solids) of already demisted gas when a very high liquid removal efficiency is required.

Characteristics:
- liquid removal efficiency >99%;
- very high pressure drop;
- sensitive to high liquid loading or slugs;
- sensitive to fouling by sticky material.

Recommended use:
- typically as a second-line gas–liquid separator to clean the gas stream exiting from the first-line gas–liquid separator;
- use filter candles with the flow from OUT to IN where solids are present;
- use filter candles with the flow from IN to OUT where ultimate efficiency is required and NO solids are present;

Non-recommended use:
- heavy fouling (sticky material) service;
- high liquid loading;
- slugs.

Typical process application:
- last demisting stage of natural gas prior to dispatch for sale.

4.13 Solid–liquid separators

4.13.1 Multi-cyclone scrubbers/separators

High efficiency separation of solids and liquids. For detailed applications see Table 4.8.

The multicyclone gas separator/scrubber offers an economical way to remove solid and liquid particles from a gas stream.

- **Advantages**:
    - On high pressure applications, the reduced diameter of the multicyclone is smaller than the vane type and markedly smaller than the demister type. This means thinner walls, reduced space, and most importantly reduced costs (especially on larger flow rates).
    - The multicyclone is able to handle and remove solid particles whereas the vane and demister type principally do not.

Note:
    The vane and demister type separators will remove small solids suspended in liquid.

- Like the wire mesh demister and vane type, the multicyclone separator is a high-capacity separator designed to operate with a low pressure drop.
    - Tables 4.8 and 4.9 also represent a separator type selection guide for other types of separators for any reference.
- The multicyclone is self cleaning and needs only a periodic blowdown for removal of collected material.
- It is highly efficient over a wide range of operating conditions.

Table 4.8 Separator Type Selection Guide

Application	Vertical Separator with Vane	Demister	Inline Vane	Cyclone	Separator Type		
					Multi-cyclone	Filter	Filter Separator
Gas Transmission							
Station scrubber	×	×			×		×
Fuel gas	×				×	×	×
City gates/ metering	×	×			×	×	×
Pressure reduction	×	×	×		×		×
Compressor station	×	×	×		×		
Process							
Glycol dehydrators		×	×	×	×		
Amine contactors		×	×	×	×		
Compressors			×		×		×
Gathering	×	×		×			
Pressure reduction	×	×	×		×		×
Wellhead	×				×		
"×" means it is recommended.							

Table 4.9 Gas Separator Types along with Typical Separation Performance

Separator Type	Slug Handling Capacity	Solid Removal Efficiency If Any (μ = microns (μm))	Liquid Removal Efficiency
Multicyclone scrubbers	Yes	100% 8 μm & larger 99% 5–8 μm	100% 5 μm & larger
Vertical gas separator with vane	Yes	No	100% 8 μm & larger
In-line vane	No	No	100% 8 μm & larger
Vertical separator with wire mesh demister	Yes	No	98% 8 μm & larger
Filter separators with vane	Yes	100% 3 μm & larger 99% 0.5–3 μm	100% 3 μm & larger
Dry gas filters	No	100% 10 μm & larger 99% 1–3μm	

4.13.2 Removal efficiency

Based on Figure 4.20 generally, the efficiency of the separator is dependent on the following factors:

- Particle size
- Distribution of particles
- Liquid loading

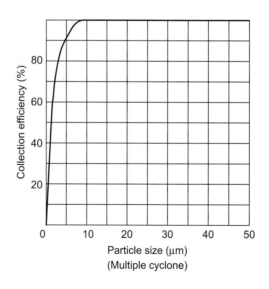

FIGURE 4.20

Efficiency curve for multi-cyclone separator solid particle removal.

- **Liquid efficiency**

The outlet gas shall contain less than 1 m³/74.81 × 10⁶ m³ liquids to gas through the scrubber (based on a liquid to gas ratio of less than or equal to 5%). In addition, based on the same inlet assumption, all liquid particles 5 μm and larger shall be removed.

- **Solids**

Based on an assumed solid loading of 22.88 g/m³. The multicyclone separator will generally remove 100% of 8 μm and larger solid particles (see Figure 4.20).

- **Operation**

Figure 4.21 shows the general detail of multicyclone scrubber/separator.

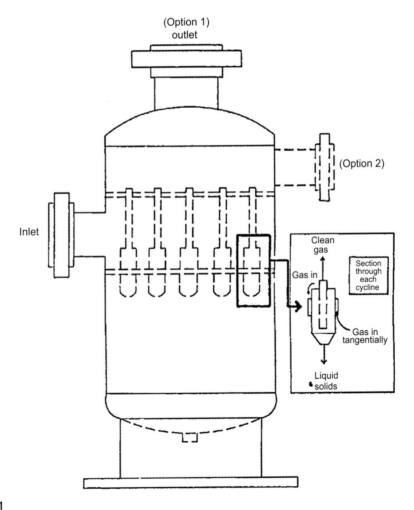

FIGURE 4.21

Sketch of multi-cyclone scrubber/separator.

4.13.3 Material specifications and construction

Multicyclone scrubbers are designed and manufactured in accordance with the main international codes, ASME VIII Div. 1, BS-5500 Cat. 1 & 2, etc. Materials of construction are as customer's specification, from standard carbon steel to stainless steel, and other steel alloys. Cyclone tubes are manufactured as standard from high-alloy cast steel.

4.13.4 Liquid–liquid separators

For liquid–liquid separators, the following requirements and criteria should be considered:

1. The vessels should be sized in such a way that the settling time for each liquid phase from the other is less than its residence time in the vessel itself.
2. Settling velocity for the dispersed droplets is calculated using Stokes', Newton's, or an intermediate law, according to the field of application.
3. A maximum settling value of 250 mm/min shall be considered for light hydrocarbons. It should be verified that the hold-up time necessary for any phase for settling satisfies process hold-up requirements.

For normal liquid droplet separation, droplet diameters ranging from 500 to 1000 μm are assumed as a guide for the design for vertical separators, and droplet diameter of about 100 μm are assumed as a guide for the design of horizontal separators.

4.13.5 Vertical separator

1. Liquid velocities

The terminal velocity V_t of any droplet moving in a medium can be calculated form Stokes' law. The settling rate of each phase from the other phase should be calculated to decide which phase is limiting.

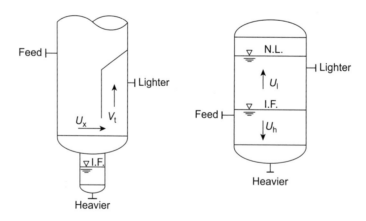

FIGURE 4.22

Vertical liquid–liquid separator.

2. Construction

Typical construction of a vertical liquid–liquid separator is shown in Figure 4.22. Feed nozzles shall be so arranged that the liquid–liquid interface may not be disturbed.

U_x is assumed to be equal to terminal velocity V_t, i.e. $U_x = V_t$.

U_h or U_l, whichever is the larger, is made equal to the terminal velocity V_t.

4.13.6 Horizontal separator

1. Liquid velocity

The liquid velocity can be determined by the equation given below, where D_p is assumed to be 100 μm:

2. Construction

Typical construction of horizontal liquid–liquid separator, is shown in Figures 4.23 and 4.24.

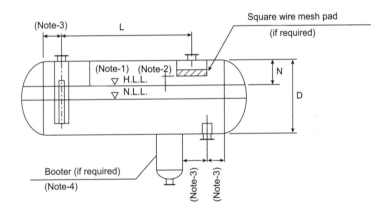

FIGURE 4.23

Horizontal separator.

Notes:

1. 20% of drum diameter of 300 mm (12 in), whichever is the greater.
2. Minimum 150 mm (6 in).
3. Minimum.
4. Boots.
   a. Boots shall be sized for a minimum residence time of 5 min as a guideline and their diameters shall be the same as the commercial pipe sizes as far as possible. The height-diameter ratio shall be 2:1–5:1.
   b. Boot diameters shall be 250 mm φ minimum, because of good operability.
   c. Maximum boot diameters shall be one-third of the drum inside diameter.

FIGURE 4.24

Horizontal liquid–liquid separator.

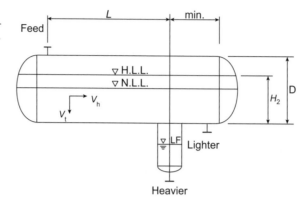

4.14 Typical equations, which can be used for terminal velocity calculation

4.14.1 Terminal velocity of a spherical fine particle settling under the influence of gravity

The terminal velocity of a spherical fine particle settling under the influence of gravity can be determined by Stokes' law:

$$V_t = \sqrt{\frac{4gD_p}{3C_D}} \sqrt{\frac{\rho_L - \rho_v}{\rho_v}} \tag{4.46}$$

$$V_t = k\frac{\rho_1 - \rho_v}{\rho_v} \tag{4.47}$$

where:

V_t is terminal velocity, in (m/s)
g is acceleration of gravity, in (m/s²)
D_p is diameter of fine particle, in (m)
C_D is drag coefficient
ρ_l is density of liquid, in (kg/m³)
ρ_v is density of vapor, in (kg/m³)
k is entrainment coefficient, in (m/s)

Based on theory and experimental results, C_D can be expressed approximately as follows. It should be noted that C_D is given as a function of Reynolds number, Re, where the velocity is the relative velocity of the fine particle to its surrounding fluid.

$$C_D = 24/Re \text{ for } Re < 2 \text{ (Stokes' law)} \tag{4.48}$$

$$C_D = 10/Re \text{ for } 2 < Re < 500 \text{ (Allen's law)} \tag{4.49}$$

$$C_D = 0.44 \text{ for } Re > 500 \text{ (Newton's law)} \tag{4.50}$$

Substituting these for C_D in Eqn (4.47).

$$V_t = \frac{gD_p^2(\rho_L - \rho_v)}{18\mu} \tag{4.51}$$

$$V_t = \left[\frac{4}{225}\frac{(\rho_L - \rho_v)^2 g^2}{\mu\rho_v}\right]^{\frac{1}{3}} D_p \tag{4.52}$$

$$V_t = \sqrt{3gD_p\frac{\rho_L - \rho_v}{\rho_v}} \tag{4.53}$$

4.14.2 Application to vertical vapor-liquid separator

Several types of wire mesh demisters are manufactured. The most widely used ones are given in Table 4.10:

"k" Value for wire mesh demisters, based on disengaging height (height between bottom of mesh, and liquid surface) of 300 mm minimum should be considered as in Table 4.11:

For variation of "k" values with disengaging height, see recommended values as in Table 4.12:

Disengaging heights must be a minimum of 460 mm in cases where wire mesh is provided. This height must be corrected for the vapor velocity in each individual case.

Provision of baffles must be considered. The reason for this is that the duty of vapor–liquid separation in the disengaging zone can be reduced by directing the liquid flow downward. Also, consideration has been given to avoid the unfavorable effects on wire mesh pad due to the conning of feed stream in cases where no baffle is provided. The same thought has been followed even in cases where no wire mesh is provided.

4.14.3 Application of these equations to vertical liquid–liquid separators

Using Allen's and Stokes' equations, the terminal velocity of the heavier particles can be calculated. Typical application of these equations to the separation of water from oil is given in Tables 4.13 and 4.14.

4.14.4 Horizontal separators

The following equation is based on Stokes' equation:

$$V_t = gD_p^2(\rho_h - \rho_L)\Big/18\mu$$

where:

g = acceleration of gravity, in 9.8 m/s^2

D_p = diameter of the heavier particle, in m

ρ_h (*rho*) = density of the lighter fraction, in kg/m^3

Table 4.10 Typical Identification of Wire Mesh Types

Density (kg/m³)	Surface Area (Note 1) (m²/m³)	Void Fraction	Wire Diameter (mm)	Manufacture's Type					Application
				York	Divment	Metex	Nihon Mesh	Naniwa Special Mesh	
192	36	0.977	0.275	421	4210	Xtradense standard	SL	3311	High efficiency, relatively clean, moderate velocity, use in 100 mm pad
144 (Note 2)	279	0.982	0.275 (Note 3)	431	4310	Nu-standard	N	3383	Standard efficiency, general purpose, use in 100 mm pad
128	459	0.984	0.15	326	3260	—	SN	—	Fine mist removal, use in 100–150 mm pad
80	151	0.990	0.275 (Note 3)	931	9310	HI-throughput	H	3346	High throughput or low density, for service containing solids or "dirty" material, use in 150 mm pad
219	902	0.972	5 strand each 0.122	—	—	—	T	4060	Very high efficiency, 250 and 300 mm recommended for special applications such as fine mists, oil vapor mist

Notes
(1) If the mesh is made of nickel, monel or copper, multiply the density values by 1.13, referenced stainless steel.
(2) Normally, 144 kg/m³.
(3) Some manufacturer fabricates with 0.254 mm wire.

Table 4.11 "k" Value for Wire Mesh Demisters

Service Condition	"k" Value	Mesh Type
Clean fluids, moderate liquid load, fits 90% of process situation	0.35–0.36 0.35 0.25	Standard High efficiency Very high efficiency
High viscosity, dirty suspended solids	0.4	• Low density or Herringbone, high throughput
Vacuum Operations		
• 50 mm Hg (abs.) • 400 mm Hg (abs.) • Corrosive chemicals	0.20 0.27 0.21	Standard or • High efficiency • Plastic-coated wire, or plastic strain

Table 4.12 Variation of "k" Values with Disengaging Height

Disengaging Height above Mesh, (mm)	Allowable "k" Value
75	0.12
100	0.15
125	0.19
150	0.22
175	0.25
200	0.29
225	0.32
250	0.35
275	0.38
300	0.40
325	0.42
350	0.43

Table 4.13 $D_p = 400{-}500$ μm

Physical Properties			
Relative Density (Sp. Gr.)	μ (cP)	Oil Fraction	Maximum Liquid Velocity, m/s
0.70	0.5	Naphtha	0.3
0.80	1.5	Kerosene	0.015
0.85	5.0	Gas–oil	0.003
0.90	10.0	Vac. Gas–oil	0.0015

Table 4.14 $D_p = 1000$ μm (1 mm)

Physical Properties			Maximum Liquid
Relative Density (Sp. Gr.)	μ (cP)	Oil Fraction	Velocity, m/s
0.70	0.5	Naphtha	0.8
0.80	1.5	Kerosene	0.04
0.85	5.0	Gas–oil	0.02
0.90	10.0	Vac. Gas–oil	0.005

ρ_L (*rho*) = density of the heavier fraction, in kg/m^3
μ (*mu*) = viscosity of the lighter fraction, in kg/m s

Substituting 9.8 m/s^2 for g and 100 μm for D_p, converting the densities to relative densities (specific gravities), and expressing the viscosity in centipoise:

$$V_t = \frac{gD_p^2(\rho_h - \rho_l)}{18\mu}$$

$$= 5.45 \times 10^{-3}\frac{S_h - S_l}{\mu} \ (m/s)$$

- **Crude oil separator**

Process design of crude oil separator shall be in accordance with requirements and criteria set forth in the following:

1. Sizing
   a. In three-phase separators, the following main functions should be considered:
      - Provision of enough residence time for the liquid so that degassing can occur.
      - Provision of enough free cross-sectional area above the liquid so that the gas velocity is low enough to allow liquid particles to be separated.
      - Provision of enough residence time for the liquid/liquid separation to occur.
   b. A separator should be either liquid or gas controlling, the gas/oil will determine which. Liquid residence time for a two-phase separator will depend on the rate of vapor breakout from the liquid. In this regard the following is recommended:
   Minimum for light gravity crudes (greater than 35 API) 2 min.
   20-35 API 4 min.
   15-20 API 6 min.
   c. When degassing water, a residence time of only 1 min is required. Recommended residence time for a three-phase separator is 3 min minimum. This can be more depending on inlet temperature, API gravity, inlet oil quality, and final oil quality required.
   d. Special considerations for sizing
      - Degassing of very light >40 API dry oil.
      - Degassing and dehydration of waxy crude.
      - Handling of foaming crude.

2. Other special separators
   a. Free water knock-outs are usually specified downstream of the degassing or first-stage separator. They are normally employed to remove large amounts of water from oil streams, i.e. 70–90% water.
      The free water knock-out is essentially a separator for which the sizing is completely controlled on liquid–liquid separation rather than gas.
      The inlet fluid enters and impinges on an inlet diverter as with a standard separator.
3. Free water knock-out special considerations
   a. The free water knock-out, as its name implies, is designed to remove free water, not emulsions, from oil.
   b. The use of demulsifying chemicals is usually a must to aid with separation, and even then outlet quality depends on the extent and tightness of the emulsion.
   c. As a guide, the following typical outlet ppm qualities of oil in water and water in oil emulsions will give an indication of achievable separation.

Oil in water emulsion

Heavy crude, <30 API-150 ppm outlet
Medium crude, 30 API to 45 API-100 ppm outlet
Light crude, >45 API-50 ppm outlet

Water in oil emulsion

Heavy crude, <30 API-5/6%
Medium crude, 30 API to 45 API-3/5%
Light crude, >45 API-2/3%
• The free water knock out (FWKO) is usually sized on a residence time between 15 and 30 min

4.15 Vessels

4.15.1 Code, regulations, and standards

All design and construction shall be in accordance with the latest edition of ASME Section VIII, Division 1, "Boiler and Pressure Vessel Code."

All pressure vessels shall be inspected and "U" stamped in accordance with ASME Code and also in accordance with other applicable Codes subject to Company's approval. Applicable Registry Certificates shall be provided for each vessel.

In all cases where more than one code or standard applies to the same conditions, the most stringent shall be followed.

4.15.2 Design

• **Pressure and temperature**

The design pressure shall be 350 kPa or 10%, whichever is greater, in excess of the maximum working pressure, except in special cases approved by the Company or when the Company specifies the design pressure. In addition, where process fluid static head or other appropriate loads significantly increase the internal pressure, the design pressure shall be increased accordingly for the vessel section concerned.

Vessels not designed for full vacuum to meet normal operating conditions but which may be subjected to partial vacuum when contents are being emptied or during steaming out shall be suitable for partial vacuum. As a minimum, vessel shall be designed for 50 kPa (ga) external pressure using ASME Code.

Where the above restrictions create excessive cost, vessels may be provided with an automatic vacuum relief, subject to written approval by the Company.

Vessels shall be designed so that each part subject to pressure will have strength equivalent to the new and cold working pressure of the shell and heads or integral external piping connections, whichever is lower.

The maximum allowable working pressure stamped on the nameplate shall be calculated for the "as built" vessel and shall be limited by the lesser of the shell, head, or flange rating.

The design temperature shall be at least 15 °C above the maximum working temperature. For cold service, the design temperature shall be 6 °C below the minimum operating temperature unless otherwise specified.

4.15.3 Corrosion

The minimum corrosion allowance shall be 1.6 mm for sweet service and 3.2 mm for sour service, unless specified otherwise.

No corrosion allowance shall be provided in high-alloy vessels.

Corrosion allowance shall be cladding or lining thickness for alloy clad or lined vessels.

Unless otherwise specified, vessel corrosion allowance shall be provided to all exposed surfaces of nonremovable internal parts and half this amount to surface of removable parts (except demister wire, column packing, etc.).

4.15.4 Design load

- **Loadings**

Loadings to be considered in designing a vessel shall be per the ASME Code Section VIII and shall include cyclic conditions and erection loadings.

- **Additional loadings**

Vessels containing circulating suspensions of solids in fluids (fluid–solids processes) shall be subjected to additional horizontal loadings.

- **During erection, start-up, or operation**

During erection, start-up, or operation, all applicable loads shall be considered as acting simultaneously, including either wind or earthquake, whichever governs.

- **During hydrostatic testing**

During hydrostatic testing, wind load (wind pressure) equivalent to 58 km/h (16 m/s) wind speed shall be considered acting simultaneously with the hydrostatic test load.

4.15.5 **Materials**

Materials shall be per ASME Code Section VIII or as specified on the Vessel Data Sheet. Proposals to use any other materials shall be submitted to purchaser for approval by Owner's Engineer.

Material production shall be by the electric furnace, basic oxygen, or open hearth processes.

Proposals to use materials having a specified maximum tensile strength greater than 620 MPa at room temperature shall be submitted to purchaser for approval by the Company.

For vessels in hydrogen service, external welded attachments, and at least a 450 mm course of all skirts, shall be of the same nominal chemistry as the material used for the vessel. Proposals to use alternative materials shall be submitted to purchaser for approval by the Company.

All carbon steel and low-alloy material used in sour service shall comply with, but not be limited to, the requirements of NACE Standard MR-01-75, (Material Requirements-Sulphilde Stress Cracking Resistant Material for Oilfield Equipment, latest Revision).

4.15.6 **Documentation**

A manufacturer's data report shall be furnished, and shall contain the same information as required by form U-1 of the ASME Code Section VIII, Division 1. If the vessel is constructed for use at a location where the ASME Code is not mandatory (if approved by the Company), it shall be noted on the form that the vessel does not carry an ASME Code symbol.

4.15.7 **Internals**

All gas scrubbers, contactors, and separators shall be provided with 316 stainless steel wire mesh mist eliminator(s) of 150 mm minimum thickness, adequately supported and fastened to prevent displacement under surge conditions.

All demister pads shall be fabricated in sections to permit their removal through manways, where applicable.

The following criteria should be considered in vessels, tower's internal design:

1. The minimum thickness of removable tray plates shall be 2 mm. Any corrosion allowance specified for the tray parts under consideration shall be added to this thickness. Active devices may be less than 2 mm thick for process reasons.
2. Each tray shall be designed for an upward pressure differential and shall be specified in the data sheet.
3. Valve type proprietary tray designs shall have been tested by Fractional Research Incorporated and shall be of a type approved by the Company.

Trays with the capped perforated areas are not acceptable.

4. Thickness
   a. Minimum thickness for 13 Cr. Monel and 18/8 stainless steel assemblies shall be as follows:
      Tray and related components 2 mm
      Support ring, downcomer bars 6 mm when alloy 10 mm and when
      parts welded to vessel carbon steel
      Other parts 3 mm

b. Minimum thickness for carbon steel assemblies shall be as follows:

Tray and related components 2 mm
Support rings, downcomer bars and 10 mm
Parts welded to vessel
Other parts 3 mm

5. Corrosion allowance

a. No corrosion allowance is required for Monel or stainless steel assemblies.

b. The corrosion allowance for all surfaces of floor beams trusses or other support members of carbon steel assemblies shall be 1 mm minimum.

6. Loading

a. The design live load for tray assemblies at operating temperatures shall be based on a liquid height of 50 mm above weirs at a density of 800 kg/m^3, with a minimum of 100 kg/m^2. For areas under downcomers, design live load shall be 100 kg/m^2.

b. For maintenance purposes, all assemblies shall be designed for a concentrated load of 135 kg at any point.

Trays in columns 900 mm inside diameter and greater shall incorporate manways.

All internal manways shall be designed to be opened from either side of the tray and shall be gasketed to prevent leakage of liquids.

Tray spacing is recommended to be 610 mm.

Trays shall be bolted to support rings and not welded to the vessel wall and shall be in bolted sections to facilitate their removal through vessel manways.

For columns less than 900 mm inside diameter, cartridge trays shall be employed and vessel break flanges shall be provided to facilitate tray package removal.

Impingement plates attached to shell or baffles shall be provided where severe erosion may occur, such as opposite inlet connections. These and all internal supports shall be attached with full fillet seal welds.

Vortex breakers shall be provided on liquid bottom outlet nozzles from vessels as follows:

1. Pump suction connection;

2. Vessels where two liquid phases may be present.

4.15.9 Miscellaneous requirements

All nozzles over DN 40 (1-½ in) shall be flanged. Connections DN 40 (1-½ in) and smaller may be made with forged steel couplings. Such connections shall be limited to vessels for which the design pressure and temperature is less than 42 bar (ga) (600 psig) and 232 °C (450 °F), respectively. Couplings shall be 420 bar (ga) (6000 psig) rating for DN 40 (1-½ in) and smaller connections. Couplings shall not be used in lined portions of alloy lined vessels, on bottom heads of vertical vessels. Threaded fittings or tapped holes are not permitted. The minimum size of nozzles shall be DN 25 (1 in), except that for alloy lined nozzles the minimum size is DN 40 (1-½ in). For vessels in hydrogen service, minimum size connection shall be DN 25 (1 in) and all connections shall be flanged.

All vessels 900 mm inside diameter (ID) or greater shall be provided with at least one 450 mm inside diameter manhole opening. Davits or hinges shall be provided for handling manhole covers.

Vessels less than 900 mm inside diameter shall be provided with two 168.3 mm outside diameter (OD) hand holes unless otherwise specified.

Drain connections shall be flushed with the bottom of the vessel. These and internal siphon drains shall as a minimum be Schedule 160 pipe.

Provide one manway in single cross-flow trays, two manways in double cross-flow trays, removable from above and below. Manways shall be as close to tray centerline as possible and as nearly aligned as practicable.

All trays are equipped with one manway top and bottom removable for each liquid flow as per 7.9.5 above, except where the free space between tray and beams is less than 400 mm. In this case manways are to be provided on both sides of beams.

Specification sheets of reactors shall contain, but not be limited to, the following information:

1. Capacity.
2. Space velocity (normal and design conditions).
3. Conversion per pass (normal and design conditions).
4. Design-temperature and pressure.
5. Recycling (normal and design conditions).
6. Reactor pressure drop (SOR & EOR).
7. Reactor bed design and bed life.
8. Catalyst and its characteristics.
9. Reactor inlet and outlet conditions (normal and design conditions).
10. Reactor bed temperature profile, etc. (normal and design conditions).
11. Reactor lining.
12. Corrosion allowance.
13. Process fluid complete physical properties (normal and design conditions).
14. Operating and design conditions.
15. Stress relieving and insulation.
16. Materials of construction.
17. Stress relieving.
18. Insulation requirements.
19. Type of trays.
20. Tray numbers and details for columns.
21. Pedestal height.
22. Details of special internals such as pans, distributors, etc.
23. Mist eliminators, supports, mesh, or packing, etc.
24. Basic recommendation for spares for commissioning and 2 years of operation.
25. Instrumentation requirements.

4.15.10 Internals for fixed bed reactors

Table 4.15 lists the basic practices and standards that may be used as an acceptable typical material for fixed bed reactor internals.

Design requirements are as follows:

1. Vessel internals that contribute to the total reactor height shall be designed for minimum height. Similar parts shall be interchangeable where possible.
2. All internals, except shrouds, shall be removable with vertical thermowells in place, and shall be designed to pass through the nearest manhole above their level.

Table 4.15 Typical Materials

| Materials | ASTM Standards | | | | |
	Plate	Sheet	Strip	Bars	Bolts & Nuts
Carbon steel	A283, A285, A36	A414, A569	A570	A675	A307, Gr B
Low and intermediate alloy steels C-1/2 Mo	A204	As specified for high alloy steels			
1-1/4 Cr-1/2 Mo through 5 Cr-1/2 Mo	A387				
High alloy steels					A193 B6, and A194 Gr 6F (with Selenium), or Gr 8
12 Cr: welded components	A176 and A240, types 405 and 410S			A276 Type 405	
12 Cr: non-welded components	A176 and A240, types 405 and 410S			A276 Type 405 Or 410	
18 Cr 8 Ni: Types 304, 316, 321, 347	A167 and A240			A276	A193 B8 and A194 Gr 8
Non-ferrous nickel copper (monel)	B265 Gr2			B164	B164
Titanium	B265 Gr2			B348 Gr 2	B348 Gr 4

Vendor's proposals to use materials or thickness alternative to those specified shall be submitted to purchaser for approval by the Company.

3. All removable internals shall be designed to permit installation and removal from the top side.
4. Screens shall be attached on top of grids and catalyst support hardware to prevent inerts and catalyst from:
   a. Falling through holes or slot openings.
   b. Blocking clearances in support hardware that are required for thermal expansion.

Loads are as follows:

1. Reactor internals shall be designed to support their own mass plus specified design live loads. For trays and decks, this design shall be based on a corroded thickness of 1.5 mm.
2. Maximum deviation from the horizontal for liquid distributor trays under loaded conditions shall not exceed 1/900 of the reactor diameter.
3. Maintenance loads.

Support members as defined in Table 4.16 shall be designed for a concentrated live load of 135 kg at any point based on the allowable stress. This design shall be based on the corroded thickness of the support members; i.e. total thickness excluding corrosion allowance.
   Minimum metal thickness and corrosion allowance as follows:

1. For internals fabricated from sheet, plate, or strip, and for internal piping, the minimum acceptable total metal thickness (including corrosion allowance) is given in Table 4.16 for the specified

Table 4.16 Metal Thickness of Internals, mm (in)

	Vessel Materials [1]		Corrosion Design Category	Vessel Shell Corrosion Allowance, mm or Metal Lining or Cladding Thickness, mm					
	Shell (interior surface)	Internals		0.25	0.8	1.6	3.0	4.5	6.0
Non-supporting members, such as: trays, decks, integral minor beams, liquid redistributor	CS	CS	10 year	–	–	2.0	3.5	4.5	5.5
			15 year			2.8	4.5	5.5	5.5
			5 year			2.0	4.5	← 2.0 Alloy →	
	CS	Alloy	All	–	–	← 2.0 →			
	Alloy	Alloy [5]	10 year	2.0	2.0	2.8	3.5	4.5	Notes (2), (4)
			15 year			2.8	4.5	5.5	
			5 year			2.0	4.5		
Tray accessories such as: bubble caps, chimneys, weirs, baffles	CS	CS	10 year	–	–	1.6	2.8	3.5	5.5
			15 year			1.6	3.5	4.5	5.5
			5 year			1.6	3.5	4.5	5.5
	CS	Alloy	All	–	–	← 1.6 →			
	Alloy	Alloy [5]	10 year	1.6	1.6	1.6	2.8	4.5	5.5
			15 year	1.6	1.6	1.6	2.8	4.5	5.5
			5 year	1.6	1.6	1.6	2.8	4.5	5.5
Internal piping (non-pressure)	CS	CS	All	–	–	← Schd STD →		← Schd XS →	
	CS	Alloy		–	–	← Schd 10 →			
	Alloy	Alloy [5]		← Schd 10 ® →		← Schd STD →			Schd XS

Notes:
(1) Abbreviations used in the above table are.
CS designates carbon, low alloy and intermediate alloy steels. Alloy designates stainless steels (300 and 400 series), nickel alloys, copper alloys; aluminum alloys. TCA total corrosion allowances.
(2) Trays, Decks, etc. shall be at least 2.0 mm thick.
(3) Tray Accessories shall be at least 1.6 mm thick.
(4) Material and metal thickness to be specified.
(5) Internals shall be of the same material as the interior surface of the vessel shell.
(6) Metric conversion of metal thickness of internals (ex. internal piping) and Corrosion Allowance values in Table 4.17 shall be as follows.
(7) Vessel shell corrosion allowance for 5, 10 and 15 year, design categories are based on the maximum values for a given range of corrosion rates.
(8) Thickness specified for each vessel shell corrosion allowance are designed so that successively lower corrosion rates are used for each (5, 10, 15 year) corrosion design category.

Corrosion Design Category and Vessel Shell Corrosion Allowance (or Metal Lining or Cladding Thickness). Unless otherwise specified, Corrosion Design Category "10 year" shall be used.

2. Supporting members formed as an integral part of the tray deck shall have the same total thickness as the tray.

3. The Total Corrosion Allowance (TCA) to be added to the design thickness of support members (major beams, supporting, etc.) is given in Table 4.16 for the specified Corrosion Design Category and Vessel Shell Corrosion Allowance (or Metal Lining or Cladding Thickness).

Distributor trays, quench decks, splash decks, and bed support grids as:

1. The nominal diameter of trays, decks, and grids shall be determined to the nearest 5 mm (¼ in) per the following:

$$\text{Diameter} = \text{Vessel ID} - [1\% \text{ Vessel ID} + 19 \text{ mm } (3/4 \text{ in})]$$

Table 4.17 Metric Conversion of Metal Thickness of Internals (ex. Internal Piping) and Corrosion Allowance Values

Corrosion Allowance				Metal Thickness					
mm	inch	mm	in	mm	in	mm	in	mm	in
0.25	(0.010)	1.5	1/16	0.25	0.01	2.8	0.105	1.6	1/16
0.8	(0.030)	3	1/8	0.8	0.03	3.5	0.134	3	1/8
1.5	(0.060)	4.5	3/16	1.6	0.06	4.5	0.179	4.5	3/16
		6	1/4	2	0.075	5.5	0.224	6	1/4
		9	3/8						
		12	1/2						

2. A minimum of 19 mm (¾ inch) overlap under the most adverse operating conditions shall be provided between the support ring and the OD of trays, decks, and grids.
3. Bolt hole spacing around the edge of Tray, deck, or grid sections shall not exceed 177 mm.
4. Access through grids, decks, and trays shall be provided either by split construction or by use of manways, as follows:
    a. Manways shall provide a minimum rectangular opening of 380 × 460 mm except as provided in Sub-par. d.2.
    b. Access through liquid distributor trays shall be at least 600 mm wide, and of sufficient length to permit catalyst leveling and arrangement of the inert ball layer at the top of each bed by a man lying on the tray. Assume a maximum reach 900 mm from the edge of any opening.

Internal pipes and thermowells

- Thermowells shall be designed to resist collapse due to pressure in the fully corroded condition.
- Clearance between thermowell and thermowell nozzle shall be minimized.
- Side-entering thermowells shall be structurally supported within the vessel.
- Internal distribution pipes shall have flanged and gasketed connections and shall have their ends blanked.

Expansion guides shall be provided for vertical thermowell installation.

- Tray guides shall be provided to minimize liquid leakage at points where thermowells or catalyst dump tubes pass through trays. Highest liquid level will be specified.

Reference

[1] Gas processors and suppliers (GPSA) engineering data book. 12th ed. 2004. Tulsa (OK, USA).

Further reading

American Petroleum Institute Spec. Specifications for oil and gas separators. 7th ed.; October 1989.
American Petroleum Institute, RP 521. Guide for pressure relieving and depressuring systems. 4th ed.; March 1997. p. 64.

American Petroleum Institute. Manual on disposal of refinery wastes. 6th ed., vol. 1; 1959. 18–20, and private industry data.

API (American Petroleum Institute) API RP 521. Guide for pressure-relieving and depressuring systems. 2nd ed; 1982.

API (American Petroleum Institute) API Spec. 12J. Specification for oil and gas separators. 7th ed.; October 1989.

API. Glossary of terms used in petroleum refining. 2nd ed; 1962.

ASME (American Society of Mechanical Engineers) ASME Section VIII, Div. 1. Boiler and pressure vessel, code.

Bahadori A, Mokhatab S. Optimizing multistage separators pressure set points maximizes oil recovery. World Oil 2007;228(6):101–5.

Bahadori A, Mokhatab S. Simple methodology predicts optimum pressures of multistage separators. Pet Sci Technol 2009;27(3):315–24.

Bahadori A, Vuthaluru HB, Mokhatab S. Optimizing separators pressures in the multistage crude oil production unit. Asia-Pac J Chem Eng 2008;3(4):380–6.

BSI (British Standards Institution) BS 5500 Cat. 1 & 2.

Dickey GD. Filtration. New York: Reinhold Publishing Corporation; 1981.

Fabian P, Cusack R, Hennessey P, Newman M. Demystifying the selection of mist eliminators. Part I. Chem Eng; November 1993.

Fewel Jr KJ, Kean JA. Computer modelling and separator retrofit. Oil Gas J; July 6, 1992:76–80.

Fewel Jr KJ, Kean JA. Vane separators in gas/liquid separation. Pipeline Eng; 1992:46.

Gerunda A. How to size liquid-vapor separators. Chem Eng May 4, 1981;91(7):81–4.

Heinze AJ. Pressure vessel design for process engineers. Hydrocarb Process; May 1979:181–91.

Hreiz R, Lainé R, Wu J, Lemaitre C, Gentric C, Fünfschilling D. On the effect of the nozzle design on the performances of gas-liquid cylindrical cyclone separators. Int J Multiph Flow 2014;58:15–26.

IPC (The Institute of Petroleum Constructors, Calgary Canada) Technical Bulletin for Multicyclone separators, Technical Bulletin for Crude oil separators.

Kim KS, Kim MH, Park J-C. Dynamic coupling between ship motion and three-layer-liquid separator by using MPS (moving particle simulation). In: Proceedings of the international offshore and polar engineering conference; 2013. pp. 366–71.

Ludwig EE. Applied process design for chemical and petrochemical plants, vol. 1. Houston (TX): Gulf Publishing Co; 1964. p. 126–159.

Mahsakazemi. Optimization of oil and gas multi stage separators pressure to increase stock tank oil. Orient J Chem 2011;27(4):1503–8.

Mott Jr S, Brown GG. Design of fractionating columnsdentrainment and capacity. Ind Eng Chem January 1934; 265(1):98.

NACE (National Association of Corrosion Engineers) NACE Standard MR 0175, Material requirements-sulphide stress cracking resistant material for oilfield equipment, Latest revision.

Pearce RL, Arnold JL. Glycol-hydrocarbon separation variables. In: Proceedings gas conditioning conference. University of Okalahoma; 1964.

Perry JH, editor. Chemical engineers' handbook. 3rd ed. New York: McGraw-Hill; 1950. pp. 1013–50. Section 15 dust and mist collection by Lapple CE.

Perry Jr D. What you should know about filters. Hydrocarb Process April 1966;45(4):145–8.

Perry RH, editor. Chemical engineers' handbook. 3rd ed. McGraw-Hill Book Company; 1950. p. 1019.

Perry RH, editor. Chemical engineers' handbook. 5th ed. McGraw-Hill Book Company; 1973 [Chapter 5], p. 5–61–65.

Powers ML. Analysis of gravity separation in free-water knockouts. In: SPE 18205, 63rd annual technical conference; October 1988. Houston, TX.

Reid LS. Sizing vapor liquid separators. In: Proceedings gas conditioning conference. University of Oklahoma; 1980. p. J-1–J-13.

Schweitzer PA. Handbook of separation techniques for chemical engineers. McGraw-Hill; 1979.

Sivalls CR. Technical Bulletin No. 133. Odessa, Texas: Sivalls, Inc; 1979.

Swanborn RA, Koene F, de Graauw J. New separator internals cut revamping costs. J Pet Technol; August 1995: 688–92.

York OH, Poppele EW. Wire mesh mist eliminators. Chem Eng Prog June 1963;59(6):45–50.

York OH. Performance of wire mesh demisters. Chem Eng Prog August 1954;50(8):421–4.

Gas Compressors

5

This chapter covers the minimum requirements, basic reference data, and necessary formulas for process calculations and proper selection of compressors to be used in the oil and gas processing (OGP) industries. Compressors are dealt within four groups; axial, centrifugal, reciprocating, and rotary, and each are covered in separate sections.

Compressors are generally divided into three major types: dynamic, positive displacement, and thermal, as shown in Figure 5.1.

Positive displacement types fall into two basic categories: reciprocating and rotary. The reciprocating compressor consists of one or more cylinders each with a piston or plunger that moves back and forth, displacing a positive volume with each stroke. The diaphragm compressor uses a hydraulically pulsed flexible diaphragm to displace the gas. Rotary compressors cover lobe-type, screw-type, vane-type, and liquid ring type, each having a casing with one or more rotating elements that either mesh with each other, such as lobes or screws, or that displaces a fixed volume with each rotation [1].

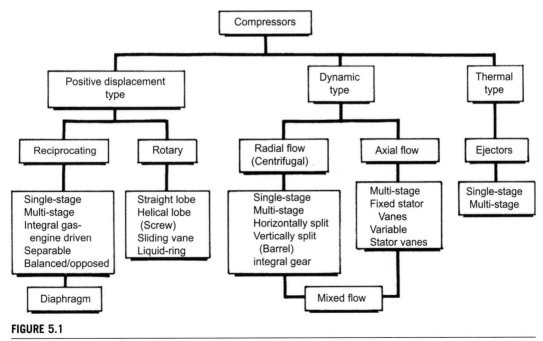

FIGURE 5.1

Compressors types.

Ref. [1].

Natural Gas Processing. http://dx.doi.org/10.1016/B978-0-08-099971-5.00005-2
Copyright © 2014 Elsevier Inc. All rights reserved.

The dynamic types include radial flow (centrifugal), axial flow, and mixed flow machines. They are rotary continuous flow compressors in which the rotating element (impeller or bladed rotor) accelerates the gas as it passes through the element, converting the velocity head into static pressure, partially in the rotating element and partially in stationary diffusers or blades [1].

Ejectors are "thermal" compressors that use a high-velocity gas or steam jet to entrain the inflowing gas, then convert the velocity of the mixture to pressure in a diffuser [1].

The type of compressor to be used shall be the most suitable for the duty involved. See the compressor coverage chart in Figure 5.2.

Figure 5.2 covers the normal range of operation for compressors of the commercially available types. The advantages of a centrifugal compressor over a reciprocating machine are:

1. Lower installed first cost where pressure and volume conditions are favorable
2. Lower maintenance expense
3. Greater continuity of service and dependability
4. Less operating attention
5. Greater volume capacity per unit of plot area
6. Adaptability to high-speed low-maintenance-cost drivers

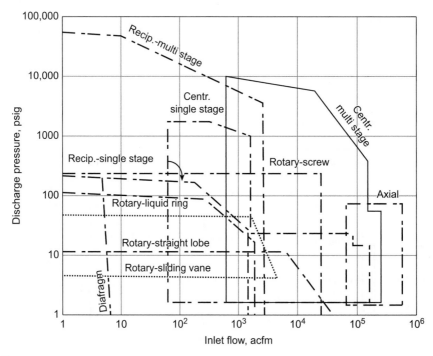

FIGURE 5.2

Compressor coverage chart.

Ref. [1].

- **The advantages of a reciprocating compressor over a centrifugal machine are:**

1. Greater flexibility in capacity and pressure range
2. Higher compressor efficiency and lower power cost
3. Capability of delivering higher pressures
4. Capability of handling smaller volumes
5. Less sensitive to changes in gas composition and density

Adequate knock out facilities, including demister pads, where necessary, shall be provided to prevent damage by liquid carry over into the compressor.

Compressors handling SO_2, HCl, or other gases, which are corrosive in the presence of water, shall not use water as a cooling medium unless the water circuit is positively isolated from the gas side (e.g., by separate water jackets). It is not sufficient to rely on gaskets or seals for isolation.

Similar restrictions shall apply to the use of glycol as a coolant for machines handling corrosive gases plus hydrogen, as the hydrogen can react with glycol to form water. The use of oil as a cooling medium will be acceptable as an alternative in special cases.

Rotodynamic compressors are to be provided with anti-surge equipment. The response time for the control equipment shall be such as to prevent surge during any anticipated process condition, due consideration being given to the speed at which process changes or upsets can move the compressor operation toward surge.

For the more complicated installations with multiple stages and side streams, or multiple units (in series or parallel), or variable-speed units, an analysis of the stability of the anti-surge control system is also necessary.

5.1 Type selection criteria

The choice of the type of compressor, whether axial, centrifugal, reciprocating, or rotary, depends primarily on the required flow to be compressed, the density of the gas in conjunction with the total head (for a given gas, this is the compression ratio), and the duty that has to be performed. Table 5.1 outlines the compression limits for the four types of compression equipment.

Table 5.1 General Compressor Limits

Compressor Type	Approx. Max. Commercially Used Disch. Press. kPa	Approx. Max. Compression Ratio per Stage	Approx. Max. Compression Ratio per Case or Machine
Reciprocating	240,000–345,000	10	As required
Centrifugal	20,600–34,500	3–4.5	8–10
Rotary displacement	690–896	4	4
Axial flow	550–896	1.2–1.5	5–6.5

5.2 Centrifugal compressors

Providing a centrifugal compressor can handle the required flow with a reasonable efficiency, then this type is the preferred choice because it has the potential to operate continuously for long periods, if properly designed and assembled. If the flow at discharge conditions is 300 m^3/h or more, than the possibility of using centrifugal compressor to be investigated.

Table 5.2 gives an approximate idea of the flow range that a centrifugal compressor will handle. A multi-wheel (multi-stage) centrifugal compressor is normally considered for inlet volumes between 850 and 340,000 m^3/h inlet volume. A single-wheel (single-stage) compressor would normally have application between 170 and 255,000 m^3/h inlet volume. A multi-wheel compressor can be thought of as a series of single-wheel compressors contained in a single casing.

Most centrifugal compressors operate at speeds of 3000 rpm or higher, a limiting factor being impeller stress considerations as well as velocity limitation of 0.8–0.85 Mach number at the impeller tip and eye. Recent advances in machine design have resulted in production of some units running at speeds in excess of 40,000 rpm.

Centrifugal compressors are usually driven by electric motors, steam, or gas turbines (with or without speed-increasing gears), or turbo-expanders.

The operating characteristics must be determined before an evaluation of compressor suitability for the application can be made.

The centrifugal compressor approximates the constant head-variable volume machine, while the reciprocating is a constant volume-variable head machine. The axial compressor, which is a low head, high flow machine, falls somewhere in between. A compressor is a part of the system, and its performance is dictated by the system resistance. The desired system capability or objective must be determined before a compressor can be selected.

Figure 5.3 is a typical performance map that shows the basic shape of performance curves for a variable-speed centrifugal compressor. The curves are affected by many variables, such as desired compression ratio, type of gas, number of wheels, sizing of compressor, etc.

Table 5.2 Centrifugal Compressor Flow Range

Nominal Flow Range (Inlet m^3/h)	Average Polytropic Efficiency	Average Isentropic Efficiency	Speed to Develop 30,000 (N·m)/kg head/wheel
170–850	0.7	0.67	20,500
850–12,700	0.8	0.78	10,500
12,700–34,000	0.86	0.83	8200
34,000–56,000	0.86	0.83	6500
56,000–94,000	0.86	0.83	4900
94,000–136,000	0.86	0.83	4300
136,000–195,000	0.86	0.83	3600
195,000–245,000	0.86	0.83	2800
245,000–340,000	0.86	0.83	2500

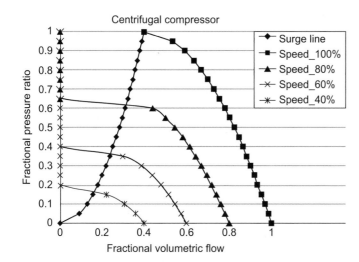

Centrifugal compressor

- ◆ Surge line
- ■ Speed_100%
- ▲ Speed_80%
- ✕ Speed_60%
- ✳ Speed_40%

FIGURE 5.3

Typical performance curve.

With variable speeds, the centrifugal compressor can deliver constant capacity at variable pressure, variable capacity at constant pressure, or a combination variable capacity and variable pressure.

Basically, the performance of the centrifugal compressor, at speeds other than design, is such that the capacity will vary directly as the speed, the head developed as the square of the speed, and the required horsepower as the cube of the speed.

As the speed deviates from the design speed, the error of these rules, known as the affinity laws or fan laws, increases. The fan laws only apply to single stages or multi stages with very low compression ratios or very low Mach numbers.

Fan laws:

$$Q \propto N; \text{i.e.,} \frac{Q_{110}}{N_{110}} = \frac{Q_{100}}{N_{100}} = \frac{Q_{90}}{N_{90}} \tag{5.1}$$

$$H \propto N^2; \text{i.e.,} \frac{H_{110}}{(N_{110})^2} = \frac{H_{100}}{(N_{100})^2} = \frac{H_{90}}{(N_{90})^2} \tag{5.2}$$

$$\text{Bhp} \propto N^3; \text{i.e.,} \frac{\text{Bhp}_{110}}{(N_{110})^3} = \frac{\text{Bhp}_{100}}{(N_{100})^3} = \frac{\text{Bhp}_{90}}{(N_{90})^3} \tag{5.3}$$

where:

Bhp = Brake horse power
H = head, kN·m/kg
N = speed, rpm
Q = inlet capacity (inlet volume rate) m³/h

By varying speed, the centrifugal compressor will meet any load and pressure condition demanded by the process system within the operating limits of the compressor and the driver. It normally accomplishes this as efficiently as possible, because only the head required by the process is developed by the compressor. This compares to the essentially constant head developed by the constant-speed compressor.

5.2.1 Calculating performance

When more accurate information is required for the compressor head, gas horsepower, and discharge temperature, the equations in this section should be used.

All values for pressure and temperature in these calculation procedures are the absolute values. Unless otherwise specified, volumes of flow in this section are actual volumes. To calculate the inlet volume:

$$Q = \frac{(w)(8.314)(T_1)(Z_1)}{(M)(P_1)} \tag{5.4}$$

where:

M = molecular mass, kg/kg mol
P = pressure, kPa (abs)
Q = inlet capacity (inlet volume rate) m³/h
T = absolute temperature, K
Z = compressibility factor

If we assume the compression to be isentropic (reversible adiabatic, constant entropy), then:

$$H_{is} = \frac{ZRT}{M(k-1)/k}\left[\left(\frac{P_2}{P_1}\right)^{(k-1)/k} - 1\right] \tag{5.5}$$

where:

H = head, kN·m/kg
is = isentropic process
k = isentropic exponent, C_p/C_v
M = molecular mass, kg/kg mol
P = pressure, kPa (abs)
R = universal gas constant = 8.314 kJ/(kmol·K)
T = absolute temperature, K
Z = compressibility factor

Because these calculations will not be wheel-by-wheel, the head will be calculated across the entire machine. For this, use the average compressibility factor:

$$Z_{avg} = \frac{Z_1 + Z_2}{2} \tag{5.6}$$

where:

Z_{avg} = average compressibility factor = $(Z_s + Z_d)/2$.

The heat capacity ratio, k, is normally determined at the average suction and discharge temperature.

- **Isentropic Calculation**

To calculate the head:

$$H_{is} = \frac{Z_{avg}RT_1}{M(k-1)/k}\left[\left(\frac{P_2}{P_1}\right)^{(k-1)/k} - 1\right]$$ (5.7)

where:

H = head, kN·m/kg
is = isentropic process
k = isentropic exponent, C_p/C_v
M = molecular mass, kg/kg mol
P = pressure, kPa (abs)
R = universal gas constant = 8.314 kJ/(kmol·K)
T = absolute temperature, K
Z_{avg} = average compressibility factor = $(Z_s + Z_d)/2$

Which can also be written in the form:

$$H_{is} = \frac{8.314}{M}\frac{Z_{avg}T_1}{(k-1)/k}\left[\left(\frac{P_2}{P_1}\right)^{(k-1)/k} - 1\right]$$ (5.8)

The gas horsepower can now be calculated from:

$$Ghp = \frac{(w)(H_{is})}{(\eta_{is})(3,600,000)}$$ (5.9)

where:

Ghp = gas power, actual compression power, excluding mechanical losses, kW
H = head, kN·m/kg
is = isentropic process
w = work, N·m
η = efficiency, expressed as a decimal

The approximate theoretical discharge temperature can be calculated from:

$$\Delta T_{ideal} = T_1\left[\left(\frac{P_2}{P_1}\right)^{(k-1)/k} - 1\right]$$ (5.10)

$$T_2 = T_1 + \Delta T_{ideal}$$ (5.11)

The actual discharge temperature can be approximated:

$$\Delta T_{actual} = T_1\frac{\left[\left(\frac{P_2}{P_1}\right)^{(k-1)/k} - 1\right]}{\eta_{is}}$$ (5.12)

$$T_2 = T_1 + \Delta T_{actual}$$ (5.13)

tags22">

the2">ief

5.2.2 Polytropic calculation

Sometimes, compressor manufacturers use a polytropic path instead of isentropic. Polytropic efficiency is defined by:

$$\frac{n}{(n-1)} = \left[\frac{k}{(k-1)}\right]\eta_p \tag{5.14}$$

where:

k = isentropic exponent, C_p/C_v
n = polytropic exponent or number of moles
η = efficiency, expressed as a decimal

The equations for head and gas horsepower based upon polytropic compression are:

$$H_p = \frac{Z_{avg}RT_1}{M(n-1)/n}\left[\left(\frac{P_2}{P_1}\right)^{(n-1)/n} - 1\right] \tag{5.15}$$

H_p = polytropic head, kN·m/kg
M = molecular mass, kg/kg mol
n = polytropic exponent or number of moles
P = pressure, kPa (abs)
R = universal gas constant = 8.314 kJ/(kmol·K)
T = absolute temperature, K
Z_{avg} = average compressibility factor = $(Z_s + Z_d)/2$

Which also can be written in the form:

$$H_p = \frac{8.314}{M}\frac{Z_{avg}T_1}{(n-1)/n}\left[\left(\frac{P_2}{P_1}\right)^{(n-1)/n} - 1\right] \tag{5.16}$$

$$Ghp = \frac{(w)(H_p)}{(\eta_p)(3,600,000)} \tag{5.17}$$

Polytropic and isentropic head are related by:

$$H_p = \frac{H_{is}\eta_p}{\eta_{is}} \tag{5.18}$$

The approximate actual discharge temperature can be calculated from:

$$T_2 = T_1\left(\frac{P_2}{P_1}\right)^{\left(\frac{n-1}{n}\right)} \tag{5.19}$$

Centrifugal compressors shall be designed in accordance with American Petroleum Institute (API) Standard No. 617.

The centrifugal (radial flow) compressor is well established for the compression of gases and vapors. It has proven its economy and uniqueness in many applications, particularly where large volumes are handled at medium pressures.

Centrifugal compressors shall conform to API Standard No. 617 for all services handling air or gas, except machines developing less than 35 kPa (0.35 bar) from atmospheric pressure, which may be classified as fans or blowers.

Compressors shall be guaranteed for head, capacity, and satisfactory performance at all specified operating points and further shall be guaranteed for power at the rated point.

The volume capacity at the surge point shall not exceed the specified percentage of normal capacity at normal speed, and normal (unthrottled) suction conditions. The rise in pressure ratio from normal capacity to the surge point at normal speed shall not be less than that specified.

The head developed at 115% of normal capacity at normal speed shall be not less than approximately 85% of the head developed at the normal operating point.

The head capacity characteristic curve shall rise continuously from the rated point to the predicted surge. The compressor, without the use of a bypass, shall be suitable for continuous operation at any capacity at least 10% greater than the predicted approximate surge capacity shown in the proposal.

For variable-speed compressors, the head and capacity shall be guaranteed with the understanding that the power may vary ±4%.

For constant-speed compressors, the specified capacity shall be guaranteed with the understanding that the head shall be within ±5% and −0% of that specified; the power shall not exceed stated power by more than 4%. These tolerances are not additive.

The compressor manufacturer shall be responsible for checking the "k" (ratio of specific heats) and "Z" (compressibility factor) values specified against the gas analysis specified.

Compressor Mach numbers shall not exceed 0.90 when measured at any point.

5.3 Design criteria

This section covers information necessary to select centrifugal compressors and to determine whether the selected machine should be considered for a specific job.

An approximate idea of the flow range that a centrifugal compressor will handle is shown in Table 5.3. A multi-stage centrifugal compressor is normally considered for inlet volumes between 850 and 340,000 Im^3/h. A single-stage compressor would normally have applications between 170 and 255,000 Im^3/h. A multistage compressor can be thought of as a series of single stage compressors contained in a single casing.

5.3.1 Effect of speed

With variable speed, the centrifugal compressor can deliver constant capacity at variable pressure, variable capacity at constant pressure, or a combination of variable capacity and variable pressure.

Basically, the performance of the centrifugal compressor, at speeds other than design, follows the affinity (or fan) laws.

By varying speed, the centrifugal compressor will meet any load and pressure condition demanded by the process system within the operating limits of the compressor and the driver.

Table 5.3 Centrifugal Compressor Flow Range

Nominal Flow Range (Inlet m³/h)	Average Polytropic Efficiency	Average Isentropic Efficiency	Speed to Develop 3048 m Head/Wheel
170–850	0.63	0.6	20,500
850–12,743	0.74	0.7	10,500
12,743–34,000	0.77	0.73	8200
34,000–56,000	0.77	0.73	6500
56,000–93,400	0.77	0.73	4900
93,400–135,900	0.77	0.73	4300
135,900–195,400	0.77	0.73	3600
195,400–246,400	0.77	0.73	2800
246,400–340,000	0.77	0.73	2500

5.3.2 Performance analysis

- **Determination of properties pertaining to compression**
  Compressibility factor (Z factor), ratio of specific heats (C_p/C_v or k value) and molecular mass are three major physical properties for compressors, which must be clarified.
- **Determination of suction conditions**
  The following conditions at the suction flange should be determined:
  - Temperature
  - Pressure
    In case of air taken from the atmosphere, corrections should be made for elevation. Air humidity should also be considered.
  - Flow rate
    All centrifugal compressors are based on flows that are converted to inlet or actual conditions (Im³/h or inlet cubic meters per hour). This is done because centrifugal compressor is sensitive to inlet volume, compression ratio (i.e., head), and specific speed.
  - Fluctuation in conditions
    Because fluctuations in inlet conditions will have large effects on the centrifugal compressor performance, owing to the compressibility of the fluid, all conceivable condition fluctuations must be taken into consideration in determination of design conditions.
- **Flow limits**
  Two conditions associated with centrifugal compressors are surge (pumping) and stone-wall (choked flow).
- **Surge point**
  At some point on the compressors operating curve, there exists a condition of minimum flow/maximum head where the developed head is insufficient to overcome the system resistance. This is the "surge point." When the compressor reaches this point, the gas in the discharge piping back-flows into the compressors. Without discharge flow, discharge pressure drops until it is within the compressor's capability, only to repeat the cycle.

The repeated pressure oscillations at the surge point should be avoided because it can be detrimental to the compressor. Surging can cause the compressor to overheat to the point the maximum allowable temperature of the unit is exceeded. Also, surging can cause damage to the thrust bearing due to the rotor shifting back and forth from the active to the inactive side.

- **Stone-wall or choked flow**
  Stone-wall or choked flow occurs when sonic velocity is reached at any point in the compressor. When this point is reached for a given gas, the flow through the compressor cannot be increased further without internal modifications.

5.3.3 Interstage cooling

Multi-stage compressors rely on intercooling whenever the inlet temperature of the gas and the required compression ratio are such that the discharge temperature of the gas exceeds about $150\,^{\circ}C$. Performance calculations indicate that the head and power are directly proportional to the absolute gas temperature at each impeller.

The gas may be cooled within the casing or in external heat exchangers. In the case of diaphragm water cooling system, API Standard No. 617, Section 2.1.4 should be followed for design parameters.

- **Interstage cooling**

There are certain processes that require a controlled discharge temperature. For example, the compression of gases, such as oxygen, chlorine, and acetylene, requires that the temperature be maintained below $100\,^{\circ}C$.

The thermal stress within the horizontal bolted joint is the governing design limitation in a horizontally split compressor case. The vertically split barrel-type case, however, is free from the thermal stress complication.

Substantial power economy can be gained by precooling the gas before it enters the interstage impellers. Performance calculations indicate that the head and the horsepower are directly proportional to the absolute gas temperature at each impeller.

The gas may be cooled within the casing or in external heat exchangers.

Two methods of cooling within the casing are used—water cooled diaphragms between successive stages and direct liquid injection into the gas.

Diaphragm cooling systems include high-velocity water circulation through cast jackets in the diffuser diaphragms. The diaphragm coolers are usually connected in series [1].

Liquid injection cooling is the least costly means of controlling discharge temperatures. It involves injecting and atomizing a jet of water or a compatible liquid into the return channels. In refrigeration units, liquid refrigerant is frequently used for this purpose. Injected liquid also functions as a solvent in washing the impellers free of deposits. Nevertheless, the hazards of corrosion, erosion, and flooding present certain problems resulting in possible replacement of the compressor rotor.

External intercoolers are commonly used as the most effective means of controlling discharge temperatures. The gas is discharged from the compressor casing after one or more stages of compression and, after being cooled, is returned to the next stage or series of stages for further compression [1].

Intercoolers usually are mounted separately. When there are two or more compressor casings installed in series, individual machines may or may not be cooled or have intercoolers. In some cases, it may be advantageous to use an external cooler to precool gas ahead of the first wheel [1].

5.3.4 Control systems

Centrifugal compressor controls can vary from the very basic manual recycle control to the elaborate ratio controllers. The driver characteristics, process response, and compressor operating range must be determined before the right controls can be selected.

In cases where constant-speed drivers are used, the inlet gas density can be reduced by throttling or by adjusting the compressor guide vanes. Different features of the centrifugal compressor control systems are described in Gas Processors and Suppliers Association (GPSA), "Engineering Data Book," volume 2, 1987.

5.3.5 Anti-surge control

It is essential that all centrifugal compressor control systems be designed to avoid possible operation in surge which usually occurs below 50–70% of the rated flow.

The surge limit line can be reached by reducing flow or decreasing suction pressures and/or increasing discharge to suction pressure ratio. An anti-surge system senses conditions approaching surge, and maintains the unit pressure ratio below the surge limit by recycling some flow to the compressor suction. A volume-controlled anti-surge system is shown in Figure 5.4. As the flow decreases to less than the minimum volume set point, a signal will cause the surge control valve to open, to keep a minimum volume flowing through the compressor.

A pressure-limiting anti-surge control system is shown in Figure 5.5. A process pressure increase over the pressure set point will cause the blow-off valve to open. The valve opens as required to keep the pressure limited to a minimum of gas or air flowing through the compressor.

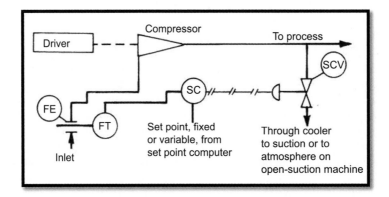

FIGURE 5.4

A volume-controlled anti-surge system. FE, Flow element; FT, Flow transmitter; SC, Set-point computer; SCV, Set-point control valve.

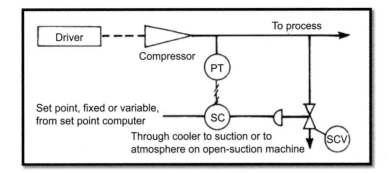

FIGURE 5.5

A pressure-limiting anti-surge control system. PT, Pressure transmitter; SC, Set-point computer; SCV, Set-point control valve.

5.4 Reciprocating compressors

Where the required flow is too small for a centrifugal compressor, or where the required head is so high that an undesirably large number of stages would be necessary, then, generally, the choice should be a reciprocating compressor.

As a reciprocating compressor cannot fulfill the minimum requirement of continuous uninterrupted operation for a 2-year period, due to fairly high maintenance requirements, a full-capacity spare shall be provided as a general rule for reciprocating compressors in critical services. Alternatively, three half-capacity machines may be specified, two running in parallel, with the third unit as a spare. Reciprocating compressors shall be in accordance with API Standard No. 618.

The reciprocating compressor is a positive displacement unit with the pressure on the fluid developed within a cylindrical chamber by the action of a moving piston. It may consist of one or more cylinders each with a piston or plunger that moves back and forth, displacing a positive volume with each stroke.

Reciprocating compressors shall conform to API Standard No. 618 for all services except portable air compressors, and standard utility air compressors of 400 kW or less with not more than 900 kPa (9 bar) discharge pressure. This latter group will generally be purchased as packaged units.

Reciprocating compressors normally should be specified for constant-speed operation to avoid excitation of torsional and acoustic resonances. When variable-speed drivers are used, all equipment shall be designed to run safely to the trip speed setting.

When considering the use of a single frame for cylinders on different services, particular attention shall be given to the means of independently controlling the different process streams. Care shall also be taken to ensure that the frame, transmission, and driver can accept the wide variety of loadings that occur during all operating modes including start-up and shut-down.

5.4.1 Speed ranges

Low-speed unit up to 330 r/min (rpm).
Medium-speed unit 330–700 r/min (rpm).

High-speed unit over 700 r/min (rpm).

Generally, high-speed units are preferred for units under 1865 kW. For larger units, the choice is between low and medium speed.

5.4.2 Capacity control

Capacity control for constant-speed units normally will be obtained by suction valve unloading (depressors or lifters), clearance pockets, a combination of both pockets and unloaders, or bypass. Operation of controls shall be automatic. Unless stated otherwise, five-step unloading shall provide capacities of 100, 75, 50, 25, and 0%; three-step unloading shall provide capacities of 100, 50, and 0%; and two-step unloading shall provide capacities of 100 and 0%.

Capacity control on variable-speed units generally is by speed control.

Clearance pockets may be either the two-position type (pocket either open or closed) or the variable-capacity type. If not specified, the vendor shall propose on the data sheet the type recommended for the purchaser's or company's operating conditions.

When unloading for start-up is necessary, unloading arrangement shall be stated on the data sheet or shall be mutually agreed upon between the purchasers and/or the companies and the vendor.

5.4.3 Design criteria

This section covers information necessary for process engineers to determine the approximate power required to compress a certain volume of gas at some intake conditions to a given discharge pressure and estimate the capacity of an existing reciprocating compressor under specified suction and discharge conditions.

Reciprocating compressors are furnished either single-stage or multi-stage. The number of stages is determined by the overall compression ratio.

On multi-stage machines, intercoolers may be provided between stages. Such cooling reduces the actual volume of gas going to the high-pressure cylinders, reduces the power required for compression, and keeps the temperature within safe operating limits.

Reciprocating compressors should be supplied with clean gas as they cannot satisfactorily handle liquids and solid particles that may be entrained in the gas.

In evaluating the work of compression, the enthalpy change is the best way. If a P–H diagram is available, the work of compression should always be calculated by the enthalpy change of the gas in going from suction to discharge conditions.

The k value of a gas is associated with adiabatic compression or expansion. The change in gas properties at different states is related by:

$$P_1 \cdot V_1^k = P_2 \cdot V_2^k = P_3 \cdot V_3^k \tag{5.20}$$

For a polytropic compression, the actual value of "n" (polytropic exponent) is a function of the gas properties, such as specific heats, degree of external cooling during compression, and operating features of the cylinder. Usual reciprocating compressor performance is evaluated using adiabatic C_p/C_v.

"Power rating" or kilowatt rating of a compressor frame is the measure of the ability of the supporting structure and crankshaft to withstand torque and the ability of the bearings to dissipate frictional heat.

"Rod loads" are established to limit the static and internal loads on the crankshaft, connecting rod, frame, piston rod, bolting, and projected bearing surfaces.

5.4.4 Performance calculation

- **Determination of properties pertaining to compression.**

Compressibility factor (Z factor), ratio of specific heats (C_p/C_v or k value), and molecular mass are three major physical properties for compression that must be clarified. Mollier diagrams should be used if available.

The k value may be calculated from the ideal gas equation:

$$k = \frac{C_p}{C_v} = \frac{M \cdot C_p}{MC_p - R} \tag{5.21}$$

where:

$MC_p =$ is molar heat capacity at constant pressure in (kJ/kmol·K)
$R =$ is gas constant, 8.3143 kJ/kmol·K

Method presented in Tables 5.4 and 5.5 can be used for calculation of k value of hydrocarbon gases and vapors.

5.4.5 Estimating compressor horsepower

Equation 5.21 is useful for obtaining a quick and reasonable estimate for compressor horsepower. It was developed for large slow-speed (300–450 rpm) compressors handling gases with a relative density of 0.65 and having stage compression ratios above 2.5.

Due to higher valve losses, the horsepower requirement for high-speed compressors (1000 rpm range, and some up to 1800 rpm) can be as much as 20% higher, although this is a very arbitrary value. Some compressor designs do not merit a higher horsepower allowance and the manufacturers should be consulted for specific applications.

$$\text{Brake power} = (0.014)\left(\frac{\text{ratio}}{\text{stage}}\right)(\text{\# of stages})(m^3/h)(F) \tag{5.22}$$

where:

$m^3/h =$ compressor capacity referred to 100 kPa (abs) and intake temperature
$F = 1.0$ for single-stage compression
$1.08 =$ for two-stage compression
$1.10 =$ for three-stage compression

Equation 5.21 will also provide a rough estimate of horsepower for lower compression ratios and/or gases with a higher specific gravity, but it will tend to be on the high side. To allow for this tendency, use a multiplication factor of 0.013 instead of 0.014 for gases with a specific gravity in the 0.8–1.0 range; likewise, use a factor in the range of 0.010–0.012 for compression ratios between 1.5 and 2.0.

Table 5.4 Calculation of k

Example Gas Mixture		Determination of Mixture Mol. Mass		Determination of MC_p Molar Heat Capacity		Determination of Pseudo Critical Pressure and Temperature			
Component Name	Mole Faction (y)	Individual Component Mol. Mass (M)	$y \times M$	Individual Component MC_p at 70°C	$y \times MC_p$ at 70°C	Component Critical Pressure P_c, kPa (abs)	$y \times P_c$	Component Critical Temperature T_c (K)	$y \times T_c$
Methane	0.9216	16.04	14.782	37.471	34.533	4640.4	4276.5	190.6	175.6
Ethane	0.0488	30.07	1.467	58.39	2.850	4944.4	241.3	305.6	14.9
Propane	0.0185	44.10	0.816	82.85	1.533	4256.4	78.7	370.0	6.8
i-Butane	0.0039	58.12	0.227	109.397	0.427	3749.0	14.6	406.9	1.6
n-Butane	0.0055	58.12	0.320	109.497	0.602	3658.6	20.1	425.2	2.3
i-Pentane	0.0017	72.15	0.123	134.379	0.228	3333.2	5.7	460.9	0.8
Total	1		$M =$ 17.735	$MC_p =$	40.173	$P_c =$	4036.9	$T_c =$	202.1

$MC_v = MC_p - 8.3143 = 31.859$

$k = MC_p/MC_v = 40.173/31.859 = 1.261$

Table 5.5 Molar Heat Capacity MC_p (Ideal-Gas State) kJ/(kmol·k)

Gas	Chemical Formula	Mol wt	0 °F	50 °F	60 °F	100 °F	150 °F	200 °F	250 °F	300 °F
							Temperature			
Methane	CH_4	16.043	8.23	8.42	8.46	8.65	8.95	9.28	0.64	10.01
Ethyne (Acetylene)	C_2H_2	26.038	9.68	10.22	10.33	10.71	11.15	11.55	11.90	12.22
Ethene (Ethylene)	C_2H_4	28.054	9.33	10.02	10.16	10.72	11.41	12.09	12.73	13.41
Ethane	C_2H_6	30.070	11.44	12.17	12.32	12.95	13.78	14.63	15.49	16.34
Propene (propylene)	C_3H_6	42.081	13.63	14.69	14.90	15.75	16.80	17.85	18.88	19.89
Propane	C_3H_8	44.097	15.65	16.88	17.13	18.17	19.52	20.89	22.25	23.56
1-Butene (Butlyene)	C_4H_8	56.108	17.96	19.59	19.91	21.18	22.74	24.26	25.73	27.16
Cis-2-butane	C_4H_8	56.108	16.54	18.04	18.34	19.54	21.04	22.53	24.01	25.47
Trans-2-butene	C_4H_8	56.108	18.84	20.23	20.50	21.61	23.00	24.37	25.73	27.07
Iso-butane	C_4H_{10}	58.123	20.40	22.15	22.51	23.95	25.77	27.59	29.39	31.11
n-Butane	C_4H_{10}	58.123	20.80	22.38	22.72	24.08	25.81	27.55	29.23	30.90
Iso-pentane	C_5H_{12}	72.150	24.94	27.17	27.61	29.42	31.66	33.87	36.03	38.14
n-Pentane	C_5H_{12}	72.150	25.64	27.61	28.02	29.71	31.86	33.99	36.08	38.13
Benzene	C_6H_6	78.114	16.41	18.41	18.78	20.46	22.45	24.46	26.34	28.15
n-Hexane	C_6H_{14}	86.177	30.17	32.78	33.30	35.37	37.93	40.45	42.94	45.36
n-Heptane	C_7H_{16}	100.204	34.96	38.00	38.61	41.01	44.00	46.94	49.81	52.61
Ammonia	NH_3	17.0305	8.52	8.52	8.52	8.52	8.52	8.53	8.53	8.53
Air		28.9625	6.94	6.95	6.95	6.96	6.97	6.99	7.01	7.03
Water	H_2O	18.0153	7.98	8.00	8.01	8.03	8.07	8.12	8.17	8.23
Oxygen	O_2	31.9988	6.97	6.99	7.00	7.03	7.07	7.12	7.17	7.23
Nitrogen	N_2	28.0134	6.95	6.95	6.95	6.96	6.96	6.97	6.98	7.00
Hydrogen	H_2	2.0159	6.78	6.86	6.87	6.91	6.94	6.95	6.97	6.98
Hydrogen sulfide	H_2S	34.08	8.00	8.09	8.11	8.18	8.27	8.36	8.46	8.55
Carbon monoxide	CO	28.010	6.95	6.96	6.96	6.96	6.97	6.99	7.01	7.03
Carbon dioxide	CO_2	44.010	8.38	8.70	8.76	9.00	9.29	9.56	9.81	10.05

Example: Compress 2000 m³/h of gas at 101.325 kPa (abs) and intake temperature through a compression ratio of nine in a two-stage compressor. What will be the power?
Solution steps

$$\text{Ratio per stage} = \sqrt{9} = 3$$

From Eqn (5.21), we find the brake power to be:

$$(0.014)(3)(2)(2000 \text{ m}^3/\text{h})(1.08) = 181 \text{ kW}$$

5.4.6 Capacity

Most gases encountered in industrial compression do not exactly follow the ideal gas equation of state but differ in varying degrees. The degree in which any gas varies from the ideal is expressed by a compressibility factor, Z, which modifies the ideal gas equation:

$$PV = nRT \tag{5.23}$$

$$PV = nZRT \tag{5.24}$$

where:

P = pressure, kPa (abs)
V = specific volume, m³/kg
n = polytropic exponent or number of moles
R = universal gas constant, 8.314 kJ/kmol·K

For the purpose of performance calculations, compressor capacity is expressed as the actual volumetric quantity of gas at the inlet to each stage of compression on a per minute basis (IVR).
From standard volumetric rate (SVR):

$$Q = \text{SVR}\left(\frac{101.325}{288}\right)\left(\frac{T_1 Z_1}{P_1 Z_L}\right) \tag{5.25}$$

SVR = standard volumetric rate m³/h measured at 101.325 kPa (abs) and 15 °C.
Q = inlet capacity (IVR) m³/h. (IVR is inlet volumetric rate m³/h, usually at suction conditions.)
L = standard conditions used for calculation or contract.
P = pressure, kPa (abs)
T = absolute temperature, K
Z = compressibility factor

From weight flow (w, kg/h):

$$Q = \frac{8.314}{\text{MW}}\left(\frac{w T_1 Z_1}{P_1 Z_L}\right) \tag{5.26}$$

From molar flow (Nm, mols/min):

$$Q = 8.314 \left(\frac{N_m T_1 Z_1}{P_1 Z_L} \right) \tag{5.27}$$

From these equations, inlet volume to any stage may be calculated by using the inlet pressure P_1 and temperature T_1. Moisture should be handled just as any other component in the gas.

In a reciprocating compressor, effective capacity may be calculated as the piston displacement (generally in m^3/h) multiplied by the volumetric efficiency.

The piston displacement is equal to the net piston area multiplied by the length of piston sweep in a given period of time. This displacement may be expressed for a single-acting piston compressing on the outer end only:

$$PD = \frac{(\text{stroke})(N)(D^2)\pi(60)}{(4)\cdot 10^9}$$
$$= 5(10^{-8})(\text{stroke})(N)(D^2) \tag{5.28}$$

where:

D = cylinder inside diameter, mm
N = speed, rpm
PD = piston displacement, m^3/h
stroke = length of piston movement, mm

For a single-acting piston compressing on the crank end only:

$$PD = \frac{(\text{stroke})(N)(D^2 - d^2)\pi(60)}{(4)\cdot 10^9}$$
$$= 5(10^{-8})(\text{stroke})(N)(D^2 - d^2) \tag{5.29}$$

For a double-acting piston (other than tail rod type):

$$PD = \frac{(\text{stroke})(N)(2D^2 - d^2)\pi(60)}{(4)\cdot 10^9}$$
$$= 5(10^{-8})(\text{stroke})(N)(2D^2 - d^2) \tag{5.30}$$

where:

d = piston rod diameter, mm
D = cylinder inside diameter, mm
N = speed, rpm
PD = piston displacement, m^3/h
stroke = length of piston movement, mm

- **Volumetric efficiency**

In a reciprocating compressor, the piston does not travel completely to the end of the cylinder at the end of the discharge stroke. Some clearance volume is necessary and it includes the space between the end of the piston and the cylinder head when the piston is at the end of its stroke.

   It also includes the volume in the valve ports, the volume in the suction valve guards, and the volume around the discharge valve seats.

   Clearance volume is usually expressed as a percent of piston displacement and referred to as percent clearance, or cylinder clearance, C.

$$C = \frac{\text{clearance volume, m}^3}{\text{piston displacement, m}^3} (100) \tag{5.31}$$

For double-acting cylinders, the percent clearance is based on the total clearance volume for both the head end and the crank end of a cylinder. These two clearance volumes are not the same due to the presence of the piston rod in the crank end of the cylinder. Sometimes additional clearance volume (external) is intentionally added to reduce cylinder capacity. The term "volumetric efficiency" refers to the actual pumping capacity of a cylinder compared to the piston displacement.

   Without a clearance volume for the gas to expand and delay the opening of the suction valve(s), the cylinder could deliver its entire piston displacement as gas capacity. The effect of the gas contained in the clearance volume on the pumping capacity of a cylinder can be represented by:

$$VE = 100 - r - C\left[\frac{Z_s}{Z_d}\left(r^{1/k}\right) - 1\right] \tag{5.32}$$

where:

   $r =$ compression ratio, P_2/P_1
   $C =$ cylinder clearance as a percent of cylinder volume
   $VE =$ volumetric efficiency, percent
   $Z_s =$ suction compressibility factor
   $Z_d =$ discharge compressibility factor

One method for accounting for suction and discharge valve losses is to reduce the volumetric efficiency by an arbitrary amount, typically 4%, thus modifying Eqns (13) and (14) as follows:

$$VE = 96 - r - C\left[\frac{Z_s}{Z_d}\left(r^{1/k}\right) - 1\right] \tag{5.33}$$

where:

   $r =$ compression ratio, P_2/P_1
   $C =$ cylinder clearance as a percent of cylinder volume
   $k =$ isentropic exponent, C_p/C_v
   $VE =$ volumetric efficiency, percent
   $Z_s =$ suction compressibility factor
   $Z_d =$ discharge compressibility factor
   $r =$ compression ratio, P_2/P_1

- **Equivalent capacity**

The net capacity for a compressor, in cubic feet per day @ 100 kPa (abs) and suction temperature, may be calculated by Eqn (5.34), which is shown in dimensioned form:

$$m^3/h = \frac{PD(VE)P_s Z_s}{100} \qquad (5.34)$$

which can be simplified to Eqn (5.35) when Z100 is assumed to equal 1.0.

$$m^3/h = \frac{PD \cdot VE \cdot P_s}{Z_s \cdot 100} \qquad (5.35)$$

For example, a compressor with 425 m³/h piston displacement, a volumetric efficiency of 80%, a suction pressure of 517 kPa (abs), and suction compressibility of 0.9 would have a capacity of 1953 m³/h. If compressibility is not used as a divisor in calculating m³/h, then the statement "not corrected for compressibility" should be added.

- **Discharge temperature**

The temperature of the gas discharged from the cylinder can be estimated from Eqn (5.35), which is commonly used but not recommended. (Note: the temperatures are in absolute units, °R or K.)

$$T_d = T_s\left(r^{(k-1)/k}\right) \qquad (5.36)$$

The discharge temperature determined from Eqn (5.36) is the theoretical value. While it neglects heat from friction, irreversibility effects, etc., and may be somewhat low, the values obtained from this equation will be reasonable field estimates.

- **Detailed horsepower calculation**

A more detailed calculation of reciprocating compressor power requirements can be performed using the following equation:

$$BP/stage = 2.78 \times 10^{-4} \times Z_{avg} \times (Q_g T_s/E)(K/(K-1))\left(\frac{P_L}{T_L}\right)\left[\left(\frac{P_d}{P_s}\right)^{\frac{k-1}{k}} - 1\right] \qquad (5.37)$$

where:

BP = brake power, kW
Q_g = gas flow rate, S m³/h
T_s = suction temperature, K
$Z_{avg} = (Z_s + Z_d)/2$
Z_s = suction compressibility factor
Z_d = discharge compressibility factor
E = overall efficiency
High-speed reciprocating units—0.82
Low-speed reciprocating units—0.85
K = ratio of specific heats, C_p/C_v
P_s = suction pressure, kPa (abs)

P_d = discharge pressure, kPa (abs)
P_L = standard pressure kPa (abs)
T_L = standard temperature, K

The total power for the compressor is the sum of the power required for each of the stages that are utilized. For multi-stage machines, an allowance should be made for the interstage pressure drop associated with piping, cooler, scrubber, etc., typically 35–70 kPa.

- **Procedure**
  - Calculate overall compression ratio $r_t = P_{dfinal} - P_s$
  - Calculate the compression ratio per stage, r, by taking the s root of r_t, where s is the number of compression stages. The number of stages, s, should be increased until the ratio per stage, r, is $< \sim 4$. This should generally result in stage discharge temperatures of $<150\,°C$ depending on the interstage cooler outlet temperature assumed
  - Multiplying r by the absolute suction pressure of the stage being considered will give you a discharge pressure of the stage
  - Calculate the horsepower required for the stage
  - Substract the assumed interstage pressure loss from the discharge pressure of the preceding stage to obtain the suction pressure for the next stage
  - Repeat steps four and five until all stages have been calculated
  - Sum the stage horsepowers to obtain the total compressor power required

Example: Compress 2250 m³/h of gas measured at 101.325 kPa (abs) and 15 °C. Intake pressure is 690 kPa (abs), and intake temperature is 38 °C. Discharge pressure is 6210 kPa (abs). The gas has a relative density of 0.80 (23 mol mass and a gas with relative density of 0.8 at 66 °C would have an approximate k of 1.21).

What is the required brake power assuming a high-speed compressor?

1. Compression ratio is

$$\frac{6210 \text{ kPa (abs)}}{690 \text{ kPa (abs)}} = 9$$

This would be a two-stage compressor; therefore, the ratio per stage is $\sqrt{9}$ or 3.

2. 690 kPa (abs) × 3 = 2070 kPa (abs) (first stage discharge pressure) 2070 − 35 = 2035 kPa (abs) Where the 35 kPa (abs) represents the pressure drop between first-stage discharge and second-stage suction.

$$\frac{6205 \text{ kPa (abs)}}{2035 \text{ kPa (abs)}} = 3.05 (\text{compression ration for 2nd stage})$$

It may be desirable to recalculate the interstage pressure to balance the ratios. For this sample problem, however, the first ratios determined will be used.

3. For most compression applications, the 66 °C curve will be adequate. This should be checked after determining the average cylinder temperature

4. Discharge temperature for the first stage can be obtained. For a compression ratio of 3, discharge temperature = approximately 104 °C. Average cylinder temperature = 71 °C

5. In the same manner, discharge temperature for the second stage (with $r = 3.05$ and assuming interstage cooling to 49 °C) equals approximately 118 °C. Average cylinder temperature = 84 °C
6. Estimate the compressibility factors at suction and discharge pressure and temperature of each stage
   1st stage: $Z_s = 0.98$
   $Z_d = 0.97$
   $Z_{avg} = 0.975$
   2nd stage: $Z_s = 0.94$
   $Z_d = 0.92$
   $Z_{avg} = 0.93$
7. Calculate the horsepower required for the first and second stages: Total BP required = 98.5 + 98.9 = 197.4 kW

5.4.7 Reciprocating compressor control devices

Output of compressors must be controlled (regulated) to match system demand. Compressor capacity, speed, or pressure may be varied in accordance with the requirements. The nature of the control device will depend on the regulating variables; whether pressure, flow, temperature, or some other variable; and on the type of compressor driver.

- **Capacity control**

The most common requirement is regulation of capacity. Many capacity controls, or unloading devices as they are usually termed, are actuated by the pressure on the discharge side of the compressor.

A common method of controlling the capacity of a compressor is to vary the speed. This method is applicable to steam-driven compressors and to units driven by internal-combustion engines.

On reciprocating compressors up to about 75 kW, two types of control are usually available. These are automatic-start and stop control and constant-speed control.

- **Step control**

Motor-driven reciprocating compressors above 75 kW in size are usually equipped with a step control. This is in reality a variation of constant-speed control in which unloading is accomplished in a series of steps, varying from full load down to no load.

- **Manual control**

Although control devices are often automatically operated, manual operation is satisfactory for many services.

- **Gas pulsation control**

Pulsation is inherent in reciprocating compressors because suction and discharge valves are open during only part of the stroke.

Pulsation must be damped (controlled) in order to:

- Provide smooth flow of gas to and from the compressor
- Prevent overloading or underloading of the compressors, and
- Reduce overall vibration

• **Pulsation dampeners (snubbers)**

A pulsation dampener is an internally baffled device. The design of the pulsation dampening equipment is based on acoustical analog evaluation which takes into account the specified operating speed range, conditions of unloading, and variations in gas composition. Detailed discussion of recommended design approaches for pulsation suppression devices is presented in API Standard No. 618.

5.5 Axial compressors

Axial compressors (Figure 5.6) can handle large volume flow and are more efficient than centrifugal compressors. However, centrifugals are less vulnerable and, hence, more reliable, have wider operating ranges, and are less susceptible to fouling.

Axial compressors should be considered only for air, sweet natural gas, or noncorrosive gases.

Axial compressors are basically high-flow, low-pressure machines, in contrast to the lower flow, high-pressure centrifugal compressors (the axial compressors used in gas turbines are often designed for higher pressures and compression ratios).

Axial compressors are generally smaller and significantly more efficient than comparable centrifugal compressors. The characteristic feature of an axial compressor, as its name implies, is the axial direction of flow through the machine. An axial flow compressor requires more stages than a centrifugal due to the lower pressure rise per stage. In general, it takes approximately twice as many stages to achieve a given pressure ratio as would be required by a centrifugal. Although the axial compressor

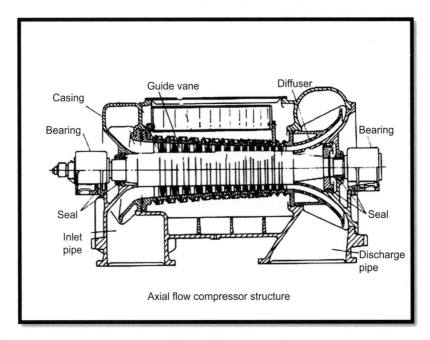

Axial flow compressor structure

FIGURE 5.6

Axial flow compressor.

requires more stages, the diametral size of an axial is typically much lower than for a centrifugal. The axial compressor's capital cost is usually higher than that of a centrifugal, but may be justified based on efficiency and size. The axial compressor utilizes alternating rows of rotating and stationary blades to transfer the input energy from the rotor to the gas in order to generate an increase in gas pressure.

A multi-stage axial flow compressor has two or more rows of rotating blades operating in series on a single rotor in a single casing. The casing contains the stationary vanes (stators) for directing the air or gas to each succeeding row of rotating blades. These stationary vanes, or stators, can be fixed or at a variable angle, or a combination of both.

Axial compressor is usually a single inlet, uncooled machine consisting essentially of blades mounted on a rotor turning between rows of stationary blades mounted on the horizontally split casing.

5.5.1 Performance guarantee

1. Compressors shall be guaranteed for head, capacity, and satisfactory performance at all specified operating points and further shall be guaranteed for power at the normal operating point
2. For variable-speed compressors, the head and capacity shall be guaranteed with the understanding that the power may vary $\pm 4\%$
3. For constant-speed compressors, the specified capacity shall be guaranteed with the understanding that the head shall be specified for 100.0 and 105.0%; the power consumption shall not exceed stated power by more than 4%. These tolerances are not additive
4. For compressors handling side loads or for two or more compressors driven by a single drive, the required performance guarantee for each compressor "section" shall be agreed upon by the company and the vendor

5.5.2 Design criteria

The minimum head rise to surge of an axial machine should be specified. The normal operating point shall be at least 10% removed in flow from the surge point.

- **Gas velocities**

General guideline for good design practice indicates an axial velocity for air of 91–137 m/s. For other gases, the axial velocity range is in direct proportion to the speed of sound of the gas compared to air. The internal shape of the machine is usually arranged to give constant gas velocity as the gas travels through.

- **Volume**

The size is determined by the inlet volume. The lower volume limit is approximately 8500 m^3/h but the upper limit practically does not exist, units have been built to handle well above 1,700,000 m^3/h.

5.6 Screw compressors

Screw compressors, also known as helical lobe compressors, fall into the category of rotary positive displacement compressors.

Rotary screw compressors are available in oil-free (dry) or oil-injected designs. Oil-free compressors typically use shaft mounted gears to keep the two rotors in proper mesh without contact. Applications for oil-free compressors include all processes that cannot tolerate contamination of the compressed gas or where lubricating oil would be contaminated by the gas [1].

Oil-injected screw compressors are generally supplied without timing gears. The injected lubricant provides a layer separating the two screw profiles as one screw drives the other. Oil-injected machines generally have higher efficiencies and utilize the oil for cooling as well, which allows for higher compression ratios in a single screw compressor stage.

Although originally intended for air compression, rotary screw compressors are now compressing a large number of gases in the hydrocarbon processing industries. In particular, screw compressors are widely used in refrigeration service and are gaining in popularity in the gas production business in booster and gas gathering applications [1].

Gas compression is achieved by the intermeshing of the rotating male and female rotors. Power is applied to the male rotor and as a lobe of the male rotor starts to move out of mesh with the female rotor, a void is created and gas is taken in at the inlet port. As the rotor continues to turn, the intermesh space is increased and gas continues to flow into the compressor until the entire interlobe space is filled. Continued rotation brings a male lobe into the interlobe spacing compressing and moving the gas in the direction of the discharge port.

The volume of gas is progressively reduced as it increases in pressure. Further rotation uncovers the discharge port and the compressed gas starts to flow out of the compressor. Continued rotation then moves the remaining trapped gas out while a new charge is drawn into the suction of the compressor into the space created by the unmeshing of a new pair of lobes as the compression cycle begins again.

Screw compressors are usually driven by constant-speed motors, with capacity control normally achieved via an internal regulating device known as a slide valve. By moving the slide in a direction parallel to the rotors, the effective length of the rotors can be shortened. This provides smooth control of flow from 100% down to 10% of full compressor capacity. Rotary screw compressors in use today cover a range of suction volumes from 1300 to 60,000 m^3/h, with discharge pressures up to 4000 kPa (ga). Typical adiabatic efficiency will be in the range of 70–80% [1].

5.7 Rotary compressors

Rotary compressors are positive displacement gas (or vapor) compressing machines. Rotary compressors cover lobe-type, screw-type, vane-type, and liquid ring type, each having a casing with one or more rotating elements that either mesh with each other, such as lobes or screws, or that displace a fixed volume with each rotation.

Rotary compressors shall conform to API Standard No. 619 for all services handling air or gas, except those machines that this section does not cover.

The compressor shall be guaranteed for satisfactory performance at all specified operating conditions.

Compressor performance shall be guaranteed at the rated point unless otherwise specified. At this point, no negative tolerance is permitted on capacity and power may not exceed 104% of the quoted power.

5.8 Compressor cooling water jacket

The compressor cooling water jacket shall be designed for the specified cooling water pressure but not less than 618 kPa (absolute). The maximum pressure drop shall be 70 kPa and provisions shall be included for complete draining and venting of the jackets (modification to API Standard No. 619). The cooling water design conditions shall be in accordance with API Standard No. 619.

Rotary compressor shall be considered only where there is proven experience of acceptable performance of this type of compressor in the duty concerned and only where there are advantages over a reciprocating compressor.

The application of oil flooded screw compressors for instrument air and of dry running rotary screw compressors, sliding vane compressors, and rotary-lube compressors for process duties, requires the explicit approval.

Rotary-type positive displacement compressors shall be in accordance with API Standard No. 619.

5.9 Atmospheric pressure

The absolute pressure of the atmosphere at the site should be considered as the "absolute pressure" in the compressor calculations. The value of the absolute pressure is taken as 101.325 kPa at sea level and declines with increasing altitude, as shown in Table 5.6.

5.10 Specification sheets

Process information required to complete specification sheets for compressors are presented. The minimum required information to be included in the process specification sheet shall be as follows:

1. Process requirements
   a. Flow at 1.013 bar (abs) and 0 °C, m^3/h (normal and design)
   b. Flow at suction conditions, m^3/h (normal and design)
   c. Suction temperature, °C (normal and design)
   d. Suction pressure, kPa (abs) (normal and design)
   e. Discharge pressure, kPa (abs) (normal and design)
   f. Discharge temperature limitation (if any) (normal and design)
   g. Compression ratio (normal and design)
   h. Approximate C_p/C_v (at suction) (normal and design)
   i. Compressibility factor (at suction) (normal and design)
   j. Mass flow molecular mass, kg/h (normal and design)
   k. Estimated polytropic head, meters (normal and design)
   l. Estimated Bhp required, kW (normal and design)
   m. Estimated gear loss, kW (normal and design)
   n. Recommended driver, kW (normal and design)
   o. Compressor speed, r/min (normal and design)

Table 5.6 Atmospheric Pressure versus Elevation

Altitude (Meters)	Average Atmospheric Pressure (kPa (abs))
0	101.325
100	99.97
200	98.84
300	97.93
400	96.60
500	95.44
600	94.54
700	93.49
800	92.04
1000	90.03
1200	87.77
1400	85.51
1600	83.42
2000	79.41
2500	74.58
3000	70.06
3500	65.54
4000	61.4
4500	57.71
5000	54.31

2. Service
   a. Approximate gas composition (vol% or mol%)
   b. Average molecular mass
   c. Relative density (specific gravity)
   d. Corrosiveness and remarks
3. Site informations
   a. Elevation of plant site from sea level, meters
   b. Minimum winter temperature, °C
   c. Maximum summer temperature, °C
   d. Normal barometer, kPa
   e. Relative humidity for process design, %
4. Available utilities
   a. Cooling water: maximum inlet temperature, °C
   b. Maximum outlet temperature, °C
   c. Pressure, kPa
   d. Fouling factor
   e. Instrument air pressure, kPa
   f. Electric power for instruments, volts, phase, Hz

5. Compressor location (outdoor, indoor)
6. Instrument graduation system
7. Remarks on control system

B.2 process specification sheet for positive displacement compressors
The minimum required information to be included in the process specification sheet shall be as follows:

1. Process requirements
  a. Flow at 1.013 bar (abs) and 0 °C, m^3/h, normal
  b. Flow at 1.013 bar (abs) and 0 °C, m^3/h, rated
  c. Flow at suction conditions, m^3/h, rated
  d. Suction temperature, °C
  e. Suction pressure, kPa (abs)
  f. Discharge pressure, kPa (abs) normal
  g. Discharge pressure, kPa (abs) rated
  h. Differential pressure, kPa, rated
  i. Compression ratio
  j. Approximate C_p/C_v (at suction)
  k. Compressibility factor (at suction)
  l. Mass flow molecular mass, kg/h, normal
  m. Mass flow molecular mass, kg/h, rated
  n. Estimated power required, kW, normal
  o. Estimated power required, kW, rated
  p. Estimated gear loss, kW
  q. Recommended driver Bhp, kW
  r. Compressor speed limitation (if any), r/min
  s. Piston speed limitation (if any), m/s
2. Service
  a. Approximate gas composition (vol% or mol%)
  b. Average molecular mass
  c. Relative density (specific gravity)
  d. Corrosiveness and relevant remarks
3. Site informations
  a. Elevation of plant site from sea level, meters
  b. Minimum winter temperature, °C
  c. Maximum summer temperature, °C
  d. normal barometer, kPa
  e. Relative humidity for process design, %
4. Available utilities
  a. Cooling water: maximum inlet temperature, °C
  b. Maximum outlet temperature, °C
  c. Pressure, kPa
  d. Fouling factor
  e. Instrument air pressure, kPa

> **f.** Fuel gas for engine, molecular mass: relative density
> **g.** Calorific value, kJ/m^3, gross pressure, kPa
> **h.** Electric power for instruments, volt, phase, Hz

5. Compressor location, outdoor, indoor
6. Instrument graduation system
7. Remarks on control system

5.11 Material for axial and centrifugal compressors and expander-compressors

This section gives technical specifications and general requirements for the purchase of "Axial and Centrifugal Compressors and Expander-Compressors for Petroleum, Chemical and Gas Industry Services" and is based on API Standard No. 617, seventh edition, July 2002, and shall be read in conjunction with that document.

Engineers shall pay due attention in specifying all the possible process variations, such as changes in pressure, temperature, molecular weight, etc., during the service life. In upstream applications, such as gas lift, gas gathering, and gas re-injection, the possible variation of the gas molecular weight shall be consulted with the reservoir specialist. If a variable speed unit is used, the speed required to achieve the specified rated operating conditions shall be indicated.

For reference purposes, the physical properties of the process gas (C_p/C_v, compressibility factor, molecular weight, etc.) may be given on the data/requisition sheets in order to facilitate the compressor calculations. However, the manufacturer shall base the performance calculations on the gas composition, as stated in the data/requisition sheets.

The manufacturer shall calculate the performance of the proposed compressor using the standard methods for computing the physical properties of the specified gases and shall take full responsibility for any design features affected. The manufacturer shall affirm the equation of state, mixing rule, interaction coefficients, and values of all physical properties of the process gas used in the computations with the proposal.

The equipment (including auxiliaries) shall be designed and constructed for a minimum service life of 25 years and at least 5 years of uninterrupted operation.

For offshore installations, the vendor who has unit responsibility shall notify the company in the proposal of the area required for the maintenance and repair of the entire equipment train to be supplied.

Unless otherwise specified, compressors and auxiliaries shall be suitable for outdoor installation in the climatic zone specified.

Compressors shall be designed and manufactured to minimize the generation of noise and shall not exceed the noise limits given in Table 5.7, at any measuring location not less than 1 m from the equipment surface.

Table 5.7 Sound Pressure Limit in dB

Compressor	87 dB
Compressor + Driver	90 dB

If the equipment produces impulsive noise, the above limits shall be taken 5 dB lower, thus, 82 dB for compressor and 85 dB for the compressor + driver.

The above requirements apply in the absence of reverberation and background noise from other sources, and for all operating conditions between the minimum flow and the rated flow.

Where excessive noise from equipment cannot be eliminated by low noise design, corrective measures preferably should take the form of acoustic insulation for pipes, gearbox, etc., where noise hoods are proposed, prior approval of the purchaser shall be obtained regarding construction, materials, and safety requirements. Noise control measures shall cause no hindrance to operations or any obstruction to routine maintenance activities.

When specified, the manufacturer should be present during initial installation and initial start-up. An outline of initial start-up procedures shall be prepared by the manufacturer and shall be agreed upon by all parties involved before operations are commenced.

All electrical components and installations shall be suitable for the area classification, gas grouping, and temperature classes specified by the purchaser on the data sheets, and shall meet the requirements of API Standards.

Throughout the service life specified, spare parts for all components of the unit shall be available for purchase and all manufacturing drawings shall be retained.

Note: For some applications, such as offshore, when the specified process gas is not accessible in the platform fabrication yard, the compressor may be field run on air for the purpose of pre-commissioning.

Compressor ratings shall not exceed the limits of the vendor's design but shall be well within the manufacturer's actual experience. Only equipment that has proven its reliability is acceptable.

A unit responsible representative shall be present in Hazard and Operability Analysis meeting, as requested by the company.

A detailed statement for supervision of installation, pre-commissioning, start-up, and commissioning of the equipment at the jobsite, including daily rates, terms, and conditions, shall be indicated in the proposal.

5.11.1 Material

The shaft shall be of forged steel material, and shall be ultrasonically examined prior to machining.

Radiographic inspection procedures shall be applied wherever possible. Butt welded joints of pressure casing shall be 100% radiographed.

Copper or copper alloys shall not be used for parts of machines or auxiliaries in contact with process fluid. Nickel–copper alloy (UNS N04400), bearing Babbitt, and precipitation-hardened stainless steels are excluded from this requirement.

All carbon and low alloy steel pressure-containing components, including nozzles, flanges, and weldments shall be impact tested in accordance with the requirements of Section VIII, Division 1, Sections UCS-65 through 68, of the American Society of Mechanical Engineers (ASME) Code standard.

High-alloy steels shall be tested in accordance with Section VIII, Division 1, Section UHA-51, of the ASME Code standard. For materials and thicknesses not covered by Section VIII, Division 1, of the ASME Code standards, the purchaser will specify requirements.

Note: Impact testing of a material may not be required depending on the minimum design metal temperature, thermal, mechanical and cyclic loading, and the governing thickness. Refer to requirements of Section VIII, Division 1, Section UG-20F of the ASME Code, for example. Governing thickness used to determine impact testing requirements shall be the greater of the following:

1. The nominal thickness of the largest butt-welded joint
2. The largest nominal section for pressure containment, excluding:
   a. Structural support sections, such as feet or lugs
   b. Sections with increased thickness required for rigidity to mitigate shaft deflection
   c. Structural sections required for attachment or inclusion of mechanical features, such as jackets or seal chambers
3. One fourth of the nominal flange thickness, including parting flange thickness, for axially split casings (in recognition that the predominant flange stress is not a membrane stress)

The results of the impact testing shall meet the minimum impact energy requirements of Section VIII, Division 1, Section UG-84, of the ASME Code standard.

5.11.2 Inter-stage diaphragms and inlet guide vanes

• **Pressure-containing casings**

All butt-welds on fabricated steel casings shall be 100% x-rayed after post-weld heat treatment. Further magnetic particle or dye penetrant examination shall occur after final machining. The purchaser and vendor shall agree prior to fabrication which welds will be inaccessible upon completion of fabrication for quality control assessment of the welds.

Cast iron casings are not acceptable. Radial split (barrel) casings and their end covers shall be forged steel if the design pressure exceeds 6900 kPa (69 Bar) or any molecular weight specified is less than 15.

The manufacturer shall provide drawings showing the location of all welds inaccessible for nondestructive testing (NDT), in particular, stressed welds on hydrocarbon containing components of the compressors.

• **Material inspection of pressure-containing parts**

All pressure welds shall be 100% radiographically examined. If radiography is not feasible, magnetic particle examination shall be performed.

Casing drains shall be individually valved and, when specified, manifolded, supported, and piped up to a single nozzle at the base-plate edge. Blocked valves shall be easily accessible. The number and locations of drains offered shall be indicated in the proposal.

Main process connections shall be flange-type, unless otherwise specified.

Socket welded connections shall not exceed DN50.

5.11.3 Rotating elements

Electrical and mechanical run-out shall be determined by rolling the rotor in V-block at the journal center lines, while measuring run-out with a noncontacting vibration probe and a dial indicator at the same shaft location. Readings shall be recorded at the probe location and one probe tip width to both sides. Metalizing or plating to reduce electrical run-out shall be permitted on precipitation-hardened steel shafts only, and the shaft must be flame or plasma sprayed aluminum applied by a vendor with proven experience in this process.

If the radial combined total electrical and mechanical run-out exceeds the 5 µm limit of this paragraph, the vendor shall first ensure that mechanical run-out at the probe locations does not exceed 2.5 µm. After achieving this mechanical run-out, two attempts shall be made to burnish to meet the 5 µm total run-out limit. After these attempts, the purchaser shall be contacted to evaluate this deviation.

Impellers shall be designed to limit the maximum stress at the maximum continuous speed to a value of 70% of the material yield strength.

The design for shaft-impellers and shaft-seal sleeves or shaft-seal labyrinth assembly shall not create distortion of the rotor assembly at rest, or bore growth, deteriorating concentricity and balance at trip speed. Finite element analyses must support the interference fit design for the wheels carrying full torques during the worst case trip speed conditions. Keys shall be the cylindrical type when used.

- **Impellers**
  - Forged rather than cast impellers are preferred
  - Cast impellers shall be inspected after machine finishing as well
  - Cast iron crack defect repair is not acceptable

5.11.4 Dynamics

The effect on rotor stability of high-density gas shall be taken into consideration.

Unless otherwise specified, the compressor vendor shall be unit responsible for undesirable speed calculations and shall furnish satisfactory verification of undesirable speed calculations, including all driver train components prior to submittal of the certified composite outline drawing.

Both torsional and lateral vibration analyses are the complete responsibility of the compressor vendor. The vendor shall furnish a copy of this analysis and related data for review.

- **Lateral analysis**

Undesirable speeds (including critical speeds) and their associated amplification factors shall be determined by means of a damped unbalanced rotor response analysis.

The location of all undesirable (including critical speeds) below the trip speed shall be confirmed on the test stand during the mechanical running test. The accuracy of the analytical model shall be demonstrated.

A train lateral analysis, considering the effect of other equipment in the train on the damped un-balance response analysis, shall be quoted as an option by the vendor who is assigned unit responsibility.

Every attempt shall be exercised to remove any undesirable speeds with amplification factor near 2.5 from operating speeds ranges.

- **Torsional analysis**

For motor-driven units and units including gears, units comprising three or more coupled machines (excluding any gears), the vendor having unit responsibility shall ensure that a torsional vibration analysis of the complete coupled train is carried out and shall be responsible for directing any modifications necessary to meet the requirements of 2.6.7.2 through 2.6.7.6.

- **Vibration and balancing**

Rigid rotors shall have a minimum of three balancing planes. The dynamic balance of the rotor shall be verified after installation of the coupling hub. All balance reading shall be recorded.

The vendor shall record the balance readings after the initial balance for the contract rotor. The rotor shall then be disassembled and reassembled. The rotor shall be check balanced after reassembly to determine the change in balance due to disassembly and reassembly.

Completely assembled rotating elements shall be subject to operating-speed (at speed) balancing in lieu of a sequential low-speed balancing. When the vendor's standard balance method is by operating-speed balancing in lieu of a sequential low-speed balancing and operating-speed balancing is not specified, it may be used with the purchaser's approval.

A rotor that is to be operating-speed balanced shall, when specified, first receive a sequential low-speed balance.

Any correction action (if needed), such as demagnetization after rotor balancing, shall be performed.

5.11.5 Bearings and bearing housings

Each radial bearing shall have two thermocouples or resistance temperature detectors (RTDs) embedded in the bottom of the most highly loaded shoe. Two thrust bearing pads on each side, 180° apart, shall be furnished with double-head thermocouples or RTDs suitable for connection to indication and alarm instruments.

All sensors shall be permanently cabled up to the local junction box at the skid edge.

The indicating device shall have two levels: one for the alarm and one for tripping. They shall be set individually and connected to the sensing points.

The number of connections in cabling shall be the minimum possible that facilitates assembly and disassembly.

5.11.6 Shaft end seals

Unless seal manufactured by main vendor, the manufacturer shall provide full technical details with the proposal, including a reference list with at least two examples of similarly sized seals each having achieved a minimum of 2 years continuous operation in similar services.

Shaft seals in inert gas service shall be of the labyrinth type. Shaft seals in toxic gas services shall be of the liquid film type, unless otherwise specified.

Unless otherwise specified, the seal design shall have provision for buffer gas injection.

When buffer gas injection is required for the compressor, the supplier shall state the gas requirements and limitations, including temperature and dew point in his proposal, and furnish the complete system.

The system for continuous buffer gas injection shall include but is not limited to dual appropriate mesh size strainers, automatic differential pressure controller, low pressure alarm and buffer gas pressure gage, and cooler, if required. Piping material for downstream of the strainer shall be stainless steel.

Various supplemental devices shall be provided to ensure sealing when the compressor is pressurized but not running and the seal oil system is shut-down. The vendor shall state the means in the proposal.

- **Clearance seals**

The labyrinth seals should be stationary parts of the machine.

- **Self-acting dry gas seal**

The dry gas seal must be able to withstand seal pressure fluctuations down to and under atmospheric conditions. If backward rotation can occur, it is required to select a seal that has the ability to rotate in both directions.

- **Lubrication and sealing systems**

Lubricating oil and seal oil systems shall comply with API Standard No. 614, as amended/supplemented.

5.11.7 Accessories

- Drivers

Steam turbine drivers shall conform to API Standard No. 611 or with API Standard No. 612, whichever is applicable.

Steam turbine drivers shall be sized to deliver continuously not less than 110% of the maximum power required by the machine train when operating at any of the specified operating conditions and specified normal steam conditions.

Electric motor drivers shall be rated with a 1.0 S.F. The motor rating shall be at least 110% of the greatest power required (including gear and coupling losses) for any of the specified operating conditions. Consideration shall be given to the starting conditions of both the driver and driven equipment and the possibility that these conditions may be different from the normal operating conditions.

Gas turbine drivers shall conform to API Standard No. 616. There shall be a power output margin of at least 7% between the demand of the driven equipment and the power of the gas turbine at the site when in a new and clean condition. Note that the power extracted by the auxiliaries, directly driven from the gas turbine, is not always included in the vendor's standard information sheets.

Having established the site rating for the gas turbine, the ISO* rating of the gas turbine (the ISO rating of a gas turbine is its rating at 15 °C ambient temperature, at 1013.25 mbar, and 60% relative humidity, with zero inlet and exhaust pressure losses) can be calculated to serve as a guide for comparing the available makes and models of the gas turbine type suitable for the application being considered.

- **Coupling and guards**

All hubs shall be fitted to the finish machined shaft ends before factory testing and before delivery.

- **Base plate**

Welded non-slip decking plates shall be provided for the top of the base plate. They shall cover all walk and work areas. The plates shall be fastened and shall allow access to inside components.

The supplier shall provide the method for leveling to facilitate use of optical, laser, or other instruments.

All auxiliary piping and accessories shall be mounted within the base-plate perimeter.

Unless otherwise specified, a common base plate extended as practically as possible to support the driver, other compressors, and gears shall be provided.

5.11.8 Controls and instrumentation

The vendor shall provide sufficient machine performance data (in accordance with Section 5) to enable the purchaser to properly design a control system for start-up operation for all specified operating conditions and for surge prevention. The vendor shall review the purchaser's overall machine control system for compatibility with vendor-furnished control equipment.

The instrument and control system shall permit local start-up of the equipment and subsequent transfer of control to the remote-control room panel. The equipment shall then normally be operated from the remote-control room panel.

Instrumentation shall be designed for minimum manning and operator attention. Alarms for equipment and process functions, equipment and process initiated shut-down annunciation, and all critical operation information shall be displayed on the remote-control room panel and be provided with suitable signals to repeat this information on the local panel. The compressors and drivers shall be suitable for automatic unattended start-up.

The instrumentation and installation shall conform to the requirements of API Standard No. 614, Chapter 1.

Compressor and driver monitoring panels shall be mounted in the control room, unless otherwise specified.

- **Control systems**

An anti-surge device shall be utilized if system requirements indicate that the compressor may operate in surge for intended periods.

- **Instrument and control panels**

Minimum instrumentation and process controls shall be furnished as specified and listed in Table 5.8. Any additional instrumentation and controls, as deemed necessary, for the smooth and safe operation of the unit under all specified operating conditions shall be provided.

Compatibility of the overall compressor control system with the furnished instrumentation and controls shall be ensured. The nominal supplies and location of each instrument written in the table shall be indicated by the following coded notes:

- Locally mounted (LM)
- Local panel mounted (LP)
- Unit control pane (UCP)

A free standing local control panel shall be supplied, completely enclosed and sealed, and suitable for pressurizing to keep out dust.

The panel shall include all the applicable items listed, together with alarm lights suitably screened to be easily visible in bright sunlight and other process instruments as required.

Table 5.8 Controls and Instrumentation

Instruments	Supply and Location
a) Pressure and level gages, pressure controllers, control valves, thermometers, pressure, differential pressure and temperature switches, and relief valves at the compressor for separate lube oil systems.	LM
b) Pressure and level gages, level controllers, pressure controllers, control valves, thermometers, pressure, differential pressure, level and temperature switches, flow meters or indicators, and relief valves, for seal oil system.	LM
c) Start and stop push button station with pilot lights for lube oil pump motor and seal oil pump motor.	LP
d) Speed indicator for compressor.	LP
e) Pressure gages for compressor suction and discharge.	LP
f) Pressure gage for lube oil pump discharge.	LM
g) Pressure gage for seal oil pump discharge.	LM
h) Pressure gage for reference gas line (if applicable).	LM
i) Pressure gage for balance gas line (if applicable).	LM
j) Pressure gage on air supply for flow regulator to seals.	LP
k) Pressure gage for lube oil to compressor bearings.	LP
l) Sight glasses on seal and lube oil out-let lines.	LM
m) Differential pressure gage for seal oil.	LM
n) Gage glass for seal oil overhead tank.	LM
o) High seal oil return temperature.	LP
p) Sight glass for lube oil reservoir.	LM
q) Temperature indicator for discharge gas.	LP

LM, locally mounted; LP, local panel mounted.

Access for easy maintenance to this panel shall be provided, and the location of the panel shall be so as to facilitate easy control of the equipment.

Consideration may also be given to the installation of a separate ground mounted panel to cover auxiliary equipment mounted on the console if easier operation would be achieved.

- **Instrumentation**

Alarms and interlock/shut-down circuits shall be generally "fail safe" (i.e., normally energized) contacts closed in the healthy state.

- **Alarms and shut-down**

Refer to API Standard No. 614 for details on instrument and control panels.

As a minimum, following the mentioned items in Table 5.9 shall be furnished. Local indication lights shall indicate green for normal operation, yellow for warning alarm, and red, for shut-down. The alarm system shall be independent of shut-down devices.

Table 5.9 Alarms and Shut-downs

Service	Alarm Lights	Alarm Switches	Shut-down Device
a) Low lube oil pressure.	LP, UCP	LM	UCP
b) Low seal oil differential pressure.	LP	LM	None
c) High seal oil differential pressure.	LP	LM	UCP
d) High lube oil temperature (after oil cooler).	LP, UCP	LM	UCP
e) High compressor discharge temperature.	LP	LM	UCP
f) Low level in lube oil reservoir.	LP	LM	None
g) Low level in seal oil reservoir.	LP	LM	None
h) Axial movement of compressor shaft.	LP	LM	None
i) High vibration of compressor.	LP	LM	UCP
j) Main seal oil pump failure.	LP, UCP	LM	UCP
k) Main lube oil pump failure.	LP, UCP	LM	None
l) Start standby lube oil pump.	LP, UCP	LM	None
m) Start standby seal oil pump.	LP	LM	None
n) High temperature in lube oil reservoir (if applicable).	LP	LM	None
o) High temperature in seal oil reservoir (if applicable).	LP	LM	None

LM, Locally mounted; LP, Local panel mounted; UCP, Unit control panel.

The alarm and shut-down systems involving equipment, environmental, and/or personnel safety shall be so designed as they cannot be bypassed during operation. The alarm units shall, however, have 20% spare lines.

- **Vibration, position, and bearing temperature detectors**
  - Unless otherwise specified, vibration and axial displacement measurement probes shall be installed in removable, externally mounted probe holders. Each holder shall be shouldered so that the probe location is maintained when the probe is removed and reinstalled. Both the probes and the holders shall be securely locked in place. Mountings shall permit removal and installation during compressor operation with a minimum spillage of oil
  - The shaft, in the region of the radial probes, shall present an unplated ground surface having a total run-out not exceeding 25% of the specified test level or 6 microns, whichever is greater. The shaft shall be demagnetized prior to the installation of the probes
  - Readout equipment provided for continuous monitoring of the vibration/axial displacement shall match the characteristics of the detector-probe system and shall not require field calibration
  - Standardized cable lengths shall be used to interconnect the detector probe and readout equipment. Radial vibration readout equipment shall be suitable for combining the two signals from H/V probes in the right phase and shall indicate the real value of the maximum peak-to-peak displacement. Readout equipment shall be mounted on the local panel. Each monitor

shall be provided with jack connectors, one per probe, to permit connection of portable analysis equipment (oscilloscope, etc.)

- The supplier shall indicate the settings for the alarm and shut-down in case of excessive radial vibration and axial displacement. Alarm and shut-down signals shall be individually connected to the alarm and shut-down alarm annunciation system of the compressor unit

As a minimum and unless otherwise specified, the following instrumentation shall be provided:

1. All thrust bearing temperatures (50% of pads on active side and two pads on inactive side) shall be monitored as follows:
   a. Each thermocouple point high temperature alarm
   b. If specified, a temperature recorder shall be provided for thermocouple points
2. Permissive start facility shall be provided for lube oil, seal oil, and control oil so that the driver cannot be started unless the oil system is functioning correctly
3. Compressor discharge high gas temperature alarm shall be provided (shut-down shall be provided if there is a possibly of compressor discharge temperature exceeding maximum case design temperature)
4. Diaphragm cooling water system, low water pressure alarm, and stand-by water pump cut-in alarm (when applicable)

5.11.9 Piping and appurtenances

All interconnecting piping between the twin units, the lube and seal oil console(s), and the various equipment groupings shall also be provided.

All piping for utilities (cooling water, steam, instrument air, purge gas, buffer gas, etc.) shall be provided and arranged in a way that will permit single inlet/outlet connections.

When a buffer gas manifold is specified, the required components, such as valves, flow meter, check valves, pressure indicators, throttle valves, differential pressure indicators, controllers, and control valves, shall be furnished by the vendor.

Process piping, if furnished, shall be in accordance with 2.4 of Chapter 1 of API Standard No. 614.

For radially split units, the vendor shall furnish a cradle or similar device for ease of removal of the compressor rotor and diaphragm.

5.11.10 Inspection, testing and preparation for shipment

Auxiliary equipment, such as drivers, gears, and oil systems, shall be inspected and tested in accordance with the specified API standards for the equipment.

The purchaser's representative shall have the right to reject any parts of the equipment that do not conform to the purchase order.

- **Inspection**

Any portion of the oil system furnished shall meet the cleanliness requirements of the relevant standard.

Final inspection for cleanliness of the compressor piping and auxiliary equipment shall be performed.

The purchaser shall be informed by the vendor of any defects noticed during manufacturing. The supplier shall also informed the purchaser and obtain permission before proceeding with any repairs that may affect equipment operation, integrity, of interchangeability. The repair procedure shall be approved by the purchaser before rectification.

- **Testing**

Acceptance of a shop test does not constitute a waiver of equipment to field performance under specified operating conditions nor does inspection relieve the vendor of his responsibilities.

The vendor shall notify the purchaser not less than 15 days before the date that the equipment will be ready for test.

- **Overspeed test**
  - Impeller Overspeed Test

Each impeller shall be subjected to an overspeed test at not less than 115% of maximum continuous speed for a minimum duration of 3 min.

- **Mechanical running test**

The vendor shall make tape recordings of all real-time vibration data starting with the initial shop run and given to the purchaser, even if not witnessed or observed.

Seal leakage rates per seal shall not exceed the guaranteed maximum rates stated in the vendor's proposal.

During the mechanical running test, the lubricating oil and seal oil temperatures shall be held for a minimum period of 30 min at the temperature corresponding to the minimum allowable viscosity and per 30 min at the temperature corresponding to the maximum allowable viscosity.

If the compressor is designed with variable stator geometry, the linkage, servo-motor, and stator blade seal shall be checked during the mechanical test.

- **Assembled machine gas leakage test**

During the leak test, the rotor shall be turned manually to check the seal adjustment. The gas leakage test of the assembled unit with seals installed shall be performed on the maximum discharge pressure. The leak test shall be the last test to be performed before dispatch. If a casing joint has to be remade after this test, the test shall be performed again.

- **Post-test inspection of compressor internals**

The compressor shall be dismantled, inspected, and reassembled after satisfactory completion of the mechanical running test. The purchaser will specify whether the gas test shall be performed before or after the post-test inspection. However, for toxic gas compressors, the leak test shall be conducted after assembly.

Note: The merits of the post-test inspection of the compressor internals should be evaluated against the benefits of shipping a unit with proven mechanical assembly and casing joint integrity.

- **Full-pressure/full-load/full-speed test**

Following the mechanical run test, the compressor shall be opened for internal inspection. The leak test shall be conducted after assembly.

- **Optional tests**
  - Performance test

If a performance test is specified, it shall be conducted in accordance with ASME PTC-10, and the following information shall be furnished prior to the test:

– Identification of the class of test (I, II, or III) required to meet objectives and selection of operating conditions to satisfy the limitations outlined in ASME PTC-10
– The nature of the test gas, if the design or specified gas cannot be used, and the means for establishing its physical and thermodynamic properties
– Procedures to be used for adjusting test results to specify operating conditions with respect to test classifications I, II, and III
- **Spare-parts test**

When spare rotors are supplied, they shall be dynamically balanced to the same tolerances as the main rotor.

- **Preparation for shipment**

The preparation shall make the equipment suitable for 12 months of outdoor storage from the time of shipment.

The paint for all exterior surfaces shall be suitable for the environment specified.

Unless otherwise specified, the rust preventive applied to unpainted exterior machined surfaces shall be of a type: (1) to provide protection during outdoor storage for a period of 12 months exposed to a normal industrial environment, and (2) to be removable with mineral spirits or any standard solvent.

The height of shanks shall be at least 35 mm.

Unless otherwise specified, separate shipment of the material is not permitted.

Unless otherwise stated in the purchase order, all packaged pipe work shall be fully fitted and assembled prior to dispatch. Pipe work between main packages and off package auxiliaries shall not be fully fitted, but shall have sufficient spare lengths for field welding, any loose items should be packed separately.

The exposed shaft end should be protected against physical damage.

5.11.11 Guarantee and warranty

- **Mechanical**

Unless exception is recorded by the vendor in the proposal, it shall be understood that the vendor agrees to the following guarantees and warranties:

- All equipment and component parts shall be guaranteed by the vendor against defective materials, design, and workmanship for 1 year after being placed in service or 18 months after the date of shipment
- If any malfunction or defects occur during the guarantee and warranty period, the vendor shall make all necessary alterations, repairs, and replacements free of charge, free on board (f.o.b.) factory. Field labor charges, if any, shall be subject to negotiation between the vendor and the purchaser

- **Performance**

Each compressor casing shall be guaranteed for head with no negative tolerance, capacity, and behavior satisfying all points mentioned on the data sheet. Power consumption shall also be guaranteed. If several compressor sections/casings are on the same shaft line, the power guarantees shall cover the total power.

The rotation speed and the guaranteed head and flow rate shall not vary from the indicated theoretical-value by more than ±3%.

If the compressor consists of several casings, or several sections in a casing handling different flow rates, each section shall satisfy its individual guaranteed performance.

Intermediate, outlet, and inlet pressures shall not differ from the guaranteed values by more than ±2% with the compressor operating under the following conditions:

- Flow rates, pressures, temperature, and nature of gas at inlet, corresponding to guarantee
- Outlet pressure corresponding to guarantee

5.12 Centrifugal and axial compressors

This section, in conjunction with previous parts of this chapter, covers minimum requirements for centrifugal and axial compressors for use in refinery services, gas, chemical, and petrochemical plants and where applicable in production and new ventures.

Compliance by the vendor with the provisions of this section specification does not relieve him of responsibility of furnishing properly designed equipment, mechanically and electrically suited to meet operating conditions.

The compressors shall be the product of a manufacturer regularly engaged in the manufacture of the machine at least for 3 years.

Unless specific exception is accompanied by a description of the proposed substitute that is recorded under the heading "exception" in the manufacturer's proposal, it shall be mutually understood that the proposal complies strictly with the requirement of this section.

5.12.1 Basic design

- **Casings**
- **Pressure casing connections**

The supplier shall advise in the proposal all possible locations for borescope inspection points and the advantages any such inspection points would have.

- **Guide vanes, stators, and stationary internals**

The supplier shall advise in the proposal whether such a feature is required or of benefit.

Where AIGVs are supplied, the supplier shall also provide a valve positioner with local position indicator, unless otherwise specified.

AIGVs shall only be supplied of a type where the identical design has been demonstrated to operate in a validly similar duty, for a period of 3 years without any operational problems. In particular, no blade linkage or bearing wear or failure has been observed due to high frequency or resonant vibration.

Design of diaphragms shall be such that they can be removed without disassembly of the rotating element, unless otherwise specified.

- **Rotating elements**
  - **Thrust balancing**

Unless otherwise stated, the pressure tapping shall be provided.

  - **Centrifugal compressor impeller construction**

The impellers preferably should be the closed type.

Riveted-type impellers are not acceptable.

Brazed impellers would be acceptable providing that references are submitted for similar impellers and services.

The impellers components preferably should be forged.

  - **Hydrodynamic radial bearings**

The use of non-split design is forbidden unless approved by the purchaser.

- **Bearing housings**

If separate lube oil and seal oil systems or labyrinth seals are used, bearing ends shall be separated from the seals by a distance of at least 20 mm. The possibility of installing a buffer gas feed (air or inert gas) between the bearings and the seals must be demonstrated.

If the process gas is liable to cause deterioration of the lube oil, the bearings shall be isolated from the shaft seal by a ventilated space and labyrinths or other seals to prevent contamination or interchange of the seal oil, process gas, and lube oil. The space shall have provision for external or inert gas purge.

- **Shaft end seals**

Shaft seals in inert gas service shall be of the labyrinth type. Shaft seals in toxic gas services shall be of the liquid film type, unless otherwise specified.

Shaft end seals and, unless other specified, shaft sleeves shall be accessible for inspection and for replacement without removing the top half of the casing for an axially split compressor or the heads of a radially split unit. (Note: This requirement may not be feasible for overhung designs.)

The guaranteed figure for total waste seal oil leakage for both seals (per case) shall not exceed 50 l/day. No waste oil shall be returned to the oil system unless all contaminating components are removed.

- **Lubrication and sealing systems**

Unless otherwise specified, the supplier shall propose the type and properties of oil to be used.

For compressors driven by electric motor or steam turbine, the compressor supplier shall have the overall responsibility for the entire unit, comprising the compressor, driver, power transmission, controls, instrumentation, and all associated auxiliary equipment. Unless otherwise specified, for compressors driven by gas turbines, the turbine vendor shall have the overall responsibility.

The supplier having the overall responsibility shall coordinate and resolve any engineering or contractual problems for the complete unit.

When compressor flushing is specified, its effect on power requirements shall be considered in sizing the driver.

For air compressors, the effect of minimum design ambient temperature on the compressor's power requirements shall be considered in sizing the driver.

Unless otherwise specified, the supplier shall state in the proposal the standard type of control signal, which will be mutually agreed upon between the purchaser and the vendor prior to order placement.

5.12.2 Piping and appurtenances

- **Inlet system for process air compressors**

If specified, the vendor shall supply the inlet system for the process air compressors per the applicable portions of API Standard No. 616. In addition, the following requirements apply:

- The free area of the inlet screen shall be at least 2½ times the cross-sectional area of the air compressor inlet flange. The inlet screen shall be 5 × 5 mesh, 3.5 mm avgerage opening, 1.6 mm wire type 304 stainless steel. The free area of this screen arrangement is 47% of the total actual area; therefore, the actual area of the inlet screen shall be at least 2.5/0.47 or 5.3 times the compressor inlet area
- Louvers, if used, shall have a free area at least equal to the free area of the screen
- Provide a flanged opening diameter nominal (DN) 150 (6 in. nominal pipe size (NPS)) minimum diameter on the inlet ducting to permit future on-stream cleaning of the compressor blading

The vendor furnishing the discharge blow-off silencers for air compressors (downstream side of blow-off valves) shall submit cross-sectional drawings and at least one example of prior satisfactory experience to the purchaser for approval. The experience case(s) submitted shall be for the same diameter and flow rate range as the proposed silencer.

The vendor furnishing the discharge blow-off silencers for air compressors (downstream side of blow-off valves) shall submit cross-sectional drawings and at least one example of prior satisfactory experience to the purchaser for approval. The experience case(s) submitted shall be for the same diameter and flow rate range as the proposed silencer.

5.12.3 Inspection, testing, and preparation for shipment

- **Testing**
  - **Mechanical running test**

Oil system components downstream of the filters shall meet the cleanliness requirements of available standards before any test is started.

  - **Assembled compressor gas leakage test**

Casings for compressors handling gas containing 30 mol-percent or higher of hydrogen, shall be subject to a helium leakage test, at not less than the casing maximum allowable working pressure.

- **Optional tests**
  - **Performance test**

Air shall not be used for closed loop testing. For all other gases, open loop air performance tests shall be performed, providing the ASME Power Test Code 10 permissible tolerances are complied with; otherwise, a closed loop performance test shall be performed. The Machine Reynolds number correction to performance test data allowed by ASME Power Test Code 10 shall not be used.

Any guarantees shall be based on the purchaser's stated guarantee point, which may not be the normal operating point.

The required guarantees for the complete compressor shall not be affected by those for the intermediate pressures.

5.12.4 Vendor's data

The following additional data shall be identified on transmittal (cover) letters and in title blocks or title pages.

1. Expected variation in bearing clearances and oil viscosity
2. Input data required for an axial compressor performance check:
  a. Number of stages
  b. Presence of inlet guide vanes (IGVs) and exit guide vanes (EGVs)
  c. Number of blades in each blade row (rotor, stator, IGV, and EGV)
  d. Inlet and outlet blade angles as a function of blade height for rotor, stator, IGV, EGV
  e. Stator blade angles for alternate operating setting if machine has variable stator geometry
  f. Design incidence angle and loss coefficients at mean radius for each rotor and stator blade row
  g. Rotor blade area ratio (tip profile area/base profile area)
  h. Hub and tip radii for inlet and exit to each blade row (rotor, stator, IGV, and EGV)
  i. For each rotor and stator blade row, the chord length as a function of blade height and the solidity (chord/pitch ratio) at the mean radius
  j. Type of airfoil (subsonic or transonic)
  k. Statement if free-vortex flow is assumed
  l. Tangential velocity of gas at rotor blade tip
  m. Rotor and stator blade thickness as a function of chord length
  n. Young's modulus, density, and material for rotor blades.
  o. Any additional mass (i.e., shrouds, lacing, wire, etc.) and radius of application.
  p. Total temperature rise per stage.
  q. Polytrophic efficiency for each stage (rotor and stator, IGV + rotor + stator and rotor + stator + EGV).
  r. Minimum flow at design speed and alternate speed conditions.

5.13 Integrally geared compressors

This section, in conjunction with Section 5.1 of this chapter, covers the minimum requirements for integrally geared compressors for use in refinery services, gas, chemical, and petrochemical plants and, where applicable, in production and new ventures.

Equipment offered by the supplier shall satisfy the following minimum service and manufacturing experience requirements:

1. Compressors shall be identical or validity similar in power rating, speed discharge pressure, suction capacity, mechanical design, and materials or rotor dynamics, as compared with at least one unit produced by the supplier at the proposed manufacturing plant. The compressor must have at least 1 year's satisfactory operation.
2. Corresponding requirements of items 1 shall also apply to the driver, gear, and auxiliary equipment.

The torsional resonances of the package shall be at least 10% above trip or 10% below any operating speed for any possible excitation frequency.

5.13.1 Bearings and bearing housings

- **Bearing housings**

If separate lube oil and seal oil systems or labyrinth seals are used, bearing ends shall be separated from the seals by a distance of at least 20 mm. The possibility of installing a buffer gas feed (air or inert gas) between the bearings and the seals must be demonstrated.

If the process gas is liable to cause deterioration of the lube oil, the bearings shall be isolated from the shaft seal by a ventilated space and labyrinths or other seals to prevent contamination or interchange of the seal oil, process gas, and lube oil. The space shall have provision for external or inert gas purge.

- **Gears**
  - Gearboxes

Gears shall have at least 80% tooth contact at full load.

Vibration monitoring shall be provided for all shafts with hydrodynamic bearings, including the driver.

5.13.2 Piping and appurtenances

All oil piping (supply and return) shall be stainless steel.

If specified, the vendor shall supply the inlet system for process air compressors per the applicable portions of API Standard No. 616. In addition, the following requirements apply:

- **Process piping and accessories**
  - The free area of the inlet screen shall be at least 2½ times the cross-sectional area of the air compressor inlet flange. The inlet screen shall be 5 × 5 mesh, 3.5 mm avgerage opening, 1.6 mm wire type 304 stainless steel. The free area of this screen arrangement is 47% of the total actual area; therefore, the actual area of the inlet screen shall be at least 2.5/0.47 or 5.3 times the compressor inlet area
  - Louvers, if used, shall have a free area at least equal to the free area of the screen
  - Provide a flanged opening DN 150 (6 in. NPS) minimum diameter on the inlet ducting to permit future on-stream cleaning of compressor blading

The vendor furnishing the discharge blow-off silencers for air compressors (downstream side of blow-off valves) shall submit cross-sectional drawings and at least one example of prior satisfactory experience to the purchaser for approval. The experience case(s) submitted shall be for the same diameter and flow rate range as the proposed silencer.

If specified, the vendor shall supply the inlet system for process air compressors, the following requirements apply:

5.13.3 Inspection, testing, and preparation for shipment

- **Testing**

Unless otherwise specified, a seal oil system shall be subjected to a separate package functional test and shall, in addition, be fully functionally tested as part of the complete unit test, if this test is required.

Unless otherwise specified, the supplier shall provide all necessary vibration monitoring equipment for test, where contract equipment is not utilized.

The gear contact pattern shall not be less than 80% of the effective width of the gear mesh.

All bearings shall be removed, inspected, and re-assembled after completion of the running test.

5.14 Expander-compressors

This section covers the minimum requirements for expander-compressor use in refinery services, gas, chemical, and petrochemical plants and, where applicable, in production and new ventures.

Compliance by the vendor with the provisions of this section does not relieve him of responsibility of furnishing properly designed equipment, mechanically and electrically suited to meet operating conditions.

The expander-compressors shall be the product of a manufacturer regularly engaged in the manufacture of the machine at least for 3 years.

Unless specific exception accompanied by a description of the proposed substitute that is recorded under the heading "exception" in the manufacturer's proposal, it shall be mutually understood that the proposal complies strictly with the requirement of API Standards.

5.14.1 Basic design

The normal point of operation shall be the guarantee point and be as close as possible to the optimum efficiency, unless otherwise specified. When other operating points are specified at higher flow, higher head, or both, the vendor shall achieve a power balance by adjusting the unit's speed and shall state any negative tolerance on head at these other flows and shall obtain company approval.

Expanders and compressors shall be designed for 110% of the largest flow that can be expected during continuous operation and start-up.

Expanders and compressors shall be capable of operating at 50% of the normal flow rate without using recycles on the compressors.

5.14.2 **Materials**

All parts shall be suitable for the lowest expected temperature.

Other components of the machinery train should also be evaluated for the prevention of brittle fracture due to materials exhibiting change from ductile to brittle fracture as temperatures are reduced.

The purchaser will specify the minimum design metal temperature and concurrent pressure used to establish the impact test and other material requirements.

Note: Normally, this will be the lower of the minimum surrounding ambient temperature or the minimum fluid pumping temperature, however, the purchaser may specify a minimum design metal temperature based on properties of the pumped fluid, such as auto-refrigeration at reduced pressures.

5.14.3 **Casings**

- **Pressure-containing casings**

The purchaser shall provide system protection for pressure that may develop by compressors when operating at trip speed. The vendor shall indicate the maximum allowable working pressure for each casing.

- **Inlet guide vanes, variable nozzles, and heat shields**

IGVs shall be capable of handling at least 125% of the design mass flow rate at normal pressures and temperatures, or at least 110% of the maximum specified mass flow at the minimum specified inlet pressure and maximum specified inlet temperature, whichever gives the highest actual flow.

Actuating devices shall be pneumatic, with integral positioners and guide vanes incorporating a pressure balanced design to limit the nozzle forces and provide smooth, accurate, and reliable operation.

IGVs shall incorporate pressure actuated sealing rings to minimize side leakage and prevent galling. The nozzle design shall allow for:

a. Rotational movement about nozzle pins only
b. Minimum travel of nozzle segments
c. Floating pins in nozzle segments to allow lapping to provide equal sealing of segments

5.14.4 **Rotating elements**

- **Shafts**

The vendor shall provide specific details for coating or overlays on journals of precipitation-hardened stainless steel shafts. The use of lube oil additives to mitigate bearing wire wooling shall be subject to the purchaser's approval.

- **Impellers**

Proposed impellers shall have proven performance on identifiably similar units, preferably supported by the manufacturer's performance tests on expanders and compressors.

- **Thrust balancing**

Thrust equalizing valves shall incorporate built in bias to prevent hunting.

5.14.5 Dynamics

- **Vibration balancing**

Rotors shall be a stiff-shaft design with no undesirable running speeds between zero and the trip speed, as proven by the manufacturer's experience.

Rotors shall be assembled and the balance verified. Rotors failing to meet the criteria shall be balance corrected by repeating the component balance and not by trim balancing the assembly.

A residual unbalance check shall be performed on the assembled rotor directly following the balance verification and before the assembled rotor is removed from the balancing machine.

A damped unbalanced rotor response analysis shall verify that the first bending critical speed is at least 125% above the trip speed. For new rotor designs, a complete lateral analysis shall be performed. The undamped analysis shall identify the first three system natural frequencies.

A torsional analysis shall be performed, if not already available, for a similar torsional model or when specified by the purchaser.

The first torsional mode shall be at least 125% above the trip speed.

5.14.6 Bearings and bearing housings

- **Hydrodynamic radial bearings**

The bearing design shall feature high stiffness to suppress oil whirl (half speed gyration) or oil film resonance over the operating speed range.

- **Hydrodynamic thrust bearings**

Each half of the hydrodynamic thrust bearing should preferably be combined with a journal bearing housing to limit the overall length of the assembly.

- **Expander-compressor shaft seals**

For application (using the pressurized oil reservoir) that uses buffer gas as sealing media, provision shall be provided by the vendor to prevent the build-up of condensate and contamination of the oil system.

- **Lubrication and sealing systems**

When approved by the purchaser, a pressurized integral oil system may be provided for a closed loop. In this case, a continuous indicator of actual lubricant viscosity at the bearing inlet shall be required.

Reference

[1] Gas Processors and Suppliers Association (GPSA). Engineering databook. 12th ed. 2004. Tulsa (OK, USA).

Further reading

Abbaspour M, Krishnaswami P, Chapman KS. Transient optimization in natural gas compressor stations for linepack operation. J Energy Resour Technol Trans ASME 2007;129(4):314–24.

Adams D. DATUM technology extended to natural gas pipeline compressors. Gas Turbine World 2004;34(4): 18–20.

American Petroleum Institute Standards: API 11P–"Specifications for Packaged Reciprocating Compressors for Oil and Gas Productive Services". API 614–"Lubrication, Shaft-Sealing and Control Oil Systems for Special-Purpose Applications". API 617–"Axial and Centrifugal Compressors and Expander Compressors for Petroleum, Chemical and Gas Industry Services". API 618–"Reciprocating Compressors for General Refinery Services". API 619–"Rotary Type Positive Displacement Compressors for, Petroleum, Chemical and Gas Industry Services". API 670–"Non-Contacting Vibration and Axial Position Monitoring, System". API 678–"Accelerometer Based Vibration Monitoring Systems".

Anon. Operating experience with gas turbine compressors in natural gas processing plant. Diesel Gas Turbine Worldwide 1988;20(9). 38, 40.

Anon. Valve, compressor contracts awarded for western hemisphere projects. Oil Gas J 1998;96(3):57.

Bergmann D, Mafi S. Selection guide for expansion turbines. Hydrocarbon Process; August, 1979.

Bloch HP. A practical guide to compressor technology. New York (NY): McGraw-Hill Book Co., Inc.

Bosman M. Availability analysis of a natural gas compressor plant. Reliab Eng 1985;11(1):13–26.

Botros KK, Golshan H, Rogers D, Sloof B. Evaluation of gas turbine outboard bleed air on overall engine efficiency and CO2e emission in natural gas compressor stations. J Eng Gas Turbines Power 2013;135(10). art. no. 101201.

Botros KK, Golshan H, Sloof B, Samoylove Z, Rogers D. Natural gas compressor operation optimization to minimize gas turbine outboard bleed air. In: Proceedings of the Biennial International Pipeline Conference. IPC 2012;vol. 1:681–9.

Brezonick Mike. New high capacity natural gas compressor package from hurricane. Diesel Progress Engines Drives 1995;61(10):3.

Brown RN. Control systems for centrifugal gas compressors. Chem Eng; Feb. 1964.

Cherednichenko I, Khodak E, Kirillov AI, Zabelin N. Optimal running conditions of cooling systems of the gas-main pipeline compressor stations. Heat Transf Res 2009;40(4):293–304.

Cierniak S, Knebel T. Repair, revamp or replacement of natural gas reciprocating compressors for underground storage. Oil Gas Eur Mag 2005;31(4):183–6.

Criqui AF. Rotor dynamics of centrifugal compressors. San Jose (CA): SolarTurbines International.

Hafaifa A, Laaouad F, Laroussi K. Fuzzy logic approach applied to the surge detection and isolation in centrifugal compressor. Autom Control Comput Sci 2010;44(1):53–9.

Hansen J, Curry B, Bequette M. New loading technique reduces compressor interruptions. Oil Gas J 2010; 108(33):140–5.

Hayes Charles R, Teach Ernesto F. Nolen Don. High-speed engine, compressor sets help pipeline meet growth. Oil Gas J 1995;93(47):66–7.

ISR Mechanalysis, Inc., 6150 Huntley Road, Columbus, Ohio 43229.

Jiang Y, Chen M. Failure analysis on the exhaust steel tube explosion of natural gas compressor. Adv Mater Res 2012;418–420:940–3.

Jopp K. Large-scale air separation compressor solutions for GTL plants. Erdoel Erdgas Kohle 2003;119(6):83–4.

Kind RH. El Paso automates main line compressor stations. Oil Gas J 1989;87(52):127–32.

Lagus PL, Flanagan BS, Peterson ME, Clowney SL. Tracer-dilution method indicates flow-rate through compressor. Oil Gas J; 89(8):36–9.

Leimgruber W. Sour- and acid-gas compressors in Oman. Tech Rev–Mitsubishi Heavy Ind 2001;38(3):8–11.

Marriott A, Ryrie J. Dominique Gilon. Mopico compressor for gas pipeline stations. Sulzer Tech Rev 1991;73(1): 30–4.

Miftakhov AA, Voronov GF. Experimental trial of end stages for unified centrifugal compressors and superchargers of natural gas. Khimicheskoe I Neftegazov Mashinostr 1992;(3):3–5.

Moore JJ, Ransom DL, Viana F. Rotordynamic force prediction of centrifugal compressor impellers using computational fluid dynamics. Proc ASME Turbo Expo 2007;5:293–304.

Moore JJ, Ransom DL. Centrifugal compressor stability prediction using a new physics based approach. Proc ASME Turbo Expo 2009;6(part B):1023–31.

Neerken RF. Compressor selection for the chemical process industries. Chem Engg 1975;82:78.

Novecosky T. Axial inlet conversion to a centrifugal compressor with magnetic bearings. J Eng Gas Turbines Power 1994;116(1):152–5.

Nystad BH, Rasmussen M. Remaining useful life of natural gas export compressors. J Qual Maint Eng 2010;16(2): 129–43.

Ovsienko AG, Petrusenko MI, Krinitskii EV. Determination of extent of operational observations required for assessing reliability of gas compressor units. Chem Pet Eng 1990;25(11–12):704–5.

Palmer Graham. Synthetic lubricant developments influence compressor selection. Ind Lubr Tribol 1989;41(2): 7–9.

Perry RH, Chilton CH. Chemical engineers handbook. 5th ed. New York (NY): McGraw-Hill Book Co., Inc. [Section 6].

Raheel M, Engeda A. Performance characteristics of regenerative flow compressors for natural gas compression application. J Energy Resour Technol Trans ASME 2005;127(1):7–14.

Reid CP. Application of transducers to rotating machinery, monitoring and analysis. Noise Control Vib Reduct; January, 1975.

Rogers LE. Design stage acoustic analysis of natural gas piping systems in centrifugal compressor stations. J Eng Gas Turbines Power 1992;114:727–36.

Scheel LF. Gas and air compression machinery. New York (NY): McGraw-Hill Book Co., Inc.

Schneider M, Mann J. Noise control technology with reference to natural gas compressor stations under the aspect of investment costs. J Eng Gas Turbines Power 1992;114(4):737–9.

Schneider TN, Brogelli R. Optimum compressor choice boosts Brent gas production. Oil Gas J 2006;104(38): 38–41. 35, 36.

Shade N. New gas compressor "takes off". Diesel Prog North Am Ed 2003;69(11):32–7.

Shen Y, Zhang B, Xin D, Yang D, Peng X. 3-D finite element simulation of the cylinder temperature distribution in boil-off gas (BOG) compressors. Int J Heat Mass Transfer 2012;55(23–24):7278–86.

Shin MW, Shin D, Choi SH, Yoon ES, Han C. Optimization of the operation of boil-off gas compressors at a liquified natural gas gasification plant. Industrial Eng Chem Res 2007;46(20):6540–5.

Song M. Optimal design for conficuration of natural gas compressor unit. Nat Gas Ind 2008;28(1):122–4. A17.

Swearingen JS. Turboexpanders and expansion processes for industrial gas. Los Angeles (CA): Rotoflow Corp.

Zeckendorf A, Altena JW. Design, simulation create low surge, low cost gas-injection compressor. Oil Gas J 1995; 93(3):57–62.

Blow-Down and Flare Systems

In designing safeguarding systems for process plants, facilities should be provided for handling, directing, and ultimately disposing of voluntary and involuntary gases and liquids. Several options are available to the process design engineer as to the selection of disposal systems. Once a specific disposal system is selected, detail design is undertaken.

This chapter covers process design and evaluation and selection of relief systems for oil, gas, and petrochemical process plants. It includes network and related ancillary installations that are to handle and direct fluids discharged as a result of overpressure, and/or operational requirements of a safe disposal system. This chapter is primarily concerned with the selection of disposal systems, sizing of relief headers, sizing of flare systems, and burning pits.

Although the various systems for the disposal of voluntary or involuntary vapor or liquid are mentioned subsequently, the actual selection of a disposal system should be conducted in accordance with the expected frequency, duration of operation, required capacity, and fluid properties.

6.1 Blow-down system for vapor relief stream

Systems for the disposal for voluntary and involuntary vapor discharges are:

1. To the atmosphere
2. To a lower-pressure process vessel or system
3. To a closed-pressure relief system and flare
4. Acid gas flares

6.1.1 Vapor discharge to atmosphere

Vapor relief streams should be vented directly into the atmosphere if all of the following conditions are satisfied (for a complete discussion on the subject, see Ref. [1]):

1. Such disposal does not conflict with current regulations concerning pollution and noise
2. The vapor is effectively nontoxic and non-corrosive
3. Vapor is lighter than air, or vapor of any molecular mass is nonflammable, non-hazardous, and non-condensable
4. There is no risk of condensation of flammable or corrosive materials
5. There is no chance of simultaneous release of liquid, apart from water
6. Relief of flammable hydrocarbons directly into the atmosphere should be restricted to cases in which it can be ensured that they will be diluted with air to below the lower flammable limit. This should occur well before they can come into contact with any source of ignition

Copyright © 2014 Elsevier Inc. All rights reserved.

This condition can most easily be met if the vapors to be released have a density less than that of air. However, with proper design of the relief vent, adequate dilution with air can be obtained in certain cases with higher-density vapors. Methods for calculation are given in Ref. [1] section 4.3.

- **Exceptions:**

1. Vapor from depressuring valves should be discharged to a closed-pressure relief system
2. Vapor that contains $\geq 1\%$ H_2S by volume should be discharged to a closed-pressure relief system

6.1.2 Vapor discharge to lower-pressure process vessel or system

Individual safety/relief valves may discharge to a lower-pressure process system or vessel capable of handling the discharge.

Although this type is rarely used, it is effective for discharges that contain materials that must be recovered.

6.1.3 Vapor discharge to closed-pressure relief system and flare

In all cases in which atmospheric discharge or release of vapor to a lower-pressure system is not permissible or practicable, vapor should be collected in a closed-pressure relief system that terminates in a flare, or a flare system. Where the concentration of H_2S is such that condensation of acid gas is probable, provision for a separate line, heat traced, should be considered.

In all cases, the installation of a closed-pressure relief system should result in a minimum of air pollution and the release of combustion products.

6.1.4 Acid gas flare

In process plants in which H_2S-free and H_2S-containing streams are to be flared, consideration should be given to the installation of a separate header and flare stack assembly for the H_2S-containing streams. The following provisions should be studied for the acid gas flare assembly:

1. Automatic injection of fuel gas downstream of the H_2S pot to make the combustion stable
2. Steam injection for smokeless operation should not be considered for the H_2S flare tip
3. A common pilot igniter should be used to ignite all flare stacks, including the acid flare
4. The H_2S flare header and subheaders may be heat traced to prevent condensation acid gas

6.2 Blow-down system for liquid relief stream

Systems for the disposal of voluntary and involuntary liquid discharges are:

1. To an onsite liquid blow-down drum
2. To a lower-pressure process vessel or system
3. To oily water sewers only if the material will not cause hazardous conditions
4. To pump suction if the pump will not overheat and can withstand the expected temperature rise
5. To a burning pit
6. To a vaporizer

Thermal expansion relief valves may discharge small quantities of volatile liquid or vapor into the atmosphere, provided the valve outlet is in a safe location.

6.2.1 Liquid discharge to onsite liquid blow-down drum

The liquid should be discharged to an onsite liquid blow-down drum that is capable of retaining the liquid discharged at the required liquid relief rate for a period of 20 min. This drum should have a vapor discharge line to the closed-pressure relief system.

6.2.2 Liquid discharge to lower-pressure process vessel or system

The liquid should be discharged to a lower-pressure process vessel or system that is capable of handling the required liquid relief rate plus any flashed vapor.

6.2.3 Liquid discharge to oily water sewer

Liquid discharge to an oily water sewer should be nonvolatile and nontoxic. The required liquid relief rate should be within the oil removal capability of the oily water treating system.

6.2.4 Liquid discharge to pump suction

Required liquid relief should discharge to an upstream liquid reservoir from which the pump takes suction. The liquid relief may discharge directly to the pump suction line if sufficient cooling is provided to prevent a temperature rise of the liquid recycled through the pump when the safety/relief valve opens or when a constant displacement pump is used.

6.2.5 Liquid discharge to burning pit

Liquid relief or voluntary liquid blow-down that needs not be returned to the process or discharged to an oily water sewer should be discharged to a burning pit, if environmentally acceptable.

6.2.6 Liquid discharge to vaporizer

The liquid should be discharged to a vaporizer that is capable of vaporizing a liquid relief of no more than 5000 kg/h.

6.3 Design of disposal system components

Depending on the process plant under consideration, a disposal system may consist of a combination of the following items: piping, knockout drum, quench drum, seal drum, flare stack, ignition system, flare tip, and burning pit.

- **Piping**

In general, the design of disposal piping should conform to the requirements of *American National Standards Institute/American Society of Mechanical Engineers*(ANSI/ASME) B31.3. Installation details should conform to those specified in Ref. [2], Part II.

- **Inlet piping**

The design of inlet piping should be in accordance with [1], Section 5.4.1.

- **Vapor relief header**

The sizing should be in accordance with [1], Section 5.4.1.2, in conjunction with [2], Part I, Section 7.

6.3.1 Vapor relief header sizing

- **Calculate N (number of velocity heads):**

$$N = \frac{fL}{D} + \sum K \tag{6.1}$$

where:
  f = Moody factor (dimensionless)
  L = length of equivalent pipe, m
  D = internal diameter of pipe, m
  K = factor representing friction resistance to flow, dimensionless (Tables 6.1 and 6.2).
- **Calculate P_3/P_1 or P_2/P_1**
  where:
  P_3 = pressure in the reservoir into which pipe discharges, 101 kPa absolute with atmosphere discharge
  P_1 = pressure at upstream low-velocity source, kPa absolute
  P_2 = pressure in the pipe at the exit or at any point distance L downstream from the source, kPa absolute.
- **Calculate Gci**

$$Gci = 6.7p_l \left(\frac{MW}{zT_l}\right)^{0.5} \tag{6.2}$$

where:
  Gci = critical mass flux, kg/s/m^2
  MW = molecular weight of vapor
  T_1 = temperature at upstream low-velocity source, Kelvin
  z = compressibility factor.
- From P_3/P_1 or P_2/P_1 and N, read G/Gci from Figure 6.1
- Calculate G in kilograms per second per square meter
- Calculate w (actual flow in kilograms per second)
  where:
  w = Gx (cross-sectional area of pipe, in square meters)

6.3.2 Liquid blow-down header

To reduce relief header loads and prevent surges resulting from two-phase gas/liquid flow as much as possible, it is advisable to direct all disposable liquids into a separate blow-down network.

Table 6.1 Typical K_f (Factor Representing Frictional Resistance to Flow, Dimensionless) Values for Pipe Fittings

Fitting	K_f (Factor Representing Frictional Resistance to Flow, Dimensionless)	Fitting	K_f
Globe valve, open	9.7	90° double-miter elbow	0.59
Typical depressuring valve, open	8.5	Screwed tee through run	0.5
Angle valve, open	4.6	Fabricated tee through run	0.5
Swing check valve, open	2.3	Lateral through run	0.5
180° close-screwed 1.95 return	1.95	90° triple-miter elbow	0.46
Screwed or fabricated tee through branch	1.72	45° single-miter elbow	0.46
90° single-miter elbow	1.72	180° welding return	0.43
Welding tee through branch	1.37	45° screwed elbow	0.43
90° standard-screwed elbow	0.93	Welding tee through run	0.38
60° single-miter elbow	0.93	90° welding elbow	0.32
45° lateral through branch	0.76	45° welding elbow	0.21
90° long-sweep elbow	0.59	Gate valve, open	0.21

Contraction or enlargement			**Ratio of diameters**		
	0	0.2	0.4	0.6	0.8
Contraction (ANSI)	–	–	0.21	0.135	0.039
Contraction (sudden)	0.5	0.46	0.38	0.29	0.12
Enlargement (ANSI)	–	–	0.9	0.5	0.11
Enlargement (sudden)	1	0.95	0.74	0.41	0.11

Table 6.2 Typical Friction Factors and Conversion Factors for Clean Steel Pipe (Based on Equivalent Roughness of 0.046 mm)

	Conversion Factor for Equivalent Length per Unit of K_f	
Diameter Nominal Pipe Size (mm)	**Moody Friction Factor (f)**	**Meters**
DN 50 schedule 40	0.0195	2.7
DN 80 schedule 40	0.0178	4.36
DN 80 schedule 100	0.0165	6.25
DN 80 schedule 150	0.0150	10.2
DN 200-6 mm wall	0.0140	14.7
DN 250-6 mm wall	0.0135	19.2
DN 300-6 mm wall	0.0129	24.0
DN 350-6 mm wall	0.0126	27.3
DN 400-6 mm wall	0.0123	31.88
DN 500-6 mm wall	0.0119	41.45
DN 600-6 mm wall	0.0115	56.67
DN 750-6 mm wall	0.0110	67.85
DN 900-6 mm wall	0.0107	83.33

Note: The above friction factors and conversion factors apply at high Reynolds numbers: namely, above 1×10^6 for DN 600 and larger, scaling down to 2×10^5 for DN 50

Once maximum load and back pressure in each segment have been established, standard pipe sizing procedures are used.

In determining back pressure, the following should be taken into consideration:

1. Flashing of liquid at relief/safety valve discharge or along the network as a result of pressure drop and/or warm-up to ambient temperatures should be analyzed
2. Solids formation resulting from auto-refrigeration and the presence of high–melting point liquids should be determined
3. If flashing and auto-refrigeration are possible, a temperature profile along the network should be established so that proper piping material selection and construction practices are undertaken
4. The network should be self-draining and should not include pockets
5. The network should be continuously purged by natural gas controlled through an orifice
6. High liquid velocities should be watched for within the network

- **Drainage**

Disposal system piping should be self-draining toward the discharge end. Pocketing of discharge lines should be avoided. Where pressure relief valves handle viscous materials or materials that can solidify as heat cool to ambient temperature, the discharge line should be heat traced. A small drain pot or drip leg may be necessary at low points in lines that cannot be sloped continuously to the knockout or blow-down drum. The use of traps or other devices with operating mechanisms should be avoided.

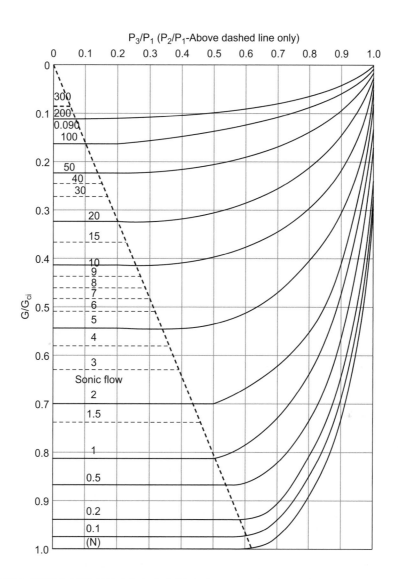

FIGURE 6.1

Isothermal flow of compressible fluid through pipes at high-pressure drops.

- **Details**

1. Safety/relief valve connection to the header

Normally, the laterals from individual relieving devices should enter a header from above.

2. Safety/relief valves connection when installed below the relief header

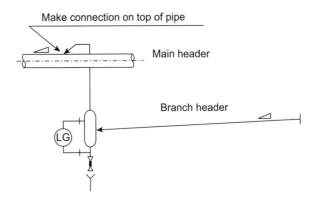

FIGURE 6.2

Drain pot.

Laterals leading from individual valves located at an elevation above the header should drain to the header. Locating a safety valve below the header elevation in closed systems should be avoided. Laterals from individual valves that must be located below the header should be arranged to rise continuously to the top of the header entry point. However, means should be provided to prevent liquid accumulation on the discharge side of these valves.

In this regard, the following should be taken into consideration:

1. For the branch header that must be connected to the main header from a level lower than the main header (e.g., sleeper flare piping), a drain pot must be installed. This is shown in Figure 6.2
2. If a safety/relief valve must be installed below the flare header, the outlet line leading to the flare header should be heat traced from the safety/relief valve to its highest point. However, the arrangement of the safety/relief valve must be reviewed; an arrangement is not permitted for safety/relief valves that discharge a medium that can leave a residue. The heat tracing can be omitted if the safety/relief valve in question handles only products that vaporize completely or do not condense at all at the lowest ambient temperature.

• **Drain pot size:**

Diameter is the larger value between twice the cross-sectional area of the branch header and pipe of DN 250. The height is a minimum of 800 mm.

1. Purge point of gas for dry seal
   a. A continuous fuel gas purge should be installed at the end of the main header and the end of any major subheader. The fuel gas purge should be controlled by means of a restriction orifice
   b. The purge gas volume should be determined so that a positive pressure is maintained and air ingress is prevented
2. Insulation of flare line

Normally, insulation of a flare line (including the outlet line of the safety/relief valve) is not required except for personnel protection.

However, to avoid hydrate formation, ice accumulation, and so forth, within the flare line, the use of insulation or heat tracing should be considered.

3. Location of safety/relief valve

More than one piece of equipment may be protected by a common safety/relief valve, provided they are connected by a line of sufficient size and that no block valve exists on the connecting lines.

4. Valves on inlet/outlet line of safety/relief valve

Unless otherwise specified by the company, all safety relief valves must have block valves on the inlet and outlet to facilitate maintenance. The block valves must be full bore and locked open. Safety valves discharging to the atmosphere should not have block valves on the outlet. A bypass line with a valve should be provided for each safety valve.

5. Provision for installation of drain holes

Where individual valves are vented to the atmosphere, an adequate drain hole (a nominal pipe size of DN 15 is usually considered suitable) should be provided at the low point to ensure that no liquid collects downstream of the valve. The vapor flow that occurs through this hole during venting is not generally considered significant, but each case should be checked to see whether the drain connection should be piped to a safe location. Vapors escaping from the drain hole must not be allowed to impinge against the vessel shell, because accidental ignition of such vent streams can seriously weaken the shell.

6. Angle entry into the relief header

The use of an angle entry at 45° (0.79 radian) or even 30° (0.52 radian) to the header axis for laterals is much more common in relieving systems than in most process piping systems.

7. Installation of valves and blinds in relief headers

Means (valve and blind) must be provided to isolate each unit from the flare system for safety and maintenance.

Extreme caution must be exercised in their use, to ensure that equipment that is operating is not isolated from its relieving system. Valves in the header system, if used, should be mounted so that they cannot fail in the closed position (for example, a gate falling into its closed position).

8. Slope of flare header

A slope of 1 m in 500 m is suggested for the flare header.

9. Absorption of thermal expansion in headers by looped pipes
   a. As a rule, headers should be designed so that thermal expansion generated in headers can be absorbed by the bent parts of the headers. In other words, the piping route of headers should incorporate several bends
   b. If thermal expansion cannot be absorbed by this method, absorption by looped pipes should be considered. Looped parts should have no drain pocket.

10. Absorption of thermal expansion by expansion joints
   a. As a rule, no expansion joints should be used. The use of expansion joints is limited to cases in which thermal expansion cannot be absorbed by pipes alone because of a short route (e.g., the route between the seal drum, or knock-out drum, and the flare stack)
   b. Drain pipes should be installed at bellows or other concave parts where drain is likely to remain
   c. The conditions for selecting bellows (design condition or materials) should be specified clearly
11. Solids formation

The possibility of solids forming within the disposal system must be studied, considering all related aspects such as hydrate formation, water or heavy hydrocarbon presence, auto-refrigeration, and so forth. Consideration should be given to a separate disposal system so that the possibility of the formation of solids is eliminated.

6.4 Sizing a knock-out drum

Sizing a knock-out drum is generally a trial-and-error process. First, the drum size required for liquid entrainment separation is determined. Liquid particles will separate when the residence time of the vapor or gas is equal to or greater than the time required to travel the available vertical height at the dropout velocity of the liquid particles, and the vertical gas velocity is sufficiently low to permit the liquid droplet to fall. This vertical height is usually taken as the distance from the liquid surface. The vertical velocity of the vapor and gas must be low enough to prevent large slugs of liquid from entering the flare. Because the flare can handle small liquid droplets, the allowable vertical velocity in the drum may be based on that necessary to separate droplets from 300 to 600 μm in diameter. The dropout velocity of a particle in a stream is calculated as follows:

$$U_d = 1.15 \sqrt{\frac{g d_P (\rho_L - \rho_V)}{\rho_V C_D}} \qquad (6.3)$$

where:
C_D = drag coefficient
g = acceleration due to gravity, 9.8 m/s^2
U_d = particle dropout velocity, m/s
d_p = particle diameter, m
ρ_L = density of liquid at operating conditions, kg/m^3
ρ_V = density of vapor at operating conditions, kg/m^3

This basic equation is widely accepted for all forms of entrainment separation.

The second step in sizing a knock-out drum is to consider the effect any liquid contained in the drum may have on reducing the volume available for vapor/liquid disengagement. This liquid may result from (1) condensate that separates during a vapor release or (2) liquid streams that accompany a vapor release. It is suggested that the volume occupied by the liquid be based on a release lasting

20–30 min. Any accumulation of liquid retained from a prior release (pressure relief valves or other sources) must be added to the liquid indicated in Items (1) and (2) to determine the available vapor-disengaging space. However, for situations in which the knock-out drum is used to contain large liquid dumps from pressure relief valves on other sources where there is no significant flashing and the liquid can be removed promptly, it would not usually be necessary to consider these volumes relative to vapor disengaging.

The economics of vessel design should be considered when selecting a drum size, and may influence the choice between a horizontal and a vertical drum. When large liquid storage is desired and the vapor flow is high, a horizontal drum is often more economical. Split entry of exit decreases the size of the drum for large flows.

As a rule, drum diameters over 3.3 m should apply split-flow arrangements for to be most economical. Horizontal and vertical knock-out drums are available in many designs; the main differences consist of how the path of the vapor is directed. The various designs include the following:

1. A horizontal drum with the vapor entering one end of the vessel and exiting at the top of the opposite end (no internal baffling)
2. A vertical drum with the vapor inlet nozzle on a diameter of the vessel and the outlet nozzle at the top of the vessel's vertical axis. The inlet stream should be baffled to direct the flow downward
3. A vertical vessel with a tangential nozzle
4. A horizontal drum with the vapor entering at each end on the horizontal axis and a center outlet
5. A horizontal drum with the vapor entering in the center and exiting at each end on the horizontal axis
6. A combination of a vertical drum in the base of the flare stack and a horizontal drum upstream to remove the bulk of the liquid entrained in the vapor. This combination permits the use of larger values for the numerical constant in the velocity equation

The following sample calculations were limited to the simplest of the designs, Items 1 and 2. The calculations for Items 4 and 5 are similar, with one-half the flow rate determining one-half the vessel length. The normal calculations would be used for Item 3 and will not be duplicated here.

- **Assume the following conditions:**

A single contingency results in the flow of 25.2 kg/s of a fluid with a liquid density of 496.6 kg/m^3 and a vapor density of 2.9 kg/m^3, both at flowing conditions. The pressure is 13.8 kPa gage, and the temperature is 149 °C. The viscosity of the vapor 0.01 cP.

Also, the fluid equilibrium results in 3.9 kg/s of liquid and 21.3 kg/s of vapor. In addition, 1.89 m^3 of storage for miscellaneous draining from the units is desired. The droplet size selected as allowable is 300 μm in diameter. The vapor rate is determined as follows:

$$\text{Vapor rate} = \frac{21.3 \text{ kg/s}}{2.9 \text{ kg/m}^3} = 7.34 \text{ m}^3/\text{s} \tag{6.4}$$

The drag coefficient, C_D, is determined from Figure 6.3 as follows:

$$C_D(R_e)^2 = \frac{0.13 \times 10^8(2.9)\left(300 \times 10^{-6}\right)^3(496.6 - 2.9)}{(0.01)^2} = 5025 \tag{6.5}$$

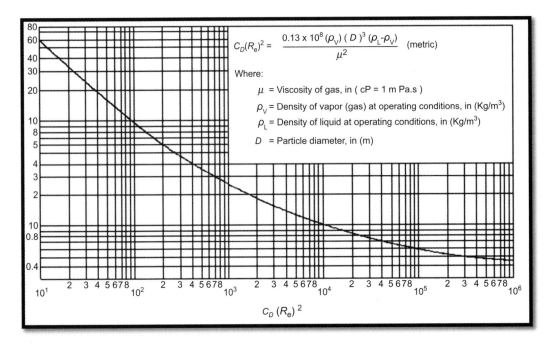

FIGURE 6.3

Determination of drag coefficient.

From Figure 6.3, $C = 1.3$.

The dropout velocity, U_c is calculated as follows:

$$U_d = 1.15 \left[\frac{9.8\left(300 \times 10^6\right)(496.6 - 2.9)}{2.9 \times 1.3} \right]^{0.5} = 0.71 \qquad (6.6)$$

Assume a horizontal vessel with an inside diameter, Di and a cylindrical length, L. This gives the following total cross-sectional area:

$$A_t = \frac{\pi}{4}(\mathrm{Di})^2 \qquad (6.7)$$

Liquid holdup for 30-min release from the single contingency, in addition to the slop and drain volume, is desired.

The volume in the heads is neglected for simplicity. The liquid holdup required is therefore calculated as follows:

1. The slop and drain volume of 1.89 m^3 will occupy a bottom segment as follows:

$$A_{L1}(\text{square meters}) = (1.89 \text{ cubic meters})/L \qquad (6.8)$$

where:

A_{L1} = vessel segment area occupied by slops and drain, m^2
L = Flare knock-out drum length, m.

2. A total of 3.9 kg/s of condensed liquids with a density of 496.6 kg/m^3 accumulated for 30 min will occupy a cross-sectional segment (see above) as follows:

$$A_{L2} = \frac{(3.9 \text{kg/s})}{496.6 \text{kg/m}^3} \times \left(\frac{60 \text{ s}}{\text{min}}\right) (30 \text{ min}) \left(\frac{1}{1 \text{ m}}\right) \tag{6.9}$$

where:

A_{L2} = the vessel segment area occupied by condensed liquid, m^2

The cross-sectional area remaining for the vapor flow is as follows:

$$A_v = A_t - (A_{L1} + A_{L2}) \tag{6.10}$$

where:

A_t = the total vessel cross-sectional area, m^2
A_v = the vessel cross-section area available for vapor flow, m^2

The vertical depths of the liquid and vapor spaces are determined using standard geometry, where h_{L1} = depth of slops and drains, $h_{L1} + h_{L2}$ = depth of all liquid accumulation, and h_v = remaining vertical space for the vapor flow.

The total drum diameter is calculated as follows:

$$h_t = h_{L1} + h_{L2} + h_v \tag{6.11}$$

where:

h_v = vertical space for vapor flow, cm
h_{L1} = vessel depth occupied by slops and drain, m
h_{L2} = vessel depth occupied by condensed liquid, m

The adequacy of the vapor space is verified as follows: The vertical drop available for liquid dropout is equal to h_v. The liquid dropout time is determined as follows:

$$\theta = \left(\frac{1}{U_c \text{ m/s}}\right) \left(\frac{h_v}{100 \text{ cm/m}}\right) \tag{6.12}$$

θ = liquid particle dropout time, s

The velocity of the vapor, based on one vapor pass, is determined as follows:

$$U_v = \left(\frac{7.34 \text{ m}^3/\text{s}}{N \text{ vapor passes}}\right) \left(\frac{1}{A_v \text{ m}^2}\right) \text{m/s} \tag{6.13}$$

where:

U_v = vapor velocity, m/s (Eqn (6.13))

The drum length required is determined as follows:

$$L_{min} = (U_v \text{ m/s})(\theta \text{ s}) \times (N \text{ vapor passes) m} \tag{6.14}$$

L_{min} = flare knockout drum minimum length required, m

L_{min} must be less than or equal to the assumed cylindrical drum length, L; otherwise, the calculation must be repeated with a newly assumed cylindrical drum length.

Table 6.3 summarizes these calculations for horizontal drums with various inside diameters to determine the most economical drum size. Drum diameters in 15-cm increments are assumed, in accordance with standard head sizes.

Now consider a vertical vessel. The vapor velocity is equal to the dropout velocity, which is 0.71 m/s. The required cross-sectional area of the drum is determined as follows:

$$\text{Cross} - \text{sectional area} = \frac{7.34 \text{ m}^3/\text{s}}{0.71 \text{ m/s}} = 10.3 \text{ m}^2 \tag{6.15}$$

The drum diameter is determined as follows:

$$D = \sqrt{10.3 \times 4/\pi} = 3.6 \text{ m} \tag{6.16}$$

Thus, a vertical drum is not a logical choice for this example, unless layout considerations dictate differently.

- **Details**
  - If the knock-out drum is disproportionally large, adoption of the vane type knock-out drum should be considered
  - The 20- to 30-min residence time is based on the release of maximum liquid quantity for the liquid space in the knock-out drum between a high-level alarm and minimum pump-out level
  - The pump installed to empty the drum should be sized to do so in 2 h
  - The header leading to the knock-out drum should be sloped toward it. The header from the drum to the flare stack should slope continuously back to the drum

6.5 Quench drum

A quench drum is provided as a means of preventing liquid hydrocarbon condensation in the flare system, to reduce flare capacity requirements, or to prevent discharge of condensable hydrocarbons into the atmosphere. In some cases, it serves the additional purpose of reducing the maximum temperature of flare gases, and hence minimizing thermal expansion problems in the mechanical design of flare headers.

The quench drum functions by means of a direct contact water spray arrangement that condenses entering heavy hydrocarbon vapors. Condensed hydrocarbons and effluent water are discharged through a seal to the sewer or pump-out to slop tankage. On the other hand, uncondensed hydrocarbon vapors are vented to the flare or into the atmosphere. Figure 6.4 presents a typical quench drum.

Table 6.3 Optimizing the Size of Horizontal Knockout Drum (SI Units)

Trial No.	Assumed Drum Inside Diameter D_i, m	Assumed Drum Cylindrical Length, L, m	Cross-Sectional Area, m²				Vertical Depth of Liquid and Vapor Spaces, cm				Liquid Dropout Time, q, s	Vapor Velocity, U_v, m/s	Required Drum Length L min, m
			A_1	A_{L1}	A_{L2}	A_v	h_{l1}	$h_{l1} + h_{l2}$	h_v	di			
1	2.24	5.79	4.67	0.33	2.45	1.9	30	140	104	224	1.45	3.9	5.6
2	2.29	6.25	4.10	0.3	2.27	1.53	29	137	91	229	1.28	4.8	6.2
3	2.13	6.86	3.57	0.28	2.07	1.23	28	133	81	213	1.13	6	6.7
4	1.98	7.62	3.08	0.25	1.86	0.98	27	128	70	198	0.98	7	7.4

Note: The data in this table are in accordance with the example given in text.
The following conclusions can be drawn from this table:
(1) All of the drum sizes above would fulfill the design requirements.
(2) The most suitable drum sizes should be selected according to the design pressure, material requirements, corrosion allowance, layout, transportation, and other considerations.
(3) The choice of two-pass flow, as shown in Figure 6.4, is optional.

6.5.1 Details

1. The quench drum should have a design pressure capable of withstanding the maximum back pressure. Minimum design pressure is 350 kPa gage
2. Water requirements are normally based on reducing gas and liquid outlet temperatures to about 50 °C. Selection of the optimum temperature is based on considerations of temperature and composition of entering streams, and the extent to which subsequent condensation of effluent vapors downstream of the drum can be tolerated.

It is generally assumed that no more than 40–50% of the liquid fed will be vaporized. The water supply should be taken from a reliable water system. If a recirculating cooling water system is used, the circulating pumps and cooling water basin must have adequate capacity to supply the maximum quench drum requirements for 20 min.

The seal height in the liquid effluent line (assuming 100% water) is sized for 175% of the maximum operating pressure, or 3 m, whichever is greater.

3. If the quenched hydrocarbons are of a sour nature, provision should be made for a proper disposal system and due consideration should be given to material specification.

Sizing a seal drum and design details should be in accordance with [1], Sections 5.4.2.2 and 5.4.2.4; and APIRP-2001, Section 3.14.3.

6.6 Flares

A gas flare, alternatively known as a flare stack, is a gas combustion device used in industrial plants such as petroleum refineries, chemical plants, and natural gas processing plants, and at oil or gas production sites with oil wells, gas wells, offshore oil and gas rigs, and landfills.

Flare systems provide for the safe disposal of gaseous wastes. Depending on local environmental constraints, these systems can be used for:

1. Extensive venting during start up or shutdown
2. Venting of excess process plant gas
3. Handling emergency releases from safety valves, blow-down, and depressuring systems

Designs will vary considerably, depending on the type of connected equipment and the complexity of the overall system. A flare system generally consists of an elevated stack, means to maintain burning conditions at the top of stack, and means to prevent flashback within the system.

Whenever industrial plant equipment items are overpressurized, the pressure relief valves provided as essential safety devices on the equipment automatically release gases and sometimes liquids. Those pressure relief valves are required by industrial design codes and standards, as well as by law.

The released gases and liquids are routed through large piping systems called flare headers, to a vertical elevated flare. The released gases are burned as they exit the flare stacks. The size and brightness of the resulting flame depend on the flammable material's flow rate in terms of joules per hour (or btu per hour).

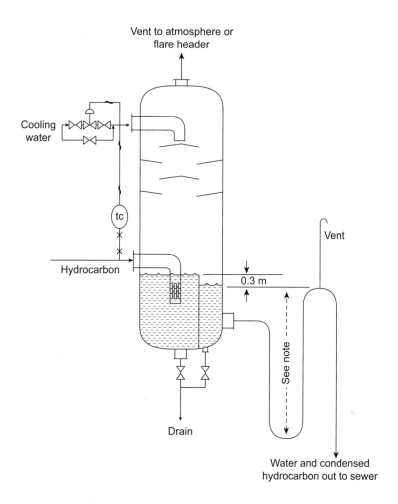

Vent to atmosphere or
flare header

Cooling
water

tc

Hydrocarbon

0.3 m

Vent

See note

Drain

Water and condensed
hydrocarbon out to sewer

FIGURE 6.4

Quench drum.

Notes:

1. It is suggested that the sewer seal be designed for a minimum of 175% of the drum's maximum operating pressure

2. Proper destination of liquid effluent should be investigated in case it contains toxic or hazardous materials

3. Criteria for venting to atmosphere should be considered

Most industrial plant flares have a vapor–liquid separator (also known as a knockout drum) upstream of the flare to remove any large amounts of liquid that may accompany the relieved gases.

Steam is often injected into the flame to reduce the formation of black smoke. To keep the flare system functional, a small amount of gas is continuously burned, like a pilot light, so that the system is always ready for its primary purpose as an overpressure safety system.

The adjacent flow diagram depicts the typical components of an overall industrial flare stack system (Figure 6.5):

- A knockout drums to remove any oil and/or water from the relieved gases
- A water seal drum to prevent any flashback of the flame from the top of the flare stack
- An alternative gas recovery system for use during partial plant startups and/or shutdowns as well as other times, when required. The recovered gas is routed into the fuel gas system of the overall industrial plant
- A steam injection system to provide an external momentum force used for efficient mixing of air with the relieved gas, which promotes smokeless burning
- A pilot flame (with its ignition system) that burns all the time so that it is available to ignite relieved gases whenever needed
- The flare stack, including a flashback prevention section at the upper part of the flare stack

6.6.1 Sizing

The sizing of flares requires determination of the required stack diameter and the required stack height.

Because the flare tip is open to the atmosphere, high gas velocities are expected at this point. Very high tip velocities cause a phenomenon known as blow-off, in which the flame front is lifted and could

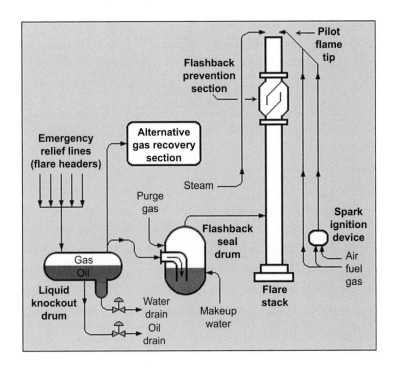

FIGURE 6.5

Typical installation of flare system.

eventually turn into a blowout. Very low velocities could damage the flare tip owing to high heat intensities and smoking. In this case, ingress of air in the system and creation of a flammable mixture are possible. Therefore, determination of the right flare diameter is important as far as operation of the system is concerned.

The location and height of flare stacks should be based on the heat release potential of a flare, the possibility of personnel exposure during flaring, and the exposure of surrounding plant equipment. There are exposure limitations set forth that must be taken into consideration. In effect, this fixes the distance between the flame and the object. If there are limitations regarding the location (distance), the stack height can be calculated; otherwise, an optimum tradeoff between height and distance should be applied.

Wind velocity, by tilting the flame, changes the flame distance and heat intensity. Therefore, its effect should be considered in determining the stack height.

If the flare is blown out (extinguished), or if there are environmental hazards associated with the flare output, the possibility of a hazardous situation downwind should be analyzed.

1. Diameter

Flare stack diameter is generally sized on a velocity basis, although pressure drop should be checked. Depending on the volume ratio of maximum conceivable flare flow to anticipated average flare flow, the probable timing, frequency, and duration of those flows, and the design criteria established for the project to stabilize flare burning, it may be desirable to permit a velocity of up to 0.5 Mach for a peak, short-term, infrequent flow, with 0.2 Mach maintained for the more normal and possibly more frequent conditions. Smokeless flares should be sized for the conditions under which they are to operate smokelessly. The formula relating velocity (as a Mach number) to flare tip diameter can be expressed as follows:

$$\text{Mach} = (3.23)(10^{-5}) \frac{W}{P_2 D^2} \sqrt{\frac{T \cdot z}{KMW}} \tag{6.17}$$

where:

  Mach = Mach number
  W = flow rate
  P_2 = pressure in the reservoir into which the pipe discharges, 101 kPa (absolute) with atmospheric discharge
  T = temperature, K
  Z = compressibility factor
  D = diameter
  K = ratio of specific heats, C_p/C_v for the vapor being relieved
  MW = Molecular weight

Pressure drops as large as 14 kPa have been satisfactorily used at the flare tip. Too low a tip velocity can cause heat and corrosion damage. The burning of the gases becomes slow, and the flame is greatly influenced by the wind. The low-pressure area on the downwind side of the stack may cause the burning gases to be drawn down along the stack for ≥ 3 m. Under these conditions, corrosive materials in the stack gases may attack the stack metal at an accelerated rate, even though the top 2.4–3 m of the flare is usually made of corrosion-resistant material.

2. Height

The flare stacks height is generally based on the radiant heat intensity generated by the flame.

The following equation may be used to determine the distance required between a location of atmospheric venting and a point of exposure where thermal radiation must be limited:

$$D = \sqrt{\frac{\tau F Q}{4 \pi K}} \tag{6.18}$$

The factor F allows for the fact that not all of the heat released in a flame can be released as radiation. Measurement of radiation from flames indicates that the fraction of heat radiated (radiant energy per total heat of combustion) increases toward a limit, similar to the increase in the burning rate with increasing flame diameter. Data from the U.S. Bureau of Mines for radiation from gaseous-supported diffusion flames are given in Table 6.4.

These data apply only to the radiation from a gas. If liquid droplets of hydrocarbon >150 μm in size are present in the flame, the values in Table 6.4 should be somewhat increased.

The fraction of heat intensity transmitted, τ, is used to correct the radiation impact. It can be calculated from the following relationship:

$$\tau = 0.79 \left(\frac{100}{r}\right)^{1/16} \left(\frac{30.5}{D}\right)^{1/16} \tag{6.19}$$

Table 6.4 Radiation from Gaseous Diffusion Flames

Gas	Burner Diameter, cm	(Radiate Output)/ (Thermal Output) × 100
Hydrogen	0.51	9.5
	0.91	9.1
	1.9	9.7
	4.1	11.1
	8.4	15.6
	20.3	15.4
	40.6	16.9
Butane	0.51	21.5
	0.91	25.3
	1.9	28.6
	4.1	28.5
	8.4	29.1
	20.3	28
	40.6	29.9
Methane	0.51	10.3
	0.91	11.6
	1.9	16
	4.1	16.1
	8.4	14.7
Natural gas 20.30	19.2	-
95% CH_4	40.6	23.2

Table 6.5 Recommended Design Flare Radiation Levels Excluding Solar Radiation	
Permissible Design Level, KW/m²	**Conditions**
15.77	Heat intensity on structures and in areas where operators are not likely to be performing duties and where shelter from radiant heat is available: for example, behind equipment
9.46	Value of at design flare release at any location to which people have access: for example, at grade below the flare or on a service platform of a nearby tower. Exposure must be limited to a few seconds, sufficient for escape only
6.31	Heat intensity in areas where emergency actions lasting up to 1 min may be required by personnel without shielding but with appropriate clothing
4.73	Heat intensity in areas where emergency actions lasting several minutes may be required by personnel without shielding but with appropriate clothing
1.58	Value of at design flare release at any location where personnel are continuously exposed

Note: On towers or other elevated structures where rapid escape is not possible, ladders must be provided on the side away from the flare, so the structure can provide some shielding when radiation intensity is >6.31 kW/m².

where:

D = distance, m
r = radius, m

This equation is strictly applicable under the following conditions: luminous hydrocarbon flame radiating at 1227 °C, 27 °C dry bulb ambient temperature, relative humidity >10%, distance from the flame between 30 and 150 m; however, it can be used to estimate the order of magnitude under a wide range of conditions. This equation should prove adequate for most flare gases, except H_2 and H_2S, which burn with little or no luminous radiation.

As for the effect of radiation level on humans, the allowable radiation level is a function of length of exposure. Table 6.6 gives exposure times necessary to reach the pain threshold.

Table 6.6 Exposure Times Necessary to Reach the Pain Threshold	
Radiation Intensity, KW/m²	**Time to Pain, KW/Threshold m² (s)**
1.74	60
2.33	40
2.9	30
4.73	16
6.94	9
9.46	6
11.67	4
19.87	2

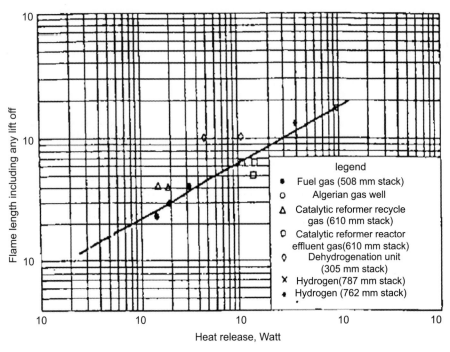

FIGURE 6.6

Heat release, watts flame length vs heat release: industrial size and releases.
Note: Multiple points indicate separate observations or different assumptions of heat content.

The correction for the location of the flame center will be significant when radiation levels are examined.

Information on this subject is limited and is usually based on visual observations in connection with emergency discharges to flares. Figure 6.6 gives flame length versus heat release.

3. Wind effect

Another factor to be considered is the effect of wind in tilting the flame, thus varying the distance from the center of the flame, which is considered to be the origin of the total radiant heat release, with respect to the plant location under consideration. A generalized curve for approximating the effect of wind is given in Figure 6.7.

4. Dispersion

Where there is concern about the resulting atmospheric dispersion, if the flare were to be extinguished, reference should be made to the *API Manual on Disposal of Refinery Wastes, Volume on Atmospheric Emissions*, to calculate the probable concentration at the point in question.

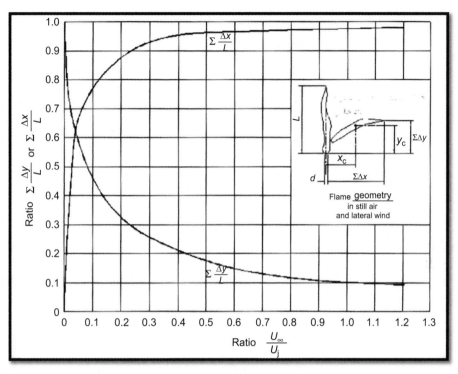

FIGURE 6.7

Approximate flame distortion resulting from lateral wind on jet velocity from flare stacks.

Notes:

U_∞ = Lateral wind velocity

U_j = Exit gas velocity from stack

6.6.2 Design details

1. Smokeless flares

Smoke-free operation of flares can be achieved by various methods, including steam injection, injection of high-pressure waste gas, forced draft air, operation of flares as a premixed burner, or distribution of the flow through many small burners.

The most common type of smokeless flare involves steam injection.

The assist medium mass requirements are low for steam and fuel gas because of their high velocity relative to the flare gas. Typical values for steam or fuel gas are from 0.20 to 0.50 kg of assist gas per kilograms of hydrocarbon flow. The following equation predicts steam use for a given hydrocarbon molecular mass (weight) gas to be burned in a smokeless flare:

$$W_{steam} = W_{HC}[0.68 - (10.8/M)] \tag{6.20}$$

2. Flashback protection

The most common method of preventing propagation of flame into the flare system caused by entry of air is the installation of a seal drum, as depicted in Figure 6.8. Flame arresters are occasionally used for flashback protection; however, they are subject to plugging, and their application is limited.

Alternatively, continuous introduction of purge gas can be used to prevent flashback. A safe condition exists in situations involving hydrocarbon air mixtures if a positive flow of oxygen free gas is maintained, allowing the oxygen concentration to be no greater than 6% at a point 7.6 m from the flare tip. The injection rate should be controlled by a fixed orifice to ensure that supply remains constant and is not subject to instrument malfunction or maladjustment. Molecular seals can be used to minimize purge gas rates.

3. Ignition

To ensure ignition of flare gases, a continuous pilot with a means of remote ignition is recommended for all flares. The most commonly used type of igniter is the flame-front propagation type, which uses a spark from a remote location to ignite a flammable mixture.

Pilot igniter controls are located at a safe distance from the base of elevated flares and at least 30 m from ground flares.

It is recommended that a low-pressure alarm for the pilot gas be provided so that the operator in the control room becomes aware of pilot blowout.

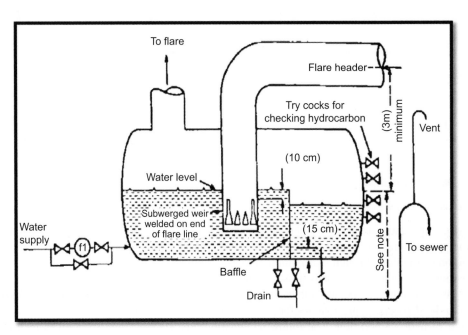

FIGURE 6.8

Flare stack seal drum.

Note: It is suggested that the sewer seal be designed for a minimum of 175% of the drum's maximum operating pressure.

Reliable pilot operation under all wind and weather conditions is essential. Flaring operations are for the most part intermittent and unscheduled. The flare must be instantly available for full emergency duty to prevent any possibility of a hazardous or environmentally offensive discharge to the atmosphere. Windshields and flame retention devices may be used to ensure continuous piloting under the most adverse conditions.

4. Fuel system

Fuel gas supply to the pilots and igniters must be highly reliable. Because normal plant fuel sources may be upset or lost, it is desirable to provide a backup system connected to the most reliable alternative fuel source, with provision for automatic cut-in on low pressure. Use of waste gas with low-energy content or with unusual burning characteristics should be avoided. Parallel instrumentation for pressure reduction is frequently justifiable. The flare fuel system should be carefully checked to ensure that hydrates cannot present a problem. Because of small lines, long exposed runs, large vertical rises up the stack, and pressure reductions, use of a liquid knockout pot or scrubber after the last pressure reduction is frequently warranted. If at all feasible in terms of distance, relative location, and cost, it is considered good practice to install a low-pressure alarm on the fuel supply after the last regulator or control valve so that operators will be warned of any loss of fuel to the pilots.

5. Fired or endothermic flares

When low heating value gases are to be sent to a flare stack, fired or endothermic flares are used (sulfur plant tail gas presents an example).

Generally, if the heating value of the gas to be flared is $<4280 \, kJ/m^3$, a fired flare with a high heating value assist gas may be required for complete combustion.

6. Location

The location of flares in the vicinity of tall refinery equipment should be examined. Flames or hot combustion products can be carried by the wind, which could cause problems and create hazards to personnel working on these elevated structures at the time of a flare release. As discussed in the section on sizing, flare height and distance depend on radiation intensity. When either the height or the distance from the plant of a flare is fixed, the other can be determined. Usually, there are constraints on the distance; therefore stack height is calculated.

If there are no constraints on the distance and flare height is to be determined, the following guideline is recommended.

For stack heights <23 m a distance of 91 m, and for stack heights >23 m a distance of 61 m from the plant is considered.

7. Due consideration should be given to installation of flow-measuring equipment on the flare system. Specifically, subheaders handling continuous relief loads from individual units should be provided with proper flow elements.

6.7 Burning pits

Burning pit flares can handle flammable liquids or gases, or mixtures of the two. The burning pit is simply a shallow earth or concrete surfaced pool area enclosed by a dike wall, a liquid/vapor inlet pipe through the wall, and provided with pilot and igniter. Although the design basis flow is adequate for handling emergency releases, a more conservative approach is recommended for continuous flaring services, incorporating up to twice the calculated pit area.

6.7.1 Burning pit flare sizing

The burning pit area is sized to provide sufficient surface to vaporize and burn liquid at a rate equal to the maximum incoming liquid rate. The calculation procedure is as follows:

1. Determine the linear regression rate of the liquid surface (i.e., the rate at which the liquid level would fall as a result of vaporization by radiant heat from the burning vapor above it, assuming no addition of incoming liquid):

$$S_R = K_1 \frac{Q}{Q_v \cdot q_1} \tag{6.21}$$

where:
   S_R = linear regression rate of liquid surface, mm/min
   K_1 = dimensional constant equal to 0.076 mm/min
   Q = heat, kJ/kg
   Q_v = heat required to vaporize liquid, kJ/kg
   q_1 = rate of vaporization and burning of liquid, in (kg/s) (selected as equal to the rate of flashed liquid entering the pit)

2. Determine the pit area necessary to vaporize and burn liquid at a rate equal to the liquid input rate:

$$A_p = K_2 \frac{\rho_L}{S_R \cdot \rho_L} \tag{6.22}$$

where:
   A_p = pit area required to vaporize and burn liquid, m^2
   K_2 = unit conversion factor equal to 60,000
   S_R = linear regression rate of liquid surface, mm/min
   ρ_L = density of liquid at operating conditions, kg/m^3

3. The dike wall height above the water level is selected to provide hollow capacity for the largest liquid release resulting from a single contingency during 30 min, plus 460 mm free board. The liquid rate is based on the actual flashed liquid entering the pit, assuming no burning or further vaporization in the pit. However, the height of the dike wall above the water level should not be <1.20 m

6.7.2 Spacing of burning pit flares

Spacing is based on radiant heat consideration at maximum heat release, using a simplified calculation procedure, as follows:

$$D = \sqrt{\frac{\tau F Q}{4 \pi K}} \tag{6.23}$$

This equation has been described in the section on flare sizing. In this equation, absorption of radiation by surrounding air is neglected. Also, the fraction of heat radiated, as shown in Table 6.4, refers to light gases, but in burning pits combustion of liquid is under consideration. Therefore, good engineering judgment should be exercised in evaluating the effect of this factor when determining the distance.

The center of the flame is assumed to be 1.5 pool diameters from the center of the pool, in the direction of the point where radiant heat density is being considered. This assumption is used to allow for flame deflection by wind.

Although permissible radiant heat densities are given in Table 6.5, its value at the property line must not exceed 1.60 kW/m^2. In addition, the following minimum spacing applies to burning pits:

- 150 m from property lines, roadways, or any process or storage facilities
- 60 m from any source of ignitable hydrocarbons, such as separators, or floating roof tanks

Valves in the inlet, seal water, and pilot gas lines should be located according to permissible radiant heat densities for personnel. Piping to the burning pit should be suitably protected against flame impingement (e.g., by installation below grade).

In designing the burning pit, all personnel and equipment safety precautions should be observed.

6.8 Determination of liquid level in a horizontal vessel

One way the liquid depth in a horizontal cylindrical vessel can be calculated is the following (volume owing to heads is neglected for simplicity):

Liquid volume = (Segment area)(Vessel length)

The segment area is given by:

$$A = r^2 \cdot \cos^{-1}\left(\frac{r-h}{r}\right) - (r-h)\sqrt{2rh - h^2} \qquad (6.24)$$

where:

A = cross-sectional area, m^2
r = radius, m in Figure 6.9
h = height in m, refer to Figure 6.9

r and h must have similar units. Obviously, the arc cosine term must be calculated in radians. This equation is applicable even if $h > r$, i.e., the vessel is more than half full.

6.9 Sample calculation for sizing a flare stack

This section presents an example for sizing a flare stack based on the effect of radiation. The effect of dispersion if the flame is extinguished is not analyzed.

6.9.1 Basic data

Hydrocarbon vapor flow rate = 12.6 kg/s
Average molecular mass of vapor = 46.1

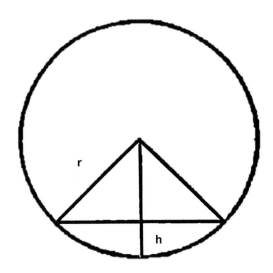

FIGURE 6.9

Liquid depth in a horizontal cylindrical vessel.

Flowing temperature $= 422$ K
Heat of combustion $= 50,000$ kJ/kg
Ratio of specific heats $= 1.1$
Flowing pressure at flare tip $= 101.3$ kPa (absolute)
Design wind velocity $= 8.9$ m/s

6.9.2 Calculation of flare diameter

For Mach $= 0.2$, the flare diameter is calculated as follows:

$$\text{Mach} = (3.23)(10^{-5}) \frac{W}{P_2 D^2} \sqrt{zT/k \cdot \text{MW}}$$

$$0.2 = (3.23)(10^{-5}) \frac{45445}{101.3\,D^2} \sqrt{422/(1.1)(46.1)}$$

$$d = 0.46 \text{ m}$$

6.9.3 Calculation of flame length

The heat liberated, φ, in kilowatts, is calculated as follows:

$$Q = (12.6)(50000) = 6.3 \times 10^5 \text{kW}$$

From Figure 6.6, the flame length, L_f, is 52 m.

6.9.4 Calculation of flame distortion caused by wind velocity

The vapor flow rate is determined as follows:

$$\text{Flow} = (12.6)(22.4/46.1)(422/273) = 9.46 \text{ actual m}^3/\text{s}$$

The flame distortion caused by wind velocity is calculated as follows:

$$U_\infty/U_j = \text{Wind velocity/Flare tip velocity}$$

The flare tip exit velocity, U_j, may be determined as follows:

$$U_j = \text{Flow}/(\pi d^2/4)$$

$$U_j = \frac{9.46}{\left[\pi(0.46)^2\right]/4} = 56.9 \text{ m/s}$$

$$U_\infty/U_j = 8.9/56.9 = 0.156$$

From Figure 6.7:

$$\Sigma\Delta Y/L = (0.35) = 18.2 \text{ m}$$
$$\Sigma\Delta X/L = (0.85) = 44.2 \text{ m}$$

Calculation of required flare stack height (for dimensional references, see Figure 6.10).
The design basis is as follows:
Fraction of heat radiated, F, is 0.3.
Maximum allowable radiation, $\varnothing$, at 45.7 m from the flare stack, is $6.3 \times 10^5 \text{kW/m}^2$.
Assume $\tau = 1.0$, then:

$$D = \sqrt{\frac{\tau F Q}{4\pi K}} = \sqrt{\frac{(1)(0.3)(6.3 \times 10^5)}{4(3.14)(6.3)}} = 48.9 \text{ m}$$

$$H' = H + \left(\tfrac{1}{2}\right)(\Sigma\Delta Y)$$

$$R' = 45.7 - \tfrac{1}{2}(44.2) = 23.7 \text{ m}$$

$$R' = R - \left(\tfrac{1}{2}\right)(\Sigma\Delta X)$$

$$D^2 = R'^2 + H'^2$$

$$48.9^2 = 23.7^2 + H'^2$$

$$H' = 42.8 \text{ m}$$

$$H = 42.8 - \tfrac{1}{2}(18.2)$$

$$H = 33.7 \text{ m}$$

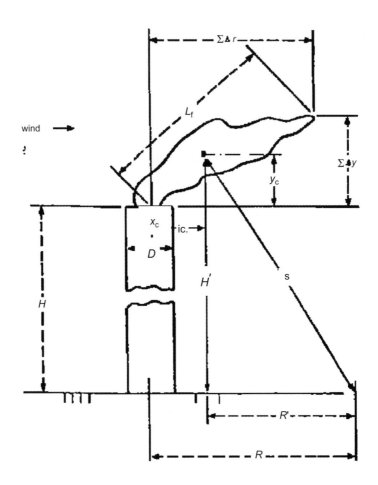

FIGURE 6.10

Dimensional references for sizing a flare stack.

6.10 Process design of emergency measures

Conceptual design of emergency measures for processing units in refineries and other chemical plants covers requirements of installation of protective devices such as depressuring valves and equipment isolation valves, tripping factors of units/equipment, relevant failure actions, and so forth.

Design of emergency measures is usually supplemented by the licensors' instructions and related regulations and requirements for licensed units to ensure complete plant protection from emergency situations.

This section is intended to cover minimum requirements and guidelines for process engineers to specify the proper type of emergency measures for probable emergency situations.

6.10.1 Design criteria

• **Isolation**

The following paragraphs describe the installation standards and design practice for valves used for equipment isolation in emergency cases, including vessels, furnaces, and compressors. Installation points of the isolation valves should be clearly indicated on piping and instrumentation diagram/drawing (P&ID), highlighting the distance from the equipment to be protected.

• **Vessels**

If the liquid volume in the vessel exceeds 10 m^3 (calculated at a normal liquid level with the addition of a tray and reboiler inventory in case of towers, and neglecting line inventory) and one or more of the following conditions exist, emergency isolation valves should be provided on vessel outlet line below the normal operating liquid level.

1. Liquid should conform to national fire protection association (NFPA) No. 30
   a. Class IA: Liquids should include liquids that have flashpoint below 23 °C and boiling point below 38 °C
   b. Class IB: Liquids should include liquids that have flashpoint below 23 °C and boiling point at or above 38 °C
2. Liquids that are heated above their flashpoint
3. Temperature is 260 °C or higher
4. Pressure is 1960 kPa(ga) or greater

The valve should be located no farther than 9 m measured horizontally from the side of the vessel. The total pipe length from the nozzle to the valve should not exceed 15 m.
   Valves should be operable from grade or platform as follows:

1. Access to a manually operated valve should be considered acceptable if the valve can be operated from a platform no more than 6 m above grade and access to the platform is by stairway. Access to the platform by ladder is not permitted
2. Valves with sizes DN 200 (8 in) and smaller may be manually operated and may be fitted with extension spindles, angle drives, and so forth, to fulfill the criteria of operability from grade
3. Valves with sizes DN 250 (10 in) and larger should be electrically or pneumatically operated and controls should be located in a place at grade safe from the danger of fire

If liquid on a side-draw flows into the bottom of a second vessel such as a stripper and the total liquid in the draw-off pan plus that in the bottom of the second vessel exceeds 10 m^3, and one or more of the four conditions noted earlier exists, an emergency isolation valve should be installed on the bottom outlet line of the second vessel.
   Where a vessel outlet, above is divided in a manifold system of branches (each with a valve), an emergency isolation block valve is not required if no more than two valves, meeting the requirements of standard, are normally open; an emergency isolation block valve is required upstream of the manifold if three or more valves are normally open.

- **Furnaces**

Fuel lines to process furnaces and steam boilers should be provided with remotely operated emergency valves. These solenoid-operated bubble tight shut-off valves should be installed in each main furnace fuel line adjacent to the control valve. Operation should be remote, manual or automatic on closing and manual only opening. Loss of fuel or atomizing steam pressure should also automatically close these valves.

In addition to this, a manually operated block valve should be provided in each fuel line. This includes the pilot gas supply line if it is a separate line. These valves should be located at least 15 m horizontally from the furnace or boiler being protected. In some instance, a plant battery limit valve may be used to meet this requirement.

Atomizing steam lines to furnaces should not be cut off automatically, but should be stopped by closing the pressure control valve on this line.

Emergency valve on the snuffing steam line should be located at least 15 m horizontally from the furnace and operable from grade.

Burner isolation valves from the main fuels and steam should not be located under the heater and should be arranged to be within arms length of the peepholes, enabling burner flames to be seen.

The regulated pilot gas, where possible, should be from an independent sweet gas supply or from a separate take-off on the fuel gas main (upstream of the main fuel gas control valve) with its own spaded block valve. If continuous pilots are specified, a solenoid operated shutoff valve should also be installed in the pilot gas line operated by an emergency shut-down (S/D) switch only. A low-pressure alarm on the pilot gas line should also be fitted.

- **Compressors**

Emergency block valves should be provided in the suction and discharge lines of all compressors with a driver of >150 kW. When the discharge goes to two different locations, block valves should be installed on both discharge lines.

Compressors <750 kW may have hand-operated emergency block valves in sizes up to and including DN 200 (8 in). For larger valve sizes, remote power-operated valves should be used. If a hand-operated emergency valve is used, it should be located at least 9 m horizontally from the compressor. Remotely operated emergency shutoff valves may be located closer than 9 m; however, the control station should be installed at least 15 m from the compressor.

Compressors of ≥750 kW should have remotely operated emergency block valves. A control station should be located in the area of the compressor, at least 15 m away in a readily accessible location that is not likely to be exposed to fire. A second control station should be installed in the control room.

- **Depressurizing**

The following paragraphs describe the installation standards and design practice for depressuring valves, which are used to protect units operating at high pressures from an emergency situation by depressuring.

It is understood that the temperature of the system may fall excessively as a result of depressuring, thus requiring other equipment to be checked for corresponding temperature and pressure changes and the chance of hydrate formation. Also, it is conceivable that any sudden opening of the valves may import shock to the flare header; hence, this should be checked, depending on individual cases. The installation points of these valves should be clearly indicated on the P&IDs.

6.10.2 Design practice

Vapor depressuring system should be installed on equipment if one of the following conditions is encountered.

1. Operating pressure is above 1800 kPa (ga)
2. Process equipment contains more than 2 ton of liquid (C4 or more volatile) under normal conditions

Depressuring valves should be located near the equipment to be protected. They should be operated from the control room or a remote accessible location at grade. The valve, electric motor, and the portion of their supply lines located within the fire area should be fireproofed.

The initial pressure is commonly taken as the operating pressure; however, the equipment design pressure should be used as the initial pressure, considering operating practices and operator response time. Unless otherwise specified, the pressure of the equipment should be reduced to 50% of the vessel design pressure within 15 min while vapor is being generated at a rate corresponding to the following:

1. Vapor generated from liquid by heat input from a fire, plus
2. Density change of the vapor in the equipment during pressure reduction, plus
3. Liquid flashing during pressure reduction

The valves should be spring-loaded, pneumatic, and diaphragm-operated without a valve positioner, and should have tight shut-off and failure open functions. The minimum size of the depressuring valves should be DN 25 (1 in). They should not be provided with a hand wheel. Locked-open block valves should be provided to isolate them. Locked-closed bypass valves should be provided at grade with steam tracing.

The method of calculating the total vapor load for a system to be depressured is given in API RP 521 Section 3.19.

6.10.3 Trip sequence

The following paragraphs summarize the tripping factor for units and equipment, and also trip action under the following categories:

1. Furnaces
2. Pumps including hydraulic power recovery turbine
3. Compressors

The trip sequence should be summarized on P&IDs. It is necessary, however, to prepare the sequence logic diagram at the same time. Also, it is desirable to explain the outline of trip sequence, which should be explained in the unit operating manual.

In design of emergency S/D systems, the following features should be considered:

1. Test of the system should not cause S/D devices to operate
2. Bypassing an emergency device in a system should give a lasting light indication in the control room
3. First-out and failsafe features should be provided
4. All emergency switches should be secured against accidental activation
5. The visual alarm should be used where possible, with one working as pre-alarm

- **Furnaces**

The furnaces should be provided with an S/D system that should cut out the fuel oil and fuel gas (including any waste gas to be burnt in furnace, such as non-condensable gases) supply to the burners and will be actuated by the following cases:

a. Low-low flow of the total feed to the heater or low-low flow of passes (two passes per furnace cell, all combined with "AND" logic for furnaces having four passes per cell; implementation of similar requirements for other cases should be studied separately), with pre-alarm at low flow (for catalytic reactor heaters, licensor's design practice is to be followed)
b. Low-low flow of recycle gas (for catalytic reactor heaters), with pre-alarm at low flow
c. Low-low pressure of the pilot gas, with pre-alarm at low pressure
d. Emergency push button (PB) installed in the central control room
e. High-high pressure of the combustion chamber of the heater, with pre-alarm at high pressure (for heaters for which an air-preheating system is foreseen)
f. Low-low pressure of the fuel gas
g. Low-low pressure of the fuel oil
h. High-high flue gas temperature (upstream and downstream of APH, combined with "OR" logic), if specifically specified by the company
i. Low-low atomizing steam pressure or low-low differential pressure of atomizing steam with regard to fuel oil pressure

In case of heater S/D, any waste gas fired in the heater should be diverted to the foreseen alternative destination.

Table 6.7 shows a summary of the S/D system of furnaces, including devices to initiate block sequences and pertinent services of the action devices.

- **Pumps including hydraulic power recovery turbine.**

The following failure action as shown in Table 6.8 should be applied to the main pump of each unit.

- **Compressors**

The following failure action as shown in Table 6.9 should be applied to compressors.

Table 6.7 Shutdown System of Furnaces

| | | Devices to Initiate Block Sequences | | | | | | Block and Pertinent Action Devices | |
| | | | Trip No. | | | | | | |
No.	S/D Switch	Causes	1	2	3	4	5	Trip No.	Services
1	FSLL	Low-low flow of total	x		x	x		1	Close fuel gas to heater
		Feed to the heater or low-low flow of passes						2	Close pilot gas to heater
		(Two passes per cell, all combined with "AND" logic), excluding catalytic reactor heater						3	Close fuel oil to heater
								4	Close waste gas (off-gas) to heater
								5	Open stack control valve (PV) and open air preheater bypass (if applicable)
2	FSLL	Low-low flow of recycle gas (for catalytic reactor heaters)	x		x	x			
3	PSLL	Low-low pressure of pilot gas	x	x	x	x			
4	PB	Emergency manual push button (in CCR)	x	x	x	x			
5	PSHH	High-high pressure of combustion chamber (heater equipped with APH)	x		x	x			
6	FSLL/PSLL/ FDF fail switch	Low-low flow of combustion air or low-low pressure at FDF discharge (or downstream of APH) or FDF failure	x		x	x			
7	PSLL	Low-low pressure of fuel gas	x			x			
8	PSLL	Low-low pressure of fuel oil			x				
9	TSHH	High-high fuel gas temperature (upstream/ downstream of APH combined with "OR" logic), if specified by company	x		x	x			
10	PSLL/PDSLL	Low-low atomizing steam pressure or low-low differential pressure of atomizing steam with regard to fuel oil pressure			x				
11	IDF fail switch	IDF failure					x		

Table 6.8 Shutdown System of Main Pump of Unit

Failure\Action	Pump S/D System	Minimum Flow Bypass Open
Manual shutdown	Yes	No
Shutdown factors (Note 1)	Yes	No
Process flow low (Note 2)	No	Yes

Notes.
(1) Requirements for this failure will be provided by vendors, typical examples of which are low-low lube oil pressure, high-high speed of turbine (for turbine-driven pumps), high-high bearing temperature, etc.
(2) This failure applies only to severe turn-down service.

Table 6.9 Shutdown System of Compressors

Failure\Action	Compressor S/D System	Minimum Flow Bypass Open
Manual shutdown	Yes	No
Shutdown factors (Note 1)	Yes	No
Knockout drum level high	Yes	No
Gas flow low	No	Yes (Note 2)

Notes.
(1) Requirements for this failure will be provided by vendors, typical examples of which are low-low lube oil/seal oil pressure, high-high speed of turbine (for turbine driven compressors), high-high bearing temperature, etc.
(2) This failure action applies to the centrifugal compressors only.

References

[1] API RP 521—Guide for pressure-relieving and depressuring systems. American Petroleum Institute, 1220 L Street, NW, Washington, DC 20005.
[2] API RP 520—Recommended practice for the design (part I) and installation (part II) of pressure relieving systems in refineries. American Petroleum Institute, 1220 L Street, NW, Washington, DC 20005.

Further reading

175 West Jackson Blvd Recommendations and guidelines—Gasoline plants, pamphlet 301. Chicago (Illinois): Oil Insurance Association; August 1971. 60604.
Akeredolu FA, Sonibare JA. A review of the usefulness of gas flares in air pollution control. Manag Environ Qual 2004;15(6):574–83.
Akpojivi RE, Akumagba PE. Effects of gas flaring on soil fertility JPT. J Pet Technol 2005;57(12):39–40.
ANSI B31.3—Chemical plant and petroleum refinery piping. The American Society of Mechanical Engineers (ASME), 345 East 47th Street, New York, NY 10017.
ANSI B31.8—Gas transmission and distribution piping. The American Society of Mechanical Engineers (ASME), 345 East 47th Street, New York, NY 10017.

API 2510—Design and construction of LPG installations at marine and pipeline terminals, natural-gas-processing plants, refineries, petrochemical plants, and tank farms. American Petroleum Institute, 1220 L Street, NW, Washington, DC 20005.

API Standard 527—Commercial seat tightness of safety relief valves with metal-to-metal seats, American petroleum Institute, 1220 L Street, NW, Washington, DC 20005.

ASME boiler and pressure vessel code, Section I and Section VIII. American Society of Mechanical Engineers (ASME), New York, NY.

Brzustowski TA. Flaring in the energy industry. In: Process energy combustion science, vol. 2. Great Britain: Pergamon Press; 1976. 129–144.

Busina M, Raventos M, Albrecht P. Continuous zero emissions in a teg dehydration unit, Brazilian Petroleum, Gas and Biofuels Institute [IBP]. In: Rio Oil and Gas Conference; 2010 (Rio de Janeiro, Brazil, 9/ 13–16/2010) [Techn].

Chiu CH. Apply depressuring analysis to cryogenic plant safety. Hydrocarbon Process; November 1982: 255–64.

Cook DK, Fairweather M, Hammonds J, Hughes DJ. Size and radiative characteristics of natural gas flares. Part I–field scale experiments. Chem Eng Res Des 1987;65(4):310–7.

Cook DK, Fairweather M, Hammonds J, Hughes DJ. Size and radiative characteristics of natural gas flares. Part 2- empirical model. Chem Eng Res Des 1987;65(4):318–25.

Gogolek PEG, Hayden ACS. Performance of flare flames in a crosswind with nitrogen dilution. J Can Pet Technol 2004;43(8):43–7.

Johnson GA. Energy relations of gas estimated from flare radiation in Nigeria. Int J Energy Res 2001;25(1):85–91.

Johnson MR, Kostiuk LW. A parametric model for the efficiency of a flare in crosswind. Proc Combust Inst 2002; 29(2):1943–50.

Paul Kandell. Program sizes pipe and flare manifolds for compressible flow. Chem Eng; June 29, 1981:89–93.

Min TC, Fauske HK, Patrick M. Ind Eng Chem Fundam; 1966:50–1.

Mohammed Ali AAK, Jassim RK. Using swirl flow to improve efficiency of flares, proceedings of the institution of mechanical engineers, part A. J Power Energy 2006;220(7):731–6.

Mourad D, Ghazi O, Noureddine B. Recovery of flared gas through crude oil stabilization by a multi-staged separation with intermediate feeds: a case study. Korean J Chem Eng 2009;26(6):1706–16.

NFPA 30, 58, 59 & 59A. National Fire Protection Association (NFPA), Batterymarch Park, Quincy, Massachusetts 02269.

Odigure JO, Abdulkareem AS, Odeniyi OD. Computer simulation of soil temperature due to heat radiation from gas flaring. Model Meas Control B 2003;72(5–6):1–9.

Overa Sverre J. Strange Ellen, Salater Per. Determination of tempertures and flare rates during depressurization and fire. San Antonio (Texas): GPA Convention; March 15–17, 1993.

Powell WW, Papa DM. Precision valves for industry. Houston (TX): Anderson, Greenwood Company; 1982. pp. 52–61.

Rahimpour MR, Jokar SM. Feasibility of flare gas reformation to practical energy in farashband gas refinery: no gas flaring. J Hazard Mater 2012;209–210:204–17.

Schwartz Robert E, White Jeff W. Predict radiation from flares. Chem Eng Prog July 1997;93:42–9.

Snow N. Nigeria group to use flare gas for electricity. Oil Gas J 2006;104(43):29–30. þ32.

Straitz III JF. Flaring for safety and environmental protection. Drilling-DCW; November 1977.

Straitz III JF. Make the flare protect the environment. Hydrocarbon Process; October 1977.

Straitz III JF. Nomograms "Determining proper flame tip diameter and height". Tulsa (OK): Oil Gas and Petroleum Equipment. July and; August, 1979.

Straitz III JF. Solving flare-noise problems. San Francisco: Inter. Noise 78; May 8–10, 1978. 1–6.

Tan SH. Flare systems design simplified. Hydrocarbon Process.

Tite JP, Greening K, Sutton P. Explosion hazard of air ingress into flare stacks. Chem Eng Res Des 1989;67(4): 373–80.

Torres VM, Herndon S, Allen DT. Industrial flare performance at low flow conditions. 2. Steam- and air-assisted flares. Ind Eng Chem Res 2012;51(39):12569–76.

Torres VM, Herndon S, Kodesh Z, Allen DT. Industrial flare performance at low flow conditions. 1. Study overview. Ind Eng Chem Res 2012;51(39):12559–68.

Van Boskirk BA. Sensitivity of relief valves to inlet and outlet line lengths. Chem Eng; August 23, 1982:77–82.

Zadakbbar O, Vatani A, Karimpour K. Flare gas recovery in oil and refineries. Oil Gas Sci Technol 2008;63(6): 705–11.

Safety Relive Valves Design

Generally speaking, safety relief valves acting as a "last resort," this fully mechanical valve is designed to open based on an overpressure situation within a process pressure system, thus, not only protecting life but safeguarding the investment and plant itself. This chapter covers the minimum requirements for process design and engineering of pressure relieving devices in OGP Industries excluding cryogenic services.

Safety relief valves shall be provided to protect all equipment subject to overpressure and under certain other conditions as specified herein.

Unless otherwise specified, the safety devices shall be provided when the overpressure exceeds the design pressure of the equipment.

Pressure/vacuum relief requirement and relief load capacity for atmospheric and low-pressure storage tanks shall be evaluated based on API-Standard 2000.

Snuffing steam shall be provided for safety valves discharging to the atmosphere in the following services:

1. Material above its auto-ignition point
2. Light hydrocarbon service at locations listed in the job specification where hazard of exposure to lightning is prevalent due to high rate of electrical storms

When snuffing steam is required, provide a DN 25 (one inch) snuffing steam connection approximately 300 mm from outlet end of the discharge piping. The steam line to this connection shall be run to grade, provided with a double valve and bleeder at grade, and a steam condensate trap upstream of the double valves.

7.1 Provisions of pressure safety relief valves

Pressure safety or relief valve shall be provided for all cases as specified herein below.

7.1.1 Vessels

When designed in accordance with American Society of Mechanical Engineers (ASME) Section VIII, Unfired Pressure Vessel Code and the overpressure exceeds the Design Pressure.

Special attention shall be made to the cases where process conditions are changed during the engineering stage or after initial start-up of the plant and require increased operating pressure.

When designed in accordance with ASME, Section I, Power Boiler Code and the overpressure exceeds the Maximum Allowable Working Pressure as defined in that code.

Natural Gas Processing. http://dx.doi.org/10.1016/B978-0-08-099971-5.00007-6
Copyright © 2014 Elsevier Inc. All rights reserved.

7.1.2 Pumps

On discharge of positive displacement pumps.

On discharge of centrifugal pumps to protect downstream equipment from overpressure based on pump shutoff pressure.

On discharge of all pumps where under blocked discharge, the horsepower rating (in kilowatts) of the electric motor drive may be exceeded.

On pump suction lines from a "bottled in" system where overpressure can be imposed on suction piping by backflow through the pump or through a control valve bypassing from pump discharge to suction.

7.1.3 Compressors

Where each stage of reciprocating compressors to protect interstage, intercooler, compressor frame, or cylinder.

On suction lines where overpressure may occur on suction lines or frame, or overload electric motor driver before interstage or discharge safety valve opens.

On discharge where under blocked discharge the design torque of the driver coupling or keyway may be exceeded due to an oversized motor or turbine.

7.1.4 Steam boilers and superheaters

The following equipment falling under the jurisdiction of ASME, Section I, Power Boiler Code when the overpressure exceeds the Maximum Allowable Working Pressure as defined by that Code:

1. Steam drums
2. Superheater outlet
3. Externally fired superheater coils installed in a process heater

7.1.5 Fired process heaters

To prevent overpressure due to heat input resulting from an action blocking the lines at downstream of the heater, where check valves or other valves upstream of the heater are closed by the same action blocking the upstream line(s), except for the condition covered under (1) below. The safety valve may be located anywhere between the upstream and downstream blocking valves:

1. A safety valve will not be required to protect the heater if the only block valve(s) between a fired heater and a tower are operable manually and intended to be used to prevent backflow from the tower to the heater in case of heater tube failure

7.1.6 Turbines and surface condensers

Condensing or non-condensing steam turbine cases where the exhaust outlet may be blocked, thus, preventing the turbine from discharging to the atmosphere, a low pressure steam header, or a surface condenser.

Gas turbine cases, where exhaust outlet is blocked, thus, preventing the turbine from exhausting to the atmosphere, or to waste-heat generators, etc.

Exhaust outlet where the design pressure of an expansion joint is less than the turbine casing design pressure.

Surface condensers for condensing turbines.

7.1.7 Piping and connected equipment

For protection of piping, heat exchangers and other equipment served by the piping against overpressure under the following conditions:

1. Downstream of steam reducing valves (including protection to steam engine and turbine drivers in this case)
2. On the exhaust steam header leaving a Unit regardless of overpressure considerations. Use a chain operated multiport relief valve located on the Unit side of the block valve where the header connects into the plant exhaust steam system
3. Downstream of restriction orifices or manually operated valves in steam make-up or other services where closing a valve would result in overpressure
4. Fuel inlet to gas engine drivers
5. Pressure line to pressure balanced valves used to control fuel supply where the steam breaking point, diaphragm bursting point, or diaphragm case breaking point may be exceeded
6. Downstream of steam control or regulating valves
7. Downstream or upstream of all control valves when the piping or equipment would be subjected to overpressure assuming that the control valve will fail in the open or closed position
8. On pedestal or gland water systems to pumps and turbines where overpressure is caused by water pump shutoff pressure. One safety valve may be provided on the supply line to a group of pumps, turbine glands, or pedestals in lieu of a safety valve on the line to each piece of equipment

7.1.8 Cold side blockage and tube failure in exchangers

To prevent overpressure due to heat input resulting from an action blocking the line(s) downstream of the cold side of the exchanger, where check valves or other valves upstream of the exchanger are closed by the same action blocking the upstream line(s). A safety valve is required where heating the cold fluid within the exchanger when the hot side inlet temperature will raise the pressure of the fluid contained between the upstream and downstream blockages to more than 1.5 times of the design pressure of any item of equipment (excluding piping) in the contained system.

7.1.9 Pressure safety relief valves not required

Pressure safety relief valves shall not be provided on the following systems:

1. Interconnected vessels (excluding those falling under the requirements of ASME, Section I, Power Boiler Code), if they meet the following conditions:
   a. The vessel that is the source of pressure shall be equipped with a safety valve sized to protect all of the interconnected vessels or requirement and the interconnected piping (including heat exchanger equipment) shall be sized or checked for pressure drop to insure that the design

pressure on the downstream vessel or vessels is not exceeded by more than the percentage accumulations allowed in available standards

    b. At least one interconnecting piping system between the protected vessel and any other vessel must be free of:

        – Any equipment that may fail or stop in a closed position

        – Any block valves, control, or check valves

        – Any orifices, or similar restrictions to flow

2. Where a lower pressure piping system, such as pumpout, is routed to an offsite slop or emergency tank with overshot connections (connections entering the top of the tank without block valves)

3. Depressuring systems routed to a flare, if all valves at the Unit limit, cooling boxes, or downstream depressuring valves are locked open

7.2 Provisions of temperature safety valves

Temperature safety valves shall be provided for all cases as required herein below:

7.2.1 Heat exchangers

Temperature safety valves shall be provided on the cold side of heat exchange equipment, including compressor jackets operating full of liquid and subject to being blocked off, where heat may be applied internally or externally, and other forms of pressure relief valves are not provided.

7.2.2 Piping

Provide temperature safety valves for the following:

1. Steam traced lines

For sections of pipe that are steam traced and contain liquid and may be blocked in.

2. Lines containing cold solvent or refrigerant

For blocked insections of pipe that contain cold liquid and may develop excessive pressure under ambient temperature.

3. Exposed liquid-filled lines

For blocked in sections of pipe that may be subjected to excessive pressure due to rays of sun when the exposed length of line is 30 m or greater.

7.2.3 Temperature safety valves not required

On pedestal or gland water piping to pumps, turbines, compressors, blowers, etc., when the improbability that blocking off both the inlet and outlet valves and heating of the glands would occur simultaneously,a temperature safety valve is not required.

7.3 **Provisions of vacuum safety valves**
7.3.1 **Pressure vessels**

Vessels in the following services shall be provided with vacuum breaking device, if they cannot withstand full vacuum:

1. Vessels where stripping steam is used and overhead condensing systems are provided
2. Vessels operating under partial vacuum
3. Vessels operating full of liquid having a vapor pressure less than atmospheric pressure at the operating temperature
4. Vessels on suction to compressors
5. Vessels such as hot process softeners, deaerating heaters, hot water storages where cold water is heated bydirect contact with steam and the vessel is not open to the atmosphere

Before vacuum safety valves are provided for vessels where explosive mixtures may occur due to bleeding air into the vessel through a vacuum safety valve and if the set pressure of the vacuum relief valve is so low that special consideration is needed to hold leakage to a minimum, consideration should be given to the alternate method of bleeding gas into the vessel by means of a regulator or control valve using the bleed gas as the operating medium instead of instrument air.

Vacuum relief valves allowing air to enter a hydrocarbon system shall not be used without prior approval of the Company. The Contractor shall provide adequate backup documents to support the design applied to the system.

7.3.2 **Spheres and spheroids**

The necessity for vacuum relief valves on spheres and spheroids shall be investigated for each case, depending on the vapor pressure and corrosivity of the fluid handled and design of such equipment in regard to vacuum.

7.4 **Provisions of rupture disks**
7.4.1 **Rupture disk alone**

A rupture disk alone shall not be used to protect vessels or equipment but it may be used in combination with safety valves as a second relief device.

7.4.2 **Rupture disk on inlet to safety valve**

A rupture disk shall be provided at the inlet to a safety valve in the following instances:

1. Where it is difficult to keep safety valves from leaking either due to high pressure involved or to a combination of high pressure and fluctuating pressure, such as high-pressure reciprocating compressors
2. Where toxic material is being handled and contamination of the atmosphere is to be avoided due to possible leakage

3. Where the fluid handled is expensive and the possibility of safety valve leakage is to be avoided
4. Where material that will make the safety valve inoperative is to be kept out of the safety valve
5. Where corrosive material is handled and a safety valve cannot be obtained of satisfactory material to resist corrosion
6. Where corrosive material is handled and the cost of a safety valve to resist corrosion is prohibitive

7.4.3 Rupture disk on outlet of safety valve

Rupture disks shall be provided on the outlet of safety valves in the following instances:

1. To keep corrosive products out of a safety valve connected into a vent system
2. Where it is necessary to confine safety valve leakage and there is insufficient space available on the safety valve inlet

7.4.4 Rupture disks used in conjunction with safety valves

Rupture disks shall be provided in parallel with safety valves that discharge to the atmosphere or to a vent system under the following conditions:

1. Where exothermic reactions may develop abnormally high and uncontrollable pressure conditions

7.5 Spare safety valves

The necessity for provision of spare safety valves shall be based upon the following considerations:

1. On process Units where the required time interval between safety valve inspection periods is less than the time interval between designated inspection and test periods of the Unit, or is less than the normally anticipated frequency of Unit shut-downs for other reasons such as clean up or catalyst change, and the vessel or item of equipment cannot be taken out of service without shutting down the Unit.

When a spare safety valve is required, the size of the safety valves shall be such as to provide the least number of safety valves that will have a total capacity equal or greater than the calculated required relieving capacity.

A spare relief valve is not required when two or more pressure or safety relief valves are required for the calculated relieving capacity on spheres or spheroids. However, where only one valve is needed for the required capacity, a spare valve shall be provided for spheres or spheroids.

When a vacuum relief valve or valves are required on spheres or spheroids, no spare relief valve is required.

Spare temperature safety valves are not required.

7.5.1 Emergency vapor depressuring requirements

In addition to (not in lieu of) the pressure relief facilities described in this section, a depressuring system shall be provided for reducing the pressure under fire conditions. To this end, all process equipment, except as mentioned below, containing more than two metric ton of liquid C4 or more volatile liquid under normal operating conditions, shall be provided with remotely operated vapor depressuring valves discharging into a closed system, preferably the relief system, leading away from the probable fire areas.

Platformate separators in platforming Units and high-pressure separators in hydrotreating and hydrocracking Units shall always be provided with depressuring facilities.

Facilities for remote operation of the depressuring valves shall be located in the control room.

7.5.2 Provisions of fire relief valves

Fire relief valves shall be considered for liquid containing equipment (e.g., vessels, columns, and settlers) if all of the following circumstances apply:

1. If the equipment is located in an area where a sizable fire may occur
2. If the equipment under consideration can be blocked in while the Unit remains in operation

7.6 Selection of type
7.6.1 Conventional (unbalanced) safety relief valves

Use the conventional safety relief valve where the service is:

1. Clean and noncorrosive
2. Corrosive, with provision of corrosion-resistant materials

Do not use this valve where service is:

1. Corrosive and corrosive materials may damage the guide and disk, or guide and spindle, or spring and bonnet
2. Such that the variable back pressure is greater than 10% where 10% accumulation is allowed or greater than 20% where 20% accumulation is allowed under fire conditions
3. Such that the differential pressure when the valve is relieving compared to the normal differential pressure across the protected equipment is less than 10%. This particularly applies when starting up a Unit where the initial back pressure on a valve is zero and the differential pressure across the valve may be below the intended operating pressure or closer than 10% above the normal operating pressure
4. Such that the material relieved may contain coke in suspension or slurries containing particles that may clog the guiding surfaces

Figure 7.1 shows conventional (unbalanced) safety relief valves.

7.6.2 Balanced bellows safety relief valves

Use the balanced bellows safety relief valve where:

1. The relieving pressure is to be independent of the back pressure
2. The bellows is used to prevent clogging of the guiding surfaces with coke or other particles in suspension

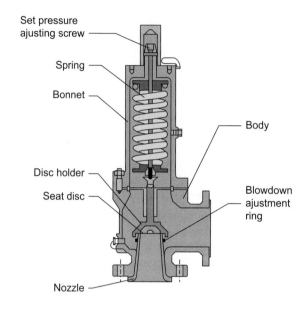

Set pressure
ajusting screw

Spring

Bonnet

Body

Disc holder

Seat disc

Blowdown
ajustment
ring

Nozzle

FIGURE 7.1

Conventional (unbalanced) safety relief valves.

3. The bellows is used to prevent corrosive products from damaging the guiding surfaces, spring, or associated pieces
4. Back pressure, as specified above, exists and either liquid or vapor relief is to be handled
5. Savings in discharge piping and size of flare header may be realized through use of higher back pressure in the flare header

Where the balanced bellows safety valve is used in a closed flare system and the variable back pressure is above 30% of the relieving pressure, check with the manufacturer on the decrease in the capacity of the particular valve when sizing. Do not use this type of safety valve where the liquid being discharged could accumulate and set up in the convolutions or bellows and make the safety valve inoperative.

Superimposed back pressure due to operation of emergency depressuring valves shall be disregarded when evaluating the necessity of balanced type valves, because simultaneous operation of relief and depressuring valves should not normally occur. As variable back pressure, the highest pressure shall be taken that can be generated in the outlet of a valve by the valve itself, and possibly other valves at the same time of blowing.

Bellows balanced valves shall not be used for services involving materials with their pour point at or above the lowest ambient temperature (e.g., materials containing wax), or where coking may be expected. Balanced valves with a piston only shall be used in such cases.

Figure 7.2 shows balanced bellows safety relief valves.

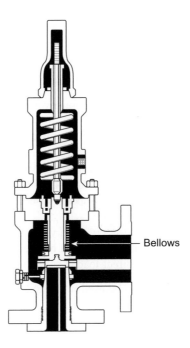

Bellows

FIGURE 7.2

Balanced bellows safety relief valves.

7.6.3 **Balanced bellows plus balanced piston safety valves**

The balanced piston added to a balanced bellows safety valve is to be provided where it is important that the valve operate properly in case of bellows failure. This type of valve is to be used only on approval of the Company.

7.6.4 **Multiport relief valves**

Use the multiport relief valve on exhaust steam systems where it is desired to open the valve manually to dump the exhaust steam to the atmosphere, or to provide protection for process Unit exhaust steam headers.

Multiport relief valves may also be used in vacuum service to protect surface condensers, etc. Such valves are to be water sealed.

7.6.5 **Pilot operated safety relief valves**

Types of pilot operated pressure safety valves:

1. With integral pilots
   a. External Pressure Sensing-Pilot Exhaust Atmosphere
   b. Integral Pressure Sensing-Pilot Exhaust Atmosphere

    c. Integral Pressure Sensing-Pilot Exhaust to Relief Valve Outlet
2. With external pilot
    a. External Pressure Sensing-Pilot Exhaust Atmosphere

Pilot operated safety valves are used in the following cases:

1. On spheres, spheroids, and other low-pressure storage vessels
2. In clean, noncorrosive services
3. Where premium tightness is desired. These valves are tight to set pressure wherever conventional valves are tight to 90–95% of set pressure. Therefore, for equipment that operates between 90% and 95% of the set pressure, utilization of pilot operated valves shall be considered
4. Where maximum capacity with least overpressure is desired
5. Where the built-up back pressure and variable superimposed back pressure does not exceed 50% of the set pressure
6. When the pressure drop to the inlet of the safety relief valve is greater than 3% of the set pressure (in this case, the remote pressure pickup type should be used)
7. Where the valve size is so large that a direct loaded safety relief valve would be unsuitable

Pilot operated safety relief valves are not to be used in the following cases:

1. In corrosive service
2. In coking or slurry service
3. In closed flare systems unless back pressure effects have been investigated and envisaged

Figure 7.3 shows a schematic of a pilot operated safety relief valve.

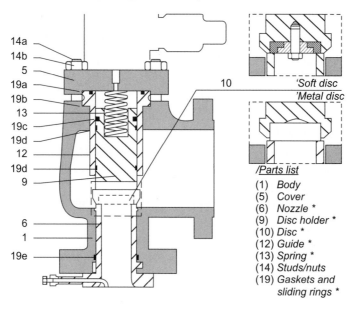

Parts list
(1) Body
(5) Cover
(6) Nozzle *
(9) Disc holder *
(10) Disc *
(12) Guide *
(13) Spring *
(14) Studs/nuts
(19) Gaskets and
       sliding rings *

FIGURE 7.3

Pilot operated safety relief valves.

7.6.6 Open spring type valves

Open spring type valves are permitted in the following services:

1. Steam service, such as boilers, waste heat boilers, etc., where ASME Power Boiler Code applies
2. Steam service, such as turbine cases, etc.,that are located inside shelters or buildings

Open spring type valves located outdoors shall have a specially constructed metal hood over the spring to prevent water seepage into the relief valve interior or accumulation of foreign particles, such as catalyst dust.

7.7 Closed spring type valves

Closed spring type valves are required in the following services:

1. Steam service, such as turbine cases exposed to the weather, where ASME Power Boiler Code does not apply
2. Hydrocarbon service
3. Chemical service
4. Air service
5. Multiport relief valves

A DN 20 (¾ inch) minimum size vent connection shall be provided in the closed bonnet of all safety valves to vent the spring housing to the atmosphere. This is especially needed on bellows type safety valves to prevent buildup of back pressure.

7.8 Safety valves with lifting devices

7.8.1 Lifting levers shall be furnished for the following valves

1. Steam safety valves where ASME, Section I, Power Boiler Code applies
2. Safety valves for air receivers
3. Open spring safety valves where the lifting lever is standard construction
4. Other safety valves when specified for special applications

When lifting levers are required, standard lifting levers shall be furnished except as follows:

1. Packed lifting levers shall be provided on safety valves with closed bonnets when specified in the job specifications where no leakage of vapor at the lifting lever can be tolerated

7.9 Temperature safety relief valves

7.9.1 Water service

Use top guided standard relief valve or puppet type spring load check valve with resilient seats.

7.9.2 Hydrocarbon or chemical service

Top guided standard relief valves with size and type connection required by the applicable piping material service specification shall be provided where valve discharges to a drain or sewer, unless otherwise accepted by the Company.

7.10 Safety valve caps

All closed bonnet safety valves shall be furnished with screwed or bolted caps.

7.11 Safety valve drains

Drain connections shall be provided on all safety valves used in condensable vapor service (see also Article 11.4 hereinafter).

7.12 Rupture disc types

The following types shall be used for all applications:

1. Insert type rupture disc preassemblies for installation inside bolt circle of two flanges
2. Multiple component discs with plastic or metallic seal and vacuum or back pressure support

7.13 Safety valve bonnet

Conventional safety relief valves for closed relief systems shall have a closed bonnet.

Safety relief valves of the balanced type shall have a vented bonnet that shall be vented to the atmosphere in such a manner as to prevent the ingress of rain and dirt. Personnel hazard should be avoided while, on the other hand, ready observation of any flow from a vent line should remain possible.

7.14 Set pressure

Unless otherwise specified in this Section, safety relief valves shall be set to relieve initially at the design pressure of the equipment, within the limitations of the allowable blowdown and accumulation specified in this Section.

Unless otherwise specified in this Section, the stated set pressure is the initial relieving pressure.

In general, Set Pressures (SPs) and Maximum Relief Pressures (MRPs) of safety/relief valves, expressed in relation to the Design Pressure (DP) of the protected equipment, all expressed in gage pressure, shall not exceed the values given in Table 7.1.

Note:

For the pressure setting of the safety/relief valve instead of the DP, the Maximum Allowable Working Pressure (MAWP) is also used.

Table 7.1 Set Pressure and Maximum Relieving Pressure

Description	Set Pressure (SP)		Maximum Relieving Pressure (MRP)	
	Non-fire Conditions	Fire Conditions	Non-fire Conditions	Fire Conditions
Single valve	100% of DP	100% of DP	110% of DP	121% of DP
Multiple valves	One valve 100% of DP the others at 105% of DP(for set pressures below 1000 kPa (10 bar), staggering of set pressure becomes impracticable because of the difference between the set pressure tolerance of 3% (according to ASME, VIII UG 134) and the value of 5% of the DP becomes too small.	110% of DP (relief valves for fire protection may only be set at 110% of DP if they are installed in addition to adequate relief protection of the process equipment against non-fire situations).	110% of DP	121% of DP

The above shall also apply to safety/relief valves discharging liquid and flashing liquid.

Note: For safety valves, protecting vessels, or other equipment falling under the jurisdiction of ASME, Section I, Power Boiler Code, see Article 8.2.4.

7.15 Pressure safety or relief valve set pressure

Maximum pressure accumulations above the set pressure (initial relieving pressure) of safety or relief valves shall be limited as shown for the following services and equipment.

1. 6% for:

Steam boilers and steam systems governed by ASME, Section I, Power Boiler Code.

2. 10% for:
  a. Air receivers
  b. Compressors
  c. Pumps
  d. Turbine cases
  e. Piping
  f. Heat exchanger units
  g. Vessels and fired heaters other than those governed by ASME, Section I, Power Boiler Code

3. 20% covers fire conditions for:
  a. Air receivers
  b. Piping
  c. Heat exchanger units
  d. Vessels other than those governed by ASME, Section I, Power Boiler Code

The following allowable blowdown shall be specified for pressure safety relief valves:

1. 3% Blowdown for all steam service
2. 4% Blowdown for all vapor and gas service other than steam.

7.15.1. Equipment and piping safety relief valve set pressure

- **Vessels**

1. ASME, Section VIII, Code Vessels: Set at Design Pressure.
2. ASME, Section VIII, Code Vessels (special conditions only): For special cases when the overpressure exceeds the DP, the set pressure shall be the MAWP as defined in applicable Code.
3. ASME, Section I, Code Vessels: Set at MAWP as defined in ASME, Section I, Code.

- **Pumps & compressors**

On discharge where under blocked discharge, the horse-power rating (in kilowatts) of the electric motor drive may be exceeded, set at a pressure plus the overpressure that does not exceed the pressure developed at the maximum safe horse-power rating (in kilowatts) recommended by the pump or compressor manufacturer.

- **Steam boiler systems**

Set at a minimum of 10% or 100 kPa (ga) (whichever is greater) above the normal operating pressure, but not to exceed the set pressure of any safety valve protecting other lower pressure boilers feeding into the system.

- **Compressor and pump couplings**

To protect against overpressure on discharge under conditions defined in 6.2.4, set at manufacturer's recommendation.

- **Turbines**

1. When turbines exhaust into a separate steam system, set turbine case safety valve at a minimum of 70 kPa (ga) above the normal operating pressure of the system into which the turbines exhaust, but not to exceed turbine case design pressure
2. For safety valves to protect gas turbine cases, set at turbine manufacturer's recommendation

Cold side blockage and tube failure in exchangers set at 1.5 times Lowest Design Pressure for any item of equipment (excluding piping) in the system contained between the blockages.

- **Piping and connected equipment**

For safety valves required to protect piping and connected equipment (including fired heater tubes), set at the lower of the following:

1. Design pressure of equipment or fired heater tubes
2. The short time pressure of the piping as determined by the classification and limits of short time conditions defined in the relevant ASME/ANSI Code B.31.3

For safety valves required to protect piping only (no connected equipment involved), set at pressure specified in (2) above.

For safety valves required to protect piping and connected equipment in pulsating service, verify that the design pressure of the piping or equipment has been established on the basis of the maximum pressure anticipated under pulsation.

For steam service, the following additional requirements shall be noted:

1. For safety valves used to protect piping or equipment downstream of a steam pressure reducing valve, set at a minimum of 70 kPa (ga) above the normal operating pressure of the system on the low pressure side but not to exceed settings of any turbine safety valves exhausting into the same system or the design pressure of other equipment to be protected
2. For safety valves provided to protect piping and equipment in steam service, set at a pressure consistent with the setting of safety valves protecting any turbine exhausting into the same system

7.15.2 Multiple safety valves

Safety valve settings for multiple safety valves shall conform to the following:

1. For safety valves protecting vessels or other equipment falling under the jurisdiction of ASME, Section I, Power Boiler Code, the first valve shall be set at or below the MAWP as defined by that code. The other safety valve(s) may be set at a pressure not to exceed the MAWP by 3%. The range of settings between the first and other safety valves shall not be more than 10% of the setting of the highest set valve
2. For safety valves protecting vessels falling under the jurisdiction of ASME, Section VIII, Unfired Pressure Vessel Code, the first valve may be set at or below the DP. The other safety valve(s) may be set at pressures that do not exceed the DP by more than 5%. For fire conditions, set pressure of all valves shall not exceed 110% of DP

7.16 Temperature safety valve set pressure
7.16.1 Valves discharging to the atmosphere

1. Heat exchangers and equipment excluding piping: Set at DP
2. Piping only: Set at the Maximum Allowable Short Time Pressure, permitted under the applicable class of short time conditions defined in the relevant ASME/ANSI Code B.31.3, for the anticipated blocked in temperature
3. Where piping and other equipment are involved in a common system to be protected: Set at DP of governing equipment or at pressure defined in "2" above, whichever is lower

7.16.2 Valves discharging to a closed system

In a consecutive system of piping (including other equipment) containing block valves, check valves, pumps, or other closed devices, it may be desirable to install a series of temperature safety valves so that each valve relieves to the section of pipe protected by the next valve, with the last safety valve in series relieving pressure to the destination of the low pressure end of the system. In such a consecutive system, the set pressure shall be determined as follows:

1. Determine the Limiting System Allowable Pressure (LSAP; e.g., design pressure of equipment in the system) or Short Time Pressure of the piping, whichever is smaller.
2. The set pressure of each safety valve must be checked to insure that the summation of the set pressures of all the safety valves in series in the system does not exceed the LSAP defined in "1" above. If the summation of the set pressures exceeds the LSAP, each safety valve set pressure shall be reduced to bring the summation within the prescribed limits.

7.17 Rupture disc set pressure

The set pressure for rupture disc is determined the same as for pressure safety valves, but the differential between the normal operating pressure and the set pressure of the rupture disc depends on whether a single component rupture disc alone is used, whether a multiple component disc with plastic or metallic seal is used, and whether the normal operating pressure is constant or pulsating.

7.17.1 Rupture disc (single component)

Where the pressure is fluctuating or pulsating, the burst pressure of the single component rupture disc shall not be closer than 33% above the normal operating pressure for a metallic disc and 25% for a nonmetallic disc.

7.17.2 Multiple component rupture discs

1. Multiple component rupture discs with plastic pressure seal and vacuum or back pressure support shall be used where the pressure is constant or fluctuating. The burst pressure of the disc shall be no closer than 20% above the normal operating pressure.
2. A similar rupture disc with metallic pressure seal and vacuum or back pressure support shall have a bursting pressure not closer than 10% above the normal operating pressure.

7.18 Vacuum relief valve set pressure

Vacuum relief valves are set as close to atmospheric pressure as possible and still not be affected by normal upsets. This is to allow as much accumulation as possible to keep the valve size to a minimum.

Vacuum relief valves are allowed to accumulate to the maximum external design pressure.

7.19 Sizing

7.19.1 Pressure safety relief valves

Safety relief valve capacity formulas used for calculation of the required orifice area shall be in accordance with API Recommended Practice 520 Part I, Section 4 "Procedures for Sizing."

The standard effective orifice areas and the corresponding letter designations shall be according to API Standard 526.

The thermal expansion valves shall be DN 25 × DN 25 (1 × 1 inch) with flange orifice area of 38.7×10^{-6} or 71.0×10^{-6} square meters as required.

All safety relief valves shall have flanged inlet and outlet connections of 300# raised face (RF) and 150# RF, respectively, unless the service requires a higher rating or a different type of facing.

When estimating the normal fluid mass inflow to the system at blocked outlet conditions, credit shall be given that the vessel under emergency is at relieving conditions (i.e., at pressure higher than the normal operating one). When evaluating relieving requirements, it is assumed that any automatic control valve, which is not the cause of upset, will remain in the normal position. Credit may therefore be taken for the normal capacity of these valves, corrected for relieving conditions, and limited to the flow rates that downstream equipment can safely handle.

A load summary table including the following information and data shall be provided for each safety relief valve:

1. Item number
2. P & ID number
3. Protected equipment
4. Size and type
5. Set pressure (bar (ga) or kPa (ga))
6. Discharge to...............
7. Emergency failure (see Paragraph 9.1.2 for the emergency failure cases, all applicable emergency causes shall be tabulated). For each applicable emergency cause, the following information/data shall be specified:
   a. MW for vapor and kg/m^3 at flowing conditions for liquid
   b. kg/h for vapor and m^3/h at flowing conditions for liquid
   c. °C
   d. V or L
   e. Area (only for fire case).
8. Remarks

7.19.2 Relief load requirements

- **Double jeopardy**
  - Safety valves shall be sized on the basis of a single involuntary occurrence.
  - The sizing shall not be on the basis of double or multiple occurrences that are completely independent of one another, except where one of the occurrences happens frequently. Where multiple occurrences are dependent on one another and the probability of the occurrences are likely, the condition shall be considered for sizing.

For each safety relief valve, the maximum of the relief conditions shall be determined for each of the applicable causes cited in Table 7.1 of API-RP 520, Part I "Bases for Relief Capacities Under Selected Conditions," with the following modifications/reservations:

1. For new designs, no corrections shall be made for the difference between operating and relieving pressure, except and subject to the Company's approval where there is an appreciable difference between maximum operating pressure and maximum allowable working pressure (e.g., a vacuum column or a near atmospheric column with an MAWP of 600–800 kPa(ga)).
2. For new designs, the cooling effect of an air cooling heat exchanger resulting from natural drought in case of fan failure shall not be taken into account.
3. In case of lean oil failure to an absorber, the quantity to be released shall be taken as the total incoming vapor, plus the quantity generated therein under normal conditions.

In addition to the above, the quantities to be discharged in the event of each emergency case, as specified in this Section, shall be determined for each valve. The conditions resulting in a maximum relief quantity found under all the possible emergency cases shall be evaluated. The size of the relief valve and of its outlet pipe up to the relief header shall be determined on the basis of the largest of the relief quantities.

The following emergency cases should be evaluated for each system. Each system shall be considered as an equipment or a group of equipment that may be isolated by control valves or other valves, but without any other isolation in between.

- **Utilities failures**
  - Power failure
  - Cooling water failure (main system)
  - Cooling water failure (within process Unit limits)
  - Cooling water plus power failure (simultaneous failure)
  - Instrument air failure
  - Steam failure
  - Fuel oil/fuel gas failure
  - Air fan coolers failure
- **Operating failures**
  - Blocked outlet
  - Control valve failure
  - Reflux failure
  - Side stream failure/pump around failure
  - Loss of fractionator or column or liquid gas separator drum condensing duty
  - Simultaneous failure of reflux and condensing duty
- **Mechanical failures**
  - External fire
  - Heat exchanger tube failure
  - Hydraulic expansion
  - Other failures

- **Utilities failures**

Impact of utility failures shall be evaluated based on API RP 521 Section 2 Clause 2.3.5 and the criteria set forth herein below. The utilities failures can be plant-wide or local.

- **Power failure**

Power failure effects on Unit operation and safety valve sizing shall be considered and shall include the following:

1. Complete Unit-wide loss of low and medium volt supply while high volt supply power source, steam drivers, gas engine drivers, etc., are still operable
2. Complete Unit-wide loss of all types of electricity while steam drivers (if any used), gas engine drivers (if any used), etc., are still operable
3. Consideration must be given to any automatic shut-down of drivers caused by power failure affecting controls.

- **Cooling water failure (main system)**

1. Complete Unit-wide or partial cooling water loss shall be considered if the cooling water pumps are motor driven with zero to partial steam turbine spare drivers that automatically are started upon power failure
2. Each service requiring cooling water shall be considered as losing the same percentage of cooling water where there is only a partial failure
3. Complete loss of cooling water due to a break in the main cooling water header or due to main cooling water header block valve being closed shall not be considered.

- **Cooling water failure within process Unit limits**

Every condenser or cooler using cooling water shall be investigated for the following:

1. One hundred percent of cooling water in any given riser taken off from the main cooling water supply line. All condensers and coolers being fed from a single riser shall be considered to lose water simultaneously due to a possible break in the riser above grade or the block valve in the riser being inadvertently closed:
   a. If there is a single riser from or to the main cooling water headers above grade, then all branches must be considered as having failed
   b. It is permissible to use parallel risers that are distinctly separate, feeding parallel condensers or coolers to prevent 100%cooling duty loss
2. Credit for cooling from air fan exchangers may be taken if used prior to a cooling water condenser or cooler when cooling water is lost and the air fan is still under normal operation
3. Cooling or condensing duty shall be calculated in the same manner as set forth in the paragraphs below.

- **Cooling water plus power failure (simultaneous failure)**

This emergency case is associated with steam failure when the electrical power generators are linked to the situation caused by steam failure. In case of steam failure, the cooling water pumps will also be failed if the pumps drivers are steam or electrical driven.

Relief load for each system shall be evaluated based on possible overpressure source when no steam is available. No credit shall be taken for running of steam turbines or heat generation by steam in the reboilers and steam heaters.

- **Instrument air failure**

The relief load corresponding to this emergency shall be calculated based on the consequence of this failure such as:

1. Blocking the vapor or liquid outlet, when fail close type control valve is applied on vapor or liquid outlet
2. Cutting out the heat input to the system (e.g.,when fail close type control valve on steam to reboiler has been utilized)
3. Others.

- **Steam failure**

Plant-wide steam failure is also related to the simultaneous failure of cooling water and electric power. Local failure shall also be considered based on the consequence of steam loss.

- **Fuel oil/fuel gas failure**

The impact of this contingency on the Unit and the associated equipment shall be investigated in the light of the equipment affected. For example, the loss of fuel gas/oil may lead to loss of heater duty.

- **Air fan coolers failure**

1. Air fan condensers and coolers shall be considered as completely losing their duty if the air fans fail
2. Where cooling water exchangers follow air fan exchangers, cooling duty credit can be taken to the extent of cooling water available per the subclause above:
   a. This cooling or condensing duty shall be calculated on the basis of entering pressure and temperature at relieving conditions and a constant water rate that will not exceed normal design rates, if available
   b. The design heat transfer coefficient based on normal conditions can be assumed but should be recalculated if obviously different at relieving conditions (e.g., a liquid subcooler transformed to condensing service relieving conditions).

- **Operating failures**

The potential operating contingencies shall be considered for evaluating overpressure in the system. The most possible cases are as follows:

- **Blocked outlet**

1. The protected system shall be included in this study (i.e., tower plus overhead condensing system and accumulator) provided there is no control valve or other restrictions between the tower and accumulator
2. No credit shall be allowed for operation of any control valve in overhead, bottom, or feed system

3. Only those sources of fluid having sufficient pressure to open the safety valve shall be considered:

    a. Where the design pressure of the vessel is greater than the stalling pressure of the positive displacement pump, the safety valve on the pump provides the protection from this source of pressure

    b. Where the discharge pressure of a centrifugal pump is above the design pressure of the vessel, the flow quantity is the pumped quantity at the accumulated relieving pressure.

4. Liquid flow

    a. The flow quantity shall be the quantity available at the accumulated pressure

    b. The liquid flow quantity and relative density (specific gravity) shall be determined at the temperature existing at the accumulated pressure

    c. Circulating streams pumped from a vessel and back into the same vessel shall be excluded when determining the flow quantity

    d. Reflux flow quantities shall be excluded except when these flows are furnished from tankage outside the Unit or where there is a control valve or other restriction in the overhead line before the reflux drum.

5. Vapor flow

    a. Only the vapor input needs to be considered if the vessel has a high level alarm plus at least 10 min liquid residence time after the alarm

    b. Large vessels that have over 30 min liquid residence time above high liquid level need not have a high level alarm and vapor input only is considered

    c. Where only vapor input is considered, the flow quantity shall be the vapor generated at the controlled temperature at a pressure equal to the safety valve accumulated relieving pressure.

- **Control valve failure**

This contingency should include control valve failure either in the open or closed position. Special attention shall be given to the inadvertent opening of a control valve on the high pressure source, such as steam that may lead to overpressure in the system.

- **Reflux failure**

1. In towers where the source of heat is in the charge to the tower, special attention shall be given to the location of safety valve and charge flowrate at the relieving conditions:

    a. A flash curve shall be calculated to determine the vapor quantity at relieving conditions using heat input if feed is pressured into the tower

    b. The available feed rate should be investigated using the pump characteristics at the relieving conditions

    c. Stripping steam if used in the tower should be added to the hydrocarbon vapor if the stripping steam pressure is above the safety valve relieving pressure.

2. For towers where reboilers, heaters, or feed preheaters are used, the heat input is calculated as follows:

    a. Heat input via fired heaters:

        – The heater outlet shall be on temperature recorder controller (TRC) control with the fuel gas or fuel oil rate being adjusted by the outlet temperature

        – Simultaneous failure of the TRC and failure of reflux shall not be considered

    – Special consideration shall be given for a multiservice heater where the entire heat input to a stream is in the convection section.

  b. Heat input via exchangers:

    – The charge rate shall be the normal design rate unless the rate is limited by the charge pump flow characteristics or flow from a higher pressure vessel at the relieving pressure

    – The heating medium fluid throughput shall be the design rate, even if on TRC control, unless the relieving conditions affect the normal design rate

    – After determining the new hot and cold fluid rates, the heat exchange shall be calculated using inlet and outlet temperatures at the relieving conditions. The heat transfer coefficient used can be the design heat transfer coefficient, unless it is obviously different at the relieving conditions.

  c. Total heat input:

    – Total heat input shall generally be the input via fired heaters, plus exchangers, plus heat inherently in the feed or side-cut streams

    – Sensible heat can normally be deducted from the total heat input where the feed enters the vessel below its bubble point at relieving conditions (i.e., sensible heat from feed inlet temperature to its bubble point at relieving conditions)

    – Sensible heat represented by the net tower bottoms and/or side-cut streams at their relieving temperature can also be deducted from the total heat input

    – Credit for condensing duty shall not be taken unless the accumulator drum has a high liquid level alarm plus at least 20min residence time at the relieving condensing rate after the alarm is actuated and there is no build-up of vapor into accumulator to prevent flow through the condensers at the relieving conditions.

- **Side stream reflux/pumparound failure**

The relief load for this contingency shall be calculated according to API RP 521.

- **Loss of fractionator column or liquid gas separator drum condensing duty**

1. Maximum loss of condensing duty shall be determined by Paragraph 9.1.2.4.8 above

2. Heat input can be considered as vaporizing liquid of the same composition as residing in the top tray:

  a. Latent heat shall be at the top tray temperature at the relieving pressure

  b. Feed vapors calculated at the relieving conditions shall be added to the vapor calculated at the top section

  c. Stripping steam, if available at the relieving conditions, shall also be added, if not condensed at the relieving conditions.

3. A portion of the vapors generated at the relieving condition may be considered as condensing. However, no partial condensing credit may be taken if vapor build-up in the accumulator or separator can prevent flow through the condensers.

- **Simultaneous failure of reflux and condensing duty**

1. This emergency case shall be considered if caused by power failure that affects both reflux and condensing duty

2. In most cases, relieving capacity is the same as for reflux failure.

- **Mechanical failure**

Failure of electrical or mechanical equipment, such as fans, pumps, compressors, etc., shall be evaluated as an emergency case. Occurrence of such mechanical failures can be evaluated as part of operating failures or a partial utility failure.

- **External fire**

The external fire shall be evaluated based on API RP 520, unless otherwise specified in this Section and the following criteria.

1. The surface area effective in generating vapor when exposed to fire is limited to areas wetted by internal liquid contents
2. No credit shall be taken for insulation (i.e., F = 1.0)
3. For establishing the fire areas, each Unit shall be divided into probable fire areas of approximately 460 m^2
4. Air coolers shall be assumed under fire even if installed above 8 m. Wetted surface area of air fin cooler shall be based on API 521, Paragraph 3.15.4.1, depending on services.

- **Vertical distance covered by fire**

1. The total wetted surface included within a height of 8 m above grade shall be used
2. Where the vessel is on a platform or flat roof, the roof or platform shall be considered grade level if it can sustain a fire
3. For spheres or spheroids, the vertical distance covered by fire is the elevation of the maximum horizontal diameter or a height of 8 m, whichever is the greater.

- **Area covered by fire**

1. For horizontal accumulators, the fire covers the area in the two heads plus the horizontal wetted surface within the height limitation above grade. The horizontal wetted surfaces depends on the internal arrangement of the vessel and the maximum liquid level covered by the level control instrument
2. For vertical accumulators, the fire covers the area in the bottom head plus the vertical wetted surface up to the height limitation. The maximum liquid level covered by the level instrument governs the vertical wetted surface
3. In settlers operating full of liquid, the total surface area up to the height limitation shall be used as the wetted surface
4. In charge drums or surge drums, the total surface area up to the height limitation shall be used as the wetted surface,because it is assumed these vessels may be operated full
5. For vessels in pits, the fire is assumed to cover the entire surface area
6. For fractionators, strippers, and absorbers, the area of the bottom head exposed to fire plus the vertical area within height limitation shall be considered as wetted surface
7. For spheres and spheroids, only the vessel surface above grade level and to the equator of the vessel shall be considered as wetted surface
8. The area covered by a fire is considered to be a circle 24 m in diameter around a given tower. The presence of firewalls to contain liquid shall also be considered
9. Storage vessels are assumed to contain their maximum working volume.

- **Heat input due to fire**

1. Heat input shall be calculated according to the procedure established in API RP 520 and the criteria set forth in this Section.

- **Quantity to be released from liquid storage**

1. Heat input due to fire is assumed to be going into vaporizing the stored products at the relieving pressure
2. The relieving temperature shall correspond to the boiling point of the material corrected to the absolute relieving pressure
3. Where material is vaporizing from a constant boiling mixture, the temperature shall be the boiling point of the mixture at the absolute relieving pressure
4. Where water, caustic, or amine solution or other liquids exist in the presence of another liquid, the temperature shall correspond to the temperature at which the partial pressure of the different components add up to the relieving pressure.

- **Heat exchanger tube failure**

1. Heat exchanger tube failure shall be evaluated based on API RP 521, Section 2 Clause 2.3.13 and Section 3 Clause 3.18 for relief load estimation
2. Flow through the ruptured tube shall be determined according to CRANE Technical Paper No. 410 Section 3.4.

- **Hydraulic expansion**

1. The hydraulic expansion shall be considered as per API RP 521 Section 3 Clause 3.14. Requirement of thermal relief valve for the process piping shall be evaluated based on pipe length and diameter.

7.19.3 Equipment safety relief valve capacity

- **Reciprocating compressors**

Compressor safety valves shall be sized for the maximum capacity of each stage at the recommended speed.

- **Positive displacement pumps:**

1. Safety valves on positive displacement pumps shall be sized for the maximum pumping capacity at recommended speed
2. Where the pump and equipment pumped into are designed for the stalling pressure of the pump under normal conditions, the safety valve shall be sized for 25%of the pump rated capacity under continuous operating conditions.

- **Sizing for protection of electric motors on positive displacement pumps:**

1. Safety valves for protecting electric motors shall be sized for the quantity that can be pumped at the relief valve setting without exceeding the horsepower rating (in kilowatts) recommended by

the motor manufacturer. Particular consideration shall be given to effects of viscosity on horsepower (in kilowatts) requirements.

- **Sizing for protection of gas turbines:**

1. Safety valves to protect the casing of gas turbines or expansion joints on turbines shall be sized for the quantity of effluent gases developed by the turbine at the safety valve relieving pressure. The quantity shall be checked with the equipment manufacturer. Consideration shall be given to pressure drop in effluent gas header before safety valve.

- **Sizing for protection of shell and tube heat exchangers:**

1. Safety valves shall be sized to protect shell and tube equipment for the pump capacity at the relieving pressure
2. Safety valves shall be sized for the wide open capacity of the control valve at its normal inlet pressure and control valve outlet pressure shall be equal to the relieving pressure of safety valve
3. Cold side blockage safety valves shall be sized based on the assumed heat input equal to the design exchanger duty. The temperature and composition of the relieving fluid and the required safety valve capacity will depend upon the location of the safety valve relative to the exchanger (e.g., a safety valve located on the exchanger will be relieving heated vapors; a safety valve in the piping 15 m upstream of the exchanger will be relieving cold fluid). Each situation must be individually examined.

- **Fired heaters**

Safety valves on fired heaters (if required) shall be sized on the basis that no less than 50%of the normal heat input shall be used due to residual heat input from the hot firebox. However, as cracking will often occur, each situation will require careful evaluation of the individual system conditions to determine the required relieving rate.

- **Vessels/towers**

Safety valves for towers and other types of pressure vessels shall be checked by each of the following methods and sized for the most severe condition:

1. Vessel blocked outlet (see 9.1.2.5.1 above)
2. Failure of condensing duty (see 9.1.2.5.5 above)
3. Failure of reflux (see 9.1.2.5.3 above)
4. Simultaneous failure of reflux and condensing duty (see 9.1.2.5.6 above)
5. Fire (see 9.1.2.7 above)
6. Electrical or electrical plus cooling water failure (this case may be covered through the above 2, 3, and 4).

7.19.4 Steam safety valve sizing

- **Safety valves on steam boilers:**

Refer to ASME, Section I, Power Boiler Code for sizing of these valves.

- **Safety valves on process Unit steam superheater coils:**

Refer to ASME, Section I, Power Boiler Code for sizing of these valves.

- **Safety valves downstream from pressure reducing and automatic control valves:**

The wide open capacity of the pressure reducing valve, make up valve, or automatic control valve, with normal operating pressure at the inlet of the valve and outlet pressure equal to the safety valve relieving pressure, is to be used for sizing safety valve downstream of automatic control valves.

- **Safety valves for protect: ion of turbine cases:**

1. Non-condensing turbines

The flowrate shall be based on the capacity of the turbine steam nozzles. Data from turbine manufacturer shall be obtained.

2. Condensing steam turbines

The flowrate shall be based on the water rate at a back pressure equal to the relieving pressure (water rate rises with increase of back pressure).

3. Vent exhaust steam header safety valves

When a safety valve is placed on the common exhaust steam header from all turbine cases, the steam consumption of all operating turbines and other operating steam-driven equipment exhausting into the header shall be used to size this valve.

7.20 Vacuum relief valve sizing

Vessels operating full of liquid shall be sized to provide protection against loss of liquid input depending on the rate of withdrawal.

Protection against loss of stripping steam is determined by whether the vessel is bare or insulated and the maximum possible condensation that may occur in a rain or snow storm.

Where a column or vessel during start-up, abnormal operating conditions, heating medium failure, etc., may possibly be subjected to vacuum, it should normally be designed to withstand vacuum conditions. Only when this is very costly or impractical, means for introducing fuel gas or inert gas at a sufficient rate may be considered upon approval of the Company. Steaming-out shall not be considered in this context.

7.21 Temperature safety valve sizing

Temperature safety valves are not sized. They are generally selected on the basis of the minimum size connection permitted on piping for the particular service. For cases where expansion relief is required around check valves, block valves, or pumps, sizes down to DN 15 (½ inch) may be acceptable provided that the applicable piping material service specification permits screwed connections for small sizes and is approved by the Company.

7.22 **Rupture disc sizing**

Rupture discs shall be sized using the manufacturer's charts.

Orifice formula may be used for sizing rupture discs using a coefficient discharge of 0.6 and picking pipe size with internal area equal to or greater than the calculated area.

7.23 **Emergency vapor depressuring systems**

Sizing of depressuring systems shall be based on the assumption that during the fire all input and output streams to and from the system have been ceased, while all internal heat sources have been shut off.

The pressure of the equipment shall be reduced to 700 kPa(ga) or 50% of the MAWP, whichever is lower, within 15 min while vapor is being generated at a rate corresponding to the following:

1. Vapor generated from liquid by heat input from the fire, plus:
2. Density change of the vapor in the equipment during pressure reduction, plus:
3. Liquid flashing during pressure reduction.

Vessels and columns need not be provided with emergency vapor depressuring facilities, if as the result of a fire, the pressure in the equipment will not rise to 700 kPa (ga) or 50% of the MAWP within 30 min.

For the purpose of sizing depressuring systems, each Unit area shall be divided into probable fire areas as outlined in (3) above. It shall be assumed that the fire is present throughout the depressuring period.

To determine the vapor quantities contributed by the factors mentioned above, it is necessary to establish the liquid inventory and vapor volume of the system. This shall include all facilities located in the fire area and all liquid and vapor in facilities outside the fire area that, under normal operating conditions, are in open conditions with the facilities located in the fire area.

Subject to corrections made necessary by the final plant design, the following assumptions may be made for estimating purposes:

1. The liquid inventory of fractionating columns can be estimated as the normal column bottom and draw off tray capacity plus normal tray liquid hold-up (total tray liquid hold-up shall be based on the pressure drop over the column)
2. The liquid inventory of accumulators, flash drums, knock-out vessels, and the like may be based on normal operating levels
3. For shell and tube heat exchangers, it may be assumed that 1/3 of the total shell volume is occupied by the tube bundle.

Depressuring valves shall be located near the vessel or vapor lines to be protected.

The valves shall be spring-loaded, pneumatic diaphragm operated, tight shut-off, spring to open, except valves for ethylene-producing Units, platformate, separators, and high-pressure separators (see 6.7 above), that shall spring to close.

The minimum size of the valves shall be DN 25 (1 inch).

Block valves shall be provided to isolate the depressuring valves.

Air for depressuring valves shall be supplied from the instrument air system. They shall not be provided with a handwheel.

7.24 Arrangement of safety relief valves

Safety devices shall be so arranged that their proper functioning is not hampered by the nature of the vessel's contents. If necessary, the use of protecting devices, such as rupture discs, swan-neck seals, or purge arrangements, is permissible, but these shall not interfere with the proper functioning of the safety device.

If a rupture disc is used in combination with a safety relief valve, a pressure gage preceded by a block valve shall be provided between the rupture disc and the safety relief valve.

The pressure drop in the connection between the equipment and the safety relief valve should not exceed 3% of the valve set pressure.

Safety relief valves discharging to the atmosphere shall, in general, be located at maximum practical elevation to economize on discharge piping, considering also ease of maintenance.

Safety relief valves connected to a closed relief system shall be located slightly above the relief header, if possible. For instance, valves protecting tall columns shall be put in a suitable position on the overhead vapor line.

If the valves must be below the header, the outlet lines leading to the header shall be steam traced from the safety relief valve to their highest point. However, agreement by the Company shall be made for lowering the valve in such a way. Branches should be connected to the headers so that the latter cannot drain back into them, even with the header full of liquid.

7.25 Location on vessels

Where a safety valve is provided for a vessel, the connection for the safety valve shall be provided on the vessel and not on the vapor line or discharge line from the vessel except as follows:

1. When access to the safety valve or discharge piping arrangement can be improved by locating the safety valve on vessel piping, such location is permissible providing the maximum pressure drop permitted between the vessel and safety valve inlet is not exceeded by the allowable figure per 10.1.2 above and provided the safety valve can be properly supported with full consideration to the reaction forces.

Safety valve nozzles shall be vertical when placed in the top head of vessels, except as follows:

1. In those cases where the vessel diameter is small enough to prohibit vertical nozzles, the safety valve nozzle should be attached at a 45° angle from the top head with a 45° ell so that the flange face is horizontal and above the nozzle.

If a safety valve inlet nozzle cannot be located in the top head of vessel, as outlined in above, due to constructions or limited working space, then the safety valve nozzle should be attached at a 45° angle from the vessel shell, with a 45° ell so that the flange face is horizontal and above the nozzle.

7.26 Location of safety valve nozzles to minimize turbulence

Where safety valves are installed to protect piping and are located downstream of control valves, pressure reducing stations, orifice plates, flow nozzles, or pipe fittings, such as elbows, which may cause turbulence, the safety valve shall be located at a sufficient distance downstream of these devices to avoid improper operation of the safety valves caused by whatever turbulence may be present.

The minimum number of straight pipe diameters that must be provided between the source of turbulence and the safety valve shall be according to API RP 520-Part II.

7.27 Location of safety valve nozzles to minimize pulsation

Where there are pressure fluctuations at the pressure source (discharge of reciprocating pumps or compressors) that peak close to the set pressure of the safety valve, it is beneficial to locate the safety valves farther from the pressure source in a more stable pressure region.

7.28 Inlet piping of safety relief valves

The piping from the vessel to the safety valve inlet shall be, as a minimum, the same size as the safety valve inlet connection with a pressure drop, including that in any block valve or fittings, of 3% of the set pressure or less.

1. Where this requirement cannot be met, means must be found to reduce the pressure drop by rounding the entrance connection, using block valves (when required) with full area ports or enlarging the inlet piping.

The inlet piping or nozzle to safety valves on vessels shall be kept as short and direct as possible, preferably with a single nozzle for each valve. Where more than one safety valve is involved, avoid mounting them on a common header tee type nozzle to prevent turbulence and excessive pressure drop.

7.29 Discharge piping of safety relief valves

In the design of the outlet piping, the effect of superimposed or build-up back pressure on the particular type of valve and its service shall be considered.

In addition to any specific requirements, the determination of discharging pressure and safety valves to the atmosphere or to a closed system shall be governed by the philosophy covered in API RP 520–Part I.

- **Permissible pressure drop in closed flare system**

1. The permissible pressure drop from the outlet of a safety valve to a flare system shall be as given for the standard safety valve or balanced bellows safety valve stated in 10.6.1.2 above

2. The pressure drop permissible shall be limited to a great extent by the lowest set safety valve in the system because this determines the pressure permissible at the point where this safety valve ties into the flare header

3. Further, from the permissible pressure drop, the pressure drop in any emergency cooler, and the pressure required at the Unit flare knock-out drum must be subtracted.

A minimum of 35 kPa(ga) shall be used for the pressure in the Unit flare knock-out drum. The type of safety valve used on the lowest set safety valve shall be such that permits this back pressure to exist.

- **Closed header design considerations**

1. A single maximum discharge occurrence for one safety valve or manual depressuring valve shall be used to determine the header size

2. The header shall be sized based on the maximum relief load, as outlined in 9.1 above

3. Consideration shall be given to utilizing balanced bellows safety valves where pressures available are limiting

4. Unless further restricted by job specifications, hot streams that flow into the closed above ground header shall be limited to 232 °C maximum temperature downstream of the safety valve. An emergency cooler shall be provided to cool to 232 °C all relief streams that exceed this maximum temperature, except:

    a. Relief stream where a temperature higher than 232 °C will be reached under fire conditions, and where the safety valve size is not set by fire conditions, and need not have an emergency cooler, unless it is required for other reasons. However, a cooler is required for such streams when the safety valve size is set by fire conditions

    b. Under the conditions permitted in "a" above, the piping and flare appurtenances shall be designed for the temperatures involved.

7.29.1 Vapor safety valve piping

- **Sizing**

Sizing of discharge piping for vapor safety valves depends on whether the valve is relieving to the atmosphere or into a closed relieving system, and whether a standard or bellows type safety valve is used.

- **Permissible pressure drop**

1. Conventional (standard) safety valve

The permissible pressure drop in discharge piping from the safety valve in vapor service is 10% of the set pressure under normal relief where 10% accumulation is used and 20% of the set pressure under emergency fire conditions where 20% accumulation is used.

2. Balanced bellows safety valve

The permissible pressure drop in discharge piping for balanced bellows safety valves in vapor service is 25–45%of the valve set pressure. For that particular valve, the manufacturer's curve shall be used.

- **Liquid relief valve piping**

1. Because the pressure drop in the outlet piping affects the set pressure of a conventional safety relief valve and also affects the capacity of both a conventional and a bellows relief valve, these impacts shall be evaluated when determining the size of the outlet piping
2. Most liquid relief valves have short discharge lines, such as relieving into suction of pump, where the discharge line relieves back to tankage or a tower that considerable pressure drop may be developed, its effect must be considered
3. The discharge piping from safety valves in liquid service shall be no smaller than the outlet connection on the safety valve and shall be increased where a pressure drop significantly affects the set pressure or capacity of the relief valve.

7.30 Block valves

1. When spare safety valves are required, in accordance with standards, block valves shall be provided as follows to permit removal of all of the valves being spared and the spare safety valve:
   a. Block valves shall be provided only on the inlet of the safety valve where each safety valve discharges independently to the atmosphere
   b. Block valves shall be provided on the inlet and outlet of a safety valve that discharges into a system operating under pressure or into a common discharge header
   c. Block valves, in accordance with the applicable piping material service specification, shall be provided in the pressure sensing piping to the pilot valves and in the inlet piping to the controlled relief valves when spare safety valves are required, as per Article 6.6.
2. Although, it is not permitted,except as noted in (a) above, provision of block valves and bypass line may be requested in the job specification for all safety relief valves provided for protection of the vessel(s) in order to make possible removal of the safety relief valve for maintenance purpose while the Unit/system is under normal operation. In such cases, due considerations shall be made to the requirements of safety features around the subject equipment/system. The inlet and outlet block valves shall be locked open in any case.

7.30.1 Block valves on inlet of safety valves

1. Block valves in accordance with the applicable inlet piping material service specifications shall be used
2. Block valves for the inlet of safety valves shall be, as a minimum, the same size as the safety valve inlet. They shall be increased in size where necessary to reduce the pressure drop in the inlet piping to that essential for proper operation of the safety valve.

7.30.2 Block valves on outlet of safety valves

1. Block valves in accordance with the Piping Material Service Specification for the service involved at the discharge may be used

2. Block valves for the inlet of safety valves shall be, as a minimum, the same size as the outlet of the safety valve. They shall be increased in size as necessary to reduce the pressure drop in the outlet piping to that essential for proper operation at the capacity required.

7.30.3 Locking methods

1. Suitable devices shall be provided on all block valves at the inlet or the outlet of safety valves that will allow locking of the valves by authorized persons in an open or closed position. The job specification will specify whether the devices are to be suitable for a padlock and chain or a car seal, depending on plant preference
2. A list of all safety valves with block valves that must be locked shall be furnished by the Contractor.

7.31 Discharge piping support

1. If guided laterally, a maximum of 3 m of piping may be supported by the welding neck flange and long radius elbow on safety valve outlet. If moved,then 3 m of piping is used or if high pressures are involved, mass and reaction forces should be supported as close to valve discharge as possible using a free support
2. Long piping should be anchored to the equipment as close to the valve discharge as possible, providing the expansion between the connection and anchor is taken care of and guided from that point up a tower, steel structure, or into the closed header piping
3. Stops and guides shall be provided to support piping independent of the safety valve when the safety valve is removed
4. Discharge piping shall be designed to prevent undesirable temperature and mechanical stresses being placed on the valve body and inlet connection
5. Supports shall be provided for discharge piping to compensate for the reaction force or load caused by discharging of the valve.

7.32 Position

1. All safety relief valves, except expansion relief valves, shall be mounted with the stem in a vertical position
2. Temperature safety valves shall be mounted in either the vertical or horizontal position to permit installation of discharge piping without pocketing.

7.33 Discharge piping of temperature safety valves
7.33.1 Water service

Temperature safety valves in water service may discharge to grade or to a sewer.

• **Hydrocarbon service**

1. Temperature safety valves in hydrocarbon service on process Units shall be piped to the nearest safe location, such as drain or sewer hub, with the end of the discharge pipe visible

2. Temperature safety valves located in National Electrical Code, Division two electrical classification areas shall have discharge piped to an open drain or sewer, with the end visible, outside the Division two area
3. Temperature safety valves in storage or product loading areas shall not discharge to the atmosphere but shall have discharge piped, using a spring loaded popped type check valve around block valves, check valves, pumps, etc., back to storage.

- **Chemical service**

1. The procedure for handling discharge from temperature safety valves in chemical service shall be specified in the job specification.

7.34 Venting and draining philosophy

Type of hydrocarbons to be vented to the atmosphere or closed drain flare system shall be according to the criteria set forth in API RP 520-Part I. Venting to a safe zone shall be in accordance with the specifications outlined herein below.

7.35 Vapor venting
7.35.1 Venting to the atmosphere

1. To determine safe zones for venting hydrocarbon vapor safety valves to the atmosphere, the following requirements shall be proceeded:
   a. Establish truncated cones on the highest platform of each vessel
   b. The top horizontal plane of the truncated portion shall be a minimum of 3.2 m above the platform. The lines forming the sides of the cone shall start at the circumference of a circle circumscribing the periphery of the platform and extend downward at an angle of 30° from the horizontal until it forms a base of the cone at a point 15 m above grade.
2. Hydrocarbon vapor safety valves may be vented to the atmosphere anywhere outside of the space encompassed by a single truncated cone or any similar intersecting truncated cones from other platforms, but not less than 15 m above grade
3. All flooring of platforms or walkways shall be a minimum of 3.2 m vertically below the top or sides of any one of the established truncated cones
4. Six millimeters drip holes shall be provided at safety valve discharge for drainage of liquids, unless otherwise prohibited by the operating personnel safety aspects (see 11.4.1(b) below).

7.35.2 Venting compressor safety valves

1. Safety valves for compressors shall discharge back to the suction except where back pressure limitation of the valve prevents its use

7.35.3 Venting steam safety valves

1. Steam safety valves may be vented to the atmosphere but the discharge must be run to a spot where there is no hazard to operating personnel
2. Discharge piping for steam safety valves in buildings shall be run outside the building and directed away from operating passageways or platforms.

7.36 Liquid venting
7.36.1 Pump relief valve

1. Relief valves on displacement type pumps shall discharge into the suction of the pump except where back pressure limitations of the valve prevent their application
2. Where the pump safety valve cannot be piped back to the suction line, the subject material shall be routed to the pump suction vessel or tank.

7.36.2 Venting acid or corrosive liquid

Special consideration shall be given to the location of safety valves discharging acid or corrosive liquids and also the destination of the discharge.

7.37 Safety valve bonnet venting
7.37.1 Conventional safety valves

In conventional safety valves, the bonnet shall be vented internally into the outlet line.

7.37.2 Balanced bellows valves

1. In balanced bellows safety valves, the bonnet shall be vented to the atmosphere or to a closed system
2. When it is not permissible to vent the bonnet to the atmosphere, the bonnet should be vented through an independent piping system to a safe location taking care not to impose a back pressure on the safety valve bonnet as a result of pressure drop in the bonnet vent line.

7.37.3 Pilot operated valves

1. The pilot may be vented to the atmosphere if the slight discharge during operation is acceptable. The quantity normally vented shall be checked with the manufacturer
2. When the pilot discharge to the atmosphere is not permissible, the pilot should be vented through an independent piping system to a safe location taking care not to impose back pressure in the pilot.

7.37.4 Sizing bonnet vent lines

The nominal size of the bonnet vent piping shall be as large or larger than the nominal size of the bonnet vent connection.

7.38 Safety valve draining

Drain provision shall be made for draining the low point of the safety valve outlet piping according to the following requirements for all condensable vapor services:

1. All safety valves discharging to the atmosphere shall be provided with a 6 mm drain hole in the bottom of the discharge line
2. In addition to the drain hole, as required in "1" above, one DN 15 (one inch) drain connection with bleeder shall be provided for all safety valves discharging to the atmosphere
3. Such drain holes and drain connections shall be piped where necessary to direct the drainage away from the operating platforms or operating areas and avoid draining on the vessel insulation.

7.39 Sizing for gas or vapor relief

The rate of flow through a relief valve nozzle is dependent on the absolute upstream pressure (as indicated in Eqn (7.1)–(7.3)) and is independent of the downstream pressure as long as the downstream pressure is less than the critical-flow pressure. However, when the downstream pressure increases above the critical-flow pressure, the flow through the relief valve is materially reduced (e.g., when the downstream pressure equals the upstream pressure, there is no flow).

The critical-flow pressure (PCF) may be estimated by the perfect gas relationship shown in Eqn (7.5). As a rule of thumb, if the downstream pressure at the relief valve is greater than one-half of the valve inlet pressure (both pressures in absolute units), then the relief valve nozzle will experience subcritical flow.

7.39.1 Critical flow

Safety valves in gas or vapor service may be sized by use of one of these equations:

$$A = \frac{131.6W\sqrt{T_1(Z)}}{(C_1)(K_d)(P_1)(K_b)(K_c)\sqrt{MW}} \tag{7.1}$$

$$A = \frac{5.875Q_v\sqrt{T_1(MW)(Z)}}{(C_1)(K_d)(P_1)(K_b)(K_c)} \tag{7.2}$$

$$C_1 = 520\sqrt{k\left(\frac{2}{k+1}\right)^{\frac{k+1}{k-1}}} \tag{7.3}$$

where:

A = required discharge area of the valve, cm^2. Use valve with the next larger standard orifice size/area
MW = molecular mass of gas or vapor

C_1 = coefficient determined by the ratio of specific heats of the gas or vapor at standard conditions

P_1 = upstream relieving pressure, kPa (abs). This is the set pressure plus the allowable overpressure plus the atmospheric pressure

W = flow, kg/h

T_1 = gas temperature, K, at the upstream pressure

Z = compressibility factor at flowing conditions

k = specific heat ratio, C_p/C_v

K_b = capacity correction factor due to back pressure

K_c = combination correction for rupture disk = 0.9 = 1.0 no rupture disk installed

K_d = coefficient of discharge, obtainable from the valve manufacturer.

Q_v = flow through valve, m³/h at standard conditions (101.325 kPa (abs), 0 °C)

C_1 and K_b can be obtained from Table 7.2, Table 7.3, Figure 7.4, and Figure 7.5. For final design, K should be obtained from the valve manufacturer. A value for K of 0.975 may be used for preliminary sizing.

Table 7.2 Values of Coefficient C_1 [1]

K	C_1
0.4	216.9274
0.5	238.8252
0.6	257.7858
0.7	274.5192
0.8	289.494
0.9	303.0392
1.0	315.37[a]
1.1	326.7473
1.2	337.2362
1.3	346.9764
1.4	356.0604
1.5	364.5641
1.6	372.5513
1.7	380.0755
1.8	387.1823
1.9	393.9112
2.0	400.2962
2.1	406.3669
2.2	412.1494

[a]*Interpolated values, because C₁ becomes indeterminate as k approaches 1.00.*

Table 7.3 Values of C_1 for Gases [1]

	Mol Mass	k	C_1		Mol Mass	k	C_1
Acetylene	26	1.28	345	Hydrochloric acid	36.5	1.40	356
Air	29	1.40	356	Hydrogen	2	1.40	356
Ammonia	17	1.33	351	Hydrogen sulfide	34	1.32	348
Argon	40	1.66	377	Iso-butane	58	1.11	328
Benzene	78	1.10	327	Methane	16	1.30	346
Carbon disulfide	76	1.21	338	Methyl alcohol	32	1.20	337
Carbon dioxide	44	1.28	345	Methyl chloride	50.5	1.20	337
Carbon monoxide	28	1.40	356	N-butane	58	1.11	328
Chlorine	71	1.36	352	Natural gas	19	1.27	345
Cyclohexane	84	1.08	324	Nitrogen	28	1.40	356
Ethane	30	1.22	339	Oxygen	32	1.40	356
Ethylene	28	1.20	337	Pentane	72	1.09	325
Helium	4	1.66	377	Propane	44	1.14	331
Hexane	86	1.08	324	Sulfur dioxide	64	1.26	342

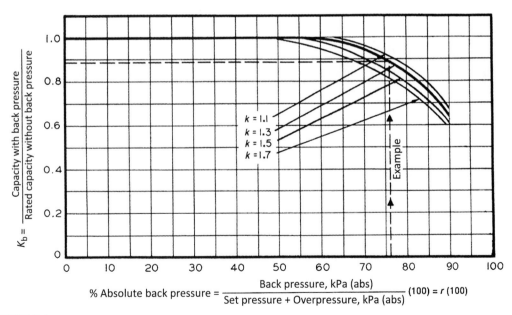

FIGURE 7.4

Constant back pressure sizing factor, K_b, for conventional safety-relief valves (vapors and gases only)

Ref. [1].

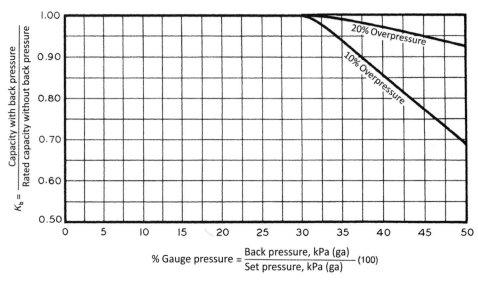

FIGURE 7.5

Variable or constant back-pressure sizing factor, K_b, for balanced bellows safety-relief valves (vapors and gases)

Ref. [1].

7.39.2 Subcritical flow

For downstream pressures, P_2, in excess of the PCF, the relief valve orifice area can be calculated from:

$$A = \frac{0.179W\sqrt{ZT_1}}{(F_2)(K_d)(K_c)\sqrt{MW(P_1)(P_1 - P_2)}} \qquad (7.4)$$

where:

A = required discharge area of the valve, cm². Use valve with the next larger standard orifice size/area

F_2 = coefficient for subcritical flow

k = specific heat ratio, C_p/C_v

K_c = combination correction for rupture disk = 0.9 = 1.0 no rupture disk installed

K_d = coefficient of discharge, obtainable from the valve manufacturer

MW = molecular mass of gas or vapor

P_1 = upstream relieving pressure, kPa (abs). This is the set pressure plus the allowable overpressure plus the atmospheric pressure

P_2 = downstream pressure at the valve outlet, kPa (abs)

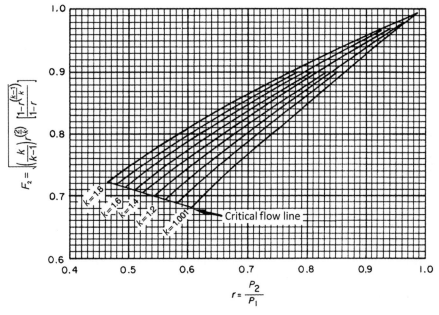

FIGURE 7.6

Values of F_2 for subcritical flow.

Ref. [1].

$T_1 =$ gas temperature, K, at the upstream pressure
$W =$ flow, kg/h
$Z =$ compressibility factor at flowing conditions

F_2 is taken from Figure 7.6.

$$P_{CF} = P_1 \left(\frac{2}{k+1}\right)^{\frac{k}{k-1}} \tag{7.5}$$

where:

$k =$ specific heat ratio, C_p/C_v
$P_1 =$ upstream relieving pressure, kPa (abs). This is the set pressure plus the allowable overpressure plus the atmospheric pressure
$P_{CF} =$ critical-flow pressure, kPa (abs)

Balanced pressure relief valves should be sized using Eqn (7.1) or Eqn (7.2) and the back pressure correction factor supplied by the valve manufacturer.

7.40 Sizing for liquid relief
7.40.1 Turbulent flow

Conventional and balanced bellows relief valves in liquid service may be sized by use of Eqn (7.6).

Pilotoperated relief valves should be used in liquid service only when the manufacturer has approved the specific application.

$$A = \frac{(7.07)(V_1)\sqrt{G}}{(K_d)(K_c)(K_w)(K_v)\sqrt{(P_1 - P_b)}} \tag{7.6}$$

where:

A = required discharge area of the valve, cm^2. Use valve with the next larger standard orifice size/area

G = relative density of gas referred to air = 1.00 at 15 °C and 101.325 kPa (abs); or, if liquid, the relative density of liquid at flowing temperature referred to water = 1.00 at 15 °C

K_c = combination correction for rupture disk = 0.9 = 1.0 no rupture disk installed

K_d = coefficient of discharge, obtainable from the valve manufacturer

K_v = capacity correction factor due to viscosity

MW = molecular mass of gas or vapor

P_1 = upstream relieving pressure, kPa (abs). This is the set pressure plus the allowable overpressure plus the atmospheric pressure

P_b = back pressure, kPa (ga)

K_w = capacity correction factor due to back pressure (Figure 7.7)

V_l = flow rate, liters/s at flowing temperature and pressure

7.40.2 Laminar flow

For liquid flow with Reynolds numbers less than 4000, the valve should be sized first with $K_v = 1$ in order to obtain a preliminary required discharge area, A. From manufacturer standard orifice sizes, the next larger orifice size, A', should be used in determining the Reynolds number, Re, from the following relationship:

$$Re = \frac{(V_1)(112654)(G)}{\mu\sqrt{A'}} \tag{7.7}$$

where:

A' = discharge area of the valve, cm^2, for valve with next standard size larger than required discharge area

G = relative density of gas referred to air = 1.00 at 15 °C and 101.325 kPa (abs); or, if liquid, the relative

Re = Reynolds number (dimensionless)

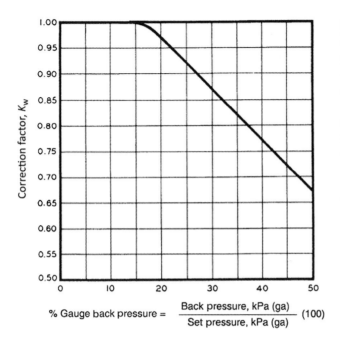

FIGURE 7.7

Variable or constant back-pressure sizing factor K_w for 25 percent overpressure on balanced bellows safety-relief valves (liquids only).

Note: The figure represents a compromise of the values recommended by a number of relief-valve manufacturers. This curve may be used when the make of the valve is not known. When the make is known, the manufacturer should be consulted for the correction factor.

Ref. [1].

$$\% \text{ Gauge back pressure} = \frac{\text{Back pressure, kPa (ga)}}{\text{Set pressure, kPa (ga)}} (100)$$

V_1 = flow rate, liters/s at flowing temperature and pressure
μ = viscosity at flowing temperature, mPa s

After the Reynolds number is determined, the factor K_v is obtained from Figure 7.8. Divide the preliminary area (A') by K_v to obtain an area corrected for viscosity. If the corrected area exceeds the standard orifice area chosen, repeat the procedure using the next larger standard orifice.

7.40.3 Sizing for mixed phase relief

When a safety relief valve must relieve a liquid and gas, it may be sized by:

- Determining the rate of gas and the rate of liquid that must be relieved
- Calculating the orifice area required to relieve the gas, as previously outlined
- Calculating the orifice area required to relieve the liquid, as previously outlined
- Summing total areas calculated for liquid and vapor to obtain the total required orifice area

When the material to be relieved through a safety valve is a flashing liquid, it is difficult to know what percentage of the flashing actually occurs in the valve in order to estimate the total required orifice area using the above four steps. A conservative approach is to assume an isenthalpic expansion and that all the vapor is formed in the valve. Then, using enthalpy values, calculate the amount of vapor formed

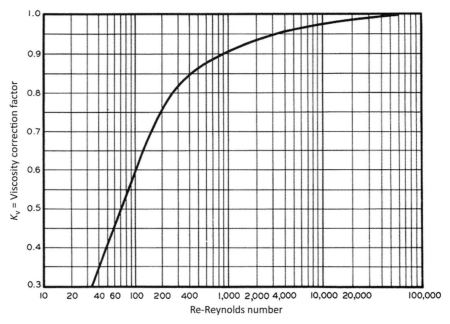

FIGURE 7.8

Capacity correction factor due to viscosity

Ref. [1].

using Eqn (7.8). Once the amounts of vapor and liquid assumed present in the valve are determined, the total required orifice area can becalculated.

$$\text{Weight Percent Flashing} = \left[\frac{h_{L1} - h_{L2}}{h_{G2} - h_{L2}} \right] (100) \tag{7.8}$$

where:

h_{G2} = enthalpy of vapor at downstream pressure, kJ/kg
h_{L1} = enthalpy of saturated liquid at upstream pressure, kJ/kg
h_{L2} = enthalpy of saturated liquid at downstream pressure, kJ/kg

7.41 Material and engineering for pressure and vacuum relief devices

Pressure and vacuum safety relief devices are normally used to terminate an abnormal internal or external rise in pressure above a predetermined design value in boilers, pressure vessels, and related piping and process equipment. Pressure relief valves or rupture discs may be used independently or in combination to provide the required protection against excessive overpressure. As used in this section, the term "pressure relief valves" includes safety valves, relief valves, and safety relief valves. Scope of this section is:

- This section covers the minimum requirements for design, material, fabrication, inspection, and shop test of pressure and vacuum safety relief devices
- Pressure and vacuum relieving devices include pressure relief valves, rupture disks, vacuum relief valves, and single units composed of both pressure and vacuum relief sides
- For installation, recommendations on pressure relieving devices, reference is made to Part two of API Recommended Practice RP-520

7.41.1 Selection of type

Conventional safety relief valve shall be provided:

1. When the built-up back pressure and variable superimposed back pressure does not exceed 10% of the set pressure
2. When constant superimposed back pressure exists

Balanced-bellows safety relief valve shall be provided:

1. When the built-up back pressure and variable superimposed back pressure does not exceed 50% of the set pressure if the back pressure is incorporated in the safety relief valve sizing
2. When the fluid-flow is a corrosive service (service in which the corrosion allowance is 6 mm or more or in which stainless steel or alloy material is used to prevent corrosion from occurring)

Pilot operated safety relief valve shall be used primarily in the following services:

1. When the built-up back pressure and variable superimposed back pressure does not exceed 50% of the set pressure
2. For equipment that operate between 90% and 95% of the set pressure
3. When the pressure drop to the inlet of the safety relief valve is greater than 3% of the set pressure (in this case, the remote pressure pickup type should be used)
4. Where the valve size is so large that a direct loaded safety relief valve would be unsuitable

All pressure relief valves (except thermal relief valves) with the inlet nozzle size of DN 25 (1 inch) and larger should be flanged, spring loaded, high lift, and high capacity type with a top guided disc. Pressure relief valves in services other than steam, hot water, and air should not be provided with a lifting device. Pressure relief valves with the inlet nozzle size under DN 25 may be of the screwed type connection.

All safety relief valves shall be provided with pressure tight bonnets except bellows type valves.

7.41.2 Dimensions

Center-to-face dimensions shall be in accordance with API Standard 526. Flange facings and dimensions shall be in accordance with ASME/ANSI B16.5.

Inlet pressure limits shall be governed by the inlet termination or the manufacturer's spring design limits, whichever is the smaller. Inlet flanges shall be capable of withstanding reaction forces due to the

Table 7.4 Standard Effective Orifice Areas and Letter Designations

Orifice Letter Designation	Effective Area (Square mm)
D	71
E	126.5
F	198
G	324.5
H	506.5
J	830.5
K	1186
L	1840.5
M	2322.5
N	2800
P	4116
Q	7129
R	10,322.5
T	16,774

valve discharge in addition to the internal pressure and are therefore generally suitable for pressure and temperature lower than the ANSI ratings. Outlet pressure limits shall be determined by the valve design.

7.41.3 Determination of orifice area

The required orifice area shall be determined in accordance with API Recommended Practice 520 Part1.

The standard effective orifice areas and the corresponding letter designations are listed in Table 7.4. Springs are usually designed and fabricated to the manufacturers own standards.

7.42 Design of rupture disks

Rupture disks are recommended in the following cases:

In services where the operation of a pressure relief valve may be affected by corrosion or corrosion products, or by the deposition of material that may prevent the valve from lifting in service. Rupture disks can be used instead of or in conjunction with a pressure relief valve.

With highly toxic or other materials where leakage through a pressure relief valve cannot be tolerated, the rupture disk should be upstream of any pressure relief valve.

For low positive set pressure, where pressure relief valves tend to leak, a rupture disk can be used in place of the valve.

For the relief of a pressure rise that is too fast for conventional pressure relief valves.

Calculation of relief area shall be in accordance with API Recommended Practice 520 Part I.

Protecting a vessel from overpressure that results from an internal explosion is a function of the vapor volume of the vessel, the area of the rupture disk, and the allowable pressure rise.

Because no generally accepted method exists by which a calculation can be made, selecting the best means of protecting a vessel against an internal explosion depends solely on the designer's judgment. The following are two possible design approaches:

By applying a suitable safety factor to the normal operating pressure, the vessel can be designed to withstand an internal explosion. However, this may become impractical when the safety factor is considered.

By applying a rupture disk based on the contained tank volume, the vessel can be protected. The basis for the sizing of the rupture disk must then be arbitrary. For cases involving an explosion pressure rise factor of 8, this choice varies from 3.3 to 6.6 m^2 of relief area per 100 m^3 of vapor volume. For normal refinery applications, a general rule of 6.6 m^2 per 100 m^3 will provide an adequate area to protect against the explosion of an air hydrocarbon mixture. Until more work is done in this area, an authoritative guideline cannot be recommended.

7.43 Material

The material of construction shall be compatible with the process fluid and the adjoining components and the environment in which the relief devices is to be used. Material for sour service shall be in accordance with NACE MR-01-75 [2]. Proposed material for pressure and vacuum relieving devices shall be approved by the Company. For pressure and vacuum relieving valves, the following materials are proposed.

7.43.1 Body and its relative parts

• **Bodies and bonnets**

Bodies and bonnets, or yokes, of pressure relief valves shall be manufactured from either:

1. Cast or forged material, listed in Table 7.5, or equivalent grades of plate or bar; or,
2. Materials other than those listed, providing they comply with a standard or specification that ensures control of chemical and physical properties and quality, appropriate to the end use

7.43.2 Disk, nozzle, and body seat ring

Material for these components shall be capable of withstanding the corrosive and erosive effects of the particular service conditions. If a resilient insert is used the material and design shall be such that it will not become distorted under operating conditions or adhere to the body seat/disk so as to change the discharge or operating characteristics of the valve. Cast iron shall not be used.

7.43.3 Materials intended to be welded

The chemical composition, by ladle analysis, of carbon, carbon manganese, and carbon-molybdenum steels intended to be welded shall have a maximum carbon content of 0.25%.

Table 7.5 Materials for Pressure Containing Components

Material	Comparable American Society for Testing and Materials (ASTM)
Castings gray iron	A126: Class B
Copper alloy	-
Carbon steel	A216: WCB
1¼% chromium, ½% molybdenum	A217: WC6
2¼% chromium, 1% molybdenum	A217: WC9
Austenitic chromium nickel	A351: Grade CF8
Austenitic chromium nickel 2½% molybdenum	A351: Grade CF8M
Carbon steel	A352: LCB
3½% nickel	A352: LC3
Forgings carbon steel	A105
Austenitic chromium nickel	A182: Grade F304
Austenitic chromium nickel molybdenum	A182: Grade F316

7.43.4 Carbon steel for sub-zero service

Steels for sub-zero service shall comply with the impact and other requirements as specified by the Company.

7.43.5 Pressure and/or temperature limitations

The following limitations shall apply:

1. Cast iron

When cast iron is used for bodies, bonnets, or caps, the set pressure shall not exceed 13 bar gage nor the design temperature exceed 220 °C or be below 0 °C.
   Note:
   Cast iron shall not be used for service with hydrocarbon vapors or for other flammable or toxic materials or in an area where a fire risk exists.

2. Carbon, low, and high alloy steels

The minimum and maximum temperature limitations shall be in accordance with the relevant material/ application standard.

3. Copper alloy

Copper alloy pressure containing parts shall not be used in locations or service where a fire risk exists, due to the relatively low melting point of the alloy.

- **Internal Parts**

The material of the guiding surfaces shall be compatible with the service conditions and shall be selected to reduce the possibility of galling or seizure.

- **Bolting**

Bolting for pressure-containing joints shall be in accordance with American society for testing and materials (ASTM) A-193 and A-194 or the Purchaser's specification sheet.

7.44 Inspection and shop tests
7.44.1 Inspection

Each pressure and vacuum relieving device shall be checked against the manufacturer's documents to see that it conforms to the design requirements.

Complete data of all parts shall be checked by the inspector prior to any test. Valve parts may be required to be dismantled for this purpose at the discretion of the inspector.

7.44.2 Shop tests

All pressure-measuring devices fitted to test equipment shall be tested and calibrated to ensure the required accuracy during testing.

Each pressure relief valve shall consecutively be subjected to the following tests, as minimum.

Unspecified details shall be in accordance with the manufacturer's test procedures, approved by the Company.

Body and other pressure parts subject to inlet pressure (primary pressure zone) shall be hydro-statically tested at a pressure of at least 1.5 times the design pressure of the parts.

A test shall be applied to the discharge side of those pressure relief valves fitted with bellows to test the pressure tightness of the bellows and its joints. The bonnet vent that shall be open shall have a soapy water film placed across it and there shall be no visible leakage. The test shall be carried out using air or nitrogen at a pressure not less than the maximum specified back pressure. The duration of the test shall be as for the seat tightness test.

For closed bonnet pressure relief valves, secondary pressure zone (parts subjected to outlet or discharge pressure) shall be tested according to the applicable clause of ASME Section I or ASME Section VIII.

Each pressure relief valve shall be tested to demonstrate its popping or set pressure in accordance with the applicable clause of ASME Section I or ASME Section VIII.

Test medium for valves on gas or vapor services shall be air, unless otherwise specified by the Company.

The set pressure tolerances shall be in accordance with ASME, as appropriate.

Blow-down shall be adjusted according to ASME Section I, where applicable.

Each valve shall be subjected to a seat tightness test, according to API Standard 527.

One sample from each group of rupture disks of the same size and material should be tested in a holder of the same form and dimensions as that with which the disk is to be used.

The disk shall burst within +5% of its specified bursting pressure.

Vacuum relief valves (such as vacuum breakers) shall be tested in accordance with the manufacturer's standard procedure. The following tests shall be carried out as minimum:

1. Pressure test of body
2. Vacuum (and pressure) setting test(s)
3. Sealing test to check blow down for vacuum (and pressure) side(s)

7.45 Marking, documentation, and preparation for shipment
7.45.1 Body marking

Each pressure relief valve shall bear legible and permanent marking on the body, as follows:

1. The inlet nominal size
2. The material designation of the body
3. The manufacturer's name and/or trade mark
4. An arrow showing the direction of flow, where the inlet and outlet connections have the same dimensions or the same nominal pressure rating
5. Ring joint number (where applicable) to be marked on the flange

7.45.2 Nameplate

The following information shall be marked on a stainless steel nameplate:

These tests shall be conducted after all machining operations on the parts have been completed and prior to any painting that may be applied.

There shall be no visible sign of leakage.

1. Limiting operating temperature for which the valve has been designed, where applicable
2. Set pressure
3. Back pressure
4. Cold differential test pressure
5. Nominal size (inlet by outlet)
6. Certified discharge capacity
7. Flow area in square millimeters
8. Manufacturer's type reference
9. Tag number
10. Serial number
11. Pressure range of spring or manufacturer's serial number
12. If required, the phrase "sour service" is stamped
13. Any other information required by the Company

The nameplate shall be permanently attached to the body or bonnet of the valve.

Rupture disk nameplate shall bear the following information:

1. The size, net inside diameter of opening leading to the flange or holding arrangement for the disk in mm
2. Material of construction

3. Temperature for continuous operation and at burst pressure
4. Relief capacity
5. Rupture pressure
6. Test pressure
7. Manufacturer's name or identifying trademark
8. Any other information required by the company

For pressure and/or vacuum relief valves, marking requirements shall be in accordance with API Standard 2000.

7.45.3 Documentation

The following documents (including records) shall be submitted for each item, along with proposal, before/after placing order, as appropriate:

1. Along with proposal:
   a. completed data sheet
   b. calculation sheet
   c. manufacturer's catalog
   d. overall dimensional drawing and list of materials
   e. overall weight of item, weight of pallets for pressure, and/or vacuum relief valves,if any
   f. recommended spare parts list
   g. list of special tools for maintenance and testing
2. After placing order:
   a. material quality certificates, as required by purchase documents
   b. shop test certificates
   c. records of inspection, any repairs, and corrections
   d. certificate of conformance
   e. any other document and instruction required for parts ordering, site handling and storage, site inspection and tests, maintenance and repair of all parts, and any other document

7.45.4 Preparation for shipment

After test and inspection, all exterior surfaces, except flange facings, shall be painted. Corrosion resistant materials need not be painted.

Unless otherwise specified, safety relief valves shall be painted with the following color identification:

1. Conventional valves: canary yellow
2. Bellows type valves: body canary yellow; bonnet-red

Machined or threaded exterior surfaces shall be protected from corrosion during shipment and subsequent storage by coating with rust preventive.

Inlet and outlet flanges shall be protected to prevent damage from or entrance of foreign material during shipment.

Threaded openings shall be plugged with suitable protective devices. Temporary plugs should be readily distinguishable from permanent metal plugs.

Valves shall be so packaged as to minimize the possibility of damage during transit or storage. Instructions shall be provided for removing devices used for temporary protection.

7.46 General specification for springs of pressure relief valves

Under normal operating conditions, springs used in pressure relief valves shall be of helical coil design made of material complying with one of the specifications given in Table 7.6.

Table 7.6 Spring Materials

Material		Recommended Limiting Section (Diameter)	Recommended Temperature Limit at the Spring	
Type	Specification		Min.	Max.
Carbon steels	BS 2803 094A65 HS	Up to 12.5 mm	−20	+150
	BS 5216 HS2 and HS3	Up to 13 mm	−20	+130
	BS 970: Part 5 070A72	10−20 mm	−20	+150
	BS 970: Part 5 070A78	10−20 mm	−20	+150
Alloy steels	BS 970: Part 4302 S25	Above 10 up to 30 mm	−200	+250
	BS 970: Part 4316 S16	Above 10 up to 30 mm	−200	+250
	BS 970: Part 5 250A58	10−40 mm	−20	+150
	BS 970: Part 5 250A61	10−40 mm	−20	+150
	BS 970: Part 5 527A60	10−40 mm	−20	+150
	BS 970: Part 5 735A50	10−80 mm	−20	+175
	BS 970: Part 5 805A60	10−80 mm	−20	+150
	BS 2056 302S26	Up to 10 mm	−200	+250
	316S42	Up to 10 mm	−200	+250
	BS 2803 685A55 HS	Up to 12.5 mm	−20	+175
	735A50 HS	Up to 12.5 mm	−20	+175
	BS 4659 BH12	Up to 50 mm	−20	+370
	BH13	Up to 50 mm	−20	+370
	BH21	Up to 50 mm	−20	+370
	British Standard 2S143	-	-	-
	British Standard 2S144	-	-	-
Alloy steels	British Standard 2S145	Up to 50 mm	−90	+350
	DTD 5086:1969§	Up to 10 mm	−90	+300
	ASTM A638:1982 Grade 660	Up to 80 mm	−200	+400
Non-ferrous	BS 3075 NA13	Up to 10 mm	−40	+200
	NA14	Up to 10 mm	−200	+370
	N18	Up to 8 mm	−200	+230
	British Standard 2HR 501	Up to 8 mm	−200	+540
	British Standard 2HR 502	Up to 8 mm	−200	+540
	SAEAMS 5698C: 1953 (Reaffirmed 1977)	Up to 12.5 mm	-	+540

Where operating conditions require an alternative material, this should be agreed between the manufacturer and the Purchaser.

The allowable stresses shall be based on previous satisfactory experience and the current understanding of the behavior of spring materials taking into consideration the temperature of the spring, the environment and the amount of relaxation that is permissible in service.

The material selected shall comply with the limitations on temperature range given in Table 7.6 and shall have corrosion resistant properties for the duty specified. The material shall be of circular section.

The use of protective coatings is not covered by this section and, if they are necessary, their use should be agreed between the manufacturer and the Purchaser.

7.46.1 Dimensions

- **Proportion**

The proportion of the unloaded length to the external diameter of the spring shall not exceed four to one.

- **Spring index**

The spring index (i.e., the mean diameter of the coil, D, divided by the diameter of the section, d) shall be within the range 3–12.

- **Spacing of coils**

The spacing of the coils shall be such that when the valve head is at the lift corresponding to its certified discharge capacity, the space between coils shall not be less than 1 mm. The total of these clearances for the spring as a whole shall not be less than 20% of the deflection of the spring from the free length to the solid length.

- **Spring plates/buttons**

The spring plates/buttons shall have a locating spigot length (excluding any radius or chamber) of at least 0.75 of the spring wire/bar diameter. The maximum clearance between the outside diameter of the locating spigot and the inside diameter of the spring shall not exceed 0.7 mm and shall rotate freely (Figure 7.9).

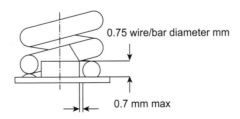

0.75 wire/bar diameter mm

0.7 mm max

FIGURE 7.9

Spring plate.

7.46.2 Stress

The corrected shear stress, q, shall be determined from the following equation:

$$q = \frac{8WDKA}{\pi D^3}$$

$$A = \frac{\delta_1 + \delta_2}{\delta_1}$$

where:

$q =$ is the corrected shear stress (N/mm^2)
$W =$ is the force at set pressure (N)
$D =$ is the mean diameter of the coil (mm)
$d =$ is the diameter of the section (mm)
$K =$ is the stress correction factor for curvature (see Table 7.7)
$\delta_1 =$ is the axial deflection due to force W (mm)
$\delta_2 =$ is the lift (in mm) of the valve, at certified discharge capacity

7.46.3 Number of working coils

The number of working or free coils in a spring, n, shall be determined from the following equation:

$$n = \frac{d^4 G \delta_1}{8 D^3 W}$$

where:

$G =$ is the shear modulus (N/mm^2)
$n =$ is the number of working coils
Other symbols used in the equation are defined in the above sections

- **Handing of coils**

Where springs are nested, the adjacent springs shall be opposite handed. Single springs may be coiled either right hand or left hand.

7.47 Testing and dimensional checks

7.47.1 Permanent set

The permanent set of the spring (defined as the difference between the free length and the length measured 10 min after the spring has been compressed solid three additional times at room temperature) shall not exceed 0.5% of the free length.

Table 7.7 Stress Correction Factor for Curvature (K)

D/d	K	D/d	K	D/d	K
3	1.6	6.1	1.235	9.1	1.148
3.1	1.571	6.2	1.231	9.2	1.146
3.2	1.545	6.3	1.226	9.3	1.145
3.3	1.522	6.4	1.222	9.4	1.143
3.4	1.500	6.5	1.218	9.5	1.141
3.5	1.480	6.6	1.214	9.6	1.140
3.6	1.462	6.7	1.211	9.7	1.138
3.7	1.444	6.8	1.207	9.8	1.136
3.8	1.429	6.9	1.203	9.9	1.135
3.9	1.414	7	1.200	10	1.133
4	1.400	7.1	1.197	10.1	1.132
4.1	1.387	7.2	1.194	10.2	1.130
4.2	1.375	7.3	1.19	10.3	1.129
4.3	1.364	7.4	1.188	10.4	1.128
4.4	1.353	7.5	1.185	10.5	1.126
4.5	1.343	7.6	1.182	10.6	1.125
4.6	1.333	7.7	1.179	10.7	1.124
4.7	1.324	7.8	1.176	10.8	1.122
4.8	1.316	7.9	1.174	10.9	1.121
4.9	1.308	8	1.171	11	1.120
5	1.3	8.1	1.169	11.1	1.119
5.1	1.293	8.2	1.167	11.2	1.118
5.2	1.286	8.3	1.164	11.3	1.117
5.3	1.279	8.4	1.162	11.4	1.115
5.4	1.273	8.5	1.160	11.5	1.114
5.5	1.267	8.6	1.158	11.6	1.113
5.6	1.261	8.7	1.156	11.7	1.112
5.7	1.255	8.8	1.154	11.8	1.111
5.8	1.25	8.9	1.152	11.9	1.110
5.9	1.245	9	1.150	12	1.109
6	1.24				

7.47.2 Dimensional checks

Following the above test, each spring shall then be subjected to the following minimum checks.

1. Measurements corresponding to:
   a. load/length at 15% of the calculated total deflection of the spring
   b. load/length at the maximum compression at which the spring will be used, or the spring rate over a given range above 15% of the calculated total deflection

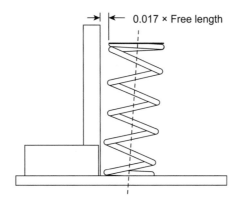

FIGURE 7.10

End-squareness.

2. Dimensional check of coil diameter and free length
3. Dimensional check for end-squareness, by standing the spring on a surface plate against a square and measuring the maximum deviation between the top end coil and the square. This shall be repeated with the spring reversed end for end (see Figure 7.10)
4. Dimensional check for end parallelism, by standing the spring on a surface plate and measuring the difference between the levels of the highest and lowest points of the surface of the upper ground end. These measurements shall be repeated with the spring reversed end for end (see Figures 7.10 and 7.11)

Where maximum hardness is specified, hardness tests shall be carried out for each spring.

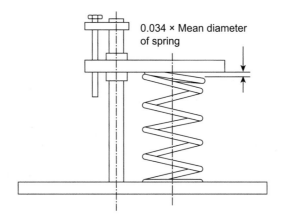

FIGURE 7.11

End-parallelism.

Table 7.8 Load/Length Tolerances

Number of Working Coils	Load Tolerances %	
	Wire/Bar Diameter up to and Including 10 mm	**Wire/Bar Diameter above 10 mm**
Less than 4	±7.5	+7.5 −5
4 to 10 inclusive	±5	+6 −4
More than 10	±3	+6 −4

7.47.3 Tolerances

- **Load/length**

The tolerances on load/length requirements shall be as given in Table 7.8.

7.47.4 Spring rate

The tolerances on spring rate shall be as given in Table 7.9.

7.47.5 Coil diameters

Assembly and machined part considerations determine whether the inside or outside diameter of the spring is critical and the spring specification shall indicate the limits on the spring diameters within which the spring shall be supplied. Inside diameter tolerances are given in Table 7.10.

7.47.6 Free length

Prior to determination of free length, springs shall, at room temperature, be compressed to the nominal free length less 85% of the average total deflection. After a 10 min wait, in an unloaded condition, the free length shall then be determined by placing a straight-edge across the top of the spring and

Table 7.9 Spring Rate Tolerances

Number of Working Coils	Spring Rate Tolerances %	
	Wire/Bar Diameter up to and Including 10 mm	**Wire/Bar Diameter above 10 mm**
Less than 4	±3.0	+8.5 −6.0
4 or More	±3.0	+7.5 −5.0

Table 7.10 Inside Diameter Tolerances

Inside Diameters, mm	Tolerance, mm
25 diameter and smaller	+0.8 −0.0
Over 25–50	+1.5 −0.0
Over 50–100	+2.5 −0.0
Over 100	+3.0 −0.0

measuring the perpendicular distance from the plate on which the spring stands to the bottom of the straight-edge at the approximate center of the spring. The measured free length shall be within the tolerance given in Table 7.11.

- **End-squareness**

The maximum deviation from end-squareness shall not exceed $0.017 \times$ free length.

- **End-parallelism**

The maximum deviation from end-parallelism shall not exceed $0.034 \times$ the mean diameter of the spring.

- **Condition of Material**

Bars or wires used in the unmachined condition shall be limited to the following:

1. Surface defects: 1% of bar diameter or 0.25 mm, whichever is the greater
2. Decarburization: 2% of bar diameter or 0.30 mm, whichever is the greater. No more than 1/3 of the total affected depth shall be complete decarburization.
3. Machined bars shall be free from all surface defects, there shall be no complete decarburization and partial decarburization shall not exceed 0.13 mm in depth.

Table 7.11 Free Length Tolerances

Nominal Free Length, mm	Tolerance, mm
Up to 75	±0.8
Over 75–165	±1.5
Over 165–250	±2.5
Over 250–360	±3
Over 360–560	±4

- **Marking**

Identification marks that involve stamping or etching shall be confined to the inactive coils and be located between the tip and 180° of the circumference from the tip.

- **Spring test certificate**

When required, the pressure relief valve manufacturer shall request a test certificate stating that the spring(s) has been made from the specified material and has been tested in accordance with Standards. This certificate shall be submitted to the Company along with other documents.

References

[1] Gas Processors and Suppliers Association (GPSA). Engineering databook. 12th ed. 2004. Tulsa (OK, USA).
[2] NACE (National Association of Corrosion Engineers) MR-01–75. Sulfide stress cracking resistant metallic materials for oil field equipment.

Further reading

ANSI (American National Standard Institute) ANSI/ASME. Chemical plant and petroleum refinery piping. B.31.3; 1990.
ANSI B31.3—Chemical plant and petroleum refinery piping. The American Society of Mechanical Engineers (ASME), 345 East 47th Street, New York, NY 10017.
ANSI B31.8—Gas transmission and distribution piping. The American Society of Mechanical Engineers (ASME), 345 East 47th Street, New York, NY 10017.
API (American Petroleum Institute) RP 520. Recommended practice for the design and installation of pressure relieving systems in refineries. 5th ed. Part I–Sizing and Selection; July 1990.
API RP 521—Guide for pressure-relieving and depressuring systems. American Petroleum Institute, 1220 L Street, NW, Washington, DC 20005.
API RP520—Recommended practice for the design (Part I) and installation (Part II) of pressure relieving systems in Refineries. American Petroleum Institute, 1220 L Street, NW, Washington, DC 20005.
API Standard 527—Commercial seat tightness of safety relief valves with metal-to-metal seats. American Petroleum Institute, 1220 L Street, NW, Washington, DC 20005.
API2510—Design and Construction of LPG Installations at Marine and Pipeline Terminals, natural-Gas-Processing plants, Refineries, Petrochemicalplants, and tank Farms. American Petroleum Institute, 1220 L Street, NW, Washington, DC 20005.
ASME (American Society of Mechanical Engineers) ASME boiler and pressure vessel code section I: "Power boilers" section VIII: "Pressure vessels" Section VIII: "UG 134".
ASME boiler and pressure vessel code, Section I and Section VIII, American Society of Mechanical Engineers (ASME), New York (NY).
ASTM (American Society for Testing and Materials) ASTM. Materials Standard.
Brzustowski TA. Flaring in the energy industry. In: Process energy combustion science, vol. 2. Great Britain: Pergamon Press; 1976. pp. 129–144.
BSI (British Standard Institution) BS. Materials Standard.
Cristea NN, Smith D. Making relief load estimates match reality. In: 28th Center for chemical process safety International Conference 2013, CCPS—Topical Conference at the 2013 AIChE Spring Meeting and 9th Global Congress on Process Safety 2; 2013. pp. 645–55.

Dempster W, Elmayyah W. Two phase discharge flow prediction in safety valves. Int J Pressure Vessels Piping 2013;110:61–5.

Hagey E. The PSV that did not fail-misconceptions about PSVs. Process Saf Prog 2013;32(1):84–9.

Khan MS. Take a quicker approach to staggered blowdown. Hydrocarbon Process 2013;(4):92.

Melhem GA. Estimate vibration risk for relief and process piping. In: 28th center for chemical process safety international conference 2013, CCPS—Topical Conference at the 2013 AIChE Spring Meeting and 9th Global Congress on Process Safety 2; 2013. pp. 533–59.

Min TC, Fauske HK, Patrick M. Ind Eng Chem Fundam; 1966:50–1.

NFPA30, 58, 59& 59A, National fire Protection Association (NFPA), Batterymarch Park, Quincy, Massachusetts 02269.

Oil Insurance Association. Recommendations and guidelines—Gasoline plants. Pamphlet August 1971;301:175. West Jackson Blvd., Chicago, Illinois 60604.

Overa Sverre J, Strange Ellen, Salater Per. Determination of temperures and flare rates during depressurization and fire. San Antonio (TX): GPA Convention; 15–17 March, 1993.

Schwartz Robert E, White Jeff W. Predict radiation from flares. Chem Eng Prog July 1997;93:42–9.

Smith D. An engineering method to mitigate the impact of regulatory focus on relief system installations by prioritizing risk. In: 28th Center for chemical process safety International Conference 2013, CCPS—Topical Conference at the 2013 AIChE Spring Meeting and 9th Global Congress on Process Safety 2; 2013. pp. 637–44.

Straitz III JF. Nomograms "Determining proper flame tip diameter and height". Tulsa (OK): Oil Gas and Petroleum Equipment; July and August, 1979.

Van Boskirk BA. Sensitivity of relief valves to inlet and outlet line lengths. Chem Eng; August 23, 1982:77–82.

Sizing of Valve and Control Valve

8

Valves are the components in a fluid flow or pressure system that regulates either the flow or the pressure of the fluid. This duty may involve stopping and starting flow, controlling flow rate, diverting flow, preventing back flow, controlling pressure, or relieving pressure.

The equations of this section are used to predict the flow rate of a fluid through a valve when all the factors, including those related to the fluid and its flowing condition are known, when the equations are used to select a valve size, it is often necessary to use capacity factors associated with the fully open or rate condition to predict an approximate required valve flow coefficient (C_v).

The valves discussed here are manually operated valves for stop and starting flow, controlling flow rate, and diverting flow. The manual valves are divided into four groups according to the way the closure member moves into the seat. The many types of check valves are likewise divided into groups according to the way the closure member moves onto the seat. The basic duty of these valves is to prevent back flow. Predicting the flow of compressible and incompressible fluids through control valve, and cavitation are covered as parts of this Engineering Standard Specification.

This Engineering chapter is intended to cover minimum process requirements for manual valves, and control valves, as well as field of application, selection of types, design considerations (e.g., cavitations), and control valve sizing calculations.

8.1 Manual valves

Manual valves serve three major functions in fluid handling systems:

1. stopping and starting flow;
2. controlling flow rate;
3. diverting flow.

8.1.1 Grouping of valves by method of flow regulation

Manual valves may be grouped according to the way the closure member moves onto the seat. Four groups of valves are thereby distinguishable:

- **Closing-down valves**

A stopper-like closure member is moved to and from the seat in direction of the seat axis.

- **Slide valves**

A gate-like closure member is moved across the flow passage.

Natural Gas Processing. http://dx.doi.org/10.1016/B978-0-08-099971-5.00008-8
Copyright © 2014 Elsevier Inc. All rights reserved.

- **Rotary valves**

A plug or disc-like closure member is rotated within the flow passage, around an axis normal to the flow stream.

- **Flex-body valves**

The closure member flexes the valve body.

8.1.2 Valve guides

The main parameters concerned in selecting a valve or valves for a typical general service are:

1. Fluid to be handled

This will affect both type of valve and material choice for valve construction.

2. Functional requirements

Mainly affecting choice of valve.

3. Operating conditions

Affecting both choice of valve type and constructional materials.

4. Flow characteristics and frictional loss

Where not already covered by (b), or setting additional specific or desirable requirements.

5. Size of valve

This again can affect choice of type of valve (very large sizes are only available in a limited range of types); and availability (matching sizes may not be available as standard production in a particular type).

6. Any special requirements

In the case of specific services, choice of valve type may be somewhat simplified by following established practice or selecting from valves specifically produced for that particular service.

Table 8.1 summarizes the applications of the main types of general purpose valves.

Table 8.2 carries general selection a stage further in listing valve types normally used for specific services.

Table 8.3 is a particularly useful expansion of the same theme relating the suitability of different valve types to specific functional requirements.

8.1.3 Selection of valves

1. Valves for stopping and starting flow

Such valves are slide valves, rotary valves, and flex-body valves.

2. Valves for control of flow rate

Table 8.1 Applications of Valve Types

Valve Category	General Application(s)	Actuation	Remarks
Screw-down stop valve	Shut-off or regulation of flow of liquids and gases (e.g., steam)	1. Handwheel. 2. Electric motor. 3. Pneumatic actuator. 4. Hydraulic actuator. 5. Air motor.	1. Limited applications for low pressure/low volume systems because of relatively high cost. 2. Limited suitability for handling viscous or contaminated fluids.
Cock	Low pressure service on clean, cold fluids (e.g., water, oils, etc.).	Usually manual.	Limited application for steam services.
Check valve	Providing flow in one direction.	Automatic.	1. Swing check valves used in larger pipelines. 2. Lift check valves used in smaller pipelines and in high pressure systems.
Gate valve	Normally used either fully open or fully closed for on-off regulation on water, oil, gas, steam, and other fluid services.	1. Hand wheel. 2. Electric motor. 3. Pneumatic actuator. 4. Hydraulic actuator. 5. Air motor.	1. Not recommended for use as throttling valves. 2. Solid wedge gate is free from chatter and jamming.
Parallel slide valve	Regulation of flow, particularly in main services in process industries and steam power plant.	—	1. Offers unrestricted bore at full opening. 2. Can incorporate venture bore to reduce operating torque.
Butterfly valve	Shut-off and regulation in large pipelines in waterworks, process industries, petrochemical industries, hydroelectric power stations, and thermal power stations.	1. Handwheel. 2. Electric motor. 3. Pneumatic actuator. 4. Hydraulic actuator. 5. Air motor.	1. Relatively simple construction. 2. Readily produced in very large sizes (e.g., up to 5.5 m (18 ft) or more).
Diaphragm valve	Wide range of applications in all services for flow regulation.	1. Handwheel. 2. Electric motor. 3. Pneumatic actuator. 4. Hydraulic actuator. 5. Air motor.	1. Can handle all types of fluids, including slurries, sludges, etc., and contaminated fluids. 2. Limited for steam services by temperature and pressure rating of diaphragm.

Continued

Table 8.1 Applications of Valve Types—cont'd

Valve Category	General Application(s)	Actuation	Remarks
Ball valve	Wide range of applications in all sizes, including very large sizes in oil pipelines, etc.	1. Handwheel. 2. Electric motor. 3. Pneumatic actuator. 4. Hydraulic actuator.	1. Unrestricted bore at full opening. 2. Can handle all types of fluids. 3. Low operating torque. 4. Not normally used as a throttling valve.
Pinch valve	Particularly suitable for handling corrosive media, solids in suspensions, slurries, etc.	1. Mechanical. 2. Electric motor. 3. Pneumatic actuator. 4. Hydraulic actuator. 5. Fluid pressure (modified design).	1. Unrestricted bore at full opening. 2. Can handle all types of fluids. 3. Simple servicing. 4. Limited maximum pressure rating.

Table 8.2 Valve Types for Specific Services

Service	Main	Secondary
Gases	• Butterfly valves • Check valves • Diaphragm valves • Lubricated plug valves • Screw-down stop valves	• Pressure control valves • Pressure-relief valves • Pressure-reducing valves • Safety valves • Relief valves
Liquids, clear up to sludges and sewage	• Butterfly valves • Screw-down stop valves • Gate valves • Lubricated plug valves • Diaphragm valves • Pinch valves	—
Slurries and liquids heavily contaminated with solids	• Butterfly valves • Pinch valves • Gate valves • Screw-down stop valves • Lubricated plug valves	—
Steam	• Butterfly valves • Gate valves • Screw-down stop valves • Turbine valves	• Check valves • Pressure control valves • Presuperheated valves • Safety and relief valves

3. Valves for diverting flow

Such valves are plug valves and ball valves.

4. Valves for fluids with solids in suspension

The valves best suited for this duty have a closure member that slides across a wiping motion.

8.1.4 Globe valves

The sealing of these valves is high.
Applications
Duty:

• Controlling flow.
• Stopping and starting flow.
• Frequent valve operation.

Service:

• Gases essentially free of solids.
• Liquids essentially free of solids.
• Vacuum.
• Cryogenic.

Table 8.3 Valve Type Suitability

Valve Type	Service or Function										
	On-off	Throttling	Diverting	No Reverse Flow	Pressure Control	Flow Control	Pressure Relief	Quick Opening	Free Draining	Low Pressure Drop	Handling Solids Suspension
Ball	Suitable	Maybe suitable	Suitable	–	–	–	–	Suitable	–	Suitable	Limited suitability
Butterfly	Suitable	Suitable	–	–	–	Suitable	–	Suitable	Suitable	Suitable	Suitable
Diaphragm	Suitable	Maybe suitable	–	–	–	–	–	Maybe suitable	Maybe suitable	–	Suitable
Gate	Suitable	–	–	–	–	–	–	Suitable	Suitable	Suitable	–
Globe	Suitable	May be suitable	–	–	–	Maybe suitable	–	Suitable	Suitable	Suitable	–
Plug	Suitable	May be suitable	Suitable	–	–	Maybe suitable	–	Suitable	Suitable	Suitable	Limited suitability
Oblique (Y)	Suitable	May be suitable	–	–	–	Maybe suitable	–	–	–	–	–
Pinch	Suitable	Suitable	–	–	–	Suitable	–	–	Suitable	Suitable	Suitable
Slide	–	May be suitable	–	–	–	Maybe suitable	–	Maybe suitable	Suitable	Suitable	Suitable
Swing check	–	–	–	Suitable	–	–	–	–	–	Suitable	–
Tilting disc	–	–	–	Suitable	–	–	–	–	–	Suitable	–
Lift check	–	–	–	Suitable	–	–	–	–	–	–	–
Piston check	–	–	–	Suitable	–	–	–	–	–	–	–
Butterfly check	–	–	–	Suitable	–	–	–	–	–	–	–
Pressure relief	Suitable	–	–	–	–	–	Suitable	–	–	–	–
Pressure reducing	–	–	–	–	Suitable	–	–	–	–	–	–
Sampling	Suitable	–	–	–	–	–	–	–	–	–	–
Needle	–	Suitable	–	–	–	–	–	–	–	–	–

8.1.5 **Piston valves**

Applications
Duty:

- Controlling flow.
- Stopping and starting flow.

Service:

- Gases.
- Liquids.
- Fluids with solids in suspension.
- Vacuum.

8.1.6 **Gate valves**

Applications
Duty:

- Stopping and starting flow.
- Infrequent operation.

Service:

- Gases.
- Liquids.
- Fluids with solids in suspension.
- Knife gate valve for slurries, fibers, powders, and granules.
- Vacuum.
- Cryogenic.

8.1.7 **Wedge gate valves**

Wedge shape is to introduce a high supplementary seating load against high but also low fluid pressures.
Duty:

- Stopping and starting flow.
- Infrequent operation.

Service:

- Gases.
- Liquids.
- Rubber-seated wedge gate valves without bottom cavity for fluids carrying solids in suspension.
- Vacuum.
- Cryogenic.

8.1.8 Plug valves (cocks)

Duty:

- Stopping and starting flow.
- Moderate throttling.
- Flow diversion.

Fluids:

- Gases.
- Liquids.
- Non-abrasive slurries.
- Abrasive slurries for lubricated plug valves.
- Sticky fluids for eccentric and lift plug valves.
- Sanitary handling of pharmaceutical and food stuffs.
- Vacuum.

8.1.9 Ball valves

Duty:

- Stopping and starting flow.
- Moderate throttling.
- Flow diversion.

Service:

- Gases.
- Liquids.
- Non-abrasive slurries.
- Vacuum.
- Cryogenic.

8.1.10 Butterfly valves

Butterfly valves are available for wide range of pressures and temperatures based on variety of sealing principles.
Duty:

- Stopping and starting flow.
- Controlling flow.

Service:

- Gases.
- Liquids.
- Slurries.

- Powder.
- Granules.
- Sanitary handling of pharmaceuticals and food stuffs.
- Vacuum.

8.1.11 Needle valves

Small sizes of globe valves fitted with a finely tapered plug are known as needle valves.
   Three basic configurations are:

1. is a simple screw down valve;
2. is an oblique version, offering a more direct flow path;
3. is another form where the controlled outlet flow is at right angles to the main flow (and may be distributed through one or more passages).

8.1.12 Pinch valves

Pinch valves are flex-body valves consisting of a flexible tube that is pinched either mechanically, or by application of a fluid pressure to the outside of the valve body.
Duty:

- Stopping and starting flow.
- Controlling flow.

Service:

- Liquids.
- Abrasive slurries.
- Powders.
- Granules.
- Sanitary handling of pharmaceuticals and food stuffs.

8.1.13 Diaphragm valves

Diaphragm valves are flex-body valves in which the body flexibility is provided by a diaphragm. Diaphragm valves fall into two main types:

- Weir-Type Diaphragm valves that are designed for a short stroke between the closed and fully open valve positions.
- Straight-Through Diaphragm valves that have a relatively long stroke that requires more flexible construction materials for the diaphragm.

Duty:
   For weir-type and straight-through diaphragm valves:

- Stopping and starting flow.
- Controlling flow.

Service:

For weir-type diaphragm valves:

- Gases, may carry solids.
- Liquids, may carry solids.
- Viscous fluids.
- Leak-proof handling of hazardous fluids.
- Sanitary handling of pharmaceuticals and food stuffs.
- Vacuum.

Service for straight-through diaphragm valves:

- Gases, may carry solids.
- Liquids, may carry solids.
- Viscous fluid.
- Sludges.
- Slurries may carry abrasives.
- Dry media.
- Vacuum (consult manufacturer).

8.2 Check valves

Check valves are automatic valves that open with forward flow and close against reverse flow. They are also known as non-return valves. Check valves shall operate in a manner that avoids:

- The formation of an excessively high surge pressure as result of the valve closing.
- Rapid fluctuating movements of the valve closure member.

The type and operating characteristics of which can influence the choice of check valve type. Suitable combinations are:

- Swing check valve used with ball, plug, gate, or diaphragm control valves.
- Tilting disc check valves similar to swing-type check valve but with a profiled disc.
- Lift check valve used with globe or angle valves.
- Piston check valve used with globe or angle valves.
- Butterfly check valve used with ball, plug, butterfly, diaphragm or pinch valves.
- Spring-loaded check valves used with globe or angle valves.
- Diaphragm check valves the closure member consists of a diaphragm that deflects from or against the seat.

8.2.1 Lift check valves

Lift check valves may be subdivided into:

1. disc check valves;
2. piston check valves;
3. ball check valves.

8.2.2 Swing check valves

- Dirt and viscous fluids cannot easily hinder the rotation of the disc around the hinge.

8.2.3 Tilting-disc check valves

- Potentially fast closing.
- Being more expensive.
- More difficult to repair.

8.2.4 Diaphragm check valves

- Are not as well-known as other check valves.
- Is well suited for applications in which the flow varies within wide limits.
- The pressure differential is limited to 1 MPa.
- Operating temperature is limited to 70 °C.
- Sizes as small as DN3 (NPS 1/8 in) and as large as DN 3000 (NPS 120 in).

8.2.5 Foot valves

- Is basically a check valve.
- Often includes a strainer.
- Are fitted to the end of a suction pipe.
- Prevents the pump emptying when it stops.

8.2.6 Poppet lift check valves

The travel of the poppet is controlled by a stop on the end of the poppet legs acting as supports for the return spring shouldered onto a washer.

8.2.7 Ball foot valves

- It is particularly suitable for use with contaminated waters or more viscous fluids.

8.2.8 Membrane foot valves

- Consist of a cylindrical rubber membrane fitted inside a steel strainer.

8.2.9 Spring-loaded check valves

- Spring-loaded for more positive shut-off action.
- More rapid response cessation of flow.
- Work in any position, inclined, upward, or downward flow.

8.2.10 Dashpots

- The most important application of dashpots is in systems in which flow reverses very fast.
- A dashpot designed to come into play during the last closing movements can considerably reduce the formation of surge pressure.

8.2.11 Selection of check valves

Most check valves are selected qualitatively by comparing the required closing speed with the closing characteristic of the valve. This selection method leads to good results in the majority of applications.

8.2.12 Check valves for incompressible fluids

These are selected primarily for the ability to close without introducing an unacceptably high surge pressure due to the sudden shut-off of reverse flow. Selecting these for a low pressure drop across the valve is normally only a secondary consideration.

8.2.13 Check valves for compressible fluids

Check valves for compressible fluids may be selected on a basis similar to that described for incompressible fluids. However, valve-flutter can be a problem for high lift check valves in gas service, and the addition of a dashpot may be required.

8.3 Control valves

A valve selected as optimum for a level control process might not be the best selection for a flow control system. Also, the best valve for one flow control system might not be optimum for a system utilizing a different primary element or flow measurement means. Control valves are used in many applications, including liquid flow control, gas pressure reduction, steam flow to heaters, etc.

Control valves can be classified according to body design. The selection of a valve for a particular application is primarily a function of the process requirements, and no attempt will be made herein to cover this subject. Some of the more common types of control valve bodies are covered.

8.3.1 Globe body valve

One of the principle advantages is a balancing feature that reduces required actuator forces. In this design two options are available:

- A single-seat construction for minimum leakage in the closed position.
- A more simplified construction where greater leakage in the closed position can be tolerated.

The valve trim may be replaced without removing the valve body from the line. The globe valve design for a double-seated type has a higher leakage rate in the closed position than a single-seat type.

Another variation is the split body valve that is available both in globe and angle-type patterns. In this valve, the seat ring is clamped between the two body sections that makes it readily removable for replacement. This design is a single-seat type and does minimize leakage in the closed position.

The split body valve is used extensively in chemical processes due to (1) its availability in alloy materials and (2) the feature of separable flanges that allows the flanges to be manufactured from less expensive materials.

8.3.2 Butterfly valve

The butterfly valve is a rotating-vane, high-pressure recovery type of valve used in applications where high-capacity and low-pressure drops are required. Although they are not normally used on minimum leakage applications.

8.3.3 Ball valve

The ball control valve is a rotating-stem, high-pressure recovery type of valve, in which the flow of fluid is restricted by using a full-type or partial-type ball in the valve body. This valve has a high flow coefficient and may be used to control many types of fluids.

8.3.4 Three-way valve

The three-way valve is a special type of valve primarily used for splitting (diverting) or mixing (combining) service. The most common applications are through or around exchangers to control the heat transferred or in the controlled mixing of two streams.

8.3.5 Flashing

If the cavitation process could be halted before the completion of the second stage so that vapor persists downstream of the region where bubble collapse normally occurs, the process would be known as flashing. Flashing, like cavitation, can cause physical damage and decreased valve efficiency. Manufacturers should be consulted for recommendations.

8.3.6 Rangeability

The rangeability required for the control valve should be considered during valve selection. Although many control valves are available with published ranges of 50 to 1 and even greater, remember that these are at constant pressure drop, a condition that rarely exists in an actual plant. The requirement for rangeability is that the valve must handle the maximum flow at the minimum pressure drop available down to the minimum flow at maximum pressure drop. Sizing calculations should be checked at both extremes to assure controllability over the entire range of flow rates and pressure drops.

8.4 Control valve sizing

Having obtained the control valve's pressure drop allocation from the pump head available, the further step is to size the valve. The other factors involved are flow rate and liquid relative density (specific gravity).

Valve sizing shall be based on maximum sizing capacity of 1.3 times the normal maximum flow or 1.1 times the absolute maximum flow, whichever is greater.

The valve should be selected such that the opening of the valve at C_v calculated should not be greater than 75% of total travel. For the exceptional cases, the approval of the company shall be obtained.

Basic steps for sizing and selecting the correct valve include calculating the required C_v. ANSI/ISA sizing equations and procedures, have been programed and they are making computer-aided sizing available on IBM-PC or compatible computers. These programs permit rapid control valve flow capacity calculations and valve selection with minimal effort. The programs also include exit velocity, noise prediction, and actuator sizing calculations.

These instructions are designed to expose the user to the different aspects of valve sizing. The step-by-step method outlined in this section is the most common method of sizing.

8.4.1 Flow capacity

The valve sizing coefficient most commonly used as a measure of the capacity of the body and trim of a control valve is C_v. One C_v is defined as one U.S. gallon per minute of 60 °F water that flows through a valve with a 1 psi pressure drop. The general equation for C_v is as follows:

$$C_v = \text{flow} \sqrt{\frac{\text{specific gravity at flowing temperature}}{\text{pressure drop}}} \qquad (8.1)$$

When selecting a control valve for an application, the calculated C_v is used to determine the valve size and the trim size that will allow the valve to pass the desired flow rate and provide stable control of the process fluid.

8.4.2 Pressure profile

Fluid flowing through a control valve obeys the basic laws of conservation of mass and energy, and the continuity equation. The control valve acts as a restriction in the flow stream. As the fluid stream approaches this restriction, its velocity increases in order for the full flow to pass through the restriction. Energy for this increase in velocity comes from a corresponding decrease in pressure.

Maximum velocity and minimum pressure occur immediately downstream from the throttling point at the narrowest constriction of the fluid stream, known as the vena contracta. Downstream from the vena contracta, the fluid slows and part of the energy (in the form of velocity) is converted back to pressure. A simplified profile of the fluid pressure is shown in Figure 8.1. The slight pressure losses in the inlet and outlet passages are due to frictional effects. The major excursions of pressure are due to the velocity changes in the region of the vena contracta.

8.4.3 Allowable pressure drop

The capacity curve shown in Figure 8.2 shows that, with constant upstream pressure, flow rate, q, is related to the square root of pressure drop through the proportionality constant C_v. The curve departs from a linear relationship at the onset of "choking" described using the F_i factor. The flow rate reaches

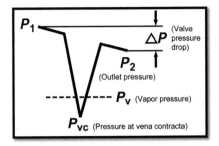

FIGURE 8.1

Pressure profile of fluid passing through a valve.

a maximum, q_{max}, at the fully choked condition due to effects of cavitation for liquids or sonic velocity for compressible fluids.

The transition to choked flow may be gradual or abrupt, depending on valve design. ANSI/ISA liquid sizing equations use a pressure recovery factor, F_L, to calculate the DP_{ch} at which choked flow is assumed for sizing purposes. For compressible fluids, a terminal pressure drop ratio, x_T, similarly describes the choked pressure drop for a specific valve.

When sizing a control valve, the smaller of the actual pressure drop or the choked pressure drop is always used to determine the correct C_v. This pressure drop is known as the allowable pressure drop, DP_a.

8.4.4 Cavitation

In liquids, when the pressure anywhere in the liquid drops below the vapor pressure of the fluid, vapor bubbles begin to form in the fluid stream. As the fluid decelerates, there is a resultant increase in pressure. If this pressure is higher than the vapor pressure, the bubbles collapse (or implode) as the

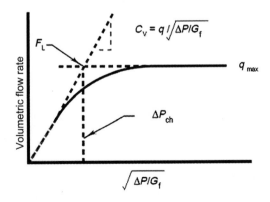

FIGURE 8.2

Choked pressure drop.

vapor returns to the liquid phase. This two-step mechanism, called cavitation, produces noise, vibration, and causes erosion damage to the valve and downstream piping.

The onset of cavitation, known as incipient cavitation, is the point when the bubbles first begin to form and collapse. Advanced cavitation can affect capacity and valve performance, which begins at a DP determined from the factor, F_i. The point at which full or choked cavitation occurs (severe damage, vibration, and noise) can be determined. Under choked conditions, "allowable pressure drop," is the choked pressure drop.

8.4.5 Liquid pressure recovery factor, F_L

The liquid pressure recovery factor, F_L, predicts the amount of pressure recovery that will occur between the vena contracta and the valve outlet. F_L is an experimentally determined coefficient that accounts for the influence of the valve's internal geometry on the maximum capacity of the valve. It is determined from capacity test data like that shown in Figure 8.2.

F_L also varies according to the valve type. High recovery valves, such as butterfly and ball valves, have significantly lower pressures at the vena contracta and, hence, recover much farther for the same pressure drop than a globe valve. Thus, they tend to choke (or cavitate) at smaller pressure drops than globe valves.

8.4.6 Liquid critical pressure ratio factor, F_F

The liquid critical pressure ratio factor, F_F, multiplied by the vapor pressure, predicts the theoretical vena contracta pressure at the maximum effective (choked) pressure drop across the valve.

8.4.7 Flashing

If the downstream pressure is equal to or less than the vapor pressure, the vapor bubbles created at the vena contracta do not collapse, resulting in a liquid–gas mixture downstream of the valve. This is commonly called flashing. When flashing of a liquid occurs, the inlet fluid is 100% liquid that experiences pressures in and downstream of the control valve that are at or below vapor pressure. The result is a two phase mixture (vapor and liquid) at the valve outlet and in the downstream piping. Velocity of this two phase flow is usually very high and results in the possibility for erosion of the valve and piping components.

8.4.8 Choked flow

Choked flow occurs in gases and vapors when the fluid velocity reaches sonic values at any point in the valve body, trim, or pipe. As the pressure in the valve or pipe is lowered, the specific volume increases to the point where sonic velocity is reached. In liquids, vapor formed as the result of cavitation or flashing increases the specific volume of the fluid at a faster rate than the increase in flow due to pressure differential. Lowering the downstream pressure beyond this point in either case will not increase the flow rate for a constant upstream pressure. The velocity at any point in the valve or downstream piping is limited to sonic (Mach = 1). As a result, the flow rate will be limited to an amount that yields a sonic velocity in the valve trim or the pipe under the specified pressure conditions.

8.4.9 Reynolds number factor, F_R

The Reynolds Number Factor, F_R, is used to correct the calculated C_v for non-turbulent flow conditions due to high viscosity fluids, very low velocities, or very small valve C_v.

8.4.10 Piping geometry factor, F_P

Valve sizing coefficients are determined from tests run with the valve mounted in a straight run of pipe that is the same diameter as the valve body. If the process piping configurations are different from the standard test manifold, the apparent valve capacity is changed. The effect of reducers and increasers can be approximated by the use of the piping geometry factor, F_P.

8.4.11 Velocity

As a general rule, valve outlet velocities should be limited to the following maximum values in Table 8.4.

The above figures are guidelines for typical applications. In general, smaller sized valves handle slightly higher velocities and large valves handle lower velocities. Special applications have particular velocity requirements; a few of which are provided below.

Liquid applications, where the fluid temperature is close to the saturation point, should be limited to 30 ft/s to avoid reducing the fluid pressure below the vapor pressure. This is also an appropriate limit for applications designed to pass the full flow rate with a minimum pressure drop across the valve.

Valves in cavitating service should also be limited to 30 ft/s to minimize damage to the downstream piping. This will also localize the pressure recovery that causes cavitation immediately downstream from the vena contracta.

In flashing services, velocities become much higher due to the increase in volume resulting from vapor formation. For most applications, it is important to keep velocities below 500 ft/s. Expanded outlet style valves, such as the Mark One-X, help to control outlet velocities on such applications. Erosion damage can be limited by using chrome-moly body material and hardened trim. On smaller valve applications that remain closed for most of the time, such as heater drain valves, higher velocities of 800–1500 ft/s may be acceptable with appropriate materials.

Gas applications where special noise attenuation trim are used should be limited to approximately Mach 0.33. In addition, pipe velocities downstream from the valve are critical to the overall noise level. Experimentation has shown that velocities around Mach 0.5 can create substantial noise even in a straight pipe. The addition of a control valve to the line will increase the turbulence downstream, resulting in even higher noise levels.

Table 8.4 Maximum Valve Outlet Velocities

Liquids	50 ft/s
Gases	Approaching Mach 1.0
Mixed gases and liquids	500 ft/s

Expansion factor, Y

The expansion factor, Y, accounts for the variation of specific weight as the gas passes from the valve inlet to the vena contracta. It also accounts for the change in cross-sectional area of the vena contracta as the pressure drop is varied.

Ratio of specific heats factor, F_k

The ratio of specific heats factor, F_k, adjusts the equation to account for different behavior of gases other than air.

Terminal pressure drop ratio, x_T

The terminal pressure drop ratio for gases, x_T, is used to predict the choking point where additional pressure drop (by lowering the downstream pressure) will not produce additional flow due to the sonic velocity limitation across the vena contracta. This factor is a function of the valve geometry and varies similarly to F_L, depending on the valve type.

Compressibility factor, Z

The compressibility factor, Z, is a function of the temperature and the pressure of a gas. It is used to determine the density of a gas in relationship to its actual temperature and pressure conditions.

8.5 Calculating C_v for liquids

Introduction

The equation for the flow coefficient (C_v) in non-laminar liquid flow is:

$$C_v = \frac{q}{F_P} \sqrt{\frac{G_f}{\Delta P_a}} \qquad (8.2)$$

where:

C_v = Valve sizing coefficient
F_P = Piping geometry factor
q = Flow rate, gpm
ΔP_a = Allowable pressure drop across the valve for sizing, psi
G_f = Specific gravity at flowing temperature

The following steps should be used to compute the correct C_v, body size and trim number (Figures 8.3–8.5):

Step 1: Calculate actual pressure drop

The allowable pressure drop, ΔP_a, across the valve for calculating C_v, is the smaller of the actual ΔP from Eqn 3.2 and choked ΔP_{ch} from Eqn 3.3.

$$\Delta P = P_1 - P_2 \qquad (8.3)$$

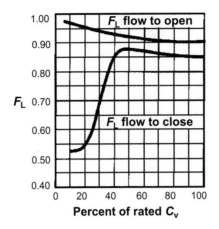

FIGURE 8.3

Globe valve F_L values chart.

Step 2: Check for choked flow, cavitation, and flashing

Use Eqn 3.3 to check for choked flow:

$$\Delta P_{ch} = F_L^2(P_1 - F_F P_V) \qquad (8.4)$$

where:

F_L = Liquid pressure recovery factor
F_F = Liquid critical pressure ratio factor
P_V = Vapor pressure of the liquid at inlet temperature, psia

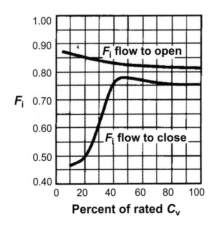

FIGURE 8.4

Globe valve F_i values chart.

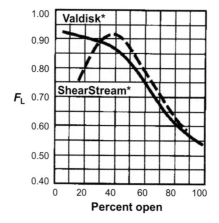

FIGURE 8.5

Rotary disc valve F_L values chart.

Table 8.5 Typical Valve Recovery Coefficient and Incipient Cavitation Factors

Valve Type	Flow Direction	Trim Size	F_L	F_i	x_T	F_d
Globe	Over seat	Full area	0.85	0.75	0.70	1.0
	Over seat	Reduced area	0.80	0.72	0.70	1.0
	Under seat	Full area	0.90	0.81	0.75	1.0
	Under seat	Reduced area	0.90	0.81	0.75	1.0
Valdisk	60° open	Full	0.76	0.65	0.36	0.71
Rotary disc	90° open	Full	0.56	0.49	0.26	0.71
ShearStream	60° open	Full	0.78	0.65	0.51	1.0
Rotary ball	90° open	Full	0.66	0.44	0.30	1.0
CavControl	Over seat	All	0.92	0.90	N/A	$0.2/\sqrt{d}$
MegaStream	Under seat	All	~1.0	N/A	~1.0	$(n_s/25d)^{2/3}$[b]
ChannelStream	Over seat	All	~1.0	0.87 to 0.999	N/A	0.040[a]
Tiger-tooth	Under seat	All	~1.0	0.84 to 0.999	~1.0	0.035[a]

Note: Values are given for full-open valves. See charts below for part-stroke values.
[a]Typical.
[b]n_s = number of stages.

$P_1 =$ Upstream pressure, psia
See Table 8.5 for F_L factors for both full-open and part stroke values.
F_F can be estimated by the following relationship:

$$F_F = 0.96 - 0.28\sqrt{\frac{P_V}{P_C}} \qquad (8.5)$$

where:

$F_F =$ Liquid critical pressure ratio
$P_V =$ Vapor pressure of the liquid, psia
$P_C =$ Critical pressure of the liquid, psia (see Table 8.6)

If ΔP_{ch} (Eqn (8.4)) is less than the actual ΔP (Eqn (8.3)), use ΔP_{ch} for ΔP_a in Eqn (8.2) (Figure 8.6).

It may also be useful to determine the point at which substantial cavitation begins. The following equation defines the pressure drop at which substantial cavitation begins:

$$\Delta P(\text{cavitation}) = F_i^2(P_1 - P_V) \qquad (8.6)$$

In high pressure applications, alternate analysis may be required; verify analysis with factory if $\Delta P > \Delta P$ (cavitation) > 300 psi (globe valves) or 100 psi (rotary valves).

where:

$F_i =$ Liquid cavitation factor (Typical values for F_i are given in Table 3.1).
$P_1 =$ Upstream pressure, psia.
$P_V =$ Vapor pressure of the liquid, psia.

Table 8.6 Critical Pressures

Liquid	Critical Pressure (psia)	Liquid	Critical Pressure (psia)
Ammonia	1636.1	Hydrogen chloride	1205.4
Argon	707.0	Isobutane	529.2
Benzene	710.0	Isobutylene	529.2
Butane	551.2	Kerosene	350.0
Carbon dioxide	1070.2	Methane	667.3
Carbon		Nitrogen	492.4
Monoxide	507.1	Nitrous oxide	1051.1
Chlorine	1117.2	Oxygen	732.0
Dowtherm A	547.0	Phosgene	823.2
Ethane	708.5	Propane	615.9
Ethylene	730.5	Propylene	670.3
Fuel oil	330.0	Refrigerant 11	639.4
Fluorine	757.0	Refrigerant 12	598.2
Gasoline	410.0	Refrigerant 22	749.7
Helium	32.9	Sea water	3200.0
Hydrogen	188.1	Water	3208.2

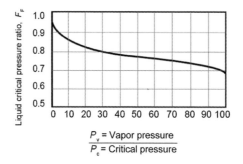

FIGURE 8.6

Liquid critical pressure ratio factor curve.

The required C_V for flashing applications is determined by using the appropriate ΔP allowable (ΔP_{ch} calculated from Eqn (8.4)).

Step 3: Determine specific gravity

Specific gravity is generally available for the flowing fluid at the operating temperature.

Step 4: Calculate approximate C_V

Generally the effects of nonturbulent flow can be ignored, provided the valve is not operating in a laminar or transitional flow region due to high viscosity, very low velocity, or small C_V. In the event there is some question, calculate the C_V, from Eqn (8.2), assuming $F_P = 1$, and then proceed to steps 5–7. If the Reynolds number calculated in Eqn (8.7a) is greater than 40,000, F_R can be ignored (proceed to step 8 after step 5.)

Step 5: Select approximate body size based on C_V

From the C_V tables in Section 8.4, select the smallest body size that will handle the calculated C_V.

Step 6: Calculate valve Reynolds number Re_V and Reynolds number factor F_R

Use Eqn 3.6a to calculate valve Reynolds number factor:

$$Re_V = \frac{N_4 F_d q}{v\sqrt{F_L C_V}} \left(\frac{F_L^2 C_V^2}{N_2 d^4} + 1 \right)^{1/4} \tag{8.7a}$$

Use Eqn 3.6b to calculate valve Reynolds number factor F_R if $Re_V < 40,000$, otherwise $F_R = 1.0$:

$$F_R = 1.044 - 0.358 \left(\frac{C_{vs}}{C_{vt}} \right)^{0.655} \tag{8.7b}$$

where:

$C_{vs} =$ Laminar flow C_V

$$C_{vs} = \frac{1}{F_s}\left(\frac{q\mu}{N_s\Delta P}\right)^{2/3} \tag{8.7c}$$

C_{vt} = Turbulent flow C_v (Eqn (8.2)).
F_s = streamline flow factor

$$F_s = \frac{F_d^{2/3}}{F_L^{1/3}}\left(\frac{F_L^2 C_v^2}{N_2 d^4} + 1\right)^{1/6} \tag{8.7d}$$

where:

d = Valve inlet diameter, inches
F_d = Valve style modifier (Table 3.1)
F_s = Laminar, or streamline, flow factor
q = Flow rate, gpm
N_2 = 890 when d is in inches
N_4 = 17,300 when q is in gpm and d in inches
N_s = 47 when q is in gpm and DP in psi
m = absolute viscosity, centipoise
n = kinematic viscosity, centistokes = m/G_f
Step 7: Recalculate C_v using Reynolds number factor

If the calculated value of F_R is less than 0.48, the flow is considered laminar; and the C_v is equal to C_{vs} calculated from Eqn 3.6c. If F_R is greater than 0.98, turbulent flow can be assumed (F_R = 1.0); and C_v is calculated from Eqn 3.1. Do not use the piping geometry factor F_P if F_R is less than 0.98. For values of F_R between 0.48 and 0.98, the flow is considered transitional; and the C_v is calculated from Eqn (8.7e):

$$C_v = \frac{q}{F_R}\sqrt{\frac{G_f}{P_1 - P_2}} \tag{8.7e}$$

For laminar and transitional flow, note the ΔP is always taken as P_1-P_2.

Step 8: Calculate piping geometry factor

If the pipe size is not given, use the approximate body size (from step 5) to choose the corresponding pipe size. The pipe diameter is used to calculate the piping geometry factor, F_P, which can be determined by Tables 8.7 and 8.8. If the pipe diameter is the same as the valve size, F_P is one and does not affect C_v.

Step 9: Calculate the final C_v

Using the value of F_P, calculate the required C_v from Eqn (8.2).

Step 10: Calculate valve exit velocity

The following equation is used to calculate entrance or exit velocities for liquids:

$$V = \frac{0.321q}{A_v} \tag{8.8}$$

Table 8.7 Piping Geometry Factors for Valves with Reducer and Increaser, F_P Versus C_v/d^2

C_v/d^2	d/D				
	0.50	0.60	0.70	0.80	0.90
4	0.99	0.99	1.00	1.00	1.00
6	0.98	0.99	0.99	1.00	1.00
8	0.97	0.98	0.99	0.99	1.00
10	0.96	0.97	0.98	0.99	1.00
12	0.94	0.95	0.97	0.98	1.00
14	0.92	0.94	0.96	0.98	0.99
16	0.90	0.92	0.95	0.97	0.99
18	0.87	0.90	0.94	0.97	0.99
20	0.85	0.89	0.92	0.96	0.99
25	0.79	0.84	0.89	0.94	0.98
30	0.73	0.79	0.85	0.91	0.97
35	0.68	0.74	0.81	0.89	0.96
40	0.63	0.69	0.77	0.86	0.95

Note: The maximum effective pressure drop (DP choked) may be affected by the use of reducers and increasers. This is especially true of Valdisk valves. Contact factory for critical applications.

Table 8.8 Piping Geometry Factors for Increaser Only on Valve Outlet, F_P Versus C_v/d^2

C_v/d^2	d/D				
	0.50	0.60	0.70	0.80	0.90
4	1.00	1.00	1.00	1.00	1.00
6	1.01	1.01	1.01	1.01	1.01
8	1.01	1.02	1.02	1.02	1.01
10	1.02	1.03	1.03	1.03	1.02
12	1.03	1.04	1.04	1.04	1.03
14	1.04	1.05	1.06	1.05	1.04
16	1.06	1.07	1.08	1.07	1.05
18	1.08	1.10	1.11	1.10	1.06
20	1.10	1.12	1.12	1.12	1.08
25	1.17	1.22	1.24	1.22	1.13
30	1.27	1.37	1.42	1.37	1.20
35	1.44	1.65	1.79	1.65	1.32
40	1.75	2.41	3.14	2.41	1.50

where: d = Valve port inside diameter in inches; D = Internal diameter of the piping in inches. (See Tables 8.9 and 8.10).

Table 8.9 Valve Outlet Areas

Valve Size (Inches)	Valve Outlet Area, A_v (Square Inches)						
	Class 150	Class 300	Class 600	Class 900	Class 1500	Class 2500	Class 4500
½	0.20	0.20	0.20	0.20	0.20	0.15	0.11
¾	0.44	0.44	0.44	0.37	0.37	0.25	0.20
1	0.79	0.79	0.79	0.61	0.61	0.44	0.37
1½	1.77	1.77	1.77	1.50	1.50	0.99	0.79
2	3.14	3.14	3.14	2.78	2.78	1.77	1.23
3	7.07	7.07	7.07	6.51	5.94	3.98	2.78
4	12.57	12.57	12.57	11.82	10.29	6.51	3.98
6	28.27	28.27	28.27	25.97	22.73	15.07	10.29
8	50.27	50.27	48.77	44.18	38.48	25.97	19.63
10	78.54	78.54	74.66	69.10	60.13	41.28	28.27
12	113.10	113.10	108.43	97.12	84.62	58.36	41.28
14	137.89	137.89	130.29	117.86	101.71	70.88	50.27
16	182.65	182.65	170.87	153.94	132.73	92.80	63.62
18	233.70	226.98	213.82	194.83	167.87	117.86	84.46
20	291.04	283.53	261.59	240.53	210.73	143.14	101.53
24	424.56	415.48	380.13	346.36	302.33	207.39	143.14
30	671.96	660.52	588.35	541.19	476.06	325.89	
36	962.11	907.92	855.30				
42	1320.25	1194.59					

Note: To find approximate fluid velocity in the pipe, use the equation $V_P = V_v A_v/A_P$ where: V_P = Velocity in pipe; A_v = Valve Outlet area from Table 3.8; V_v = Velocity in valve outlet; A_P = Pipe area from Table 3.7.
To find equivalent diameters of the valve or pipe inside diameter use: $d = \sqrt{4A_v/\pi}, D = \sqrt{4A_p/\pi}$.

where:

V = Velocity, ft/s
q = Liquid flow rate, gpm
A_v = Applicable flow area, in^2 of body port (Table 8.9)

After calculating the exit velocity, compare that number to the acceptable velocity for that application. It may be necessary to go to a larger valve size.

Step 11: Recalculate C_v if body size changed

Recalculate C_v if the F_P has been changed due to selection of a larger body size.

Step 12: Select trim number

First identify if the valve will be used for on/off or throttling service. select the appropriate trim number for the calculated C_v and body size selected. The trim number and flow characteristic

(Section 8.9) may be affected by how the valve will be throttled. When cavitation is indicated, refer to evaluate special trims for cavitation protection.

8.6 Liquid sizing examples
8.6.1 Example one

Given:	
Liquid	Water
Critical pressure (P_C)	3206.2 psia
Temperature	250 °F
Upstream pressure (P_1)	314.7 psia
Downstream pressure (P_2)	104.7 psia
Specific gravity	0.94
Valve action	Flow-to-open
Line size	4 in (Class 600)
Flow rate	500 gpm
Vapor pressure (P_V)	30 psia
Kinematic viscosity (ν)	0.014 cSt
Flow characteristic	Equal percentage

Step 1: Calculate actual pressure drop using Eqn (8.3).

$$\Delta P = 314.7 \text{ psia} - 104.7 \text{ psia} = 210 \text{ psi}$$

Step 2: Check for choked flow. Find F_L using Table 8.5. Looking under "globe, flow-under," find F_L as 0.90. Next, estimate F_F using Eqn (8.5):

$$F_F = 0.96 - 0.28\sqrt{\frac{30}{3206.2}} = 0.93$$

Insert F_L and F_F into Eqn (8.4):

$$\Delta P_{ch} = (0.90)^2[314.7 - (0.93)(30)] = 232.3 \text{ psi}$$

Since the actual ΔP is less than ΔP_{ch}, the flow is not choked; therefore, use the smaller (or actual ΔP) to size the valve. At this point, also check for incipient cavitation using Eqn (8.6) and Table 8.5:

$$\Delta P(\text{cavitation}) = (0.81)^2(314.7 - 30) = 187 \text{ psi}$$

Since ΔP (actual) exceeds ΔP (cavitation), substantial cavitation is occurring, but flow is not choked. Special attention should be paid to material and trim selection.

Step 3: The specific gravity for water is given as 0.94.

Step 4: Calculate the approximate $C_v F_P$ using Eqn (8.8). and assuming F_P is 1.0:

$$C_v = 500\sqrt{\frac{0.94}{210}} = 33.4$$

Step 5: From the C_v tables (Mark One, flow-under, equal percentage, Class 600) select the smallest body size for a C_v of 33.4, which is a 2-inch body.

Step 6: Calculate the Reynolds number factor, F_R, using Eqn (8.7a and e), as required.

$$Re_v = \frac{(17.300)(1)(500)}{(0.014)\sqrt{(0.90)(33.4)}}\left(\frac{(0.90)^2(33.4)^2}{(890)(2)^4} + 1\right)^{1/4} = 114 \times 10^6$$

Step 7: Since $Re_v > 40{,}000$, $F_R = 1.0$ and the recalculated $C_v F_P$ remains as 33.4.

Step 8: Using the 2-inch body from step 5, determine the F_P using Table 8.7, where:

$$d/D = 2/4 = 0.5 \quad \text{and} \quad C_v/d^2 = 33.4/2^2 = 8.35$$

Therefore, according to Table 8.7, the F_P is 0.97.

Step 9: Recalculate the final C_v:

$$C_v = \frac{500}{0.97}\sqrt{\frac{0.94}{210}} = 34.5$$

Step 10: Using Eqn (8.8), the velocity for a 2-inch body is found to be nearly 51 ft/s. Since this application is cavitating, damage may result in a 2-inch valve.

A 3-inch body reduces velocity to about 23 ft/s which is a more acceptable value. However, special trim may be required to eliminate cavitation damage.

Note: In this example, a 2 × 4-inch Mark One-X might also be chosen. It is less costly than a 3-inch valve and the larger outlets will lower the velocities. It will also be less costly to install in a 4-inch line.

Step 11: Since the body size has changed, recalculate the C_v by following steps 8 and 9. The F_P for a 3-inch body is nearly 1.0, and the final C_v is 33.4.

Step 12: Referring to the C_v tables, a C_v 33, 3-inch valve would require at least a trim number of 1.25. Trim number 2.0 may also suffice and have no reduced trim price adder. Refer to Section 8.14 on special trims for cavitation protection.

8.6.2 Example two

Given:	
Liquid	Ammonia
Critical pressure (P_C)	1638.2 psia
Temperature	20 °F
Upstream pressure (P_1)	149.7 psia
Downstream pressure (P_2)	64.7 psia
Specific gravity	0.65
Valve action	Flow-to-close
Line Size	3-inch (Class 600)
Flow rate	850 gpm
Vapor pressure (P_V)	45.6 psia
Kinematic viscosity (v)	0.02 cSt
Flow characteristic	Linear

Step 1: Calculate actual pressure drop using Eqn 3.2.

$$\Delta P = 149.7 \text{ psia} - 64.7 \text{ psia} = 85 \text{ psid}$$

Step 2: Check for choked flow. Find F_L using Table 3.1 Looking under "globe, flow-over," find F_L as 0.85. Next, estimate F_F using Eqn (8.5):

$$F_F = 0.96 - 0.28\sqrt{\frac{45.6}{1638.2}} = 0.91$$

Insert F_L and F_F into Eqn (8.4):

$$\Delta P_{ch}(\text{choked}) = (0.85)^2[149.7 - (0.91)(45.6)] = 78.2 \text{ psi}$$

Since the actual ΔP is more than ΔP_{ch}, the flow is choked and cavitating; therefore, use the ΔP_{ch} for ΔP_a to size the valve. Since the service is cavitating, special attention should be made to material and trim selection.

Step 3: The specific gravity for ammonia is given as 0.65.
Step 4: Calculate the approximate C_V using Eqn (8.2):

$$C_V = 850\sqrt{\frac{0.65}{78.2}} = 77.5$$

Step 5: From the C_V tables (Mark One, flow-over, linear, Class 600) select the smallest body size for a C_V of 77.5, which is a 3-inch body.

Steps 6 and 7: Turbulent flow is assumed, so Reynolds number factor is ignored, $F_R = 1.0$.
Step 8: With the 3-inch body and 3-inch line, $F_P = 1$.
Step 9: Since $F_P = 1$, the final C_v remains as 77.5.
Step 10: Using Eqn (8.8), the velocity for a 3-inch body is found to be over 38 ft/s. Since this application is cavitating, this velocity may damage a 3-inch valve. However, since the size is restricted to a 3-inch line, a larger valve size cannot be chosen to lower the velocity. Damage problems may result from such a system. A cavitation control style trim should be suggested.
Step 11: If cavitation control trim is not selected, C_v recalculation is not necessary since the body size or trim style did not change.
Step 12: Referring to the C_v tables, a C_v 77.5, 3-inch valve would use a trim number of 2.00 or the full size trim number 2.62. Use of this trim, however, could result in cavitation damage to body and trim.

8.6.3 Flashing liquids velocity calculations

When the valve outlet pressure is lower than or equal to the saturation pressure for the fluid temperature, part of the fluid flashes into vapor. When flashing exists, the following calculations must be used to determine velocity.

Flashing requires special trim designs and/or hardened materials. Flashing velocity greater than 500 ft/s requires special body designs. If flow rate is in lb/h:

$$V = \frac{0.040}{A_v} w \left[\left(1 - \frac{x}{100\%}\right) v_{f2} + \frac{x}{100\%} v_{g2} \right]$$ (8.9)

if the flow rate is given in gpm, the following equation can be used:

$$V = \frac{20}{A_v} q \left[\left(1 - \frac{x}{100\%}\right) v_{f2} + \frac{x}{100\%} v_{g2} \right]$$ (8.10)

where:

V = Velocity, ft/s
w = Liquid mass flow rate, lb/h
q = Liquid flow rate, gpm
A_v = Valve outlet flow area, in^2, see Table 3.8
v_{f2} = Saturated liquid specific volume (ft^3/lb at outlet pressure)
v_{g2} = Saturated vapor specific volume (ft^3/lb at outlet pressure)
x = % of liquid mass flashed to vapor

Calculating % flash.
The % flash (x) can be calculated as follows:

$$x = \left(\frac{h_{f1} - h_{f2}}{h_{fg2}}\right) \times 100\%$$ (8.11)

where:

$x = \%$ of liquid mass flashed to vapor
$h_{f1} =$ Enthalpy of saturated liquid at inlet temperature
$h_{f2} =$ Enthalpy of saturated liquid at outlet pressure
$h_{fg2} =$ Enthalpy of evaporation at outlet pressure

For water, the enthalpies (h_{f1}, h_{f2} and h_{fg2}) and specific volumes (v_{f2} and v_{g2}) can be found in the saturation temperature and pressure tables of any set of steam tables.

8.6.4 Flashing liquid example

Assume the same conditions exist as in Example one, except that the temperature is 350 °F rather than 250 °F. By referring to the saturated steam temperature tables, you find that the saturation pressure of water at 350 °F is 134.5 psia, which is greater than the outlet pressure of 105 psia (90 psia). Therefore, the fluid is flashing. Since a portion of the liquid is flashing, Eqns (8.10) and (3.11) must be used. x (% flashed) can be determined by using Eqn (8.11):

$h_{f1} = 321.8$ BTU/lb at 350 °F (from saturation temperature tables)
$h_{f2} = 302.3$ BTU/lb at 105 psia (from saturation pressure tables)
$h_{fg2} = 886.4$ BTU/lb at 105 psia (from saturation pressure tables)

$$x = \left(\frac{321.8 - 302.3}{886.4}\right) \times 100\% = 2.2\%$$

Therefore, the velocity in a 3-inch valve can be determined by using Eqn (8.10):

$v_{f2} = 0.0178$ ft^3/lb at 105 psia (from saturation pressure tables)
$v_{g2} = 4.234$ ft^3/lb at 105 psia (from saturation pressure tables)

$$V = \frac{(20)(500)}{7.07}\left[\left(1 - \frac{2.2\%}{100\%}\right)0.0178 + \left(\frac{2.2\%}{100\%}\right)4.234\right]$$

Flashing velocity is less than 500 ft/s, which is acceptable for Mark One bodies. Hardened trim should also be considered.

8.7 Calculating C_v for gases

Because of compressibility, gases and vapors expand as the pressure drops at the vena contracta, decreasing their specific weight. To account for the change in specific weight, an expansion factor, Y, is introduced into the valve sizing formula. The form of the equation used is one of the following, depending on the process variables available.

$$w = 63.3 F_P C_v Y \sqrt{x P_1 \gamma_1} \tag{8.12}$$

$$Q = 1360 \, F_P C_v P_1 Y \sqrt{\frac{x}{G_g T_1 Z}} \tag{8.13}$$

$$w = 19.3 \, F_P C_v P_1 Y \sqrt{\frac{xMW}{T_1 Z}} \tag{8.14}$$

$$Q = 7320 \, F_P C_v P_1 Y \sqrt{\frac{x}{MW T_1 Z}} \tag{8.15}$$

where:

$w =$ Liquid mass flow rate, lb/h
$F_P =$ Piping geometry factor
$C_v =$ Valve sizing coefficient
$Y =$ Expansion factor
$x^T =$ Pressure drop ratio
$\gamma 1 =$ Specific weight at inlet conditions, lb/ft^3
$Q =$ Gas flow in standard ft^3/h (SCFH)
$G_g =$ Specific gravity of gas relative to air at standard conditions
$T_1 =$ Absolute upstream temperature
$°R = (°F + 460 \, °F)$
$Z =$ Compressibility factor
$MW =$ Molecular weight
$P_1 =$ Upstream absolute pressure, psia

Note: The numerical constants in Eqns (8.12–8.15) are unit conversion factors.
The following steps should be used to compute the correct C_v, body size and trim number:

Step 1: Select the appropriate equation.

Based on the information available, select one of the four Eqns (8.12–8.15).

Step 2: Check for choked flow.

Determine the terminal pressure drop ratio, x_T, for that particular valve by referring to Table 8.10.
Next, determine the ratio of specific heats factor, F_k, by using the equation below:

$$F_k = \frac{k}{1.40} \tag{8.16}$$

Table 8.10 Pressure Drop Ratios, x_T

Valve Type	Flow Direction	Trim Size	x_T
Globe	Flow-to-close	Full area	0.70
	Flow-to-close	Reduced area	0.70
	Flow-to-open	Full area	0.75
	Flow-to-open	Reduced area	0.75
High performance	60° open	Full	0.36
Butterfly	90° open	Full	0.26
Multi-stage	Under seat	All	~1.00
Ball	90° open	Full	0.30

Table 8.11 Gas Physical Data

Gas	Critical Pressure (psia)	Critical Temperature (°R)	Molecular Weight (MW)	Ratio of Specific Heats (k)
Air	492.4	227.1	28.97	1.40
Ammonia	1636.1	729.8	17.0	1.31
Argon	707.0	271.1	39.9	1.67
Carbon dioxide	1070.2	547.2	44.0	1.29
Carbon monoxide	507.1	238.9	28.0	1.40
Ethane	708.5	549.4	30.1	1.19
Ethylene	730.6	508.0	28.1	1.24
Helium	32.9	9.01	4.00	1.66
Hydrogen	188.2	59.4	2.02	1.40
Methane	667.4	342.8	16.04	1.31
Natural gas	667.4	342.8	16.04	1.31
Nitrogen	492.4	226.8	28.0	1.40
Oxygen	732.0	278.0	32.0	1.40
Propane	615.9	665.3	44.1	1.13
Steam	3208.2	1165.1	18.02	1.33

where:

F_k = Ratio of specific heats factor
k = Ratio of specific heats (taken from Table 8.11)

Calculate the ratio of actual pressure drop to absolute inlet pressure, x, by using Eqn (8.17):

$$x = \frac{\Delta P}{P_1} \tag{8.17}$$

where:

x = Ratio of pressure drop to absolute inlet pressure
ΔP = Pressure drop ($P_1 - P_2$)
P_1 = Inlet pressure, psia
P_2 = Outlet pressure, psia

Choked flow occurs when x reaches the value of $F_k x_T$. Therefore, if x is less than $F_k x_T$, the flow is not choked. If x is greater, the flow is choked. If flow is choked, then $F_k x_T$ should be used in place of x (whenever it applies) in the gas sizing equations.

Step 3: Calculate the expansion factor.
The expansion factor, Y, may be expressed as:

$$Y = 1 - \frac{x}{3F_k x_T} \tag{8.18}$$

Note: If the flow is choked, use $F_k x_T$ for x.

Step 4: Determine the compressibility factor.

To obtain the compressibility factor, Z, first calculate the reduced pressure, P_r, and the reduced temperature, T_r:

$$P_r = \frac{P_1}{P_C} \tag{8.19}$$

where:

P_r = Reduced pressure
P_1 = Upstream pressure, psia
P_C = Critical Pressure, psia (from Table 8.12)

$$T_r = \frac{T_1}{T_C} \tag{8.20}$$

where:

T_r = Reduced temperature
T_1 = Absolute upstream temperature
T_C = Critical absolute temperature (from Table 8.11) Using the factors P_r and T_r, find Z in Figures 8.7 or 8.8

Step 5: Calculate C_v.

Using the above calculations, use one of the four gas sizing equations to determine C_v (assuming F_P is 1).

Step 6: Select approximate body size based on C_v.

From the C_v tables, select the smallest body size that will handle the calculated C_v.

Step 7: Calculate piping geometry factor.

If the pipe size is not given, use the approximate body size (from step 6) to choose the corresponding pipe size. The pipe size is used to calculate the piping geometry factor, F_P. If the pipe diameter is the same as the valve size, F_P is one and is not a factor.

Step 8: Calculate the final C_v.

With the calculation of the F_P, figure the final C_v.

Step 9: Calculate valve exit Mach number.

Eqns (8.21–8.23) or (8.24) are used to calculate entrance or exit velocities (in terms of the approximate Mach number). Use Eqn (8.21) or (8.22) for gases, Eqn (8.23) for air, and Eqn (8.24) for steam. Use downstream temperature if it is known, otherwise use upstream temperature as an approximation.

$$M(\text{gas}) = \frac{Q_a}{5574 \, A_v \sqrt{\frac{kT}{MW}}} \tag{8.21}$$

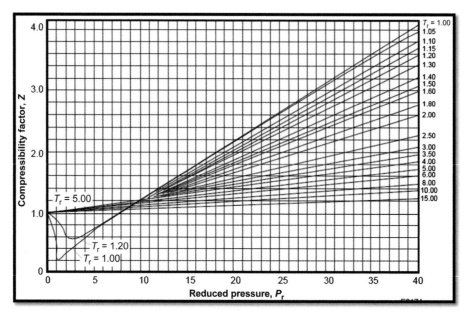

FIGURE 8.7

Compressibility factors for gases with reduced pressures from 0 to 40.

Reproduced from charts of L.C. Nelson and E.F. Obert, Northwestern Technological Institute.

$$M(\text{gas}) = \frac{Q_a}{1036\, A_v \sqrt{\frac{kT}{G_g}}} \tag{8.22}$$

$$M(\text{air}) = \frac{Q_a}{1225\, A_v \sqrt{T}} \tag{8.23}$$

$$M(\text{steam}) = \frac{wv}{1514\, A_v \sqrt{T}} \tag{8.24}$$

where:

M = Mach number
Q_a = Actual flow rate, ft³/h (CFH, not SCFH)
A_v = Applicable flow area, in², of body port (Table 8.9)
T_1 = Absolute temperature °R, (°F + 460°)
w = Liquid mass flow rate, lb/h
v = Specific volume at flow conditions, ft³/lb
G_g = Specific gravity at standard conditions relative to air
MW = Molecular weight
k = Ratio of specific heats

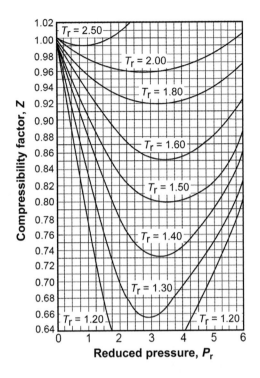

FIGURE 8.8

Compressibility factors for gases with reduced pressures from 0 to 6.

Reproduced from charts of L.C. Nelson and E.F. Obert, Northwestern Technological Institute.

Note: To convert SCFH to CFH use the equation:

$$\frac{(P_a)(Q_a)}{T_a} = \frac{(P_s)(Q)}{T_s} \tag{8.25}$$

where:

P_a = Actual operating pressure
Q_a = Actual volume flow rate, CFH
T_a = Actual temperature, °R (°F + 460°)
P_s = Standard pressure (14.7 psi)
Q = Standard volume flow rate, SCFH
T_s = Standard temperature (520° Rankine)

After calculating the exit velocity, compare that number to the acceptable velocity for that application. Select a larger size valve if necessary. Refer to Section 8.13 to predict noise level.

Caution: Noise levels in excess of 110 dBA may cause vibration in valves/piping resulting in equipment damage.

Step 10: Recalculate C_v if body size changed.

Recalculate C_V if F_P has changed due to the selection of a larger body size.

Step 11: Select trim number.

Identify if the valve is for on/off or throttling service. Using the C_V tables, select the appropriate trim number for the calculated C_V and body size selected.
    The trim number and flow characteristic may be affected by how the valve is throttled.

8.7.1 Gas sizing examples

• **Example one**

Given:	
Gas	Steam
Temperature	450 °F
Upstream pressure (P_1)	140 psia
Downstream pressure (P_2)	50 psia
Flow rate	10,000 lb/h
Valve action	Flow-to-open
Critical pressure (P_C)	3206.2 psia
Critical temperature (T_C)	705.5 °F
Molecular weight (MW)	18.026
Ratio of specific heats (k)	1.33
Flow characteristic	Equal percentage
Line size	2-inch (Class 600)
Specific volume	10.41

Step 1: Given the above information, Eqn (8.14) can be used to solve for C_V.
Step 2: Referring to Table 8.10, the pressure drop ratio, x_T, is 0.75. Calculate F_k using Eqn (8.16) and x using Eqn (8.17):

$$F_k = \frac{1.33}{1.40} = 0.95$$

$$x = \frac{140 - 50}{140} = 0.64$$

Therefore, $F_k x_T$ is (0.95)(0.75) or 0.71. Since x is less than $F_k x_T$, flow is not choked. Use x in all equations.
    Step 3: Determine Y using Eqn (8.18):

$$Y = 1 - \frac{0.64}{3(0.71)} = 0.70$$

Step 4: Determine Z by calculating P_r and T_r using Eqns (8.19–8.20):

$$P_r = \frac{140}{3208.2} = 0.04$$

$$T_r = \frac{450 + 460}{705.5 + 460} = 0.78$$

Using Figure 8.7, Z is found to be 1.0,

Step 5: Determine C_v using Eqn (8.14) and assuming F_P is 1:

$$C_v = \frac{10,000}{(19.3)(140)(0.70)} \sqrt{\frac{(910)(1.0)}{(0.64)(18.02)}} = 47.0$$

Step 6: From the C_v tables (Mark One, flow-under, equal percentage, Class 600), select the smallest body size for a C_v of 47, which is a 2-inch body.
Steps 7 and 8: Since the pipe size is the same as the body, F_P is one and is not a factor. Therefore, the C_v is 47.
Step 9: The gas is steam, calculate the Mach number using Eqn (8.24). Assume a constant enthalpy process to find specific volume at downstream conditions; from steam tables, $v = 10.41$ ft^3/lb at $T_2 = 414\,°$F:

$$M = \frac{(10,000)(10.41)}{1515(3.14)\sqrt{414 + 460}} = 0.74$$

This is greater than Mach 0.5 and should be reviewed for excessive noise and use of noise reducing trim.

Step 10: If body size does not change, there is no impact on C_v calculation.
Step 11: Referring to the C_v tables, a C_v 47, 2-inch Mark One would use a trim number of 1.62. If noise is a consideration.
- **Example two**

Given:	
Gas	Natural gas
Temperature	65 °F
Upstream pressure (P_1)	1314.7 psia
Downstream pressure (P_2)	99.7 psia
Flow rate	2,000,000 SCFH
Valve action	Flow-to-open
Critical pressure (P_C)	672.92 psia
Critical temperature (T_C)	342.8 °R

Continued

Molecular weight (MW)	16.042
Ratio of specific heats (k)	1.31
Flow characteristic	Linear
Line size	Unknown (class 600)

Step 1: Given the above information, Eqn (8.15) can be used to solve for C_v.
Step 2: Referring to Table 8.10, the pressure drop ratio, x_T, is 0.75 by assuming a Mark One flow-under. Calculate F_k using Eqn (8.16) and x using Eqn (8.17):

$$F_k = \frac{1.31}{1.40} = 0.936$$

$$x = \frac{1314.7 - 99.7}{1314.7} = 0.92$$

Therefore, $F_k x_T$ is (0.94)(0.75) or 0.70. Since x is greater than $F_k x_T$, flow is choked. Use $F_k x_T$ in place of x in all equations.

Step 3: Determine Y using Eqn (8.18):

$$Y = 1 - \frac{0.70}{3(0.70)} = 0.667$$

Step 4: Determine Z by calculating P_r and T_r using Eqns (8.19–8.20):

$$P_r = \frac{1314.7}{667.4} = 1.97$$

$$T_r = \frac{65 + 460}{342.8} = 1.53$$

Using Figure 8.8, Z is found to be about 0.86.

Step 5: Determine C_v using Eqn (8.15) and assuming F_P is 1:

$$C_v = \frac{200,000}{(7320)(1314.7)(0.667)} \sqrt{\frac{(16.04)(525)(0.86)}{(0.70)}} = 31.7$$

Step 6: From the C_v tables (Mark One, flow-under, linear, Class 600), select the smallest body size for a C_v of 31.7, which is a 1½-inch body.
Steps 7 and 8: Since the pipe size is unknown, use 1 as the F_P factor. Therefore, the C_v is 31.7.
Step 9: Since the gas is natural gas, calculate the Mach number using Eqn (8.21):

$$M = \frac{(297,720^*)}{5574(1.77)\sqrt{\frac{(1.31)(65 + 460)}{16.04}}} = 6.61$$

Note: To convert SCFH to CFH, use Eqn (8.25).

Step 10: Mach numbers in excess of sonic velocity at the outlet of the valve are not possible. A larger valve size should be selected to bring the velocity below the sonic level. To properly size the valve, select a size to reduce the velocity to less than Mach 1.0.

Step 11: Using Eqn (8.21), solve for the recommended valve area required for Mach 0.5 velocity:

$$0.5\ M = \frac{297{,}720\ \text{CFH}}{5574A\sqrt{\dfrac{1.31(65+460)}{16.04}}} = A_v = 16.3\ \text{in}^2$$

Solve for the valve diameter from the area by:

$$A_v = \pi d^2 \quad \text{or} \quad d = \sqrt{\frac{4A_v}{\pi}} = \sqrt{\frac{4(16.3)}{\pi}} = 4.6\ \text{in}$$

Thus a 6-inch valve is required.

Step 12: Referring to the C_v tables, a C_v of 31.7, 6-inch Mark One would use a trim number of 1.62. Since the flow is choked, noise should be calculated and special trim may be selected.

8.8 Calculating C_v for two phase flow

The method of C_v calculation for two phase flow assumes that the gas and liquid pass through the valve orifice at the same velocity. The required C_v is determined by using an equivalent density for the liquid gas mixture. This method is intended for use with mixtures of a liquid and a non-condensable gas. To size valves with liquids and their own vapor at the valve inlet will require good engineering judgment.

- **Nomenclature:**

    A_v = flow area of body port (Table 8.9)
    ΔP_a = allowable pressure drop
    q_f = volumetric flow rate of liquid, ft^3/h
    q_g = volumetric flow rate of gas, ft^3/h
    w_f = liquid mass flow rate, lb/h
    w_g = liquid mass flow rate, lb/h
    G_f = liquid specific gravity at upstream conditions
    G_g = gas specific gravity at upstream conditions
    T_1 = upstream temperature (°R)

Step 1: Calculate the limiting pressure drop.

First it must be determined whether liquid or gas is the continuous phase at the vena contracta. This is done by comparing the volumetric flow rate of the liquid and gas. Whichever is greater will be the limiting factor:

If $q_f > q_g$, then $\Delta P_a = \Delta P_a$ for liquid
If $q_g > q_f$, then $\Delta P_a = \Delta P_a$ for gas

The ΔP_a for liquid or gas is either P_1-P_2 or the choked pressure drop of the dominating phase if the valve is choked. (See the gas and liquid choked pressure equations.)

Step 2: Calculate the equivalent specific volume of the liquid-gas mixture.

$$v_e = \frac{\left(f_g v_g\right)}{Y^2} + f_f v_f \tag{8.26}$$

$$f_g = \frac{w_g}{w_g + w_f} \tag{8.27}$$

$$f_f = \frac{w_f}{w_g + w_f} \tag{8.28}$$

$$v_g = \frac{T_1}{\left(2.7P_1 G_g\right)} \tag{8.29}$$

$$v_f = \frac{1}{\left(62.4 G_f\right)} \tag{8.30}$$

where:

Y = gas expansion factor (Eqn (8.18))

Step 3: Calculate the required C_v of the valve.

$$C_v F_P = \left(\frac{w_g + w_f}{63.3}\right)\sqrt{\frac{v_e}{\Delta P_a}} \tag{8.31}$$

Use the smaller of P_1-P_2 and ΔP_{ch} for P_a.

Step 4: Select body size based on C_v.

From the C_v tables, select the smallest body size that will handle the calculated C_v.

Step 5: Calculate piping geometry factor.

If the pipe size is not given, use the approximate body size (from step 6) to choose the corresponding pipe size. The pipe size is used to calculate the piping geometry factor, F_P, which can be determined by Tables 3.7 or 3.8. If the pipe diameter is the same as the valve size, F_P is 1.

Step 6: Calculate final C_v.

With the calculation of the F_P, figure the final C_v.

Step 7: Calculate the valve exit velocity.

Where:

$$\text{Velocity} = \frac{\left(q_f + q_g\right)}{A_v} \tag{8.32}$$

$$q_f = \frac{w_f}{62.4G_f} \tag{8.33}$$

$$q_g = \frac{w_g T_1}{2.7 G_g P_2} \tag{8.34}$$

After calculating the exit velocity, compare that number to the acceptable velocity for that application. Select a larger valve size if necessary. Recommended two phase flow velocity limits are similar to those for flashing when the gaseous phase is dominant. If liquid is the dominant phase, velocity of the mixture should be less than 50 ft/s in the body.

Step 8: Recalculate C_v if body size changed.

Recalculate C_v if F_P has been changed due to the selection of a larger body size.

Step 9: Select trim number.

Identify if the valve will be used for on/off or throttling service. Using the C_v tables in Section 8.4, select the appropriate trim number for the calculated C_v and body size selected. The trim number and flow characteristic (Section 8.9) may be affected by how the valve is throttled. Special trim and materials may be required if high noise levels or cavitation are indicated (Table 8.12).

8.9 Engineering and material for control valves

This engineering and materials section covers the minimum requirements for control valve bodies, actuators and accessories, designed, constructed and materially tested in accordance with the References outlined herein.

The following description is primarily intended to indicate the general and minimum requirement of the control valves body design criteria, to be used in oil and gas Industries. Control valves have inherent operation characteristics that hinder precise positioning under varying operating conditions. Factors such as pressure differential across the valve seat, over tightening of packing, and viscous or fouling service can create additional forces preventing the valve from assuming the position called for by the controller.

Valve types shall be selected by taking into account such factors as operating and design conditions, fluids being handled, range ability required, cost, allowable leakage, noise and other special requirements.

Pneumatic control valves in electronic control loops shall be equipped with a 24 V d.c. device (convertor or transducer) to convert 4–20 mA, d.c. electronic instrument signal to a 0.2–1 barg pneumatic signal.

Control valves shall be furnished with pneumatic type (or d.c. electric-pneumatic) positioners in the following applications:

1. Temperature control valve other than minor applications.
2. Valves 4 in body size and over.
3. More than one valve on a single controller.
4. Critical pressure drop service for valve trim 1 in and larger.

Table 8.12 Pipe Flow Areas, A_p (Square Inches)

Nominal Pipe Diameter	Schedule												
	10	20	30	40	60	80	100	120	140	160	STD	XS	XXS
½				0.30		0.23				0.17	0.30	0.23	0.05
¾				0.53		0.43				0.30	0.53	0.43	0.15
1				0.86		0.72				0.52	0.86	0.72	0.28
1½				2.04		1.77				1.41	2.04	1.77	0.95
2				3.36		2.95				2.24	3.36	2.95	1.77
3				7.39		6.61				5.41	7.39	6.61	4.16
4				12.73		11.50		10.32		9.28	12.73	11.50	7.80
6				28.89		26.07		23.77		21.15	28.89	26.07	18.83
8		51.8	51.2	50.0	47.9	45.7	43.5	40.6	38.5	36.5	50.0	45.7	37.1
10		82.5	80.7	78.9	74.7	71.8	68.1	64.5	60.1	56.7	78.9	74.7	
12		117.9	114.8	111.9	106.2	101.6	96.1	90.8	86.6	80.5	113.1	108.4	
14	143.1	140.5	137.9	135.3	129.0	122.7	115.5	109.6	103.9	98.3	137.9	132.7	
16	188.7	185.7	182.6	176.7	169.4	160.9	152.6	144.5	135.3	129.0	182.6	176.7	
18	240.5	237.1	230.4	223.7	213.8	204.2	193.3	182.7	173.8	163.7	233.7	227.0	
20	298.6	291.0	283.5	278.0	265.2	252.7	238.8	227.0	213.8	202.7	298.0	283.5	
24	434	425	411	402	382	365	344	326	310	293	425	415	
30	678	661	649	663	602	574	542	513					
36	975	956	938	914	870	830	782						
42	1328	1302	1282	1255	1187	1132	1064						

STD, Standard; XS, Extra-strong; XXS, double extra-strong.

5. Line pressure exceeding 20 bar.
6. Extension bonnets (includes radiation fins and bellows seals).
7. Butterfly valves (rotary action).
8. Sounders patent bodies.
9. Three-way valves.
10. Slow system, such as mixing, thermal, or level process.
11. For fast systems, detailed stability analysis should be presented and if required boosters will be used instead.
12. On duties where the controlled liquid will vaporize across the ports.

Filter Regulators and Positioners or Boosters shall be factory mounted and tubbed. All connecting tubing in instrument air service shall be plastic coated copper with brass compression type fittings, unless otherwise specified.

Single seated bodied valves shall be top guided, double seated bodied valves shall be top and bottom guided construction. Unless otherwise as specified in data sheet.

Control valves shall have removable trims and sufficient clearance shall be allowed for access and removal.

Butt-welding valves should not be used, however, if line specification calls for butt-welding, consideration shall be given to the welding of control valves.

For flashing conditions, the type and size and additionally the flashing condition of the control valve shall be specified in data sheet and/or agreed with the user.

For control valves intended for operating high temperature, particular attention shall be paid to the clearance between plug and guide bushing to avoid valve sticking when the valve is hot.

In gas pressure let down stations when high differential pressure is present across the valve, special low noise valve shall be used. In this case the noise level should not exceed certain limits as specified in standards (Noise and vibration control standard) or through the requirement of data sheets.

The action of valves on failure of the operating medium shall be determined by process requirements with regard to safe operation and emergency shut-down requirements.

Where cage guided control valves are specified, balanced trim should be considered for large sized valves.

For control valves on vacuum services, special provisions should be considered for prevention and detection of leakage.

Where temperature of control fluid is below zero degree Celsius a bonnet extension shall be used.

Extension bonnet or finned also shall be provided on services above 200 °C, in order to maintain the temperature of stuffing box within the limits specified in accordance with the manufacturer's recommendations.

Air operated diaphragms and springs shall be selected to optimize a bench setting range of 0.2–1 barg for the specified maximum upstream pressure with the downstream pressure of zero bar. The "Bench Setting Range" and the "In Service Stroking Range" shall be specified on the control valve data plates. Air operated control valves with an in-service stroking range other than 0.2–1 barg may be used if so dictated by availability of standard operators, and user's approval.

Manual Loading Type Hand Operators shall be considered in lieu of a side mounted handwheel in relatively low pressure/pressure drop applications where block and bypass valves are not provided and a handwheel may cause a hazardous condition for automatic start-up or shutdown of the related equipment. These hand operators shall consist of a three-way air switch and a handwheel operated air regulator. The handle and ports shall be clearly marked as, MANAUTO. The regulators may be common to other components.

For globe body control valves, the trim construction shall be either single-seated with heavy duty top guiding for the plug, Double-seated with top and bottom guiding for the plug, or cage type. For liquid services with a high pressure drop (i.e., boiler feed water), and gas service (pressure let down), cage trims shall be specified to have the plug supported at the critical area.

Balance type control valve in place of single seat valve in high pressure service shall be considered.

Control valves for steam heated reboilers shall be located in the steam lines and not in the condensate lines, unless otherwise agreed by the user.

Where control valves are liable to freezing due to operating or ambient conditions, they shall be insulated or heat traced.

8.9.1 Control valve materials selection

Since the majority of control valve applications are relatively non-corrosive at reasonable pressure and temperature, cast iron and carbon steel bodies are the most common valve body materials used in the oil industries.

Most control valve materials can be placed in two categories:

The pressure containment materials for valve body, bonnet, bottom flange, and bolting.
The valve trim materials for valve plug, seat ring, cage, valve stem, guide bushing, and packing box parts.

For oxygen services, body and trim materials shall be AISI-316 stainless steel. Body casting shall internally be completely machined to a smooth surface to remove any casting imperfections.

For material selection of body, bolts, nuts, etc., the relevant piping class or any other information shall be adhered to for the particular application.

Control valve material shall be as specified in data sheets or shall be selected from ANSI-B16.5 specifications and applicable sections of the codes and standards.

Supplier shall comply with the pressure and temperature ratings of more common materials established by the ANSI-B16.5.

In case, corrosive condition would require very exotic materials, consideration may be given to a composite construction, such as internal metallic lining of the body.

For very severe erosive services the small fluid impact area inside the valve body shall be covered with a hard facing.

The minimum requirement for the body material is that the valve shall have a cast steel body, and the trim, consist of plug, seat ring and stem, shall have stainless steel 316, unless otherwise specified by the nature of process fluid being handled and/or requested through relevant data sheet.

When valves are used for chlorine service or other fluids that become corrosive when in contact with a moist atmosphere, suitable valve stem material must be chosen or other precautions taken. For chlorine services neoprene diaphragm valves is recommended.

For extremely erosive-and-corrosive services the hard facing material made of two disks of tungsten carbide material in angle pattern body can be used. This material is especially useful in oil production where severe sand erosion exists.

Hardened plug and seat rings shall be selected for the following applications:

1. Erosive service.
2. Wet gas or wet steam service with a pressure drop above 5 bar, other services when the pressure drop is above 10 bar, at design condition.

Small-sized valves for erosive services shall have their plug and seat rings made for solid satellite No. 6. For economic reasons hardened stainless steel 440 °C may be used as trim material if this is suitable for the particular process conditions.

When tight shut off is required, a ball or plug valve, a single seated globe body valve shall be selected. The seats shall be of soft material, such as glass fiber filled polytetrafluoroethylene (PTFE), the selection shall be based on suitability for the specified process conditions. The selected material shall be suitable for temperature at least 50 °C above the maximum process design conditions. The soft seat ring shall be properly clamped between metal parts.

When valves are used for sour gas services the trim and bolting material construction shall comply with the recommendation of National Association of Corrosion Engineers (NACE) MR-01-75 latest revision.

Packing glands shall be equipped with flange style gland followers with bolted constructions. A lubricator with steel isolating valve shall be provided where packing lubrication is required.

Guide bushing shall be a corrosion resistant material. It is preferred that the guide bushing material be a minimum of 125 brinell harder than the trim (i.e., 17-4 PH [Precipitation Hardened] stainless steels or better).

Stainless steel bellows seals may be considered for services with dangerous and or poisoning fluids such as Tetra Ethyl Lead (TEL) or Tetra Methyl Lead (TML ,) but should be avoided wherever possible. A purge with suitable pressure shall be used (monitored for purge) as an alternative method of sealing.

Butterfly valves material shall be as specified in data sheet for the related service conditions or shall be at manufacturer's option and in accordance with the applicable standard such as British Standard (BS)-5155.

Butterfly valves body material shall be selected from those listed in Table 8.1, if not specified in data sheets.

Butterfly valves trim material shall be suitable for specified service conditions and compatible with the piping material.

Butterfly valves trim material including disks, shafts, bushings, body, and/or disk seating surfaces, internal keys and pins and screws when in contact with the contained fluid shall be selected from Table 8.13, if not specified in the data sheet.

Seats in the body and on the disk may be separate or integral. Seat facings may be applied to valve bodies and/or disks as deposited metal, integral metal, mechanically retained metal, or resilient materials.

8.9.2 Control valve bodies

A control valve consists of two major sub-assemblies, a valve body sub-assembly and an actuator. The valve body sub-assembly is the portion that actually controls the passing fluid. It consists of a housing, internal trim, bonnet, and sometimes a bottom flange (Figure 8.9).

Table 8.13 Basic Materials for Butterfly Valves

1	2	3
Component	**Material**	**BS Reference**
Body Body with integral seat Disk Hand wheel Disk with integral seat Rings fitted to body or disk for sealing, seating, or retaining purposes	Cast iron Austenitic cast iron Spheroidal graphite iron Carbon steel Stainless steel Gunmetal Aluminum bronze	1452 3468 2789 1501.151 1503.221 1504.161 1501: Part 3 1503 1504, 3100 1504 1400 1400
	Rings of deposited metal or resilient + material	
Shaft	Carbon steel Stainless steel Aluminum bronze Nickel copper alloy	970: Part 1 970: Part 4 2672 or 2874 3076
Shaft bearings seals (when fitted)	**No requirement in this section**	
Internal fastenings	Carbon steel Stainless steel Phosphor bronze Aluminum bronze Nickel copper alloy	 2870, 2873 2872, 2874, 2875 3076

Body sub-assemblies occur in many shapes and working arrangements depending upon the individual service conditions and piping requirements. Each type has certain advantages and disadvantages for given service requirements and should, therefore, be selected with care.

Control valves operate by one of two primary motions: Reciprocating (sliding stem) motion or rotary motion. The selection of a valve for a particular application is primarily a function of the process requirements. Some of the more common types of control valve bodies are discussed in the following sections.

• **Globe body control valves.**

The most common control valve body style is in the form of a globe; (Figure 8.10), such a control valve body can be either single or double-seated.

A single-seat construction, for minimum leakage in the close position shall be used.

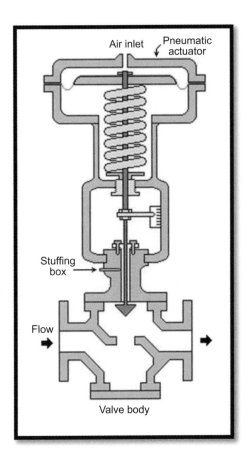

FIGURE 8.9

Typical control valve assembly.

A double-seat or balance construction when requiring less actuator force, but allowing some leakage in the close position, shall be used.

Single-seated valves shall have a top guided construction. The valve plug is guided within the lower portion of the valve bonnet (Figure 8.10).

Double-seated valves shall be top and bottom guided construction (Figure 8.11).

Three-way valves are a design extension of a typical double-ported globe valve. They are used for diverting services and mixing or combining services (Figure 8.12).

Control valve with a globe body shall be considered for all applications (throttling or on-off control) except where adverse operating conditions, such as high pressure drops or high capacities make other types more suitable.

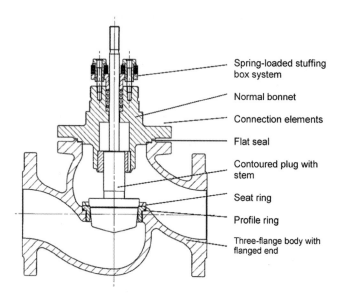

FIGURE 8.10

Globe body valve, top-guided.

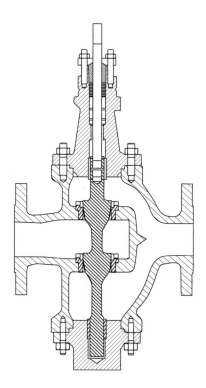

FIGURE 8.11

Globe body valve, top, and bottom-guided.

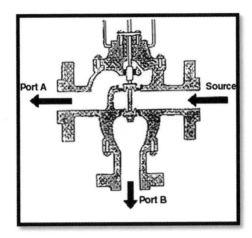

FIGURE 8.12

Three-way valve.

- **Angle body valves.**

Angle body valves should be considered for hydrocarbon services with a tendency for high pressure drop or coking and erosive services such as slurries and applications where solid contaminants might settle in the valve body (Figure 8.13).

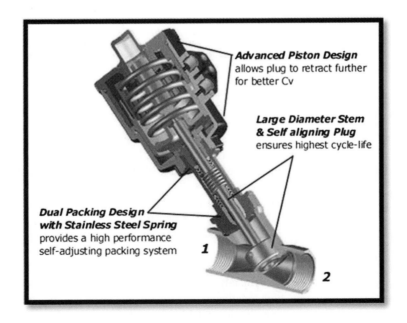

FIGURE 8.13

Angle body valve.

FIGURE 8.14

The weir type (a) and straight-through type
(b) diaphragm valves.

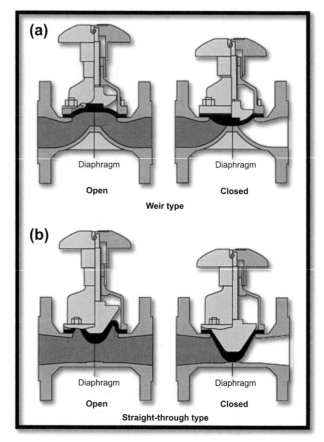

(a)

Diaphragm   Diaphragm

Open   **Closed**

Weir type

(b)

Diaphragm   Diaphragm

Open   **Closed**

Straight-through type

- **Diaphragm valves.**

Diaphragm valve may be considered for simple services and applications where the body lining in a standard valve becomes economically unattractive. When used for throttling service a characterized positioner may be required for obtaining the required valve-characteristic (Figure 8.14).

- **Cage guided valves.**

Top entry or cage guided valves have the advantages of easy trim removal. Valves of this type usually have stream lined body passages to permit increased flow capacity (Figure 8.7) (Figure 8.15).

- **Rotary Type control valves.**

All types of rotary valves share certain basic advantages and disadvantages among the advantages are low weight, simplicity of design, high relative C_v, more reliable, and friction-free packing, and generally low initial cost. They are generally not suitable in size, below 2 in and pressure-drop ratings are limited.

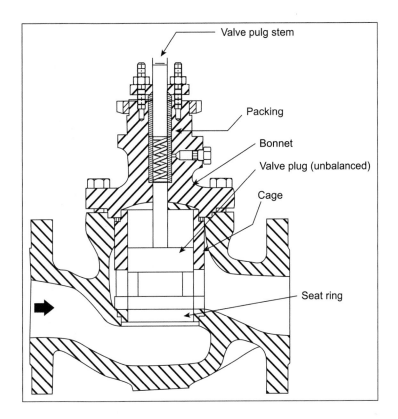

FIGURE 8.15

Globe body valve, cage-guided.

- **Butterfly valves.**

The most common type of rotary valve, is the butterfly valve. Butterfly valves shall be considered for high capacity low pressure drops and where no tight shut-off is required (Figure 8.16). Although not normally used in minimum leakage applications, it is available with piston ring, pressurized seat or various types of elastomer seating surfaces if minimum leakage is required.

Heavy pattern butterfly valves shall be used where they are practical and economical. They shall normally be furnished with diaphragm or piston actuators with positioners. Where hand wheel is required, the shaft mounted declutch able type is preferred. Long stroke position actuators shall be used where practical.

- **Ball valves.**

Ball valves may be considered for on-off and throttling services under moderate operating conditions (Figure 8.17).

Characterized ball valves may be used for fluids containing suspended solids or fluids likely to polymerize or crystallize.

FIGURE 8.16

Butterfly valves. http://www.arm-tex.com/sure-seal-butterfly-valves.html.

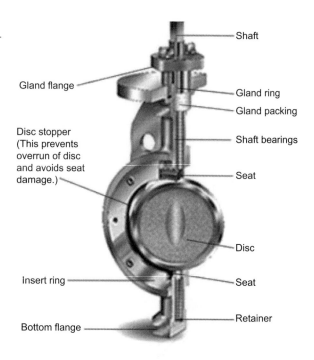

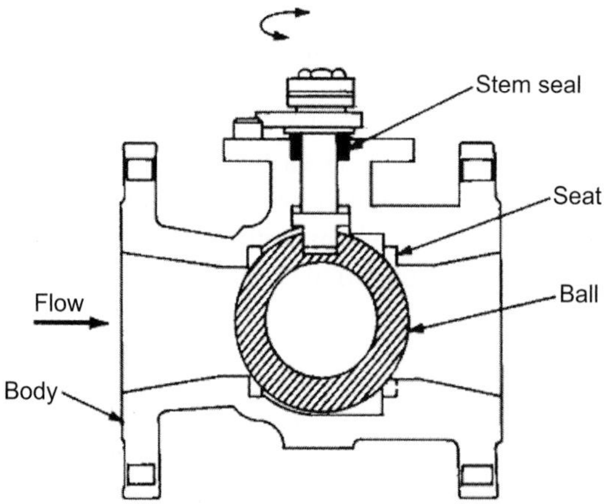

Gland flange

Disc stopper
(This prevents
overrun of disc
and avoids seat
damage.)

Insert ring

Bottom flange

Shaft

Gland ring

Gland packing

Shaft bearings

Seat

Disc

Seat

Retainer

FIGURE 8.17

Ball valves.

Stem seal

Seat

Ball

Flow

Body

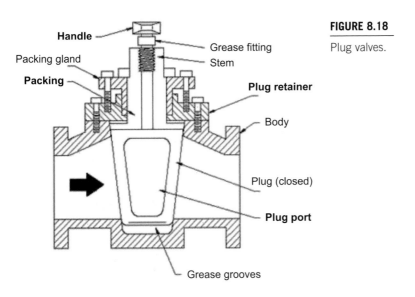

Handle
Packing gland
Packing
Grease fitting
Stem
Plug retainer
Body
Plug (closed)
Plug port
Grease grooves

FIGURE 8.18

Plug valves.

- **Emergency shut-off valves.**

For emergency shut-off valves on fuel service Ball Valve shall be used for temperature up to 150 °C. Above this temperature single seated tight shut-off globe valves shall be used.

- **Plug valves.**

Plug valves may be considered for special applications such as throttling control on slurry services in chemical plants (Figure 8.18).

- **Eccentric rotating plug valves.**

Eccentric rotating-plug valves are general substitute for globe body control valves provided that the application allows the use of long bolting.

- **Special type control valves.**

Special body types, such as angle, split body, low noise low flow valves shall be considered where the process fluid may be erosive, viscous or carrying suspended solids and or high differential pressure is required. Flushing connection shall be provided on slurry services.

- **Low noise valves.**

For services at high pressure drops, the application of a conventional valve trim often results in very high fluid velocities and unacceptable high noise levels. Where this would be the case, the fluid velocity must be controlled by using a valve trim having specially designed multiple orifices in series and/or in parallel, or having a tortuous path forcing the fluid to change the direction continuously, causing high turbulence friction.

- **Low flow or miniature valves.**

Where control valves with a very low capacity factor (C_v) are required, these may be of the miniature valve type with flanged or threaded connections and a needle trim.

8.10 Control valve body size and flange rating
8.10.1 Globe body valves

- **Body sizes.**

Nominal body sizes for the globe body, shall be selected from the following series; (Inches) 1, 1½, 2, 3, 4, 6, 8, 10, 12 in etc. The use of odd sizes such as 1¼, 2½, 5, 7, 9 in etc., shall be avoided. 1½ and 3 in valves are less common in petroleum industries.

The minimum globe control valve body size to be used shall be 1 in screwed, unless flange type is specified, and the internal trim size shall be in accordance to the requirements as specified in data sheet.

Body sizes smaller than 1 in may be used for special applications, and pressure regulation services. For valve sizes smaller than 1 in, reduced trim in 1 in size bodies normally will be preferable.

Flange rating-Globe control valves shall normally have flanged ends, but flangeless bodies may be considered for special applications.

The flange rating shall generally be in accordance with the piping class, but for carbon steel bodies the flange rating shall be class 300 minimum.

For pressure-temperature rating of globe body control valves reference should be made to relevant standards.

All globe body control valve manifolds and by pass valves shall follow the piping class and ratings. The dimensions however, shall be in accordance with the recognized standard such as ANSI/ISA RP-75.06.

8.10.2 Butterfly body valves

Lug-type and wafer-type butterfly valves shall have body pressure-temperature ratings for the selected American Society of Testing and Materials (ASTM) material specification in accordance with the applicable ANSI B-16 standard.

The wafer type butterfly valves, other than lugged type, shall be provided with or without holes for the passage of bolts securing the connecting flanges dependent upon valve design.

Lugged type, wafer valves shall be supplied with threaded or drilled holes the lugs to the size, nominal pressure rating, and type of connecting flange.

The end flanges of double flanged steel butterfly control valves shall be cast or forged integral with the body.

For other types of valves such as eccentric rotating plug valves or Butterfly valves flanges shall be wafer type (i.e., suitable for installation between flanges).

Butterfly valves shall be one of the following types, with metal or resilient seating or linings:

1. Double flanged: A valve having flanged ends for connection to pipe flanges by individual bolting.
2. Wafer: A valve primarily intended for clamping between pipe flanges using through bolting:
   a. single flange;
   b. flangeless;
   c. u-section.

8.11 **Control valve characteristics**

Control valve sizing is necessary to optimize operation, provide sufficient range ability, and minimize cost. The key to correct control valve sizing is the proper determination of the required valve capacity coefficient (C_v). By definition (C_v) is the number of gallon per minute of water at 15 °C that will pass through a given flow restriction with a pressure drop of 1 lb/in^2. For example, a control valve that has a maximum flow coefficient (C_v) of 12 has an effective port area in the full open position such that it passes 12 gallons per minute of water with a pressure drop of 1 lb/in^2.

In using these methods, full knowledge of actual flowing conditions is essential. The primary factors that should be known for accurate sizing are:

- The upstream and downstream pressures at the flow rates being considered.
- The generic identity of process fluid.
- The temperature of the fluid.
- The fluid phase (gas, liquid, slurry, and so forth).
- The density of the fluid (specific gravity, specific weight, and molecular weight).
- The viscosity (liquids).
- The vapor pressure (liquids).

Valve sizing shall be based on a maximum sizing capacity of 1.3 times the normal maximum flow or 1.1 times the absolute maximum flow, whichever is greater. The sizing pressure drop (ΔP sizing) shall be sufficient to obtain good regulation at the normal maximum case, maintain maximum quantity as well as the normal minimum quantity within the rangeability of the selected valve.

If in primary design stage maximum flow is not available, then valves shall be selected to have twice the C_v required for normal design flow at specified conditions.

Control valves with inherent high pressure-recovery characteristics can cause cavitation when fluid pressure and temperature conditions would indicate. Valves with low pressure recovery, special trim should be used to minimize or prevent cavitation.

Flashing, like cavitation, can cause physical damage and decreased valve capacity. Manufacturers should be consulted for recommendations.

The pressure drop across the control valve at maximum process flow shall be at least 20% of the pressure drop across the control valve at normal flow.

1. The control valve shall be sized such that the C_v value of the control valve for maximum process flow with the pressure drop across the control valve at maximum process flow is approximately 80% of the maximum C_v value for that control valve. Furthermore, the control valve shall never have less than 25% lift for minimum process flow at the specified pressure drop.

If neither a maximum nor a minimum process flow is stated, these flows shall be assumed to be 120% and 80%, respectively, of the normal process flow.

Sizing calculations should be checked for at both extremes to assure controllability over the entire range of the flow rates and pressure drop.

Butterfly valves shall be sized for maximum angle operating of 60°. Proposals to use angles greater than 60° shall be submitted to the purchaser for approval.

Shafts of rotary actuated valves shall be sized for pressure drop equal to maximum upstream pressure.

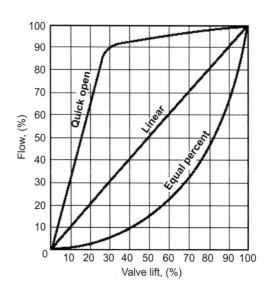

FIGURE 8.19

Percent of valve opening representative inherent flow characteristic curves.

Control valve body size normally shall not be less than half that of normal line size.

Control valve flow characteristics are determined principally by the design of the valve trim. The three inherent characteristics available are quick opening, linear, and equal percentage. These are shown in Figure 8.19. A modified percentage characteristic generally falling between the linear and equal percentage characteristics is also available. The three inherent characteristics can be described as follows:

1. Quick opening.

As the name implies, this characteristic provides a large opening as the plug is first lifted from the seat, with lesser flow increase as the plug opens further. This type is most commonly used where the valve will be either open or closed with no throttling of flow required.

2. Linear.

Linear trim provides equal increases in C_V for equal increases in stem travel. Thus the C_V increase is linear with plug position throughout its travel.

3. Equal percentage.

Equal percentage trim provides equal percentage increases in C_V for equal increments of stem travel. This is accomplished by providing a very small opening for plug travel near the seat and very large increases toward the more open position. As a result, a wide rangeability of C_V is achieved.

Characteristic of the inner valve shall normally be equal percentage except where system characteristics indicate otherwise. Linear and quick opening characteristics shall be used where required. In general linear trim shall be used only for Split-Range service or where control valve pressure drop remains constant over the range of 10–100% of flow capacity.

Shut off valves should normally have quick closing or equal percentage characteristic, but another characteristic (such as modified equal percentage) may be required for special cases (e.g., to avoid or reduce the consequence of hydraulic shock).

Characteristics of valves may change due to particular requirements.

Butterfly and angle valves and characterized ball valves ("V-Ball") shall normally have equal percentage characteristics.

Three-way valves in control services shall normally have linear characteristics.

Valves with shut-off function shall be single seated. The pressure drop across the control valve at maximum flow shall be at least 25% of the pressure drop across the control valve at normal flow.

Three-way valve shall be capable of operating against the maximum differential pressure that can exist across a single port. Each 3-way valve shall be specified as flow-mixing or flow splitting in accordance with the intended application.

The action of the valves on failure of the operating medium shall be determined by process requirements with regard to safe operation and emergency shut-down.

Extension bonnets shall be provided on services above 232 °C and below −6.7 °C or in accordance with the manufacturer's recommendation.

Pressure balanced valves of the double diaphragm type shall be considered for use on fuel gas to heaters in temperature control systems. When a single diaphragm type is used, a pneumatic ratio relay shall be installed in the control air line with the input–output ratio as required.

Control valves installed in pipe lines should normally be at least one pipe size smaller than the computed line size. This is to allow margin for future expansion and a better controllability of the process.

Where it is necessary to reduce from line size to control valve size, swaged reducers shall be used between the block valves and the control valve. Sufficient spacing between block valves shall allow for installation of larger size control valves.

Oversized bodies with reduced trims shall be used for valves in severe flashing or cavitating service. Angle type or multiple seat type valves may be considered for this service.

Valves used in pairs, as three-way valves, including rotary actuated valves such as Ball or Butterfly types, shall have linear characteristics. Characterized positioners may be used to meet this requirement. In this case calibration for the required characterization must be done by the valve manufacturer.

Gas compressor recycle control valves shall have linear characteristics.

Valves in pressure reducing service, where the pressure drop is constant, shall have linear characteristics.

8.12 **Control valve manifold design**

Control valves and bypass valves are mostly manifolded in piping systems to allow manual manipulation of the flow through the systems in those situations when the control valve is not in service.

For application information and guidance reference shall be made to ISA Handbook of control valves, API RP 550, or other relevant publications.

Dimensions for flanged globe control valves shall be used as per ANSI/ISA-S75.03 standard.

Control valve body nominal size covered by the designs are 1, 1½, 2, 3, 4, and 6 in.

Reference should be made to ISA RP 75.06 "Control Valve Manifold Design" for additional information and dimensions for all ANSI classes.

8.13 Control valve block and bypass valves

Where significant future expansion is not anticipated, a less flexible but more economical approach that gives a minimum acceptable design is to make the block valves one size larger than the control valve (but not larger than line size).

The bypass line and valve should normally have a capacity at least equal to the calculated or required C_v of the control valve, but not greater than twice the selected C_v of the control valve.

Bypass valves in sizes of 4 in or less are usually globe valves that allow throttling. For larger sizes, because of cost, gate valves are normally used.

Where block valves are provided, vent valves shall be fitted between them so that pressure may be relieved and the control valve drained when the block valves are closed. Suitable drain lines shall be provided where necessary.

Vent and drain connections shall not be less that ¾ in nominal bore.

A bypass connection and valve shall be installed around each control valve unless other means are available for manual control when the control valve is out of service.

Consideration shall be given to the elimination of by-pass and block-valves around control valves sizes 2 in and over, but this shall be by agreement with the user.

Block and by-pass valve assemblies should be avoided in the following instances:

- On hydrogen service.
- Around three-way valves.
- Around self-acting steam pressure reducing valves.
- Around control valves forming part of a protective system.

Block and bypass valve assemblies shall be provided in the following instances:

- Where a valve controls a service common to a number of plants.
- Where valves are in continuous operation and there is not sufficient assurance of reliability over the anticipated period between plant overhauls (e.g., on erosive or corrosive service or where the temperature is below 0 °C or above 180 °C). The cost of a failure shall also be taken into account.

Where failure of the control valve would necessitate continuous operator attention (e.g., on the fuel control to heaters).

Where bypass valves are not provided, a permanent side-mounted hand wheel shall be fitted to the control valve. Where the cost of the hand wheel is greater than the cost of block and bypass valves, the latter shall be provided except on hydrogen service and protective service.

Where block and bypass valves are not fitted initially adequate space should be allowed for possible future installation.

When control valves are placed in pre-stressed lines they shall be in a bypass assembly to the main pipeline.

8.14 Control valve packing and sealing

All valves shall be drilled and tapped to accept a gland lubricator except when otherwise specified in data sheet.

The bottom flange or the bottom of the body of a control valve shall not be drilled and tapped.

For special duties as specified in data sheet (e.g., toxic, control valve stems should be bellows sealed, with an independent gland seal, the enclosed space being monitored for bellows leakage).

When sealing by bellows is not possible, a purge should be used, monitored for flow failure. Bellows seals may also be required to prevent leakage of penetrating liquids.

On clean fluids, interlocking self-lubricating gland packings with spring followers may be used.

On higher temperature duties or where carbon or other deposits may settle on the stem, special packing should be used.

If dangerous fluids are encountered, horizontal lines shall be fitted with suitable drains on the underside. This does not replace the vent.

Packing materials for butterfly valves shall be suitable for the specified service conditions.

Where the controlled liquid contains particles or materials that would damage the valve guide, stem or packing, a purge system shall be considered.

8.15 Control valve noise and vibration caused by sonic flow

Sonic flow occurs when the velocity of the fluid reaches the speed of sound in that medium.

At subsonic velocities the flow is characterized by turbulent mixing and this is responsible for the noise produced. This noise best described as a "hiss" for small jets or as a roar for larger jets has no discrete dominating frequency. Its spectrum is continuous with a single, rather flat maximum. As pressure ratio increases past the critical ratio and the fluid reaches its sonic velocity, the sound emanating undergoes a fundamental change, while the roaring noise due to the turbulent mixing is still present, it may be almost completely dominated by a very powerful "whistle" or "serooch" of a completely different character. This noise is rather harsh and of a confused nature, becoming much more like a pure note.

At sonic flow, vibration can be caused in various, frequency bands due to vertical/horizontal movement of control valve components (20-80 KHz), impingement of fluid on control valve internals at high velocities (400–1600 Hz) aerodynamic noise from shock waves by the sonic velocities (1200–4000 Hz), internal components vibrating at their natural frequencies (3000–6000 Hz), and high pressure drop gas services (above 8000 Hz). For guidance in specifications the permissible noise exposure, is noted below Table 8.14:

8.15.1 Cause of noise and vibration

High pressure drop gives rise to sonic flow. Sonic flow generates shock waves which in turn produce high frequency noise and vibration (1.2–4.8 KHz). The noise has a characteristic whistle or scream at its peak frequency, is directional in nature when discharged into the atmosphere and even more dependent upon fluid jet pressure than the turbulent mixing noise of subsonic flow.

The most dangerous vibration occurs in frequency band 3–7 KHz and is the result of resonance by the valve parts. This can lead to failure due to metal fatigue.

The most effective method of solution is to remove the cause that is the high pressure drop. Values of safe pressure drop may be taken from:

$$\text{for subsonic conditions} \quad \Delta P < 0.5 C_f^2 (P_1) \tag{8.35}$$

Table 8.14 Permissible Noise Exposure Rate

Duration Hours per Day	Sound Level dB Slow Response
8	90
6	92
4	95
3	97
2	100
1.5	102
1	105
0.5	110
0.25 or less	115

$$\text{for sonic conditions,} \quad \Delta P \geq 0.5 C_f^2 (P_2) \tag{8.36}$$

where:

ΔP = Pressure drop across valve $(P_1 - P_2)$ bar
P_1 = upstream pressure, bar Abs
P_2 = downstream pressure, bar Abs
C_f = critical flow factor for valve, dimensionless (obtained from valve manufacturer)

The absence of sonic flow means an absence of its effects of noise and vibration. In cases where pressure drop must remain high, a special type of "low-noise" control valve is recommended.

If the calculated sound pressure level (SPL) value of a reducing valve under maximum load exceeds the stated limit by only 5–10 dB, then one of the following simple cures must be considered:

1. Increase the pipe wall thickness downstream (doubling the wall thickness will decrease the SPL by 5 dB).
2. Use acoustical isolation downstream. This will reduce SPL by 0.2–0.5 dB/mm of insulation, depending on the density of the insulating material.

If the valve noise is 10 dB above the selected limit, then one must choose a different approach such as the use of downstream, in line silencers. The silencers generally attenuate between 10 and 20 dB depending on the frequency range.

The silencers must be installed directly adjacent to the valve body and that the valve outlet velocity is below sonic; (say 1/3 match). Otherwise the silencer will act as a pressure-reducing device, for which it is not suitable.

The use of expansion plates downstream of valves is recommended. The primary function of these plates is not to attenuate the valve noise, but to absorb some of the pressure reduction over the whole system. In this way the pressure across the control valve can be kept below critical. In a typical installation the expansion plate downstream flow area must be increased to compensate for the changes in the density due to pressure drop.

8.16 **Control valve actuators**

For the purpose of this section the following definitions apply:

- **Actuator:**

Any device designed for attachment to a general purpose industrial valve in order to provide the operation of valve. The device is designed to operate by using motive energy which can be electrical, pneumatic, hydraulic, etc., or a combination of these. The movement is limited by travel, torque or thrust.

- **Multi-turn actuator:**

An actuator that transmits to the valve a torque for at least one revolution and shall be capable to withstand the thrust.

- **Torque:**

A turning moment transmitted through the mounting flanges and couplings.

- **Thrust:**

An axial force transmitted through the mounting flanges and couplings. Control valve actuators should be selected, so that; on failure of the operating medium, the valve will automatically take a position (open, closed, or locked) that will result in the safest configuration for the operating unit.

8.16.1 **Actuator travel limit**

In the opening direction, a stop shall be engaged before the valve plug reaches its travel limit. The stop shall have sufficient contact area to absorb any force transmitted to it. In the closing direction, the valve plug shall seat before the actuator reaches its travel limit. Valve connection to the actuator shall be adjustable, with positive locking of the adjustment.

Rotary actuated valves (such as butterfly, ball) shall have shaft keyways that allow the action of the valve to be changed.

For Rotary actuated valves in cryogenic service, shaft thrust bearings shall be provided. Standard spring range shall be 0.2–1 barg.

Hand wheels, when specified, shall be mounted and designed to operate in the following manner:

- For globe valves, hand wheels shall be mounted on the yoke, arranged so that the valve stem can be jacked in either direction.
- Neutral position shall be clearly indicated.
- Handwheel operation shall not add friction to the actuator.
- Clutch/Linkage mechanisms for handwheels on rotary valves shall be designed such that control of valve position is not lost when engaging the handwheel.

Pneumatic actuators should have as low as an operating pressure is practicable in order to minimize the need for spare capacity in the instrument air system. In no case may the operating pressure exceed 4 bar.

For shut-off valves, the actuators shall be capable of opening the valve against the full upstream pressure, with the downstream pressure assumed to be atmospheric.

For butterfly valves, the actuators shall have sufficient force for coping with all operating conditions from the fully closed position to the fully open position, and for coping with all pressure drop and torque requirements.

The stroking time of control valves shall be evaluated on the basis of the process control requirements. For critical analogue control systems such as surge control of compressors, the stroking time shall be less than 5 s. For other analogue control systems, longer stroking times may be acceptable.

For valves having spring-to-close action, the stroking time is determined by the spring force, the diameter of air exhaust ports in the actuator and solenoid valve, and the mechanical inertia of moving parts.

8.16.2 Types of control valve actuators

There are many types of actuators for stroking control valves. These actuators may be classified into four (4) general types:

- Pneumatically operated diaphragm actuators.
- Piston (cylinder) actuators.
- Electro-hydraulic actuators.
- Electro-mechanical (motor) operated actuators.

- **Pneumatically operated diaphragm actuators**

Control valves shall normally be operated by pneumatic diaphragm actuators. The actuator shall normally be operated between 0.2–1.0 bar and 0.4–2.0 bar may be used as specified for full stroke. Where the control signal is electric, the electro-pneumatic convertor shall be used.

There are two types of diaphragm actuators:

- Direct acting.
- Reverse acting.

The actuator shall be designed to provide dependable on-off or throttling operation of automatic control valve.

Reverse acting diaphragm actuators using seals or glands are permitted only for those applications where the direct acting type of actuator is unsuitable (typical spring diaphragm actuators are shown in Figures. 8.20 and 8.21).

The air pressure required for stroking the valve may vary from the operating spring range. The maximum air pressure allowable on spring diaphragm actuators shall not exceed 4 bar.

- **Pneumatic and hydraulic cylinder or piston actuators**

Cylinder actuators shall be used where a long stroke and high force is required, such as for dampers and louvers in large ducting for combustion air or flue gas services.

Piston actuators may be pneumatically or hydraulically operated.

The cylinders shall be connected directly to the valve as an integral part. Actuator may also be purchased separately and mounted at site.

For throttling applications the cylinder actuator shall be provided with a positioner mechanism and, where necessary, with a position transmitter. A pair of oil filter with isolating valves shall be installed as close as possible to the positioner.

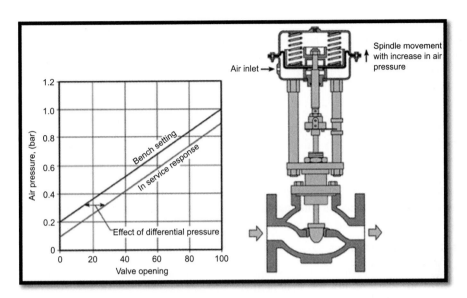

FIGURE 8.20

Direct acting actuator, air-to-close, direct acting valve–normally open. http://www.spiraxsarco.com/resources/
steam-engineering-tutorials/control-hardware-el-pn-actuation/control-valve-actuators-and-positioners.asp

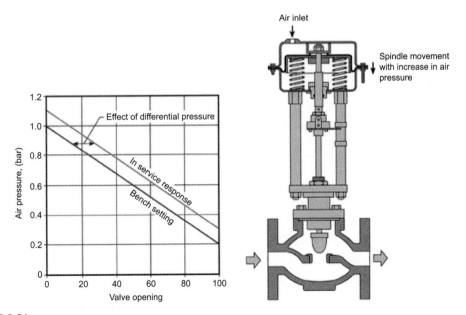

FIGURE 8.21

Reverse acting actuator, air-to-close, direct acting valve–normally open. http://www.spiraxsarco.com/
resources/steam-engineering-tutorials/control-hardware-el-pn-actuation/control-valve-actuators-and-
positioners.asp

Cylinder actuators on valves with provisions for manual (local) control shall be provided with external bypass valves if these are not integral with the actuator. Four way valves are often require in the piping of the cylinder to allow local operation.

For hydraulic cylinders, the following points should be considered:

1. If the hydraulic manifold is rigidly piped, it should be connected to the hydraulic fluid supply and return headers by flexible metallic hose.
2. To assure a continuous supply of hydraulic to actuators, it is advisable to provide both an oil filter or strainer and a spare suitably valved and piped so that either unit may be removed and cleaned without shutting off the supply.
3. Vent valve should be provided at high points in the hydraulic fluid system.
4. Depending upon whether the valve served by the actuator will move if the hydraulic oil pressure is lost, it may be necessary to use automatic fluid trapping valves that lock the hydraulic fluid in the cylinders upon failure of the hydraulic system.

- **Electro-hydraulic actuators**

Electro-hydraulic actuators may be used to operate a large rotary and sliding stem valves. Electrical signal of 4–20 mA or 10–50 mA d.c. may be used.

Electro-hydraulic actuators shall be used for an electronic control loop, where fast stroking speeds, high thrust, and long stroke are required.

Electro-hydraulic actuators may be used at locations where a suitable air supply is not available.

- **Electro-mechanical actuators**

The electro-mechanical valve actuator has essentially the same advantages as the electro-hydraulic actuator with respect to field use. It is capable of being used over long distances with only inconsequential signal transmission delays. It is also immune from the pneumatic system problem of freeze-up in extremely cold ambient conditions. Electromechanical actuators are still, however, generally more expensive, although more efficient, than electro-hydraulic units.

An electro-mechanical valve actuator is composed of a motorized gear train and screw assembly that drives the valve stem or rotary shaft valves. The varying input signal, whose magnitude corresponds to the required position of the inner valve stem, is fed into the positioner (usually a differential amplifier) and produces a voltage to actuate the motorized gear train and screw.

The resultant movement of the stem and the take-off attached to it and to a potentiometer or linear differential transformer, produces a voltage that increases with stroke, and is sent into the positioner. When the input signal voltages and feedback voltages are equal, the output voltage to the motor goes to zero and the motor stops with the valve stem at the required position. Conversely, when the voltages are not equal, the motor is run in the direction to make them equal.

- **Motor operated actuators**

The actuator shall consist of a motor driven, reduction gearing, thrust bearing where applicable, handwheel and local position indicator, together with torque and limit switches, space heaters terminals, control power transformer, and integral motor starter controls, all furnished as a self-contained, totally enclosed unit.

Motor starter for remote mounting shall be specified. The motor shall be sized for torque requirement according to the valve size, operating differential pressure and temperature, and speed operation.

The actuator terminal box should be double-sealed in such a way that when the actuator terminal box cover is removed for the connection of incoming cables, the remaining electrical components are still protected by the watertight enclosure.

The valve and actuator mounting bracket must be capable of withstanding the stall torque of the actuator, with torque and limit switches disconnected.

The actuator shall provide both torque and position limitation in both directions. An automatic override shall be provided to prevent the torque switches from tripping the motor on initial valve unseating. The actuator shall be capable of operating at any mounting angle. The actuator shall be designed so that there will be no release of high stem thrust of torque reaction spring forces when covers are removed from actuator gear box, even when the valve is under full line pressure conditions.

Failure of the motor, power, or motor gearing shall not prevent manual operation of the valve through the use of the handwheel. When the motor drive is declutched, the handwheel drive shall be engaged safely, with the motor running or stopped.

A means of locking the actuator in either the manual or motor condition shall be provided. If not locked in either position, starting of the motor shall automatically restore the power drive. The actuator shall be capable of functioning within the ambient temperature range as specified in the job specification, but in any case not less than $-15\,°C$ to $+80\,°C$.

Clockwise rotation of the actuator handwheel shall close the valve. The handwheel drive must be mechanically independent of the motor drive, and gearing should be such as to permit emergency manual operation in a reasonable time.

The motor shall be of sufficient size to open and close the valve against maximum cold working pressure when voltage to motor terminals is within 10% of rated voltage. Motor nameplate rating shall be 380 V, 3 phase, 50Hz, unless otherwise specified.

Local pushbuttons shall be provided for "Open-Stop-Close" control of valve, with a lockable selector switch providing positions "local control only," "local and remote," "remote plus local stop," and "off" position.

Provision shall have made for the addition of extra sets of limit switches in each actuator. Each set shall be adjustable to any point of valve position. Each unit shall be provided with auxiliary contacts, rated 5 A at 110 V.a.c. for remote position indication. They shall be adjustable continuously in the range of open to close positions.

Actuators shall have an output speed of 0.4 rev/s unless otherwise specified on the data sheets (e.g., for high speed emergency isolation). No valve shall require more than 2 min for full operation, i.e., from closed to fully open or open to fully closed.

In the case of high speed actuators a pulse timer should be included that will give variable slow down times for fast closing and fast opening, normally set to operate over the last 25% of the valve closure.

Provision shall also be made for remote operation through interposing relays supplied with d.c. power of 24 V unless otherwise specified. Mechanical dial indication of valve position shall be incorporated in the actuator. Indication shall be continuous if valve is specified to be in regulating device.

Position limit switches shall be provided at each end of travel for remote indication and sequencing. Torque and limit switches shall be easily adjustable without special tools or removal of switch assembly from the actuator. Repeatability of switch actuation shall be ±5% of set point.

Control power transformers when used, shall have fuse protection on the secondary. Fuses shall be readily accessible for replacement or deactivation at the terminal board.

An electrically and mechanically interlocked motor starter shall be provided in the actuator housing, unless the motor starter is specified for remote mounting.

All electrical components shall be prewired by the actuator vendor to a legibly marked terminal strip. Power and control wiring shall be segregated and insulated from each other. All wiring shall be identified by the Vendor, and access for maintenance provided.

Motor overload protection shall be provided. One or more winding temperature detectors embedded in the motor winding, or three thermal overload relays in the motor controller are acceptable. Either must be capable of being deactivated at the terminal board.

All motor operators shall be explosion proof approved in accordance with the requirements of the National Electrical Code latest edition (NFPA No. 70) for use in the hazardous area classification, or unless otherwise specified in individual data sheets.

All electrical equipment and motors shall be totally enclosed for outdoor services and should be water tight to International Equipment Components (IEC) 34.5 and IEC 144-IP 67.

All electrical equipment used in hazardous area, shall also meet the electrical area classification requirements as per standards.

Vendor shall supply the valve actuator compatible with the valve. All information required for sizing the actuator shall be obtained from the purchaser and/or Valve Supplier.

Torque requirements of valve and torque characteristics of the actuator shall be supplied to the purchaser for approval.

The actuating unit shall include a 3-phase electric motor, reduction gearing geared limit switches, and torque switches, space heaters, terminals, together with a handwheel for manual operation with declutching lever and valve position indicator.

All gearing shall be totally enclosed and continuously lubricated. All shafts shall be mounted on ball or roller bearings. Limit switch drive shall be stainless steel or bronze.

Power terminals shall be of stud type, segregated by an insulating cover. Four conduit entrance taps shall be provided as a minimum. Each tap will be provided with standard electrical connections in metric type, such as M20.

The actuator terminal box should be double sealed in such a way that when the terminal box cover is removed for the connection of incoming cables the remaining electrical components are still protected by the watertight enclosure.

The actuator shall have an integral motor starter, local controls and lamp indication. Provide phase discriminator relay.

The starter shall include a mechanically and electrically interlocked reversing contactor, with control transformer having a grounded screen between primary and secondary windings. The common point of contactor coils and secondary winding shall also be grounded, so that any ground fault will cause contactors to drop-out. Terminals for remote controls shall be provided.

The starter components shall be readily accessible for inspection without disconnecting external cables. Internal wiring shall be number-identified at both ends.

Lamp indication of "close" (Green), intermediate (White) and "open" (Red) positions shall be provided.

Open and close torque and/or position limit switches, plus two auxiliary limit switches at each end of travel shall be provided. Switch ratings shall be 5 A at 115 V.a.c or as specified in data sheet.

Internal control wiring of 5 A tropical grade polyvinyl chloride (PVC) cable shall be provided, terminating in a separately sealed housing with stud terminals. The 3-phase leads of the motor shall be brought to separately stainless studs.

The motor shall be pre-lubricated and all bearings shall be of anti-frictions type.

The motor shall be sized for the torque requirements according to the valve size, operating pressure and temperature, and speed operation.

The motor shall have class "B" insulation, short time rated, with burn-out protection provided.

The motor shall be of sufficient size to open and close the valve against maximum differential pressure when voltage to motor terminals is within 10% of rated voltage. Motor name plate rating shall be 380 V, 3-phase, 50 Hz.

8.17 Actuator construction materials

Materials of construction shall be manufacturer's standard for the specified environmental exposure. The material of diaphragm housing shall be steel, unless otherwise specified. For piston type actuators aluminum housing are acceptable except for valve on depressurizing or emergency shut-off services. In special cases such as for the larger sizes of butterfly valves, consideration may be given to (long-stroke) cylinder actuators.

The enclosure housing the electrical components of a valve shall be made of iron, steel, brass, bronze, aluminum, or an alloy containing not less than 85% aluminum. A metal such as zinc or magnesium or other alloys shall not be used.

Copper shall not be used for an enclosure for use in Class I group A locations. A copper alloy shall not be used for an enclosure unless it is coated with tin nickel or other acceptable coating, or unless the copper content of the alloy is not more than 30%.

Construction material of actuators may be considered and selected according to the requirements. The following materials shall be considered for different parts of actuators:

- Diaphragm casing: Steel, cast iron, or cast aluminum.
- Diaphragm: Nitrile on nylon or nitrile on polyester.
- Diaphragm plate: Cast iron, cast aluminum, or steel.
- Actuator spring: Alloy steel.
- Spring adjuster: Steel.
- Spring seat: Steel or cast iron.
- Actuator stem: Steel.
- Travel indicator: Stainless steel.
- O-rings: Nitrile.
- Seat bushing: Brass.
- Stem connector: Steel zinc plated.
- Yoke: Iron or steel.

Further reading

Ancrum R. Control valve diagnostics: ready for prime time. Control (Chicago, IL) 1996;9(3):60–2.

Anon. Correct sizing of control valves. Process Control Eng 1993;46(10):52.

Baltz N. Flow control with three-way valves. Chem Eng (NY) 1996;103(4):4.

Baumann HD. Viscosity flow correction for small control valve trim. J Fluids Eng. Transactions ASME 1991; 113(1):86–9.

Boccardi G, Bubbico R, Piero Celata G, Mazzarotta B. Two-phase flow through pressure safety valves. Experimental investigation and model prediction. Chem Eng Sci 2005;60(19):5284–93.

Bollinger R. Sizing sliding gate valves for steam service. Plant Eng (Barrington, IL) 1995;49(14):47–8.

Buzzetta P, Hall J, Heider R. Control optimization saves money in distillation. Hydrocarb Process 2007;86(6): 57–60.

Carlson B. Avoiding cavitation in control valves. ASHRAE J 2001;43(6):58, 60, 62–63.

Darby R. On two-phase frozen and flashing flows in safety relief valves: recommended calculation method and the proper use of the discharge coefficient. J Loss Prev Process Ind 2004;17(4):255–9.

DeRose D. Sizing and selecting FRLs. Fluid Power J 2004;11(8):29–33.

Diener R, Kiesbauer J, Schmidt J. Improved control valve sizing for multiphase flow. Hydrocarb Process 2005; 84(3):59–64.

Diener R, Schmidt J. Sizing of throttling device for gas/liquid two-phase flow Part 1: safety valves. Process Saf Prog 2004;23(4):335–44.

Diener R, Schmidt J. Sizing of throttling device for gas/liquid two-phase flow Part 2: control valves, orifices, and nozzles. Process Saf Prog 2005;24(1):29–37.

Fisher Controls International, Control Valve Handbook.

Foy M. Right-sizing pneumatic valves. Mach Des 2003;75(8):80, 82.

George JA. Sizing and selection of low flow control valves. InTech 1989;36(11):46–8.

Hagglund P, Nielsen T. Valve condition monitoring simplifies evaporator operation. Process Control News (Pulp Pap Ind.) 2001;21(7):8.

Harold D. Control valves. Match size with application. Control Eng (North Am Ed) 1999;46(9):93–8.

Harrold D. Select and size control valves properly to save money. Control Eng (North Am Ed) 1999;46(10): 55–60.

Hegberg MC. Control valve selection for hydraulic systems. ASHRAE J 2000;42(11):33, 35–40.

ISA/ANSI (Instrument Society of America/American National Standards Institute). Flow equations for sizing control valves, ANSI/ISA-75.01.01–2002.

Kohan D. Criteria given for sizing and selection of control valves. Oil Gas J 1985;83(50):76–80.

Lam S. HVAC control valve selection. Hong Kong Eng 1989;17(9):12–3. 15.

Lipták B. How to select control valves, Part 3. Control (Chicago, IL) 2006;19(11):55–8.

Liptak BG, editor. Instrument engineers' handbook. 4th ed. CRC Press; 2003.

Luft G, Broedermann J, Scheele T. Pressure relief of high pressure devices. Chem Eng Technol 2007;30(6): 695–701.

Luyben WL. A rational approach to control valve sizing. Ind Eng Chem Res 1990;29(4):700–3.

McLin R. Compressor control retrofits. Hydrocarb Eng 2005;10(5):85–9.

Monsen JF. Spreadsheet sizes control valves for liquid/gas mixtures. InTech 1990;37(12):38–41.

Montana WM. Butterfly valves for control. Chem Eng (NY) 1986;93(5):123.

O'Neill M. Right-sizing valves for downsizing utilities. Power Eng (Barrington, IL) 1997;101(12):105–6, 108, 110.

Perandinou-Petersen A. Careful selection is key to control valve operation. Chem Process 1994;57(8):53–7.

Perry RH, Green DW, editors. Perry's chemical engineers' handbook. 8th ed. McGraw-Hill; 2007.

Pezzinga G, Pititto G. Combined optimization of pipes and control valves in water distribution networks. J Hydraul Res 2005;43(6):668–77.

Pokki J-P, Hurme M, Aittamaa J. Dynamic simulation of the behaviour of pressure relief systems. Comput Chem Eng 2001;25(4–6):793–8.

Rahimi Mofrad S. Instrumentation/safety: relief rate calculation for control valve failure. Hydrocarb Process 2008;87(1):105–6, 108–109.

Singh J. Sizing relief vents for runaway reactions. Chem Eng (NY) 1990;97(8):104–11.

Singleton EW. Development of a high-performance choke valve with reference to sizing for multiphase flow. Meas Control 1991;24(9):273–81.

Wollerstrand J. Flexible sizing and its influence on DH network operating conditions. Euroheat Power/ Fernwarme Int 2002;31(10):52–7.

Natural Gas Dehydration

9

Water in natural gas can create problems during transportation and processing, of which the most severe is the formation of gas hydrates or ice that may block pipelines, process equipment, and instruments. Corrosion of materials in contact with natural gas and condensed water is also a common problem in the oil and gas industry.

Natural gas reservoirs always have water associated with them. Thus, gas in the reservoir is water saturated. When the gas is produced, water is produced as well. Some of this water is produced water from the reservoir directly. Other water produced with the gas is water of condensation formed because of the changes in pressure and temperature during production.

In the sweetening of natural gas, the removal of hydrogen sulfide and carbon dioxide, aqueous solvents are usually used. The sweetened gas, with the H_2S and CO_2 removed, is saturated with water. In addition, the acid gas byproduct of the sweetening is also saturated with water. Furthermore, water is an interesting problem in the emerging technology for disposing of acid gas by injecting into a suitable reservoir–acid gas injection.

In the transmission of natural gas, further condensation of water is problematic. It can increase pressure drop in the line and often leads to corrosion problems. Thus, water should be removed from the natural gas before it is sold to the pipeline company.

For these reasons, the water content of natural gas and acid gas is an important engineering consideration.

Dehydration is the process used to remove water from natural gas and natural gas liquids (NGLs), and is required to [1]:

- prevent formation of hydrates and condensation of free water in processing and transportation facilities,
- meet a water content specification,
- prevent corrosion.

Three of the most common methods for dehydration of natural gas are physical absorption using glycols, adsorption on solids (e.g., molecular sieve/silica gel), and condensation by a combination of cooling and chemical injection (ethylene glycol/methanol). Triethylene glycol (TEG) dehydration is the most frequent method used to meet pipeline sales specifications. Adsorption processes are used to obtain very low water vapor concentration (0.1 ppm or less) required in low temperature processing, such as deep NGL extraction and liquefied natural gas (LNG) plants.

Some relatively new processes for dehydration involve applying isentropic cooling and separation using high centrifugal forces in supersonic gas flow. To estimate the limits of such techniques, it is important to have models that can calculate water vapor concentration in equilibrium with hydrate, ice, and water in natural gas at operational temperature and pressure.

Natural Gas Processing. http://dx.doi.org/10.1016/B978-0-08-099971-5.00009-X
Copyright © 2014 Elsevier Inc. All rights reserved.

A distinct difference between the chemical based (e.g., glycol absorption) and the nonchemical based (e.g., adsorption) dehydration techniques is that the chemical based techniques will saturate the gas with chemicals at operational conditions. Both techniques can, in principle, remove almost all the water from the gas, but the phase behavior of the natural gas leaving the processes will be different. Even though chemicals used for absorption, in general, will have low vapor pressure and relatively small amounts will condense per cubic meter gas, the effect of condensation has to be considered in design of pipelines and process equipment.

9.1 Phase behavior of dehydrated natural gas

Adsorption of water on solid surfaces, such as molecular sieves and silica gel, are commonly used to remove water in natural gas to very low levels. Such solid adsorbent based processes can also be designed to remove traces of chemicals such as TEG and mono-ethylene glycol (MEG) from the natural gas.

In the treated natural gas, the content of chemicals will normally be too low to influence the precipitation of water. For such a case, water will precipitate directly as liquid water (with insignificant traces of chemicals), ice, or natural gas hydrate.

Dependent on temperature, pressure, and water vapor concentration, the maximum water precipitation temperature will correspond either to the water dew point, the frost point, or the hydrate point. In a conservative design, the specification of water vapor concentration should be based on this maximum precipitation temperature, and not the traditional water dew point temperature. Accurate conversion between water vapor concentration and precipitation temperature is therefore crucial.

In Figure 9.1, the typical phase behavior of natural gas is illustrated, where water has been removed in an adsorption process. The water content in the gas is 40 ppm (mole) and does not have any traces of chemicals. For this gas, the hydrate saturation line appears at higher temperatures than the equivalent lines for ice and subcooled water. In a cooling process though, ice and liquid water (subcooled) can be formed before hydrates because of the relative slow kinetics of hydrate formation. It is often discussed if hydrates can form directly from the gas phase. Even though the process is expected to be kinetically slow, direct hydrate formation can occur from a thermodynamic point of view.

As we can see from Figure 9.1, natural gas hydrate will generally be the first thermodynamically stable phase that can precipitate in the whole pressure range. If the water content in the gas is higher, we can have pressure regions where water or ice can precipitate at higher temperature than gas hydrate.

Sufficient dehydration is commonly achieved by contacting the natural gas with a TEG solution at high pressure and relatively low temperature (typically 50–100 bar and 20–40 °C). The treated natural gas from a TEG contactor will be at its aqueous dew point downstream from the glycol absorber; hence it is saturated with TEG and water.

It is a good approximation that the treated gas is in thermodynamic equilibrium with the lean TEG. A typical lean TEG composition is 99 wt% TEG and 1 wt% water. If the treated gas temperature decreases, a solution of water and TEG will start to condense from the natural gas. At absorber operating pressure, the first droplet that is formed will have the same composition as the lean TEG. The aqueous dew point at glycol absorber operation pressure is therefore equal to the temperature of the gas leaving the absorber. TEG will work as a hydrate inhibitor in the condensed phase.

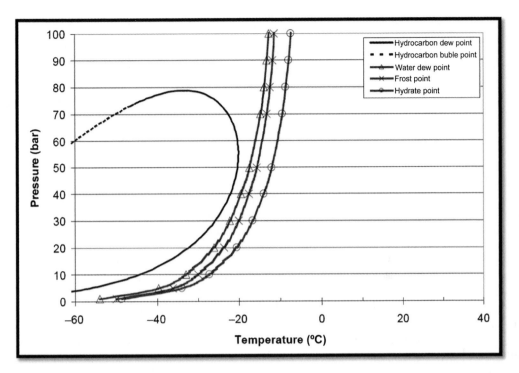

FIGURE 9.1

Phase behavior of natural gas with traces of water (40 ppm (mole)), NG composition (mole): 85% C_1, 10% C_2, 4% C_3, 0.5% nC_4, 0.5%iC_4.

9.2 Water content of natural gases

A number of different water dew point specifications exist for natural gas. The water dew point specification is specified in sales gas contracts or given by requirements for transport, processing, or storage. For example, the water dew point specification for gas transported through pipelines to Europe is typically 8 °C at 70 bar. For LNG production, the water specification needs to be more stringent, and a specification of <0.1 ppm (mole) water in the gas is normally used. Due to the low solubility of water in hydrocarbon liquids, the water specification for liquefied petroleum gas (LPG) products need to be low. Table 9.1 summarizes typical water specification used for natural gas products.

As can be seen, specifications are given both as a dew point at a given pressure and as water vapor concentration. Accurate tools are necessary for converting between the various specifications. Online dew point analyzers are normally calibrated to report ppm (mole) of water in natural gas.

A large number of methods have been developed for the estimation of water content and water dew point of natural gas. Only a few of them will be examined here. The methods range from simple methods, such as generalized charts for direct reading of water content to advanced numerical demanding thermodynamic models (e.g., modern models based on equations of state). We can say that

Table 9.1 Typical Specification for Water in Natural Gas Products	
Natural Gas Type	**Water Dew Point/ Content Specification**
Pipeline sales gas	−8 °C at 70 bar
LNG	0.1 ppm(mole)
Propane	No free
LPG	No free

the simple methods will have a limited range of validity when it comes to gas compositions, pressure, and temperature range. The more advanced models will generally be more accurate and be applicable for a larger span of variables, such as gas composition, pressure, and temperature.

Most of the simple methods based on charts and empirical correlations are only developed to predict the water dew point, and will not be able to predict hydrate precipitation, ice, or the aqueous dew point. Some of the more recently developed models based on fundamental thermodynamics are developed to predict all mentioned dew points.

9.3 Gas water content prediction using generalized charts

The estimation of water content of natural gas based on reading from charts is a common method to use. The charts are generally based on experimental data or thermodynamic models. Normally, the water content is plotted on a semi-logarithmic scale against the dew point temperature in form of isobaric lines. A large number of such diagrams have been made and published in reference literature in gas processing. The diagrams that are most used in the gas industry are the diagrams from McKetta and Wehe [2]. These diagrams are based on experimental data measured in the 1940s.

In 1958, McKetta and Wehe [2] published a chart for estimating the water content of sweet natural gas. This chart has been modified slightly over the years and has been reproduced in many publications, most notably the *GPSA Engineering DataBook* [1]. To obtain the values in this study, the original chart was photo-enlarged to two times its original size. Even so, it is difficult to read the chart to an accuracy of more than two significant figures. Therefore, the values reported here are only two significant figures.

Figure 9.2 has been widely used for many years in the design of "sweet" natural gas dehydrators. It was first published in 1958 and was based on experimental data available at that time. The gas gravity correlation should never be used to account for the presence of H_2S and CO_2 and may not always be adequate for certain hydrocarbon effects, especially for the prediction of water content at pressures above 10,000 kPa (abs). The hydrate formation line is approximate and should not be used to predict hydrate formation conditions.

Figure 9.2 Plotted on this graph is an equilibrium curve of hydrate formation, which should be a function of gas composition.

Determination of water content by this chart produces an error not exceeding 4%, which is acceptable for engineering purposes. As seen in Figure 9.2, the water content of a natural gas increases with the increase in temperature and decreases with decrease in pressure. Moreover, the water content of natural gases drops with an increase in their molecular weight and with an increase in the water salinity. The two auxiliary graphs shown in Figure 9.2 are for finding the correction factors for the molecular weight (gas density), Cg, and water salinity (Cs).

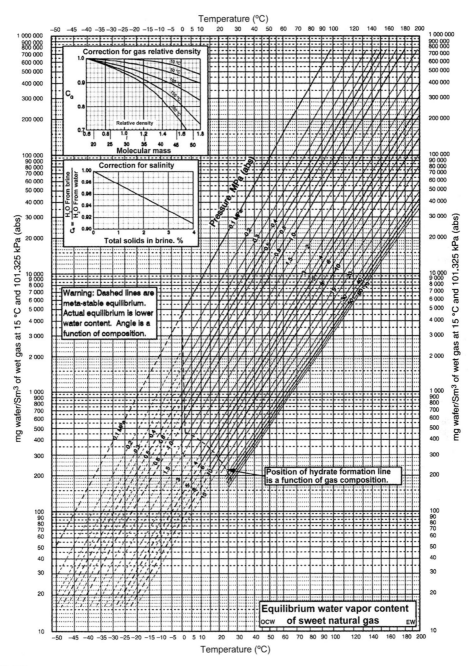

FIGURE 9.2

Water content of hydrocarbon gas.

Reprinted with permission from [1].

The McKetta-Wehe chart [2] is not applicable to sour gas, as will be clearly demonstrated here. Fortunately, most engineers who work in the natural gas industry are aware of this limitation.

The saturated water content of a gas depends on pressure, temperature, and composition. The effect of composition increases with pressure and is particularly important if the gas contains CO_2 and/or H_2S.

For lean, sweet natural gases containing over 70% methane and small amounts of heavy hydrocarbons, generalized pressure-temperature correlations are suitable for many applications.

One of the limitations in accuracy of using these methods is that water content of natural gasses is generally dependent on the gas composition, and such composition dependency can, in general, not easily be implemented in chart based methods. The compositional dependency is normally handled by developing charts for various types of gas mixtures (e.g., sweet gas and sour gas). Another limitation of such methods is that it cannot be used to estimate hydrate, ice, or the aqueous dew point. The area in the charts where the most stable phase is ice or natural gas hydrate is normally plotted with dotted lines indicating equilibrium water content with meta-stable liquid water. Acid gas components, such as CO_2 and H_2S, can have a large influence of water solubility in the gas phase, and generalized charts for such gas mixtures are difficult to develop.

For the chart based method of McKetta and Wehe, a correction factor that is a function of relative gas density can be applied to correct for deviations in gas composition. Correction factors for taking into account the effect of salinity of water have also been presented in this publication. Use of these correction factors is expected to improve the results of the method, but will, in general, be too simple to expect accurate results.

Even though there are many weaknesses in chart based methods for estimating water content of natural gasses, because of the simplicity of this method, these methods must be expected to be used by the gas industry also in the future. When using such methods, it is important to have in mind the weaknesses of the methods, and always look for charts developed for gases with similar compositions.

The following examples are provided to illustrate the use of Figure 9.2:

EXAMPLE 9.1:

Determine the saturated water content for a sweet lean hydrocarbon gas at 66 °C and 6900 kPa (abs). From Figure 9.2,

$W = 3520$ mg/Sm^3

For a 26 molecular mass gas, Cg $= 0.98$ (Figure 9.2)

$W = (0.98)(3520) = 3450$ mg/Sm^3

For a gas in equilibrium with a 3% brine,

Cs $= 0.93$ (Figure 9.2)
$W = (0.93)(3520) = 3270$ mg/Sm^3.

Both H_2S and CO_2 contain more water at saturation than methane or sweet natural gas mixtures. The relative amounts vary considerably with temperature and pressure.

An accurate determination of water content requires a careful study of the existing literature and available experimental data. In most cases, additional experimental data is the best way to verify predicted values. Even the most sophisticated equation of state (EOS) techniques may give results of questionable reliability.

Figure 9.3 is used to estimate the ratio of water content sour gases over water content of sweet gases. The CO_2 is converted to equivalent H_2S as for the second method, and the factor is 70%.

$$\%H_2S \text{ Equivalent} = (H_2S \text{ mole pernet} + 0.70 \, CO_2 \text{ mole percent}) \tag{9.1}$$

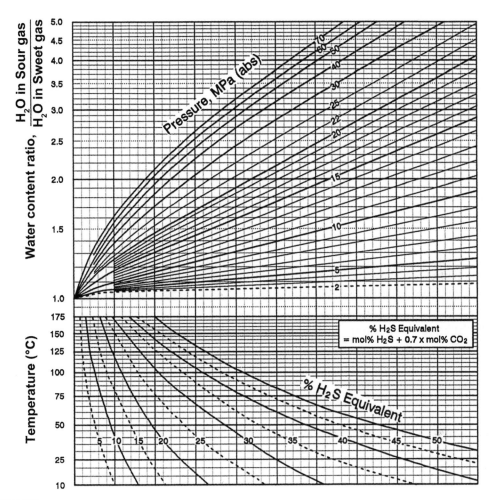

FIGURE 9.3

A chart to estimate the ratio of water content sour gases over water content of sweet gases.

Reprinted with permission from GPSA [1].

9.4 Gas water content prediction using empirical methods

Various simple empirical models have been developed for the calculation of water content of natural gas. The simplest models are based on functions fitted to the experimental data for the vapor pressure of pure water. In an ideal gas, the water content will be directly given by the vapor pressure of water and the total pressure. However, such models will generally be invalid for pressures higher than typically 10 bar. The maximum pressure will depend on how ideal the gas mixture behaves.

For gases with high solubility in water (e.g., gases with high CO_2 or H_2S content), the ideal method is not valid even at low pressures. Some empirical models correct for the non-ideality of the gas by fitting the model to high pressure experimental data. Such models can give reasonable results at higher pressures, but will, in general, be limited to gases with similar composition as what was as experimental basis. A popular empirical correlation used in the natural gas industry was developed by Bukacek. This method is published as a standard for defining the relation between water content and water dew point of natural gas (ASTM D1142-95).

9.4.1 Ideal model

In this model, the water content of a gas is assumed equal to the ratio of the vapor pressure of pure water divided by the total pressure of the system. This yields the mole fraction of water in the gas in pounds per MMscf.

$$W = 47484 \frac{Pv}{P} \tag{9.2}$$

This model is reasonably good at very low pressures. This equation can be used with reasonable accuracy for sweet natural gas and pressures up to about 200 psia.

9.4.2 Bukacek's method

Correlation permitting determination of water content of natural gases (lb/MMscf) for pressures up to 10,000 psia and for temperature ranged from-40 to 230 °F. The following expression is used for calculating gas water content:

$$W = \frac{A}{P} + B \tag{9.3}$$

where:

A = Coefficient equal to the water content of ideal gas
B = Coefficient dependent on the gas composition

The values of A and B are given in Table 9.2 and can be calculated by regression analysis, as it is shown in the computer program.

Bukacek suggested a relatively simple correlation for the water content of sweet gas. The water content is calculated using the following equations:

$$W = 47,484 \frac{P_{\text{water}}^{\text{sat}}}{P_{\text{total}}} + B \tag{9.4}$$

$$\log B = \frac{-3083.87}{459.6 + T} + 6.69449 \tag{9.5}$$

This correlation is reported to be accurate for temperatures between 60 and 460 °F and for pressures from 15 to 10,000 psia.

Where W is the water content (lb/MMscf), P is the total pressure (psia), A is a constant proportional to the vapor pressure of water, and B is a constant depending on temperature and

Table 9.2 Coefficients for Eqn (9.2)

Temperature, °F	A	B	Temperature, °F	A	B
−40	0.1451	0.00347	89.6	36.1	0.1895
−36.4	0.178	0.00402	93.2	40.5	0.207
−32.8	0.2189	0.00465	96.8	45.2	0.224
−29.2	0.267	0.00538	100.4	50.8	0.242
−25.6	0.3235	0.00623	104	56.25	0.263
−22	0.393	0.0071	107.6	62.7	0.285
−18.4	0.4715	0.00806	111.2	69.25	0.31
−14.8	0.566	0.00921	114.8	76.7	0.335
−11.2	0.6775	0.01043	118.4	85.29	0.363
−7.6	0.8909	0.01168	122	94	0.391
−4	0.966	0.0134	125.6	103	0.422
−0.4	1.144	0.0151	129.2	114	0.454
3.2	1.35	0.01705	132.8	126	0.487
6.8	1.59	0.01927	136.4	138	0.521
10.4	1.868	0.021155	140	152	0.562
14	2.188	0.0229	143.6	166.5	0.599
17.6	2.55	0.0271	147.2	183.3	0.645
21.2	2.99	0.03035	150.8	200.5	0.691
24.8	3.48	0.0338	154.4	219	0.741
28.4	4.03	0.0377	158	238	0.793
32	4.67	0.0418	161.6	260	0.841
35.6	5.4	0.0464	165.2	283	0.902
39.2	6.225	0.0515	168.8	306	0.965
42.8	7.15	0.0571	172.4	335	1.023
46.4	8.2	0.063	176	363	1.083
50	9.39	0.0696	179.6	394	1.148
53.6	10.72	0.0767	183.2	427	1.205
57.2	12.39	0.0855	186.8	462	1.25
60.8	13.94	0.093	190.4	501	1.29
64.4	15.75	0.102	194	537.5	1.327
68	17.87	0.112	197.6	582.5	1.327
71.6	20.15	0.1227	201.2	624	1.405
75.2	22.8	0.1343	204.8	672	1.445
78.8	25.5	0.1453	208.4	725	1.487
82.4	28.7	0.1595	212	776	1.53
86	32.3	0.174	230	1093	2.62

gas composition. The effect of gas composition is indirect corrected for by multiplying the B factor with a term dependent on gas gravity. The B factor has also been estimated based on experimental data for water content in sweet natural gas mixtures and values for the constant are tabulated in the original publication.

Bukacek's method is fitted to water content in natural gas in equilibrium with liquid water. The method can, thus, not be expected to be able to estimate water content in equilibrium with hydrate or ice.

The simple empirical models cannot be expected to be as accurate as more advanced thermodynamic models. Extrapolation in gas composition, temperature, and pressure must be done with caution. Due to their simplicity and low numerical requirements, such methods are widely used in the industry. Many of the online water dew point analyzers use these empirical correlations to estimate water dew point based on water content analysis.

9.4.3 Maddox correlation

Maddox proposed a method for calculating the water content of sour gas. This correlation assumes that the water content of sour gas is the sum of three terms: (1) a sweet gas contribution, (2) a contribution from CO_2, and (3) a contribution from H_2S. Figures 9.4 and 9.5 are provided to estimate the contributions for CO_2 and H_2S. The chart for CO_2 is for temperatures between 80 and 160 °F, and the chart for H_2S is for temperatures between 80 and 280 °F, as it is shown in Figures 9.7 and 9.8.

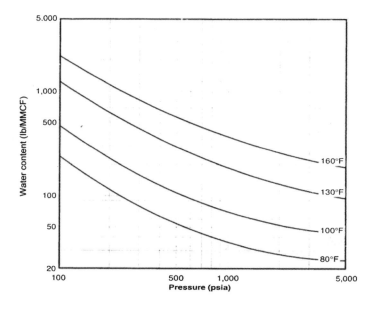

FIGURE 9.4

Maddox correction for the water content of CO_2.

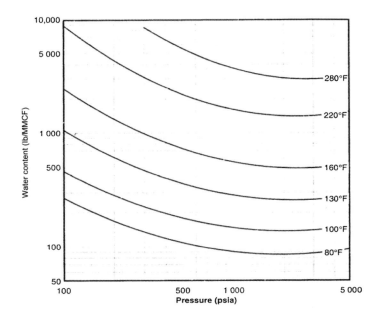

FIGURE 9.5

Maddox correction for the water content of H_2S.

The water content of both CO_2 and H_2S were correlated as a function of the pressure using only the following relation:

$$\log(W) = a_0 + a_1 \log P + a_2 (\log P)^2 \tag{9.6}$$

where W is the water content in lb/MMscf and a set of coefficients, a_0, a_1, and a_2 was obtained for each isotherm. The coefficients are listed in Table 9.3.

Table 9.3 Values of the Correlation Coefficients in Eqn (9.5)

Temperature(°F)	a_0	a_1	a_2
Carbon Dioxide			
80	6.0901	−2.5396	0.3427
100	6.1870	−2.3779	0.3103
130	6.1925	−2.0280	0.2400
160	6.1850	−1.8492	0.2139
Hydrogen Sulfide			
80	5.1847	−1.9772	0.3004
100	5.4896	−2.0210	0.3046
130	6.1694	−2.2342	0.3319
160	6.8834	−2.4731	0.3646
220	7.9773	−2.8597	0.4232
280	9.2783	−3.3723	0.4897

Table 9.4 Values of Constants in Eqn (9.9)

Constants	Value
c_1	2.8910758E + 01
c_2	−9.668146E + 03
c_3	−1.663358E + 00
c_4	−1.308235E + 05
c_5	2.0353234E + 02
c_6	3.8508508E − 02

9.4.4 Sloan correlation

Sloan fitted the water content of natural gas versus both temperature and pressure. His equation is valid for temperatures between −40 and 120 °F and for pressures from 200 to 2000 psia.

$$W = EXP\left\{c_1 + \frac{c_2}{T} + c_3 \ln(P) + \frac{c_4}{T^2} + \frac{c_5 \ln(P)}{T} + c_6(\ln(P))^2\right\} \tag{9.7}$$

where:

$T =$ Temperature in °R

$c_1–c_6 =$ Constants are given in Table 9.4

9.5 Methods based on EOS

Thermodynamic models based on EOS for calculating water dew point and water content of natural gas can be relatively complex and computers have to be utilized in doing efficient calculations. However, many of the developed models have been shown to give accurate predictions of water dew point for a large number of gas compositions and total pressures. Some of the popular classic EOS, like the SRK- and PR-EOS often used in the oil and gas industry, have traditionally had problems in handling polar components like water.

The traditional way of fixing this limitation of modeling polar components correctly has been to use a modified attractive term in the EOS to reproduce the vapor pressure of polar components more accurately. Still, we have seen a limitation of such models in that the density of the liquid phase was predicted badly and that relatively large and binary interaction parameters between water and hydrocarbons had to be used. Scientific development of EOS during the last decades has more or less removed this limitation by adding explicit terms for modeling the effect of the hydrogen bonding between polar components.

Most modern EOS are developed by fitting parameters (e.g., binary interaction parameters) to experimental data for both pure components and mixtures. This is the case for the EOS used in ISO 18,453 that has been developed for converting between water content and water dew point of natural gas. Some of the more recent models have been shown to be able to predict mixture properties based on knowledge of only pure component properties. It was demonstrated that such a predictive model could be applied for the calculation of the water content of natural gas mixtures with high accuracy. The model was also developed to predict hydrate and ice precipitation from natural gas, and is also suited to estimate the aqueous dew point in gases with traces of production chemicals. This model is normally referred to as the CPA-EOS and is described in more detail in this chapter.

The advantages of methods based on fundamental thermodynamic models are that they are expected to cover a larger range of gas compositions, temperatures, and pressures. The models used in this work are based on EOS.

9.6 Hydrates in natural gas systems

Gas hydrates are representatives of a class of compounds known as clathrates or inclusion compounds. Natural gas and crude oil normally reside in reservoir in contact with connate water. Water can combine with low-molecular weight natural gases to form a solid, hydrate, even if the temperature is above the water freezing point.

Hydrates are considered a nuisance because they block transmission lines, plug blowout preventers, jeopardize the foundation of deep-water platforms and pipelines, cause tubing and casing collapses, and foul process heat exchangers, valves, and expanders.

Hydrates act to concentrate hydrocarbons; one cuft of hydrates may contain as much as 180 scf of gas. Large natural reserves of hydrocarbons exist in hydrated form, both in deep oceans and in the permafrost. Evaluation of these reserves is highly uncertain, yet even conservative estimates indicate that there is perhaps twice as much energy in the hydrated form as in all other hydrocarbon sources combined.

This chapter is intended to provide the basic information needed for engineering purposes about hydrates. Figure 9.6 provides initial hydrate-formation estimation for natural gases based on gas gravity.

In the last 50 years, several studies have been conducted on the measurement and prediction of hydrate conditions for various gas mixtures.

Because it was impractical to measure the pressure and temperature of hydrate formation for every gas composition, estimation of the hydrate formation conditions for natural gases are carried out using K-value and correlations developed with gravity method. These correlations can be classified into four major methods, such as the Vapor–solid equilibrium ratio method, Modified K-factor method Gas Gravity Method [3], and Empirical correlations method. Various empirical correlations have been presented in the literature for predicting the hydrate formation conditions.

9.6.1 Kobayashi et al. correlation

Kobayashi et al. (1987) [4] developed an empirical equation that predicts the hydrate-forming temperatures at given pressures for systems including only hydrocarbons in limited range of temperatures, pressures, and gas specific gravities. Table 9.5 gives the adjusted parameters for the Kobayashi et al. correlation.

$$
\begin{aligned}
T = 1 / \Big[& A_1 + A_2 (\ln \gamma_g) + A_3 (\ln P) + A_4 (\ln \gamma_g)^2 + A_5 (\ln \gamma_g)(\ln P) \\
& + A_6 (\ln P)^2 + A_7 (\ln \gamma_g)^3 + A_8 (\ln \gamma_g)^2 (\ln P) \\
& + A_9 (\ln \gamma_g)(\ln P)^2 + A_{10} (\ln P)^3 + A_{11} (\ln \gamma)^4 \\
& + A_{12} (\ln \gamma_g)^3 (\ln P) + A_{13} (\ln \gamma_g)^2 (\ln P)^2 \\
& + A_{14} (\ln \gamma_g)(\ln P)^3 + A_{15} (\ln P)^4 \Big]
\end{aligned}
\tag{9.8}
$$

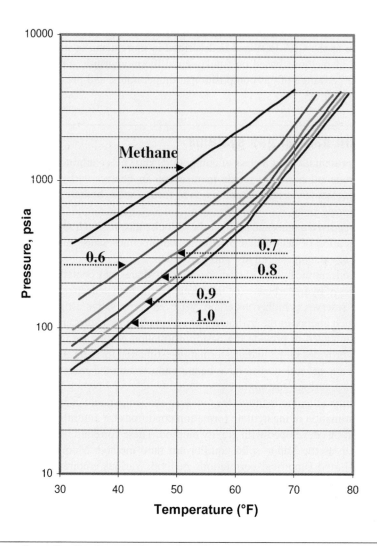

FIGURE 9.6

Initial hydrate-formation estimation for natural gases based on gas gravity.

Ref. [1].

Table 9.5 Adjusted Parameters for Kobayashi et al. Correlation

Const	Values	Const	Values
A_1	25.9987×10^0	A_9	16.0621×10^0
A_2	9.6153×10^{-1}	A_{10}	-2.2886×10^{-2}
A_3	-11.2419×10^0	A_{11}	-7.0447×10^{-3}
A_4	9.0125×10^{-2}	A_{12}	7.6032×10^0
A_5	18.7886×10^0	A_{13}	2.2559×10^{-1}
A_6	-1.6364×10^{-1}	A_{14}	10.3107×10^0
A_7	-2.1699×10^0	A_{15}	2.4159×10^{-1}
A_8	7.1141×10^{-3}	—	—

where:

T = temperature, °F
P = Pressure, psia
γ_g = Gas specific gravity

9.6.2 Berge correlation

The equations developed by Berge (1986) [5] are temperature explicit(i.e., temperature is calculated directly for a given pressure and specific gravity of the gas). The equations for predicting hydrate temperatures are:

For $0.555 \leq \gamma_g \leq 0.58$: $T = -96.03 + 25.37$

$$\times \ln P - 0.64 \times (\ln P)^2 + \frac{\gamma_g - 0.555}{0.025} \times \left[80.61 \times P + 1.16 \times \frac{10^4}{(P + 599.16)} \right]$$

$$-\left(-96.03 + 25.37 \times \ln P - 0.64 \times (\ln P)^2 \right) \tag{9.9}$$

For $0.58 \leq \gamma_g \leq 1.0$

$$T = \frac{\left\{ 80.61 \times P - 2.1 \times 10^4 - 1.22 \times \frac{10^3}{(\gamma_g - 0.535)} - \left[1.23 \times 10^4 + 1.71 \times \frac{10^3}{\gamma_g - 0.509} \right] \right\}}{\left[P - \left(-260.42 - \frac{15.18}{\gamma_g - 0.535} \right) \right]} \tag{9.10}$$

where:

T = temperature, °F
P = Pressure, psia
γ_g = Gas specific gravity

9.6.3 Motiee correlation

A regression method was used to determine six coefficients that would correlate temperature, pressure, and specific gravity. Table 9.6 gives the adjusted parameters for the Motiee et al. correlation. The equations are:

$$\text{Log}(P) = a_1 + a_2 T + a_3 T^2 + a_4 \gamma_g + a_5 \gamma_g^2 + a_6 T \gamma_g$$

$$T = b_1 + b_2 \text{Log}(P) + b_3 (\text{Log}(P) + b_4 \gamma_g + b_5 \gamma_g^2 + b_6 \gamma_g \text{Log}(P) \tag{9.11}$$

Table 9.6 Adjusted Parameters for Kobayashi et al. Correlation

Mottie Correlation (Pressure Explicit From)			
Constants	**Values**	**Constants**	**Values**
a_1	−152.9132	a_4	162.8287
a_2	32.2076	a_5	−55.6539
a_3	−1.1256	a_6	−8.0644

Table 9.7 Adjusted Parameters for Hammerschmidt Correlation

Hammerschmidt	
Constants	**Values**
α	14.7593
β	0.2101

9.6.4 Hammerschmidt correlation

Hammerschmidt gives the following relationship for initial hydrate forming temperature below:

$$T = \alpha P^\beta \tag{9.12}$$

By transforming Eqn (9.6) to pressure explicit form, the equation for initial pressure calculation becomes:

$$P = \left(\frac{T}{\alpha}\right)^{\frac{1}{\beta}} \tag{9.13}$$

Table 9.7 shows adjusted parameters for the Hammerschmidt correlation.

9.7 Thermodynamic model for the hydrate phase

The hydrate phase is modeled based on the van der Waals and Platteeuw model (van der Waals and Platteeuw, 1959 [6]; Sloan and Koh, 2007 [7]), combined with a fugacity approach as proposed by Chen and Guo (1996) [8] and Klauda and Sandler (2000) [9]. The equilibrium criteria of the hydrate–forming mixture are based on the equality of fugacity of the specified component i in all phases that coexist simultaneously:

$$f_i^H = f_i^L = f_i^V (= f_w^I) \tag{9.14}$$

In this equation, H is the hydrate phase either structure I, II, or H, L is the liquid phase such as the water-rich or guest-rich liquid phases, V is the vapor phase, and i is the ice phase. Each of the possible hydrate structures I, II, or H has to be considered as one independent phase because the mixed or double hydrates can possibly exist for the systems containing one or more guest components. The fugacity of water in the hydrate phase $f_w^H(T,P)$ is given by:

$$f_W^H(T,P) = f_W^{MT}(T,P)\exp\left(\frac{-\Delta\mu_w^{MT-H}}{RT}\right) \tag{9.15}$$

Here, $f_W^{MT}(T,P)$ represents the fugacity of water in the hypothetical empty hydrate cavity and $\Delta\mu_w^{MT-H}$ is the chemical potential difference of water in the empty hydrate lattice, μ_w^β and the hydrate phase, $\mu_w^H \cdot \Delta\mu_w^{MT-H}$ is calculated from the following equation:

$$\Delta\mu_w^{MT-H} = RT \sum_m v_m \ln\left(1 - \sum_j \theta_{mj}\right) \tag{9.16}$$

where v_m is the number of cavities of type m per water molecule in the hydrate phase (which are $v_1 = 1/23$: and $v_2 = 3/23$ for structure I hydrate and $v_1 = 2/17$ and $v_2 = 1/17$ for structure II hydrate) and θ_{mj} the fraction of cavities of type m occupied by the molecules of component j. This fractional occupancy is determined by a Langmuir-type expression:

$$\theta_{mj} = \frac{C_{mj}f_j^V}{1 + \sum_k C_{mk}f_k^V} \tag{9.17}$$

In Eqn. (7.31) C_{mj}, is the Langmuir constant of component j in a cavity of type m and f_j^V is the fugacity of component j in the vapor phase in equilibrium with the hydrate. In the present work, the Langmuir constants C_{mj} are estimated by assuming the following temperature dependence:

$$C_{mj} = \frac{A_{mj}}{T} \exp\left(\frac{B_{mj}}{T}\right) \tag{9.18}$$

The values of the parameters A_{mj} and B_{mj} are given by Munck et al., (1988) [10] and Delahaye et al., (2006) [11] and are tabulated in Table 9.8.

The fugacity of water in the empty hydrate lattice, $f_W^{MT}(T,P)$, is calculated from:

$$f_w^{MT} = P_w^{sat,MT} \phi_w^{sat,MT}(T)\exp\left(\frac{v_w^{MT}(P - P_w^{sat,MT}(T))}{RT}\right) \tag{9.19}$$

With $P_w^{sat,MT}$ and $\phi_w^{sat,MT}$ respectively as the saturation vapor pressure and fugacity coefficient of the hypothetical empty hydrate lattice at temperature T.

Based on X-ray diffraction data obtained by Tse (1987) [12] and Hirai et al. (2000) [13], Klauda and Sandler (2000) [9] expressed the molar volume of the empty hydrate lattice, v_w^{MT} for each structure as a function of the temperature and pressure as:

$$v_{w,I}^{MT}(T,P) = \left(11.835 + 2.217 \times 10^{-5}T + 2.242 \times 10^{-6}T^2\right)^3 \frac{10^{-30}N_A}{46}$$
$$-8.006 \times 10^{-9}P + 5.448 \times 10^{-12}P^2 \tag{9.20}$$

$$v_{w,II}^{MT}(T,P) = \left(17.13 + 2.249 \times 10^{-4}T + 2.013 \times 10^{-6}T^2 + 1.009 \times 10^{-9}T^3\right)^3 \frac{10^{-30}N_A}{136}$$
$$-8.006 \times 10^{-9}P + 5.448 \times 10^{-12}P^2 \tag{9.21}$$

Table 9.8 The Optimized A_{mj} and B_{mj} Parameters for the Calculation of Langmuir Constants

		CO$_2$		THF	
		A(K/Pa)	B(1/K)	A(K/Pa)	B(1/K)
Structure I	Large cavities	4.19E-07	2813	–	–
	Small cavities	2.44E-09	3410	–	–
Structure II	Large cavities	8.40E-06	2025	6.5972	1003.22
	Small cavities	8.34E-10	3615	–	–

The saturation vapor pressure $P_w^{sat,MT}$ is modeled based on Ruzicka and Majer (1996) [14]. This expression is chosen because of its accuracy in extrapolation. The expression is shown below:

$$\ln(P_w^{sat,MT}) = A \ln(T) + \frac{B}{T} + C + DT \tag{9.22}$$

where A, B, C, and D are adjustable parameters and their values are regressed from phase equilibrium data. The adjustable parameter C is fixed at 2.7789. This is done to enable a comparison with the data of Klauda and Sandler (2000) [9].

9.8 Hydrate predictions for high CO_2/H_2S content gases

The Katz method of predicting hydrate formation temperature gives reasonable results for sweet normal paraffin hydrocarbon gases. The Katz method should not be used for gases containing significant quantities of CO_2 and/or H_2S despite the fact that K-values (K the *value* of the *equilibrium constant*) values are available for these components.

Hydrate formation conditions for high CO_2/H_2S gases can vary significantly from those composed only of hydrocarbons. The addition of H_2S to a sweet natural gas mixture will generally increase the hydrate formation temperature at a fixed pressure.

A method by Baille andWichert for predicting the temperature of high H_2S content gases is shown in Figure 9.7. This is based on the principle of adjusting the propane hydrate conditions to account for the presence of H_2S as illustrated in the following example.

EXAMPLE 9.2:

Estimate the hydrate formation temperature at 4200 kPa (abs) of a gas with the following analysis using 9.7.

Component	Mol%
N_2	0.30
CO_2	6.66
H_2S	4.18
C_1	84.27
C_2	3.15
C_3	0.67
iC_4	0.20
nC_4	0.19
C_{5+}	0.40
$M = 19.75$	$\gamma = 0.682$

Solution steps:
1. Enter left side of Figure 9.7 at 4200 kPa (abs) and proceed to the H_2S concentration line (4.18 mol%).
2. Proceed vertically to the relative density of the gas ($\gamma = 0.682$).
3. Follow the diagonal guide line to the temperature at the bottom of the graph ($T = 17.5\,°C$).

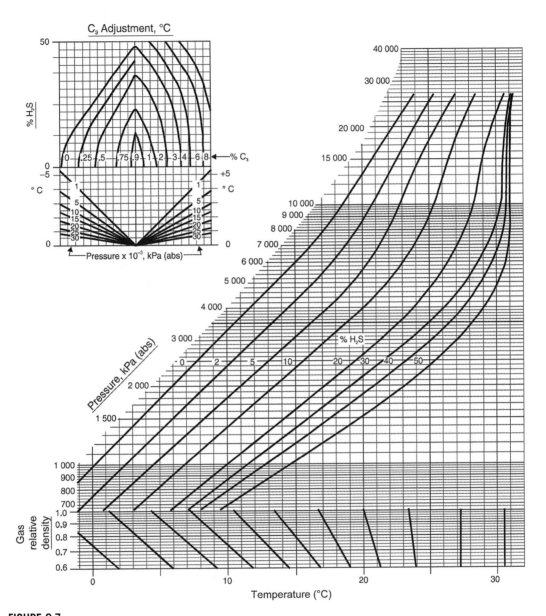

FIGURE 9.7

Hydrate chart for gases containing H_2S.

Ref. [1].

---

4. Apply the C_3 correction using the insert at the upper left. Enter the left hand side at the H_2S concentration and proceed to the C_3 concentration line (0.67%). Proceed down vertically to the system pressure and read the correction on the left hand scale ($-1.5\,°C$).

   Note: The C_3 temperature correction is negative when on the left hand side of the graph and positive on the right hand side.

   $TH = 17.5 - 1.5 = 16\,°C.$

Figure 9.7 was developed based on calculated hydrate conditions using the Peng-Robinson EOS. It has proven to be quite accurate when compared to the limited amount of experimental data available. It should only be extrapolated beyond the experimental data base with caution.

9.9 Hydrate inhibition

The formation of hydrates can be prevented by dehydrating the gas or liquid to eliminate the formation of a condensed water (liquid or solid) phase. In some cases, however, dehydration may not be practical or economically feasible. In these cases, chemical inhibition can be an effective method of preventing hydrate formation. Chemical inhibition utilizes injection of thermodynamic inhibitors or low-dosage hydrate inhibitors (LDHIs).

Thermodynamic inhibitors are the traditional inhibitors (i.e., one of the glycols or methanol), which lower the temperature of hydrate formation. LDHIs are either kinetic hydrate inhibitors (KHIs) or anti-agglomerants (AAs). They do not lower the temperature of hydrate formation, but do diminish its effect. KHIs lower the rate of hydrate formation, which inhibits its development for a defined duration. AAs allow the formation of hydrate crystals but restrict them to submillimeter size.

9.9.1 Thermodynamic inhibitors

Inhibition utilizes injection of one of the glycols or methanol into a process stream where it can combine with the condensed aqueous phase to lower the hydrate formation temperature at a given pressure.

Both glycol and methanol can be recovered with the aqueous phase, regenerated, and reinjected. For continuous injection in services down to $-40\,°C$, one of the glycols usually offers an economic advantage versus methanol recovered by distillation.

At cryogenic conditions (below $-40\,°C$) methanol usually is preferred because glycol's viscosity makes effective separation difficult.

Ethylene glycol (EG), diethylene glycol (DEG), and TEG glycols have been used for hydrate inhibition. The most popular has been EG because of its lower cost, lower viscosity, and lower solubility in liquid hydrocarbons.

To be effective, the inhibitor must be present at the very point where the wet gas is cooled to its hydrate temperature. For example, in refrigeration plants, glycol inhibitors are typically sprayed on the tube-sheet faces of the gas exchangers so that it can flow with the gas through the tubes. As water condenses, the inhibitor is present to mix with the water and prevent hydrates. Injection must be in a manner to allow good distribution to every tube or plate pass in chillers and heat exchangers operating below the gas hydrate temperature.

The inhibitor and condensed water mixture is separated from the gas stream along with a separate liquid hydrocarbon stream. At this point, the water dew point of the gas stream is essentially equal to the separation temperature. Glycol-water solutions and liquid hydrocarbons can emulsify when agitated or when expanded from a high pressure to a lower pressure(e.g., JT expansion valve). Careful separator design will allow nearly complete recovery of the diluted glycol for regeneration and reinjection.

The regenerator in a glycol injection system should be operated to produce a regenerated glycol solution that will have a freezing point below the minimum temperature encountered in the system.

This is typically 75–80 wt%. The minimum inhibitor concentration in the free water phase may be approximated by Hammerschmidt's

$$d = \frac{K_H X_I}{MW_I(1 - X_I)} \tag{9.23}$$

$$X_I = \frac{d MW_I}{K_H + d MW_I} \tag{9.24}$$

where

$$K_H(\text{glycols}) = 2335 - 4000 \text{ and } K_H(\text{methanol}) = 2335$$

where:

d = depression of the water dew point or the gas hydrate freezing point, °C
X_I = mass fraction of inhibitor in the liquid phase
I = Inhibitor
MW_I = molecular mass of inhibitor

The K_H range of 2335–4000 for glycols reflects the uncertainty in the value of this parameter. At equilibrium, such as for a laboratory test, 2335 is applicable.

In some field operations, however, hydrate formation has been prevented with glycol concentrations corresponding with K_H values as high as 4000. This is because hydrate suppression with glycols depends on the system's physical and flow characteristics etc., as well as the properties of the gas and the glycol. Therefore, in the absence of reliable field-test data, a system should be designed for a K_H of 2335. Once the system is operating, the glycol concentration can be reduced to tolerable levels. Equations (9.22) and (9.23) should not be used beyond 20–25 wt% for methanol and 60–70 wt% for the glycols. For methanol concentrations up to about 50%, the Nielsen–Bucklin equation provides better accuracy:

$$d = -72.0 \ln(X_{H_2O}) \tag{9.25}$$

where:

d = depression of the water dew point or the gas hydrate freezing point, °C
X = mole fraction in the liquid phase

Note that "xH$_2$O" in Eqn (9.24) is a mole fraction, not a mass fraction.

Once the required inhibitor concentration has been calculated, the mass of inhibitor required in the water phase may be calculated from 9.25

$$m_I = \frac{X_R \cdot m \, H_2O}{X_L - X_R} \tag{9.26}$$

where:

I = Inhibitor
m = the mass of inhibitor required in the water phase
X = mass fraction in the liquid phase
L = lean inhibitor
R = rich inhibitor

The amount of inhibitor to be injected not only must be sufficient to prevent freezing of the inhibitor water phase, but also must be sufficient to provide for the equilibrium vapor phase content of the inhibitor and the solubility of the inhibitor in any liquid hydrocarbon. Figure 9.8 shows methanol vaporization loss.

The vapor pressure of methanol is high enough that significant quantities will vaporize. Glycol vaporization losses are generally very small and are typically ignored in calculations.

Inhibitor losses to the hydrocarbon liquid phase are more difficult to predict. Solubility is a strong function of both the water phase and hydrocarbon phase compositions. Figure 9.9 shows the solubility

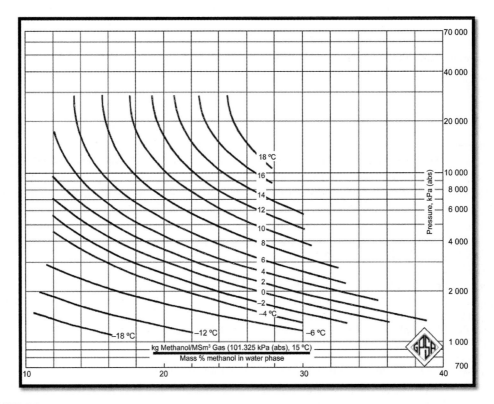

FIGURE 9.8

Methanol vaporization loss chart.

Ref. [1].

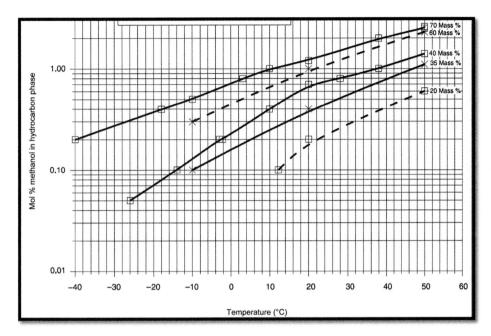

FIGURE 9.9

Methanol solubility in hydrocarbon liquid phase.

Ref. [1].

of methanol in a paraffinic hydrocarbon liquid as a function of temperature and methanol concentration. Methanol solubility in naphthenic hydrocarbons is slightly less than paraffinic, but solubility in aromatic hydrocarbons may be four to six times higher than in paraffins.

Solubility of EG in the liquid hydrocarbon phase is extremely small. A solubility of 40 g/m^3 of NGL is often used for design purposes. However, entrainment and other physical losses may result in total losses significantly higher than this.

EXAMPLE 9.3:

2.83×10^6 Sm^3/day of natural gas leaves an offshore platform at 38 °C and 8300 kPa (abs). The gas comes onshore at 4 °C and 6200 kPa (abs). The hydrate temperature of the gas is 18 °C. Associated condensate production is 56 m^3/Standard million cubic meters. The condensate has a density of 778 kg/m^3 and a molecular mass of 140. Calculate the amount of methanol and 80 mass% EG inhibitor required to prevent hydrate formation in the pipeline.

Solution steps:
Methanol

1. Calculate the amount of water condensed per day from Figure 9.2
   $W_{in} = 850$ mg/Sm^3.
   $W_{in} = 152$ mg/Sm^3.
   $\Delta W = 698$ mg/Sm^3.
   Water condensed $= (2.83 \times 10^6)(698) = 1975(10^6)$ mg/d $= 1975$ kg/d.

2. Calculate required methanol inhibitor concentration:
   $d = 14\,°C$.
   $M = 32$.
   Solving for X_I,
   $X_I = 0.255$.
   $X_I = 0.275$ (use this value in subsequent calculations).
3. Calculate mass rate of inhibitor in water phase (assume 100% methanol is injected).

$$m_1 = \frac{X_R \cdot m_{H_2O}}{X_L - X_R} = \frac{(0.275)(1975)}{(1 - 0.275)} = 749 \text{ kg/d}$$

4. Estimate vaporization losses from Figure 9.8 at 4 °C and 6200 kPa (abs).

$$\text{losses} = 16.8 \left(0^{-6}\right) \frac{\text{kg/m}^3}{\text{wt\% MeOH}}$$

$$\text{daily losses} = \left(1.68 \times 10^{-6}\right)\left(2.83 \times 10^6\right)(27.5) = 1310 \text{ kg/d}$$

5. Estimate losses to hydrocarbon liquid phase from Figure 9.9 at 4 °C and 27.5 wt% MeOH, xMeOH ≈ 0.2 mol% lb mols of condensate per day.

$$\left(\frac{2.83 \times 10^6 \text{ Sm}^3}{d}\right)\left(\frac{56 \text{ m}^3}{10^6 \text{ Sm}^3}\right)\left(\frac{778 \text{ kg}}{\text{m}^3}\right)\left(\frac{1 \text{ kg mol}}{140 \text{ kg}}\right) = 881 \text{ kg mol/d}$$

kg mol methanol $= (881)(0.002) = 1.76$ kg mols/d.
kg methanol $= (1.76)(32) = 56$ kg/d.
Total methanol injection rate $= 749 + 1310 + 56 = 2115$ kg/d.

Methanol left in the gas phase can be recovered by condensation with the remaining water in downstream chilling processes. Likewise, the methanol in the condensate phase can be recovered by downstream water washing.

80 wt% EG

1. Calculate required inhibitor concentration.
   $d = 14\,°C$ $M = 62$.
   Solving for X_I, $X_I = 0.28$.
2. Calculate mass rate of inhibitor in water phase.

$$m_I = \frac{(0.28)(1975)}{(0.8 - 0.28)} = 1063 \text{ kg/d}$$

9.9.2 Low dosage hydrate inhibitors

LDHIs can provide significant benefits compared to thermodynamic inhibitors including:

- Significantly lower inhibitor concentrations and therefore dosage rates. Concentrations range from 0.1 to 1.0 mass percent polymer in the free water phase, whereas alcohols can be as high as 50%
- Lower inhibitor loss caused by evaporation, particularly compared to methanol.
- Reduced capital expenses through decreased chemical storage and injection rate requirements; and no need for regeneration because the chemicals are not currently recovered. These are especially appropriate for offshore where weight and space are critical to costs.

- Reduced operating expenses in many cases through decreased chemical consumption and delivery frequency.
- Increased production rates, where inhibitor injection capacity or flow line capacity is limited.
- Lower toxicity.

9.9.3 Kinetic hydrate inhibitors

KHIs were designed to inhibit hydrate formation in flow lines, pipelines, and downhole equipment operating within hydrate-forming conditions, such as subsea and cold-weather environments.

Their unique chemical structure significantly reduces the rate of nucleation and hydrate growth during conditions thermodynamically favorable for hydrate formation, without altering the thermodynamic hydrate formation conditions (i.e., temperature and pressure). This mechanism differs from methanol or glycol, which depress the thermodynamic hydrate formation temperature so that a flow line operates outside hydrate-forming conditions.

9.9.4 KHIs compared to methanol or glycols

KHIs inhibit hydrate formation at a concentration range of 0.1–1.0 mass percent polymer in the free water phase. At the maximum recommended dosage, the current inhibition capabilities are $-2\,°C$ of subcooling in a gas system and $-7\,°C$ in an oil system with efforts continuing to expand the region of effectiveness. For relative comparison, methanol or glycol typically may be required at concentrations ranging 20–50 mass percent, respectively, in the water phase.

9.10 Natural gas dehydration methods

Dehydration of natural gas is the removal of the water that is associated with natural gases in vapor form. The natural gas industry has recognized that dehydration is necessary to ensure smooth operation of gas transmission lines. Dehydration prevents the formation of gas hydrates and reduces corrosion. Unless gases are dehydrated, liquid water may condense in pipelines and accumulate at low points along the line, reducing its flow capacity. Several methods have been developed to dehydrate gases on an industrial scale.

The three major methods of dehydration are (1) direct cooling, (2) adsorption, and (3) absorption. Molecular sieves (zeolites), silica gel, and bauxite are the desiccants used in adsorption processes. In absorption processes, the most frequently used desiccants are diethylene and TEGs. Usually, the absorption/stripping cycle is used for removing large amounts of water, and adsorption is used for cryogenic systems to reach low moisture contents.

Natural gas usually contains water, in liquid or vapor form, at source or as a result of sweetening with an aqueous solution. Operating experience and thorough engineering have proved that it is necessary to reduce and control the water content of gas to ensure safe processing and transmission. Removal of the water vapor that exists in solution in natural gas requires a more complex treatment. This treatment consists of dehydrating the natural gas, which is accomplished by lowering the dew point temperature of the gas at which vapor will condense from the gas. There are several methods of dehydrating natural gas. The most common of these are liquid desiccant (glycol) dehydration and solid desiccant dehydration.

In summary, techniques for dehydrating natural gas, associated gas condensate, and NGLs include:

- Absorption using liquid desiccants,
- Adsorption using solid desiccants,
- Dehydration with $CaCl_2$,
- Dehydration by refrigeration,
- Dehydration by membrane permeation,
- Dehydration by gas stripping, and
- Dehydration by distillation.

Among various gas dehydration processes, absorption is the most common technique, where the water vapor in the gas stream becomes absorbed in a liquid solvent stream. Glycols are the most widely used absorption liquids as they approximate the properties that meet commercial application criteria. Several glycols have been found suitable for commercial application. TEG is by far the most common liquid desiccant used in natural gas dehydration as it exhibits most of the desirable criteria of commercial suitability.

Absorption dehydration involves the use of a liquid desiccant to remove water vapor from the gas. Although many liquids possess the ability to absorb water from gas, the liquid that is most desirable to use for commercial dehydration purposes should possess the following properties:

1. High absorption efficiency.
2. Easy and economic regeneration.
3. Non-corrosive and non-toxic.
4. No operational problems when used in high concentrations.
5. No interaction with the hydrocarbon portion of the gas, and no contamination by acid gases.

The glycols, particularly EG, DEG, TEG, and tetraethylene glycol come closest to satisfying these criteria to varying degrees. Water and the glycols show complete mutual solubility in the liquid phase due to hydrogen-oxygen bonds, and their water vapor pressures are very low. One frequently used glycol for dehydration is TEG.

The flow sheet of a TEG dehydration unit is shown in Figure 9.10.

In those situations where inhibition is not feasible or practical, dehydration must be used. Both liquid and solid desiccants may be used, but economics frequently favor liquid desiccant dehydration when it will meet the required dehydration specification.

Liquid desiccant dehydration equipment is simple to operate and maintain. It can easily be automated for unattended operation; for example, glycol dehydration at a remote production well. Liquid desiccants can be used for sour gases, but additional precautions in the design are needed due to the solubility of the acid gases in the desiccant solution. At very high acid gas content and relatively higher pressures, the glycols can also be "soluble" in the gas.

Glycols are typically used for applications where dew point depressions of the order of 15°–49 °C are required. DEG, TEG, and tetraethylene glycol (TREG) are used as liquid desiccants, but TEG is the most common for natural gas dehydration.

Good practice dictates installing an inlet gas scrubber, even if the dehydrator is near a production separator. The inlet gas scrubber will prevent accidental dumping of large quantities of water (fresh or salty), hydrocarbons, treating chemicals, or corrosion inhibitors into the glycol contactor. Even small

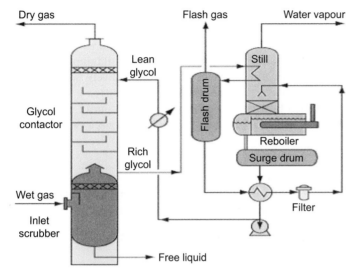

FIGURE 9.10

Typical TEG-natural gas dehydration system.

quantities of these materials can result in excessive glycol losses due to foaming, reduced efficiency, and increased maintenance. Integral separators at the bottom of the contactor are common.

Summarizing the flow path of natural gas and TEG in a typical TEG dehydration unit, wet natural gas first enters an inlet separator to remove all liquid hydrocarbons from the gas stream. Then the gas flows to an absorber (contactor) where it is contacted counter-currently and dried by the lean TEG. TEG also absorbs volatile organic compounds (VOCs) that vaporize with the water in the reboiler. Dry natural gas existing the absorber passes through a gas/glycol heat exchanger and then into the sales line. The wet or rich glycol exiting the absorber flows through a coil in the accumulator where it is preheated by hot lean glycol. After the glycol–glycol heat exchanger, the rich glycol enters the stripping column and flows down the packed bed section into the reboiler. Steam generated in the reboiler strips absorbed water and VOCs out of the glycol as it rises up the packed bed [2,3]. The water vapor and desorbed natural gas are vented from the top of the stripper. The hot regenerated lean glycol flows out of the reboiler into the accumulator (surge tank) where it is cooled via cross exchange with returning rich glycol; it is pumped to a glycol/gas heat exchanger and back to the top of the absorber. Figure 9.1 illustrates a typical TEG-Natural Gas Dehydration System [15].

Glycols are very good absorbers for water because the hydroxyl groups in glycols form similar associations with water molecules [16]. The contact between a wet gas and glycol can be made in any gas–liquid contact device. This is mainly an absorption/stripping type process, similar to the oil absorption process. The wet gas is dehydrated in the absorber, and the stripping column regenerates the water-free TEG. The glycol stream should be recharged constantly because some TEG may react and form heavy molecules, which should be removed by the filter shown in Figure 9.10 or by distillation of a slip stream.

Evaluation of a TEG system involves first establishing the minimum TEG concentration required to meet the outlet gas water dew point specification. Figure 9.11 shows the water dew point of a natural gas stream in equilibrium with a TEG solution at various temperatures and TEG concentrations. Figure 9.11 can be used to estimate the required TEG concentration for a particular application or the theoretical dew point depression for a given TEG concentration and contactor temperature.

Actual outlet dew points depend on the TEG circulation rate and number of equilibrium stages, but typical approaches to equilibrium are 6–11 °C. Equilibrium dew points are relatively insensitive to pressure and Figure 9.11 may be used up to 10 300 kPa (abs) with little error.

Figure 9.11 combines the results published by Parrish et al. covering TEG purity up to 99.99 mass percent and those reported by Bucklin and Won in covering TEG purity up to 99.999 mass percent.

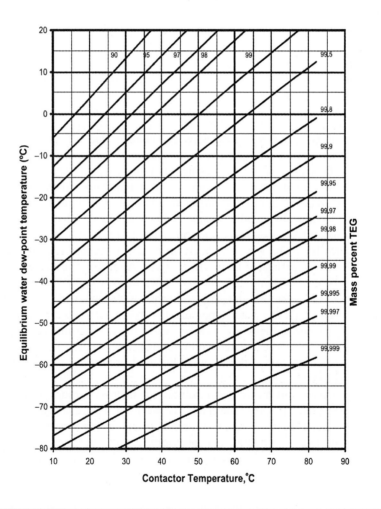

FIGURE 9.11

Equilibrium H$_2$O dew point vs temperature at various TEG concentrations.

Ref. [1].

9.10.1 Simple methodology for sizing of absorbers for TEG gas dehydration systems

Please note that the equilibrium water dew points on the ordinate of Figure 9.11 are based on the assumption the condensed water phase is a metastable liquid. At low dew points the true condensed phase will be a hydrate. The equilibrium dew point temperature above a hydrate is higher than that above a metastable liquid. Therefore, Figure 9.11 predicts dew points that are colder than those that can actually be achieved. The difference is a function of temperature, pressure, and gas composition but can be as much as 8–11 °C.

When dehydrating to very low dew points, such as those required upstream of a refrigeration process, the TEG concentration must be sufficient to dry the gas to the hydrate dew point.

Once the lean TEG concentration has been established, the TEG circulation rate and number of trays (height of packing) must be determined. Most economical designs employ circulation rates of about 15–40 l TEG/kg H_2O absorbed.

The relationship between circulation rate and number of equilibrium stages uses the absorption calculation techniques. This has been done for TEG systems with the results presented in Figures 9.12–9.17.

The graphs in these figures apply only if the feed gas is water saturated. They are based on feed-gas and therefore contactor temperatures of 27 °C, but are essentially independent of temperature.

Although the K values in the absorption factors (i.e., L/VK) do increase with temperature, the required TEG rates also increase, and this tends to compensate for the increasing K values and keep the absorption factors fairly constant.

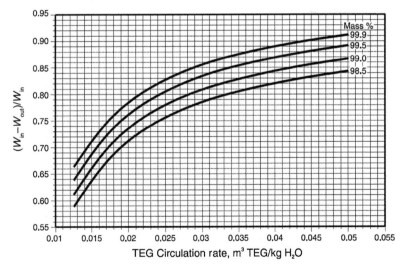

FIGURE 9.12

Water removal vs TEG circulation rate at various TEG concentrations (N = 1.0)

Ref. [1].

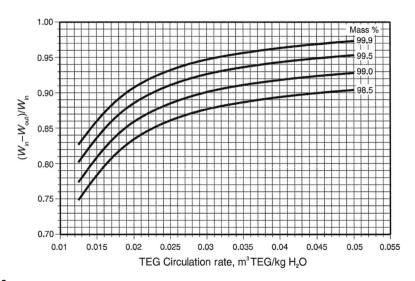

FIGURE 9.13

Water removal vs TEG circulation rate at various TEG concentrations (N = 1.5).

Ref. [1].

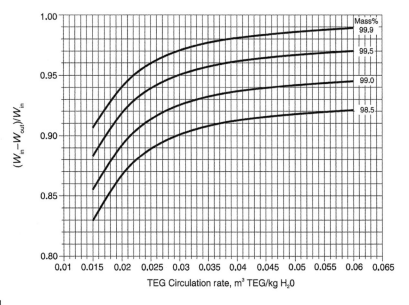

FIGURE 9.14

Water removal vs TEG circulation rate at various TEG concentrations (N = 2.0).

Ref. [1].

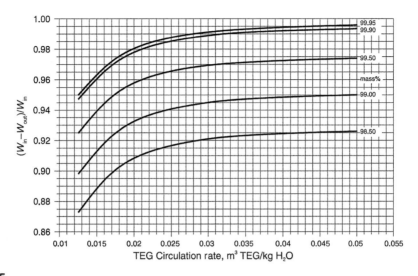

FIGURE 9.15

Water removal vs TEG circulation rate at various TEG concentrations (N = 2.5).

Ref. [1].

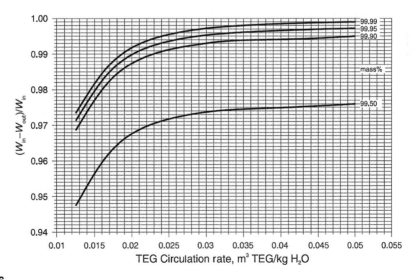

FIGURE 9.16

Water removal vs TEG circulation rate at various TEG concentrations (N = 3.0).

Ref. [1].

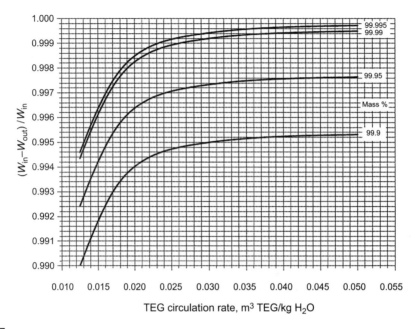

FIGURE 9.17

Water removal vs TEG circulation rate at various TEG concentrations (N = 4.0).

Ref. [1].

Conversion from equilibrium stages to actual trays can be made assuming an overall tray efficiency of 25–30%.

For random and structured packing, Height of Packing Equivalent to a Theoretical Plate (HETP) varies with TEG circulation rate, gas rate, gas density, and packing characteristics but a value of about 1.5 m is usually adequate for planning purposes. When the gas density exceeds about $100\,kg/m^3$ (generally at very high pressures), the above conversions may not provide sufficient packing height and number of trays. Also, when the contactor temperature is less than about 15 °C, the increased viscosity of the TEG can reduce mass transfer efficiency, and temperatures in this range should be avoided.

Typical tray spacing in TEG contactors is 0.6 m. Therefore, the total height of the contactor column will be based on the number of trays or packing required plus an additional 1.8–3 m to allow space for vapor disengagement above the top tray, inlet gas distribution below the bottom tray, and rich glycol surge volume at the bottom of the column. Bubble cap trays have historically been used in glycol contactors due to the low liquid rates versus gas flow, but structured packing is widely used. Generally, structured packing allows a significantly smaller contactor diameter and a slightly smaller contactor height. In 1991, ARCO published a report on its testing and application of structured packing to TEG dehydration. Contactor diameter is set by the gas velocity.

Structured packing vendors frequently quote an F_s value for sizing glycol contactors, where F_s is defined in Eqn (9.26).

$$F_s = v\sqrt{\rho_v} \qquad (9.27)$$

where:

F_s = sizing parameter for packed towers
ρ_v = Vapor density,$\frac{kg}{m^3}$
v = vapor velocity, m/s

Values of $F_s = 2.5–3.0$ will generally provide a good estimate of contactor diameter for structured packing.

EXAMPLE 9.4:

0.85 million Sm^3/d of a 0.65 relative density natural gas enters a TEG contactor at 4100 kPa (abs) and 38 °C. Inlet water content is 1436 mg H_2O/Sm^3 and outlet water content specification is 110 mg H_2O/Sm^3. The TEG circulation rate is 25 l TEG/kg H_2O. Estimate the contactor diameter and number of bubble cap trays or height of structured packing required to meet this requirement. Consider $z = 0.92$, at $T = 38$ °C lean TEG concentration ≈ 99.0 mass%, H_2O Dew point $= -4$ °C, which is equivalent to a water content of 110 mg H_2O/Sm^3 at 4100 kPa (abs).

Solutions steps:
Estimate required TEG concentration from 9.11 H_2O dew point $= -4$ °C, which from Figure 9.2 is equivalent to a water content of 110 mg H_2O/Sm^3 @ 4100 kPa (abs) Assume a 6 °C approach to equilibrium @ $T = 38$ °C, lean TEG concentration ≈ 99.0 mass%.

- Estimate number of theoretical stages. Calculate water removal efficiency.

$$\frac{W_{in} - W_{out}}{W_{in}} = \frac{1436 - 110}{1436} = 0.922$$

From Figure 9.13 and $N = 1.5$ at 25 l TEG/kg H_2O and 99 mass% TEG:

$$R = \frac{W_{in} - W_{out}}{W_{in}} = 0.885$$

From Figure 9.14 ($N = 2.0$) at 25 l TEG/kg H_2O and 99.0 mass% TEG

$$R = \frac{W_{in} - W_{out}}{W_{in}} = 0.925$$

So, use $N = 2$.

- theoretical stages are approximately eight bubble cap trays at 0.6 m tray spacing
- theoretical stages are approximately 3 m of structuring packing
  - Size the contactor:

Bubble caps, 0.6 m tray spacing, so mass velocity (G) is calculated:

$$G = C_b \sqrt{\rho_v(\rho_L - \rho_v)}$$

$$G = 176\sqrt{32(1119.7 - 32)} = 32800 \frac{kg}{m^2 h}$$

- Calculate mass flow rate:

$$m = \left(\frac{0.85 \times 10^6 Sm^3}{day}\right)\left(\frac{1\ kmol}{23.64}\right)\left(\frac{0.65 \times 28.97\ kg}{kmol}\right)\left(\frac{1\ day}{24\ h}\right) = 28211 \frac{kg}{h}$$

- Then, calculate cross-sectional area:

$$A = \frac{m}{G} = \frac{28211}{32800} = 0.86 \text{ m}^2$$

- Bubble cap diameter is calculated:

$$D_b = \sqrt{\frac{4A}{\pi}} = \sqrt{\frac{4 \times 0.86}{3.14}} = 1.05 \text{ m}$$

- Diameter for structured packing will be:

$$D_P = \sqrt{\frac{C_b}{C_P}} \times D_b = \sqrt{\frac{176}{366}} \times 1.05 = 0.70 \text{ m}$$

- TEG will typically absorb about 8×10^{-3} Sm^3 of sweet natural gas per liter of glycol at 6900 kPa (abs) and 38 °C. Solubilities will be considerably higher if the gas contains significant amounts of CO_2 and H_2S.

Below is nomenclature for above equations:

C_b = Coefficient for bubble cap column
C_p = Coefficient for packed column
D_b = Bubble Cap Diameter, m
D_p = Bubble Cap Diameter, m
G = mass velocity, kg/(m^{2}h)
$\dot{m}$ = mass flow rate, kg/h
N = number of theoretical stages
R = Water removal efficiency
W = water content of gas, mg/(Sm^3)
X = Triethylene glycol purity, Mass fraction
ρ_v = Vapor density, kg/m^3
ρ = Liquid density, kg/m^3

9.11 Adsorption of water by a solid

There are several solid desiccants that possess the physical characteristic to adsorb water from natural gas. These desiccants generally are used in dehydration systems consisting of two or more towers and associated regeneration equipment. Figure 9.18 shows a simple two-tower system. One tower is on-stream adsorbing water from the gas while the other tower is being regenerated and cooled. Hot gas is used to drive off the adsorbed water from the desiccant, after which the tower is cooled with an unheated gas stream. The towers are switched before the on-stream tower becomes water saturated. In this configuration, part of the dried gas is used for regeneration and cooling, and is recycled to the inlet separator.

Solid desiccant units generally cost more to buy and operate than glycol units. Therefore, their use is typically limited to applications such as high H_2S content gases, very low water dew point requirements, simultaneous control of water and hydrocarbon dew points, and special cases

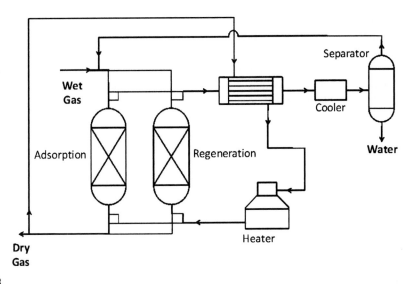

FIGURE 9.18

Scheme of the temperature swing adsorption dehydration process.

such as oxygen containing gases, etc. In processes where cryogenic temperatures are encountered, solid desiccant dehydration usually is preferred over conventional methanol injection to prevent hydrate and ice formation. Solid desiccants are also often used for the drying and sweetening of NGL liquids.

The second dehydration method is adsorption of water by a solid desiccant. In this method, water is usually adsorbed on a mole sieve, on a silica gel, or on alumina.

Adsorption (or solid bed) dehydration is the process where a solid desiccant is used for the removal of water vapor from a gas stream. The solid desiccants commonly used for gas dehydration are those that can be regenerated and, consequently, used over several adsorption–desorption cycles.

The mechanisms of adsorption on a surface are of two types; physical and chemical. The latter process, involving a chemical reaction, is termed "chemisorptions." Chemical adsorbents find very limited application in gas processing. Adsorbents that allow physical adsorption hold the adsorbate on their surface by surface forces. For physical adsorbents used in gas dehydration, the following properties are desirable.

1. Large surface area for high capacity. Commercial adsorbents have a surface area of 500–800 m^2/g
2. Good "activity" for the components to be removed, and good activity retention with time/use
3. High mass transfer rate(i.e., a high rate of removal)
4. Easy, economic regeneration
5. Small resistance to gas flow, so that the pressure drop through the dehydration system is small
6. High mechanical strength to resist crushing and dust formation. The adsorbent also must retain enough strength when "wet."
7. Cheap, non-corrosive, non-toxic, chemically inert, high bulk density, and small volume changes upon adsorption and desorption of water.

Some materials that satisfy these criteria, in the order of increasing cost are; bauxite ore, consisting primarily of alumina ($Al_2O_3 \cdot xH_2O$); alumina; silica gels and silica-alumina gels; and molecular sieves.

Activated carbon, a widely used adsorbent, possesses no capacity for water adsorption and is therefore not used for dehydration purposes, although it may be used for the removal of certain impurities. Bauxite also is not used much because it contains iron and is thus unsuitable for sour gases.

9.11.1 Alumina

A hydrated form of aluminum oxide (Al_2O_3), alumina is the least expensive adsorbent. It is activated by driving off some of the water associated with it in its hydrated form ($Al_2O_3 \cdot 3H_2O$) by heating. It produces an excellent dew point depression values as low as $-100\,°F$, but requires much more heat for regeneration.

Also, it is alkaline and cannot be used in the presence of acid gases, or acidic chemicals used for well treating. The tendency to adsorb heavy hydrocarbons is high, and it is difficult to remove these during regeneration. It has good resistance to liquids, but little resistance to disintegration due to mechanical agitation by the flowing gas.

9.11.2 Silica gel and silica-alumina gel

Gels are granular, amorphous solids manufactured by chemical reaction. Gels manufactured from sulfuric acid and sodium silicate reaction are called silica gels, and consist almost solely of silicon dioxide (SiO_2). Alumina gels consist primarily of some hydrated form of Al_2O_3. Silica-alumina gels are a combination of silica and alumina gel. Gels can dehydrate gas to as low as 10 ppm, and have the greatest ease of regeneration of all desiccants. They adsorb heavy hydrocarbons, but release them relatively more easily during regeneration. Because they are acidic, they can handle sour gases, but not alkaline materials such as caustic or ammonia. Although there is no reaction with H_2S, sulfur can deposit and block their surface. Therefore, gels are useful if the H_2S content is less than 5–6%.

9.11.3 Molecular sieves

These are a crystalline form of alkali metal (calcium or sodium) alumina-silicates, very similar to natural clays. They are highly porous, with a very narrow range of pore sizes, and very high surface area.

Manufactured by ion-exchange, molecular sieves are the most expensive adsorbents. They possess highly localized polar charges on their surface that act as extremely effective adsorption sites for polar compounds such as water and hydrogen sulfide. Molecular sieves are alkaline and subject to attack by acids. Special acid-resistant sieves are available for very sour gases.

Because the pore size range is narrow, molecular sieves exhibit selectivity toward adsorbates on the basis of their molecular size, and tend not to adsorb bigger molecules such as the heavy hydrocarbons. The regeneration temperature is very high. They can produce a water content as low as 1 ppm. Molecular sieves offer a means of simultaneous dehydration and desulfurization and are therefore the best choice for sour gases.

Solid desiccants or absorbents are commonly used for dehydrating gases in cryogenic processes. The use of solid adsorbent has been extended to the dehydration of liquid. Solid adsorbents remove water from the hydrocarbon stream and release it to another stream at higher temperatures in a regeneration step.

In a dry desiccant bed, the adsorbate components are adsorbed at different rates. A short while after the process has begun, a series of adsorption zones appear. The distance between successive adsorption zone fronts is indicative of the length of the bed involved in the adsorption of a given component. Behind the zone, all of the entering component has been removed from the gas; ahead of the zone, the concentration of that component is zero. Note the adsorption sequence: C_1 and C_2 are adsorbed almost instantaneously, followed by the heavier hydrocarbons, and finally by water that constitutes the last zone.

Almost all the hydrocarbons are removed after 30–40 min and dehydration begins. Water displaces the hydrocarbons on the adsorbent surface if enough time is allowed. At the start of dehydration cycle, the bed is saturated with methane as the gas flows through the bed. Then ethane replaces methane, and propane is adsorbed next. Finally, water will replace all the hydrocarbons.

For good dehydration, the bed should be switched to regeneration just before the water content of outlet gas reaches an unacceptable level. The regeneration of the bed consists of circulating hot dehydrated gas to strip the adsorbed water, then circulating cold gas to cool the bed down.

References

[1] Gas Processors and Suppliers Association (GPSA). In: Engineering databook. 12th ed. 2004. Tulsa (OK, USA).

[2] McKetta JJ, Wehe AH. Use this chart for water content of natural gases. Pet Refin Hydrocarbon Process August 1958;37(8):153.

[3] Kobayashi R, Song KY. RR-120 "Water content values of a CO_2–5.31 mol percent methane mixture. Gas Processors Association; January 1989.

[4] Kobayashi R, Song K, Sloan E. Phase Behavior of Water/Hydrocarbon Systems. In: Bradley HB, Gipson Fred W, editors. Petroleum Engineering Handbook. Richardson, TX, USA: Society of Petroleum Engineers; 1987.

[5] Berge BK. Norsk Hydro. Hydrate Prediction on A Microcomputer. SPE 15306, presented at the Society of Petroleum Engineers Symposium on Petroleum Industry Applications. Colorado, USA: Silvercreek; 1986. 18–20 June.

[6] van der Waals JH, Platteeuw JC. Clathrate solutions. Adv Chem Phys 1959;21(2):1–57.

[7] Sloan Jr ED, Koh CA. Clathratehydrates of natural gas. 3rd ed. Florida: CRC Press; 2007.

[8] Chen G, Tianmin G. Thermodynamic modeling of hydrate formation based on new concepts. Fluid Phase Equilibr 1996;122:43–65.

[9] Klauda JB, Sandler SI. A fugacity model for gas hydrate phase equilibria. Ind Eng Chem Res 2000;39(9): 3377.

[10] Munck J, Skjoldjorgensen S, Rasmussen P. Computations of the formation of gas hydrates. Chem Eng Sci 1988;43(10):2661.

[11] Delahaye A, Fournaison L, Marinhas S, Chatti I, Petitet J-P, Dalmazzone D, Fürst W. Effect of THF on Equilibrium Pressure and Dissociation Enthalpy of CO_2 Hydrates Applied to Secondary Refrigeration. Ind Eng Chem Res 2006;45:391–7.

[12] Tse JS. Thermal expansion of the clathrate hydrates of ethylene oxide and tetrahydrofuran. J Phys 1987; 48(CI):543–9.

[13] Hirai H, Kondo T, Hasegawa M, Yagi T, Yamamoto Y, Komai T, et al. Methane hydrate behavior under high pressure. J Phys Chem B 2000;104:1429–33.

[14] Ruzicka K, Majer V. Simple and controlled extrapolation of vapor pressures towards the triple point. AIChE J 1996;42:1723–40.

[15] Katz DL. Prediction of conditions for hydrate formation in natural gases. Trans AIME 1945;160:140.

[16] Brady CJ, Cunningham JR, Wilson GM. RR-62-"Water-hydrocarbon liquid-liquid-vapor equilibrium measurements to 530°F". Gas Processors Association; September 08, 1982.

Further reading

Al-Marhoun MA. PVT correlations for MiddleEast crude oils. J Pet Technol; May 1988:650.

AL-Tolaiby MA, Jamieson G. Determination of moisture content of sour hydrocarbons in upstream petroleum production using gas chromatography. Saudi Aramco J Technol; Summer; 1997:45–51.

Aoyagi Keichi, Song Kyoo Y, Kobayashi Riki, Sloan E Dendy, Dharmawardhana PB. RR-45, "I. The water content and correlation of the water content of methane in equilibrium with hydrates. II. The water content of a high carbon dioxide simulated prudhoe bay gas in equilibrium with hydrates". Gas Processors Association; December 1980.

Bahram A, John MC. Solubility of gaseous hydrocarbon mixtures in water. Soc Pet Eng J February 1972;253: 21–7.

Baillie C, Wichert E. Chart gives hydrate formation temperature for natural Gas. Oil Gas J April 6, 1987;85(14):37.

Barreau L. Experimental studies in liquid/liquid equilibrium of methanol/water/hydrocarbons. IFP Report No. April 1992;39652.

Behr WR. Correlation eases absorber equilibrium line calculations for TEG natural gas dehydration. Oil Gas J; November 1983:96–8.

Blanc C, Tournier-Lasserve J. "Controlling hydrates in high pressure flowlines", World Oil, vol. 211, No. 5, p. 63.

Bukacek J. quoted in McCain W.D. The properties of petroleum fluids. 2nd ed. Tulsa (OK): Penn Well Books; 1990.

Bukacek RF. Equilibrium moisture content of natural gases, vol. 8. Chicago: Research Bulletin IGT; 1959. 198–200.

Burden RL, Faires JD. Numerical analysis. 5th ed. Boston (USA): PWS Kent Publishing Company; 1993. 437–449.

Campbell JM. Gas conditioning and processing. 8th ed., vols. I & II. Oklahoma (USA): Campbell Petroleum Series; 2004.

Carroll JJ. Natural gas hydrates. New York (USA): Gulf Professional Publishing. Elsevier Science; 2003:232–53.

Carroll JJ. The water content of acid gas and sour gas from 100 to 220F and pressures to 10,000psia. Texas (USA): Gas Liquids Engineering Ltd; November 2002.

Carson DB, Katz DL. Natural gas hydrates. Petroleum Transactions, AIME 1942;146,:150–8.

Carson DB, Katz DL. Natural gas hydrates. Trans AIME 1942;146:150.

Chapoy A, Mohammadi AH, Chareton A, Tohidi B, Richon D. Experimental measurement and phase behavior modeling of hydrogen sulfide–water binary system. Ind Eng Chem Res 2005;44(19):7567–74.

Chapoy A, Mohammadi AH, Chareton A, Tohidi B, Richon D. Measurement and modeling of gas solubility and literature review of the, properties for the carbon dioxide–water system. Ind Eng Chem Res 2004;43(7): 1794–802.

Chapoy A, Mohammadi AH, Chareton A, Tohidi B, Richon D. Measurement and modeling of gas solubility and literature review of the properties for the hydrogen sulfide–water system. Ind Eng Chem Res 2005;43(20). 1802.

Clark MA, Wichert E. Designing an optimized injection strategy for, acid gas disposal without dehydration. In: Proceedings annual convention gas, processors association. Calgary Univ; 1998:49–56.

Clark MA. Experimentally obtained saturated water content phase behavior and density of acid gas mixtures. M.Sc. Thesis. Calgary. (Alberta): University of Calgary; July 1999.

Défontaines AD. Experimental studies in hydrate formation in methanol hydrocarbons at very low temperatures. IFP Report No. July 1995;42324.

Dobrynin VM, Korotajev Yu P, Plyuschev DV. In: Meyer RF, editor. Long-term Energy resources. Boston: Pitman; 1981. p. 727.

Dodson CR, Standing MB. Pressure-volume-temperature and, solubility relations for natural gas water mixtures. In: API Drilling and Production Practice; 1944. p. 173.

Douglas ED. Determination of water vapor and hydrocarbon dew point in natural Gas. In: Proceeding of international school of hydrocarbon measurement. Houston: Texas; 1991. pp. 476–7. USA.

Draper NR. Applied regression analysis. 2nd ed. New York (USA): John Wiley & Sons; 1981. 57.

Engineering data book. 12th ed., vols. I & II. Tulsa: FPS, Gas Processors Suppliers, Association; 2004.

Erbar JH, et al. RR-42, "Predicting synthetic gas and natural Gas thermodynamic properties using a modified Soave-Redlich-Kwong equation of state". Gas Processors Association; August 1980.

Ganapathy V. Compute dew point of acid gases. Hydrocarbon Process; February 1993:93–8.

Gudmundsson J-S, Parlaktuna M. Gas-in-ice concept evaluation. University of Trondheim, The Norwegian Institute of Technology; 1991.

Guo B, Ghalambor A. Natural gas engineering handbook. Houston (USA): Gulf Publishing Company; 2005.

Hammerschmidt EG. Formation of gas hydrates in natural gas transmission lines. Ind Eng Chem 1934;26: 851.

Haridy AT, Abdel Fattah KhA, Awad M, Kenawy F. New model estimates water content in saturated natural gas. Oil Gas J; April 2002:50–3.

Herskowitz M, Gottlieb M. Vapor-liquid equilibrium in aqueous solutions of various glycols and poly (ethylene) glycols, triethylene glycol. J Chem Eng Data 1984;29:173.

Ismail A, Monnery WD, Svrcek WY. Model predicts equilibrium water content of high-pressure acid gases. Oil Gas J; July 2005:48–54.

Jacoby RH. Vapor-liquid equilibrium data for the use of methanol in preventing gas hydrates. In: Proceedings of gas hydrate control conference. Univ. of Oklahoma; May 5–6, 1953.

Katz DL, Kobayashi R. Handbook of natural gas engineering. New York (USA): McGrawHill; 1959.

Kazim FM. Quickly calculate of the water content of natural gas. Hydrocarbon Process; March 1996:105–8.

Kvenvolden KA. Methane hydrate-amajor reservoir of carbon in the shallow geosphere. Chem Geol 1988;71:41.

Leinonen PT, Mackay D, Phillips CR. A correlation for the solubility of hydrocarbons in water. Can J Chem Eng April 1971;49:288–90.

Lekvam K. Kinetics of natural gas hydrates. Ph.D. Dissertation. University of Aalborg and Hogskolen at Stavanger; August 1995.

Maddox RN, et al. Estimating water content of sour gas mixtures. Gas Conditioning Conference. Norman (OK): Univ. of Oklahoma; March 1988.

Maddox RN, Lilly LL, Moshfeghian M, Elizondo E. Estimating water content of sour natural gas mixtures. Norman (OK): Laurance Reid Gas Conditioning, Conference; March 1988.

Maddox RN, et al. "Predicting hydrate temperature at high inhibitor concentration", Proceedings 1991 gas conditioning and processing conference, Univ. of Oklahoma, Norman (OK).

Makogon YF. Gazov Prom-st 1965;5:14.

Makogon YF. Natural gas hydrates: the state of study in the USSR and perspectives for its use. In: Paper presented at the third Chemical Congress of North America. Toronto: Canada; June 5–10; 1988.

Makogon YF. Hydrates of hydrocarbons. Tulsa (OK, USA): Penn Well Publishing Company; 1997. 56.

Maria AB, Graciela M. Evaluation of hydrate formation and inhibition using equations of state. Pet Eng Int; January 1999:65–70.

McCain Jr WD. The properties of petroleum fluids. 2nd ed. Tulsa: PennWell Books; 1989. pp. 525–528.

McCain WD. The properties of petroleum fluids. 2nd ed. Tulsa (OK, USA): Penn Well Publishing Company; 1990.

McCleod HO, Campbell JM. Natural Gas hydrates at pressures to 10,000psia. Trans AIME 1961;222:590.

McIver RD. In: Meyer RF, Olson JC, editors. Long-term Energy resources, vol. 1. Boston: Pitman; 1981. p. 713.

McKetta Jr JJ, Wehe Katz DL. Phase equilibra in the methane–butane-water system. Ind Eng Chem 1948;40:853.

Mehta AP, Sloan ED. Structure H hydrates: the state of-the-art. In: Proceedings 75th GPA annual convention; 1996. Denver, CO.

Mehta AP, Sloan Jr ED. Structure H Hydrates: the state-of-the-art. In: Proceedings 2nd international conference on natural gas hydrates. Toulouse: France; 1996. pp. 1–9.

Meyer RF. In: Meyer RF, Olson JC, editors. Long-term Energy resources. Boston: Pitman; 1981. p. 49.

H–J Ng, C–J Chen, Robinson DB. RR-92, "The effect of ethylene glycol or methanol or hydrate formation in systems containing ethane, propane, carbon dioxide, hydrogen sulfide or a typical gas condensate". Gas Processors Association; September 1985.

H–J Ng, C–J Chen, Robinson DB. RR-106, "The influence of high concentrations of methanol or hydrate formation and the distribution of glycol in liquid-liquid mixtures". Gas Processors Association; April 1987.

H–J Ng, C–J Chen, Robinson DB. RR-117, "The solubility of methanol or glycol in water–hydrocarbon systems". Gas Processors Association; March 1988.

Nielsen RB, Bucklin RW. Hydrocarbon Process; April 1983:71.

Nielsen RB, Bucklin RW. Why not use methanol for hydrate control? Hydrocarbon Process April 1983;62(4):71.

Ning Y, Zhang H, Zhou G. Mathematical simulation and program for water content chart of natural gas. Chem Eng Oil Gas 2000;29:75–7.

Noaker LJ, Katz DL. Gas hydrates of hydrogen sulphide-methane mixtures. Trans AIME 1954;201:237.

Olds RH, Sage BH. Composition of the dew point of methane water system. Ind Eng Chem October 1942;34(10):1223–7.

Otakar J, Lee M. Particulate probe answers water content questions for Alabama gas pipeline. Oil Gas J; September 2000:68–73.

Parrish WR, Won KW, Baltatu ME. Phase behavior of the triethylene glycol-water system and dehydration/regeneration design for extremely low dew point requirements. In: Proceedings 65th GPA annual convention; 1986. p. 202. San Antonio, TX.

Pedersen KS, Aa Fredenslund, Thomasson P. Properties of oils and natural gases. Houston (Texas): Gulf Publishing Company; 1989.

Poettmann FH, Sloan ED, Mann SL, McClure LM. Vapor-solid equilibrium ratios for structure I and II natural gas hydrates. In: Proceedings 68th annual GPA convention, March 13–14. San Antonio: Texas; 1989.

Poettmann FH. Here's butane hydrates equilibria. Hydrocarbon Process June 1984;63(6):111.

Polderman LD. The glycols as hydrate point depressants in natural Gas systems. Norman (OK): Proc. Gas Conditioning Conference; 1958.

Ripmeester JA, Tse JS, Ratcliffe CI, Powell BM. Nature 1987;325:135–6. 40.

Robinson DB, Ng HL. Improve hydrate predictions. Hydrocarbon Process December 1975;54(12):95.

Robinson JM, et al. Estimation of the water content of sour natural gases. Trans AIME August 1977;263:281.

Robinson JN, Moore RG, Heidemann RA. Estimation of water content of sour natural gases. Soc Pet Eng J; Aug. 1977:281.

Robinson JN, Wichert G. Charts estimate H2O content of sour gases. Oil Gas J; February 1978:76–8.

Rosman A. Water equilibrium in the dehydration of natural Gas with triethylene glycol. Trans AIME 1973;255:297.

Scauzillo FR. Equilibrium ratios of water in the water-triethylene glycol-natural gas system. Trans AIME 1961; 222:697.

Schroeter JP, Kobajashi Riki, Hildebrand HA. TP-10-"Hydrate decomposition conditions in the system hydrogen sulfide, methane and propane". Gas Processors Association; December 1982.

Selleck FT, Carmichael LT, Sage BH. Phase behavior in the hydrogen sulfide-water system. Ind Engr Chem Sept 1952;44(9):2219.

Sharma S, Campbell JM. Predict natural Gas water content with total gas usage. Oil Gas J; August 1969:136–7.

Skinner W. The water content of natural gas at low temperatures. Norman: University of Oklahoma; 1948. M.Sc. Thesis.

Sloan ED, et al. Vapor-solid equilibrium ratios for structure I & II natural gas hydrates. In: Proceedings 60th GPA annual convention. San Antonio: Tx; 1989.

Sloan ED, Khoury F, Kobayashi R. Water content of methane Gas in equilibrium with hydrates. Ind Eng Chem Fundam 1976;15(4):318–23.

Sloan ED. Clathratehydrate of natural gases. 2nd ed. New York (USA): Marcel Dekker Inc; 1998.

Sloan ED. Natural gas hydrates. J Pet Technol; December 1991:1414–2141.

Song K, Kobayashi R. Measurement and interpretation of the water content of amethane-propane mixture in the gaseous state in equilibrium with hydrate. Ind Eng Chem Fundam 1982;21(4):391–5.

Song K, Kobayashi R. Water content of CO2 in equilibrium with liquid water and/or hydrates. SPE Form Eval; December 1987:500–8.

Song KY, Kobayashi R. RR-50, "Measurement and interpretation of the water content of amethaned5.31mol% propane mixture in the gaseous state in equilibrium with hydrate". Gas Processors Association; January 1982.

Song KY, Kobayashi R. RR-80, "The water content of CO2–rich fluids in equilibrium with liquid, water, or hydrate". Gas Processors Association; May 1984.

Stange E, Majeed A, Overa S. Norsk-Hydro experimentations and modeling of the multiphase equilibrium and inhibition of Hydrates.

Takahashi S, Kobayashi R. TP-9, The water content and solubility of CO2 in equilibrium with DEG-water and TEG-Water.

Townsend FM. Vapor-liquid equilibrium data for DEG and TEG-water-natural gas system. Proceedings 1953 gas conditioning conference, Univ. of Oklahoma, Norman (OK).

Trofimuk AA, Cherskiy NV, Tsatev VP. In: Meyer RF, editor. Future Supply of nature-made petroleum and gas. New York: Pergamon Press; 1977. p. 919.

Unruh CH, Katz DL. Gas hydrates of carbon dioxide-methane mixtures. Trans AIME 1949;186:83.

Wattenbarger RA, John L. Gas reservoir engineering. TX (USA): Society of Petroleum Engineers; 1996.

Wichert G, Wichert E. Chart estimates water content of sour natural Gas. Oil Gas J; March 1993:61–4.

Wichert G, Wichert E. New charts provide accurate estimates water content of sour natural gas. Oil Gas J; October 2003:64–6.

Wiebe R, Gaddy VL. Vapor phase composition of carbon dioxide water mixtures at various temperatures and pressures to 700 atmospheres. J Am Chem Soc 1941;63:475–7.

W,I Wilcox. Carson DB, Katz DL. Natural gas hydrates. Ind Eng Chem 1941;33:662.

Worley S. Super dehydration with glycols. Proceedings 1967 gas conditioning conference, Univ. of Oklahoma, Norman, OK.

Yaws CL, et al. Hydrocarbons: water solubility data. Chem Eng April 1990;97(4):177.

Natural Gas Sweetening

The primary operation of natural gas treatment process falls into gas sweetening and dehydration by absorption, adsorption, and chemical reactions. This chapter covers the minimum process requirements, criteria, and features for process design of gas sweetening units. Describing the basic principles for process design of main equipment, piping, and instrumentation together with guidelines on present developments and process selection in the gas sweetening process are the main objectives throughout this chapter.

Acid gas constituents present in most natural gas streams are mainly hydrogen sulfide (H_2S) and carbon dioxide (CO_2). Many gas streams, however, particularly those in a refinery or manufactured gases, may contain mercaptans, carbon sulfide, or carbonyl sulfide.

The level of acid gas concentration in the sour gas is an important consideration for selecting the proper sweetening process. Some processes remove large quantities of acid gas, and other processes remove acid gas constituents to the ppm range. However, the sweetening process should meet the pipeline specification.

10.1 Chemical solvent processes

Chemical reaction processes remove the H_2S and CO_2 or both from the gas stream by chemical reaction with a material in the solvent solution. The reactions may be reversible or irreversible. In reversible reactions, the reactive material removes the gases in the contactor at high partial pressure, low temperature, or both. The reaction is reversed by high temperature or low pressure in the stripper. In irreversible processes, the chemical reaction is not reversed, and removal of the H_2S and/or CO_2 requires continuous makeup of the reacting material [1]. Table 10.1 illustrates physical properties of some gas treating chemicals.

10.2 Process selection

The selection of a solvent process depends on:

- process objectives and characteristics of the solvents, such as selectivity for H_2S, COS, HCN, etc.,
- ease of handling water content in feed gas,
- ease of controlling water content of circulating solvent,
- concurrent hydrocarbon loss or removal with acid gas removal, costs, solvent supply, chemical inertness,
- royalty cost,
- thermal stability and proper plant performance for various processing techniques.

Natural Gas Processing. http://dx.doi.org/10.1016/B978-0-08-099971-5.00010-6
Copyright © 2014 Elsevier Inc. All rights reserved.

Table 10.1 Physical Properties of Gas-Treating Chemicals

Chemical Properties	Monoethanolamine	Diethanolamine	Triethanolamine	Diglycol®-Amine	Diisopropanolamine	Selexol
Formula	$HOC_2H_4NH_2$	$(HOC_2H_4)_2NH$	$(HOC_2H_4)_3N$	$H(OC_2H_4)_2NH_2$	$(HOC_3H_6)_2NH$	Polyethylene Glycol Derivative
Molecular wt	61.08	105.14	148.19	105.14	133.19	280
Boiling point @ 760 mm Hg, °C	170.5	269	360 (Decompose)	221	248.7	270
Freezing point, °C	10.5	28	22.4	−12.5	42	−28.9
Critical pressure, kPa (abs)	5985	3273	2448	3772	3770	—
Critical temperature, °C	350	442.1	514.3	402.6	399.2	—
Density @ 20 °C, gm/cc	1.018	1.095	1.124	1.058 @ 15.6 °C	0.999 @ 30 °C	1.031 @ 25 °C
Weight, kg/m³	1016 @ 15.6 °C	1089 @ 15.6 °C	1123 @ 15.6 °C	1057 @ 15.6 °C	—	—
Relative density 20 °C/20 °C	1.0179	1.0919 (30/20 °C)	1.1258	1.0572	0.989 @ 45 °C/25 °C	—
Specific heat @ 15.6 °C, kJ/(kg·°C)	2.55 @ 20 °C	2.51	2.93	2.39	2.89 @ 30 °C	2.05 @ °C
Thermal conductivity W/(m·°C) @ 20 °C	0.256	0.22	—	0.209	—	0.19 @ 25 °C
Latent heat of vaporization, kJ/kg	826 @ 760 mmHG	670 @ 730 mmHG	535 @ 760 mmHG	510 @ 760 mmHG	430 @ 760 mmHG	—
Heat of reaction, kJ/kg of acid gas (H_2S)	—	—	−930	−1568	—	−442 @ 25 °C
Heat of reaction, kJ/kg of acid gas (CO_2)	—	—	−1465	−1977	—	−372 @ 25↑ °C
Viscosity, mPa·s	24.1 @ 20 °C	350 @ 20 °C (at 90% wt solution)	1013 @ 20 °C (at 95% wt solution)	40 @ 16 °C	870 @ 30 °C 198 @ 45 °C 86 @ 54 °C	5.8 @ 25 °C
Refractive index, Nd 20 °C	1.4539	1.4776	1.4852	1.4598	1.4542 @ 45 °C	—
Flash point, COC, °C	93	138	185	127	124	151

Chemical Properties	Propylene Carbonate	Methyldiethanolamine	Sulfolane®	Methanol	10% Sodium Hydroxide
Formula	$C_3H_6CO_3$	$(HOC_2H_4)_2NCH_3$	$C_4H_8SO_2$	CH_3OH	—
Molecular wt	102.09	119.16	120.17	32.04	19.05
Boiling point @ 760 mm Hg, °C	242	247	285	64.5	102.8
Freezing point, °C	−49.2	−23	27.6	−97.7	−10
Critical pressure, kPa (abs)	—	—	5290	7956	—
Critical temperature, °C	—	—	545	240	—
Density @ 20 °C, gm/cc	1.2057	—	—	—	—
Weight, kg/m³	—	1040	1273 @ 30 °C/30 °C	792	1109
Relative density 20 °C/20 °C	1.203	1.0418	1.268	—	1.11
Specific heat @ 15.6 °C, kJ/(kg·°C)	1.4	2.24	1.47 @ 30 °C	2.47 @ 5–10 °C	3.76
Thermal conductivity W/(m·°C) @ 20 °C	0.21 @ 10 °C	0.275	0.197 @ 38 °C	0.215	
Latent heat of vaporization, kJ/kg	484 @ 760 mm Hg	476	525 @ 100 °C	1103 @ 760 mm Hg	—
Viscosity, mPa·s	1.67 @ 38 °C $19.4 \times 10^{-6}\,m^2/s$ @ − 40 °C $1.79 \times 10^{-6}\,m^2/s$ @ − 38 °C $0.827 \times 10^{-6}\,m^2/s$ @ 99 °C	$1.3 \times 10^{-6}\,m^2/s$ @ 10 °C $0.68 \times 10^{-6}\,m^2/s$ @ 38 °C $0.28 \times 10^{-6}\,m^2/s$ @ 100 °C	10.3 @ 30 °C 6.1 @ 50 °C 2.5 @ 100 °C 1.4 @ 150 °C 0.97 @ 200 °C	0.6 @ 20 °C	1.83 @ 20 °C 0.97 @ 50 °C 0.40 @ 100 °C
Refractive index, Nd 20 °C	1.4209	1.469	1.481	1.3286	
Flash point, COC, °C	132.2	129.4	177	14	

The choice of the process solution is determined by the pressure and temperature conditions at which the gas to be treated is available; its composition with respect to major and minor constituents, and the purity requirements of the treated gas. In addition, consideration should be given to whether simultaneous H_2S and CO_2 removal or selective H_2S absorption is desired.

In general the gas treating process can affect the design of entire gas processing facilities, including the methods used for acid gas disposal and sulfur recovery dehydration, absorbent recovery, etc.

For an appropriate process selection, the following factors should be considered for evaluation and decision making as a general approach to all sour gas sweetening treatment installations:

- air pollution regulations regarding H_2S removal;
- type and concentration of impurities in sour gas;
- specification of treated gas (sweet gas);
- temperature and pressure at which the sour gas is available and at which the sweetened gas should be delivered;
- volume of the gas to be treated;
- hydrocarbon composition of sour gas;
- selectivity required for acid gas removal;
- capital cost and operating cost;
- liquid product specifications (where application).

An accurate analysis of the sour gas stream should be made showing all of the impurities including COS, CO_2, and mercaptans in the project specification. These impurities shall in particular have significant effect on process design of the gas treating and downstream facilities.

The selectivity of a sweetening agent is an indication of the degree of removal that can be obtained for one acid gas constituent as opposed to another. There are sweetening processes that display rather marked selectivity for one acid gas constituent. There are cases where no selectivity is demonstrated and all acid gas constituents will be removed.

The selectivity of sweetening agent can be made on the basis of operating conditions. Only rarely will natural gas streams be sweetened at low pressures. However, there are processes that are unsuitable for removing acid gases under low pressure conditions. Other sweetening agents may adversely be effected by temperatures, and some lose their economic advantage when large volumes of gas are to be treated.

The feasibility and desirability of sulfur recovery can also place considerable limitation on the selectivity of sweetening process. Removal of H_2S to produce contractually sweetened gas is difficult, and emphasis should be placed on those processes that have major potential application for sweetening of natural gases.

10.3 Chemical reaction processes

Chemical reaction processes remove the H_2S and/or CO_2 from the gas stream by chemical reaction with a material in the solvent solution. The alkanolamines are the most generally accepted and widely used of the many solvents available because of their reactivity and availability at low cost.

The alkanolamine processes are particularly applicable where acid gas partial pressures are low or low levels of acid gas are considered in the sweet gas. The alkanolamines are clear, colorless liquids that have a slightly pungent odor. All except triethanolamine are considered stable materials, because they can be heated to their boiling points without decomposition. Triethanolamine decomposes at below its normal boiling point.

10.3.1 Aqueous alkanolamine processes

Originally applied to gas treating in 1930 by Bottoms, seven alkanolamines have become the most widely used solvents for the removal of acid gases from natural gas streams. Triethanolamine (TEA) was the first used commercially for gas treating. It has been displaced for conventional applications by other alkanolamines such as monoethanolamine (MEA), diethanolamine (DEA), diisopropanolamine (DIPA), diglycolamine (DGA), and methyldiethanolamine (MDEA).

Table 10.2 lists approximate guidelines for a number of alkanolamine processes.

The alkanolamine (hereafter amine) processes are particularly applicable where acid gas partial pressures are low and/or low levels of acid gas are desired in the residue gas. Because the water content of the solution minimizes heavy hydrocarbon absorption, these processes are well suited for gases rich in heavier hydrocarbons. Some amines can be used to selectively remove H_2S in the presence of CO_2.

Chemistry:

The overall equilibrium reactions applicable for H_2S and CO_2 and primary and secondary amines are shown below with a primary amine. A qualitative estimation of the velocity of the reaction is given. For hydrogen sulfide removal:

$$RNH_2 + H_2S \leftrightarrow RNH_3^+ + HS^- \quad \text{Fast} \tag{10.1}$$

$$RNH_2 + HS^- \leftrightarrow RNH_3^+ + S^{--} \quad \text{Fast} \tag{10.2}$$

The overall reactions between H_2S and amines are simple since H_2S reacts directly and rapidly with all amines to form the bisulfide by Eqn (10.1) and the sulfide by Eqn (10.2). For carbon dioxide removal:

$$2RNH_2 + CO_2 \leftrightarrow RNH_3^+ + RNHCOO^- \quad \text{Fast} \tag{10.3}$$

$$RNH_2 + CO_2 + H_2O \leftrightarrow RNH_3^+ + HCO_3^- \quad \text{Slow} \tag{10.4}$$

$$RNH_2 + HCO_3^- \leftrightarrow RNH_3^+ + CO_3^{--} \quad \text{Slow} \tag{10.5}$$

Concerning the chemical reactions with CO_2, primary amines (RNH_2) such as MEA and DGA® agent, and secondary amines (such as DEA and DIPA), differ from tertiary amines such as TEA and MDEA.

10.3.2 Primary and secondary amines

With the primary and secondary amines, the predominant overall reaction (Eqn (10.3)) rapidly leads to the formation of a stable carbamate, which is slow to further hydrolyze to bicarbonate.

The other overall reactions leading to bicarbonate (Eqn (10.4)) and to carbonate (Eqn (10.5)) are slow because they have to proceed through the hydration of CO_2.

Therefore, according to Eqn (10.3), there is a theoretical limit to the chemical loading capacity of the primary and secondary amine solutions to 0.5 mol CO_2 per mole of amine, even at relatively high partial pressures of CO_2 in the gas to be treated.

Table 10.2 Approximate Guidelines for Amine Processes[a]

	MEA	DEA[i]	DGA	Sulfinol	MDEA[i]
Acid gas pick up, m^3/100 l @ 38 °C, normal range[b]	2.3–3.2	2.85–6.4	3.5–5.40	3.0–12.75	2.2–6.4
Acid gas pick up, mol/mol amine, normal range[c]	0.33–0.40	0.20–0.80	0.25–0.38	NA	0.20–0.80
Lean solution residual acid gas, mol/mol amine, normal range[d]	0.12±	0.01±	0.06±	NA	0.005–0.01
Rich solution acid gas loading, mol/mol amine, normal range[c]	0.45–0.52	0.21–0.81	0.35–0.44	NA	0.20–0.81
Solution concentration, wt%, normal range	15–25	30–40	50–60	3 components. Varies.	40–50
Approximate reboiler heat duty, kJ/l, lean solution[e]	280–335	235–280	300–360	100–210	220–250
Steam heated reboiler tube bundle, approx. Average heat flux, $Q/A = MJ/(h \cdot m^2)$[f]	100–115	75–85	100–115	100–115	75–85
Direct fired reboiler fire tube, average heat flux, $Q/A = MJ/(h \cdot m^2)$[f]	90–115	75–85	90–115	90–115	75–85
Reclaimer, steam bundle or fire tube, average heat flux, $Q/A = MJ/(h \cdot m^2)$[f]	70–90	NA[g]	70–90	NA	NA[g]
Reboiler temperature, normal operating range, °C[h]	107–127	110–127	121–132	110–138	110–132
Heats of reaction[j]; approximate: kJ/kg H_2S	1420	1290	1570	N/A	1230
kJ/kg CO_2	1540	1700	2000	N/A	1425
NA—not applicable or not available					

[a]These data alone should not be used for specific design purposes. Many design factors must be considered for actual plant design.
[b]Dependent upon acid gas partial pressures and solution concentrations.
[c]Dependent upon acid gas partial pressures and corrosiveness of solution. Might be only 60% or less of value shown for corrosive systems.
[d]Varies with stripper overhead reflux ratio. Low residual acid gas contents require more stripper trays and/or higher reflux ratios yielding larger, reboiler duties.
[e]Varies with stripper overhead reflux ratios, rich solution feed temperature to stripper, and reboiler temperature.
[f]Maximum point heat flux can reach 230–285 MJ/(h·m2) at highest flame temperature at the inlet of a direct fired fire tube. The most satisfactory, design of fire tube heating elements employs a zone by zone calculation based on thermal efficiency desired and limiting the maximum tube wall temperature as required by the solution to prevent thermal degradation. The average heat flux, Q/A, is a result of these calculations.
[g]Reclaimers are not used in DEA and MDEA systems.
[h]Reboiler temperatures are dependent on solution conc. flare/vent line back pressure and/or residual CO_2 content required. It is good practice to operate the reboiler at as low a temperature as possible.
[i]According to Total.
[j]The heats of reaction vary with acid gas loading and solution concentration. The values shown are average.

10.3.3 Tertiary amines

Unlike primary and secondary amines, the nitrogen (N) in tertiary amines has no free hydrogen (H) to rapidly form carbamate as per overall Eqn (10.3). As a consequence, the removal of CO_2 by tertiary amines can only follow the slow route to bicarbonate by Eqn (10.4) and carbonate by Eqn (10.5).

The slowness of the reaction leading to bicarbonate is the underlying reason why tertiary amines can be considered selective for H_2S removal, by playing with absorption contact time, and this attribute can be used to full advantage when complete CO_2 removal is not necessary. However, the slow route to bicarbonates theoretically allows at equilibrium a chemical loading ratio of 1 mol CO_2 per mole of amine. Furthermore, at high partial pressure, the solubility of CO_2 in tertiary amines is far greater than in the primary and secondary amines, thus further enhancing the CO_2 loading by physical solubility at high partial pressure.

Therefore, in the case of gases to be treated for bulk CO_2 removal, large amounts of CO_2 can be liberated from the rich solvent by simple flash, alleviating the thermal regeneration duty with consequent energy savings.

10.3.4 Activated tertiary amines

The use of activators mitigates the slowness of the reaction to bicarbonate for tertiary amines. Activators are generally primary or secondary amines; they are tailored to increase both the hydrolysis of the carbamate and the rate of hydration of dissolved CO_2, thus making the activated tertiary amines especially suitable for efficient and economical bulk CO_2 removal when selectivity is not required (see section on MDEA).

Process Flow—The general process flow for an alkanolamine-treating plant is shown in Figure 10.1. The basic flow varies little for different solutions, though some designs incorporate multiple feeds and contactor sections.

Sour natural gas enters through an inlet separator for the removal of liquids and/or solids. From the separator, the gas stream enters the bottom of the contactor, where it contacts the amine solution flowing down from the top of the column.

The acid gas components in the gas react with the amine to form a regenerable salt. As the gas continues to pass up the contactor, more acid gases chemically react with the amine. The sweetened gas leaves the top of the contactor and passes through an outlet separator to catch any solution that might be carried over. The sweet gas leaving the contactor is saturated with water, so dehydration is normally required prior to sale. If MEA is the sweetening agent, or the contactor is operating at an unusually high temperature, a water wash may be used to attempt to recover some of the vaporized and/or entrained amine from the gas leaving the contactor. If a water wash is used, it generally will consist of three or four trays at the top of the contactor, with makeup water to the unit being used as the wash liquid.

Rich amine solution leaving the contactor flows through a flash drum to remove absorbed hydrocarbons or skim off. From the flash drum, the rich solution passes through the rich/lean exchanger where heat is absorbed from the lean solution. The heated rich amine goes to the mid portion of the stripper. As the solution flows down the column to the reboiler, it is stripped of H_2S and CO_2. The amine solution leaves the bottom of the stripper as lean solution. This lean solution is then passed through the rich/lean exchanger and a lean cooler to reduce the lean solution temperature to approximately $5\,°C$ warmer than the inlet gas temperature, to stay above the hydrocarbon dew point.

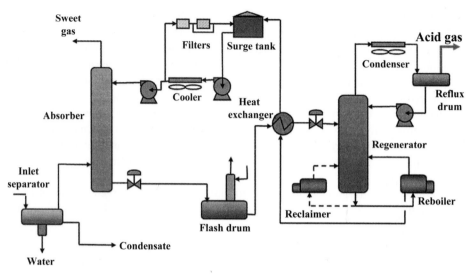

FIGURE 10.1

Schematic for a typical natural gas sweetening unit using a reversible chemical reaction process.

At this point, the lean solution is returned to the contactor to repeat the cycle. Acid gas stripped from the amine passes out of the top of the stripper. It goes through a condenser and separator to cool the stream and recover water. The recovered water is usually returned to the stripper as reflux. The acid gas from the reflux separator is either vented, incinerated, sent to sulfur recovery facilities, compressed for sale, or reinjected into a suitable reservoir for enhanced (see acid gas injection) oil recovery projects or for sequestration.

10.3.5 Reclaimer

A reclaimer is usually required for MEA and DGA amine-based systems. The reclaimer helps remove degradation products from the solution and also aids in the removal of heat-stable salts, suspended solids, acids, and iron compounds. The reclaimers in MEA and DGA systems differ. For MEA, a basic solution helps reverse the reactions. Soda ash or caustic soda is added to the MEA reclaimer to provide a pH of approximately 8–9; no addition is required for the DGA reclaimer system. Reclaimers generally operate on a side stream of 1–3% of the total amine circulation rate. Reclaimer sizing depends on the total inventory of the plant and the rate of degradation expected.

Reclaimer operation is a semi-continuous batch operation. The reclaimer is filled with hot amine solution and, if necessary, soda ash is added. As the temperature in the reclaimer increases, the liquid will begin to distill. Overhead vapors can be condensed and pumped back into the amine system, but generally the reclaimer is operated at slightly above stripper column pressure and the vapors are returned to the stripper.

The initial vapor composition is essentially water. Continued distillation will cause the solution to become more and more concentrated with amine. This raises the boiling point of the solution, and amine will begin to distill overhead. Fresh feed is continually added until the boiling point of the material in the reclaimer reboiler reaches 140–150 °C. At this point, distillation is continued for a short time, adding only water to help recover residual amine in the reclaimer reboiler. The reclaimer is then cleaned and recharged, and the cycle is repeated. Reclaimer "sludge" removed during cleaning must be handled with care. Disposal of the "sludge" must be in accordance with the governing regulations.

If needed, a reclaiming company can be contracted to remove degradation products or heat-stable salts from the amine. One type of reclaimer performs vacuum distillation on batches of spent amine mixed with sufficient caustic to neutralize the excess acidity. Another type of reclaimer uses ion exchange resin beds to remove heat-stable salts.

10.3.6 Amines used

- **Monoethanolamine**

Gas sweetening with MEA is used where there are low contactor pressures and/or stringent acid gas specifications. MEA removes both H_2S and CO_2 from gas streams. H_2S concentrations well below 4.0 ppmv can be achieved. CO_2 concentrations as low as 100 ppmv can be obtained at low to moderate pressures. COS and CS_2 are removed by MEA, but the reactions are irreversible unless a reclaimer is used. Even with a reclaimer, complete reversal of the reactions may not be achieved. The result is solution loss and build-up of degradation products in the system. Total acid gas pick-up is traditionally limited to 0.3–0.35 mol acid gas/mole of MEA, and solution concentration is usually limited to 10–20 wt%. Inhibitors can be used to allow much higher solution strengths and acid gas loadings. Because MEA has the highest vapor pressure of the amines used for gas treating, solution losses through vaporization from the contactor and stripper can be high. This problem can be minimized by using a water wash. The technical points that follow need to be take into account for MEA:

- MEA should commonly be used as a 10–20% solution in water. The acid gas loading should usually be limited to 0.3–0.4 mol acid gas per mole of amine for carbon steel equipment.
- MEA itself is not considered to be particularly corrosive. However, its degradation products are very corrosive. COS, CS_2, SO_2, and SO_3 can partially deactivate MEA, which may essentially require to be recovered with a reclaimer.
- Since MEA is primary amine, it has a high pH. This enables MEA solutions to produce gas containing less than 6 mg/S·m^3 (¼ grains H_2S per 100 S ft^3) of acid gas at very low H_2S partial pressures.
- The heat of reaction for CO_2 in MEA is about 1930 kJ/kg of CO_2 (460 kcal/kg of CO_2). The heat of reaction for all amines is a function of loading and other conditions. It varies by only 117–138 kJ/kg (28–33 kcal/kg) up to about 0.5 mol/mole of total acid gas loadings. Above this loading, the heat of reaction varies considerably and should be calculated as a function of loading.
- MEA will easily reduce acid gas concentrations to pipeline specifications (generally less than 6 mg H_2S/S m^3 gas or 0.25 grains per 100 S ft^3). By proper design and operation, the acid gas content can be reduced as low as 1.2 mg H_2S/S·m^3 gas or 0.05 grains per 100 S ft^3.

- **Diethanolamine**

This process employs an aqueous solution of DEA. DEA will not treat to pipeline quality gas specifications at as low a pressure as will MEA. Among the processes using DEA is the SNPA-DEA process developed by Societe Nationale des Petroles d'Aquitaine (today Total) to treat the very sour gas that was discovered in Lacq, France in the 1950s. The original patents covered very high acid gas loading of 0.9–1.3 mol per mole of amine.

This process is used for high-pressure, high acid gas content streams having a relatively high ratio of H_2S/CO_2. The original process has been progressively improved and Total through Prosernat is now proposing high DEA solution concentrations up to 40 wt% with the high acid gas loading together with corrosion control by appropriate design and operating procedures.

Maximum attainable loading is limited by the equilibrium solubility of H_2S and CO_2 at the absorber bottom conditions.

Although mole/mole loadings as high as 0.8–0.9 have been reported, most conventional DEA plants still operate at significantly lower loadings.

The process flow scheme for conventional DEA plants resembles the MEA process. The advantages and disadvantages of DEA as compared to MEA are as follows:

- The mole/mole loadings typically used with DEA (0.35–0.82 mol/mole) are much higher than those normally used (0.3–0.4) for MEA.
- Because DEA does not form a significant amount of nonregenerable degradation products, a reclaimer is not required. Also, DEA cannot be reclaimed at reboiler temperature as MEA can.
- DEA is a secondary amine and is chemically weaker than MEA, and less heat is required to strip the amine solution.
- DEA forms a regenerable compound with COS and CS_2 and can be used for the partial removal of COS and CS_2 without significant solution losses.

The technical points that follow need to be take into account for DEA:

- DEA is commonly used in the 25–35 mass percent range. The loading for DEA is also limited to 0.3–0.4 mol/mole of acid gas for carbon steel equipment.
- When using stainless-steel equipment, DEA can safely be loaded to equilibrium. This condition can be considered for carbon steel equipment by adding inhibitors.
- The degradation products of DEA are much less corrosive than those of MEA. COS and CS_2 may irreversibly react with DEA to some extent.
- Since DEA is a secondary alkanolamine, it has a reduced affinity for H_2S and CO_2. As a result, for some low-pressure gas streams, DEA cannot produce pipeline specification gas. However, certain design arrangement such as split flow can be considered to fulfill the specified requirement.
- Under some conditions, such as low pressure and liquid residence time on the tray (of about 2 s), DEA will be selective toward H_2S and will permit a significant fraction of CO_2 to remain in the product gas.
- The heat of reaction for DEA and CO_2 is 151 kJ/kg of CO_2 (360 kcal/kg of CO_2), which is about 22% less than for MEA.

- **Diglycolamine**

This process uses DGA brand (2-(2-aminoethoxy)) ethanol in an aqueous solution. DGA is a primary amine capable of removing not only H_2S and CO_2 but also COS and mercaptans from gas and liquid streams. Because of this, DGA has been used in both natural and refinery gas applications. DGA has been used to treat natural gas to 4.0 ppmv at pressures as low as 860 kPa (ga). DGA has a greater affinity for the absorption of aromatics, olefins, and heavy hydrocarbons than the MEA and DEA systems.

Therefore, adequate carbon filtration should be included in the design of a DGA treating unit. The process flow for the DGA treating process is similar to that of the MEA treating process. The three major differences are as follows:

- Higher acid gas pick-up per gallon of amine can be obtained by using 50–70% solution strength rather than 15–20% for MEA (more moles of amine per volume of solution).
- The required treating circulation rate is lower. This is a direct function of higher amine concentration.
- Reboiler steam consumption is reduced.

Typical concentrations of DGA range from 50% to 60% DGA by weight, while in some cases as high as 70 wt% has been used. DGA has an advantage for plants operating in cold climates where freezing of the solution could occur. The freezing point for 50% DGA solution is $-34\,°C$. Because of the high amine degradation rate, DGA systems require reclaiming to remove the degradation product. DGA reacts with both CO_2 and COS to form N, N′, bis (hydroxyethoxyethyl) urea, generally referred to as BHEEU. DGA is recovered by reversing the BHEEU reaction in the reclaimer.

The technical points that follow need to be take into account for DGA:

- DGA is generally used as 40 to 60 mass percent solutions in water. The reduced corrosion problems with DGA allow mole per mole solution loadings equivalent to MEA in most applications even with these high mass percentage.
- For gas streams with acid gas partial pressures, absorber bottom temperatures as high as $82\,°C$ and above can occur. This will reduce the possible loading.
- DGA has a tendency to preferentially react with CO_2 over H_2S. It also has a higher pH than MEA and thus can easily achieve 6 mg $H_2S/S \cdot m^3$ gas (0.25 grains per 100 S ft^3) except in some cases where large amounts of CO_2 are present relative to H_2S.
- DGA has some definite advantages over the other amines in that higher DGA concentrations in the solution result in lower freezing points. One of the primary advantages of DGA is high heats of reaction as compared with other amines and shown in Table 10.3.

- **Methyldiethanolamine**

MDEA is a tertiary amine that can be used to selectively remove H_2S to pipeline specifications at moderate to high pressure. If increased concentration of CO_2 in the residue gas does cause a problem with contract specifications or downstream processing, further treatment will be required. The H_2S/CO_2 ratio in the acid gas can be 10–15 times as great as the H_2S/CO_2 ratio in the sour gas. Some of the benefits of selective removal of H_2S include:

- Reduced solution flow rates resulting from a reduction in the amount of acid gas removed.
- Smaller amine regeneration unit.
- Higher H_2S concentrations in the acid gas resulting in reduced problems in sulfur recovery.

CO_2 hydrolyzes much slower than H_2S. This makes possible significant selectivity of tertiary amines for H_2S. This fact is used by several companies who provide process designs using MDEA for selective removal of H_2S from gases containing both H_2S and CO_2.

A feature of MDEA is that it can be partially regenerated in a simple flash. As a consequence, the removal of bulk H_2S and CO_2 may be achieved with a modest heat input for regeneration.

Table 10.3 Comparative Data Table

Amine	MEA	DEA	DGA	MDEA
Solution strength, mass %	15–20	25–35	40–60	30–50
Acid gas loading mole/mole	0.3–0.4	0.3–0.4	0.3–0.4	0.7–0.8
DH˙r for H_2S kJ/kg (kcal/kg)	1281 (306)	1189 (284)	1570 (375)	1214 (290)
DH˙r for CO_2 kJ/kg (kcal/kg)	1922 (459)	1520 (363)	1729 (413)	1398 (334)
Ability to preferentially absorb H_2S	No	Under some conditions	No	Under most conditions

˙kJ/kg (kcal/kg) and for total loadings below 0.5 mol acid gas/mole amine.

However, because MDEA solutions only react slowly with CO_2 (see chemistry), activators must be added to the MDEA solution to enhance CO_2 absorption, and the solvent is then called activated MDEA.

The technical points that follow need to be take into account for MDEA:

- MDEA is most commonly used in the 30–50 mass percent range. Due to considerably reduced corrosion problems, acid gas loadings as high as 0.7–0.8 mol/mole are practical in carbon steel equipment.
- Since MDEA is a tertiary amine, it has less affinity for H_2S and CO_2 than DEA. Thus, as in the case for DEA, MDEA can not produce pipeline specification gas for some low-pressure streams.
- MDEA has several distinct advantages over primary and secondary amines. These include lower vapor pressure, lower heats of reaction, higher resistance to degradation, fewer corrosion problems, and selectivity toward H_2S in the presence of CO_2.

- **Triethanolamine**

TEA is a tertiary amine and has exhibited selectivity for H_2S over CO_2 at low pressures. TEA was the first amine commercially used for gas sweetening. It was replaced by MEA and DEA because of its inability to remove H_2S and CO_2 to low outlet specifications.

TEA has potential for the bulk removal of CO_2 from gas streams. It has been used in many ammonia plants for CO_2 removal.

- **Diisopropanolamine**

DIPA is a secondary amine that exhibits, though not as great as tertiary amines, selectivity for H_2S. This selectivity is attributed to the steric hindrance of the chemical.

- **Formulated Solvents and Mixed Amines**

Formulated solvents is the name given to a new family of amine-based solvents. Their popularity is primarily due to equipment size reduction and energy savings over most of the other amines. All the advantages of MDEA are valid for the formulated solvents.

Some formulations are capable of slipping larger portions of inlet CO_2 (than MDEA) to the outlet gas and at the same time removing H_2S to less than 4 ppmv. For example, under conditions of low absorber pressure and high CO_2/H_2S ratios, such as Claus tail gas clean-up units, certain solvent formulations can slip up to 90% of the incoming CO_2 to the incinerator.

At the other extreme, certain formulations remove CO_2 to a level suitable for cryogenic plant feed. Formulations are also available for CO_2 removal in ammonia plants. Finally, there are solvent formulations that produce H_2S to 4 ppmv pipeline specifications, while reducing high inlet CO_2 concentrations to 2% for delivery to a pipeline. This case is sometimes referred to as bulk CO_2 removal.

This need for a wide performance spectrum has led formulated solvent suppliers to develop a large stable of different MDEA-based solvent formulations. Most formulated solvents are enhancements to MDEA. Thus, they are referred to as MDEA-based solvents or formulations. Benefits claimed by suppliers are:

- *For new plants*
  - reduced corrosion
  - reduced circulation rate
  - lower energy requirements
  - smaller equipment due to reduced circulation rates.
- *For existing plants:*
  - increase in capacity, i.e., gas throughput or higher inlet acid gas composition
  - reduced corrosion
  - lower energy requirements and reduced circulation rates.

Formulated solvents are proprietary to the specific supplier offering the product. Companies offering these products and/or processes include Huntsman Corporation, Dow Chemical Company, Shell Global Solutions and Total Fina Elf via Prosernat.

10.4 Simplified design calculations

A simplified procedure for making rough estimates of the principal parameters, for conventional MEA, DEA, and DGA amine-treating facilities when both H_2S and CO_2 are present in the gas, is given below. The procedure involves estimating the amine circulation rate and using it as the principal variable in estimating other parameters. For estimating amine circulation rate, the following equations are suggested:

- *For MEA:*

$$\text{Flow}(m^3/h) = 328(Qy/x) \tag{10.6}$$

(0.33 mol acid gas pick-up per mole MEA assumed)
- *For DEA (conventional):*

$$\text{Flow}(m^3/h) = 360(Qy/x) \tag{10.7}$$

(0.5 mol acid gas pick-up per mole DEA assumed)
- *For DEA (high loading):*

$$\text{Flow}(m^3/h) = 256(Qy/x) \tag{10.8}$$

(0.7 mol acid gas pick-up per mole DEA assumed)

Table 10.4 Estimated Heat Exchange Requirements

	Duty, kW	Area, m²
Reboiler (direct fired)	93 × (flow rate, m³/h)	4.63 × (flow rate, m³/h)
Rich-lean amine heat exchanger (HEX)	58 × (flow rate, m³/h)	4.60 × (flow rate, m³/h)
Amine cooler (air cooled)	19.3 × (flow rate, m³/h)	4.18 × (flow rate, m³/h)
Reflux condenser	38.6 × (flow rate,m³/h)	2.13 × (flow rate, m³/h)

- *For DGA*

$$\text{Flow}\left(m^3/h\right) = 446(Qy/x) \tag{10.9}$$

(0.39 mol acid gas pick-up per mole DGA assumed)
(DGA concentrations are normally 50–60% by weight)

where:

$Q =$ Sour gas to be processed, MSm³/day
$y =$ Acid gas concentration in sour gas, mol%
$x =$ Amine concentration in liquid solution, mass%

After the amine circulation has been estimated, heat and heat exchange requirements can be estimated from the information in Table 10.4. Pump power requirements can be estimated from Table 10.5.

Equations (10.6)–(10.9) normally provide conservative (high) estimates of required circulation rate. They should not be used if the combined H_2S plus CO_2 concentration in the gas is above 5 mol%. They also are limited to a maximum amine concentration of about 30% by weight.

The diameter of an amine plant contactor can be estimated using the following equation:

$$D_c = 10750 * \sqrt{Q/\sqrt{P}} \tag{10.10}$$

where:

$Q =$ MSm³/day gas to contactor
$P =$ Contactor pressure is kPa (abs)
$D_c =$ Contactor diameter in millimeters before rounding up to nearest 100 mm.

Table 10.5 Estimated Power Requirements

Parameter	Equation
Main amine solution pumps	(m³/h)kPa(ga)(0.00031) = kW
Amine booster pumps	(m³/h)(0.20) = kW
Reflux pumps	(m³/h)(0.20) = kW
Aerial cooler	(m³/h)(1.20) = kW

Similarly, the diameter of the regenerator below the feed point can be estimated using the following equation:

$$D_r = 160 * \sqrt{m^3/h}$$

(10.11)

where:

m^3/h = Amine circulation rate in gallons per minute
D_r = Regenerator bottom diameter in millimeters

The diameter of the section of the still above the feed point can be estimated at 0.67 times the bottom diameter.

EXAMPLE :

1.0 MSm^3/day of gas available at 5860 kPa (ga) and containing 0.6% H$_2$S and 2.8% CO$_2$ is to be sweetened using 20%, by mass, DEA solution. If a conventional DEA system is to be used, what amine circulation rate is required, and what will be the principal parameters for the DEA treating system?

Solution:
Using Eqn (10.7), the required solution circulation is:
   Flow (m^3/h) = 360(Qy/x) = 360(1.0·3.4/20) = 61.2m^3/h of 20% DEA solution per minute.
   Heat exchange requirements (from Table 10.4)

Reboiler:
H = (93)(61.2) = 5690 kW
A = (4.63)(61.2) = 283 m^2

Rich-lean amine exchanger:
H = (58)(61.2) = 3550 kW
A = (4.6)(61.2) = 282 m^2

Amine cooler:
H = (19.3)(61.2) = 1180 kW
A = (4.18)(61.2) = 255 m^2

Reflux condenser:
H = (38.6)(61.2) = 2360 kW
A = (2.13)(61.2) = 130 m^2
   Power requirements (Table 10.5)

Main amine pumps:
Power (kW) = (61.2)(5860)(0.00,031) = 111

Amine booster pumps:
Power (kW) = (61.2)(0.2) = 12.2

Reflux pumps:
Power (kW) = (61.2)(0.2) = 12.2

Aerial cooler:
Power (kW) = (61.2)(1.2) = 73
   Contactor diameter:

$$D_c = 10750 * \sqrt{1.0/\sqrt{5961}} = 1223 \text{ mm or } 1200 \text{ mm}$$

10.5 General considerations

Over the years of operation of amine sweetening units, a substantial amount of information has been reported, which is necessary to consider in the design stages of a sweetening unit as follows.

10.5.1 Corrosion

1. The temperature of the solution in the reboiler and the temperature of the steam should be kept as low as possible.
2. The use of high-temperature, heat-carrying media, such as oil in reboilers should be avoided to maintain the lowest possible metal skin temperature.
3. Lowest possible pressure on stripping columns and reboilers should be considered to avoid severe corrosion of reboiler tubes.
4. Inert gas blanketing facilities should be considered over all portions of the solution that could be exposed to the atmosphere. A positive pressure on the suction side of the pumps should also be ensured for excluding oxygen from the system.

10.5.2 Inlet scrubbing

The importance of proper and efficient inlet scrubbing of the sour gas is difficult to over-emphasize. Many problems in amine plant operations such as foaming, corrosion, and reboiler tube burn-out can be traced to the presence of excessive amounts of foreign material in the amine solution. Liquid hydrocarbons and entrained solids frequently can enter the plant with the sour gas stream. In addition, such foreign materials as corrosion inhibitors, drilling muds, and well acidizers can also enter with the sour gas stream, particularly on lease-type sweetening units. The inlet separation equipment should be sized and designed with these factors in mind. The fact that these materials do not enter at a steady rate but slug or surge into the plant should also be kept in mind. Instantaneous flow rates may be extremely high and the inlet separating system must be designed to handle that.

10.5.3 Amine losses

Amine losses can be very expensive. A separator on the sweet gas stream leaving the contactor is advisable. This will also help eliminate amine losses from unexpected foaming or surges. The savings in downtime and loss of production will be other advantages of the sweet gas scrubber, which may be many times the savings in amine solution lost.

10.5.4 Filtration

Proper solution is important for the maintenance of a clean, efficient amine solution. Filtration of the treating solution to remove entrained solids is essential to the successful operation of a gas treating plant. Positive removal of all solids from the system minimizes corrosion and associated problems.

In general, a two-stage filtration is often recommended for this service. Any secondary filtration may be activated charcoal and will remove degradation products, should they exist, along with some of the smaller particles of entrained solids. The filter system should be capable of handling at least 10–20% of the amine circulation rate. Doing this will permit quick cleanup of the amine solution after an upset.

With full-flow filtration, parallel filters with no bypass is recommended. The best place for filter location is on the rich amine solution at the outlet of the contactor or on the lean amine solution just before the solution enters the contactor.

Filtration rates should be as high as practical and may range from 5% of circulation to full stream. Removing particles down to 5 μm is recommended. To do this efficiently, two stages of filtration may be required.

The first stage, typically a cartridge-type or precoat filter, is designed to remove particles down to the 10 μm or less range. The second stage of filtration, typically an activated carbon filter, removes hydrocarbons and other contaminants. This is accomplished by adsorption. The carbon filter can also remove smaller particles from the amine stream.

The carbon granule size can be selected to remove particles down to the 5 μm range. The activated carbon filter should always be located downstream of the first stage filter because the deposition of solids would plug the carbon filter.

The carryover of carbon fines can be controlled by either locating a second cartridge-type filter immediately downstream of the carbon filter or using a graded carbon bed. In a graded bed, larger granules are placed at the outlet of the filter to trap fines. Large carbon granules produce fewer fines but are less efficient for adsorption.

Basic degradation products are identified by gas chromatography and mass spectrometry. Acidic degradation products are identified by ion chromatography exclusion. These tests are recommended when the amine solution appears to lose its ability to pick up acid gas. Degradation products affect the results of the conventional estimation of amine concentration by titration.

This may cause artificially high or low apparent amine concentrations Also, the carbon bed will adsorb very little strong acid degradation products. In this case, purging or reclamation of the solution is recommended.

Carbon filters can be partially regenerated with steam, which removes hydrocarbons and other adsorbed contaminants. Regeneration or bed changeout is recommended when foam tests on the inlet and outlet streams show no improvement. This indicates carbon bed saturation.

Filters can be located on the lean or rich solution side. Filtration on the rich side removes particles that are more insoluble under the rich solution conditions. It also prevents solids accumulation in the hot environment of the stripper. However, proper design and operating procedures for personnel protection during filter maintenance is mandatory when H_2S may be present.

Filtration equipment should be used continuously beginning with the first day of plant operations.

When starting up the plant, the full flow filter, even if temporary, may prove its worth by removing the scale and other solid particles and allowing much quicker and easier start-up of the plant.

10.5.5 Amine reclamation

1. The entrained solids, dissolved salts, and degradation products that cause foaming and corrosion problems are removed by reclaimer. The following points should be considered for reclaimer:

  a. An easy-to-open entry should be located on the reclaimer shell so the solids can be simply washed out at the end of the cycle. The drain line should be large enough to pass solids.

  b. The tube bundles should be raised 15 cm or more from the bottom of the reclaimer shell to provide space for sludge accumulation below the tubes, and to give better solution flow around the tubes.

  c. A packed column should be placed on top of the reclaimer to eliminate foam and entrainment from the overhead vapor stream. The installation of a glass site port in the vapor line will help keep a check on this carryover. An analytical evaluation will help pinpoint suspected entrainment.

  d. The tubes should be widely spaced for easy cleaning.

  e. Make certain the steam supply is not superheated.

  f. A recorder should be used to monitor the reclaimer temperature throughout the cycle.

  g. Amine feed to the reclaimer should be controlled by a level controller on the kettle to maintain a liquid level at least 15 cm above the tube. A temperature indicator should be provided on reclaimer outlet line.

  h. Sufficient vapor space should be allowed above the liquid layer in the reclaimer to prevent liquid carryover in the overhead vapor line.

  i. Provisions for chemical analysis of the reclaimer bottoms and the amine solution should be considered to identify the solution contaminants and determine the degree and rate of solution contamination.

10.5.6 Foaming

A sudden increase in differential pressure across a contactor or a sudden liquid level variation at the bottom of the contactor often indicates severe foaming. When foaming occurs, there is poor contact between the gas and the chemical solution. The result is reduced treating capacity and sweetening efficiency, possibly to the point that outlet specification cannot be met. Some reasons for foaming are:

- Suspended solids
- Organic acids
- Corrosion inhibitors
- Condensed hydrocarbons
- Soap-based valve greases

- Makeup water impurities
- Degradation products
- Lube oil

Foaming problems can usually be traced to plant operational problems. Contaminants from upstream operations can be minimized through adequate inlet separation. Condensation of hydrocarbons in the contactor can usually be avoided by maintaining the lean solution temperature at least 5 °C above the hydrocarbon dew point temperature of the outlet gas. Temporary upsets can be controlled by the addition of antifoam chemicals. These antifoams are usually of the silicone or long-chain alcohol type.

10.5.7 Amine–amine heat exchanger

If an intermediate flash separator is not used, contactor pressure should be maintained through the amine-amine heat exchanger. Doing this minimizes breakout of acid gases from the rich amine solution, thereby minimizing excessive corrosion of control valves, heat exchanger, and downstream piping. Operation at high pressure increases the cost of the exchangers, but this generally will be more than offset by the increased life due to lessened corrosion. The rich amine solution should be on the tube side of the exchanger, which again tends to minimize investment in the exchanger.

Linear velocities in the amine-to-amine heat exchanger should be low—in the range of 0.6–1.0 m/s. This reduces the heat transfer coefficient and increases the surface area requirement. However, again, the return on the reduced maintenance costs for the exchanger will more than offset the increased investment. Flowing amine solutions should not impinge directly on vessel surfaces; impingement baffles should be utilized in the exchangers.

10.5.8 Reboiler of regenerator

For careful and accurate design of the reboiler system in the amine regeneration section, the following points should be considered:

- Low-pressure saturated steam at approximately 300–380 kPa (135–142 °C) is commonly used to strip the amine solution. Steam temperatures above 140 °C should be avoided to prevent excessive skin temperatures on the tubes.
- Amine Units should be designed for "pressure operation". Higher operating pressures increase the bottom temperature of the regenerator and are believed to provided more complete stripping of the acid gases, especially the CO_2.
- The maximum allowable kettle temperature recommended for regeneration of the solution depends on the type of amine used to prevent amine degradation.
- The temperature controller valve should be on the steam inlet, not on the condensate outlet, to keep excessive condensate out of the tubes. When condensate partially floods the reboiler tube, the heat load concentrates in the top section of the bundle, which can cause tube failure.
- Stripping steam requirements will vary depending on the degree of sweetening required for the process stream being treated. Normally, the steam consumption should be at least 120 g of excess steam per liter of solution circulated.

- To provide good circulation of the amine solution around the tubes and to reduce fouling causes by a sludge accumulation, the tube bundle should be placed on a slide about 15 cm above the bottom of the reboiler shell.
- The amine solution should enter the reboiler at several locations to help improve the natural circulation of liquid in the reboiler shell. Several vapor exit locations will reduce the stagnant pockets of acid gases in the reboiler.
- The tube bundle should be supported to prevent tube vibration. Teflon or other protective inserts will prevent tube cutting.
- The length of the tubes should be limited to prevent condensate "logging" and water hammer. The water hammer effect can produce severe tube vibration and grooving of the tubes at the support baffles.
- A square pitch tube pattern is recommended for both reboiler and heat exchange bundles to provide easy cleaning. The tubes should be widely spaced to permit a rapid escape of liberated gases and to reduce the high velocity scrubbing action associated with two-phase flow. This scrubbing action will remove the protective film and increase corrosion.
- The reboiler should be designed to provide a liberal amount of vapor disengaging space between tubes, and with sufficient surface to produce a simmering action rather than violent boiling. In existing installations where vapor binding is a problem, some tubes can be removed to form an "X" or "V" in the center of the bundle to provide a path of low resistance to escaping vapors.
- The reboiler bundle should always be kept covered with 15–20 cm of liquid to prevent localized drying and overheating. Severe corrosion will occur if the liquid level is lowered until some of the tubes are exposed.
- An analysis of the amine solution entering and leaving the reboiler will determine the efficiency of the stripping operation. A high acid gas loading in the reboiler causes tube corrosion.

10.5.9 Amine solution selection

Many process factors should be considered when selecting an amine for a sweetening application as outlined here:

1. The initial selection should be based on the pressure and acid gas content of the sour gas as well as the purity specification for the product gas.
2. Based on the currently "accepted" operating conditions, MEA is usually not preferred for its high heat of reaction and lower acid gas carrying capacity per unit volume of solution. However, MEA is still used for plants where the inlet gas pressure is low and pipeline specification gas or total removal of the acid gases is required.
3. DEA, is used for its lower heats of reaction, higher acid gas carrying capacity, and resultant lower energy requirements. However, its potential for selective H_2S removal from streams containing CO_2 has not fully been realized.
4. DGA, despite a high heat of reaction, has a very high gas carrying capacity that usually produces very reasonable net energy requirements. DGA also has a good potential for absorbing COS and some mercaptans from gas and liquid streams, and, because of this, DGA has been used in both natural and refinery gas applications.

5. MDEA, with its some outstanding capabilities, resulting from its low heat of reaction, can be used in pressure swing plants for bulk acid gas removal. MDEA is currently best known for its ability to preferentially absorb H_2S.

10.6 Corrosion in gas sweetening plants

Corrosion is an operating concern in nearly all sweetening installations. The combination of H_2S and CO_2 with water practically ensures that corrosive conditions will exist in portions of the plant. In general, gas streams with high H_2S to CO_2 ratios are less corrosive than those having low H_2S to CO_2 ratios. H_2S concentrations in the ppmv range with CO_2 concentrations of 2% or more tend to be particularly corrosive.

Because the corrosion in sweetening plants tends to be chemical in nature, it is strongly a function of temperature and liquid velocity. The type of sweetening solution being used and the concentration of that solution have a strong impact on the corrosion rate. Increased corrosion can be expected with stronger solutions and higher gas loadings.

Hydrogen sulfide dissociates in water to form a weak acid. The acid attacks iron and forms insoluble iron sulfide. The iron sulfide will adhere to the base metal and may provide some protection from further corrosion, but it can be eroded away easily, exposing fresh metal for further attack.

CO_2 in the presence of free water will form carbonic acid. The carbonic acid will attack iron to form a soluble iron bicarbonate that, upon heating, will release CO_2 and an insoluble iron carbonate or hydrolyze to iron oxide. If H_2S is present, it will react with the iron oxide to form iron sulfide.

High liquid velocities can erode the protective iron sulfide film with resulting high corrosion rates. In general, design velocities in rich solution piping should be 50% of those that would be used in sweet service. Because of the temperature relationship to corrosion, the reboiler, the rich side of the amine–amine exchanger, tend to experience high corrosion rates. Because of the low pH, the stripper overhead condensing loop also tends to experience high corrosion rates.

Acid degradation products also contribute to corrosion. A suggested mechanism for corrosion is that degradation products act as chelating agents for iron when hot. When cooled, the iron chelates become unstable, releasing the iron to form iron sulfide in the presence of H_2S. Primary amines are thought to be more corrosive than secondary amines because the degradation products of the primary amines act as stronger chelating agents.

Treating plants normally use carbon steel as the principal material of construction. Vessels and piping should be stress relieved to minimize stress corrosion along weld seams. Corrosion allowance for equipment ranges from 1 to 6 mm, typically 3 mm. In some instances, when corrosion is known to be a problem, or high solution loadings are required, stainless steel or clad stainless steel may be used in the following critical areas:

1. Reflux condenser
2. Reboiler tube bundle
3. Rich/lean exchanger tubes
4. Bubbling area of the contactor and/or stripper trays
5. Rich solution piping from the rich/lean exchanger to the stripper
6. Bottom five trays of the contactor and top five trays of stripper, if not all.

Usually 304, 316, or 410 stainless steel will be used in these areas, even though corrosion has been experienced with 410 stainless in DEA service for CO_2 removal in the absence of H_2S. L grades are recommended if the alloys are to be welded.

Controlling oxygen content to less than 0.2 ppmw is effective in preventing chloride stress corrosion cracking (SCC) in waters with up to 1000 ppmw chloride content, at temperatures up to 300 °C. There has been an increased use of duplex stainless steels, and they have been successfully used in the water treatment industry to prevent chloride SCC in high chloride waters.

This suggests duplex stainless steels could be utilized in amine plant service where high chloride content is expected. As with any specialty steel, proper fabrication techniques and welding procedures are required.

10.7 Flash tank

Rich solution leaving the contactor may pass through a flash tank. A flash tank is more important when treating high-pressure gas. Gases entrained in the rich solution will be separated.

In addition, the amount of absorbed gas will be decreased because of the lower operating pressure of the flash tank. Using a flash tank will:

- Reduce erosion in rich/lean exchangers
- Minimize the hydrocarbon content in the acid gas
- Reduce the vapor load on the stripper
- Possibly allow the off-gas from the flash tank to be used as fuel (may require sweetening).

When heavy hydrocarbons are present in the natural gas, the flash tank can also be used to skim off the heavy hydrocarbons that were absorbed by the solution. Residence times for flash tanks in amine service vary from 3 to 10 min, depending on separation requirements. Inlet gas streams containing only methane and ethane require shorter residence times. Rich gas streams require longer times for the dissociation of gas from solution or the separation of liquid phases.

10.8 Combined physical/chemical purification processes

The newly classified physical/chemical combined purification process, i.e., alkanol amines (mono- or diethanol amine) mixed with methanol, is more successful than a single physical solvent used in many of the gas treatment plants. The main advantage of this solvent lies in the good physical absorption of a physical solvent component in combination with the chemical reaction of the amine. The combination of a chemically active amine with a low boiling point polar physical solvent such as methanol offers major advantages in the absorption of CO_2 and sulfur components:

- Very low clean gas sulfur contents of less than 0.1 ppm, which are required for synthesis gases
- Very low CO_2 contents in the purified gas
- Good absorption of trace components, such as HCN, COS, mercaptans, and higher hydrocarbons
- Low regeneration temperature, owing to the fact that the solvent methanol has a boiling point approximately 35 °C below that of water
- The solvent is non-corrosive, so carbon steel equipment can be used.

As the conventional alkanolamines such as monoethanolamine (MEA) and diethanolamine (DEA) are used as non-selective solvents, the sulfur components H_2S and COS are absorbed together with CO_2 and jointly occur in the off-gas. This is a disadvantage, because the sulfur-rich off-gas should be treated. These disadvantages have been eliminated by extensive research, and MEA/DEA are replaced with following aliphatic alkylamines:

1. diisopropylamine—DIPAM
2. diethylamine—DETA.

These new amines offer the following advantages over the conventional MEA and DEA:

1. High thermal and chemical stability so that no reclaimer is needed
2. Higher effective CO_2 and H_2S loadings
3. High selectivity of the solvent between sulfur components and CO_2, along with the very low residual sulfur contents in the clean gas
4. Good industrial availability of these amines at moderate cost
5. Smaller difference between absorption and desorption temperature
6. No foaming tendency due to low surface tension
7. No corrosion problems
8. Considerable reduction in vapor pressure of the amine neutralized by sour gases.

A particular advantage of the aliphatic alkylamines is the high selectivity of the solvent between sulfur components and CO_2, along with very low residual sulfur contents in the clean gas. Even with very low sulfur and high CO_2 contents in feed gas, it is unproblematic to produce a high sulfur off-gas containing more than 25% H_2S, which is suitable for Clause plants.

The CO_2 produced by the alkylamine process is free from COS and meets today's stringent environmental pollution requirements. CO_2 produced in this process is pure and can be used for urea synthesis.

Taking the advantage of varied properties of aliphatic alkylamine, the process lends itself mainly for the following applications:

1. Joint absorption of H_2S, COS, and CO_2 as well as trace contaminants, such as HCN, NH_3, mercaptans, thiophenes, etc., from raw gases to produce highly pure synthesis gases.
2. Selective absorption of all sulfur components from raw gases to obtain highly pure product gas and a high sulfur Clause gas. CO_2 is removed in second stage and can be recovered in pure form.

It should be kept in mind that combined physical/chemical purification processes and aliphatic alkylamine processes have predominantly chemical characteristics, and so these processes are particularly cost-effective for all gases with low to medium partial pressures of the sour gas, and if the sour gas partial pressures are high, physical gas purification processes should preferably be used.

10.9 Carbonate process

The Hot Potassium Carbonate Process has been utilized successfully for bulk removal of CO_2 from a number of gas mixtures. It has been used for sweetening natural gases containing both CO_2 and H_2S.

The Hot Potassium Carbonate Process is not suitable for sweetening gas mixtures containing little or no CO_2 since the potassium bisulfide would be very difficult to regenerate if CO_2 is not present. This process has the following advantages/disadvantages:

1. *Advantages*
   a. It is a continuous circulating system employing an inexpensive chemical
   b. It is an isotherm system in that both absorption and desorption of the acid gas are conducted at as nearly a uniform high temperature as can be obtained, thus requiring no heat exchange equipment in the fluid circulating system
   c. The desorption by stripping is accomplished with a smaller steam rate than required for an amine plant
2. *Disadvantages*
   a. It will not commercially reduce the H_2S content to pipeline specification. To accomplish this, a conventional amine plant should be used
   b. It is similar to other acid gas removal processes in that it is also prone to corrosion
   c. Like other liquid absorbents in sweetening plants, it is prone to the problems of suspended solids and foaming

10.9.1 Corrosion

In the carbonate process, corrosion will occur where the conversion to bicarbonate is high and where carbon dioxide and steam are released by pressure reduction. In addition, stress corrosion is expected because potassium carbonate is an electrolyte. This requires stress relieving of welds and vessels.

Successful corrosion inhibitors that sofar reported includes arsenic and vanadium salts as well as dichromates.

Pilot plant studies conducted by the Bureau of Mines encountered severe corrosion of carbonate in several areas. In general this corrosion occurred where the conversion to bicarbonate was occurred where carbon dioxide and steam were released by pressure reduction. In addition, stress corrosion expected because potassium carbonate is an electrolyte. This requires stress relieving of welded vessels.

Stainless steel is recommended for reboiler tubes, though carbon steel has been used successfully. Control valves and solution pumps should also be of stainless steel, particularly those parts which encounter the carbonate solution such as impellers and inner valves.

The absorber and stripper, whether using trays or packing, can be made of carbon steel. They should be stress relieved. Piping can also be carbon steel if stress relieved. Stripper columns in particular are sometimes plastic coated or gunnite lined.

When carbon dioxide is the only acid gas constituent, potassium dichromate will serve as an effective corrosion inhibitor. From 1000 to 3000 ppm will generally be effective. Where H_2S is present in the gas, chromates cannot be used. Both H_2S and COS cause reduction of the hexavalent chromium to the insoluble trivalent state. When this happens the corrosion inhibitor causes erosion and precipitation. Fortunately, if sufficient quantity of H_2S is present, formation of the potassium bisulfide inhibits corrosion to a significant extent. In cases where corrosion is a problem with H_2S present, the commercially available film-forming fatty amines are usually effective as corrosion inhibitors.

Though water vapor exerts a partial pressure over solutions of potassium carbonate, the potassium carbonate does not. For this reason, in a process unit with a properly sized vessel and good mist extractors, potassium carbonate losses should be negligible. Products of undesirable side reactions, resulting from trace gas impurities, can be maintained at satisfactory levels by a small continuous purge of carbonate solution.

The carbonate solution should be filtered through a side stream filter arrangement. There will be occasional pieces of scale and other solid materials that the filter should remove. Filtration minimizes severe erosion problems.

10.9.2 Catacarb process

The Catacarb Process employs a modified potassium salt solution containing a very active, stable, and nontoxic catalyst and corrosion inhibitor. In this process amine borate is utilized to increase the activity of the hot potassium carbonate.

Several other catalysts and inhibitors are used in this process. The choice depends on the gas compositions to be treated by the process. This process is also capable of removing trace amounts of other acid gases such as COS, CS_2, and RSH.

10.10 Physical absorption methods

Various amines, the Hot Potassium Carbonate Process, and the Catacarb Process rely on the chemical reaction to remove acid gas constituent from sour gas streams. Removal of acid gases by physical absorption is possible, and there are number of competitive processes that should be considered when:

1. The partial pressure of the acid gas in the feed is greater than 350 kPa or 3.5 bar (50 psi). For this figure reference is made to the Gas Processors Suppliers Association
2. The heavy hydrocarbon concentration in the feed gas is low
3. Bulk removal of the acid gas is desired
4. Selective removal of H_2S is desired
5. In general, physical solvents are capable of removing COS, CS_2, and mercaptans.

10.10.1 Solvent regeneration

The solvents are regenerated by:

1. multi-stage flushing to low pressure
2. regeneration at low temperature with an inert stripping gas
3. heating and stripping of solution with steam/solvent vapor.

10.10.2 Fluor process

The main characteristics desired in the Fluor Process for the right selection of solvent are as follows:

1. *Low vapor pressure at operating temperature is desirable*

This is desirable from two related standpoints. If solvent vapor pressure is appreciable, losses will be high. These losses necessitate either a complicated solvent recovery system or high operating costs due to solvent loss. For these reasons, Fluor eliminated from their consideration several high vapor pressure solvents that had good solubilities for acid gas constituents.

2. *The primary constituents in the gas stream should be only slightly, if at all, soluble in the solvent*

In the case of a natural gas stream, methane and the heavier hydrocarbons should not be appreciably soluble in the solvent. If they are soluble, then expensive and complicated procedures may be required to prevent excessive losses.

3. *The solvent should have low viscosity*

High viscosities increase pumping costs. In general they also have an adverse effect on tray efficiencies and mass transfer. Operation at sub-ambient temperatures may aggravate the viscosity problem. For these cases, some solvent satisfactory for ambient temperature operation might prove undesirable at lower temperature.

4. *Low solubility for water*

In general, increasing the water content in the circulating solvent will lower its carrying capacity for acid gases. Dissolved water will also tend to increase corrosion and solvent decomposition effects. If water is dissolved, then steps must be taken to maintain the water content of the solvent at some specified level. Again, this increases plant complexity, costs, and operational problems.

5. *The solvent should not degrade under normal operating conditions*

The solvent should not degrade chemically under the temperature and pressure conditions of normal operation. This problem can be handled by filtration, reclaiming, and so forth, but these items do increase the investment and operating costs. The solvent should be stable with regard to oxygen and other materials. Again, storage tanks can be inert gas blanketed, but this is a complicating factor to be avoided if at all possible.

6. *The solvent should not react chemically with any component in the gas stream*

This can lead to solvent degradation loss and loss of solvent effectiveness.

7. *For obvious reasons, the solvent should be non-corrosive to the common metals*

Use of carbon steel construction, preferably without necessity for stress relieving, will minimize plant investment. In the physical absorption process, conditions are usually ideal for minimizing corrosion due to carbon dioxide and hydrogen sulfide. Temperatures are low and a proper solvent should help minimize corrosion problems rather than cause them.

8. *The solvent should be readily available at reasonable cost*

An excellent solvent that, because of its cost, would have an appreciable effect on plant investment would not be desirable.

Using the above characteristics as a guideline, the following solvents are recommended for use for the removal of carbon dioxide from high-pressure gas streams:

1. Polyethylene carbonate
2. Butoxydiethylene glycol acetate
3. Glycol triacetate
4. Methoxy triethylene glycol acetate.

10.10.3 **Other miscellaneous physical methods**

There are newly commercialized physical solvent processes that should be evaluated for proper process selection in sour gas treatment projects. Most of the equilibrium data of these solvents are proprietary to the process licensors. Therefore, definitive comparative information about the solvent performance should be gained in the course of process evaluation and license selection.

Some of these physical solvent processes and their comparative data are given in Table 10.6.

Table 10.6 Typical Comparative Data of Miscellaneous Physical Solvents

Process Name	Selexol	Fluor Solvent
Solvent Name	**Selexol**	**PC**
FOB fact solvent cost $/lb	To be completed at the time of comparison	To be completed at the time of comparison
Licensor	Norton	Fluor
Viscosity @ 25 °C, cP	5.8	3.0
Relative density (specific gravity) @ 25 °C, kg/m^3	1030	1195
Molecular mass	280	102
Vapor pressure @ 25 °C, mm Hg	7.3×10^{-4}	8.5×10^{-2}
Freezing point, °C	−28	−48
Boiling point, °C @ 760 mm Hg	−	240
Thermal conductivity W/m·k (Btu/Hr/Ft^2/(°F/Ft))	0.19 (0.11)	0.21 (0.12)
Maximum operating temp., °C	175	65
Specific heat @ −4 °C (25 °F)	0.49	0.339
Water solubility @ 25 °C	Infinity	94 g/L
Solvent solubility in water @ 25 °C	Infinity	236 g/L
CO$_2$ solubility @ 25 °C m^3/l (ft^3/US Gal)	0.0036 (0.485)	0.0034 (0.455)
Number of commercial plants	32	13
Bulk CO$_2$ removal		
Synthesis gas	6	3
Natural gas	6	10
Landfill gas	3	0
Selective H$_2$S removal		
Synthesis gas	9	0
Natural gas	8	0

Continued

Table 10.6 Typical Comparative Data of Miscellaneous Physical Solvents—cont'd

Process Name	Purisol	Sepasolv MPE	Estasolvan
Solvent name	NMP	Sepasolv	TBP
Licensor	Lurgi	BASF	Uhde & IFP
Viscosity @ 25 °C, cP.	1.65	–	2.9
Relative density (specific gravity) @ 25 °C kg/m^3	1027	–	973
Mol mass	99	320	266
Vapor pressure @ 25 °C, mm Hg	0.40	0.00037	<0.01
Freezing point, °C	−24	–	−80
Boiling point, °C @ 760 mm Hg	202	320	(180 @ 30 mm Hg)
Thermal conductivity W/m.k (Btu/Hr/Ft^2/(°F/Ft))	0.16 (0.095)	–	–
Maximum operating temp., °C	–	175	–
Specific heat @ −4 °C (25 °F)	0.4	–	–
Water solubility @ 25 °C	–	–	65 g/L
Solvent solubility in water @ 25 °C	–	–	0.42 g/L
CO$_2$ solubility @ 25 °C, m^3/L (ft^3/U.S. Gal)	0.0035 (0.477)	0.0034 (0.455)	0.0025 (0.329)
Bulk CO$_2$ removal			
Synthesis gas	2	0	
Natural gas	1	0	
Landfill gas	0	0	
Selective H$_2$S removal			
Synthesis gas	1	0	
Natural gas	1	0	

10.11 Solid bed sweetening methods (batch Processes)

The use of solids for sweetening gas is based on adsorption of the acid gases on the surface of the solid sweetening agent, or reaction with some component on that surface. These types of processes are not as widely used as the liquid processes, but there are several advantages that make them worth considering.

The solids processes are usually best applied to gases containing low-to-medium concentrations of H$_2$S or mercaptans. The solids processes tend to be highly selective and do not normally remove significant quantities of CO$_2$. Consequently, the H$_2$S stream from the process is usually high purity. In addition, pressure has relatively little effect on the adsorptive capacity of a sweetening agent. Most of solids processes are of the batch type and tend to have low investment and operating costs.

10.11.1 Iron oxide (sponge) process

The iron sponge process selectively removes H_2S from gas or liquid streams. The process is limited to treating streams containing low concentrations of H_2S at pressures ranging from 170 to 8300 kPa (ga). The process employs hydrated iron oxide, impregnated on wood chips. Care must be taken with the iron sponge bed to maintain pH, gas temperature, and moisture content to prevent loss of bed activity. Consequently, injections of water and sodium carbonate are sometimes needed. H_2S reacts with iron oxide to form iron sulfide and water.

When the iron oxide is consumed, the bed must be changed out or regenerated. The bed can be regenerated with air; however, only about 60% of the previous bed life can be expected. The bed life of the batch process is dependent on the quantity of H_2S, the amount of iron oxide in the bed, residence time, pH, moisture content, and temperature.

Below are some technical points related to the iron sponge process:

The iron oxide or dry box process is one of the oldest known methods for removal of sulfur compounds from gas streams. It was introduced to England about the middle of the nineteenth century and is still widely used in many areas of special application.

This process often offers advantage when the sulfur in the gas does not exceed 7–9 ton per day and the concentration does not exceed 2400 g/100 S m^3 [1000 grains H_2S per 100 S ft^3] of gas.

This process consists of wood chips impregnated with varying amounts of hydrate iron oxide (Fe_2O_3). This reacts with the H_2S to form Fe_2S_3, which may be regenerated with air. Continuous regeneration is possible by injecting a small stream of air into the feed-gas stream, which converts the sulfide to the oxide and liberates the elemental sulfur. Regeneration is normally finished when the outlet oxygen concentration reaches 4–6% and the bed outlet temperature starts dropping. Each charge of sponge may be regenerated several times, but it gradually becomes less efficient and requires replacement.

- *Advantages of iron sponge process*
    a. Complete removal of small to medium concentrations of hydrogen sulfide without removing carbon dioxide
    b. Relatively small investment, for small to moderate gas volumes, compared with other processes
    c. Equally effective at any operating pressure
    d. Used to remove mercaptans or convert them to disulfides.
- *Disadvantages of the iron sponge process*
    The disadvantages of the process are:
    - A batch process requiring duplicate installation or flow interruption of processed gas
    - Prone to hydrate formation when operated at higher pressures and at temperatures in the hydrate-forming range
    - Effectually removes ethyl mercaptan that has been added for odorization
    - Coating of the iron sponge with entrained oil or distillate requires more frequent change out of the sponge bed.

10.11.2 Molecular sieves

Molecular sieves can be used for removal of sulfur compounds from gas streams. Hydrogen sulfide can be selectively removed to meet 4 ppmv specification. The sieve bed can be designed to dehydrate and sweeten simultaneously. In addition, molecular sieve processes can be used for CO_2 removal.

Crystalline sodium-calcium alumino silicates can be used for selective removal of H_2S and other sulfur compounds from natural gas streams. Some crystalline forms of these materials are found naturally. The common crystalline forms used in commercial adsorption are synthetically manufactured materials. The activated crystalline material is porous.

The pore openings in a given structure are all exactly the same size and are determined by the molecular structure of the crystal and the size of molecules present in the crystal. The pores are formed by driving off water of crystallization that is present during the synthesis process. The exactness of the pore size and distribution has given rise to the name molecular sieves, which is used almost universally to describe these materials.

Molecular sieves have the large surface area typical of any solid adsorbent. In addition, however, molecular sieves have highly localized polar charges. These localized charges are the reason for the very strong adsorption of polar or polarizable compounds on molecular sieves. This also results in much higher adsorptive capacities for these materials by molecular sieves than by other adsorbents, particularly in the lower concentration ranges.

In general, the concentrations of acid gas are such that cycle times are 6–8 h. To operate properly, the sieves must be regenerated at a temperature close to 315 °C for enough time to remove all adsorbed materials, usually 1 h or more. Exact arrangement of the regeneration cycle depends on process conditions.

Regeneration of a molecular sieve bed concentrates the H_2S into a small regeneration stream that must be treated or disposed of. During the regeneration cycle, the H_2S will exhibit a peak concentration in the regeneration gas. The peak is approximately 30 times the concentration of the H_2S in the inlet stream. Knowing the concentration of this stream is essential for the design of a gas treater for the regeneration gas. The problem of COS formation during processing according to the reaction:

$$H_2S + CO_2 \Rightarrow COS + H_2O \qquad (10.12)$$

has been extensively studied. Molecular sieve products have been developed that do not catalyze COS formation. The central zone in the regeneration cycle is most favorable to COS formation.

10.12 Process design

10.12.1 Design philosophy

The selected process should have given satisfactory service at process conditions and with gas compositions similar to that company's proposed treatment project.

The contractor shall pay particular attention to the heavy hydrocarbon analysis of the feed gas. If hydrocarbons condensed are absorbed in the treating solvent, severe process problems occur. The contractor's design shall incorporate features to remove or accommodate heavy hydrocarbons.

Oxygen causes oxygenation of certain treating solvents. Engineers should consider this in solvent selection if oxygen is in the feed gas.

Solvent storage tanks shall be blanketed with sweet natural gas or inert gas. Vacuum systems shall be avoided. If the solvent selected is subject to oxygen degradation, provisions shall be made in design to prevent oxygen from entering the system.

Solvent-acid gas loading shall be proportionated within accepted industry guidelines. However, for MEA, DEA, DGA, and MDEA solvents, the criteria given shall be considered respectively.

For other proprietary solvents, the solvent to acid gas loading shall rigidly based on the recommendations of the process licensor(s).

Solvent storage shall be provided with heating coils if freezing or high viscosity should prevent its normal transfer.

Solvent filters shall be provided in accordance with accepted industrial practices.

If the solvent selected, is capable of regeneration, complete regeneration facilities should be considered and designed.

If proprietary processes are used, regeneration equipment shall be designed in strict accordance with Licensor's specification.

Acid gases may be combusted in a flare or thermal oxidizer if compatible with the environmental regulations upon Company's approval.

The Company shall specify the maximum allowable concentration of H_2S, SO_2, and such compounds at ground level. The guaranteed values should be specified in project specification.

Due to considerable water vapor in treating plant, all essential determinations shall be considered in the process design, including but not limited to the followings:

1. The solvent stripping still shall be designed to prevent vacuum collapse in the event the tower is blocked in, when at a hot condition
2. All equipment should be designed for potential vacuum collapse
3. The acid gas disposal lines and facilities shall be designed so that water will not accumulate at the bottoms/lower ends
4. Particular emphasis shall be given to lines in intermittent service such as drains, instruments, gage glasses, etc., to be freeze-protected.

The following safety measures, but not limited to, shall be taken into consideration in process design of the plant:

1. For toxic solvents (if selected), safety showers, eye-wash fountains, shall be provided at strategic locations
2. All solvent and hydrocarbon tanks shall be diked to contain their contents in the event of a tank rupture
3. An Emergency Shut-Down (ESD) system shall be provided if applicable. The ESD shall be automatically and manually actuated. The ESD system shall be designed to block the feed and products gas lines of (affected) train in the treating plant. The affected train shall be vented to the flare, at a rate not to damage the equipment. Gas flow to all fired heaters shall be stopped and all rotating equipment shall be stopped
4. Plant pressure relief valves shall be separate from the ESD vents. ESD pushbuttons shall be provided at main gate, main control panel, and main entrance
5. Hydrocarbon detectors shall automatically trip the ESD system. They shall alarm at 20% of the lower explosive limit for methane and trip at 40%. They shall be provided at:

- air intake to each control building
- analyzer house
- each compressor building
- in the gas contacting area
- above each cooling tower cell
- any other locations may deem necessary for safety requirements.

6. H_2S detectors and alarms shall be provided if the gas to be treated contains more than 10 ppmv H_2S. The detector shall have solid state sensors. The detector heads shall be located in proper places and the alarm system should have a two-level warning tone. One level shall be announced at an H_2S concentration in ambient air of 10 ppmv. The other level shall be announced at 30 ppmv H_2S in ambient air.

10.12.2 Material selection for sour services

The following requirements should be considered for vessels, drums, and separator when they are used for sour services:

1. *General*
Whenever the fluid to be handled in the process system contains:
   a. >50 ppmw dissolved H_2S in the free water
   b. free water pH < 4 and some dissolved H_2S
   c. free water pH > 7.6 and 20 ppmw dissolved hydrogen cyanide (HCN) in the water and some dissolved H_2S present;
   d. >0.0003 MPa absolute (0.05 psia) partial pressure H_2S in the gas in processes with a gas phase.

Partial pressure of H_2S is determined by multiplying the mole fraction (mol. % +100) of H_2S in the gas times the system pressure.

2. *Materials*
   a. All carbon steel and low alloy steel material used in sour service shall comply with, but not be limited by, the requirements of NACE Standard MR 0103-2003. (Material Requirements-Sulfide Stress Cracking Resistant Material for Oil field Equipment, latest Revision).
   b. The use of special corrosion-resistant materials, alloys, or clad steels other than standard oil field sour service materials shall be subject to approval by the Company. Copper or copper alloys shall not be used.
   c. The following Table 10.7 lists approved materials for use in sour services, based on American Society for Testing and Materials (ASTM) specifications. The Vendor/Contractor may suggest alternate materials that are equal or better for consideration by the Company. These materials should also satisfy the physical and chemical properties requirements.

10.12.3 Piping for sour services

1. Whenever the fluid handled in the process system contains:
   a. Sour gas of 0.34 kPa (abs) or greater partial pressure hydrogen sulfide (H_2S), and if the system pressure is 450 kPa (abs) or greater

Table 10.7 ASTM Specifications	No Impact Tests Required	When Impact Tests Required
Plate	**A 516**	**A 516 (Normalized and impact tested to −45 °C maximum)**
Nozzles	A 106, Grade B	A 333, Grade 6
Flanges (weld neck only)	A 105	A 350, Grade LF 2
Fittings	A 234, Grade WPB	A 420, Grade WPL 6
Bolts	A 193, Grade B7X/M	A 320, Grade L7M
Nuts	A 194, Grade 2HX/M	A 194, Grade 2M
Gaskets	TP-304 SS or TP-316 SS	TP-304 SS or TP-316 SS

    b. Sour oil and multiphase of 70 kPa (abs) or greater partial pressure H_2S, whenever the system pressure is 1825 kPa (abs) or greater or, whenever the gas phase contains over 15 percent H_2S at any system pressure

    c. Rich sweetening agents, such as amines, sufinol, etc.

2. In general, the materials used in sour service shall be based on the following:

    a. All carbon and low alloy steel material used in sour service shall comply with, but not be limited, by the requirements of NACE Standard MR 0103-2003, latest revision

    b. The use of special corrosion-resistant materials, alloys, or cold steels other than standard oil field sour service materials shall be subject to approval by the Company. Copper and copper alloys shall not be used

    c. Stainless steels of the 400 series or other martensitic steels shall not be used unless specifically approved by the Company

    d. All carbon and low-alloy steel materials shall have the following properties:

      – The ratio of percent manganese to percent carbon (%Mn/%C) shall be greater than or equal to 3.0

      – Percent carbon plus one-quarter of the percent manganese (%C + %Mn/4) shall be less than or equal to 0.55

3. For materials used in sour service piping components such as plates, pipes, fittings, flanges, valves, gaskets, etc., the applicable provisions of the following standards shall be considered along with manufacturer's standard specification:

    a. ASTM-A 516, ASTM-A 106, Grade B, or ASTM-A 333 for plates and pipes respectively

    b. ASTM-A 234 Grade WPB for fittings unless otherwise specified

    c. ASTM-A 105 or ASTM-A 350, Grade LF2 for flanges unless otherwise specified

    d. All valves shall meet the requirement of NACE Standard MR 0103-2003

    e. All gaskets in sour service shall be TP 304 SS, TP 304 SS oval rings to ASME B 16.20 shall be used as minimum.

Reference

[1] Gas Processors and Suppliers Association (GPSA). Engineering databook. 12th ed. 2004. Tulsa (OK, USA).

Further reading

API RP. 945, Avoiding Environmental Cracking in Amine Units, para. 3.2.3. Bhatia K, Edwards M. Application of sulfa-check in the gas Industry. In: Proceedings of the gri Liquid Redox Sulfur Recovery Conference; April, 1992. p. 353. Austin Texas.

ASME (AMERICAN SOCIETY OF MATERIAL ENGINEERING) ASME B 16.20. Material gaskets for pipes, flanges, ring Joint, Spiral Wounded & Jacketed.

ASTM (AMERICAN SOCIETY FOR TESTING AND MATERIALS) ASTM A 516/A 516/M Grade B ASTM A 106, Grade B ASTM a 333 ASTM a 234, Grade WPB ASTM a 105, Grade LF2 ASTM a 350, Grade LF2.

Ballaguet JP, et al. Direct H2S removal from high pressure natural gas with the sulfint HP redox process, presented at Sulfur 99, Calgary, Alberta, Canada.

Ballaguet JP, Streicher C. Pilot operating experience with a new redox process for the direct high pressure removal of H2S. San Antonio, Texas: 80th GPA Annual Convention; March, 2001.

Ballard D. How to operate an amine plant. Hydrocarbon Process Pet Refin April, 1966;45(4):137.

Benson HE, Parrish RW. HiPure process removes CO2/H2S. Hydrocarbon Process; April, 1974:81.

Beychok MR. Aqueous wastes from petroleum and petrochemical plants. John Wiley; 1973. 113–143.

Blake RJ, Rothert KC. Reclaiming monoethanolamine solutions. In: Proceedings Gas Conditioning Conference. University of Oklahoma Extension Division; 1962.

Cameron CJ, Barthel Y, Sarrazin P. Mercury removal from wet natural gas. In: Proceedings of the 73rd GPA annual convention; March, 1994. p. 256. New Orleans.

Carroll JJ, Maddocks JR. Design considerations for acid gas injection. In: Laurance Reid Gas Conditioning Conference; March, 1999. Norman (OK).

Clarke MA, et al. Designing an optimized injection strategy for acid gas disposal without dehydration. In: Proceedings of the 78th GPA annual convention; March, 1999. p. 49. Nashville (TN).

Clem KR, Kaufman JL, Lambert MM, Brown DL. Gas treating advances with Flexsorb technology. In: Proceedings Gas Conditioning Conference. University of Oklahoma Extension Division; 1985.

Dalrymple DA, Srinivas G. CrystaSulf liquid redox and TDA gas phase H2S conversion technologies for sour gas treating. GPA 78th Annual Convention; March 1–3, 1999. Nashville (TN).

Dillon ET. Gas sweetening with a novel and selective hexahydrotriazine. In: Proceedings of the gri Liquid Redox Sulfur Recovery Conference; April, 1992. p. 373. Austin (TX).

Dingman JC, Jackson JL, Moore TF, Branson JA. Equilibrium data for the H2S–CO2 diglycolamine agent-water system. San Francisco: 62nd Gas Processors Association Annual Meeting; March 14–16, 1983.

Eickmeyer AG, Wiberg EA, Gangriwala HA. Energy savings in the gas patch using catacarb processes. Petroenergy; September 1, 1981:81. Houston (TX).

Gall GH, Sanders ES. Removal of carbon dioxide from liquid ethane using membrane technology. Dallas, Texas: 81st GPA Annual Convention; March, 2002.

Goar BG, Arrington TO. Guidelines set for handling sour Gas. Oil Gas J; June 26, 1978:160.

Goar BG. Sulfinol process has several key advantages. Oil Gas J; June 30, 1969:117.

Goddin CS. Comparison of processes for treating gases with high CO2 content. 61st Annual GPA Convention; March, 1982.

Goldstein AM, Edelman AM, Beisner WD, Ruziska PA. Hindered amines yield improved gas treating. Oil Gas J; July 16, 1984:70.

Harbison JL, Dingman JC. Mercaptan removal experiences in DGA sweetening of low pressure Gas. In: Proceedings Gas Conditioning Conference. University of Oklahoma Extension Division; 1972.

Hardison LC. Treating hydrogen sulfide: an alternative to claus. Chem Eng; January 21, 1985:62.

Hardison LC. Go from H2S to S in one unit. Hydrocarbon Process; April, 1985:70.

Hardison LC. Catalytic gas sweetening process selectively converts H2S to sulfur, treats acid gas. Oil Gas J; June 4, 1984:60.

Hegwer AM, Harris RA. Selexol solves high H2S/CO2 problem. Hydrocarbon Process April, 1970;49(4):103.

Herrin JP. Removing very fine iron sulfide particles from natural Gas. presented at the Laurance Reid Gas Conditioning Conference. University of Oklahoma; March, 1994.

Holder HL. Diglycolamineda promising new acid-gas remover. Oil Gas J 1966;64(18):83.

Jones JH, Froning JR, Claytor Jr EE. Chem Eng Data January, 1959;4:85.

Jones VW, Pearce RL. Fundamentals of Gas treating. In: Gas Conditioning Conference. Norman: Oklahoma; March 1980.

Kensell WW. How to pick a treating plant. Hydrocarbon Process; August, 1979:143.

Kent RL, Eisenberg B. Better Data for Amine Treating. Hydrocarbon Process. 1976;55(2):87.

Klen JP. Developments in sulfinol and ADIP processes increases uses. Oil Gas Int; September, 1970:34.

Kohl AL, Buckingham PA. Fluor solvent CO2 removal process. Pet Refin May, 1960;39(5):193.

Kohl AL, Reisenfeld FC. Gas purification. McGraw-Hill, Book Co., Inc; 1960. Chemical Engineering Series.

Kutsher GS, Smith GA, Greene PA. Sour gas scrubbing-allied chemical solvent process. In: Proceedings of the 46th. Houston (TX): NGPA Annual Convention; March 14–16, 1967.

Lallemand F, Minkkinen A. Highly sour gas processing in an ever-greener world. Texas: presented at the 80th Annual GPA convention in San Antonio; March 12–14, 2001.

Lee JI, Otto FD, Mather AE. Gas processing/canada; March–April, 1973:26.

Lee JI, Otto FD, Mather AE. The solubility of mixtures of CO2 and H2S in DEA solutions. Can J Chem Eng 1974; 52:125.

Leppin D, Dalrymple DA. Overview of liquid redox technology for recovering sulfur from natural Gas. In: Laurance Reid Gas Conditioning Conference; March, 2000. Norman (OK).

Lidal H, Nilsen JI, Isaksen H, Hoang-Dinh V. CO2 removal at sleipner. In: Laurance Reid Gas Conditioning Conference. University of Oklahoma; March, 1998.

Lock BW. Acid gas disposal: a field perspective. In: Proceedings of the 76th GPA annual convention; March, 1997. p. 161. San Antonio (TX).

Maddox RN. Gas and liquid sweetening. Campbell Pet Ser; April, 1977:99.

Manning WP. The chemsweet process. In: Permian basin regional meeting of the gas processors association; May, 1982.

McClure GP, Morrow DC. Amine process removes cos from propane economically. Oil Gas J; July 2, 1979: 107.

McIntush KE, et al. Status of the first commercial applications of the crystaSulf. Dallas, Texas: 81st GPA Annual Convention; March, 2002.

Menzel J, Tondorf O. A cost effective physical solvent process for acid gas removal. Scotland: presented at the GPA Europe spring meeting in Aberdeen; May 17–19; 2000.

Mick MB. Treat propane for cos removal. Hydrocarbon Process; July, 1976:137.

Mitariten M, Dolan W, Maglio A. Innovative molecular gate systems for nitrogen rejection carbon dioxide removal and NGL recovery. Dallas, Texas: 81st GPA Annual Convention; March, 2002.

NACE (NATIONAL ASSOCIATION OF CORROSION ENGINEERS) NACE Standard MR 0103–2003. Standard material requirements-material resistant to sulfide stress cracking in corrosive Petroleum refining Environments.

Newman SA, editor. Acid and sour gas treating processes. Gulf Publishing Company; 1985. pp. 112–30.

Nilsen FP, Lidal H, Linga H. Selective H2S removal in 50milliseconds. The Netherlands: presented at the GPA Europe annual conference in Amsterdam; September 26–28, 2001.

Perry CR. A new look at iron sponge treatment of sour Gas. In: Proceedings of Gas Conditioning Conference. University of Oklahoma Extension Division. 1970.

Perry CR. Activated carbon filtration of amine and glycol solutions. In: Proceedings Gas Conditioning Conference. University of Oklahoma Extension Division. 1974.

Perry RH, Chilton CH. Chemical engineer's handbook. 5th ed. NY: McGraw Hill Book Co; 1973. pp. 21–19.

Picciotti M. Optimize caustic scrubbing systems. Hydrocarbon Process; May, 1978:201.

Process industries corrosion. NACE Publication; 1986. 298.

Russell Richard M. Liquid-liquid contactors need careful attention. Oil Gas J; December 1, 1980:135.

Say GR, Heinzelman FJ, Iyengar JN, Savage DW, Bisio A, Sartori A. A new hindered amine concept for simultaneous removal of CO_2 and H_2S from gases. Chem Eng Prog; October, 1984:72.

Schell WJ, Hoernschemeyer DL. Principles of membrane gas separation. Separex Corp AICHE Symp; June, 1982.

Sterner Anthony J. Acid gas fractionation. In: Laurance Reid Gas Conditioning Conference 2001 proceedings; February 25–28, 2001. p. 17.

Stiltner J. Mercury removal from natural gas and liquid streams. In: Proceedings of the 81st GPA annual convention; March, 2002. Dallas (TX).

Stuksrud DB, Dannstrom H. Membrane gas-liquid contactor for natural gas sweetening. Aberdeen, Scotland: GPA Europe Spring Meeting; May 17–19, 2000.

Tennyson RN, Schaaf RP. Guidelines can help choose proper process for gas treating plants. Oil Gas J; January 10, 1977:78.

Turnock PH, Gustafson KJ. Advances in molecular sieve technology for natural Gas sweetening. In: Proceedings Gas Conditioning Conference. University of Oklahoma Extension Division; 1972.

Weber S, McClure G. New amine process for FCC desulfurizes light liquid streams. Oil Gas J; June 8, 1981:161.

Wendt CJ. The purisol process for acid gas treatment. In: Proceedings Gas Conditioning Conference. University of Oklahoma; 1969. Extension Division.

Wichert E, Royan T. Acid gas injection eliminates sulfur recovery expense. Oil Gas J; April 28, 1997:67.

Williams WW. How to remove mercaptans from natural Gas. Hydrocarbon Process Pet Refin July 1969;43(7). 121.

Wizig HW. An innovative approach for removing CO2 and sulfur compounds from a Gas stream. In: Gas Conditioning Conference; March 4, 1985. Norman (OK).

Woodward C. Temporary sweetening unit aids early oil, gas production. Oil Gas J; February 13, 1995:74.

Sulfur Recovery

Sulfur is an extremely useful element. Its largest application is for the manufacture of fertilizers, with other principal users including rubber industries, cosmetics, and pharmaceuticals. Sulfur is present in many raw industrial gases and in natural gas in the form of hydrogen sulfide (H_2S). The noxious hydrogen sulfide fumes that characterize many gas processing and refinery operations and petroleum production sites represent a genuine threat to our environment.

Sulfur removal facilities are located at the majority of oil and gas processing facilities throughout the world. The sulfur recovery unit does not make a profit for the operator, but it is an essential processing step to allow the overall facility to operate, as the discharge of sulfur compounds to the atmosphere is severely restricted by environmental regulations.

Concentration levels of H_2S vary significantly depending upon their source. H_2S produced from absorption processes, such as amine treating of natural gas or refinery gas, can contain 50–75% H_2S by volume or higher. Many other processes can produce H_2S with only ppm concentration, but in quantities that preclude the gases from being vented without further treatment.

Sulfur is present in natural gas principally as hydrogen sulfide and, in other fossil fuels, as sulfur-containing compounds that are converted to hydrogen sulfide during processing. The H_2S, together with some or all of any carbon dioxide (CO_2) present, is removed from the natural gas or refinery gas by means of one of the gas treating processes described in Chapter 10. The resulting H_2S-containing acid gas stream is flared, incinerated, or fed to a sulfur recovery unit. This chapter is concerned with recovery of sulfur by means of the modified Claus and Claus tail gas cleanup processes.

11.1 The Claus process

The large variations in concentrations and flows require different methods for H_2S removal and sulfur recovery. For relatively small quantities of H_2S/sulfur, scavenger processes are often used. For sulfur quantities up to approximately 5 long tons per day of sulfur, liquid reduction–oxidation (redox) processes are common. The sulfur is produced as an aqueous slurry.

Direct oxidation can sometimes be utilized for low H_2S concentrations to produce high-quality liquid sulfur. The Claus process is the most widely used process for conversion of H_2S to elemental sulfur. Variations of the Claus process, direct oxidation and liquid redox processes, can overlap other processes' applicability ranges. Figure 11.1 is a sulfur recovery process applicability chart, which presents the relative ranges of technology applications.

The Claus process was invented in 1883 by the English scientist Carl Friedrich Claus. The basic Claus process mixed hydrogen sulfide with oxygen and passed the mixture across a preheated catalyst bed. The end products were sulfur, water, and thermal energy.

Because the process performed best at 400–600°F, and the reaction heat could be removed only by direct radiation, only a small amount of H_2S could be processed at one time without overheating the

Natural Gas Processing. http://dx.doi.org/10.1016/B978-0-08-099971-5.00011-8
Copyright © 2014 Elsevier Inc. All rights reserved.

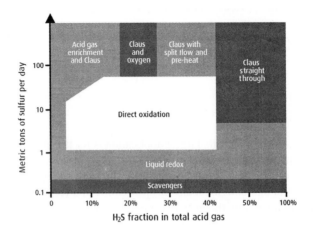

FIGURE 11.1

Sulfur recovery process applicability chart representing the relative ranges of technology applications.

reactor. The process was improved in 1938 by the addition of free-flame oxidation ahead of the catalyst bed and by revising the catalytic step. This "modified Claus process" greatly increased the sulfur yield and is the basis of most sulfur recovery units (SRUs) in use today.

11.1.1 How the process works

Feed gas for a Claus sulfur recovery unit usually originates in an acid gas sweetening plant. The stream, containing varying amounts of H_2S and CO_2, is saturated with water and frequently has small amounts of hydrocarbons and other impurities in addition to the principal components.

In a typical unit (Figure 11.2), H_2S-bearing gas enters at about 8 psig and 120 °F. Combustion air is compressed to an equivalent pressure by centrifugal blowers. Both inlet streams then flow to a burner, which fires into a reaction furnace.

The free-flame modified Claus reaction can convert approximately 50–70% of the sulfur gases to sulfur vapor. The hot gases, up to 2500 °F, are then cooled by generating steam in a waste heat boiler.

The gases are further cooled by producing low-pressure steam in a separate heat exchanger, commonly referred to as a sulfur condenser. This cools the hot gases to approximately 325 °F, condensing most of the sulfur that has formed up to this point. The resultant liquid sulfur is removed in a separator section of the condenser and flows by gravity to a sulfur storage tank. Here it is kept molten, at approximately 280 °F, by steam coils. Sulfur accumulated in this reservoir is pumped to trucks or rail cars for shipment.

The Claus process as used today is a modification of a process first used in 1883 in which H_2S was reacted over a catalyst with air (oxygen) to form elemental sulfur and water.

$$H_2S + \tfrac{1}{2}O_2 \rightarrow S + H_2O \tag{11.1}$$

Control of this highly exothermic reaction was difficult and sulfur recovery efficiencies were low. In order to overcome these process deficiencies, a modification of the Claus process was developed and introduced in 1936 in which the overall reaction was separated into: (1) a highly

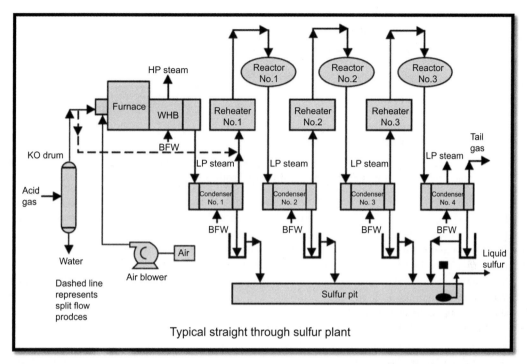

FIGURE 11.2

Typical straight-through sulfur plant.

exothermic thermal or combustion reaction section, in which most of the overall heat of reaction (from burning one-third of the H_2S and essentially 100% of any hydrocarbons and other combustibles in the feed) is released and removed; and (2) a moderately exothermic catalytic reaction section, in which sulfur dioxide (SO_2) formed in the combustion section reacts with unburned H_2S to form elemental sulfur.

The principal reactions taking place (neglecting those of the hydrocarbons and other combustibles) can then be written as follows:

Thermal or Combustion Reaction Section

$$H_2S + 1^1\!/_2O_2 \rightarrow SO_2 + H_2O \tag{11.2}$$

$$\Delta H@25\ °C = -518{,}900\ kJ$$

Combustion and Catalytic Reaction Sections

$$2H_2S + SO_2 \rightarrow \frac{3}{X}S_X + 2H_2O \tag{11.3}$$

$$\Delta H@25\ °C = -96{,}100\ kJ$$

Overall Reaction

$$3H_2S + 1^1/_2O_2 \rightarrow \frac{3}{X}S_X + 3H_2O \tag{11.4}$$

$$\Delta H @ 25 \,^\circ C = -615,000 \text{ kJ}$$

11.1.2 Simplified process description

The hot combustion products from the furnace at 1000–1300 °C enter the waste heat boiler and are partially cooled by generating steam. Any steam level from 3 to 45 barg can be generated.

- The combustion products are further cooled in the first sulfur condenser, usually by generating low-pressure steam at 3–5 barg. This cools the gas enough to condense the sulfur formed in the furnace, which is then separated from the gas and drained to a collection pit.
- In order to avoid sulfur condensing in the downstream catalyst bed, the gas leaving the sulfur condenser must be heated before entering the reactor.
- The heated stream enters the first reactor, containing a bed of sulfur conversion catalyst. About 70% of the remaining H_2S and SO_2 in the gas will react to form sulfur, which leaves the reactor with the gas as sulfur vapor.
- The hot gas leaving the first reactor is cooled in the second sulfur condenser, where low-pressure steam is again produced and the sulfur formed in the reactor is condensed.
- A further one or two more heating, reaction, and condensing stages follow to react most of the remaining H_2S and SO_2.
- The sulfur plant tail gas is routed either to a tail gas treatment unit for further processing or to a thermal oxidizer to incinerate all of the sulfur compounds in the tail gas to SO_2 before dispersing the effluent to the atmosphere.

For the usual Claus plant feed gas composition (water saturated with 30–80 mol% H_2S, 0.5–1.5 mol% hydrocarbons, the remainder CO_2), the modified Claus process arrangement results in thermal section (burner) temperatures of about 980–1370 °C. The principal molecular species in this temperature range is S_2 and conditions appear favorable for the formation of elemental sulfur by direct oxidation of H_2S Eqn (11.5) rather than by the Claus reaction Eqn (11.6).

However, both laboratory and plant measurements indicate that the more highly exothermic oxidation of H_2S to SO_2 Eqn (11.2) predominates and the composition of the equilibrium mixture therefore is determined by the slightly endothermic Claus reaction Eqn (11.6).

$$2H_2S + O_2 \rightarrow 2H_2O + S_2 \tag{11.5}$$

$$\Delta H @ 25 \,^\circ C = -314,500 \text{ kJ}$$

$$2H_2S + SO_2 \rightarrow 2H_2O + \frac{3}{2}S_2 \tag{11.6}$$

$$\Delta H @ 25 \,^\circ C = +47,500 \text{ kJ}$$

To attain an overall sulfur recovery level above about 70%, the thermal, or combustion, section of the plant is followed by one or more catalytic reaction stages. Sulfur is condensed and separated from the process gases after the combustion section and after each catalytic reaction stage in order to improve equilibrium conversion.

Gases leaving the final sulfur condensation and separation stage may require further processing. These requirements are established by local, state, or national regulatory agencies. These requirements can be affected by the size of the sulfur recovery plant, the H_2S content of the plant feed gas, and the geographical location of the plant.

11.1.3 Claus process considerations

The Claus sulfur recovery process includes the following process operations:

- Combustion—burn hydrocarbons and other combustibles and one-third of the H_2S in the feed.
- Waste heat recovery—cool combustion products. Because most Claus plants produce 1030–3450 kPa (ga) steam at 185–243 °C, the temperature of the cooled process gas stream is usually about 315–370 °C.
- Sulfur condensing—cool outlet streams from waste heat recovery unit and from catalytic converters. Low-pressure steam at 345–480 kPa (ga) is often produced and the temperature of the cooled gas stream is usually about 177 °C or 127–149 °C for the last condenser.
- Reheating—Reheat process stream, after sulfur condensation and separation, to a temperature high enough to remain sufficiently above the sulfur dew point, and generally, for the first converter, high enough to promote hydrolysis of carbonyl sulfide (COS) and carbon disulfide (CS_2) to H_2S and CO_2.

$$COS + H_2O \rightarrow CO_2 + H_2S \tag{11.7}$$

$$CS_2 + 2H_2O \rightarrow CO_2 + 2H_2S \tag{11.8}$$

- Catalytic conversion—Promote reaction of H_2S and SO_2 to form elemental sulfur Eqn (11.3).

11.1.4 Catalytic reaction

Any further conversion of the sulfur gases must be done by catalytic reaction. The gas is reheated by one of several means and is then introduced to the catalyst bed. The catalytic Claus reaction releases more energy and converts more than half of the remaining sulfur gases to sulfur vapor.

This vapor is condensed by generating low-pressure steam and is removed from the gas stream. The remaining gases are reheated and enter the next catalytic bed.

This cycle of reheating, catalytic conversion, and sulfur condensation is repeated in two to four catalytic steps. A typical SRU has one free-flame reaction and three catalytic reaction stages. Each reaction step converts a smaller fraction of the remaining sulfur gases to sulfur vapor, but the combined effect of the entire unit is to reduce the hydrogen sulfide content to an acceptable level.

11.1.5 High yields plus energy

Claus sulfur plants can normally achieve high sulfur recovery efficiencies. For lean acid gas streams, the recovery typically ranges from 93% for two-stage units (two catalytic reactor beds) up to 96% for three-stage units.

For richer acid gas streams, the recovery typically ranges from 95% for two-stage units up to 97% for three-stage units. Since the Claus reaction is an equilibrium reaction, complete H_2S and SO_2 conversion is not practical in a conventional Claus plant. The concentration of contaminants in the acid gas can also limit recovery. For facilities where higher sulfur recovery levels are required, the Claus plant is usually equipped with a tail gas cleanup unit to either extend the Claus reaction or capture the unconverted sulfur compounds and recycle them to the Claus plant.

All Claus SRUs produce more heat energy as steam than they consume. This is particularly true for those plants equipped with waste heat boilers on the incinerator.

The steam produced can be used for driving blowers or pumps, reboiler heat in the gas treating or sour water stripper plants, heat tracing, or any of a number of other plant energy requirements.

Following are some sulfur recovery and tail gas cleanup methods:

- Straight-through Claus
- Split-flow Claus
- Direct oxidation
- Acid gas enrichment
- Oxygen enrichment
- Cold bed adsorption
- Shell Claus off-gas treating

11.2 Technology overview

This section presents an overview of the technologies available through several process plants.

11.2.1 Types of plants

Claus sulfur recovery units are generally classified according to the method used for the production of SO_2 and the method used to reheat the catalyst bed feeds. The various reheat methods can be used with any SO_2 production method, whereas the technique used for the production of SO_2 is determined by the H_2S content of the acid gas feedstock.

Most sulfur recovery plants utilize one of three basic variations of the modified Claus process: "straight-through," "split-flow," or "direct oxidation." "Acid gas enrichment" can be applied ahead of the SRU to produce a richer acid gas stream and "oxygen enrichment" may be used in combination with any of these variations.

These three varieties of the modified Claus process differ in the method used to oxidize H_2S and produce SO_2 ahead of the first catalytic reactor. The first two processes use a flame reaction furnace ahead of the catalytic stages. The third process reacts oxygen directly with the H_2S in the first catalytic reactor to produce the SO_2.

11.2.2 Straight-through process

A "straight-through" unit passes all the acid gas through the combustion burner and reaction furnace. The initial free-flame reaction usually converts more than half of the incoming sulfur to elemental sulfur. This reduces the amount that must be handled by the catalytic sections and thus leads to the highest overall sulfur recovery.

The amount of heat generated in the reaction depends on the amount of H_2S available to the burner. With rich acid gas (60–100% H_2S), the reaction heat keeps the flame temperature above 2200 °F. When the gas is leaner, the flame temperature is reduced; the greater mass is heated to a lower temperature. If the temperature falls below a critical point, approximately 1800–2000 °F, the flame becomes unstable and cannot be maintained.

This point is usually reached when the acid gas has an H_2S content of 50% or less. The problem can be overcome, within limits, by preheating the acid gas and/or air before it enters the burner. However, the lower the H_2S content, the higher the preheat requirement becomes; when the gas composition falls below about 40% H_2S, this approach ceases to be practical.

11.2.3 Split-flow process

The second method of SO_2 production, known as the "split-flow" technique, is used to process leaner acid gases with 15–50% H_2S content. In these units, at least one-third of the acid gas flows into the combustion burner and the balance usually bypasses the furnace entirely.

Enough H_2S is burned to provide the necessary 2:1 ratio of H_2S to SO_2 in the catalyst beds. The flame temperature is kept above the minimum, since the constant amount of heat supplied is absorbed by a lower mass of gas. The free-flame Claus reaction is reduced or eliminated entirely by this approach, since little or no H_2S is available to react in the furnace. This results in a slight reduction of the overall sulfur recovery.

11.2.4 Split-flow process for ammonia destruction

A variation of the split-flow process is often applied in refinery SRUs that must process sour water stripper (SWS) off-gas and destroy the ammonia it contains. Efficient ammonia destruction is critical for SRUs in refineries, since ammonia can combine with the sulfur compounds in the process gas to form salts that precipitate in the lower-temperature section of the Claus unit. Accumulation of such ammonium salts would lead to unreliable operation and unacceptable maintenance costs.

All of the SWS gas is routed to the combustion burner along with a portion of the amine acid gas, so that at least one-third of the total H_2S is supplied to the burner.

This creates a high-temperature combustion zone at the inlet end of the reactor furnace where the ammonia breaks down into nitrogen and water by thermal decomposition. The remainder of the amine acid gas is injected into the middle part of the reactor furnace, where it mixes with the burner combustion products. The outlet end of the furnace provides residence time for the SO_2 produced by the burner to react with the H_2S in the bypass acid gas and form sulfur, and for any hydrocarbons in the bypass acid gas to oxidize. An optical pyrometer is typically used to monitor the temperature in the inlet section of the furnace, and is often used to adjust the amount of bypass acid gas to control the temperature at the desired value.

11.2.5 Direct oxidation process

When the H_2S concentration is below about 15% in the acid gas, the direct oxidation version of the modified Claus process may be used. Rather than using a burner to combust H_2S to form SO_2, the direct oxidation process catalytically reacts oxygen with H_2S by mixing the air and acid gas upstream of a catalytic reactor.

As SO_2 forms, it then reacts with the remaining H_2S via the Claus reaction to form sulfur. The direct oxidation is typically followed by one or more standard Claus reactors to produce and recover additional sulfur. The direct oxidation process is sensitive to catalyst deactivation by contaminants in the acid gas feed (particularly hydrocarbons), so it is not used as much as the other Claus process varieties.

11.3 Acid gas enrichment

When the acid gas produced by the gas treating system is low in H_2S concentration, it is sometimes advantageous to "enrich" the acid gas by contacting it with a second solvent. The second solvent is typically a selective solvent designed to absorb essentially all of the H_2S from the acid gas while letting most of the remainder (generally CO_2) "slip" through.

The enriching process can often raise the H_2S concentration of the SRU feed gas by a factor of five or more. Not only does this allow using smaller equipment in the SRU, but it can often allow a more reliable SRU process to be used, such as a straight-through Claus process instead of a direct-oxidation Claus process.

11.4 Oxygen enrichment

Since air is approximately 79% nitrogen and 21% oxygen, the introduction of air to supply the oxygen for combustion of H_2S to SO_2 also introduces a large quantity of nitrogen. When air is used as the oxygen source, approximately 5.6 mol of nitrogen is introduced into the gas flow for every mol of H_2S that is burned. Nitrogen does not react and the added mass of the nitrogen lowers the adiabatic flame temperature in the reaction furnace.

The nitrogen must also be heated, cooled, and reheated through the combustion, sulfur condensation, and reheat ahead of the reactors. Pure oxygen or enriched sources of oxygen can be used instead of air in the Claus process. Higher flame temperatures can be achieved with lower H_2S concentrations. In addition, the relative equipment sizes can be reduced in proportion to the amount of nitrogen that is not introduced with oxygen for combustion.

11.5 Reheat methods

Gas leaving the sulfur condenser is at its sulfur dew point temperature. Since the catalytic reaction requires a higher temperature for proper operation, the gas must be reheated before entering the reactor. This can be done directly (internally) or indirectly (externally). The method chosen is an important characteristic of any sulfur recovery unit.

11.5.1 **Direct reheat**

Two direct reheat methods are commonly employed: "inline burning" and "hot gas bypass." The first of these burns fuel or acid gas with air in an inline burner, allowing the combustion products to mix directly with the process gas flow. The second method bypasses a portion of the hot boiler outlet gas around the sulfur condenser and mixes it with the reactor feed stream.

Both methods have an adverse effect on overall sulfur recovery. In the case of the inline burner, it is difficult to maintain precise control of the overall air/H_2S ratio. Any excess of deficiency of oxygen to the inline burner can cause undesirable reactions in the catalyst beds. Too little oxygen can lead to carbon deposits on the catalyst, reducing its activity. Too much oxygen can lead to catalyst deactivation, increased corrosion, and, in extreme cases, fire in the reactor. All of these result in reduced sulfur recovery.

The "hot gas bypass" method allows a portion of the sulfur-bearing process gas to skip one or more catalytic reaction and sulfur condensing steps. When this happens, the sulfur gases have less opportunity to convert to sulfur vapor and the overall sulfur recovery drops. With both methods, the recovery loss is more pronounced in a lean acid gas unit than it is in a rich gas SRU.

11.5.2 **Indirect reheat**

The indirect reheat method uses an outside heat source to raise the temperature of the acid gases in a heat exchanger. Although this requires additional equipment, it eliminates the conversion loss problems associated with direct reheat. There are three common variations of indirect reheat: "gas–gas exchange," "fuel gas firing," and "steam reheat." Gas–gas exchangers use a sulfur condenser feed to reheat the sulfur condenser outlet gas. This exchange works well as long as the temperature of the heating gas is maintained above a minimum level. If the upstream catalyst bed has lost some of its efficiency, however, the drop in conversion will lower the outlet temperature. Less heat energy will be available for reheat. As a result, the following catalyst bed will perform less efficiently and overall sulfur recovery will decrease.

Steam reheat and fuel gas-fired heaters have none of the problems associated with the gas–gas exchange method described above. Fuel gas firing uses a conventional fired heater; the acid gases are heated in tubes and the combustion products are vented to the atmosphere. For this reason, fuel gas firing usually involves higher utility costs than the other indirect methods.

Steam reheat can usually be accomplished by utilizing a portion of the steam produced by the SRU itself, often in the same vessel with the source of steam production. Steam reheat is the method commonly preferred in better-quality sulfur plant designs.

Table 11.1 can be used as a guide in Claus process selection.

Table 11.1 Claus Plant Configurations	
Feed H$_2$S Concentration (mol%)	**Claus Variation Suggested**
55–100	Straight-through
30–55	Straight-through or straight-through with acid gas and/or air preheat
15–30	Split-flow or straight-through with feed and/or air preheat
10–15	Split-flow with acid gas and/or air preheat
5–10	Split-flow with fuel added or with acid gas and air preheat, or direct oxidation or sulfur recycle
<5	Sulfur recycle or variations of direct oxidation or other sulfur recovery processes

11.6 Combustion operation

Most Claus plants operate in the "straight-through" mode. The combustion is carried out in a reducing atmosphere with only enough air (1) to oxidize one-third of the H_2S to SO_2, (2) to burn hydrocarbons and mercaptans, and (3) for many refinery Claus units, to oxidize ammonia and cyanides. Air is supplied by a blower and the combustion is carried out at 20–100 kPa (ga), depending on the number of converters and whether a tail gas unit is installed downstream of the Claus plant.

Numerous side reactions can also take place during the combustion operation, resulting in such products as hydrogen (H_2), carbon monoxide (CO), COS, and CS_2. Thermal decomposition of H_2S appears to be the most likely source of hydrogen since the concentration of H_2 in the product gas is roughly proportional to the concentration of H_2S in the feed gas. Formation of CO, COS, and CS_2 is related to the amounts of CO_2 and/or hydrocarbons present in the feed gas. Plant tests indicate concentrations of H_2 and CO in the product gas to be approximately at equilibrium at reaction furnace temperatures; Table 11.2 indicates potential COS and CS_2 formation in the Claus furnace.

Heavy hydrocarbons, ammonia, and cyanides are difficult to burn completely in a reducing atmosphere. Heavy hydrocarbons may burn partially and form carbon, which can cause deactivation of the Claus catalyst and the production of off-color sulfur. Ammonia and cyanides can burn to form nitric oxide (NO), which catalyzes the oxidation of sulfur dioxide (SO_2) to sulfur trioxide (SO_3); SO_3 causes sulfation of the catalyst and can also cause severe corrosion in cooler parts of the unit. Unburned ammonia may form ammonium salts, which can plug the catalytic converters, sulfur condensers, liquid sulfur drain legs, etc. Feed streams containing ammonia and cyanides are sometimes handled in a special two-combustion-stage burner or in a separate burner to ensure satisfactory combustion.

Flame stability can be a problem with low H_2S content feeds (a flame temperature of about 980 °C appears to be the minimum for stable operation). The split-flow, sulfur recycle, or direct oxidation process variations often are utilized to handle these H_2S-lean feeds; but in these process schemes, any hydrocarbons, ammonia, cyanides, etc., in all or part of the feed gas are fed unburned to the first catalytic converter. This can result in the cracking of heavy hydrocarbons to form carbon or carbonaceous deposits and the formation of ammonium salts, resulting in deactivation of the catalyst and/or plugging of equipment. A method of avoiding these problems while still improving flame stability is to preheat the combustion air and/or acid gas, and to operate "straight-through."

11.7 Sulfur condenser operation

Sulfur is condensed ahead of the first catalytic converter (except in the case of split-flow operation) and following each catalytic converter in order to promote the Claus reaction.

These condensers (other than the one following the last catalytic converter) are typically designed for outlet temperatures of 166–182 °C, which results in a condensed liquid sulfur of reasonably low viscosity and a metal skin temperature (on the process gas side) above the sulfurous/sulfuric acid dew point.

The final sulfur condenser outlet temperature can be as low as 127 °C, depending on the cooling medium available. A large temperature difference between process gases and cooling medium should

Table 11.2 Potential COS and CS$_2$ Formation in Claus Furnaces

Feed Composition (mol%)				COS, CS$_2$ Formation, % of Sulfur in Feed
Hydrocarbon (as C$_3$H$_8$)	Water	CO$_2$	H$_2$S	
0	6	4	90	0.5
0	6	14	80	1.5
0	6	24	70	2.5
0	6	34	60	3.5
0	6	44	50	4.5
0	6	54	40	5.5
0	6	64	30	6.5
0	6	74	20	7.5
2	6	4	88	2
2	6	14	78	3
2	6	24	68	4.5
2	6	34	58	6
2	6	44	48	7
2	6	54	38	9
2	6	64	28	11
2	6	74	18	14
4	6	4	86	3.5
4	6	14	76	5
4	6	24	66	6
4	6	34	56	8
4	6	44	46	10
4	6	54	36	12
4	6	64	26	14
4	6	74	16	18

(a) Maximum. Actual production varies with operating temperature and pressure, residence time, burner mixing, and burner efficiency.
(b) Units feeding <30% H$_2$S may operate other than "straight-through," causing reduced COS and CS$_2$ production proportional to amount fed to main burner.

be avoided, however, because of the possible formation of sulfur fog; this is especially important for the final sulfur condenser.

11.8 Waste heat recovery operation

Most Claus plants cool the process gases, leaving the combustion section by generating steam in a fire-tube waste heat boiler. Steam pressures usually range between 1035 and 3450 kPa (ga). The waste heat boiler outlet temperature is therefore normally above the sulfur dew point of the process gases; however, some sulfur may condense, especially during partial loads, and provision should be made to drain this sulfur from the process stream (or the piping should be arranged so the sulfur will drain through downstream equipment).

Other methods of cooling the hot combustion gases include the use of glycol–water mixtures, amine solutions, circulating cooling water (no boiling), and oil baths. The utilization of one of these alternate cooling fluids can be especially advantageous at locations where good-quality boiler feed water is not available, or where steam generation is not desired.

Some small Claus units use a closed steam system. Steam is generated at 140–210 kPa (ga), condensed with air in an elevated condenser, and the steam condensate returned by gravity to the boiler as feed water.

11.9 Catalyst converter operation

The Claus reaction is exothermic at converter temperatures, and the reaction equilibrium is favored by lower temperatures. However, COS and CS_2 hydrolyze more completely at higher temperatures, as shown by Figure 11.3. The first catalytic converter is therefore frequently operated at temperatures high enough to promote the hydrolysis of COS and CS_2; the second and third catalytic converters are operated at temperatures only high enough to obtain acceptable reaction rates and to avoid liquid sulfur deposition and associated catalyst deactivation. A three-converter Claus unit will utilize inlet

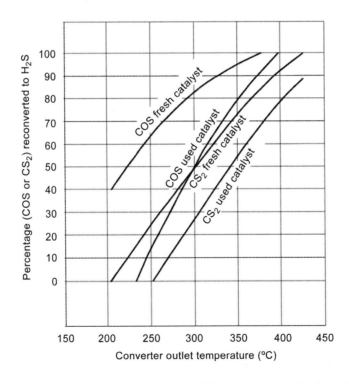

FIGURE 11.3

Hydrolysis of COS and CS_2 with activated alumina catalyst in sulfur converter (GPSA, 2004) [1].

temperatures in the following range: (1) first converter, 232–249 °C; (2) second converter, 199–221 °C; (3) third converter, 188–210 °C.

A temperature rise occurs across each catalytic converter because both the Claus and COS/CS_2 hydrolysis reactions are exothermic. The temperature rise will generally be 44–100 °C for the first converter, 14–33 °C for the second converter, and 3–8 °C for the third converter. Because of heat losses, measured temperatures for the third converter will often show a small temperature drop.

The foregoing is based on using the regular Claus catalyst in all of the converters. Regular catalyst is made of activated alumina (Al_2O_3). The primary function of activated alumina is to increase the rate of the Claus reaction and ensure full equilibrium conversion to sulfur. It also helps hydrolyze the carbon sulfides, COS and CS_2, to H_2S and CO_2 in the first converter Eqn (11.7) and (11.8); but it achieves reasonable hydrolysis only at high temperatures of 300–340 °C (Figure 11.3), which reduces the equilibrium conversion to sulfur in the Claus reaction (Figure 22.2). A catalyst made of activated titania (i.e., titanium dioxide; TiO_2) can achieve greater than 90% hydrolysis of the carbon sulfides at temperatures of 300–340 °C.

11.10 Claus tail gas treating process selection

With the sulfur content of crude oil and natural gas on the increase and tightening sulfur content in fuels, refiners and gas processors are pushed for additional sulfur recovery capacity.

At the same time, environmental regulatory agencies of many countries continue to promulgate more stringent standards for sulfur emissions from oil, gas, and chemical processing facilities. It is necessary to develop and implement reliable and cost-effective technologies to cope with the changing requirements. In response to this trend, several new technologies are now emerging to comply with the most stringent regulations.

Typical sulfur recovery efficiencies for Claus plants are 90–96% for a two- stage and 95–98% for a three-stage plant. Most countries require sulfur recovery efficiency in the range of 98.5–99.9%+. Therefore the sulfur constituents in the Claus tail gas need to be reduced further.

The key parameters effecting the selection of the tail gas cleanup process are:

- Feed gas composition, including H_2S content and hydrocarbons and other contaminants
- Existing equipment and process configuration
- Required recovery efficiency
- Concentration of sulfur species in the stack gas
- Ease of operation
- Remote location
- Sulfur product quality
- Costs (capital and operating).

Various aspects and considerations when choosing the most optimum process configuration for tail gas treating are discussed. There are several key features affecting the selection of the tail gas cleanup process that should be taken into account. When required recovery efficiency and concentration of sulfur species in the stack gas is known, selection of the tail gas process is one step closer. The first step is one of the most important criteria for the selection of the tail gas treating processes. When the required sulfur recovery is established, the selection of the tail gas process

Table 11.3 Residual Sulfur with Fresh Catalyst

Contaminant	Part per million volume (PPMV)
Carbonyl sulfide (COS)	<20
Carbon monoxide (CO)	<200
Carbon disulfide (CS_2)	0
Methyl mercaptan (CH_3SH)	0

will be limited. Table 11.3 represents the various tail gas cleanup processes and the recovery that will be achieved.

When concentration of impurities in the acid gas, such as COS and CS_2, H_2S content, and feed gas composition, and finally treated gas specifications, are established, the type of amine used for a particular application could be selected in step two. Finally the third step is the evaluation between the identical process chosen for ease of operation, capital and operating cost, and remote location. For revamp units, minimum equipment modifications and process configuration should be considered as a main key factor.

The fundamental process employed typically heats the Claus tail gas to 550–650 °F ($\sim$290–340 °C) by inline substoichiometric combustion of natural gas in a reducing gas generator (RGG) for subsequent catalytic reduction of virtually all non-H_2S sulfur components to H_2S. Conversion of SO_2 and elemental sulfur (S_x) is by hydrogenation:

$$SO_2 + 3H_2 \rightarrow H_2S + 4H_2O + \Delta H \tag{11.9}$$

$$S_X + XH_2 \rightarrow XH_2S + \Delta H \tag{11.10}$$

Conversion of COS and CS_2 is by hydrolysis:

$$COS + H_2O \rightarrow H_2S + CO_2 + \Delta H \tag{11.11}$$

$$CS_2 + 2H_2O \rightarrow 2H_2S + CO_2 + \Delta H \tag{11.12}$$

CO is essentially hydrolyzed to yield additional H_2 according to the "water gas shift" reaction:

$$CO + H_2O \rightarrow H_2 + CO_2 + \Delta H \tag{11.13}$$

A cobalt–moly catalyst, similar to hydrodesulfurization catalyst, is typically employed. As received, the catalyst is an alumina substrate impregnated with oxides of cobalt and molybdenum, which must be converted to the active sulfided state. To convert the cobalt oxide to the sulfide, a simple exchange of the oxide with H_2S is all that is necessary:

$$CoO + H_2S \rightarrow CoS + H_2O + \Delta H \tag{11.14}$$

Converting molybdenum trioxide to the active disulfide, however, requires a change in oxidation number that also requires hydrogen:

$$MoO_3 + 2H_2S + H_2 \rightarrow MoS_2 + 3H_2O + \Delta H \tag{11.15}$$

CO and H_2 naturally present in the Claus tail gas will typically satisfy up to 70% of tail gas unit (TGU) demand, with the balance generated in the RGG.

The reduced tail gas is then cooled to 90–100 °F ($\sim$30–40 °C) to condense most of the water vapor, which accounts for $\sim$35% of the stream. While it is recognized that there is potential for H_2S recovery using an alkanolamine, there was some concern about formation of heat-stable thiosulfate resulting from SO_2 breakthroughs. Consequently, the Stretford redox process could be adopted by employing an alkaline vanadium salt solution to oxidize absorbed H_2S to elemental sulfur particles, which were subsequently removed by froth flotation, filtered, and melted. The process actually had some advantages over amine absorption:

- No acid gas recycle to the Claus unit
- No steam consumption
- <5 ppm residual H_2S, obviating incineration
- Temporary high capacity for excessive Claus tail gas H_2S or SO_2 resulting from off-ratio operation.

However, these were outweighed by poor sulfur quality, high chemical make-up costs, high disposal costs from purging of by-product thiosulfate, absorber fouling, oxidizer foaming, inconsistent froth formation, troublesome filter operation, and atmospheric corrosion.

A typical (tail gas) sulfur recovery using an amine system is shown in Figure 11.4.

The recent sulfur recovery unit comprises three process steps:

- Reducing gas generation and tail gas preheat
- Hydrogenation/hydrolysis of SO_2 and other sulfur species to H_2S
- Gas cooling and waste heat recovery

Modern proprietary RGG design provides process gas reheating and reducing gas (H_2 and CO) generation in one single process unit. No external supply of hydrogen gas is required. This feature enhances the reliability of the process unit by eliminating the uncertainties associated with the availability of external hydrogen supply and the quality of hydrogen gas.

Many traditional methods use the inline burner design shown in Figure 11.5.

Recent design (Figures 11.6 and 11.7) employs a brick-lined internal combustion zone for stable combustion unaffected by downstream turbulence. Optimum outer-shell skin temperatures are easily ensured, heat loss is minimized, and potential leakage through the combustion zone wall does not result in atmospheric release. Some units have been in service for 30+ years with no major refractory repairs. The RGG is typically elevated so that minor entrained sulfur will free-drain to the reactor (and vaporize).

Industry consensus is apparently lacking with regard to the optimum air/fuel ratio. Many traditional units operate at stoichiometric air and rely on supplemental H_2 for hydrogenation of SO_2 and S_x. Perhaps contrary to intuition, equilibrium O_2 is nominally 0.6% at stoichiometric air, and only goes to zero at <90% of stoichiometric. There is experience to suggest that chronic O_2 leakage leads to catalyst sulfation, although there is disagreement within the industry on this point. Nonetheless, WorleyParsons generally recommends operating at 80% of stoichiometric to avoid, or at least minimize, O_2 leakage (and also maximize H_2 yield).

The advisability of supplemental H_2 is also a source of controversy. Many clients consider the availability of import H_2 necessary to minimize the risk of SO_2 breakthroughs, whereas in reality it is as easy to reduce Claus combustion air (with the same effect) as to increase H_2 addition. In the absence of supplemental H_2, the operator quickly learns the value of monitoring residual H_2 as a sensitive

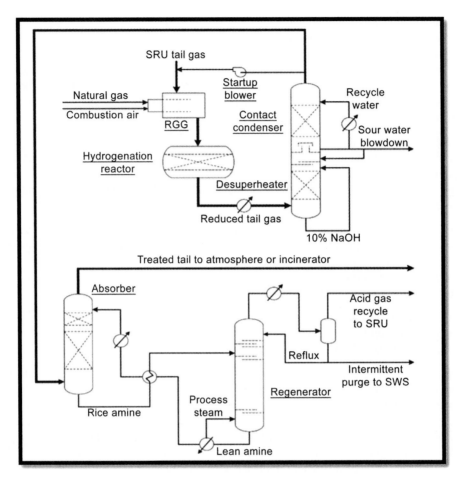

FIGURE 11.4

A typical (tail gas) sulfur recovery using an amine system.

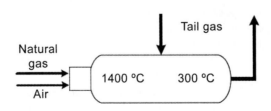

FIGURE 11.5

Common tail gas unit feed heater.

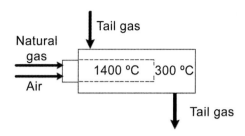

FIGURE 11.6

Modern reducing gas generator.

indicator of Claus tail gas ratio, and arguably is more likely to routinely optimize Claus air demand when constrained by a limited H_2 supply.

Three-stage Claus units clearly do not need supplemental H_2, whereas residual H_2 may be marginal with 2-stage units, in which case supplemental H_2 may be advisable to ensure ability to optimize the Claus tail gas H_2S/SO_2 ratio.

H_2 analyzers based on thermal conductivity measurement are very reliable, with minimal servicing. Where the TGU is coupled to a single Claus train, the H_2 analyzer can in fact supplant the Claus air demand analyzer. Where multiple Claus trains are coupled to a single TGU, combustion air to a Claus unit whose air demand analyzer is out of service can be temporarily adjusted based on TGU residual H_2.

Low-pressure steam injection to the burner in the nominal ratio of 1:1 lb/lb steam/fuel is generally advisable for soot inhibition when firing substoichiometrically, by virtue of the following reactions:

$$C + H_2O \rightarrow CO + H_2 - \Delta H \qquad (11.16)$$

$$C + 2H_2O \rightarrow CO_2 + 2H_2 - \Delta H \qquad (11.17)$$

While modern high-intensity burners may be operable at as low as 80% of stoichiometric air without steam injection, injection is still prudent in view of the possibly of lower air/gas ratios resulting from

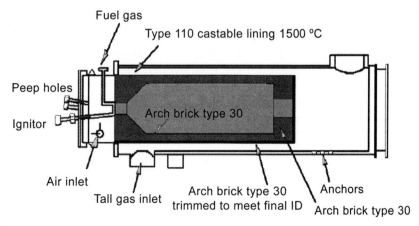

FIGURE 11.7

Details of modern reducing gas generator.

meter error or localized fuel-rich zones due to burner damage or fouling. With high-intensity burners, steam injection via a dedicated steam gun is preferred. Otherwise, injection into the combustion air is the most practical.

11.10.1 Hydrogenation reactor

With good catalyst activity and no excessive hydrocarbons in the acid gas feed to the reaction furnace, organic residuals in the absorber off-gas should be as shown in Table 11.3.

With fresh conventional catalyst, temperatures of 400–450 °F (204–232 °C) are typically required to initiate the hydrogenation reactions and 540–560 °F (282–293 °C) for hydrolysis. As the catalyst loses activity with age, progressively higher temperatures may be required. Typically, activity loss is first evidenced by (1) reduced COS, CS_2, and CO conversion, and (2) potential methyl mercaptan formed by the reaction of CS_2 and H_2, while hydrogenation of SO_2 and S_x may still be complete because of the lower initiation temperatures required.

The potential formation of methyl mercaptan at low temperature or impaired catalyst activity is perhaps not widely appreciated. In cases where the TGU tail gas is discharged without incineration, nominal mercaptan levels can result in serious nuisance odors. In Stretford units, there is reason to expect that the mercaptan is oxidized to disulfide oil, which can impair froth formation.

Excessive hydrocarbons in the SRU acid gas feed will tend to increase the carbon–sulfur compounds in the reactor effluent.

11.10.2 Low-temperature hydrogenation catalyst

Low-temperature catalysts eliminate use of the reducing gas generator; an indirect heating system could be used instead. Low-temperature TGU catalysts reportedly capable of operating at inlet temperatures of 210–240 °C (410–464 °F), achievable with steam reheat, have recently become available. The primary advantage (in a new unit) is elimination of the RGG, translating to (1) lower capital cost, (2) operating simplicity, (3) improved turndown, (4) reduced TGU tail gas volume, (5) reduced CO_2 recycle to the SRU, and (6) elimination of risk of catalyst damage by RGG misoperation.

Historically, Claus tail gas treating units (TGTU) have required reactor inlet temperatures of ∼550 °F for appreciable hydrolysis of COS, CS_2, and CO, typically requiring preheat by inline firing or heat exchange with hot oil or heat transfer fluid.

Vendor claims of energy savings are questionable since they tend to assume the plant is long on low-pressure steam and disregard the cost of high-pressure steam. Long-term performance of low-temperature catalysts is still uncertain. The following considerations should be taken into account:

- A steam reheater will limit the ability to compensate for normal catalyst activity loss with age, potentially limiting its useful life.
- A bottom layer of titania in the first Claus converter may be required for COS/CS_2 hydrolysis.
- Higher residual CO levels could mean operating the incinerator at 1500 °F (∼800 °C) instead of 1200 °F (∼650 °C).

- Incomplete CS_2 destruction, and hence methyl mercaptan formation, can result in serious nuisance odors if the TGU tail gas is discharged without incineration.

Reactor inlet temperatures are only half the story; outlet temperatures are the other half. Any catalyst will probably initiate SO_2 hydrogenation at 400–450 °F ($\sim$205–230 °C) and, with sufficient temperature rise and excess catalyst, will subsequently achieve virtually complete hydrolysis.

New catalysts require lower activation temperatures achievable by indirect reheat by 600# steam, thus reducing investment cost, operating complexity, and, in some cases, energy consumption. In addition, lower reactor outlet temperatures may obviate the downstream waste heat boiler.

While reduced investment and complexity are a given, whether the claimed energy savings is real is site-specific. Reduced feed preheat energy only constitutes a savings if the plant is already long on low-pressure waste heat steam (40–70 psig). Otherwise, incremental heat input is fully recovered. Furthermore, in the absence of a steam surplus, elimination of the waste heat boiler may have forfeited recoverable BTUs.

Relative COS, CS_2, and CO conversion efficiencies need to be compared. It is not necessarily sufficient to achieve regulatory compliance.

11.10.3 COS, CS$_2$, and CO hydrolysis using low-temperature catalyst

Relative COS, CS_2, and CO conversion efficiencies can be critical. It is not necessarily sufficient to achieve regulatory compliance.

- Regulations could become more stringent in the future.
- Some plants must also buy emission credits per pound of SO_2 discharged.
- Excessive CO residuals could require higher incinerator temperatures, or require incineration otherwise obviated in units able to achieve TGTU absorber H_2S emissions <10 ppm by the use of acid-aided methyldiethanolamine (MDEA).

Hydrolysis of COS, CS_2, and CO typically requires higher temperatures than hydrogenation of SO_2 and S_x. Perhaps accordingly, COS, CS_2, and CO conversion efficiencies are the first to suffer as conventional catalysts lose activity with age. Higher reactor inlet temperatures will tend to compensate for deactivation, thus extending catalyst life considerably. Depending on the design limits, temperatures can generally be increased by 50–150 °F (28–83 °C).

Assuming the same holds true for the low-temperature catalysts, a steam reheater will substantially limit the extent to which temperatures can be increased, in effect potentially shortening catalyst life. The lower initiation temperature of the Criterion 734 at start-of-run is thus significant, as it affords the greatest margin for increase.

At 464 °F (240 °C), hydrolysis of CO, COS, and CS_2 approaches that of conventional high-temperature catalysts. At 428 °F (220 °C), however, COS/CS_2 conversion must be accomplished in the first Claus stage by (1) supplementing the alumina bed with a bottom layer of expensive titania catalyst, or (2) increasing the inlet temperature to 550–600 °F (288–316 °C). The latter will nominally:

- reduce Claus recovery efficiency
- increase SRU tail gas rate
- increase TGTU sulfur load.

However, the first stage will not affect CO conversion.

Conventional cobalt-moly catalyst will generate minor, but significant, levels of methyl mercaptan by the reaction of CS_2 and hydrogen at 480 °F (249 °C) when in good condition, and at much higher temperatures if the catalyst is aged or damaged. Although the manufacturers claim no residual mercaptans with the low-temperature catalysts, there is some uncertainty—in the author's view—as to whether that will remain true a few years into the run.

11.10.4 Hydrogen balance using low-temperature catalyst

Compared with firing the feed heater at stoichiometric air and importing H_2, a steam reheater will of course have no impact on the H_2 balance. However, many plants avoid the need for supplemental H_2 by the use of an RGG, typically burning natural gas substoichiometrically to generate H_2 and CO.

In the absence of an RGG, the alternative is to operate the SRU more air-deficiently as necessary to maintain, say, 2% residual H_2 downstream of the TGTU reactor. This will nominally:

- reduce Claus recovery efficiency
- increase SRU tail gas rate
- increase TGTU sulfur load.

11.10.5 CO$_2$ balance using low-temperature catalyst

Eliminating the inline burner has the benefit of reducing the TGTU tail gas volume (for the assumed basis with an RGG). Assuming 85% CO_2 slip, the acid gas load on the TGTU amine is reduced.

11.10.6 Energy balance using low-temperature catalyst

A steam reheater will not only eliminate the following natural gas required by the RGG, but will also reduce incinerator fuel by virtue of the reduced tail gas rate, resulting in:

- RGG fuel savings
- incinerator fuel savings.

Assuming $H_2S/SO_2 = 2$ in the SRU tail gas, a supplemental H_2 will be required to maintain a 2% residual in the TGTU tail gas. As a rule of thumb, the value of relatively pure (nonreformer) H_2 is four times that of natural gas.

Figure 11.8 represents WorleyParsons BSR/amine with the low-temperature catalyst.

11.11 Contact condenser (two-stage quench)

Common industry practice is to cool the reduced tail gas from the reactor by the generation of low-pressure waste heat steam followed by direct quench with a recirculating water stream to cool it to 90–100 °F ($\sim$30–40 °C), thus condensing most of the water vapor, which accounts for $\sim$35% of the

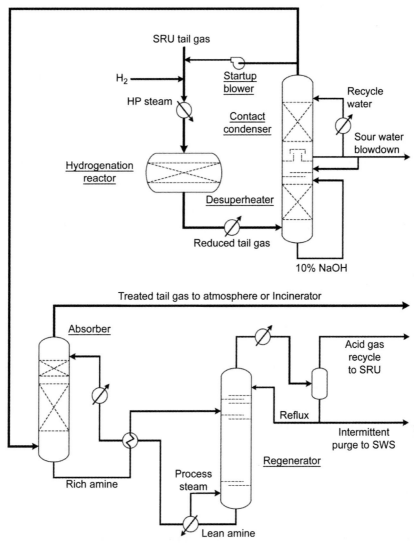

FIGURE 11.8

A sulfur recovery amine flow scheme with low-temperature catalyst.

stream. Modern design utilizes a unique two-stage tower composed of a bottom de-superheater section and top contact condenser.

- The contact condenser has two sections: the first section de-superheats the gas and scrubs any SO_2 that may break through from the hydrogenation reactor, and the second section cools the gas and condensates the water; therefore there is no need for make-up water to maintain the caustic concentration. The condensate water will provide the water to maintain the caustic concentration. We do not have continuous purge, but we provide water make-up for the water that is evaporated, just like any other quench system.

- Tail gas is de-superheated in the lower section of the contact condenser by a circulating water stream. This water is maintained alkaline to protect against any SO_2 breakthrough from the reactor. In the upper packed section of the tower, most of the water vapor in the tail gas is condensed by direct contact with a circulating stream of cooled water. A pH analyzer with a low-pH alarm is installed in the quench water circulation line and will indicate when the pH of the quench water is reducing, from either a breakthrough of SO_2 or incomplete reduction of the sulfur compounds in the gas stream from the hydrogenation reactor.

A 10%-wt NaOH solution is recirculated through the de-superheater to capture SO_2 potentially resulting from a process upset, while also cooling it to its dew point of ~ 165 °F (~ 75 °C). The only cooling is by vaporization. The gas is further cooled to 90–100 °F (~ 30–40 °C) by direct contact with an externally cooled recycle water stream in the upper contact condenser section. A recycled water slipstream is returned to the de-superheater on de-superheater level control via two bubble-cap wash trays to capture entrained caustic.

A blowdown slipstream of recycled water is purged, usually to sour water, on contact condenser-level control. While the recycle water is usually classified as sour water, the H_2S content is typically <50 ppmv by virtue of CO_2 saturation. In situations where the increased load on the plant sour water stripper is undesirable, a simple blowdown stripper is occasionally provided at the TGU. This typically involves low-pressure stripping steam injection (as opposed to a reboiler) and return of the uncondensed overhead stream to the de-superheater.

11.11.1 Start-up blower

Recent designs provide a start-up blower on the contact condenser overhead to eliminate flaring large quantities of H_2S to atmosphere and to prevent violation of the emission. For those cases that a booster blower is required, the booster blower will have dual function as a start-up blower and as a booster blower.

11.11.2 Booster blower

Many of the Claus units that are in operation do not have enough pressure to handle a new tail gas unit; in other words, the provision of operating the Claus unit at the higher pressure was not considered: if the source pressure changes, the existing amine unit requires higher reboiler duty and in most cases requires significant changes in the amine unit. WorleyParsons has been offering a booster blower in the tail gas unit to overcome the pressure limitation. Retrofit tail gas units will typically require a booster blower downstream of the contact condenser to overcome the additional pressure drop. The blower is located after the contact condenser to minimize the actual volume (by virtue of cooling and condensation) and before the absorber to take advantage of the higher pressure. With proper design and operation, booster blowers are inherently very reliable, requiring minimal maintenance. Typically, the case is cast iron or carbon steel, with an aluminum impellor. N_2-purged tandem shaft seals (typically carbon rings) eliminate process leakage to atmosphere on the discharge end as well as air aspiration into the process on the suction end, which is typically at a vacuum.

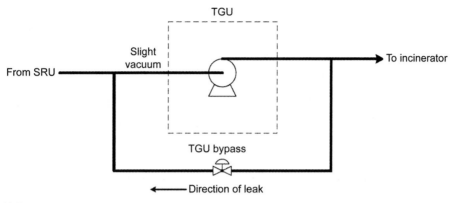

FIGURE 11.9

A reducing gas generator vacuum operation.

Though often viewed as a liability by clients, booster blowers arguably improve operability in several ways:

- By recirculating tail gas, the TGU can be started up and shut down independent of the SRUs.
- Tail gas recycle ensures process stability at high SRU turndown by: (1) avoiding undue RGG burner turndown potentially conducive to sooting due to poor mixing or air/gas flowmeter inaccuracy, and (2) diluting potentially high SO_2 levels often typical of high SRU turndown. With advance warning, tail gas recycle can avoid RGG shutdown in the event of an SRU trip.
- By routing the SRU and TGU tail gas to the incinerator via a common header, a vacuum can be maintained at the RGG (Figure 11.9) without risk of leaking air from the incinerator back into the TGU, thus potentially further increasing SRU capacity. In the event that the tail gas bypass valve leaks, clean TGU tail gas is recycled to the RGG rather than SRU tail gas bypassing the TGU (as when the RGG pressure is positive). Any such reverse flow will improve bypass valve reliability by excluding sulfur vapor, and the valve can be partially stroked periodically to verify operability without increased emissions.

In the absence of a booster blower, a single start-up blower recycle is usually provided for tail gas recycle. While these machines tend to be less sophisticated, N_2-purged tandem shaft seals are still required.

The overall configuration of using the booster blower is shown in Figure 11.10. This configuration could be used with low-temperature catalyst and indirect reheater instead of the RGG.

11.12 Solvent selection criteria in the tail gas unit

The most common solvent is 40–45%-wt MDEA (HS-101, or similar) designed for a maximum rich loading of 0.1 mol acid gas $(H_2S + CO_2)$ per mol amine with typical emission reduction to ~ 100 ppmv H_2S. Cooling of the lean amine to at least 100 °F (38 °C) is important for minimization of

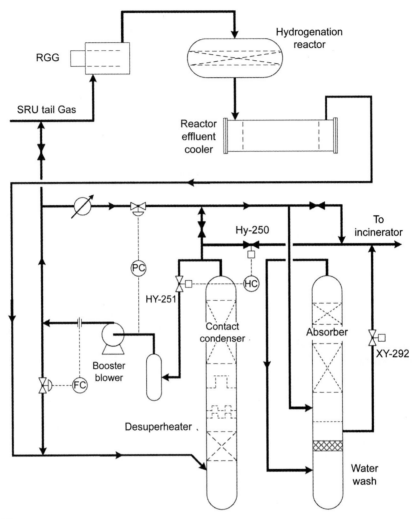

FIGURE 11.10

A process flow diagram for sulfur recovery tail gas unit with booster blower configuration.

emissions and amine circulation rate. Specialty TGU amines are essentially pH-modified MDEA to facilitate stripping to lower residual acid gases for treatment to <10 ppm H_2S, potentially obviating incineration. CO_2 slip is also improved. These products are variously marketed as:

- Dow UCARSOL HS-103
- Ineos Gas/Spec TG-10
- Huntsman MS-300.

An alternative to MDEA is ExxonMobil's Flexsorb SE, a proprietary hindered amine patented by Exxon in partnership with the Ralph M. Parsons Company. The main advantage is a 20–30% reduction in circulation rate. The solvent is much more stable than MDEA, but is also more expensive. Flexsorb SE Plus is also available for treatment to <10 ppmv H_2S. Both solvents require a license agreement with ExxonMobil.

It used to be assumed that TGU carbon filtration was not required in view of the absence of hydrocarbons. For MDEA-based solvents, at least, this has proven untrue, presumably due to the generation of surfactant amine degradation products.

Solvent applications include:

- FLEXSORB® SE: Selective removal of H_2S
- FLEXSORB® SE Plus: Selective removal of H_2S to less than 10 ppm

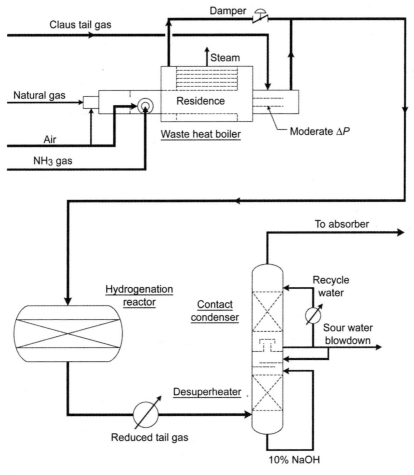

FIGURE 11.11

Ammonia destruction in tail gas unit (Rameshni ammonia conversion).

- FLEXSORB® SE Hybrid: Removal of H_2S, CO_2, and sulfur compounds (mercaptans and COS) In sulfur plant tail gas applications, FLEXSORB® SE solvents can use as little as one-half of the circulation rate and regeneration energy typically required by MDEA-based solvents. Flexsorb solvents offer other advantages compared to the other amine solvents. For instance, most of the applications require no reclaiming, have good operating experience, have low corrosion, and have low foaming due to low hydrocarbon absorption; by providing water wash of treated gas at low pressure, system amine losses are minimum.

11.13 Ammonia destruction in a TGU (RACTM)

The general industry consensus is that the amount of ammonia that can be conventionally processed in the SRU is limited to 30–35% vol on a wet basis. With what appears to be a trend toward

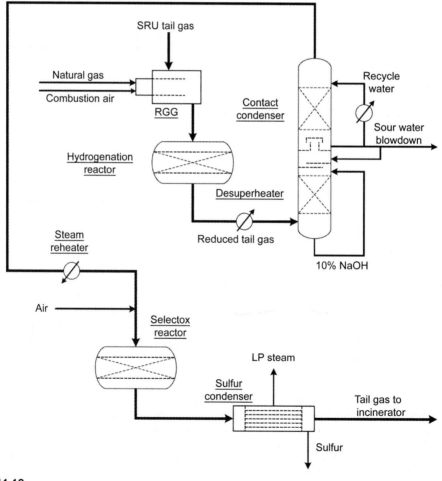

FIGURE 11.12

A process flow diagram for BSR Selectox.

higher-nitrogen crudes, refiners are increasingly faced with the need for alternative processing schemes, as well as SRU de-bottlenecking. With sour water stripper schemes such as Chevron's wastewater treatment process for separating H_2S and NH_3, producing a pure marketable NH_3 product is relatively difficult compared with bulk separation of NH_3 containing minor H_2S.

The Rameshni Ammonia Conversion (RACTM) process, for which a patent is pending, sub-stoichiometrically combusts a high-NH_3 H_2S-contaminated stream in the RGG (Figure 11.11). Typically, the NH_3–gas heat release will exceed that required to reheat the Claus tail gas, thus necessitating a waste heat boiler prior to the TGU reactor. A supplemental natural gas fire ensures process stability in the event of NH_3–gas curtailment. Sub-stoichiometric combustion of the NH_3–gas generates supplemental H_2 for the hydrogenation reactor and minimizes NO_x. Most of any NO_x that is made is reduced in the reactor. Minor unconverted NH_3 is automatically recycled to the sour water stripper via the contact condenser blowdown.

11.14 BSR Selectox

Selectox catalyst is a proprietary catalyst patented for low-temperature H_2S oxidation and Claus-reaction catalyst development by the Ralph M. Parsons Company and Unocal. Reduced tail gas from the BSR contact condenser is steam-reheated to about 400 °F ($\sim$200 °C) and combined with a stoichiometric quantity of air in the reactor to produce elemental sulfur, which is subsequently condensed (Figure 11.12). Overall recoveries of 98.5–99.5% are achievable. The reactor inlet is limited to 5% vol H_2S, above which recycle dilution (or inter-bed heat removal) is necessary to limit the exothermic.

Reference

[1] GPSA. Gas processors and suppliers association, engineering databook. Tulsa (OK, USA). 12th ed. 2004.

Futher reading

Berben PH, Borsboom J, Lagas JA. SUPERCLAUS, the Answer to claus plant Limitations. In: 38th Canadian chemical Engineering Conference; October 2–5, 1988.

Borsboom H, Clark P. New insights into the Claus thermal stage chemistry and temperatures. Banff (Alberta, Canada). In: Brimstone2002 sulfur recovery symposium; May 2–10, 2002. p. 2001.

Butwell KF, Kubek DJ, Sigmund PW. Alkanolamine treating. Hydrocarbon Processing; March 1982.

Campbell JM. Gas conditioning and processing, vol. 2. Campbell, Petroleum Series, Form@an,@Oklahoma @19/9.

Clark PD, Lesage KL, Fitzpatrick E, Davis PM. H2S solubility in liquid sulfur as a function of temperature and H2S gas phase partial pressure, Alberta Sulphur Research Ltd. Quarterly Bulletin October–December, 1997; vol. XXXIV(No. 3).

Clark PD. Alberta Sulphur Research Ltd.,. personal communication with Heigold RE; January 2002.

Calabrian Corporation. SUPER process doubles Claus capacity. Sulphur Magazine; March–April, 1999.

C Corporation, "Purification of amines with granular activated carbon", Brochure.

Fedich RB, McCaffrey DS, Stanley JF. Advanced gas treating to enhance producing and refining projects using FLEXSORBÒ SE solvents. Veracruz (Mexico). In: Encuentra y Exposicion Internacional de la Industria Petrolera, Meeting and international exposition of the petroleum industry (E EXITEP 2003); 2003.

Gamson BW, Elkins RH. Sulfur from hydrogen sulfide. Chem Eng Prog April 1953;49(4):203–15.

Goar BG, Sames JA. Tail gas clean up processesda review. In: Gas conditioning conference. University of Oklahoma; 1983.

Goar BG. Sulfur forming and degassing processes. In: Gas conditioning conference. University of Oklahoma; 1984.

Hakka LE, Parisi PJ, Cansolv Technologies Inc, Hatcher NA, Johnson JE, Black & Veatch Pritchard Inc. Integrated CANSOLV system technology into your sour gas treatment/sulfur recovery plant, Lawrence Reid gas conditioning conference. Norman: Oklahoma; March 1998.

Kelley KK. The thermodynamic properties of sulphur and its inorganic compounds. Bulletin 406, U.S. Bureau of Mines; 1937.

Kettner R, Liermann N. New Claus tail-gas process proved in German operations. Oil Gas J; January 11, 1988: 63–6.

Kobe KA, Long EG. Pet Refin, 28; February 1949. vol. 28, 11, 127-132 (November 1949); vol. 29, 1, 126-130 (January 1950).

Kohl AL, Riesenfeld FC. Gas purification. 3rd ed. Houston (TX): Gulf Publishing Co; 1979.

Kwong KV. PROClaus process: an Evolutionary Enhancement to Claus performance. In: Lawrence Reid gas conditioning conference; February/March 2000. Norman (OK).

Lagas JA. Stop emissions from liquid sulfur. Hydrocarbon Process October 1982;61:85–9.

Lewis G, Randall M. Thermodynamics and the free energies of chemical substances. McGraw Hill; 1923. 530–550.

Maddox RN, Morgan DJ. Gas and liquid sweetening. Campbell Petroleum Series; April 1998.

McIntush KE, Rueter CO, DeBerry KE, Petrinec BJ. H_2S removal and sulfur recovery options for high pressure natural gas with medium amounts of sulfur. Hydrocarbon Engineering; February 2001.

Perry Engineering Corporation. Activated-carbon filter, Brochure. Gas conditioning fact Book, Midland (MI): Dow Chemical Co., 19.

Scheirman WL. Filter DEA treating solution. Hydrocarbon Processing; August 1973.

Shuai X, Meisen A. New correlations predict physical properties of elemental sulfur. Oil Gas J; October 16, 1995.

Tonjes M, Hatcher N, Johnson J, Stevens D. Oxygen enrichment revamp checklist for sulfur recovery facilities. In: Lawrence Reid gas conditioning conference. Norman (OK): Black & Veatch Pritchard Inc; February/March 2000.

Tuller WN, editor. The sulphur data book. McGraw Hill; 1954.

Valdes AR. A new look at sulfur plants. Hydrocarbon Process Pet Refin March 1964;43:104–8. 122–124, (April 1964.

Liquefied Petroleum Gas (LPG) Recovery

12

The purpose of the liquefied petroleum gas (LPG) unit is to process light hydrocarbons (C_1–C_5) into required component streams. The LPG unit feed streams in a refinery are usually obtained from refinery units such as atmospheric crude distillation, plat former, hydro cracker, and other catalytic processing units. LPG is also produced in gas fractionation units charged by associated gas from oil reservoirs or gas from gas reservoirs. However, any LPG stream that is economically justified for recovery of the components shall be routed to the LPG recovery unit.

This chapter covers minimum process design requirements for LPG recovery and splitter units. It should be expressed that only general process requirements and the unit-specific design conditions shall be determined based on the feed analysis and final product specifications during execution of the unit conceptual design.

Raw natural gas contains valuable heavier hydrocarbons such as ethane, propane, butane, and fractions of higher hydrocarbons. These associated hydrocarbons, known as natural gas liquids (NGL), must be recovered from the gas in order to control the dew point of natural gas stream and to earn revenue by selling these components as products for different industries. Natural gas liquids are fractionated to produce LPG.

Natural gas is one of the world's favorite and promising fuels. Transportation of gas is not something easy; therefore, converting this gas into liquid simplifies and eases the transportation process. Liquefied petroleum gas, which is a super-pressurized gas stored in a liquid form in tanks or canisters, is a known type of natural gas into which we are going to look very closely.

LPG is a flammable mixture of hydrocarbon gases used as a fossil fuel closely linked to oil; almost two-thirds of the LPG that is used is extracted directly from the earth in the same way as natural gas. The rest is manufactured indirectly from petroleum drilled from the earth in wells (crude oil).

LPG is considered to be a mixture of two flammable nontoxic gases, known as propane (C_3H_8) and butane (C_4H_{10}). Propylene and butylenes are present in small concentrations too. Mainly the LPG gas is odorless, which makes it hard for people to detect the leakage if it happens, so a small amount of a pungent gas such as ethanethiol are added to help people smell potentially dangerous gas leaks.

LPG is used as fuel, especially for vehicles such as cars and motorcycles, also as an aerosol propellant and refrigerant to avoid damage to the ozone. It is an advantage to use LPG as a fuel for vehicles because it burns cleaner than petrol and diesel.

Another use is as a refrigerant. Propane gas and butane gas are used to make hydrocarbon refrigerants. Hydrocarbons are known to be more energy efficient and cheaper than other chemicals, which is why they are suitable to be used as refrigerants.

LPG can be used as a back-up or secondary fuel in generating the energy for the household. For example, in order to heat water in winter, LPG is used alongside a solar panel to provide enough energy for this purpose.

Natural Gas Processing. http://dx.doi.org/10.1016/B978-0-08-099971-5.00012-X
Copyright © 2014 Elsevier Inc. All rights reserved.

12.1 Properties

LPG is as twice as heavy as air and half as heavy as water and it is colorless and odorless. LPG can be compressed at a ratio of 1:250, which enables it to be marked in portable containers in liquid form, as mentioned above. LPG also produces less air pollutants and carbon dioxide than most other fuels; it helps to reduce the emissions of the typical house by almost 1.5 tons of CO_2 a year. LPG reduces black carbon emissions as well; these emissions are the second-biggest contributor of global warming and cause serious health hazards.

LPG has a high heating (the amount of heat released during combustion of a specified amount of it) of 12,467 kcal/m^3, which is much higher than the average heat value of most natural gas (9350 kcal/m^3). Also LPG has a very high Wobbe index (WI, an indicator of the interchangeability of fuel gases) of 73.5–87.5°MJ/Sm^3, which is a high combustion energy output.

$$WI = \frac{\text{Gross heat value}}{\sqrt{\text{Specific gravity}}} \qquad (12.1)$$

LPG can be used as an alternative fuel to natural gas (methane) in residential, commercial, and industrial applications; as an alternative to gasoline for automotive fuel purposes; and as a feedstock in petrochemical applications. Both propane and butane are gaseous hydrocarbons at normal temperatures (15 °C) and atmospheric pressure. However, they can be stored and distributed in liquid form at temperatures of under −42 °C and −2 °C for propane and butane, respectively. Table 12.1 shows the typical properties of LPG.

12.2 Natural gas liquids processing

Raw natural gas contains valuable heavier hydrocarbons when extracted from the well head. The heavier hydrocarbons that are associated with the raw natural gas are ethane, propane, butane, and natural gasoline (condensate from). These associated hydrocarbons are called natural gas liquids (NGL). These NGL components must be recovered to control the dew point of natural gas stream and

Table 12.1 Typical Properties of LPG		
Property	**Propane**	**Butane**
Liquid density	0.50–0.51	0.57–0.58
Conversion (liter/ton)	1968	1732
Gas density/air	1.40–1.55	1.90–2.10
Boiling point (°C)	−45	−2
Latent heat of vaporization	358 kJ/kg	372 kJ/kg
Specific heat (as liquid)	0.60 BTU/deg	0.57 BTU/deg
Sulfur content	0–0.02%	0–0.02%
Calorific value	2500 BTU/ft^3	3270 BTU/ft^3

also to earn revenue by selling out the separated components. Following are the different processes used to separate impurities:

- Oil and condensate removal
- Water removal
- Separation of natural gas liquids
- Sulfur and carbon dioxide removal

Our aim/objective in this report is to study the fractionation of natural gas liquids to produce LPG. We will discuss first different LPG manufacturing processes.

12.2.1 **LPG recovery processes**

Natural gas mainly contains methane and smaller amounts of ethane, propane, butane, and heavier hydrocarbons along with varying amounts of water vapors, carbon dioxide, sulfur compounds, and other non-hydrocarbons. Ethane, propane, butane, and propane are known as associated gases. The removal of these gases from raw natural gas is necessary to meet the desired consumer specifications of natural gas and to extract valuable products such as LPG from natural gas. Various techniques are used to recover LPG from natural gas/oil:

1. Recontacting–compression
2. Refrigeration
3. Absorption
4. Adsorption
5. A combination of above

- **Recontacting–compression**

This process is normally used for the recovery of LPG from the crude oil fractionator. This technique is hardly used in gas industry. The top product from a crude oil fractionator consists of lighter fractions, namely methane, ethane, propane, and butane. This top product stream is compressed, combined with top liquid product, cooled, and fed to the separator. The liquid phase from the separator is passed through the de-ethanizer and the vapor phase containing some LPG fractions is used as fuel gas. The liquid product of the de-ethanizer is LPG. The recovery of LPG by this technique is 75%.

- **Refrigeration**

This technique is more common for recovery of LPG from gas streams. The principle behind this technique is to refrigerate the gas stream and thus to obtain LPG fractions. Recovered fractions are fractionated to get the LPG components.

The technique is employed in three different processes:

- Low-temperature separation
- Expander plants
- Combined processes

- **Lean oil absorption**

This method employs the hydrocarbon oil to recover lighter fractions. This process is used in refineries and also in gas processing plants. LPG recovery by this process is 98%.

• **Adsorption**

Adsorbents are used in this process so that gas molecules are bonded to the surface. Normally silica gel, activated carbon, and alumina are used as adsorbent. The LPG recovery by this process is significantly lower than in the other processes.

12.2.2 LPG manufacturing

LPG is produced by fractionation of natural gas liquids and from crude oil by distillation, catalytic cracking, delayed cokers, and hydrocrackers. The LPG manufacturing process starts with the acid gas removal and extraction unit, then the fractionation unit, and ends with the product treatment plant. The simple process is described in Figure 12.1.

• **Acid gas removal**

Raw gas from the well head is received in knock-out drums to separate gas and liquid phases. The oil field gases contain corrosive acid gases like CO_2 and H_2S. Removal of these gases is necessary to further process the gas for LPG production or more products. These acid gases are removed by amine treatment or Benfield processes. After this, acid gases' free natural gas is sent to extraction unit.

• **Extraction unit**

The feed of the extraction unit is the combination of associated gases and condensate. The product streams are divided into three steps. One stream, having the liquid stream rich in propane, butane, and gasoline, is sent to the fractionation tower for LPG production; the other two streams go to the product gas unit for further processing.

• **Fractionation unit**

Liquid stream consisting of ethane, propane, butane, and pentane is treated in the fractionator trains to separate them and sold as LPG. A complete process flow sheet is shown in Figure 12.2. The fractionation tower consists of three columns: de-ethanizer, depropanizer, and debutanizer. The whole process description is as follows.

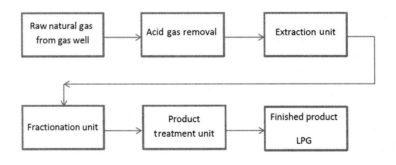

FIGURE 12.1

Block diagram for LPG manufacturing.

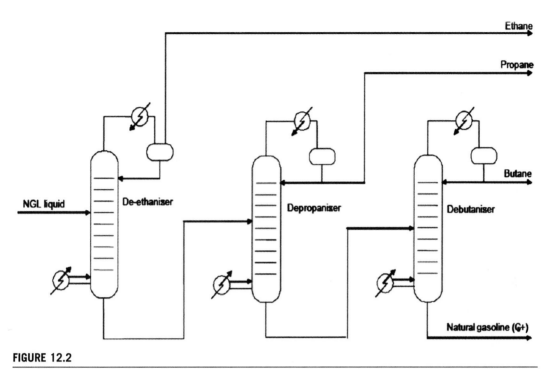

FIGURE 12.2

Typical fractionator train for natural gas liquids (NGL).

• *De-ethanizer section*

Raw gas containing associated gases is fed from the top of the de-ethanizer. The de-ethanizer operates at approximately 390 lb/in^2. We separate out ethane from this column. The overhead product is ethane in the form of vapors, which is partially condensed in the condenser by using propane at 20 °F and collected in the reflux drum. Condensed product is recycled to the de-ethanizer tower and non-condensed vapors (mainly ethane) are sent to the fuel gas system. Temperature inside the tower is maintained by supplying heat from the reboiler. The bottom product from the de-ethanizer enters into the next columns, the depropanizer.

• *Depropanizer section*

The pressure of the de-ethanizer bottom product is reduced to 290 lb/in^2 and then entered into the depropanizer. The overhead product of this column is propane rich and is condensed in the condenser by using cooling water. The condensed product is collected in the reflux drum. Some amount of this is refluxed back to the column. Heat is supplied through direct-fired heater.

• *Debutanizer section*

Depropanizer bottom product is expanded to a pressure of 110 lb/in^2 and fed to the top of the tower. Propane is separated as top product and condensed further in the condenser by using cooling water.

Table 12.2 Fractionator Types for LPG Production

Type of fractionator	Feed	Top Product	Bottom Product
Demethanizer	C_1/C_2	Methane	Ethane
De-ethanizer	LPG	Ethane	Propane plus
Depropanizer	De-ethanizer bottoms	Propane	Butanes plus
Debutanizer	Depropanizer bottoms	Butanes	Natural gasoline (pentanes plus)
Deisobutanizer	Debutanizer top	Isobutane	Normal butane

Bottom products are the heavier hydrocarbons. Fractionators of different types are commonly used in gas plants. Commonly used fractionators for LPG are listed in Table 12.2.

12.2.3 Product treatment plant

Propane and butane products separated from the fractionation plant contain some impurities as residual water, H_2S, carbon disulfide, and sulfur compounds. These impurities should be removed in order to meet the desired product specifications. The contaminants and their reasons for removal are listed in Table 12.3.

Numerous processes are available to remove contaminants, but two of them are the most important and commonly used:

1. Absorptive purification
2. Adsorptive purification

12.2.3.1 Feed specifications for NGL fractionation

Feed for NGL fractionation plants comes from upstream processing plants, which receive feed directly from gas reservoirs. Feed composition is different from different reservoirs. Feed composition is important for design considerations. The feed for NGL fractionation trains contains methane, ethane, propane, butane, and heavier components.

Table 12.3 Different Contaminants in LPG

Contaminants	Reasons for Removal
Hydrogen sulfide	Safety and environmental
Carbon dioxide	Corrosion control
Carbonyl sulfide	Product specification
Carbon disulfide	
Mercaptans	• Prevention of freeze-out at low temperatures
Organic sulfides	• Deactivation of catalyst
Nitrogen	Poisoning in downstream facilities
Water	Hydrate formation and corrosion

12.3 **Fractionation**

Fractionation is a unit operation utilized to separate mixtures into individual products. Fractionation involves separating components by relative volatility (a). The difficulty of a separation is directly related to the relative volatility of the components and the required purity of the product streams.

Virtually all gas processing plants producing natural gas liquids require at least one fractionator to produce a liquid product that will meet sales specifications. The schematic of an example fractionator in Figure 12.3 shows the various components of the system. Heat is introduced to the reboiler to produce stripping vapors. The vapor rises through the column contacting the descending liquid.

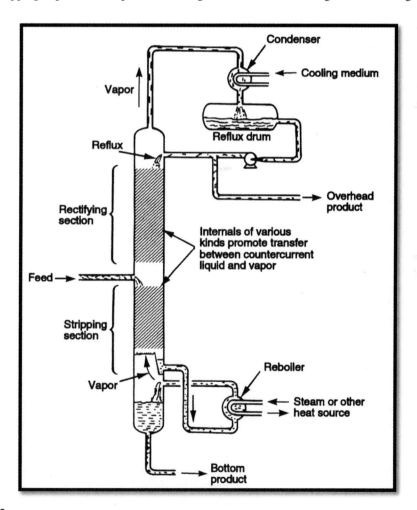

FIGURE 12.3

Fractionation schematic diagram [1].

The vapor leaving the top of the column enters the condenser, where heat is removed by some type of cooling medium. Liquid is returned to the column as reflux to limit the loss of heavy components overhead.

Internals such as trays or packing promote the contact between the liquid and vapor streams in the column. Intimate contact of the vapor and liquid phases is required for efficient separation. Vapor entering a separation stage will be cooled, which results in some condensation of heavier components.

The liquid phase will be heated, which results in some vaporization of the lighter components. Thus, the heavier components are concentrated in the liquid phase and eventually become the bottom product. The vapor phase is continually enriched in the light components, which will make up the overhead product.

The vapor leaving the top of the column may be totally or partially condensed. In a total condenser, all vapor entering the condenser is condensed to liquid and the reflux returned to the column has the same composition as the distillate or overhead product. In a partial condenser, only a portion of the vapor entering the condenser is condensed to liquid. In most partial condensers only sufficient liquid will be condensed to serve as reflux for the tower. In some cases, however, more liquid will be condensed than is required for reflux and there will actually be two overhead products: one a liquid having the same composition as the reflux and the other a vapor product that is in equilibrium with the liquid reflux.

12.3.1 Types of fractionators

The number and type of fractionators required depend on the number of products to be made and the feed composition.

Typical NGL products from a fractionation process include:

- Demethanized product (C_2^+)
- De-ethanized product (C_3^+)
- Ethane/propane mixtures (EP)
- Commercial propane
- Propane/butane mixture (LPG)
- Butane(s)
- Butane/gasoline mixtures
- Natural gasoline
- Mixtures with a vapor pressure specification

An example fractionation train used to produce three products is illustrated in Figure 12.4. The feed stream contains too much ethane to be included in the products; thus, the first column is a de-ethanizer. The overhead stream is recycled to the upstream processing plant or sent to a fuel system. The bottom product from this column could be marketed as a de-ethanized product. The second column, a depropanizer, produces a specification propane product overhead. The bottom product, a butane-gasoline mixture, is often sold to a pipeline without further processing. The third column, a debutanizer, separates the butane and gasoline products. This separation is controlled to limit the vapor pressure of the gasoline. The overhead butane product can be sold as a mixture or an additional column can be used to separate the isobutane and normal butane.

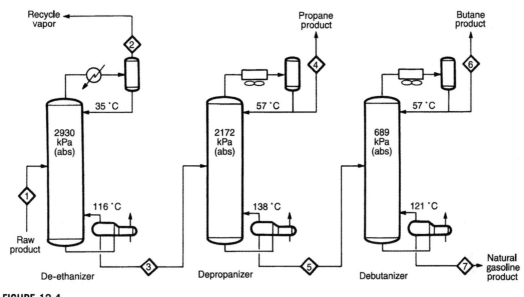

FIGURE 12.4

Fractionation train.

12.3.2 Product specifications

A material balance around the column is the first step in fractionation calculations. In order to perform this balance, assumption of the product stream compositions must be made. There are three ways of specifying a desired product from a fractionator:

- A percentage recovery of a component in the overhead or bottom stream.
- A composition of one component in either product.
- A specific physical property, such as vapor pressure, for either product.

The recovery and composition specifications can be used directly in the material balance. However, property specifications are used indirectly. For instance, if vapor pressure is a desired specification of a product, a material balance is performed with an assumed component split. The calculated vapor pressure of the resulting stream is then compared with the desired value and the material balance redone until reasonable agreement is reached.

In a multicomponent mixture, there are typically two components that are the "keys" to the separation. The light key component is defined as the lightest component in the bottom product in a significant amount. The heavy key component is the heaviest component in the overhead product in a significant amount. Normally, these two components are adjacent to each other in the volatility listing of the components. For hand calculations, it is normally assumed for material balance purposes that all components lighter than the light key are produced overhead and all components heavier than the heavy key are produced with the bottom product. By definition, the key components will be distributed between the product streams Table 12.3.

EXAMPLE:

For the given feed stream, estimate the product stream compositions for 98% propane recovered in the overhead product with a maximum isobutane content of the overhead stream of 1%.

Feed:

$C_2 = 2.4$
  $C_3 = 162.8$
  $i\text{-}C_4 = 31.0$
  $n\text{-}C_4 = 76.7$
  $C_5 = 76.5$
  Total $= 349.4$ mol

Solution steps:

For propane (light key):
Moles in overhead $= (0.98)162.8 = 159.5$.
  Moles in bottoms $= 162.8 - 159.5 = 3.3$.

For ethane:
Moles in overhead $= 2.4$ (100% to overhead). Since the isobutane (the heavy key) is 1% of the overhead stream, the sum of propane and ethane must be 99% (all $n\text{-}C_4$ and C_5^+ are in the bottoms). Thus:
  Total overhead moles $= (159.5 + 2.4)/0.99 = 161.9/0.99 = 163.5$
  Moles of $i\text{-}C_4 = 163.5 - 161.9 = 1.6$
  The overall balance is shown in Table 12.4.

In actual operation, the lighter than light key components and heavier than heavy key components will not be perfectly separated. For estimation purposes and hand calculations, perfect non-key separation is a useful simplifying assumption.

12.3.3 Trayed columns

12.3.3.1 Internals

Various types of trays are used in fractionation columns. Figure 12.5 shows the vapor flow through bubble cap trays, sieve trays, and valve trays. Due to the riser in the bubble cap, it is the only tray that can be designed to prevent liquid from "weeping" through the vapor passage. Sieve or valve

Table 12.4 Overall Balance for Solved Example

Component	Feed Moles	Overhead Moles	Overhead Mole%	Bottoms Moles	Bottoms Mole%
C_2	2.4	2.4	1.5	—	—
C_3	162.8	159.5	97.5	3.3	1.8
$i\text{-}C_4$	31	1.6	1	29.4	15.8
$n\text{-}C_4$	76.7	—	—	76.7	41.2
C_5	76.5	—	—	76.5	41.2
Total	349.4	163.5	100	185.9	100

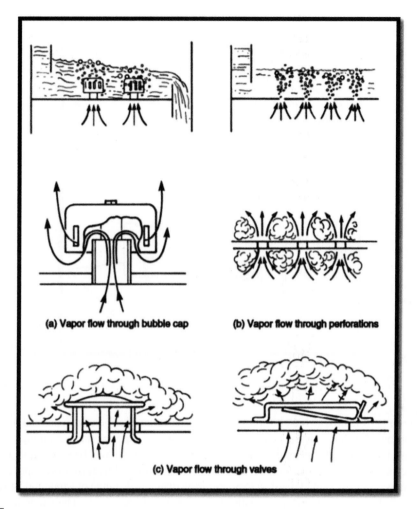

FIGURE 12.5

Flow through vapor passages (a) bubble cap (b) perforations) (c) valves [1].

trays control weeping by vapor velocity. The bubble cap tray has the highest turndown ratio, with designs of 8:1 to 10:1 ratio being common. Bubble cap trays are almost always used in glycol dehydration columns.

Valve and sieve trays are popular due to the lower cost and increased capacity over bubble cap trays for a given tower diameter. Figure 12.6 shows two valve designs. The upper drawing shows a floating valve free to open and close with varying vapor flow rates. The lower drawing shows a "caged" valve, which prevents valve loss due to erosion of the tray. Various other designs are common, such as using multiple disks and rectangular valves. Valves of assorted weights have also been used to increase flexibility.

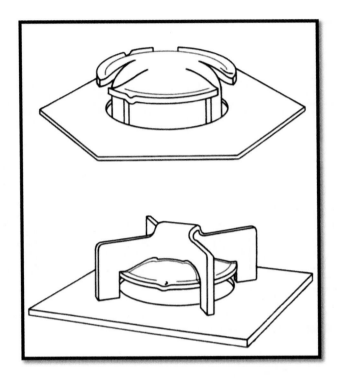

FIGURE 12.6

Valve types [1].

The sieve or perforated tray is the simplest construction of the three general types and thus is the least expensive option. The sieve tray is simply a plate with holes for vapor passage. Although the sieve tray generally has higher capacity, its main disadvantage is that sieve trays will be susceptible to "weeping" or "dumping" of the liquid through the holes at low vapor rates and its turndown capacity is limited. Trayed columns generally provide satisfactory operation over a wide range of vapor and liquid loadings.

Figure 12.7 shows operating characteristics for a representative system. The vapor and liquid rates can vary independently over a broad range and the column will operate satisfactorily. At low vapor rates unsatisfactory tray dynamics may be characterized by vapor pulsation, dumping of liquid, or uneven distribution.

At high vapor rates, the tower will eventually flood as liquid is backed up in the downcomers. At low liquid rates, poor vapor–liquid contact can result. High liquid rates can cause flooding and dumping as the liquid capacity of the downcomers is exceeded.

12.4 Packed columns

Traditionally the majority of fractionation columns in gas processing plants were equipped with trays. However, an alternative to trayed columns is to use packing. With packed columns, contact between the vapor and liquid phases is achieved throughout the column rather than at specific levels.

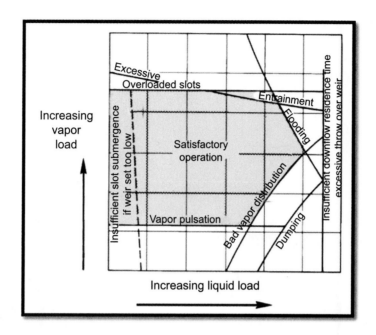

FIGURE 12.7

Limits of satisfactory tray operation for a specific set of tray fluid properties [1].

There are generally three types of packed columns:

- Random packing, wherein discrete pieces of packing are dumped in a random manner into a column shell. These packings are of a variety of designs. Each design has particular surface area, pressure drop, and efficiency characteristics. Examples of various packing types are shown in Figure 12.8.

Random packing has gone through various development phases from the first-generation packings, which were two basic shapes, the Raschig ring and the Berl saddle. Second-generation packings include the Pall ring and the Inatalox saddle, which are still used extensively today. Third-generation packings come in a multitude of geometries, most of which evolved from the Pall ring and the Intalox saddle:

- Structured packing, where a specific geometric configuration is achieved. These types of packing can either be the knitted-type mesh packing or sectionalized beds made of corrugated sheets. There are a number of commercially available packings, which differ in the angle of the crimps, the surface grooves, and the use of perforations.
- Grids, which are systematically arranged packing that use an open-lattice structure. These types of packings have found application in vacuum operation and low-pressure-drop applications. Little use of these types of packings is seen in high-pressure services. Structured packings have found application in low liquid loading applications, which are below 49 m^3/h/m^2. Structured packing has performed very well in extremely low liquid loading applications such as glycol dehydration.

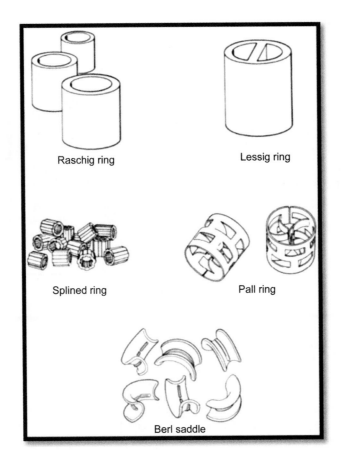

FIGURE 12.8

Various types of packing [1].

The high surface tension in glycol dehydrators also helps the structured packing to perform well. Above 49 m³/h/m², random packings are more advantageous. Structured packings have been tried in fractionators, with little success. Numerous cases of structured packing failures have been experienced in high-pressure and/or high-liquid rate services. Structured packings generally have lower pressure drop per theoretical stage than random packings. This can be important in low-pressure applications but not for high-pressure NGL fractionators. Figure 12.9 shows sample packed column internals.

12.5 Basic design requirements

The term liquefied petroleum gas (LPG or LP-Gas) as used in this specification is to be taken as applying to any material that is composed predominantly of any of the following hydrocarbons or mixtures thereof: propane, propene, normal and isobutane, butenes.

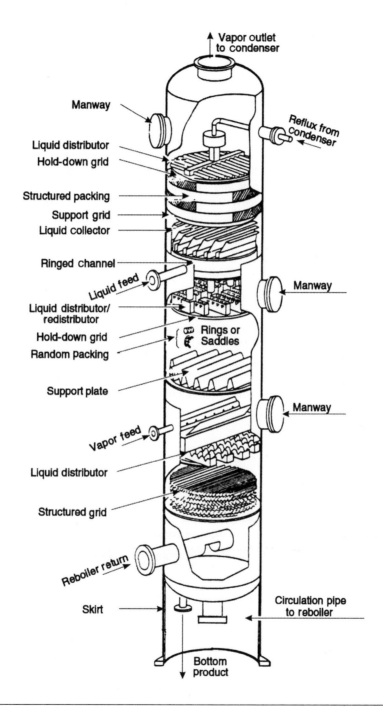

FIGURE 12.9

Sample packed column internals [1].

12.5.1 Physical properties and characteristics

The composition of a specific grade of LPG product is not normally rigidly specified, and thus product composition can vary from one particular refinery or petrochemical process plant to another; hence, for process design requirement of LPG storage and handling facilities, the physical, thermodynamic, and other properties of the product should be specified in project specification.

Table 12.5 gives a typical LP-Gas specification for different refinery products. It should be noted that whenever the words butane or propane appear hereafter, the commercial version of these products is intended. Pure and commercial products will be differentiated from one another as necessary for design and installation of relevant storage and handling systems.

Process configuration of the LPG recovery units shall be established based on the following factors and submitted for approval:

- feed composition;
- upstream unit process configurations;
- ultimate product consumption;
- product specifications;
- minimum C_3, C_4, and C_5 (if any) recovery.

Unless otherwise specified, economical study shall be practiced to justify provision of the absorption/stripping lean oil system if required by the minimum product recovery specification, as instructed by the project scope of the work for minimum C_3 recovery.

Table 12.6 shows sample LPG specifications. The final process configuration shall be approved by the company. For C_4 and C_5 the following requirements should be considered:

- Minimum C_4 recovery: 98 (vol%).
- Minimum C_5 recovery: 99.5 (vol%).

The unit product minimum specifications shall be as shown in Table 12.7, unless otherwise specified in the project specification.

Table 12.5 A typical LP-Gas Specification Based on Different Refinery Products			
		Specification	**Test Methods**
C_2 hydrocarbon	%vol	Nil	ASTM D 2163
C_3 hydrocarbon	%vol	Varies seasonally for refineries	ASTM D 2163
C_4 hydrocarbon	%vol	Varies seasonally for refineries	ASTM D 2163
C_5 hydrocarbon	%vol	2 max	ASTM D 2163
Hydrogen sulfide		Negative	ASTM D 2420
Mercaptan sulfur	mg/dm^3	0.0288 max (the limit applies to the product before addition of odorizing agent [ethyl mercaptan])	IP 10 (A)
Odorizing agent	g/m^3	12	

Table 12.6 LPG Specifications

		Specification	Test Methods
C_2 hydrocarbon	vol%	0.2 max	ASTM D 2163
C_3 hydrocarbon	vol%	Varies seasonally for refineries	ASTM D 2163
C_4 hydrocarbon	vol%	Varies seasonally for refineries	ASTM D 2163
C_5 hydrocarbon	vol%	2 max	ASTM D 2163
Hydrogen sulfide		Negative	ASTM D 2420
Mercaptan sulfur	mg/dm^3	0.23 max (the limit applies to the product before addition of odorizing agent [ethyl mercaptan])	ASTM D 3227
Odorizing agent	g/m^3	12	

LPG, liquefied petroleum gas.

Table 12.7 Unit Product Minimum Specifications

	C_3	C_4	C_5
C_2 (vol%)	0.3 max	—	—
C_3 (vol%)	Balance	0.2 max	—
C_4 (vol%)	3.5 max	Balance	1 max
C_5 (vol%)	—	1 max	Balance

If required by the feed compositions and product specifications, treating facilities shall be provided.

The unit design throughput shall be based on the sum of maximum flow rates of various feed streams to the unit when the upstream units are operating at their design capacities.

C_3 and C_4 products must be manufactured separately and each stream must be suitable for LPG blending. Drying facilities for C_3 product shall be provided.

Feed surge drums shall be provided to receive all feed streams into the LPG recovery unit. Unless otherwise specified, the unit turndown capacity should be 60% of design throughput, without loss of efficiency in fractionation while meeting the product specifications.

The unit design capacity shall be determined based on the licenser's information on the upstream licensed process units and shall take into consideration the variations resulting from the relevant process units.

One cooler shall be supplied to cool the feed gas from the crude distillation overhead compressor (if any).

The following design notes shall be taken into consideration if an LPG caustic treating section is to be supplied as per feed and product specifications:

- Caustic dissolving facilities shall be included if supply of the caustic outside of the unit battery limit is not feasible.

- Caustic regeneration facilities shall be provided, if economically justified.
- Spent caustic degassing and storage shall be provided.

Safety considerations shall be fully complied with. Special attention shall be made to the flexibility and ease of operations, equipment interchangeability, and optimization.

Maximum energy conservation shall be applied.

Kettle-type reboilers shall be provided to maintain the bottom temperature of the following towers:

- de-ethanizer;
- depropanizer;
- propane dryer;
- debutanizer.

12.6 Fractionation and system configuration

A stabilization tower shall be used where a natural gasoline or stable liquid is to be produced.

The two-tower system is most commonly used to produce an LPG mixture in the overhead and a natural gasoline product as the bottoms. In this system, the de-ethanizer must remove all methane, ethane, and other constituents not suitable in the two product streams from the second tower. Any material that enters the second tower must necessarily leave in one of the product streams.

The three-tower system most commonly produces commercial propane, commercial butane, and natural gasoline as products. In this system also, the de-ethanizer must work properly to remove all constituents that cannot be sold in one of the three products.

The sequence of fractionation following the de-ethanizer may be varied. In the second tower, an LPG mixture could be produced overhead with natural gasoline produced as bottoms. The third tower would then split the LPG into commercial propane overhead and commercial butane as bottoms. This sequence is sometimes favored where the market situation is variable and a market for LPG only exists during a portion of the year. During this period, the third tower would be shut down and not operated.

Regardless of how the fluids are removed from natural gas and/or gasoline, fractionation is necessary if products that meet any kind of rigid specification are to be made. The number of fractionating columns required depends on the number of products to be made and the character of the liquid that serves as feed. The single tower system ordinarily produces one specification product from the bottom stream, with all other components in the feed passing overhead.

The operating pressure of a fractionation tower is ordinarily fixed by the condensing temperature of the overhead product. Temperature in the condenser is ordinarily controlled by the cooling medium. Allowing for sufficient temperature difference between the cooling medium and the overhead product, condenser temperature is fixed by the designer. In the case of a liquid distillate, the bubble point pressure is then calculated; for a vapor distillate product, the dew point pressure would be calculated. This pressure is the minimum pressure at which the tower can operate at the chosen condenser temperature.

Economic investigation shall be made for selection of a total and partial condenser for a tower. At a given pressure, the dew point is always a higher temperature than the bubble point, and this tends to minimize cooling costs, where all other elements are equal.

12.6.1 **Fractionation design considerations**

If the tower involved is the first one in a fractionation system, the conditions of the feed to the column will be fixed by the separation process. A surge vessel prior to this tower might change the analysis of the feed if some vapor is withdrawn at that point. (An example of this is a rich oil flash drum situated between the absorber and the de-ethanizer.)

For a system containing several towers, the split desired in each column should be made before completing analysis of any one. This assures that the splits set up for the different towers produce satisfactory products in all streams. A perfect separation between adjacent components cannot be specified. This will lead to a situation impossible to achieve in an actual column. In producing propane, for example, it must be allowed that a small amount of ethane and butanes be present. In this case, however, the propane must still meet the purity specifications demanded.

There are three ways in which the desired operation of a fractionation column is ordinarily specified:

* A specified percentage recovery of one component in the distillate or one component in the bottom.
* A specified composition of one component in the distillate or bottom stream.
* A specified vapor pressure for either the distillate or bottom product.

The design of a fractionator shall incorporate the following considerations:

* feed composition, quantity, temperature and pressure;
* desired products and their specifications;
* a reasonable condensing temperature for the overhead stream;
* degree of separation between products.

Overall tray efficiency for the most fractionators typically are shown in Table 12.8.

Table 12.8 Typical Fractionator Parameters

Description	Operating Pressure (kPa (ga))	Number of Actual Trays	Reflux Ratio (Relative to Overhead Product (mol/mol))	Reflux Ratio (Relative to Feed, dm^3/dm^3 (gal/gal))	Tray Efficiency (%)
Demethanizer	1400–2800	18–26	Top feed	Top feed	45–60
De-ethanizer	2600–3100	25–35	0.9–2.0	0.6–1.0	50–70
Depropanizer	1700–1900	30–40	1.8–3.5	0.9–1.1	80–90
Debutanizer	500–620	25–35	1.2–1.5	0.8–0.9	85–95
Butane splitter	550–700	60–80	6.0–14.0	3.0–3.5	90–110
Rich oil fractionator (still)	900–1100	20–30	−1.75–2.0	0.35–0.40	Top 67, bottom 50
Rich oil de-ethanizer	1400–1750	40	—	—	Top 25–40, bottom 40–60
Condensate stabilizer	700–2800	16–24	Top feed	Top feed	40–60

12.7 Absorption/stripping

12.7.1 Basic requirements

The absorption/stripping process may be used in the LPG recovery unit if needed by the minimum recovery specification of the product streams, as requested in the project scope of the work.

The design should incorporate both stripper and absorber towers with all associated facilities.

The resultant rich oil shall be stripped or denuded of the absorbed materials in the stripper tower. The stripped oil shall be recirculated to the absorber tower as lean oil for absorption of the LPG components.

12.7.2 Design considerations

A temperature rise of 6–16 °C is usually designed into initial condition for the absorption process. The rise above this must be handled by intercoolers.

12.8 Control and optimization

Advanced process control (APC) and optimization may be applied to upgrade plant safety, product quality and quantity, and plant operation. The extent of application shall be as per company instructions.

The following APC loops may be incorporated in the process design as typical. The APC loops shall preferably be functioned without implementation of an online process analyzer.

1. De-ethanizer tower:
   a. Control system for matching specification of C_2 at bottom and LPG recovery at top
2. Depropanizer and debutanizer towers:
   a. C_3 quality control
   b. C_4 quality control

12.9 Storing and handling of liquefied petroleum gases (LPGs)

12.9.1 Siting

Site selection is concerned with minimizing the potential risk to adjacent property presented by the storage facility and the risk presented to the storage facility by a fire or explosion on adjacent property.

The following factors should be considered during site selection:

1. proximity to populated area;
2. proximity to public ways;
3. risk from adjacent facilities;
4. storage quantities;
5. present and predicted development of adjacent properties;
6. topography of the site, including elevation and slope;
7. access for emergency response;
8. utilities;
9. requirements for receipt and shipment of products.

- **Above-ground pressurized LPG tanks and equipment**

Pressurized LPG tanks shall not be located within the building, within the spill containment area of flammable or combustible liquid storage tanks as determined in NFPA 30, or within the spill contaminant area for refrigerated storage tanks.

Rotating equipment and pumps taking suction from the LPG tanks shall not be located within the spill contaminant area of any storage facility.

Horizontal vessels used to store LPG should be oriented so that their longitudinal axes do not point toward other containers, process equipment, control rooms, loading and unloading facilities, or flammable or combustible liquid storage facilities located in the vicinity of the horizontal vessel.

Horizontal vessels used to store LPG should be grouped with no more than six vessels in one group. Where multiple groups of horizontal LPG vessels are to be provided, each group should be separated from adjacent groups by minimum horizontal shell-to-shell distance of 15 m.

- **Layout and spacing**

Spacing and design of LPG facilities are interdependent and must be considered together. Spacing should be sufficient to minimize both the potential for small leak ignition and the exposure risk to adjacent vessels, equipment, or installations, should ignition occur.

Typical layouts are illustrated in Figure 12.10. These illustrate the slope of tank sites; the location of vessels with respect to each other; and the positioning of separation wall, when necessary, and manifolds with respect to vessels.

12.9.2 Minimum distance requirement for above-ground LPG tanks

The minimum horizontal distance between the shell of a pressurized LPG tank and the line of adjoining property that may be developed shall be as specifically given in API Standard 2510 under Clause 3.1.21.

The minimum horizontal distance between the shells of pressurized LPG tanks or between the shell of a pressurized LPG tank and the shell of any other pressurized hazardous or flammable storage tank shall be as follows:

1. If the storage is in spheres or vertical vessels, one-half the diameter of the larger sphere or vertical vessel but not less than 2 m.
2. If storage is in horizontal vessels, 2 m. Greater distances should be considered if the vessel diameter exceeds 3.5 m.
3. If the storage is in spheres and vertical or horizontal vessels, the greater of the distances given by item (1) or (2) shall be used as the spacing.

The minimum horizontal distance between the shell of a pressurized LPG tank and the shell of any other nonpressurized hazardous or flammable storage tank shall be the largest of the following:

1. If the other storage is refrigerated, three-quarters of the larger diameter.
2. If the other storage is in atmospheric tanks and is designed to contain material with a flash point of 38 °C or less, one diameter of the larger tank.
3. If the other storage is in atmospheric tanks and is designed to contain material with a flash point greater than 38 °C, one-half the diameter of the larger tank.

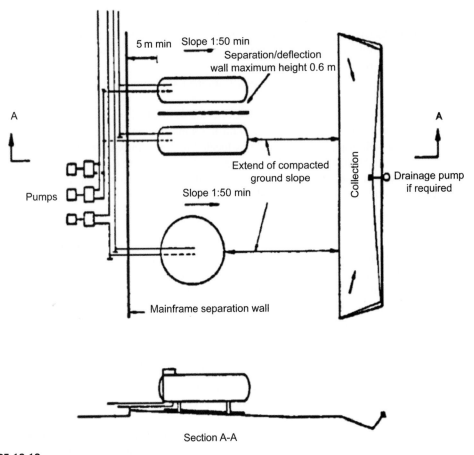

FIGURE 12.10

Typical layout of LPG pressure storage with collection pit/retaining system.

The minimum horizontal distance between the shell of an LPG tank and a regularly occupied building shall be as follows:

1. If the building is used for the control of the storage facilities, 15 m.
2. If the building is used solely for other purposes, 30 m (unrelated to control of the storage facilities).

The minimum horizontal safety distances between the shell of an LPG tank and facilities or equipment not covered in the above sections shall be as outlined in Table 12.9

The minimum distances between the LPG storage vessels and other facilities and equipment that are not covered herein shall be taken as given under Clause 3.1.2.5 of API Standard 2510.

Table 12.9 Safety Distances Between LPG Tanks and Other Equipment and Facilities

Factor	Minimum Safety Distance (m)
Between LPG storage vessels	Equal to the diameter of the largest vessel, but with a lower limit of 2.5 m and an upper limit of 10 m
Between LPG storage vessels and manifold (measured horizontally between the equator of vessel and center of separation wall)	5 m
Between LPG storage vessels and edge of refinery/depot roads and/or pipe tracks	10 m
Between LPG storage vessels and the end of fixed ending of the loading/discharging points for rail and road tank cars and cylinder filling/storage plants and areas	15 m
Between LPG storage vessels and processing units, laboratories, main offices, buildings with flammable material, other buildings where people are concentrated, and site boundary fence	30 m
Between LPG storage vessels and stationary internal combustion engines	15 m

12.9.3 Type and size of storage vessels

The type, size, and number of vessels to be used must be based on operational requirements and technical/economic considerations. The following is given as guidance for the types normally used and the size limitations generally applicable to them.

1. Horizontal vessels, which can in many instances be shop fabricated and moved to the site in one piece, are normally used for unit capacities up to 200 m^3.
2. Vertical cylinder vessels have an advantage to horizontal vessels in that they require less space for a specific capacity. They are normally limited in size to a maximum of 10 m diameter and 25 m height.
3. Spherical vessels (spheres) are normally considered if the unit capacity exceeds 400 m^3.

12.9.4 Spill containment

Spill containment should be provided in locations in which any of the following conditions exist:

1. The physical properties of the stored LPG (for example, a mixture of butane and pentane) make it likely that liquid material will collect on the ground.
2. Climatic conditions during portions of the year make it likely that liquid material will collect on the ground.
3. The quantity of material that can be spilled is large enough that any unvaporized material will result in a significant hazard.

If spill containment is to be provided, it shall be by remote impoundment of spilled material or by diking the area surrounding the vessel.

The pronounced volatility of LPG generally allows impoundment areas to be reduced and in some cases makes spill containment of LPG impractical. However, the ground and surroundings of a vessel used to store LPG shall be graded to drain any spills to a safe area away from the vessel.

All provisions under Clause 3.2 of API Standard 2510 regarding establishment of spill containment facilities for LPG storage vessels shall be considered as an integral part in design and installation of LPG storage vessels.

12.10 Design considerations
12.10.1 Storage vessels

- **Design code**
  1. 110% of the maximum operating pressure;
  2. the maximum operating pressure plus 170 kPa (1.7 kg/cm^2).

The design pressure to be used for the bottom of the vessel shall be that of above for the top of the vessel plus the static head of the content.

- **Design vacuum**

LPG storage vessel design shall consider vacuum effects. Where an LPG vessel is not designed for full vacuum, some alternatives, in order of preference, are as follows:

1. Design for partial vacuum with a vacuum relief valve and a connection to a reliable supply of inert gas. This alternative requires a means of venting inert gas that has been admitted to the storage vessel after it is no longer required for maintaining pressure.
2. Design for partial vacuum with a vacuum relief valve and a connection to a reliable supply of hydrocarbon gas. This alternative may compromise product quality.
3. Design for partial vacuum with a vacuum relief valve that admits air to the vessel. This alternative presents a hazard from air in the LPG storage vessel, and this hazard shall be considered in the design.

- **Design temperature**

Both a minimum and a maximum design temperature should be specified. In determining a maximum design temperature, consideration should be given to factors such as ambient temperature, solar input, and product rundown temperature. In determining a minimum design temperature, consideration should be given to the factors in the preceding sentence plus the autorefrigeration temperature of the stored product when it flashes to atmospheric pressure.

- **Filling and discharge line**

Only one product line shall be connected to the bottom of the vessel and this line shall be used for filling, discharge, and drainage. However, operational considerations may dictate the use of a separate top-connected filling line.

To enable complete drainage of the vessel, the connection of the bottom line to the vessel shall be made flush with the inside of the vessel.

Top-connected filling and vapor lines shall be provided with a remote-controlled fail-safe-type shut-off valve if the line extends below the maximum liquid level; otherwise, a shut-off valve plus either a non-return valve or an excess flow valve may be used.

The product line connected to the bottom of the vessel shall be provided with either of the following:

1. A remote-controlled fail-safe-type shut-off valve located at the manifold side of the separation wall between manifold and vessel. This line is to have a minimum size of diameter nominal (DN) 100 and should be of Schedule 80 for DN 100 size and Schedule 40 for DN 150 and larger. A hand-operated fire-safe valve should be provided between the remote-controlled valve and the manifold separation wall if it is considered necessary.
2. A remote-controlled fail-safe-type valve mounted internally in the vessel. Design consideration should be given to the possibility/practicability of emptying the vessel in the case of malfunctioning of the remote-controlled valve. If considered necessary, a by-pass line connected to the vessel shall be provided with a shut-off valve and shall be blanked.

Vessel design shall meet the requirements of Section 12.8 of the ASME Boiler and Pressure Vessel Code (commonly called the ASME Code), Division 1 or 2.

When complete rules and design requirements for any specific design are not given, the manufacturer, subject to the approval of the company, shall provide a design as safe as would be provided in the currently applicable ASME code cited above.

- **Design pressure**

It is assumed that the maximum operating pressure at the top of a vessel is equivalent to the vapor pressure of the product being handled at the maximum temperature that the vessel's contents may reach under prolonged exposure of the vessel to solar radiation ("the assessed temperature").

The design pressure to be used for the top of the vessel shall be equal to the greater of:

1. 110% of the maximum operating pressure;
2. the maximum operating pressure plus 170 kPa (1.7 kg/cm^2).

The design pressure to be used for the bottom of the vessel shall be that of the above for the top of the vessel plus the static head of the content.

- **Water drawing**

Water can accumulate under certain conditions in LPG storage vessels and must be removed for product quality reasons. Also, in freezing climates, ice formation in bottom connections can rupture piping and lead to major LPG releases. Thus facilities shall be provided and procedures shall be established to handle water draw-off safely.

Considering the potential risk associated with improper handling of water removal, a detailed written procedure should be prepared and rigidly followed. The procedure outlined in API Standard 2510 under Paragraphs 3.4.4.1 through 3.4.4.3 is recommended to be considered during the entire process of water removal.

- **Safety/relief valves**

LPG storage vessels shall be adequately protected by safety/relief valves directly connected to the vapor space of the vessel. Safety/relief valves shall be provided to protect against:

1. over-pressurization due to abnormal operational conditions, e.g., overfilling, high run-down temperatures, or high temperature due to solar radiation;
2. over-pressurization due to fire exposure.

Consideration should be given to the provision of a spare safety/relief valve or connection such as to facilitate servicing/maintenance of safety/relief valves.

The materials used for safety/relief valves including components, e.g., springs, valve discs, must be suitable for use with LPG and for operation at low temperatures.

Pressure relief valves installed on the LPG storage vessels shall be designed to protect the vessels during fire exposure. Other causes of tank over-pressure, such as overfilling and introduction of material with a higher vapor pressure in a common piping system, shall be considered.

Pressure relief valves shall be designed and sized in accordance with API Recommended Practice 520, Part 1, and API Recommended Practice 521.

All safety provisions given by NFPA codes, standards, recommended practice manuals, and guides in Volume 2, Chapter 6, and applicable to spring-loaded safety relief valve installation on LPG storage vessels shall be considered to the extent of process design requirements.

When a closed relief system is used, all applicable points of API 2510 A, under Paragraph 2.10.3, shall strictly be considered in process design of the system. Atmospheric relief system, if proposed in the project specification, requirements of API 2510 A, under Paragraph 2.10.2, shall essentially be taken into account in process design of such system.

Tanks that may be damaged by internal vacuum shall be provided with at least one vacuum-relieving device set to open at not less than the partial vacuum design pressure.

When a closed inner tank design is used with an outer vapor-tight shell, the outer shell shall be equipped with a pressure- and vacuum-relieving device or devices.

- **LPG tank's other accessories**

If sampling connections are required, they shall be installed on the tank piping rather than on the tank. Sampling provisions given in API Publication 2510A under Paragraph 3.5 shall, to the extent of process design requirements, be considered.

- **Accessory equipment and shut-off valves**

Accessory equipment and shut-off valves shall be designed to meet extreme operating pressure and temperature.

- **Liquid level gauging device**

Each nonrefrigerated storage vessel shall be equipped with a liquid level gauging device of approved design. If the liquid level gauging device is a float type or a pressure differential type and the vessel is a nonrefrigerated type, the vessel shall also be provided with an auxiliary gauging device, such as a fixed dip tube, slip tube, rotary gauge, or similar device.

Refrigerated LPG storage vessels shall be equipped with a liquid level gauging device of approved design. An auxiliary gauging device is not required for refrigerated storage vessels. However, in lieu of an auxiliary gauge, refrigerated vessels, if subject to overfilling, shall be equipped with an automatic device to interrupt filling of the tank when the maximum filling level is reached.

All other safety requirements relating to liquid level gauging specified in NFPA V.2, 31-69 under Paragraphs 4.4.3 through 4.4.6 shall be considered.

- **Venting noncondensibles**

Noncondensible gases, including air, can enter to an LPG storage vessel through a variety of means, including the following:
1. Dissolved or entrained gases from processing, such as sweetening.
2. Operation of vacuum breakers.
3. System leaks while under vacuum.
4. Air or inert gas in a vessel when it is put into service.
5. Vapor return lines from trucks or rail cars that contain air or inert gas prior to loading.

The gases may cause the relief valve to operate when the liquid level is subsequently raised and the noncondensible gases are thereby compressed.

Criteria should be developed to vent the compressed noncondensibles periodically when the oxygen concentration exceeds a specified value or when the head-space pressure exceeds the product vapor pressure by a specified amount.

The noncondensibles may be vented to air. If regulations require venting to a flare system, then caution is necessary, since the vented gas may contain air. In these cases precautions shall be taken to prevent sending a flammable mixture to the flare.

- **Drain facilities**

A drain connection shall be provided on the filling/discharge line at the manifold side of the first shutoff valve (manual or remote-controlled).

If in exceptional cases a drain connection on the storage vessel is unavoidable, the company's prior approval should be obtained in order to agree on an acceptable design of drainage system.

The outlet of the drain line, where flammable vapor can be released, should be discharged at a safe point, i.e., away from roads, working areas, etc.

At locations where freezing conditions can occur, the drain facilities shall be adequately traced and insulated. Insulation and, possibly, tracing of the filling/discharge line may also be necessary.

Operational rules for drainage should be given in the operating manual or as an instruction to drainage procedures.

12.10.2 Refrigeration system

- **Load**

The refrigeration load should take into consideration the following factors:
1. Heat flow from the following sources:
   a. The difference between the design ambient and storage temperatures.
   b. Maximum solar radiation.
   c. Receipt of product that is warmer than the design temperature, if such an operation is expected.

    d. Foundation heaters.

    e. Heat absorbed through connected piping.

2. Vapor displaced during filling or returned during product transfer.

3. Changes in barometric pressure.

- **Vapor handling**

The vapor load resulting from refrigeration may be handled by one or a combination of the following methods:

1. Recovery by a liquefaction system.

2. Use as fuel.

3. Use as process feedstock.

4. Disposal by flaring or another safe method.

Alternative handling methods shall be provided to dispose of vented vapors in case of failure of the normal methods. If compressors are used, castings shall be designed to withstand a suction pressure of at least 121% of the tank design pressure.

- **System accessories**

A refrigerated LPG system shall contain the following accessories:

1. An entrainment separator in the compressor suction line.

2. An oil separator in the compressor discharge line (unless the compressor is a dry type).

3. A drain and a gauging device for each separator.

4. A noncondensible gas purge for the condenser.

5. Automatic compressor controls and emergency alarms to signal:

    a. when tank pressures approach the maximum or minimum allowable tank working pressure or the pressure at which the vacuum vent will open, or

    b. when excess pressure builds up at the condenser because of failure of the cooling medium.

- **Pressure-relieving devices**

Refer to API Recommended Practice 520, Parts I and II, for the proper design of pressure-relieving devices and systems for process equipment used in liquefaction and vaporization facilities.

12.10.3 Pumps

Centrifugal or positive displacement pumps may be used for LPG service. The pumps should be able to operate at a low net positive suction head (NPSH).

In process design of pumps, reference is made to API Standard 610.

Centrifugal and rotary positive displacement pumps shall be equipped with mechanical seals. Consideration should be given to the use of auxiliary glands.

If a centrifugal pump is used, a return line connecting the discharge with the suction side of the pump should be installed. The flow through this return line should be 10–25% of the design flow of the pump at its highest efficiency.

Positive displacement pumps, if used, shall be safeguarded by a differential relief valve in a return line from discharge to suction side. The return line shall be designed for at least 100% of the designed capacity of the pump.

The return line should run back to either the suction line of the pump or the vapor space of the supplying storage vessel.

Centrifugal pumps shall be provided with a vent in order to remove any accumulated vapor before the pump is started. This vent shall be connected to the vapor space of the storage vessel or vent to atmosphere at a safe place.

12.10.4 Fire protection facilities

Provisions given in API Standard 2510 and API Publication 2510A as an amendment to API Standard 2510, in Clause 5.3 and 8.6, respectively, shall be subject to verification or modification through analysis of local conditions and used for process design of an efficient and perfectly reliable protection facility.

12.10.5 Piping

All applicable portions of API Standard 2510 under Paragraph 2.5 and Section 12.6 shall be followed in process design of piping.

Other considerations for piping process design are those relating to parts given in API Publication 2510A under Paragraph 2.7.

12.10.6 Vaporizers

Liquefied petroleum gases are used in gaseous form. A vaporizer is required when the heat transferred to the liquid is inadequate to vaporize sufficient gas for maximum demand. A steam, hot water, or direct-fired-type vaporizer may be used.

A vaporizer should be equipped with an automatic means of preventing liquid passing from vaporizer to gas discharge piping. Normally this is done by a liquid level controller and positive shut-off of liquid inlet line or by a temperature control unit for shutting off line at low-temperature conditions within the vaporizer.

Some installations operate on "flash vaporization," whereby the liquid is converted to a gas as soon as it enters the vaporizer, while others maintain a liquid level in the vaporizer.

- **Indirect vaporizers**

Indirect vaporizers shall comply with Clause 2.5.4.2(a) through (e) of NFPA Code, Standard, Vol. 2, 31–69, Ed. 1989 and the following:

1. A shut-off valve shall be installed on the liquid line to the LPG vaporizer unit at least 15 m away from the vaporizer building.
2. The heating medium lines into and leaving the vaporizer shall be provided with suitable means for preventing the flow of gas into the heat systems in the event of tube rupture in the vaporizer. Vaporizers shall be provided with suitable automatic means to prevent liquid passing from the vaporizer to the gas discharge piping.

3. The device that supplies the necessary heat for producing steam, hot water, or other heating medium shall be separated from all compartments or rooms containing liquefied petroleum gas vaporizers, pumps, and central gas mixing devices by a wall of substantially fire-resistive material and vapor-tight construction.

- **Direct-fired vaporizers**

Direct-fired vaporizers shall be designed in full conformity with requirements given in NFPA Codes, Vol. 2, Edition 1989, Chapter 5, Paragraph C.3.

12.10.7 Instrumentation

As a minimum, the requirements in API Standard 2510, Section 12.5 shall be followed. In addition, the considerations given in sections 2.9.1 through 2.9.5 of API Publication 2510A to the extent applicable to process design shall be considered.

12.11 Transfer of LPG within the off-Site facilities of oil and gas processing (OGP) plants

LPG as in liquid form is permitted to be transferred from storage vessels by liquid pump or by pressure differential.

Pressure differential may be used under certain conditions. Fuel gas or inert gas, which is at a higher pressure than LP-gas, shall be used under the following conditions:

1. Adequate precautions shall be taken to prevent liquefied petroleum gas from flowing back into the fuel gas or inert gas line or system by installing two back-flow check valves in series in these lines at the point where they connect into the liquefied petroleum gas system. In addition, a manually operated positive shut-off valve shall be installed at this point.
2. Any fuel gas or inert gas used to obtain a pressure differential to move liquid LPG shall be noncorrosive and dried to avoid stoppage by freezing.
3. Before any fuel gas or inert gas is placed in a storage vessel, permission shall be obtained from the vendor of the LPG storage vessel by considering all design requirements for introducing such vapor into the vessel.
4. Transfer operations shall be conducted by competent trained personnel.
5. Unloading piping or hoses shall be provided with suitable bleeder valves or other means for relieving pressure before disconnection.
6. Precautions shall be exercised to assure that only those gases for which the system is designed, examined, and listed are employed in its operation, particularly with regard to pressure.

12.11.1 Requirements

The transfer system shall incorporate a means for rapidly and positively stopping the flow in an emergency. Transfer systems shall be designed to prevent dangerous surge pressures when the flow in either direction is stopped.

Transfer pumps may be centrifugal, reciprocating, gear, or another type designed for handling LPG. The design pressure and construction material of the pumps shall be capable of safely withstanding the maximum pressure that could be developed by the product, the transfer equipment, or both.

All process design requirements shall be considered when centrifugal or positive displacement pumps are used.

Compressors used for liquid transfer normally shall take suction from the vapor space of the container being filled and discharge into the vapor space of the storage vessel from which the withdrawal is being made.

Provisions relating to process design requirements of LPG transfer loading and unloading facilities are given in API Standard 2510, under Paragraph 7.

All safety considerations deemed necessary to be specified in process design of LPG transfer facilities shall be in conformity with NFPA, Vol. 2, Chapter 4.

12.12 Pressure storage spheres for LPG

For storage of LPG, the principal above-ground storage methods are:

1. Pressure storage at ambient temperature
2. Fully refrigerated (at around atmospheric pressure)
3. Refrigerated pressure

For the purpose of this section, only the pressure storage at ambient temperature and refrigerated pressure storage are considered.

12.12.1 Pressure storage at ambient temperature

Because of the high vapor pressure of LPG, the liquid at ambient temperature must be stored under pressure in vessels and spheres designed to withstand safely the vapor pressure at the maximum liquid temperature.

12.12.2 Refrigerated-pressure storage

Refrigerated-pressure storage, sometimes referred to as semi-refrigeration storage of LPG, combines partial refrigeration with low or medium pressure. An attractive feature of refrigerated-pressure storage is its flexibility, making it possible for a vessel to be used at different times for butane or propane.

Thus, a storage sphere designed for pressure storage of butane at atmospheric temperature could be used for the refrigerated-pressure storage of propane by chilling the propane and insulating the vessel so that the vapor pressure does not exceed the sphere normal working pressure.

Refrigerated-pressure storage in spheres has the following advantages:

1. The evolved vapor (boil-off-for reliquefaction) comes off at a sufficient pressure to overcome line friction where the refrigeration equipment is remote from the sphere.
2. The ratio of surface area to volume is less, and therefore heat leak from the atmosphere is proportionately less.

This section covers the minimum requirements for design of pressure storage spheres. In this section, pressure storage means storage spheres with design pressure above 100 kPa (1 bar) gauge. The requirements of this section apply to both refrigerated and nonrefrigerated LPG pressure storage spheres. It is intended for use in oil refineries, chemical plants, marketing installations, gas plants, and, where applicable, in exploration, production, and new ventures.

12.13 Material selection

All materials of construction for pressure storage spheres shall meet the requirements of Section 12.2 of the ASME Boiler and Pressure Vessel Code.

The following requirements are supplementary:

- The selector of the material of construction for pressure parts and their integral attachments shall take into account the suitability of the material with regard to fabrication and to the conditions under which they will eventually operate.
- Special consideration should be given to the selection of materials for pressure storage spheres designed to operate below 0 °C. Austenitic stainless steels and aluminum alloys are not susceptible to low-stress brittle fracture and no special requirements are necessary for their use at temperatures down to −196 °C.
- Some carbon steel material for construction of pressure storage spheres for normal services are given in Table 12.10.
- Casting shall not be used as pressure components welded to the shell of pressure storage spheres.
- Materials having a specified minimum yield strength at room temperature greater than 483 mPa (70,000 psi) shall not be used without prior approval of the owner's engineer.

Table 12.10 Carbon Steel Material Specifications for Pressure Storage Spheres for Normal Services

Parts	ASTM Specification
Shell and head plates	A 285 A 442 A 516 A 537 CL.2 A 662
Flange min of PN 20 (150#)	A 105
Nozzles	A 53 GR. B seamless A 106 GR. B
Neck	For large-size nozzles and manholes neck, same material as shell plates shall be used
Bolts	A 193 GR. B7
Nuts	A 194 GR.2H

- Materials of non-pressure-retaining parts to be welded directly to pressure-retaining parts shall be of the same material as the pressure-retaining parts.
- Materials, such as lower support columns, platforms, stairways, pipe supports, and insulation support rings, shall be carbon steel of ASTM A 283 Gr. C or equivalent. External non-pressure-retaining part boltings shall be carbon steel of ASTM A 307 Gr. B or equivalent.
- The internal bolts and nuts, including U-bolts, shall be of type 410 or 405 stainless steel material.
- Material of anchor bolts shall be carbon steel of ASTM A 307 Gr. B or equivalent.

12.14 General information

While this section concerns pressure storage spheres for LPG only, it is important to have some indication of the basic differences between LPG and liquefied natural gas (LNG). LPG refers, in practice, to those C_3 and C_4 hydrocarbons, i.e., propane, butane, propylene, butylene, and the isomers of the C_4 compounds that can be liquefied by moderate pressure. Methane and mixtures of methane with ethane cannot be liquefied by pressure alone, since the critical temperatures of these gases are too low and some precooling is required.

The liquefied forms of methane/ethane are loosely referred to as liquefied natural gas (LNG). Some examples of the main gases, together with their boiling points at atmospheric pressure, are given in Table 12.11 below.

As is seen from the boiling points listed in Table 12.11 above, to liquefy these gases for ease of storage and transportation it is normally necessary to reduce the temperature to well below ambient or to pressurize them until a liquid is formed. In practice, temperature reduction by refrigeration, pressurization, or a combination of the two are commonly used to achieve liquefaction.

Commercial grades of propane and butane are not pure compounds; thus, commercial propane is mainly propane with small amounts of other hydrocarbons such as butane, butylene, propylene, and ethane, and commercial butane is mainly normal butane and isobutane, with small amounts of propane, propylene, and butylene.

Table 12.11 Formula and Boiling Points of LPG and LNG

Liquefied Petroleum Gas (LPG)			Liquefied Natural Gas (LNG)		
Name	Formula	NBP (°C)	Name	Formula	NBP (°C)
Propylene	C_3H_6	−42.7	Methane	CH_4	−161.5
Propane	C_3H_8	−42.5	Ethylene	C_2H_4	−103.7
Butylene	C_4H_8	−6.9	Ethane	C_2H_6	−88.6
Butane	C_4H_{10}	0.5			

12.15 Design of pressure storage spheres

Design of pressure storage spheres shall be in accordance with Section 12.8 of the ASME "Boiler and Pressure Vessel Code," Div. 1 or 2.

12.15.1 Design data

The following requirements shall be considered as supplementary. The shell plates and column supports of pressure storage spheres shall be designed based on the severest loading under the following two conditions:

- **Condition I (normal operating condition)**

1. Load combination shall be considered on the assumption that the following loads act simultaneously:
   a. Internal or external pressure, when necessary, at design temperature.
   b. Operating weight.
   c. Wind load or earthquake load, whichever governs.
2. The shell plate thickness shall be that corresponding to the corroded condition, that is to say, nominal thickness minus corrosion allowance.

- **Condition II (condition of hydrostatic testing at the operating position)**

1. Load combination shall be considered on the assumption that the following loads act simultaneously:
   a. Internal pressure due to hydrostatic test.
   b. Empty weight of the sphere.
   c. Weight of water for testing.
   d. One-third the wind load.
2. The shell plate thickness shall be that corresponding to the corroded condition, i.e., nominal thickness minus corrosion allowance.

12.15.2 Corrosion allowance

- Generally, a minimum corrosion allowance of 1.5 mm shall be provided for carbon steel material, unless otherwise specified. No corrosion allowance shall be provided for high-alloy or nonferrous materials.
- All pressure-retaining parts shall be provided with the specified corrosion allowance on all surfaces exposed to corrosive fluid.
- For nonremovable internal parts, one-half of specified corrosion allowance shall be added to all surfaces and one-fourth of the corrosion allowance shall be added to all surfaces of removable internal parts.
- No corrosion allowance shall be provided for external parts, unless otherwise specified.
- The pressure retaining parts of pressure storage spheres and their support columns shall be designed to be filled with water.

- Pressure storage spheres shall be supported so that the bottom is no less than 1 m above finished grade.

12.15.3 Calculation of optimum volume

The volume of liquid stored in a vessel must be limited to allow sufficient room for thermal expansion. The maximum volume (V) of liquid gas at a certain temperature (T, °C) that may be charged into a vessel is determined by the formula:

$$V = \frac{DW}{100\gamma F} \tag{12.2}$$

where:

W = Water capacity of storage vessel at 15.6 °C (60 °F)
D = Maximum filling density (Table 12.12)
γ = Specific gravity of liquid gas at 15.6 °C
F = Liquid volume correction factor from temperature T° to 15.6 °C (Table 12.13).

The filling density (D) is the percent ratio of the weight of liquid gas in a vessel to the weight of water required to fill the vessel at 15.6 °C and can be obtained from Table 12.12.

A volume correction factor (F) is necessary because the lower the temperature of the liquid below ambient at the time of filling the vessel, the greater will be the expansion when the temperature of the liquid reaches ambient. Volume correction factor (F) can be obtained from Table 12.13.

Table 12.12 Maximum Permitted Filling Density (Percent)

Specific Gravity at 15.6 °C	Above-Ground Vessels	
	Up to 5000 l	Over 5000 l
0.496–0.503	41	44
0.504–0.510	42	45
0.511–0.519	43	46
0.520–0.527	44	47
0.528–0.536	45	48
0.537–0.544	46	49
0.545–0.552	47	50
0.553–0.560	48	51
0.561–0.568	49	52
0.569–0.576	50	53
0.577–0.584	51	54
0.585–0.592	52	55
0.593–0.560	53	56

Table 12.13 Liquid Volume Correction Factors Specific Gravities at 15.6 °C/15.6 °C (60 °F/60 °F)

Observed Temperature °F	°C	0.500	Propane 0.5079	0.510	0.520	0.530	0.540	0.550	Iso-butane 0.5631	0.570	0.580	n-Butane	0.590
						Volume Correction Factors							
−50	−5.6	1.160	1.155	1.153	1.146	1.140	1.133	1.127	1.122	1.116	1.111	0.108	1.106
−45	−42.8	1.153	1.148	1.146	1.140	1.134	1.128	1.122	1.115	1.111	1.106	1.103	1.101
−40	−40.0	1.147	1.142	1.140	1.134	1.128	1.122	1.117	1.110	1.106	1.101	1.199	1.097
−35	−37.2	1.140	1.135	1.134	1.128	1.122	1.116	1.112	1.105	1.101	1.096	1.094	1.092
−30	−34.4	1.134	1.129	1.128	1.122	1.116	1.111	1.106	1.100	1.096	1.092	1.090	1.088
−25	−31.7	1.127	1.122	1.121	1.115	1.110	1.105	1.100	1.094	1.091	1.087	1.085	1.083
−20	−28.9	1.120	1.115	1.114	1.109	1.104	1.099	1.095	1.089	1.086	1.082	1.080	1.079
−15	−26.1	1.112	1.109	1.107	1.102	1.097	1.093	1.089	1.083	1.080	1.077	1.075	1.074
−10	−23.3	1.105	1.102	1.100	1.095	1.091	1.087	1.083	1.078	1.075	1.072	1.072	1.069
−5	−20.6	1.098	1.094	1.094	1.089	1.085	1.081	1.077	1.073	1.070	1.067	1.066	1.065
0	−17.8	1.092	1.088	1.088	1.084	1.080	1.076	1.073	1.068	1.066	1.063	1.062	1.061
2	−16.7	1.089	1.086	1.085	1.081	1.077	1.074	1.070	1.066	1.064	1.061	1.060	1.059
4	−15.6	1.086	1.083	1.082	1.079	1.075	1.071	1.068	1.064	1.062	1.059	1.058	1.057
6	−14.4	1.084	1.080	1.080	1.076	1.072	1.069	1.065	1.061	1.059	1.057	1.055	1.054
8	−13.3	1.081	1.078	1.077	1.074	1.070	1.066	1.063	1.059	1.057	1.055	1.053	1.052
10	−12.2	1.078	1.075	1.074	1.071	1.067	1.064	1.061	1.057	1.055	1.053	1.051	1.050
12	−11.1	1.075	1.072	1.071	1.068	1.064	1.061	1.059	1.055	1.053	1.051	1.049	1.048
14	−10.0	1.072	1.070	1.069	1.066	1.062	1.059	1.056	1.053	1.051	1.049	1.047	1.046
16	−8.9	1.070	1.067	1.066	1.063	1.060	1.056	1.054	1.050	1.048	1.046	1.045	1.044

18	−7.8	1.067	1.065	1.064	1.061	1.057	1.054	1.051	1.048	1.046	1.044	1.043	1.042
20	−6.7	1.064	1.062	1.061	1.058	1.054	1.051	1.049	1.046	1.044	1.042	1.041	1.040
22	−5.6	1.061	1.059	1.058	1.055	1.052	1.049	1.046	1.044	1.042	1.040	1.039	1.038
24	−4.4	1.058	1.056	1.055	1.052	1.049	1.046	1.044	1.042	1.040	1.038	1.037	1.036
26	−3.3	1.055	1.053	1.052	1.049	1.047	1.044	1.042	1.039	1.037	1.036	1.036	1.034
28	−2.2	1.052	1.059	1.049	1.047	1.044	1.041	1.039	1.037	1.035	1.034	1.034	1.032
30	−1.1	1.040	1.047	1.046	1.044	1.041	1.039	1.037	1.035	1.033	1.032	1.032	1.030
32	0.0	1.046	1.044	1.043	1.041	1.038	1.036	1.035	1.033	1.031	1.030	1.030	1.028
34	1.1	1.043	1.041	1.040	1.038	1.036	1.034	1.032	1.030	1.029	1.028	1.028	1.026
36	2.2	1.039	1.038	1.037	1.035	1.033	1.031	1.030	1.028	1.027	1.025	1.025	1.024
38	3.3	1.036	1.035	1.034	1.032	1.031	1.029	1.027	1.025	1.025	1.023	1.023	1.022
40	4.4	1.033	1.032	1.031	1.029	1.028	1.026	1.025	1.023	1.023	1.021	1.021	1.020
42	5.6	1.030	1.029	1.028	1.027	1.025	1.024	1.023	1.021	1.021	1.019	1.019	1.018
44	6.7	1.027	1.020	1.025	1.023	1.022	1.021	1.020	1.019	1.018	1.017	1.017	1.016
46	7.8	1.023	1.022	1.022	1.021	1.020	1.018	1.018	1.016	1.016	1.015	1.015	1.014
48	8.9	1.020	1.019	1.019	1.018	1.017	1.016	1.015	1.014	1.013	1.013	1.013	1.012
50	10.0	1.017	1.016	1.016	1.015	1.014	1.013	1.013	1.012	1.011	1.011	1.011	1.010
52	11.1	1.014	1.013	1.012	1.012	1.011	1.010	1.010	1.009	1.009	1.009	1.009	1.008
54	12.2	1.010	1.010	1.009	1.009	1.009	1.008	1.007	1.007	1.007	1.006	1.006	1.006
56	13.3	1.007	1.007	1.006	1.006	1.005	1.005	1.005	1.005	1.005	1.004	1.004	1.004
58	14.4	1.003	1.003	1.003	1.003	1.003	1.003	1.002	1.002	1.002	1.002	1.002	1.002
60	15.6	1.000	1.000	1.000	1.000	1.000	1.000	1.000	1.000	1.000	1.000	1.000	1.000
62	16.7	0.997	0.997	0.997	0.997	0.997	0.997	0.997	0.998	0.998	0.998	0.998	0.998
64	17.8	0.993	0.993	0.994	0.994	0.994	0.994	0.995	0.995	0.995	0.996	0.996	0.996
66	18.9	0.990	0.990	0.990	0.990	0.991	0.992	0.992	0.993	0.993	0.993	0.993	0.993
68	20.0	0.986	0.986	0.987	0.987	0.988	0.989	0.990	0.990	0.990	0.991	0.991	0.991
70	21.1	0.983	0.983	0.984	0.984	0.985	0.986	0.987	0.988	0.988	0.989	0.989	0.989
72	22.2	0.979	0.980	0.981	0.981	0.982	0.983	0.984	0.986	0.986	0.987	0.987	0.987
74	23.3	0.976	0.976	0.977	0.978	0.980	0.980	0.982	0.983	0.984	0.985	0.985	0.985
76	24.4	0.972	0.973	0.974	0.975	0.977	0.978	0.979	0.981	0.981	0.982	0.982	0.983
78	25.6	0.969	0.970	0.970	0.972	0.974	0.975	0.977	0.978	0.979	0.980	0.980	0.981
80	26.7	0.965	0.967	0.967	0.969	0.971	0.972	0.974	0.976	0.977	0.978	0.978	0.979

Continued

Table 12.13 Liquid Volume Correction Factors Specific Gravities at 15.6 °C/15.6 °C (60 °F/60 °F)—cont'd

Observed Temperature		0.500	Propane 0.5079	0.510	0.520	0.530	0.540	0.550	Iso-butane 0.5631	0.570	0.580	n-Butane 0.590
°F	°C	Volume Correction Factors										
82	27.8	0.961	0.963	0.963	0.966	0.968	0.969	0.971	0.973	0.974	0.976	0.977
84	28.9	0.957	0.959	0.960	0.962	0.965	0.966	0.968	0.971	0.972	0.974	0.975
86	30.0	0.954	0.956	0.956	0.959	0.961	0.964	0.966	0.968	0.969	0.971	0.972
88	31.1	0.950	0.952	0.953	0.955	0.958	0.961	0.963	0.966	0.967	0.969	0.970
90	32.2	0.946	0.949	0.949	0.952	0.955	0.958	0.960	0.963	0.964	0.967	0.968
92	33.3	0.942	0.945	0.946	0.949	0.952	0.955	0.957	0.960	0.962	0.964	0.966
94	34.4	0.938	0.941	0.942	0.946	0.949	0.952	0.954	0.958	0.959	0.962	0.964
96	35.6	0.935	0.938	0.939	0.942	0.946	0.949	0.952	0.955	0.957	0.960	0.961
98	36.7	0.931	0.934	0.935	0.939	0.943	0.946	0.949	0.953	0.954	0.957	0.959
100	37.8	0.927	0.930	0.932	0.936	0.940	0.943	0.946	0.950	0.952	0.954	0.957
105	40.6	0.917	0.920	0.923	0.927	0.931	0.935	0.939	0.943	0.946	0.949	0.951
110	43.3	0.907	0.911	0.913	0.918	0.923	0.927	0.932	0.937	0.939	0.943	0.946
115	46.1	0.897	0.902	0.904	0.909	0.915	0.920	0.925	0.930	0.933	0.937	0.940
120	48.9	0.887	0.892	0.894	0.900	0.907	0.912	0.918	0.924	0.927	0.931	0.934
125	51.7	0.876	0.881	0.884	0.890	0.898	0.903	0.909	0.916	0.920	0.925	0.928
130	54.4	0.865	0.871	0.873	0.880	0.888	0.895	0.901	0.909	0.913	0.918	0.923
135	57.2	0.854	0.861	0.863	0.871	0.879	0.887	0.894	0.902	0.907	0.912	0.916
140	60.0	0.842	0.850	0.852	0.861	0.870	0.879	0.886	0.895	0.900	0.905	0.910

12.15.4 Sample calculation for the practice engineers

As an example, the maximum volume of commercial propane (specific gravity $(G) = 0.51$) at 10 °C that may be charged into a storage vessel of 50 m^3 (water capacity at 15.6 °C) is: $D = 45$ from Table 12.12.

So, volumetric correction factor or $F = 1.016$ at 10 °C (from Table 12.13).
$W = 50$ m^3, so

$$V = \frac{DW}{100\gamma F} = \frac{(45)(50)}{(0.51)(1.016)(100)} = 43.42 \text{ m}^3 \tag{12.3}$$

Minimum required thickness of shell plate for pressure storage sphere shall be calculated per ASME "Pressure Vessel Code," Section 12.8.

12.16 Nozzles and connections

Nozzle and connection design for pressure storage spheres shall be in accordance with ASME Code Section 12.8, with the following supplementary requirements:

Attachments and shell openings, wherever practicable, shall be located so that the welds do not overlap shell seams or interfere with welds of other attachments.

All nozzles for piping shall be flanged and shall have the same P-number as the sphere shell. Corrosion protection equivalent to that of the sphere shall be provided.

Connections smaller than DN-50 (2 in) may be threaded, except for relief valve connections. Coupling shall be used for threaded connections. Additional coupling thickness shall be provided for corrosion allowance, as required. All couplings shall be installed with full penetration welds.

Nozzles and openings in the bottoms of pressure storage spheres shall be kept as minimum as possible. Connections to the storage sphere shall be positioned, as much as possible, above the maximum liquid level.

The design of nozzles pipe is governed by the following three main considerations:

- Ability to withstand design pressure. For this purpose the minimum thickness of nozzles shall be calculated in accordance with ASME Code Section 12.8 for cylindrical shells.
- Ability to withstand superimposed loading by connected pipe work or fittings. Not conforming to the minimum thickness, the nominal thickness of a nozzle intended for connection to external piping shall not be less than:
  - The value given in Table 12.14 increased by any required corrosion allowance.
  - The nominal (as-built) thickness of the main portion of the vessel shell where this is less than above.
- Suitability for the recommended forms of branch-to-shell attachment welds.

Threaded connections shall not be permitted for piping.

For nozzles fabricated from pipe, only seamless pipe shall be used.

Forged integral reinforced long-welding neck nozzles shall be used for flange rating of PN 100 (ANSI Class 600) and higher.

Table 12.14 Thickness of Branches

Branch Nominal Size		Minimum Thickness
DN	Inches	mm
15	0.5	2.4
20	0.75	2.4
25	1	2.7
32	1.25	3.1
40	1.5	3.1
50	2	3.6
65	2.5	3.9
80	3	4.7
100	4	5.4
125	5	5.4
150	6	6.2
200	8	6.9
250	10	8
300	12	8
350	14	8.8
400	16	8.8
450	18	8.8
500	20	10
600	24	10

Nozzles shall normally be designed as set-through types and be double welded, from both sides.

Necks of nozzles with sizes DN 350 (14 in) and larger and manholes may be made of plate materials. Where plate materials are used for nozzle and manhole necks, the method specified in ASTM A-672 shall be used and the material specification shall be the same as those used for pressure storage shells and full radiographic examination of longitudinal joints shall be conducted.

All nozzle flanges shall be welding-neck and raised-face-type smooth finish, except that manhole flanges may be slip-on flange type. Where slip-on flanges are used, they shall be welded both sides.

Nonstandard flanges shall be calculated per ASME code Section 12.8, Div. 1, according to the design conditions of the pressure storage sphere and external loads imposed by piping reaction.

A DN 500 (20 in nominal pipe size (NPS)) minimum manhole shall be provided in the top and bottom of each sphere. Generally, top manhole shall have a davit and bottom manhole shall have a hinge.

Minimum projection of nozzles and manholes shall normally be 150 mm for nozzle size up to DN 150 (6 in NPS), and 200 mm for DN 200 (8 in NPS) and over.

The necessity of a reinforcing pad on the shell around the nozzle opening shall be determined using the procedures outlined in ASME Code Section 12.8.

As a general rule, nozzles of DN 50 (2 in NPS) and over and manholes should be provided with a reinforcing pad and smaller-size nozzles shall be provided with a half-coupling type reinforcement except in case for nozzles fabricated from pipe.

12.17 Mountings

Pressure storage spheres, upon approval of the owner, shall be provided with the following:

- **Sample outlet**

Sample outlet shall be provided with double block and bleed valves located at places where they are convenient to the user.

- **Water draw-off**

1. If a water draw-off is specified for spheres that will operate in a freezing climate, if it is possible for water to accumulate and freeze in the bottom manhole, the draw-off pipe shall be extended down to within 75 mm of the manhole cover. A union shall be provided at the turn-down point above the manhole to facilitate access. If the sphere is to be used for storage of materials subject to polymerization or formation of peroxide, i.e., butadiene, isoprene, etc., the internal water draw-off piping shall be deleted and the draw-off connection shall be flushed with the bottom.
2. If water draw-off is specified for spheres in a non-freezing climatic location and the product will autorefrigerate below 0 °C on reduction to atmospheric pressure, the water draw-off may be located in the bottom of the sphere, bottom manhole cover plate, or on a bottom nozzle without internal riser, whichever location provides for all water removal. The water draw-off shall be at least DN 20 but shall not exceed DN 50. and shall be equipped with two valves at least 150 mm apart. The valve nearest the vessel shall be of the quick closing type, such as a plug valve (ball valves are not permitted unless certified "Fire-Safe").
3. The water draw-off line shall terminate at a minimum of 500 mm from the sphere.

- **Suction internal extension**

The hydrocarbon pump suction internal extension shall be 150 mm above the bottom tangent of the sphere or 150 mm above the water draw-off inlet, whichever is higher.

- **Instrument**

Instrument connections shall be provided as follows:

1. Pressure gauge (DN 15 coupling at the top of pressure storage sphere).
2. One internal float-type automatic ground reading level gauge (coupling and connections as required).
3. One differential-type local reading level gauge (two DN 15 couplings at the top and bottom of storage).

Table 12.15 Deluge Reed Pipe and Nozzle Dimensions

Tank Diameter, (m)	Deluge Feed Pipe Size (NPS) × Wall Thickness (mm)	Deluge Nozzle Welding Reducer Size (NPS) (mm)
13.5	Schd 40 80 × 5.49	80 × 40
>13.5–18	Schd 40 100 × 6.02	100 × 50
>18–28	Schd 40 150 × 7.11	150 × 80

4. One high-level alarm (two DN 40 couplings located at high liquid level).

5. Dial thermometer (DN 25 coupling located at low liquid level).

- **Cooling water spray and deluge system**

Pressure storage spheres shall be provided with a cooling water spray or deluge system. This will provide a minimum rate of 0.24–0.37 m^3/h/m^2. The water deluge shall cover the total surface above the maximum equator (Table 12.15).

For tanks larger than 26 m diameter the system shall be sized to deliver 0.37 m^3/h/m^2. Design of water spray and deluge systems shall be in accordance with Subsections 8.5.2 and 8.5.4 of API Standard 2510.

- **Pressure and vacuum relief valves**

1. Design of pressure safety relief and relief valves shall be in accordance with Parts UG-131, UG-133, UG-134, UG-135 and UG-136 of ASME Code Section 12.8 Div. 1. Shut-off valves, if required, shall be provided for all tank connections except safety valve connections and connections for thermometer wells.

- **The draw-off is operated as follows**
  1. To draw water, first open valve 2 wide and then throttle with valve 1.
  2. After water has been drawn, close valve 1 and open valve 3.
  3. Allow time to displace water from the draw-off line and then close valves 2 and 3.
  4. When water is not being drawn, all valves should remain closed.

12.18 Access facilities

Stairways and platforms shall be provided to allow access to operating valves and instruments. Auxiliary structures to service instruments, connections, and access openings at the bottom of the pressure storage sphere shall be provided if specified.

A stairway with handrail shall be provided from the ground to the top of the pressure storage sphere. Stairways shall have the following provisions:

1. Maximum angle with a horizontal line shall be 45°.

2. Minimum effective tread width shall be 200 mm.

3. Minimum effective width of stairways shall be 760 mm.

4. Stair landings shall not be less than 760 mm in the direction of the stairways.

Railing and toe plates shall enclose all attachments located on the top of the pressure storage sphere.

Steps and flooring plates shall be checkered type, unless otherwise specified. One drain hole of approximately 13 mm diameter shall be provided for every 1.5 m^2 of floor plate area. Holes shall be located and drilled after installation.

Supports shall be capable of supporting the sphere full of water.

Spheres, for capacities exceeding 400 m^3, shall have tubular steel leg supports welded to the shell.

Pressure storage spheres shall be grounded to provide protection against lightning.

Skirts and tubular leg supports of spheres shall be fireproofed up to the shell of the sphere irrespective of its height.

A rain deflector installed on top of the fireproofing shall prevent ingress of moisture.

12.19 Fabrication

Fabrication of pressure storage spheres shall meet the requirements of ASME Code Section 12.8. Permissible out-of-roundness for the shells shall be in accordance with the said stipulations.

All connections shall be prefabricated and welded to the shell plates, e.g., manholes, nozzles, supports, column stubs, and major structural attachments. These parts shall be post weld heat treated, if required, as a subassembly.

If the entire sphere is to be post weld heat treated, all gussets and lugs shall be welded before the heat treatment. Site erected spheres requiring full post-weld heat treatment shall have a bottom nozzle of sufficient size to introduce the heating equipment. The piping shall be connected with a reducer to this bottom nozzle.

Plate material that is cold formed shall be stress relieved if subjected to more than 5% strain at the surface during forming.

12.20 Insulation

For low-temperature storage spheres required to be insulated because of the nature of the product stored, sufficient insulation is required to minimize heat in leakage, to minimize condensation and icing effects. The requirements of this clause are to be regarded as minimal and the detailed design of the insulation system should be undertaken in cooperation with competent insulation engineers.

Before applying shell insulation, the sphere should have been satisfactorily tested and the surfaces to be insulated should be clean and free from rust and scale, and any specified painting completed.

During the period of application of shell insulation, the surfaces to be insulated should be kept dry. The work should be adequately protected against the weather and the ingress of water to the insulation should be prevented at all times.

The weatherproofing applied over shell insulation should be water and vapor tight, and care should be taken to ensure that damage to the insulation is avoided.

Reference

[1] GPSA. Engineering databook. 12th ed. Tulsa (OK, USA): Gas Processors and Suppliers Association; 2004.

Further reading

Abdel-Aal HK, Aggour M, Fahim MA. Petroleum and gas field processing. New York (Basel): Marcel Dekker Inc; 2003. 317–29.

Asian development bank. Project completion report on the LPG pipeline project (Loan 1591-IND) in India; 2003.

ASTM (American Society for Testing of Materials) ASTM-D-2163, ASTM-D-2420.

Bahnassi E, Khouri AR, Alderton P, Fleshman J. Achieving product specifications for ethane through to pentane plus from NGL fractionation plants. AIChE Fall Conference. Foster Wheeler; 2005.

Beychok MR. Aqueous wastes from petroleum and petrochemical plants. John Wiley & Sons, Ltd; 1967.

Brown GG, Souders M. Fundamental design of absorbing and stripping columns for complex vapors. Ind Eng Chem 1932;24:519.

Campbell JM. Gas conditioning and processing, vol. 2. Campbell Petroleum Series; 1978. 4.

Chen GK. Packed column internals. Chem Eng; March 5, 1984:40–51.

Chien HHY. A rigorous calculation method for the minimum stages in multicomponent distillation. Chem Eng Sci 1973;28:1967–74.

Chien HHY. A rigorous method for calculating minimum reflux rates in distillation. AIChE J; July 1978:24.

Eckert JS. Selecting the proper distillation column packing. Chem Eng Prog 1970;66(3):39.

Eckert JS. Tower packings comparative performance. Chem Eng Prog 1963;59(5):76–82.

Elvers BS. Handbook of fuels: energy sources for transportation. Wiley-VCH-Verlag GmbH & Co; 2008. p. 142–9.

Erbar JH, Maddox RN. Latest score: reflux vs. trays. Pet Refin 1961;40(5):183–8.

Fair JR, Bolles WL. Modern design of distillation columns. Chem Eng April 22, 1968;75(9):156–78.

Fair JR. What You need to design thermosiphon reboilers. Pet Refin 1960;39(2):105–23.

Fenske MR. Fractionation of straight-run Pennsylvania gasoline. Ind Eng Chem 1932;24:482–5.

Gillespie PC, Wilson GM. Vapor-liquid and liquid-liquid equilibria. GPA-RR-48; 1982.

Jamison RH. Internal design techniques. Chem Eng Prog 1969;65(3):46–51.

Katz DL, Cornell D, Kobayashi R, Poettman FH, Vary JA, Elenbaas JR, et al. Handbook of Natural Gas Engineering. New York: McGraw-Hill; 1959.

Kern DQ. Process heat transfer. McGraw-Hill; 1950. p. 453–91.

Kister H. Distillation design. McGraw-Hill; 1992 (Chapter 9).

Kister HZ. C. E. Refresher: column internals. Chem Eng; May 19, 1980:138–42. July 28, 1980, 79–83; September 8, 1980, 119–23; November 17, 1980, 283–5; December 29, 1980, 55–60; February 9, 1981, 107–9; April 6, 1981, 97–100.

Kremser A. Theoretical analysis of absorption process. Natl Pet News 1930;22(21):48.

O'Connell HE. Plate efficiency of fractionating columns and absorbers. Trans AIChE 1946;42:741–55.

Parkash S. Petroleum fuels manufacturing handbook: including specialty products and sustainable manufacturing techniques. McGraw-Hill Professional Publishing; 2009. p. 4–10.

Raaijmakers R. Offshore terminals for the transportation of liquefied petroleum gas. Bluewater Offshore Production Systems (USA), Inc.

Smith BD. Design of equilibrium stage processes. McGraw-Hill; 1963.

Underwood AJV. Fractional distillation of multicomponent mixtures. Chem Eng Prog 1948;44:603–14.

Van Winkle M. Distillation. McGraw-Hill; 1967. p. 480–645.

Vital T, Grossel S, Olsen P. Estimating separation efficiency. Hyd Proc 1984;63(11):75–8, 147–53.

Walker GJ. Design sour water strippers quickly. Hyd Proc; June 1969:121–4.

Liquefied Natural Gas (LNG)

13

Liquefied natural gas (LNG) is natural gas in a liquid form that is clear, colorless, odorless, noncorrosive, and nontoxic. It is made from natural gas and has many applications, including use as a fuel for power generation, industrial and home heating, and as a chemical feedstock.

LNG is formed when natural gas is cooled by a refrigeration process to temperatures of between −159 and −162 °C through a process known as liquefaction. During this process, the natural gas, which is primarily methane, is cooled below its boiling point, whereby certain concentrations of hydrocarbons, water, carbon dioxide, oxygen, and some sulfur compounds are either reduced or removed. LNG is also less than half the weight of water, so it will float if spilled on water.

Natural gas is transported by pipeline to its consumers, but when the distance between source and consumption is great ($\sim$1500 km by sea or 5000 km over land), then liquefaction of the gas to reduce its volume by a factor of 600 becomes economic. LNG is transported in double-hulled ships specifically designed to handle the low temperature of LNG. These carriers are insulated to limit the amount of LNG that evaporates.

LNG carriers are up to 305 m (1000 ft) long and require a minimum water depth of 12 m (40 ft) when fully loaded. LNG is about 47% as dense as water and is odorless, colorless, noncorrosive, and nontoxic. When vaporized it burns only in concentrations of 5–15% when mixed with air. Neither LNG nor its vapor can explode in an unconfined environment.

Generally, natural gas liquefaction plants consist of two main sections, pretreatment and liquefaction. In the pretreatment section, acidic gases (CO_2 and H_2S), water, mercury, and any other impurities that may solidify when natural gas is refrigerated are removed, and the liquefaction section removes sensible and latent heat from natural gas before it is expanded to atmospheric pressure.

As a liquid, LNG is not explosive. LNG vapor will only explode in an enclosed space within the flammable range of 5–15%.

Benefits of LNG in transportation applications:

- LNG is produced both worldwide and domestically at a relatively low cost and is cleaner burning than diesel fuel. Because LNG has a higher storage density, it is a more viable alternative to diesel fuel than compressed natural gas for heavy-duty vehicle applications.
- In addition, LNG in heavy-duty natural gas engines achieves significantly lower NO_x and particulate emission levels than diesel.

Natural gas entering a liquefaction plant must be pretreated to remove impurities such as water, acid gases (e.g., CO_2 and H_2S), and mercury to prevent freezing out in the process equipment, corrosion, and depositions on heat exchanger surfaces, and to control heating values in the final product.

Natural Gas Processing. http://dx.doi.org/10.1016/B978-0-08-099971-5.00013-1
Copyright © 2014 Elsevier Inc. All rights reserved.

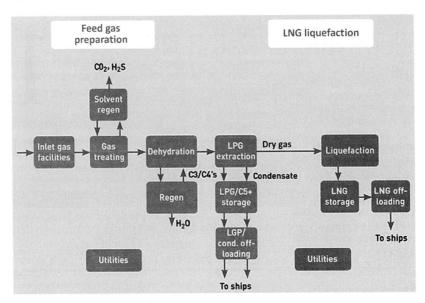

FIGURE 13.1

Gas to liquefied natural gas (LNG) block flow diagram.

Before the natural gas liquefaction process, the impurities, including carbon dioxide, sulfur compounds, heavier hydrocarbons, and water, should be eliminated by several processes (Figure 13.1). If nitrogen is present in the natural gas at high levels, it may be removed at the end of the process as it condenses at an even lower temperature than pure methane ($-196\,°C$).

When LNG is received at most terminals, it is transferred to insulated storage tanks specifically built to hold LNG. These tanks can be found above or below ground and keep the liquid at low temperature to avoid evaporation.

The liquefaction process can be designed to achieve the purify of the LNG to almost 100% methane, or to leave in more ethane and some liquefied petroleum gases (LPGs) (propane and butane) to match the pipeline gas specifications in the receiving gas system or destination.

Nitrogen is removed at end flash, whereas heavy hydrocarbons may be removed at the precooling stage because they are valuable products as natural gas liquids (NGLs), LPGs, and for refrigerant makeup. The compositions of natural gas suitable for the liquefaction process may contain a mixture of methane ($\sim$85–95%), lighter hydrocarbons, and small fraction of nitrogen.

The natural gas liquefaction processes convert pretreated natural gas into liquid suitable for transportation or storage. The LNG has a temperature of $-162\,°C$ at atmospheric pressure. The liquefaction of natural gas reduces the volume of natural gas by a factor of 600, thus enabling transportation in tanks on board specialized ship.

The process of turning gas into LNG, storing it, and loading it onto an LNG tanker requires many pump applications. In this arena, achieving low cost per ton of LNG and high reliability makes the product attractive, and improving economy of scale with larger plants and equipment continues to enhance the odds of winning. It is this reality that is driving the technology forward. At the same time, the entire industry is based on converting a gas that has to be compressed for transport into a liquid that can be pumped; pumps will continue to play an indispensable role. With a good understanding of the basic LNG chain and the forces driving innovation, the pump manufacturers and rotating equipment engineers will be able to anticipate and respond to industry demands in a manner that provides high value for their customers.

13.1 The LNG chain

The LNG chain starts with gas production, usually from offshore wells, although some plants receive gas from onshore sources. The gas produced can be from a gas field (nonassociated gas) or it may be produced along with oil (associated gas). The distinction between associated and nonassociated gas is important because associated gas must have LPG components (i.e., propane and butane) extracted to meet heating value specifications of the LNG product.

The produced gas enters the LNG liquefaction facility and goes through several steps of treating before being liquefied. The LNG leaving the liquefaction plant must be stored until a ship arrives to transport the product.

For a facility making 8 million metric tons per annum (MTPA), a 140,000 m^3 ship will arrive every 3 days. The ships are powered by steam engines and typically travel at 19 knots; thus, a round trip voyage of 8047 km (5000 miles) takes between 9 and 10 days of travel plus at least a day of turnaround at each end, for a total duration of 12 days. The time it takes to load a ship once the loading pumps are started is about 12–14 h.

13.2 The LNG liquefaction facility

The liquefaction facility is the greatest contributor to the LNG price at the receiving end, with the possible exception of shipping depending on distance to market. LNG plants produce LNG and condensate (natural gasoline) products, and in some cases LPGs (propane and butane).

The major pump services in the liquefaction unit are as follows:

- Amine circulation
- Reflux for scrub column and fractionation towers
- LNG product pumps
- Seawater pumps (if seawater cooled)
- Hot oil pumps

13.3 Liquefaction process

See the block flow diagram in Figure 13.1. The first step in the process is removal of acid gases such as CO_2 and H_2S. CO_2 would freeze at cryogenic process temperatures, and H_2S must be removed to meet the LNG product specifications. Typical specifications for acid gas removal are 50 ppm for CO_2, 4 ppm for H₂S, and total sulfur content <25 ppm. An amine solvent process is most common for acid gas removal. The process has an absorber tower where "lean" solvent contacts the natural gas and absorbs acid gas components, thus becoming "rich" solvent. The rich solvent leaves the bottom of the absorber and regenerates with a drop in pressure and heating in the stripper tower. The regenerated solvent is now "lean" again and cooled and pumped up to the absorber pressure.

13.4 LNG storage

There are three common types of LNG storage, known as "single containment," "double containment," and "full containment." In all cases, there is secondary containment in the event of a spill, and the differences between the types is mostly in the method of secondary containment. The single containment storage has a 9% nickel self-supporting inner tank and a carbon steel outer wall. There is perlite insulation between the two tanks. In the event of an inner tank leak, the outer wall fails because carbon steel is not capable of holding cryogenic materials. In this case, secondary containment is provided by a dike surrounding the tank.

The double containment tank has a posttensioned concrete outer wall capable of holding cryogenic materials, and no dike is needed because the outer wall provides the secondary containment. However, the cold vapors contacting the roof may cause the roof to fail, thus the containment is not "full containment" because vapors may be released in the event of an inner tank leak. The full containment tank is similar to double containment except that the roof is made of materials that can handle cryogenic temperatures, and if the inner tank leaks, all liquids and vapors are still contained within the outer wall and roof.

The main advantage of the single containment tank is the low cost relative to the other storage types. The main disadvantage is that the impoundment basin requires more land, and then providing enough distance between the dike and the plant fence to protect the public from heat and vapor dispersion requires even more land.

The LNG loading pumps are similar to the LNG product pumps in that they are submersed in the LNG, but instead of a separate container the pumps are inside pump columns that extend to the tank roof. The key design feature of this pumping system is that it is possible to pull the pump for maintenance while continuing to operate the storage tank. There is a foot valve at the bottom of the column that prevents LNG from entering the column when the pump is pulled. The operators purge the column with nitrogen and then remove the pump from the top of the column.

13.5 In-tank pump process objectives

The LNG loading pump capacities are usually based on filling a ship in 12 h. The liquefaction plant typically has multiple storage tanks, and two to four pumps per tank. It is common to have a total of eight pumps running during loading, each with a capacity in the 1200–2000 m³/h range and 150–160 m of head. In many plants, there is also a smaller pump in each tank in addition to the loading pumps. The purpose of this smaller pump is to recirculate LNG in the loading lines when no ship is present. The loading lines are large diameter 61–91 cm (24–36 in) and must be kept cold between ship loadings because cooling them down is a long procedure.

13.6 LNG shipping

LNG terminal layout and site selection are typically based on the following ship parameters:

- 130,000–135,000 m³ capacity, having an overall length of up to 310 m, width of 46 m, and fully loaded draft of 11.6 m. The net delivery unloading rate into the receiving terminal is approximately 10,000 m³/h. There are smaller ships (down to <60,000 m³), but the industry trend is toward larger ship sizes with designs on the drawing board for up to 250,000 m³.
- 15 m minimum water depth.

The LNG ships have two different types of pumps: the large cargo pumps for transferring LNG and the small spray pumps that provide LNG for the spray ring that helps keep the entire storage container in a cool state. The storage on the ship is usually one of two types, either self-supporting aluminum spheres or stainless-steel membrane compartments supported by the ship hull. There are either four or five spheres or compartments, and each contains two cargo pumps and one spray pump. The cargo pumps usually have a capacity of 1200–1400 m³/h and the spray pumps have a capacity of 40–50 m³/h.

13.6.1 Natural gas liquefaction options

Table 13.1 shows various natural gas liquefaction options.

13.7 Liquefaction and refrigeration

Refrigeration systems are common in the natural gas processing industry and processes related to the petroleum refining, petrochemical, and chemical industries. Several applications for refrigeration include NGL recovery, LPG recovery, hydrocarbon dew point control, reflux condensation for light hydrocarbon fractionators, and LNG plants.

Selection of a refrigerant is generally based upon temperature requirements, availability, economics, and previous experience. For example, in a natural gas processing plant, ethane and propane may be at hand, whereas in an olefins plant, ethylene and propylene are readily available. Propane or propylene may not be suitable in an ammonia plant because of the risk of contamination, whereas ammonia may very well serve the purpose. Halocarbons have been used extensively because of their nonflammable characteristics.

13.7.1 Refrigeration cycle

The refrigeration effect can be achieved by using one of these cycles:

- Vapor compression-expansion
- Absorption
- Steam jet (water-vapor compression)

By using the pressure–enthalpy (P-H) diagram, the refrigeration cycle can be broken down into four distinct steps:

Table 13.1 Natural Gas Liquefaction Options

Liquefier Type	Operating Principle	Remarks and Trade-offs
Precooled Joule–Thomson (J-T) cycle	A closed-cycle refrigerator (e.g., using Freon or propane) precools compressed natural gas, which is then partially liquefied during expansion through a J-T valve.	Relatively simple and robust cycle, but efficiency is not high.
Nitrogen refrigeration cycle (also called closed Brayton/ Claude cycle)	Nitrogen is the working fluid in a closed-cycle refrigerator with a compressor, turboexpander, and heat exchanger. Natural gas is cooled and liquefied in the heat exchanger.	Simple and robust cycle with relatively low efficiency. Using multiple refrigeration stages can increase efficiency. Used in CryoFuel systems.
Cascade cycle	Many closed-cycle refrigerators (e.g., using propane, ethylene, methane) operating in series sequentially cool and liquefy natural gas. More complex cascades use more stages to minimize heat transfer irreversibility.	High-efficiency cycle, especially with many cascade steps. Relatively expensive liquefier due to need for multiple compressors and heat exchangers. Cascade cycles of various designs are used in many large-capacity peak-shaving and LNG export plants.
Mixed refrigerant cycle (MRC)	Closed-cycle refrigerator with multiple stages of expansion valves, phase separators, and heat exchanger. One working fluid, which is a mixture of refrigerants, provides a variable boiling temperature. Cools and liquefies natural gas with minimum heat transfer irreversibilities, similar to cascade cycle.	High-efficiency cycle that can provide lower cost than conventional cascade because only one compressor is needed. Many variations on MRC are used for medium and large liquefaction plants.
Open cycles with turboexpander, Claude cycle	Classic open Claude cycle uses near-isentropic turboexpander to cool compressed natural gas stream, followed by near-isenthalpic expansion through J-T valve to partially liquefy gas stream.	Open cycle uses no refrigerants other than natural gas. Many variations, including Haylandt cycle used for air liquefaction. Efficiency increases for more complex cycle variations.

Turboexpander at gas pressure drop	Special application of turboexpander at locations (e.g., pipeline city gate), where high-pressure natural gas is received and low-pressure gas is sent out (e.g., to distribution lines). By expanding the gas through a turboexpander, a fraction can be liquefied with little or no compression power investment.	This design has been applied for peak-shaving liquefiers. Very high or "infinite" efficiency, but special circumstances must exist to use this design.
Stirling cycle (Phillips refrigerator)	Cold gas (usually helium closed-cycle using regenerative heat exchangers and gas displacer to provide refrigeration to cryogenic temperatures. Can be used in conjunction with heat exchanger to liquefy methane.	Very small-capacity Stirling refrigerators. These units have been considered for small-scale LNG transportation fuel production.
TADOPTR	TADOPTR, thermo acoustic driver orifice pulse tube refrigerator. Device applies heat to maintain standing wave that drives working fluid through Stirling-like cycle. No moving parts.	Currently being developed for liquefaction applications including LNG transportation fuel production. Progressing from small-scale to field-scale demonstration stage.
Liquid nitrogen open-cycle evaporation	Liquid nitrogen stored in dewar is boiled and superheated in heat exchanger, and warmed nitrogen is discharged to atmosphere. Counter flowing natural gas is cooled and liquefied in heat exchanger.	Extremely simple device has been used to liquefy small quantities of natural gas. More than one pound of liquid nitrogen is required to liquefy one pound of natural gas. Nitrogen is harmless to atmosphere. Economics depends on price paid for liquid nitrogen.

FIGURE13.2

Process flow diagram and pressure–enthalpy (P-H) diagram (GPSA, 2004).

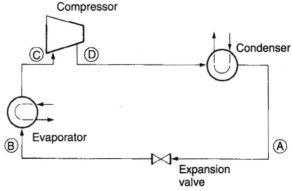

Process flow diagram

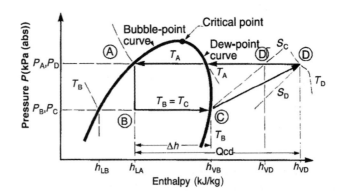

1. Expansion
2. Evaporation
3. Compression
4. Condensation

The vapor-compression refrigeration cycle can be represented by the process flow and P-H diagram shown in Figure 13.2.

13.7.2 Expansion step

The starting point in a refrigeration cycle is the availability of liquid refrigerant. Point A in Figure 13.2 represents a bubble point liquid at its saturation pressure, P_A, and enthalpy, h_{LA}. In the expansion step, the pressure and temperature are reduced by flashing the liquid through a control valve to pressure P_B. The lower pressure, P_B, is determined by the desired refrigerant temperature, T_B (point B).

At point B, the enthalpy of the saturated liquid is h_{LB}, whereas the corresponding saturated vapor enthalpy is h_{VB}. Because the expansion step (A–B) occurs across an expansion valve and no energy has been exchanged, the process is considered to be isenthalpic. Thus, the total stream enthalpy at the outlet of the valve is the same as the inlet, h_{LA}. Because point B is inside the envelope, vapor and liquid coexist.

To determine the amount of vapor formed in the expansion process, let X be the fraction of liquid at pressure P_B with an enthalpy h_{LB}. The fraction of vapor formed during the expansion process with an enthalpy h_{VB} is $(1 - X)$. Equations for the heat balance and the fraction of liquid formed are as follows:

$$(X)h_{LB} + (1 - X)h_{VB} = h_{LA} \tag{13.1}$$

$$X = \frac{(h_{VB} - h_{LA})}{(h_{VB} - h_{LB})} \tag{13.2}$$

$$(1 - X) = \frac{(h_{LA} - h_{LB})}{(h_{VB} - h_{LB})} \tag{13.3}$$

where:

X = weight fraction
h = enthalpy, kJ/kg
A, B, C, D = denote unique points of operation on P-H diagrams
L = liquid state
V = vapor state

13.7.3 Evaporation step

The vapor formed in the expansion process (A–B) does not provide any refrigeration to the process. Heat is absorbed from the process by the evaporation of the liquid portion of the refrigerant. As shown in Figure 13.2, this is a constant temperature, constant pressure step (B–C). The enthalpy of the vapor at point C is h_{VB}. Physically, the evaporation takes place in a heat exchanger referred to as an evaporator or a chiller. The process refrigeration is provided by the cold liquid, X, and its refrigerant effect can be defined as $X (h_{VB} - h_{LB})$ and substituting from Eqn (13.2), the effect becomes

$$\text{Effect} = h_{VB} - h_{LA} \tag{13.4}$$

The refrigeration duty (or refrigeration capacity) refers to the total amount of heat absorbed in the chiller by the process, generally expressed as kW. The refrigerant flow rate is given by

$$m = \frac{Q_{ref}}{(h_{VB} - h_{LA})} \tag{13.5}$$

where:

m = refrigerant flow, kW
Q = heat duty, kW
ref = refrigeration
h = enthalpy, kJ/kg

A, B, = denote unique points of operation on P-H diagrams
L = liquid state
V = vapor state

13.7.4 Compression step

The refrigerant vapors leave the chiller at the saturation pressure P_C. The corresponding temperature equals T_C at an enthalpy of h_{VB}. The entropy at this point is S_C. These vapors are compressed isentropically to pressure PA along line $C - D'$ (Figure 14.2). The isentropic (ideal) work, W_i, for compressing the refrigerant from P_B to P_A is given by:

$$W_i = m\left(h'_{VD} - h_{VB}\right) \tag{13.6}$$

where:

W = work of compression, kW
m = refrigerant flow, kW
h = enthalpy, kJ/kg
h'_{VD} = isentropic enthalpy, kJ/kg
i = isentropic
B, D = denote unique points of operation on P-H diagrams
V = vapor state

The quantity h'_{VD} is determined from refrigerant properties at P_A and an entropy of S_C. Because the refrigerant is not an ideal fluid and because the compressors for such services do not operate ideally, isentropic efficiency, i, has been defined to compensate for the inefficiencies of the compression process. The actual work of compression, W, can be calculated from

$$W = \frac{W_i}{\eta_i} = \frac{m(h_{VD} - h_{VB})}{\eta_i} = m(h_{VD} - h_{VB}) \tag{13.7}$$

where:

h = enthalpy, kJ/kg
i = isentropic
m = refrigerant flow, kW
W = work of compression, kW
B, D = denote unique points of operation on P-H diagrams
V = vapor state
η = isentropic efficiency

The enthalpy at discharge is given by

$$h_{VD} = \frac{\left(h'_{VD} - h_{VB}\right)}{\eta_i} + h_{VB} \tag{13.8}$$

The work of compression can also be expressed as

$$BP = \frac{W}{3600} \tag{13.9}$$

where:

BP = Brake power, kW
W = work of compression, kW
3600 kJ/h = 1 kW

13.7.5 Condensation step

The superheated refrigerant leaving the compressor at P_A and T_D (point D in Figure 13.2) is cooled at nearly constant pressure to the dew point temperature, T_A, and refrigerant vapors begin to condense at constant temperature.

During the desuperheating and condensation process, all heat and work added to the refrigerant during the evaporation and compression processes must be removed so that the cycle can be completed by reaching point A (the starting point) on the P-H diagram, as shown in Figure 13.2.

By adding the refrigeration duty to the heat of compression, we calculate the condensing duty, Q_{cd}, from

$$Q_{cd} = m[(h_{VB} - h_{VA}) + (h_{VD} - h_{VB})] = m(h_{VD} - h_{LA}) \tag{13.10}$$

where:

m = refrigerant flow, kW
Q = heat duty, kW
cd = condenser
ref = refrigeration
h = enthalpy, kJ/kg
A, B, D = denote unique points of operation on P-H diagrams
L = liquid state
V = vapor state

The condensing pressure of the refrigerant is a function of the cooling medium available—air, cooling water, or another refrigerant. The cooling medium is the heat sink for the refrigeration cycle.

Because the compressor discharge vapor is superheated, the refrigerant condensing curve is not a straight line. It is a combination of desuperheating and constant temperature condensing. This fact must be considered for proper design of the condenser.

Natural gas has a critical point at −80 to −900 °C, and its liquefaction cannot be done by pressurization and expansion alone.

Liquefaction of gas is done by refrigerating the gas to the temperature below its critical conditions.

The process requires a refrigeration cycle that removes energy from natural gas in the form of sensible and latent heat. Selection of best refrigeration cycle for liquefaction of natural gas can be done after thorough study of local conditions.

The thermodynamic principles of liquefaction and refrigeration process are quite similar, but the designing of the two systems are different. The refrigeration process/cycle for liquefaction of natural gas involves some equipment in which refrigerant is compressed, cooled to reject heat at ambient conditions, and expanded to produce refrigerant capacity required. In refrigeration cycles that operate as close loop, refrigerant is constantly circulating as working fluid and there is no accumulation or

FIGURE 13.3

Flow diagram for single cycle refrigeration system.

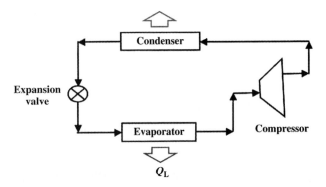

withdraw of refrigerant from the cycle. The diagram showing a simple refrigeration circuit is given in Figure 13.3. The system comprises of four components: evaporator, compressor, condenser, and throttling valve.

The temperature of the evaporator is usually near the normal boiling point of the refrigerant, so the pressure of the evaporator may be approximately atmospheric. The throttling valve maintains a pressure difference between the higher and lower side of the refrigeration cycle. Note that the work supplied to the refrigeration cycle increases with the temperature lift (difference from evaporating to condensing temperature).

The condensing pressure must be higher enough to make refrigerant condense at ambient conditions using water or air. Ambient temperature must be less than the critical temperature of the refrigerant to effect condensation using the environment as a coolant.

The refrigerant is in closed circuit and circulated by compressor. By keeping the pressure of refrigerant low in evaporator, the refrigerant boil by absorbing heat from the fluid to be cooled and at the same time it continues to remove the vaporized refrigerant and compress it to the condensing pressure.

13.7.6 The effect of natural gas pressure on liquefaction processes

Temperature–entropy (T-S) diagram of natural gas mixture with phase envelope in black and lines of constant pressure in blue is presented in Figure 13.4 below.

From the diagram it can observed that if natural gas is liquefied at low pressure, work (area W) is increased and also there is some increase on amount of heat required to be removed from natural gas (area Q). Thermodynamically, it is useful to liquefy natural gas at the highest possible pressure so that work can be saved and reduce the heat load. Practically, there are some constrains; for example, it is required that natural gas be below the critical pressure to obtain liquid/gas separation in the heavy hydrocarbon removal column. Another limitation is equipment design pressure, e.g., heat exchangers hence natural gas pressure should be within heat exchanger-designed pressure.

A block diagram showing some of the key sections of an LNG Plant is presented in Figure 13.5 below. The sections include gas reception (slug catcher), pretreatment (acid gas removal, molecular sieve dehydration, and mercury removal), and liquefaction sections.

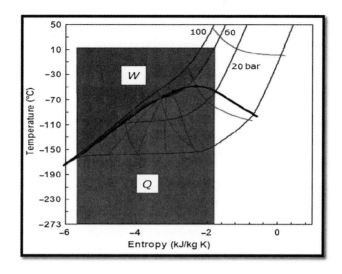

FIGURE 13.4

Temperature–entropy diagram of natural gas with area showing heat removed (Q) and ideal work for reversible liquefaction process (W).

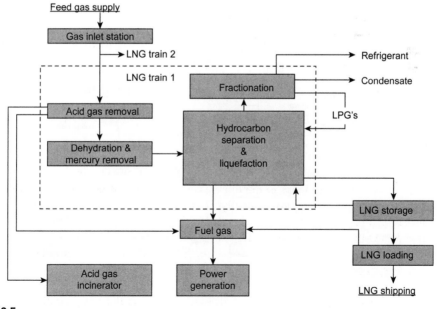

FIGURE 13.5

Principal block diagram for liquefied natural gas (LNG) plant.

A typical LNG plant is comprised of the following units:

- Feed gas compression, in case the natural gas pressure is low
- CO_2 removal, mostly by a wash process and drying or H_2O removal by an adsorber (CO_2 and H_2O would otherwise freeze and cause clogging in the downstream liquefaction equipment)
- Natural gas liquefaction
- LNG storage
- LNG loading stations
- LNG metering stations

As well as the following utilities:

- An MRC make-up and boil-off gas handling system
- Gas turbine with waste heat recovery for hot oil heating
- Other utilities

Figure 13.6 shows LNG plant block scheme.

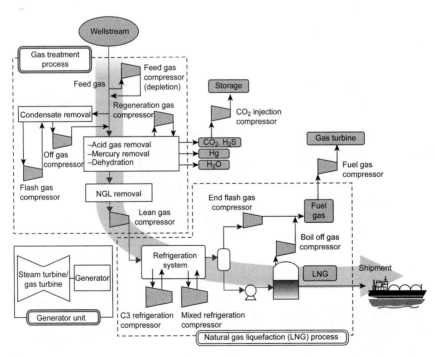

FIGURE 13.6

Liquefied natural gas (LNG) plant block scheme.

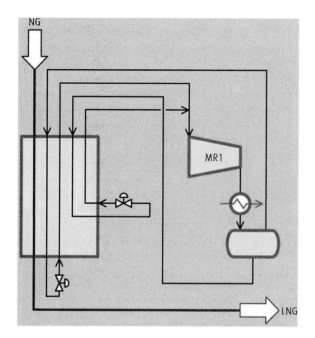

FIGURE 13.7

Basic single flow liquefied natural gas (LNG) process.

13.8 Basic single flow LNG process

The basic single flow LNG process (Figure 13.7) consists of the following:

- A plate-fin heat exchanger set in a cold box, where the natural gas is cooled to LNG temperatures by a single MRC.
- A separation vessel, where the mixed refrigerant (MR) is separated into a liquid fraction. The liquid fraction and a gas fraction provides the cold temperature after expansion in a J-T valve for the natural gas precooling and liquefaction.
- A gas reaction that provides the LNG subcooling temperature after condensation and J-T expansion at the bottom of the heat exchanger.
- Recompression of the cycle gas streams leaving the heat exchanger in the turbo compressor.
- Cooling of the compressed cycle gas against air or water.

13.9 Multistage MR process

The multistage MR process (Figure 13.8) is comprised of the following:

- A coil-wound heat exchanger (CWHE) where the natural gas is precooled, liquefied, and subcooled against various fractions of a single MRC.
- A medium-pressure refrigerant separator, from which the liquid is used to provide the precooling duty after J-T expansion to the lower section of the CWHE.

FIGURE 13.8

Multistage mixed refrigerant process.

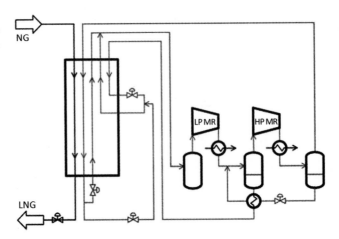

- A high-pressure refrigerant separator, from which the gas is cooled and partially condensed in the lower section of the CWHE.
- A low-temperature refrigerant separator, from which the liquid is used to provide the natural gas liquefaction duty after J-T expansion. The gaseous refrigerant stream from this separator is used to provide the subcooling duty after condensation and J-T expansion in the upper section of the CWHE.
- The combined refrigerant cycle stream from the bottom of the CWHE is compressed in a two stage compressor with intercooling and aftercooling against air or water.

13.10 Mixed fluid cascade process

The mixed fluid cascade (MFC) process (Figure 13.9) is highly efficient due to the low shaft power consumption of the three MRC compressors. The process is comprised of the following:

- Plate-fin heat exchangers for natural gas precooling.
- CWHEs for the natural gas liquefaction and LNG subcooling.
- Three separate MRCs, each with different compositions, that result in minimum compressor shaft power requirement.
- Three cold suction turbo compressors. Up to 12 MTPA LNG can be produced in a single train.

13.11 Classification of natural gas liquefaction processes

The processes for liquefying natural gas can be classified into three groups: cascade process, MR liquefaction process, and expander or turbine-based process. The classification of the mentioned natural gas liquefaction processes is shown in Figure 13.10.

The cascade process operates with the pure-component refrigerants methane, ethylene, and propane. There are few LNG plants in the world operating on the cascade process. The disadvantage of the

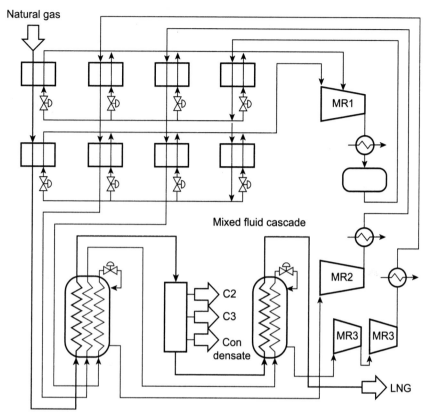

FIGURE 13.9

Mixed fluid cascade process.

cascade cycle/group is its relatively high capital cost due to the number of refrigeration compression circuits that each require their own compressor and refrigerant storage. Costs for maintenance and spares tend to be comparatively high. The main advantage of the cascade cycle is its lower power requirement compared with other liquefaction cycles, mainly because the flow of refrigerant is lower than the flow in other cycles.

It is also flexible in operation because each refrigerant circuit can be controlled separately. The mean temperature differences between the composite curves are wide relative to those of the MRC. Economies of scale dictate that the cascade cycle is most suited to very large train capacities where the low heat exchanger area and low power requirement offset the cost of having multiple machines.

The MRC uses a single MR instead of multiple pure refrigerants as the cascade cycle. The MR normally consists of nitrogen, ethane, propane, butane, and pentane. Such a mixture evaporates over a temperature trajectory instead of at a constant evaporating point, and this has large benefits for the total process.

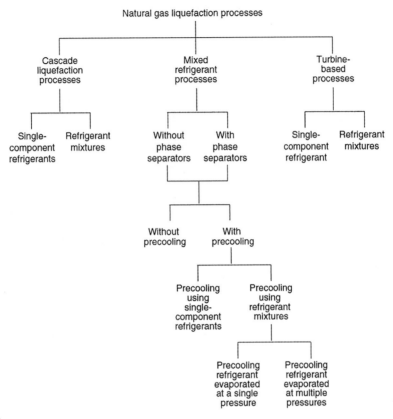

FIGURE 13.10

Classification of natural gas liquefaction processes.

The refrigeration effect will be distributed over a range of temperatures and accordingly the overall temperature difference between the natural gas and MR is small. Small driving temperature differences give operation nearer to reversibility; leading to a higher thermodynamic efficiency.

Simultaneously, the power requirement will be lower and the entire machinery smaller. Some of MRC technologies are as follows:

- Mixed fluid cascade (MFC) process,
- Single mixed refrigerant (SMR) process,
- Mixed refrigerant with propane precooling (C3MR),
- Double mixed refrigerant (DMR) process,
- AP-X and small-scale MRC.

13.11.1 MFC process

The MFC process consists of three pure refrigerants that have different boiling temperatures, such as methane, ethylene, and propane. First, natural gas is cooled to −35 °C in the propane cycle, and

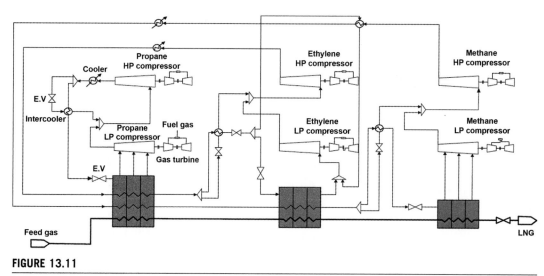

FIGURE 13.11

Schematic diagram of cascade process using liquid–gas heat exchanger.

then it is cooled to $-90\,°C$ in the ethylene cycle. Finally, it is liquefied to $-155\,°C$ in the methane cycle.

- **Cascade process with two-staged intercooler**

In this process, two-staged compression with an intercooler type is applied on the basic cascade process. Liquefied refrigerant from the condenser is bypassed and evaporated in the intercooler after being expanded. Therefore, the main refrigerant to evaporator is subcooled.

These processes involve liquid–gas heat exchangers being applied between the two cycles. One of these processes is that subcooled liquid refrigerant that is bypassed from the intercooler in the ethylene cycle and hot gaseous refrigerant from the outlet of the high-pressure compressor in the methane cycle are exchanged; the heat in the liquid–gas heat exchanger is used to liquefy the hot gaseous methane. The other process is that one more liquid–gas heat exchanger is applied between the propane and ethylene cycle in the process. Figure 13.11 shows the MFC process.

13.11.2 Single mixture process

A low-cost, highly efficient SMR process using simple, proven, and low-risk technology may benefit many future LNG projects. The process illustrated in Figure 13.12 is based on a simple, single MR cycle, but the performance is significantly enhanced by the addition of conventional combined heat and power technology and conventional industrial ammonia refrigeration.

The improved process is called the optimized single mixed refrigerant (OSMR) process. The heart of the process is a very simple single MRC that consists of a suction scrubber, compressor, aftercooler, and cold box. It uses a standard single-stage centrifugal compressor that does not require a gear box, helper motor, or interstage components as do most other LNG plants.

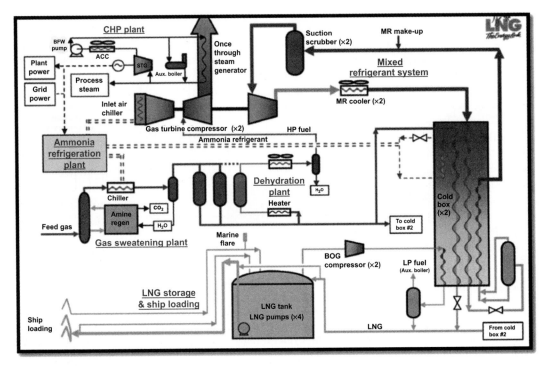

FIGURE 13.12

Optimized single mixed refrigerant (OSMR) process.

- **Process description**

Feed gas enters the LNG plant where it is sweetened in a conventional amine plant by using methyl diethanolamine to remove CO_2 and H_2S. The warm saturated gas exiting the amine contactor is cooled using ammonia refrigerant to remove the bulk of the water before being dehydrated in a conventional molecular sieve plant.

The removal of bulk water is needed to reduce the size of the dehydration plant and to allow regeneration to occur at high pressure, thus avoiding compression of fuel gas consumed by the main turbines as is normally needed in other processes.

Bulk water removal also ensures that the regeneration gas quantity is less than the fuel gas demand for the gas turbines, so no recycle compressor is required. Additional make-up fuel for the gas turbines is provided by the dry gas stream. Steam generated from gas turbine waste heat is used as the "free" heat source for the amine reboiler, molecular sieve regeneration heater, and fuel gas superheater. Prefabricated packaged sweetening and dehydration units can be used where it is desirable to reduce on-site work in remote or high-cost locations.

- **Liquefaction and boil-off gas**

Sweet dry gas enters the cold box where it is liquefied at high pressure in parallel brazed aluminum heat exchangers. The LNG exits the bottom of the cold box and flows to the LNG tank where it flashes to low pressure. No flash vessel or LNG pumps are needed. The flashed vapor and boil-off gas (BOG) are recovered from the LNG tank by two identical high-efficiency, two-stage integrally geared BOG compressors. Only one compressor operates during normal operation; the second unit is started during ship loading. LNG is sprayed into the vapor return line from the ship during loading to maintain constant vapor temperature entering the LNG tank and therefore constant suction (-150 C) and constant discharge ($-60\,°C$) temperature on the BOG compressors.

The BOG and flash vapor are compressed to only 7 bar and returned to the cold box where they are substantially reliquefied. The reliquefied BOG is separated and liquid methane returns to the LNG tank. Nitrogen concentrates in the flash gas that is used as a low British thermal unit (BTU) fuel gas for the auxiliary boiler, so this system also acts as a very effective nitrogen rejection unit. Only a small portion of the cold box refrigeration capacity is used for BOG reliquefaction; however, to avoid flaring of BOG during ship loading, the feed gas quantity to the cold box is reduced slightly. Other LNG processes require much larger high-pressure cryogenic compressors to enable the BOG to be used as gas turbine fuel gas. Flaring is also commonplace with other LNG processes during ship loading.

- **Refrigeration**

Refrigeration for the cold box is principally provided by the single MR supplemented by ammonia refrigeration at the warm end (top) of the cold box. The ammonia refrigeration plant is powered by "free waste energy" generated by the combined heat and power (CHP) plant. The sizing of the ammonia refrigeration plant is based on the spare power available from the CHP plant after all other heat and power users in the plant have been met. This ensures optimum use and balance of all available energy.

Ammonia is the most commonly used natural industrial refrigerant in the world, so safe practices are well established. A conventional ammonia refrigeration plant is used to enhance the LNG process, and this is the same type of plant that is used in thousands of applications.

The ammonia refrigeration uses a conventional industrial refrigeration process comprising motor-driven screw compressors, condensers, separator vessels, pumps, pipework, instrumentation, and control system. Alternatively, a single direct steam turbine-driven integrally geared centrifugal compressor can be used. A comparison with propane refrigeration for this application revealed substantial advantages by using ammonia, including efficiency, cost, and safety.

The ammonia refrigerant is first applied to cooling wet gas from the amine contactor and then applied to cooling inlet air to the gas turbines to increase power. The remainder is used in the cold box for precooling the MR. The result is a substantial increase in plant capacity and a substantial improvement in fuel efficiency. As an added bonus, pure water is condensed and produced when gas turbine inlet air is cooled with ammonia, and this is more than enough to feed the demineralized water plant.

Cooling required for the MR cooler, ammonia condenser, and steam condenser can be provided by water (cooling tower or once through cooling) or by air or by a combination to the two.

Several key differences between the OSMR process and the traditional processes are as follows:

- Gas turbine waste heat recovery to produce power. This has been used in the power industry for decades and poses no technical challenges or risk.

- Gas turbine inlet air cooling using ammonia. Direct cooling of gas turbine inlet air is successfully used in many power plants and has technical and economic advantages over the more traditional method using chilled water.
- Precooling of MR using ammonia. Because the cold box is a very simple design with minimal streams, the addition of ammonia to cool the MR from ambient temperature down to around 0 °C only is not technically challenging. So, these low-risk but important differences have been proven in other industries and are combined in the OSMR process to generate substantial improvements in performance. This is achieved by using fewer equipment items and a simpler configuration than that used in traditional LNG processes.

Although the single MR system is highly integrated, which it needs to be to achieve high efficiency, the overall plant availability exceeds 96%, compared with 90–92% for traditional processes. This is mainly due to the fact that if one gas turbine is down for maintenance, the plant will still run at half capacity. Also, if an ammonia compressor fails, the plant capacity simply reduces slightly.

This is far better than traditional technology such as propane-MR where the loss of one gas turbine causes a total train shutdown.

The single MR system consists of only four components:

1. compressor,
2. MR cooler,
3. cold box,
4. suction scrubber.

The gas turbine efficiency is approximately 40% (net heating value), a value that is ∼20% better than the GE Frame 7 or Frame 9 gas turbines used in conventional modern large-scale propane-MR processes. Aero-derivative gas turbines are used, and these are well proven in mechanical drive applications. The Darwin LNG project uses these turbines; however, there are many applications of aero-derivative gas turbines driving compressors in far more arduous applications than required for an LNG process (e.g., high-pressure gas injection offshore). Because this is a routine and conventional application for the compressor, a full load string test is considered unnecessary.

The compressor is a standard single-stage barrel/bundle type centrifugal with a polytropic efficiency of >86%. The compressor is directly coupled to a standard mechanical drive aero-derivative gas turbine package. No gearbox, no helper motor, no interstage cooler, no interstage scrubber, no discharge separator, no liquid mixed refrigerant pumps, and no mixed refrigerant metering is required, as they are for other processes. The MR comprises four components only, namely, methane, ethane, butane, and nitrogen. The refrigerant composition, flow rate, and pressures are carefully selected to provide an excellent match of the composite cooling curves and to allow major equipment such as the cold box to be economically sized.

- **Cold box**

The cold boxes for this capacity plant typically comprise six parallel cores manifolded together plus a common mixed refrigerant separator vessel. Only five streams are required within each core so the cold box configuration is very simple compared with alternative LNG processes and typical ethylene processes. The differential temperatures between streams and resulting thermal stresses inside the cores are within the limits required by the standards and comply with the heat exchanger manufacturer's requirements under all operating conditions. Start-up (including cool down) and shutdown procedures and control systems ensure thermal stresses are kept within limits during all operating conditions, including process upsets. The ammonia cools the high-pressure MR stream and ensures the MR suction temperature is low so that the compressor performance is much improved. Ammonia is the most commonly used natural industrial refrigerant in the world, so safe practices are well established.

- **Combined heat and power**

Proven CHP technology is used to recover the waste heat from the gas turbine so that all the heat and electric power requirements for the plant are met, including all power for the ammonia refrigeration system. Steam is generated via once through steam generators (OTSGs) that power a single pressure steam turbine generator as well as supply the required quality of steam to various process heat users.

Approximately half of the electric power is used for the ammonia compressor drives, and the remainder is consumed by various plant users. OTSGs are used to simplify the steam system design, again reducing the number of equipment items. No bypass stack or diverter damper is required, so a gas turbine(s) can continue to run and produce LNG even if the OTSG(s) is not operating.

13.11.3 MCR cycle

About 77% of LNG plants are using the propane precooled MCR cycle licensed by Air Products and Chemicals, Inc. for natural gas liquefaction, as is illustrated in Figure 13.13.

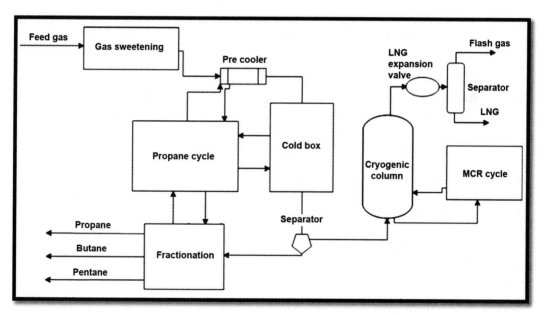

FIGURE 13.13

Schematic illustration of a typical liquefied natural gas (LNG) production process.

From Ref. [1].

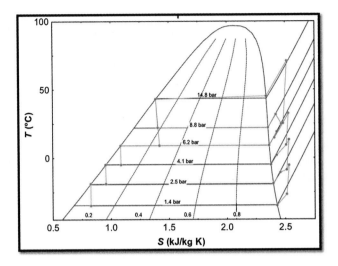

FIGURE 13.14

Temperature-entropy (T-S) diagram of the propane cycle.

From Ref. [1].

As shown in Figure 13.13, the feed gas is passed through the gas sweetening plant for the removal of H_2S, CO_2, water, and mercury. As it passes through the precooler and cold box, its temperature decreases to about 30 °C, causing certain gas components to condense. The remaining gas and condensate are separated in the separator. The condensate is then sent to the fractionation unit, where it is separated into propane, butane, pentane, and heavier hydrocarbons. The gas is further cooled in the cryogenic column to below −160 °C and liquefied.

The pressure is then reduced to atmospheric pressure by passing through the LNG expansion valve. There are two refrigeration cycles used in this complete process: the propane cycle and the MCR cycle. The first cycle (the propane cycle) provides the required cooling to the precooler, cold box, and fractionation plant. The second cycle supplies the cooling demand of the cryogenic column. The T-S diagram of the propane cycle is shown in Figure 13.14.

13.11.4 C3MR (propane mixed refrigerant) process

Figure 13.15 shows a schematic diagram of C3MR (propane mixed refrigerant) process.

This process consists of two cycles, such as the MR cycle and C3 (propane) cycle. Natural gas is precooled to about −35 °C through a C3 cooler and then liquefied at −160 °C in the MR heat exchanger.

13.11.5 Dual mixed refrigerant process

With the increased demand for natural gas, there has been an increase in the research on and development of liquefied-natural-gas floating, production, storage, and offloading (LNG FPSO) unit technologies for LNG service in place of onshore LNG plants. The DMR cycle, which precools natural gas

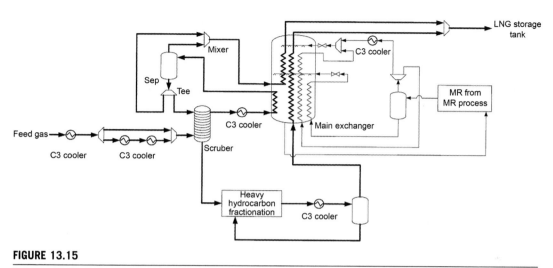

FIGURE 13.15

Schematic diagram of C3MR (propane mixed refrigerant) process.

with the MRs ethane, propane, butane, and methane and then liquefies the natural gas with another set of mixed refrigerants (nitrogen, methane, ethane, and propane), is well known for having the highest efficiency among the liquefaction cycles, and it is being examined for possible application to LNG FPSO.

The DMR cycle consists of two seawater (SW) coolers, three compressors, four heat exchangers, four valves, one tee, one phase separator, and two common headers, as shown in Figure 13.16. It consists of two cycles, makes use of a valve instead of an expander, and uses two kinds of MRs for the precooling, liquefaction, and subcooling of the natural gas.

The precooling refrigerant, which is the MR consisting of methane, ethane, propane, and butane, cools the natural gas, the main refrigerant, and itself by circulating in the precooler cold box. The main refrigerant consisting of nitrogen, methane, ethane, and propane is cooled as it passes through the precooler cold box. Then, the refrigerant liquefies and subcools the natural gas and cools itself.

The natural gas is cooled, liquefied, and subcooled by heat exchangers 1, 2, 3, and 4, with the refrigerant at −160.15 °C. The subnatural gas is expanded as it passes through valve 5, and the liquid is separated from the vapor (flash gas) and is sent to storage as LNG. The vapor (flash gas) is heated, recompressed, and used as fuel.

Most existing baseload natural gas liquefaction plants operate on the MR processes, with the C3MR process being the most widely used.

Expander cycle produces LNG by means of refrigeration generated by the isentropic expansion of gases used as refrigerant. There are various expander technologies; some of them use a single cycle, but others use a dual expansion cycle or a precooling cycle is added to improve the overall efficiency. In the industry nitrogen expansion cycles have been used for low-capacities LNG plant (0.02–0.14 MTPA), especially in peak-shaving plants and also in reliquefaction units located in very large LNG carriers.

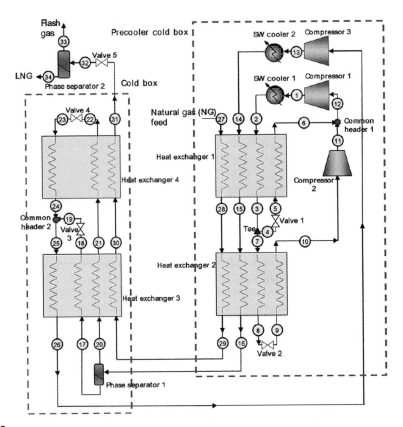

FIGURE 13.16

Configuration of the dual mixed refrigerant cycle (the number in the circle pertains to the number of streams).

From Ref. [2].

13.12 Type of LNG plants

LNG plants can be grouped into three types: base-load, peak-shaving, and small-scale plants.

- *Base-load plants*—These are large plants that are directly linked to a specific gas field development and serve to transport gas from the field. A base-load plant typically has a production capacity of >3 MTPA of LNG. The main worldwide LNG production capacity comes from this type of plants.
- *Peak-shaving plants*—These are smaller plants that are connected to a gas network. During periods of the year when gas demand is low, natural gas is liquefied and LNG is stored as a gas buffer. LNG is vaporized during short periods when gas demand is high. These plants have a relatively small liquefaction capacity (e.g., 200 tons/day) and large storage and vaporization capacity (e.g., 6000 tons/day). Especially in the United States, many such plants exist.

- *Small-scale plants*—Small-scale plants are connected to a gas network for continuous LNG production in a smaller scale. The LNG is distributed by LNG trucks or small LNG carriers to various customers with a small-to-moderate need of energy or fuel. This type of LNG plant typically has a production capacity <500,000 tons per annum (TPA). In Norway and China, several plants within this category is in operation.

13.13 Liquefaction cycle for LNG FPSO

Natural gas will play a central role in meeting the world's growing demand for energy in the upcoming decades. It is the cleaner energy alternative compared with oil and coal, and it is readily available. This is especially important in view of the rising concerns about environmental pollution and nuclear power plant hazards.

Until now, natural gas in an offshore production site has been transported by a pipeline to the onshore LNG plant, where the natural gas is liquefied, as shown in Figure 13.17(a). With the world's growing demand for natural gas, floating facilities capable of processing, liquefying, and storing natural gas are needed to develop offshore gas fields, which would otherwise entail a high cost for the development of LNG, as shown in Figure 13.17(b).

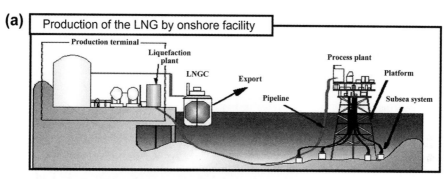

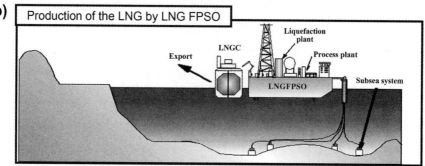

FIGURE 13.17

(a) Production of liquefied natural gas (LNG) by an onshore facility. (b) Production of LNG by liquefied-natural-gas floating, production, storage, and offloading (LNG FPSO).

© *Elsevier: From Ref. [2].*

LNG FPSO is a floating vessel with a production facility, storage tank, and offloading system for LNG, and a turret system. The LNG FPSO located at the offshore field site takes away the need for gas compression platforms; long subsea pipelines transporting the natural gas to the onshore LNG plant; and onshore construction, including onshore LNG plants, roads, storage yards, and accommodation facilities. Moreover, LNG FPSO can be redeployed to another gas field when the gas resources in the first gas field are exhausted. Therefore, LNG FPSO can accelerate LNG development.

13.13.1 Configuration of LNG FPSO

LNG FPSO consists of a hull, a turret, and a topside. The topside is divided into two parts: the process system and the utility system. The process system consists of separation, pretreatment, fractionation, and liquefaction, as shown in Figure 13.18. In the liquefaction process system, the separated and pretreated natural gas is condensed into a liquid (LNG) whose volume takes up about 1/600th the volume of natural gas.

The resulting LNG is stored in atmospheric tanks ready for export by ships. Therefore, the liquefaction process is important in the LNG FPSO topside process system and typically accounts for

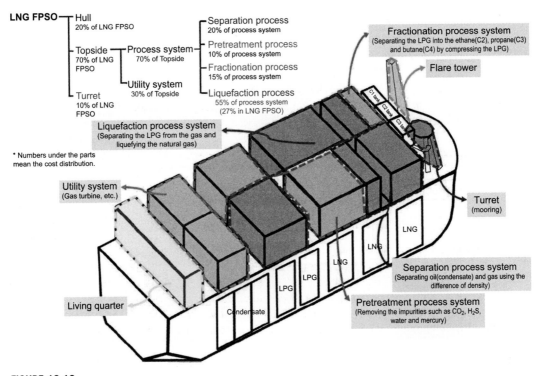

FIGURE 13.18

Configuration of liquefied-natural-gas floating, production, storage, and offloading (LNG FPSO).

© Elsevier: From Ref. [2].

Table 13.2 Gas Well Components

Components	Elementary Symbol	Boiling Point (°C)
Natural Gas (NG)		
Methane	CH_4	−161.5
Natural Gas Liquid (NGL)		
Ethane	C_2H_6	−88.6
Liquefied Petroleum Gas (LPG)		
Propane	C_3H_8	−42.1
Iso-butane	C_4H_{10}	−11.73
Normal-butane	C_4H_{10}	−0.5
Condensate		
Iso-pentane	C_5H_{12}	27.88
Normal-pentane	C_5H_{12}	36.06
Normal-hexane	C_6H_{14}	68.73

70% of the capital cost of the topside process system and 30–40% of the overall plant cost (Shukri, 2004).

The production from the gas well is separated into natural gas, LPG, and condensate. They are stored separately in the LNG FPSO. The main component of natural gas is methane (CH_4), whereas the main components of LPG are propane (C_3H_8) and butane (C_4H_{10}). If a fuel is composed of ethane (C_2H_6), propane (C_3H_8), and butane (C_4H_{10}), it is called an NGL. Because the boiling points of natural gas, LPG, and NGL at 1 atm are higher than at room temperature (21 °C), as shown in Table 13.2, they exist in the gas phase at 1 atm and room temperature. In contrast, the condensate is composed of pentane (C_5H_{12}) and hexane (C_6H_{14}).

Because the boiling point of the condensate is higher than room temperature (21 °C), as shown in Table 13.2, it exists in the liquid phase at 1 atm and room temperature, and it is called oil. The topside process system in LNG FPSO proceeds as follows. At first, as shown in Figure 13.19, the mixture of water, condensate (liquid), and gas components of NGL and natural gas is separated using the difference in density through the separation process system. The separation process system consists of a slug catcher (1-1 in Figure 13.19), a gas/liquid separator (1-2 in Figure 13.19), and a stabilizer (1-3 in Figure 13.19). The slug catcher stabilizes the slug flow from the gas well.

The gas/liquid separator then separates the mixture of water, condensate, and gas components using the difference in density. Because a part of gas components is not separated completely from the condensate in the gas/liquid separator, the stabilizer again separates the gas components from the condensate, returns them to the gas flow, and stores the condensate that is left in the condensate tank.

Second, the pretreatment process system removes impurities such as CO_2, H_2S, water, and mercury from the separated gas components. The pretreatment process system consists of the

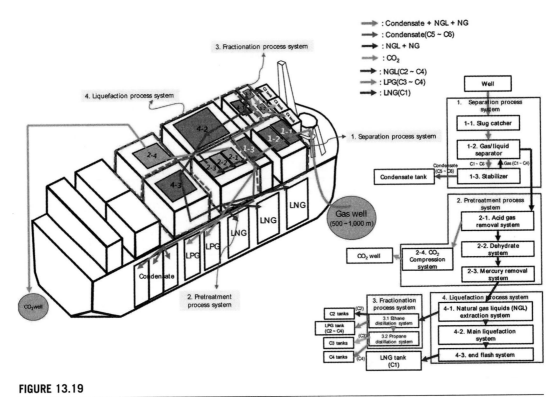

FIGURE 13.19

Topside process system of liquefied-natural-gas floating, production, storage, and offloading (LNG FPSO).

© Elsevier: From Ref. [2].

acid–gas removal system (2-1 in Figure 13.19), the dehydrate system (2-2 in Figure 13.19), the mercury removal system (2-3 in Figure 13.19), and the CO_2 compression system (2-4 in Figure 13.19). The acid–gas removal system removes consists of the NGL extraction system (4-1 in Figure 13.19), the main liquefaction system (4-2 in Figure 13.19), and the end flash system (4-3 in Figure 13.19). The gas components consisting of NGL and natural gas were separated into NGL and natural gas in the NGL extraction system by precooling such components.

The NGL was separated into ethane and LPG through the ethane distillation system. The ethane separated from the NGL was used as the refrigerant in the liquefaction system. Some of the LPG was stored in the LPG tank, and the rest was used in the propane distillation system. The propane distillation system separates LPG into propane and butane through compression and uses them as the refrigerant in the liquefaction system. The natural gas separated from the gas components in the NGL extraction system is liquefied by using the refrigerant in the main liquefaction system. At this time, the pressure of LNG is 60 bar.

To store the LNG in the LNG tank at atmospheric pressure (1.01 bar), the pressure of the LNG is reduced to the atmospheric pressure through the end flash system. Finally, the LNG is stored in the LNG tank.

13.13.2 Major considerations for the selection of the liquefaction cycle for offshore applications

Considering the limited spaces in offshore applications, the main considerations in the potential liquefaction cycles for LNG FPSO are reliability, safety, ship motion, compactness, and module layout.

- **Reliability**

All major oil companies require liquefaction cycles to be reliable based on the results of previous onshore projects. For example, because the DMR cycle was verified from the Sakhalin onshore liquefaction cycle in 2005, the world's first LNG FPSO has adopted such a cycle. Moreover, the C3MR cycle is the most commonly used in onshore applications.

This means that C3MR is the most reliable cycle in the world. The DMR cycle was also derived from the C3MR cycle, and it could be reliably applied to liquefaction cycles. Because the DMR and C3MR cycles are dual cycles, the dual cycle may be considered a reliable liquefaction model for offshore applications.

- **Safety**

To achieve a safe liquefaction cycle for LNG FPSO, many kinds of safety studies should be performed, such as hazard and operability analysis (HAZOP), hazard identification (HAZID), failure modes and effects analysis (FMEA), fault tree analysis (FTA), event tree analysis (ETA), CFD exhausts dispersion study, Helideck study, dropped object study, and explosion risk analysis. After the performance of the relevant safety studies, all the safety points derived from them should be incorporated into the engineering outputs. Especially, the critical safety issues will have a great influence on the approval of offshore projects due to the increased cost and delayed schedule.

- **Ship motion effect**

If LNG FPSO is inclined $>1.5\ ^\circ C$, its capacity for LNG production can be reduced by 10%. Therefore, the liquefaction cycle in LNG FPSO has to be designed considering the compactness, mechanical damping devices, internal turret system, and dynamic positioning system to reduce the ship motion effect. The compactness of the liquefaction cycle is particularly important in reducing the ship motion effects.

- **Compactness**

Because the available area for the liquefaction cycle for offshore applications is smaller than that for onshore plants, the compactness of the liquefaction cycle is important. The compactness of the liquefaction cycle leads to be cost saving, weight reduction, and the reduction of motion effects.

- **Module layout**

There are several modules in LNG FPSO. Among these, the liquefaction module is the largest, and it must be minimized considering the capacity of hull construction all over the world. Because the maximum feasible size of the hull is limited to 500 m, after achieving the compactness of the components, which consist of the liquefaction cycle, the size of the liquefaction module should be optimized from the viewpoint of the layout. The smaller size of the liquefaction module will also reduce the motion effect and cost.

13.13.3 Classification of the liquefaction cycles

The liquefaction cycle was classified according to three criteria: the number of cycles, the turbine-based cycle, and the use of an MR.

The number of cycles is the number of working fluids used for the precooling, liquefaction, and subcooling of the natural gas. Thus, "three cycles" means that three working fluids are used for precooling, liquefaction, and subcooling; "two cycles" means that two working fluids were used for the same purposes; and "one cycle" means that only one working fluid was used.

The turbine-based cycle is a cycle with an expander. This cycle is preferred for use in peak-shaving plants because of its simplicity and the possibility of quick startup during operation.

The amount of refrigerant that can be expended by the expander, however, is smaller than that by the valve, and the liquefied refrigerant cannot be expanded by the expander. Thus, for large-scale LNG liquefaction, the valve is used instead of the expander.

The use of an MR means that the cycle is operated with an MR instead of a pure refrigerant. The MRC can be classified according to the use of an MR for precooling, liquefaction, or subcooling. The thermodynamic efficiency of the cycle that uses an MR is high, but it is difficult to meet the proper refrigerant composition, and MRs are highly inflammable substances, giving rise to potential ventilation-related problems.

The DMR cycle consists of two cycles, makes use of a valve instead of an expander, and uses two kinds of MRs for precooling and main cooling.

Table 13.3 shows a summarized classification of the liquefaction cycles, and Figure 13.20 represents the simple schematics of these according to number of cycles.

13.14 Proposed LNG liquefaction processes for FPSO

There are several technologies for liquefaction of natural gas, but for FPSO LNG, the most proposed technologies are based on MRCs that includes SMR and DMR and expander-based cycles that include Niche LNG (CH_4 and N_2) and dual nitrogen expanders.

The proposed natural gas liquefaction cycles vary in both sophistication and power consumption. Choosing the optimum cycle for FPSO LNG is crucial and many factors are involved. Some of the major factors are:

1. LNG-FPSO liquefaction process should be light and compact due to the space and weight limitations.
2. Should be adaptable/flexible to varying natural gas compositions.
3. Rapid start-up and shutdown in a safe and controlled manner, due to unstable condition at offshore. There is high probability of having many stops due to weather.
4. Ease operation and high uptime.
5. Low requirement for handling potentially hazardous refrigerant.
6. Reliable and insensitive to the motion of LNG-FPSO (minimize weather related downtime).
7. The process should also be low-cost and easy to maintain.
8. Optimal power requirements to increase LNG production efficiency.
9. Process option to recover LPG and NGL.

Table 13.3 Classification of the Liquefaction Cycles

Number of Cycles	Use of a Turbine for Expansion	Use of a MR	Liquefaction Cycles
3 cycles	With a turbine	—	—
		Precooling	—
		Liquefaction	—
		Subcooling	—
		Precooling, liquefaction	—
		Precooling, subcooling	—
		Precooling, liquefaction, subcooling	—
	Without a turbine	—	Classical cascade cycle, optimal cascade cycle
		Precooling	
		Liquefaction	AP-X cycle
		Subcooling	
		Precooling, liquefaction	
		Precooling, subcooling	
		Precooling, liquefaction, subcooling	Multifluid cascade cycle
2 cycles	With a turbine	—	Dual independent expander refrigerant cycle
		Precooling	—
		Main cooling	—
		Precooling, liquefaction, subcooling	—
	Without a turbine	—	—
		Precooling	—
		Liquefaction, subcooling	C3MR cycle
		Precooling, liquefaction, subcooling	DMR cycle
1 cycle	With a turbine	—	(Reverse) Brayton cycle
		Precooling, liquefaction, subcooling	—
	Without a turbine	—	—
		Precooling, liquefaction, subcooling	Single MR cycle

13.14.1 Single mixed refrigerant process (PRICO® SMR TECHNOLOGY (SMR-PRICO))

The process is considered as one of the simplest and most basic processes currently in operation in the industry. The mixed refrigerant used in this process contains methane, ethane, propane, pentane, and nitrogen. Mixed refrigerant is compressed and passes through the main exchanger where it is

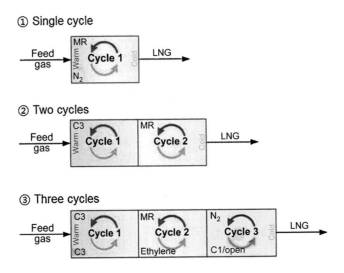

FIGURE 13.20

Simple schematics of liquefaction cycles.

© *Elsevier: From Ref. [2].*

condensed. It is then expanded across a J-T valve and evaporated as it returns counter-currently through the exchanger back to the compressor. The simplest flow diagram of SMR is shown as Figure 13.21 below. The process is simple and requires small equipment number but its capacity is limited to 1.2 MTPA per train, though the setup reduces capital costs significantly.

There is a considerable amount of refrigerant used in the process to facilitate the cooling of the natural gas which leads to a lot of compression work needed. Its low production capacity can be considered a disadvantage because more trains will be required to produce in high capacities.

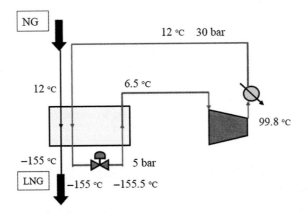

FIGURE 13.21

Principal flow diagram of SMR PRICO process.

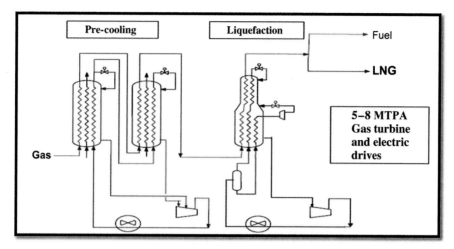

FIGURE 13.22

Diagram of dual mixed refrigerant process.

The production rate closely mirrors the capital cost not allowing for future improvement options without a total overhaul.

13.14.2 Dual mixed refrigerant process

This process contains two refrigeration cycles. The refrigerant used in the first cycle is a mixture of ethane and propane while in the second cycle is a mixture of nitrogen, methane, ethane, propane and butane. The process is licensed by Shell Global Solutions and its capacity is reported at about 4.5 MTPA. The principles of the Shell developed dual mixed refrigerant process is illustrates Figure 13.22 below.

A main argument for developing the DMR process is the need for a pre-cooling refrigerant that can cover a wider temperature range than propane, and thus give a better load distribution between the compressors. Especially in a cold or arctic climate, and for more optimal integration of heavy hydrocarbons (HHC) extraction, the minimum precooling temperature needs to be extended. This has been solved in the double mixed refrigerant (DMR) process by using a refrigerant mixture mainly based on ethane and propane. The process has been applied in the Sakhalin LNG Plant in Russia where cold climate, air cooling and large seasonal variation in temperature can be adapted to in a better way by using a mixed pre-cooling refrigerant where composition can be optimized to meet varying operating conditions.

13.14.3 Dual nitrogen expander

To eliminate the weakness of single nitrogen expander, a second stage expander was introduced on the process. A large part of refrigerant is expanded at warm temperature and only sufficient amount for sub-cooling is expanded at low temperature. There are many companies who have developed their technology based on dual nitrogen expander cycle some of them are Mustang Engineering,

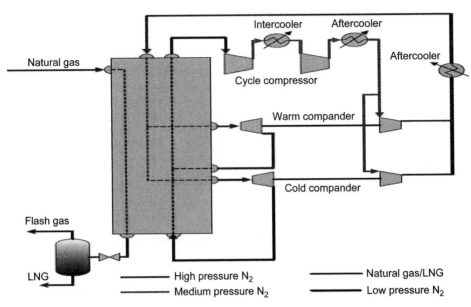

FIGURE 13.23

Diagram of dual nitrogen expander process.

Hamworthy, Petroleum Pty Ltd, Dubar (PHB), Kanfa Aragon and Statoil. The, schematic flow diagram of dual nitrogen expander cycle is given in Figure 13.23.

The industrial reference for dual nitrogen expander cycle is Kollsnes II, built by Hamworthy and the plant has an energy demand reported to be 510 kWh/ton LNG, which is a considerable reduction from the Snurrevarden plant.

13.14.4 Nitrogen and methane expander (Niche LNG)

This process is offered by CB&I Lummus and consists of two cycles. The first cycle uses methane as refrigerant or feed natural gas after heavy hydrocarbon has been removed and provides cooling at warm and moderate level. The second cycle uses nitrogen and provides refrigeration at lower temperature levels. Both cycles use turboexpanders, and their refrigerant is always in gaseous phase.

13.14.5 Exergy losses in natural gas liquefaction processes

Exergy losses in the natural gas liquefaction process are important parameters because such losses are to be compensated by more work or power input. The major losses are within the compression system (compressors), during heat transfer in heat exchangers (LNG heat exchanger and aftercoolers), and from refrigerant letdown and superheating of refrigerant (compressor discharge temperature). The losses can be categorized in three groups: heat transfer loss that includes losses in LNG heat exchangers and after collars, process losses that include letdown losses and superheating of refrigerant, and compressor losses. Temperature entropy diagram showing the different losses is presented in Figure 13.24 below.

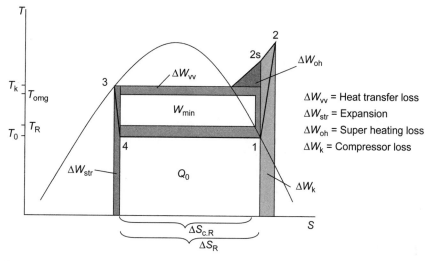

FIGURE 13.24

Temperature-entropy (T-S) diagram showing different losses in refrigeration cycle.

13.15 Storage and transfer facilities of LNG

LNG occupies about 1/600th of its gaseous volume at standard conditions. The LNG main constituents methane (CH_4) and ethane (C_2H_6), of which methane predominates, cannot be liquefied by pressure alone, because the critical temperature of these gases is well below ordinary ambient temperature, and some precooling is therefore necessary before they can be liquefied by pressure.

13.15.1 Physical properties

The properties of natural gas and LNG are similar to those of methane, although the presence of other constituents should be taken into account.

Above $-113\ ^\circ C$, methane vapor is lighter than air. The flammable range for methane is 5–15% by volume in air, a range that is wider than for most other gaseous fuels. In addition, the lower flammable limit (5%) and the ignition temperature (632 °C) are higher than for other fuels.

13.15.2 Spacing and diking

In addition to American Petroleum Institute (API) Standard 620, National Fire Protection Association (NFPA) 59A, and the governing local regulations (if any), the following factors shall also be considered for spacing, diking, and impounding of LNG storage tank(s) and other process equipment.

1. The tank(s) shall be located as far as possible from dwelling areas or locations where large numbers of people are working.
2. The tank(s) should have good access from at least two directions, so that fires can be handled if they should occur.

3. Necessary impounding and diking design and capacity should be in full conformity with Paragraph 2.2.2 of NFPA Recommended Practices Vol. 2. However, other methods of diking, including natural topography, steel structural dikes, prestressed concrete dikes, and a conventional earthen dike using an LNG collection sump within the dike area, should be consulted with the company and used upon its approval.
4. The degree to which the facilities can, within limits of practicality, be protected against forces of nature.
5. The area around natural gas installation should be kept free from vegetation and other combustible materials. Smoking should be strictly prohibited in the vicinity of the storage facility and suitable warning notices displayed.
6. Adequate arrangements for dealing with fire and larger spillages should be made instruction and emergency procedure for fire fighting should be given and regularly practiced. For fire and fire fighting reference shall also be made to BS 5429.

13.15.3 Criteria and requirements

- **Cryogenic process system**

Process system equipment containing LNG, flammable refrigerants, or flammable gases shall be either

1. installed outdoors for ease of operation, safe disposal of accidentally released liquids and gases, or;
2. installed outdoors in enclosing structures complying with NFPA Codes in Paragraphs 2.3.1 and 2.3.2 of Vol. 2, 59A.

- **Pumps and compressors**

- Pumps and compressors shall be designed for materials suitable for the temperature and pressure conditions.
- Valving shall be installed so that each pump or compressor can be isolated for maintenance. Pumps or centrifugal compressors discharge lines shall be equipped with check valves.
- Pressure-relieving device on the discharge to limit the pressure to maximum safe working pressure of the casing and downstream piping shall be provided.
- The foundations and sumps for cryogenic pumps shall be designed to prevent frost heaving.
- Pumps used for transfer of LNG at temperature below $-30\,°C$ shall be provided with suitable means for precooling to reduce effect of thermal shock.
- Compression equipment handling flammable gases shall be provided with vents from all points, including distance pieces. Vents shall be piped to a point of safe disposal.

- **Storage tanks**

Process design of storage tanks of LNG and refrigerants shall comply with requirements under API Standard 2510 and applicable portions of NFPA 59A, Paragraph 3.3. The following should essentially be determined for design:

1. purpose of the storage, i.e., peak shaving or base load;
2. volumetric capacity;
3. LNG properties;

4. operating temperature minimum and maximum;
5. operating and design pressure;
6. barometric data for sizing relief valve;
7. external loading;
8. evaporation rate.

Storage tanks may use internal or external pumps. If internal pumps are used, all connections are made in the roof. When external pumps are used, bottom connection shall be made.

To avoid LNG from falling or pouring to the bottom, a tank cooled-down line connected to a sprayrig should be inside at the top of the tank.

Liquid fill lines may be designed for installation at top, bottom, or both.

To provide positive suction head, the tank foundation should be elevated to an appropriate level.

- **Insulation**

Insulation of low-temperature installations in general and refrigerated LNG storage tanks in particular is one of the most critical areas of low-temperature storage design. The selection of any of the following methods:

1. High vacuum
2. Multiple layer
3. Power
4. Rigid foam

Should carefully be studied with specific concern to the capacity of the tank, the temperature, and economic and safety requirements.

Any exposed insulation shall be noncombustible, shall contain or inherently shall be a vapor barrier, shall be water free, and shall resist dislodgment by fire hose streams. When an outer shell is used to retain loose insulation, the shell shall be constructed of steel or concrete.

The space between the inner tank and the outer tank shall contain insulation that is compatible with LNG and natural gas and is noncombustible. The insulation shall be such that a fire external to the outer tank will not cause significant deterioration to the insulation thermal conductivity by, e.g., melting or setting.

- **Relief devices**

The LNG storage tank shall be protected when overpressured by safety relief valve(s) providing an effective rate of discharge. The minimum required rate of discharge shall be determined so as to prevent pressures exceeding those allowed by the governing code giving proper consideration to fire exposure, process upsets, or loss of product.

Sizing, locating, and installing of necessary relieving devices shall be in accordance with provision of API Standard 620. Safety requirement under Chapter four of NFPA Standard 59A shall also be considered for applicable portions in process design.

- **Instrumentation**

Each LNG storage tank shall be equipped with an adequate liquid level gaging device. Density variations shall be considered in the selection of the gaging device. Considerations shall be given to a secondary or backup gaging. At least one of these gages shall be replaceable without taking the tank out of operation.

The storage tank shall be provided with a high-liquid level alarm that shall be separate from the liquid level gaging device.

Each tank shall be equipped with a pressure gage connected to the tank at a point above the maximum intended liquid level.

Vacuum-jacketed tank shall be equipped with instruments or connections for checking the absolute pressure in the annular space.

Temperature monitoring devices shall be provided in field erected storage tanks to assist in controlling temperatures when placing the tanks into service or as a method of checking and calibrating liquid level gages.

In addition to requirements mentioned above, all instrumentation requirement given in Section B for storage tanks of LPG shall also be considered as minimum requirement for LNG tanks.

- **Other requirements**

To control corrosion rate in the specified level for the tank if the bottom of the tank rests directly on the ground, appropriate cathodic protection should be established, if required in the project specification.

All tanks in which water might accumulate under the hydrocarbon contents shall be provided with adequate drains that are suitably protected from freezing.

All openings and accessories for tanks constructed according to this section shall be installed so that, e.g., any period checking, inspection, and cleaning can readily be made.

- **Piping system design requirements**

All piping systems shall be designed in accordance with ANSI/ASME B31.4. The additional provisions as given hereunder shall be applicable to pressurized piping systems and components for LNG, flammable refrigerants, flammable liquids and gases, or low-pressure piping systems including vent lines and drain lines that handle LNG, flammable refrigerants, with service temperatures below −30 °C.

Piping systems and components shall be designed to accommodate the effects of fatigue resulting from the thermal cycling to which the systems will be subjected. Particular consideration shall be given where changes in size or wall thickness occur between pipes, fittings, valves, and components.

Provision for expansion and contraction of piping and piping joints due to temperature changes shall be in accordance with Clause 319 of ANSI B31.3.

All piping materials, including gaskets and thread compounds, shall be suitably used with the liquids and gases handled throughout the range of temperatures to which they will be subjected. The temperature limitations for pipe materials shall be as specified in ANSI B31.3.

Safety requirements for piping process design as given in Chapter six of NFPA, 59A shall be considered.

- **Transfer of LNG and refrigerants**

Transfer facilities shall comply in process design requirements and criteria with appropriate provisions elsewhere, such as those applying to siting, piping and instrumentation, and safety considerations as well as the following specific provisions: When making bulk transfers into stationary storage tanks, the LNG being transferred shall be

1. compatible in composition or temperature and density with that already in the container, or;
2. when the composition or temperature and density are not compatible, instruction shall be given in operating manual to prevent stratification which might result in "roll over" and an excessive rate of vapor evolution;

3. stratification can be prevented by means such as introducing the denser liquid above the surface of the stored liquid, introducing mechanical agitation, or introducing the LNG into the tank through an inlet nozzle designed to promote mixing. If agitation system or mixing nozzle is provided it shall be designed for sufficient energy to accomplish its purpose.

- **Piping system**

Isolating valves shall be installed so that each transfer system can be isolated at its extremities.

When power operated isolating valves are used, a design analysis should be made to determine that the closure time will not produce a hydraulic shock capable of causing line or equipment failure.

Check valves shall be provided as required in transfer systems to prevent backflow and shall be located as close as practical to the point of connection to any system from which backflow might occur.

A piping system used for periodic transfer of cold LNG, shall be provided with suitable means for precooling before use.

All applicable provisions to process design of LNG transfer system given under Chapter 8, of NFPA, 59A shall be considered as an integral part of requirements of this Section.

- **Fire protection**

Fire protection system shall be considered for all LNG facilities. In process design of the system, the extent of protection shall be evaluated with specific concern to a sound fire protection engineering principles, local conditions, process hazards, and NFPA 59A requirements.

References

[1] Mortazavi A, Somers C, Hwang Y, Radermacher R, Rodgers P, Al-Hashimi S. Performance enhancement of propane pre-cooled mixed refrigerant LNG plant. Appl Energy 2012;93:125–31.
[2] Hwang J-H, Roh M-I, Lee K-Y. Determination of the optimal operating conditions of the dual mixed refrigerant cycle for the LNG FPSO topside liquefaction process. Comput Chem Eng 2013;49:25–36.

Further reading

ASHRAE. Thermodynamic properties of refrigerants. 1791Tullie Circle N.E. Atlanta (GA) 30329.
Barclay M, Shukri T. Enhanced single mixed refrigerant process for stranded gas liquefaction. In: Proceedings of 79th annual GPA convention, Atlanta; 2000.
Cengel YA. Introduction to thermodynamics and heat transfer. 2nd ed. New York (NY, USA): McGraw-Hill; 2008.
Cha JH, Lee JC, Roh MI, Lee KY. Determination of the optimal operating condition of the Hamworthy mark I cycle for LNG-FPSO. J Soc Nav Archit Korea 2010;47(5):733–42.
Chang HM, Chung MJ, Kim MJ, Park SB. Thermodynamic design of methane liquefaction system based on reversed-Brayton cycle. Cryogenics 2009;49(6):226–34.
Chang HM, Park JH, Woo HS, Lee S, Choe KH. New concept of natural gas liquefaction cycle with combined refrigerants. In: International cryogenic engineering conference; 2010.
Douglas JM. Conceptual design of chemical processes. New York (NY, USA): McGraw-Hill; 1988.
DuPont de Nemours & Co. Bulletins, EI DuPont de Nemours & Co. Bulletins G-1, C-30, S-16, T-11, T-12, T-22, and T-114D. Wilmington (DE) 19898.
Elliott multistage compressors. Bulletin P-25A. Jeanette (PA): Elliott Co; 1975.

Finn AJ, Johnson GL, Tomlinson TR. LNG technology for offshore and mid-scale plants. In: Proceedings of the 79th annual GPA convention, Atlanta; 2000.

Form 16JB-3P Carrier hermetic absorption liquid chillers. Syracuse (NJ): Carrier Corporation; 1975.

Hwang JH, Ahn YJ, Lee GN, Kim HC, Roh MI, Lee KY, et al. Application of an integrated FEED process engineering solution to generic LNG FPSO topsides. In: Proceedings of the 19th international offshore and polar engineering conference, Osaka (Japan); 2009. pp. 136–43.

Jensen JB. Optimal operation of refrigeration cycles [Ph.D. thesis]. Norwegian University of Science and Technology; 2008.

Kaiser V, Becdelievre C, Gilbourne DM. Mixed refrigerant for ethylene. Hyd Proc; October 1976:129.

Kaiser V, Salhi O, Pocini C. Analyze mixed refrigerant cycles. Hyd Proc; July 1978:163.

Kennett AJ, Limb DI, Czarnecki BA. Offshore liquefaction of associated gas – a suitable process for the North Sea. In: Proceedings of the 13th annual OTC, Houston; 1981. pp. 31–40.

Kim H, Park C, Lee J, Kim W. Optimal degree of superheating of the liquefaction process for application of LNG FPSO. In: Proceeding of the annual spring meeting, the Society of Naval Architecture of Korea; 2010. pp. 961–5.

Lee GC, Smith R, Zhu XX. Optimal synthesis of mixed refrigerant systems for low temperature processes. Ind Eng Chem Res 2002;41:5016–28.

Lee J, Kim W, Kim H, Park C. Comparison between the LNG liquefaction plant for the ships and large LNG liquefaction plant process. In: Proceeding of the annual spring meeting, the Society of Naval Architecture of Korea; 2010. pp. 956–60.

Lee KY, Roh MI, Cho SH. Multidisciplinary design optimization of mechanical systems using collaborative optimization approach. Int J Veh Des 2001;25(4):353–68.

Lee S, Cha K, Park C, Lee C, Cho Y, Yang Y. The study on natural gas liquefaction cycle development. In: Proceedings of the 9th ISOPE Pacific/Asian offshore mechanics symposium, Busan, Korea; 2010.

Lim W, Tak K, Moon I, Choi K. Simulation comparison of liquefaction technologies for LNG offshore plant design. In: Proceedings of the 9th ISOPE Pacific/Asian offshore mechanics symposium, Busan, Korea; 2010. pp. 34–7.

Mehra YR. Refrigerant properties of ethylene, propylene, ethane and propane. Chem Eng; December 18, 1978. 97. January 15, 1979, 131; Febraury 12, 1979, 95; March 26, 1979, 165.

Mehra YR. Refrigeration systems for low-temperature processes. Chem Eng; July 12, 1982:94.

Moran MJ, Shapiro HN. Fundamentals of engineering thermodynamics. 6th ed. Hoboken (NJ, USA): John Wiley & Sons; 2008.

Nogal FD, Kim JK, Perry S, Smith R. Optimal design of mixed refrigerant cycles. Ind Eng Chem Res 2008;47: 8724–40.

Remeljej CW, Hoadley AFA. An exergy analysis of small-scale liquefied natural gas (LNG) liquefaction processes. Energy 2006;31:2005–19.

Roberts MJ, Agrawal R. Dual mixed refrigerant cycle for gas liquefaction; 2001. US Patent 6,269,655.

Sandler SI. Chemical and engineering thermodynamics. 3rd ed. Hoboken (NJ, USA): John Wiley & Sons; 1998.

Shukri T. LNG technology selection. Hydrocarbon Eng 2004;9(2).

Sibley HW. Selecting refrigerants for process systems. Chem Eng; May 16, 1983:71.

Smith JM, Van Ness HC, Abbott MM. Introduction to chemical engineering thermodynamics. 7th ed. New York (NY, USA): McGraw-Hill; 2005.

Starling KE. Fluid thermodynamic properties for light petroleum systems. Houston (TX): Gulf Publishing; 1973.

Venkatarathnam G. Cryogenic mixed refrigerant processes. New York: Springer; 2008.

Basic Engineering Design for Natural Gas Processing Projects

14

The author defines the minimum requirements for the basic engineering design package (BEDP). Before the package is prepared, a feasibility study should be conducted.

In addition, this chapter concerns the fundamental requirements for a project engineering design. It is intended as a guide to make use and follow-up convenient. The author also provide the check points to be considered by the process engineers for assurance of fulfillment of prerequisites at any stage in the implementation of process plant projects.

Before the construction of an industrial plant, engineering studies are needed that involve engineering specialties such as:

- process
- pressure vessels
- rotating equipment
- instrumentation
- electrical facilities
- computing
- piping, civil work projects
- cost control and scheduling

According to the nature of a project, engineering studies will include all or part of the following steps:

1. Basic engineering design (BED), covering:
   a. Conceptual process studies (material balances, process flow sheets, and so on) and preliminary plot plan.
   b. Preliminary piping and instrument diagrams.
   c. Definition and sizing of main equipment resulting in process specifications.
   d. Specification of effluents.
   e. Definition of control and safety devices.
   f. And, generally speaking, all the basic studies required to support a BEDP containing all data needed by a competent contractor to perform the detail engineering.
      These basic engineering studies may consist of consolidating a process package initiated by an external process licensor.
2. Front end engineering design (FEED), covering:
   a. Mechanical data sheets of the main equipment, starting from the process specifications issued during the BED and incorporating the specific requirements of codes and standards to be applied to the project in question.

   b. Thermal rating of heat exchangers.
   c. Preparation of tender packages for the main equipment.
   d. Development of process and utility piping and instrument diagrams released for detail engineering.
   e. Development of detailed plot plans and hazardous areas.
   f. Elaboration of the main piping, instrument, and electrical and civil work layouts.
   g. And, generally speaking, all the studies to be performed before ordering the main equipment.
3. Detail engineering, covering:
   a. Purchase of equipment, both main and bulk.
   b. Thermal rating of heat exchangers.
   c. Development of piping and instrument diagrams released for construction.
   d. Development of detailed piping drawings, including isometrics and stress calculations.
   e. Development of detailed drawings related to instrumentation, electrical facilities, and civil works.
   f. Management of vendor drawings.
   g. Cost and schedule control.
   h. Start-up procedures.
   i. And, generally speaking, all the studies to be performed before construction of the plant.

14.1 Contents of BEDP

The BEDP should be sufficiently comprehensive to allow a third-party contractor to carry out the detail design engineering and procurement/supply of equipment and should consist of technical data and information as required.

   The contents should essentially consist of at least the following items, although some of them can be deleted in accordance with the scope of the project and the company's requirements. The contents listed in Table 14.1 can be prepared as common for all units, whereas the contents required as per Table 14.2 can be prepared for each unit separately and independent of the other units.

   Contents of the BEDP should be arranged in sequence as shown in Tables 14.1 and 14.2 with partition papers with titles inserted for each section.

14.2 Items common for all units

The documents listed in Table 14.1 contain the items that are common for all units throughout the complex/refinery. This common information should be provided in an individual package apart from the contents shown in Table 14.2.

   However, this information can be combined in the same volume containing the individual items for each unit if only the basic design of one unit is required. The contents of each document should be as described next.

Table 14.1 Contents of Basic Engineering Design Package (Common for All Units)

Item	Contents
General design data	• General information • Basic engineering design data • Complex/plant material balance • Utility summary tables • Flare load summary tables • Effluent summary tables • Winterizing and heat conservation and insulation data • Safety • Resins, chemicals, solvents, and catalysts • Hazardous area classification • Licensor's proprietary items
Project specifications	• Fired heaters • Storage tanks • Towers, reactors, and vessels • Heat exchangers • Machinery • Electrical • Instrument • Piping • Insulation • Painting • Civil • Heating, ventilation, air conditioning cooling, and refrigeration (HVAC&R) • Miscellaneous
Manuals	Manuals
Drawings	• Flow diagram legend and general notes • Complex/refinery plot plan • Block flow diagram

14.2.1 General design data

This should include the following items where required:

- Numbering system;
- Basis of design including unit design and normal capacities;
- Specifications and properties of feed, raw materials, and products;
- Battery limit conditions for various incoming/outgoing streams;
- Set of design calculations (this information should be prepared in a separate volume and be attached to the BEDP);
- Material general specifications;
- Storage tanks, vessels, heat exchangers, heaters, etc., design criteria;
- Driver's selection philosophy;
- Control philosophy;

Table 14.2 Contents of Basic Engineering Design Package (Individual Items for Each Unit)

Item	Contents
General design data	• General information • Unit design basis • Equipment item index • Heat and material balance tables • Utility summary tables • Flare load summary tables • Effluent summary tables • Safety • Resins, chemicals, solvents, and catalysts • Hazardous area classification • Review of detailed design
Specifications and data sheets	• Fired heaters • Storage tanks • Towers, reactors, and vessels • Heat exchangers • Machinery • Electrical • Instrument • Piping • Miscellaneous
Drawings	• Preliminary unit plot plan • Process flow diagram (PFD) • Piping and instrumentation diagrams (P&IDs) • Utility distribution flow diagrams (UDFDs) • Emergency shut-down block diagram and/or logic diagram • Cause-and-effect tables/diagrams • Single-line diagrams

- Product loading, storage, and dispatch philosophy;
- Other miscellaneous general requirements.

14.2.2 Basic engineering design data

These data provide the necessary basic information required for the basic design of the facilities and should contain the following information:

- Standards and codes for design;
- Utility information;
- Site condition information;
- Regulations concerning environmental pollution (air, water disposal, noise, etc.);
- Equipment including instrumentation and electrical general design information;
- Other requirements such as general design information for buildings, insulation, painting, fire proofing, etc.

- **Complex/plant material balance**

This should provide the overall feed and products material balance in the plant. The data should also be shown on the block flow diagram that should be prepared by the basic designer.

- **Utility summary tables**

The maximum estimated utility requirements in summer and winter cases should be tabulated for whole plant. The utility services to be considered are electric power, steam, condensate, boiler feed water, potable (drinking) water, fire water, plant (service) water, raw water, cooling water, sea water, fuel gas, fuel oil, natural gas, instrument and plant air, inert gas, nitrogen, and others. Special requirements such as purging nitrogen, plant air for regeneration and decoking, etc., should be provided.

- **Flare load summary tables**

Gas and liquid flow rates to be routed to the flare or blow-down system should be tabulated for each failure and each safety relief valve.

- **Effluent summary tables**

All gaseous and liquid effluents showing quantities and qualities of impurities that are object to control (sulfur, phenol, oil, BOD_5, COD, etc.) should be tabulated to facilitate determining the environmental impact of the project.

- **Winterizing and heat conservation and insulation data**

Design basis for winterizing and heat conservation should be provided, as well as a table indicating the basis of selection of tracer size and number for various pipeline sizes and fluid temperatures in the required winterizing temperatures.

Base data to select insulation thickness such as ambient temperature, wind velocity, insulation material and thermal conductivity, etc., should be provided. Insulation thickness tables for selection of hot insulation, personnel safety, cold insulation, and other purposes should also be presented.

- **Safety**

The following requirements should be provided:

1. Specifications, fire rating, and identification of areas for fire-proofing of steel structure and equipment.
2. Recommendation for firefighting facilities required for process unit(s) and storage areas including rate of water/foam required, cooling system, deluge valve requirements, and identification of the areas that require it.
3. Recommendation for firefighting facilities as well as provision of blast-proof walls and/or roof for control room.
4. Recommendations for handling/storage of various chemicals, catalysts, hazardous materials, etc.
5. Number and location of safety showers, eye washes, fire hydrants, etc.
6. Proposed methods of fire-fighting for different types of fires.
7. Any other specific safety requirement.

- **Resins, chemicals, solvents, and catalysts**

1. The following information should be prepared for each catalyst, packing, and solid absorbent used in the process:
   a. Service
   b. Name or designation
   c. Acceptable suppliers
   d. Volume required
   e. Density
   f. Pellet or grain size and shape
   g. Design life
   h. Regeneration characteristics
2. The following information should be provided for all resins, chemicals, additives, solvents, and inhibitors used in the process:
   a. Name or designation
   b. Initial fill quantity
   c. Annual consumption
   d. Physical properties
   e. Loading, unloading, and make-up procedures
   f. Shelf life (if any)
   g. Any specific warehousing requirement

- **Hazardous area classification**

It should indicate the sources of hazards and the extent of hazardous areas, providing area classifications for the selection of electrical equipment.

- **Licensor's proprietary items**

Sufficient process design data such as heat and material balance tables, required calculations, and mechanical design data should be supplied so as to perform detail design and procure the items from the licensor's approved manufacturers.

The licensor's standard engineering specifications and drawings for the detailed design of the specific unit should be provided. The licensor should supply these specifications and drawings as applicable to particular sections of the unit. The specifications and drawings should cover details or practices not given in the engineering procedures and specifications supplied by the company.

14.2.3 Project specifications

The BEDP should include all engineering standards/specifications and drawings applicable to the various parts of the project. The engineering standards/specifications should be provided to define the basic requirements for basic design and engineering, vendor selection, procurement, manufacturing, inspection, and installation of the equipment and materials.

For the licensed units, the specific project engineering standards should be provided by the licensor together with the process design specifications. Where required, specifications provided by the licensor should be adhered to the project general specifications by the basic engineering designer.

Project specifications should cover but not be limited to:

- Fired heaters, including steam boilers, waste heat boilers, incinerators, and all kinds of furnaces.
- Storage tanks, including all types of tanks, spheres, sumps, and basins.
- Vessels, including all types of the vessels, towers, reactors, separators, and fractionators. The specifications for the internal parts of the vessels and towers such as trays, packings, distributors, etc., should also be included in this section.
- Heat exchangers, including all types of the heat transfer equipment such as process-to-process heat exchangers, coolers, double pipe heat exchangers, reboilers, chillers, plate-type heat exchangers, condensers, etc.
- Machinery, including all types of compressors, turbines, blowers, and pumps (e.g., rotary, reciprocating, centrifugal, etc.).
- Electrical, including all electric motors, cables, switches, transformers, generators, lighting, telecommunication, etc.
- Instrument, including distributed control system (DCS), and all instrumentation systems and equipment such as alarms and shut-down systems, analyzers, transmitters, differential pressure elements, thermometers, thermocouples, level gauges, pressure gauges, control valves, measuring devices, cables, computer, etc.
- Piping, including piping material, classification, layout, fittings, valves, etc.
- Insulation, including project required insulation thickness tables and general specifications.
- Painting.
- Civil, including specifications for concrete, paving, stairs, platforms, structures, buildings, site preparations, rough gradings, foundations, etc.
- HVAC&R, including air conditioning and refrigeration general specifications and basic design data (e.g., room temperature, humidity, etc.).
- Miscellaneous, including all project specifications for package equipment, welding, buildings, safety, noise, etc., not included in the above mentioned specifications.

14.3 Manuals

The basic engineering designer should prepare the following preliminary manuals if required by the company. Each manual should be provided in an individual bound volume apart from the BEDP.

- **Operating manual**

The operating manual should include an outline of start-up, shut-down, and alternative operations. It should also indicate emergency procedures covering utility failures and major operating upsets. In addition, all safety procedures, catalyst regeneration instructions, and process descriptions in all operation modes should be included. This operating manual should be reviewed and revised/completed at the end of the detail design phase.

- **Laboratory manual**

The laboratory manual should include the following:

1. Laboratory equipment
   a. Equipment list
   b. Basic specifications
2. Laboratory building
   a. Building plot plan
   b. Equipment location diagram
   c. Utility distribution diagram
3. Analytical method for
   a. Raw materials
   b. Chemicals
   c. Process control
   d. Products
   e. Catalysts
   f. Water impurities
4. Instructions for sample taking and its frequency

- **Maintenance manuals**

Maintenance manuals should include:

- particular emphasis on preventive maintenance;
- maintenance instructions for each equipment, including specific types of lubricant/grease required;
- periodicity of major shutdowns for regular overhaul/maintenance;
- any other special maintenance procedure;
- The maintenance manuals will be prepared in the detail design phase.

- **Test run procedure manual**

The manual should include details of performance test runs and other aspects of commissioning, such as:

- operating data to be recorded by the company in the log book;
- sampling method;
- analytical methods;
- methods of calculations;
- interpretation and measuring of parameters during the test;
- methods of taking operating data;
- methods of evaluating the unit performance.

14.3.5 Drawings

- **Flow diagram legend and general notes**

The flow diagram legend and general notes should be prepared to cover all legends/symbols to be used in the engineering documents. If the legends and/or symbols cannot be integrated and indicated on one flow diagram, a specific project document should be provided to cover all required legends, symbols, and general notes.

- **Complex/refinery plot plan**

A preliminary plot plan should be provided to show the layout of control room(s), process unit(s), utility unit(s), offsite facilities, tankage, loading and unloading facilities, and other buildings, basins, and so on throughout the complex/refinery. The plot plan should be prepared in accordance with the safety, operation, and maintenance requirements instructed by the company's engineering standards and approved by the company.

- **Block flow diagram**

An overall simplified schematic diagram of the all units showing all process units and utility and offsite facilities should be provided. The complex/refinery terminal material balances should also be shown.

14.4 Individual items for each unit

The BEDP for each individual unit should contain the documents listed in Table 14.2. The contents of each document should be congruent with but not limited to the following items.

14.4.1 General design data

A minimum of the following information should be included.

- Brief description of the unit's different operation modes
- Specifications and properties of feed, raw materials, and products
- Battery limit conditions for various incoming/outgoing streams
- Other miscellaneous general requirements

14.4.2 Unit design basis

The unit design basis should include but not be limited to:

- unit normal and design capacity and turn-down ratio and expected yields of the products;
- test methods and procedures;
- storage, handling, and safety aspects;
- design calculations, where required by the company;
- equipment design criteria;
- other design requirements.

- **Equipment item index**

All equipment in the unit should be listed, together with item number, name, quantity, service, and referenced P&ID.

- **Heat and material balance tables**

A heat and material balance table should be provided to indicate the stream properties (flow rates, phase, composition, temperature, pressure, operating and standard relative densities (specific gravities) or densities, enthalpies, viscosities and molecular mass (weight) for gases and vapors) for each stream number shown on the PFD.

- **Utility summary tables**

The normal and maximum estimated utility requirements for each unit including tankage and offsite facilities should be tabulated. Such figures should be consistent with a single operating case.

The services to be considered are fuel (all types), electrical power, steam and condensate (all levels), boiler feed water, water (all services), cooling water, air (instrument and plant), nitrogen, inert gas, etc. where required.

- **Review of detailed design**

This should include review and approval by the licensor or basic engineering designer for critical equipment detailed design data, information, and drawings and critical piping detailed design with respect to the licensor's basic design. The equipment and facilities to be reviewed by the licensor/basic designer should be clearly described for all services in this section.

14.5 Specifications and data sheets

Specification sheets indicating all process and mechanical basic design data required for designing the equipment should be provided. Further specific design information should be inclusively provided together with the data sheets where required.

The blank forms of the equipment data sheets should be prepared in accordance with the data sheets and requirements stipulated in the project specifications and/or standards attached to the BEDP.

The units of measurement used in the data sheets should be based on International System of units (SI) except where otherwise specified.

Equipment data sheets should include the following general requirements in additions to the items specified for each individual equipment data sheet as described next:

- Equipment item number
- Type
- Quantity
- Applicable standards and codes
- Service
- Material table
- Corrosion allowances where required
- Site conditions
- Capacity/dimensions
- Nozzles identification table including rating, size, and quantity for each nozzle

14.5.1 Fired heaters

- **Furnaces and non-reaction-type heaters**

Information that is required and included in the BEDP includes:

- Type of heater and coil arrangement
- Vaporization curves
- Fluid flow rates
- Operating and design temperatures and pressures at inlet and outlet
- Heat duty and physical properties at inlet and outlet conditions
- SOR and EOR design and allowable pressure drop and allowable differential temperature
- Maximum turn-down ratio of heater and burner considering its effect on the unit performance
- Fuel characteristics such as type, composition, viscosities, impurities, etc.
- Minimum efficiency
- Limiting fluid peak temperatures
- Limiting transfer rates or velocities
- Heat flux density (maximum allowable)
- Control specifications
- Specific design and fabrication requirements
- Whether air preheat system envisaged with forced draft fans and ducting is required
- Whether induced draft fan is required
- Material of construction for tubes
- Corrosion allowances
- Maximum allowable excess air for any type of fuel
- Type of terminal fittings
- Whether coil temperature and pressure profile is required from the vendor
- Whether steam-air decoking is required.

14.5.2 Reaction-type heaters

For this type of heater involving a thermal or catalytic reaction, the information should include:

- Conceptual furnace design
- Capacity and dimensions
- Inlet and outlet design and operating conditions and design parameters
- Radiant tube maximum allowable diameter and material of construction
- Heat flux (maximum allowable and normal)
- Heat transfer coefficient
- Duty for each operating case
- Number of cells
- Specific mechanical design where required
- Description of kinetics involved.

14.5.3 Storage tanks

The data sheet should include but not be limited to the following items:

- Type (e.g., floating roof, fixed roof, double roof, etc.)
- Operating and design temperature and pressure
- Maximum operating temperature and pressure
- Fluid properties
- Internal facilities, such as air sparger, steam heater, mixers, etc., requirements
- Requirements of insulation
- Any special surface finish requirement
- Stress-relieving and post-weld heat treating requirement
- Instrumentation requirements
- A simplified storage tank sketch showing all internals and nozzles and instrumentation

14.5.4 Towers, reactors, and vessels

Outline sketches for towers, reactors, and vessels and filters should be prepared and include:

- Diameter and height or length
- Number, type and spacing of trays (for columns)
- Operating and design temperature and pressure
- Materials and corrosion allowances
- Liquid levels and/or interface levels
- Insulation requirement
- Basic information of internals such as pans, distributors, spray nozzles, packing, demisters, etc.
- Lining or cladding requirements
- Special performance requirements for other process equipment such as coalescers, separators, etc.
- Space velocity (normal and design), conversion per pass (normal and design), reactor pressure drop, recycling (normal and design), reactor inlet and outlet conditions (normal and design), and reactor bed temperature profile (normal and design) for reactors
- Vapor/liquid loading for towers
- Foaming characteristics
- Stress-relieving requirement
- Requirements of mist eliminators, supports, mesh, or packing, etc.
- Nozzles identification table including quantity, rating, and size of all nozzles (e.g., process, instruments, manholes, etc.)
- Nozzles elevations excluding elevations of level gauge, transmitter, and switch
- Instrumentation requirements
- Skirt or leg requirement

14.5.5 Heat exchangers

The data sheets for shell and tube type heat exchangers, chillers, coolers, condensers, process heat exchangers, air coolers, and special type heat exchangers should be prepared, including the following data as minimum requirement. The rating of heat exchangers should be included in the detail engineering phase in the scope of work:

- Duty (normal and design)
- Flow rates for liquid/vapor phases separately
- Physical properties of the fluid such as specific heat, boiling temperature, viscosities, densities, thermal conductivities, etc.
- Inlet and outlet temperature and pressure
- Maximum allowable and normal operating pressure drop
- Fouling resistances
- Condensing or vaporization curve (if necessary)
- Type of TEMA design
- Mechanical design conditions
- Materials and corrosion allowances
- Limiting transfer rates and velocities where applicable
- Restrictions on combining air fin services
- Tube and baffle arrangement
- Alternative specifications for specific services
- A simplified sketch showing nozzles and general configurations for kettle-type exchangers; the sketch should also include vapor space, surge volume required, high/normal/low liquid levels, and instrumentation.

14.5.6 Machinery

The data sheets should be provided for pumps, compressors, blowers, expanders, and turbines and include the following data as a minimum requirement:

- Type
- Minimum, normal, and maximum flow rates required considering all modes of operations
- Material of construction and corrosion allowances
- Special mechanical features required
- Operating and design temperatures and pressures at suction and discharge
- Control requirements
- Physical properties of the fluid at suction and discharge conditions
- Basic recommendations for spares
- Hydraulic horsepower in kilowatts
- Sealing and lubrication requirements
- Cooling requirements
- Alternative specifications if necessary for specific services
- Spare requirements
- Flushing requirements
- NPSH where required
- Revolutions per minute (rpm)
- Inlet filter size and type

- Number of stages of turbine
- Safety devices
- Governor type for turbines

14.5.7 Electrical

The following specifications and/or data sheets should be provided if not included in the common specifications:

1. Specifications of power supply system and electrical installation
2. Specifications of emergency power supply
3. Single line diagram
4. Requirement of information for:
   a. Interlocking method with instrumentation system
   b. Fire alarm system
   c. Lighting system
   d. Consumers requiring emergency power and duration of it
   e. Any other electrical equipment/system
5. Classification of hazardous area:
   a. Electrical area classification
   b. Equipment selection criteria for areas with flammable and/or inflammable materials
   c. List of flammable and inflammable materials to be handled along with their properties such as ignition temperature, applicable gas group, etc.
6. Information for motor controls (local, central, etc.) with reference to measurement required for units as follows:
   a. Control, indication, metering, and annunciation philosophy of various drivers with basic logic diagrams
   b. Data on control requirements like remote start-up and auto start-stop for motors
7. High, medium, and low voltage requirements for each unit
8. Implications of power failure and recommended unit's emergency supply scheme for all types of electrical loads that require emergency power
9. List of drives requiring emergency power feed along with their power consumption
10. Specifications of critical drives/variable speed drives and their controls
11. Requirements of uninterrupted power supply (UPS) system such as:
    a. Total load
    b. Rated voltage and permissible variation
    c. Duration for which UPS system should be designed
    d. Maximum change-over time allowed
    e. Step load/permissible voltage dip
    f. In-rush current in worst case
    g. Load power factor
    h. Redundancy, if needed
12. Electric motor driver's data sheets

14.5.8 Instruments

The following specifications and/or data sheets should be included in this section. The following specifications should not be duplicated in case of inclusion in common part for general requirements:

1. Information for alarm signal and interlocking system
2. List and specification for special control and measuring system
3. Preliminary layout for physical arrangement of control panels in the central control room, including reference to requirements regarding ventilation, air-conditioning, etc.
4. Recommendation on wiring of instruments
5. Special requirements for distributed control system (DCS) not included in the general specification for each individual process unit
6. Requirements for advanced process control (APC) and optimization
7. General requirements of computer system
8. List and basic specifications of instruments
   The specification/data sheets should include the following items as minimum requirements:
   a. Tag number
   b. Name and service
   c. Quantity
   d. Location
   e. Applicable standards/codes
   f. Basic design data
   g. Process design data
   h. Materials of construction
   i. Alternate operation modes (if any)
9. Specification/data sheets should be provided for but not be limited to the following equipment:
   a. Board mounted instruments
   b. Flow instruments
   c. Pressure instruments
   d. Pressure switches
   e. Pressure gauges
   f. Level instrument
   g. Level switches
   h. Gauge glasses
   i. Temperature instrument and wells
   j. Thermometers and wells
   k. Thermocouples and wells
   l. Orifice flanges
   m. Control valves
   n. Safety relief valves
   o. Solenoid valves
   p. Resistance bulbs and wells
   q. Converters
   r. Analyzers
   s. Tank gaging.

14.5.9 Miscellaneous

Specifications and data sheets should be provided for the miscellaneous equipment not included in other equipment category items. Complete duty specification to be provided for each item, including all process and mechanical design data as required for the basic equipment design and operation in different modes.

14.6 Drawings

14.6.1 Preliminary unit plot plan

The plot plan in the basic design should be based on the preliminary equipment dimensions and should include preliminary layout of control room(s), buildings, equipment, and other required facilities in the unit.

For licensed units, the plot plan should be prepared based on the licensor's information and requirements for normal and emergency operation, safety, and maintenance.

The unit plot plan should be prepared with due consideration to the overall plot plan layout of the entire plant and interrelations of the unit with other units.

14.6.2 Process flow diagram

The PFD should be provided in accordance with the requirements for each unit separately.

14.6.3 Piping and instrumentation diagrams

Piping and instrumentation diagrams (P&IDs) should be prepared in accordance with the requirements. P&IDs should be provided for each unit separately and should include all facilities, piping, and equipment and interconnections between the unit and adjacent facilities. Interconnection P&IDs should also be prepared to clarify piping connection points and their tie-in between the new plant and other adjacent units.

14.6.4 Utility distribution flow diagrams

Utility distribution flow diagrams (UDFDs) should be provided for each unit separately and should include all equipment item numbers with all required pipelines and interrelations between headers/subheaders and all utility users.

The diagrams should be prepared in accordance with the requirements specified in standards:

- Emergency shut-down block diagram and/or logic diagram
- Cause-and-effect tables/diagrams
- Single line diagrams

14.7 Recommended practice for feasibility studies

The development of an industrial investment project from the stage of the initial idea until the plant is in operation can be shown in the form of a cycle comprising three distinct phases: the preinvestment, the investment, and the operational phases. The objective of this recommended practice is to explore a sound and well-established procedure for carrying out the various studies required during the pre-investment phase in general and within the feasibility studies in particular.

This recommended practice is set forth to explore the problems encountered in carrying out the various studies required during the feasibility studies of an industrial investment project.

The preinvestment phase (see Figure 14.1) comprises several stages as following:

- "Opportunity studies" (the identification of investment opportunities)
- "Prefeasibility studies" (analysis of project alternatives and preliminary project selection as well as project preparation)
- "Feasibility studies (final project selection and project appraisal investment decisions) and appraisal report"

14.7.1 Opportunity studies

Identification of investment opportunities should be made at the starting point in a series of investment-related activities as shown in Figure 14.1. Information on the newly identified and viable investment opportunities should be obtained.

This information, data, and required parameters should be generated, qualified, and used as the main instrument to develop a project idea into a proposal. Compilation of necessary information, data, and determining parameters followed with detailed analysis is termed an "opportunity study."

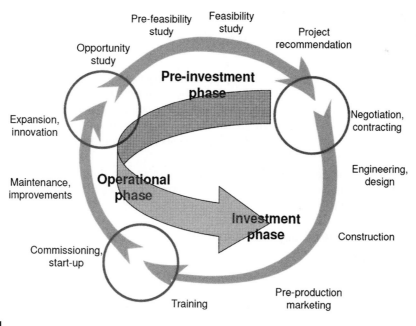

FIGURE 14.1

Preinvestment, investment, and operating phases of the project cycle.

The following objective points should typically be analyzed together with an overall look to investment potential of the company, general interest of other government authorities, domestic and foreign investors, and promotion of the project(s):

- Natural resources with potential for processing and manufacture such as hydrocarbon resources for chemical and petrochemical industries and crude oil for refinery processing.
- Possible resources of utilities such as air, steam, fuel, power, water, etc. and their transportation alternates for chemical and petrochemical or refinery plants.
- The existing industrial pattern that serves as a basis for the proposed new hydrocarbon-based industries.
- Future demand of the products that should have been created as a result of increased population or purchasing power or for newly developed consuming goods.
- Imports, in order to identify areas for import substitution.
- Environmental impact.
- Manufacturing sectors successful in other countries with similar economic background and levels of development, capital, labor, and natural resources.
- Possible interlinkage with other industries, indigenous or transnational.
- Possible extension of existing production lines by backward or forward integration, linking, for example, a downstream petrochemical industry with a refinery.
- Possibilities for diversification for example from a petrochemical complex into other commercial product manufacturing.
- Possible expansion of existing industrial capacity to attain economies of scale.
- The general investment climate.
- Industrial policies.
- Availability and cost of production factors.
- Export possibilities.

Opportunity studies should be taken as sketchy in nature, relying more on aggregate estimates rather than on the detailed analysis. Cost data usually should be taken from comparable existing projects and not from quotes from sources such as licensors, engineering companies, or equipment suppliers.

These parameters may be integrated and/or cross-checked with those already practiced in the company's planning and investment promotion departments for the purpose of adequate stimulation of the company's (as investor) response.

It should be noted that the information conveyed in a project opportunity study should not involve any substantial costs in its preparation, as it is intended primarily to highlight the principal investment aspects of a possible industrial proposition.

14.7.2 Outline of general opportunity study

- **Outline of an area study**

1. Basic features of the area: area size and leading physical features, with maps showing the main characteristics
2. Population, occupational pattern, per capita income, and socioeconomic back-ground of the area, highlighting differences in the area considered
3. Leading exports from and imports to the area
4. Basic exploited and potentially exploitable production factors
5. Structure of any existing manufacturing industry utilizing local resources

6. Infrastructural facilities, especially in the field of transport and power, fuel, and water, conductive to development of industries
7. A comprehensive checklist of industries that can be developed on the basis of the available resources and infrastructural facilities
8. A checklist revising that in Item 7 via a process of elimination, excluding the following industries:
   a. Those for which present local demand is too small and transport costs are too high
   b. Those that face too severe competition from adjoining areas
   c. Those that can be more favorably located in other areas
   d. Those that would have unacceptable environmental impacts
   e. Those that require feeder industries not available in the area
   f. Those requiring substantial export markets, if the area is located in the interior and transport to the port is difficult or freight costs are high
   g. Those for which markets are distantly located
   h. Those that are geographically not suited to the area
   i. Those that do not fit in with national plan priorities and allocations
9. Estimation of present demand and identification of opportunity for development based on other studies or secondary data, such as trade statistics, for the list of industries left after the revision referred to in Item 8
10. Identification of recommendable project objectives and suitable strategies determining the type and scope of the project, including approximate capacities of new or expanded units that could be developed
11. Estimated capital costs of selected projects (lump sum), taking into account: 1
    a. Land
    b. Technology
    c. Equipment
    d. Production equipment
    e. Auxiliary equipment
    f. Service equipment
    g. Spare parts, wear-and-tear parts, and tools
    h. Civil engineering works
    i. Site preparation and development
    j. Buildings
    k. Outdoor works
    l. Project implementation
    m. Preinvestment capital expenditures, including expenditures for preparatory investigations
    n. Working capital requirements

- **Major input requirements**

For each project, approximate quantities of essential inputs should be estimated, so as to obtain the total input requirements. Sources of inputs should be stated and classified (e.g., local, shipped from other areas of the country, or imported). Inputs should be classified as:
- Raw materials.
- Processed industrial materials and components.
- Factory supplies, such as auxiliary materials and utilities.
- Labor.

- **Further project requirements**
- Estimated production costs to be derived from Item 12.
- Estimated annual sales revenues.
- Organizational and management aspects typical for the industry.
- An indicative time schedule for project implementation.
- Estimated level of total investment contemplated in projects and peripheral activities, such as development of infrastructure.
- Projected and recommended sources of finance (estimated).
- Estimated foreign exchange requirements and earnings (including savings).
- Financial evaluation: approximate pay-off period, approximate rate of return.
- Assessment of possible enlargement of product mix, increased profitability, and other advantages of diversification (if applicable).
- A tentative analysis of overall economic benefits, and especially those related to national economic objectives, such as balanced dispersal of economic activity, estimated saving of foreign exchange, estimated generation of employment opportunities, and economic diversification.
- Indicative figures based on reference programming data, such as surveys and related studies, secondary data, and data on the performance of other similar industrial establishments should be sufficient for this purpose.

14.7.3 Outline of resource-based opportunity studies

1. Characteristics of the resource, prospective and proven reserves, past rate of growth, and potential for future growth.
2. Role of the resource in the national economy, its utilization, demand in the country, and exports.
3. Industries currently based on the resources, their structure and growth, capital used and labor engaged, productivity and performance criteria, future plans, and growth prospects.
4. Major constraints and conditions in the growth of industries based on the resource.
5. Estimated growth in demand and prospects of export of items that could use the resource.

14.7.4 Outline of other factor-based opportunity studies

1. Present size and growth rates of demand for items that are not imported and for those that are wholly or partially imported.
2. Rough projections of demand for each item.
3. Identification of the items in short supply that have growth or export potential.
4. A broad survey of the raw materials indigenously available.

14.8 Prefeasibility studies

The project idea should be elaborated in more detailed studies. However, formulation of a feasibility study that enables a definite decision to be made on the project is a costly and time-consuming task. Therefore, before assigning greater funds for such a study, a further assessment of the project idea should be made in a prefeasibility study.

A prefeasibility study should be viewed as an intermediate stage between a project opportunity study and a detailed feasibility study, the difference being in the degree of detail of the information obtained and the intensity with which project alternatives are discussed.

14.8.1 A detailed feasibility study

A detailed review of available alternatives should take place at the stage of prefeasibility study. In particular, the review should cover the various alternatives identified in the following main fields of study, such as:

* Project or corporate strategies and scope of the project.
* Market and marketing concept.
* Raw materials and supplies.
* Location, site, and environment.
* Engineering and technology.
* Organization and overhead costs.
* Human resources, in particular, managerial staff, labor costs, and training requirements and costs.
* Project implementation schedule and budgeting.

The financial and economic impacts of each of these factors should be assessed. Occasionally, a well-prepared and comprehensive opportunity study justifies bypassing the prefeasibility study stage. A prefeasibility study should, however, be conducted if the economics of the project are doubtful.

14.9 Outline of prefeasibility study
14.9.1 Project background and history

* Project sponsors.
* Project history.
* Cost of studies and investigations already performed.

14.9.2 Market analysis and marketing concept

* Definition of the basic idea of the project, objectives, and strategy.
* Demand and market.
* Structure and characteristics of the market.
* Estimated existing size and capacities of the industry (specifying market leaders), its past growth, estimated future growth (specifying major programs of development), local dispersal of industry, major problems and prospects, and general quality of goods.
* Past imports and their future trends, volume, and prices.
* Role of the industry in the national economy and the national policies, priorities, and targets related or assigned to the industry.
* Approximate present size of demand, its past growth, major determinants, and indicators.
* Marketing concept, sales forecast, and marketing budget.
* Description of the marketing concept, selected targets, and strategies.

- Anticipated competition for the project from existing and potential local and foreign producers and supplies.
- Localization of markets and product target group.
- Sales program.
- Estimated annual sales revenues from products and byproducts (local and foreign).
- Estimated annual costs of sales promotion and marketing.
- Production program required.
- Products.
- Byproducts.
- Wastes (estimated annual cost of waste disposal).

14.9.3 Material inputs

This section covers approximate input requirements, their present and potential supply positions, and a rough estimate of annual costs of local and foreign material inputs.

- Raw materials.
- Processed industrial material.
- Components.
- Factory supplies.
- Auxiliary materials and utilities.

14.9.4 Location, site, and environment

- Preselection, including, if appropriate, an estimate of the cost of land.
- Preliminary environmental impact assessment.

14.9.5 Project engineering

- Determination of plant capacity.
- Feasible normal plant capacity.
- Quantitative relationship between sales, plant capacity, and material inputs.
- Preliminary determination of scope of project.
- Technology and equipment.
- Technologies and processes that can be adopted, given in relation to capacity.
- Technology description and forecast.
- Environmental impacts of technologies.
- Rough estimate of costs of local and foreign technology.
- Rough layout of proposed equipment (major components).
- Production equipment.
- Auxiliary equipment.
- Service equipment.
- Spare parts, wear-and-tear parts, tools.
- Rough estimate of investment cost of equipment (local and foreign), classified as above.
- Civil engineering works.

- Rough layout of civil engineering works, arrangement of buildings, and short description of construction materials to be used.
- Site preparation and development.
- Buildings and special civil works.
- Outdoor works.
- Rough estimate of investment cost of civil engineering works (local and foreign), classified as above.

14.9.6 Organization and overhead costs

- Rough organizational layout
- General management
- Production
- Sales
- Administration
- Estimated overhead costs
- Factory
- Administrative
- Financial

14.9.7 Human resources

- Estimated human resource requirements, broken down into labor and staff and into major categories of skills (local/foreign).
- Estimated annual human resource costs, classified as above, including overheads on wages and salaries.

14.9.8 Implementation scheduling

- Proposed approximate implementation time schedule
- Estimated implementation costs

14.9.9 Financial analysis and investment

- Total investment costs.
- Rough estimate of working capital requirements.
- Estimated fixed assets.
- Project financing.
- Proposed capital structure and proposed financing (local and foreign) cost of finance.
- Production cost (significantly large cost items to be classified by materials, personnel, and overhead costs, as well as by fixed and variable costs).
- Financial evaluation based on the above-mentioned estimated values.
- Payback period.
- Simple rate of return.

- Break-even point.
- Internal rate of return.
- Sensitivity analysis.
- National economic evaluation (economic cost-benefit analysis).
- Preliminary tests, for example, of
  - Foreign exchange effects
  - Value-added generated
  - Absolute efficiency
  - Effective protection
  - Employment effects
  - Determination of significant distortions of market prices (foreign exchange, labor, capital)
  - Economic industrial diversification; estimate of employment-creation effect

14.10 Feasibility studies

A feasibility study should provide all data necessary for an investment decision (see Figure 14.1). The commercial, technical, financial, economic, and environmental prerequisites for an investment project should therefore be defined and critically examined on the basis of alternative solutions already reviewed in prefeasibility study.

A feasibility study should be carried out only if the necessary financing facilities, as determined by the studies, can be identified with a fair degree of accuracy. Possible project financing should be considered as early as the feasibility study stage, because financing conditions have a direct effect on total costs and thus on the financial feasibility of the project.

14.10.1 Main objectives

The main objectives to be considered, as the minimum requirement, in feasibility studies of projects are classified and covered in the following sections:

- Section I: Executive summary
- Section II: Project background and basic idea
- Section III: Market analysis and marketing concept
- Section IV: Raw materials and supplies
- Section V: Location, site, and environment
- Section VI: Engineering and technology
- Section VII: Organization and overhead costs
- Section VIII: Human resources
- Section IX: Implementation planning and budgeting
- Section X: Financial analysis and investment appraisal

Each of these sections should separately be studied on the basis of specific procedure and guidelines given herein this specification. For further detailed studies of the scope covered in these sections, specific references are given at the end of each section.

- **Executive summary (Section I)**

For convenience of preparation, the feasibility study should begin with a brief executive summary, outlining the project data (assessed and assumed) and the conclusions and recommendations that would then be covered in detail in the body of the study.

    The executive summary should concentrate on and cover all critical aspects of the study, such as the degree of reliability of data on the business environment, project input and output, the margin of error (uncertainty, risk) in forecast of the market, supply and technological trends, and project design.

    The executive summary should have the same structure as the body of the feasibility study and cover but not be limited to the following areas.

- **Summary of the project background and history (Section II)**

- Name and address of project promoter.
- Project background.
- Project (corporate) objective and outline of the proposed basic project strategy, including geographical area and market share (domestic, export), cost leadership, differentiation, market niche.
- Project location: orientation toward the market or toward resources (raw materials).
- Economic and industrial policies supporting the project.

- **Summary of market analysis and marketing concept (Section III)**

- Summarize results of marketing research: business environment, target market and market segmentation (consumer and product groups), channels of distribution, competition, life cycles (sector, product).
- List annual data on demand (quantities, prices) and supplies (past, current, and future demand and supplies).
- Explain and justify the marketing strategies for achieving the project objectives and outline the marketing concept.
- Indicate projected marketing costs, elements of the projected sales program and revenues (quantities, prices, market share, etc.).
- Describe impacts on: raw materials and supplies, location, the environment, the production program, plant capacity and technology, etc.

- **Raw materials and supplies (Section IV)**

Describe the general availability of:

- Raw materials
- Processed industrial materials and components
- Factory supplies
- Spare parts
- Supplies for social and external needs
- List annual supply requirements of material inputs
- Summarize availability of critical inputs and possible strategies (supply marketing)

- **Location, site, and environment (Section V)**

- Identify and describe location and plant site selected, including:
- Ecological and environmental impact.
- Socioeconomic policies, incentives, and constraints.
- Infrastructure conditions and environment.
- Summarize critical aspects and justify choice of location and site.
- Outline significant costs relating to location and site.

- **Engineering and technology (Section VI)**

- Outline the production program and plant capacity.
- Describe and justify the technology selected, reviewing its availability and possible significant advantages or disadvantages, as well as the life cycle, transfer (absorption) of technology, training, risk control, costs, legal aspects, etc.
- Describe the layout and scope of the project.
- Summarize main plant items (equipment, etc.) and their availability and costs.
- Describe required major civil engineering works.

- **Organization and overhead costs (Section VII)**

Describe basic organizational design and management and measures required.

- **Human resources (Section VIII)**

- Describe the socioeconomic and cultural environment as related to significant project requirements, as well as human resources availability, recruitment and training needs, and the reasons for the employment of foreign experts, to the extent required for the project.
- Indicate key persons (skills required) and total employment (numbers and costs).

- **Project implementation schedule (Section IX)**

- Indicate duration of plant erection and installation.
- Indicate duration of production start-up and running-in period.
- Identify actions critical for timely implementation.

- **Financial analysis and investment appraisal (Section X)**

- Summary of criteria governing investment appraisal
- Total investment costs
- Major investment data, showing local and foreign components
- Land and site preparation
- Structures and civil engineering works
- Plant machinery and equipment
- Auxiliary and service plant equipment
- Incorporated fixed assets
- Preproduction expenditures and capital costs

- Net working capital requirements
- Total costs of products sold
- Operating costs
- Depreciation charges
- Marketing costs
- Finance costs
- Project financing
- Source of finance
- Impact of cost of financing and dept service on project proposal
- Public policy on financing
- Investment appraisal: key data
- Discounted cash flow (internal rate of return, net present value)
- Pay-off period
- Yield generated on total capital invested and on equity capital
- Yield for parties involved, as in joint venture projects
- Significant financial and economic impact on the national economy and environmental implications
- Aspects of uncertainty, including critical variables, risks and possible strategies and means of risk management, probable future scenarios, and possible impact on the financial feasibility of the investment project
- National economic evaluation
- Conclusions
- Major advantages of the project
- Major drawbacks of the projects
- Chances of implementing the project

14.10.2 Project background and basic idea

To ensure the success of the feasibility study, it must be clearly understood how the project idea fits into the framework of general conditions and industrial development of the country. The project should be described in detail and the sponsors identified, together with a presentation of the reasons for their interest in the project.

Describe the project idea and include:

- A list of the major project parameters that served as the guiding principles during the preparation of the study.
- The project (corporate) objectives and description and analysis of proposed basic project strategy, including the:
  - Geographical area and market share (domestic, export)
  - Cost leadership
  - Differentiation
    Describe the project history and include:
- The historical development of the project (titles, authors, completion dates).
- Studies and investigations already performed.

- Conclusions arrived at and decisions taken on the basis of former studies.
  Describe the feasibility study and include:
- Author and title.
- Ordering party.
  Describe the cost of preparatory studies and related investigations and include the:
- Preinvestment studies.
- Opportunity studies.
- Prefeasibility studies.
- Feasibility studies.
- Any other supporting studies.
- Expert, consultant, and engineering fees (if their services are used).
- Preparatory investigations such as:
- land surveys.
- quantity surveys (quantification of building materials).
- quality (laboratory) test.
- ther investigations and tests (if performed).

14.10.3 Market analysis and marketing concept

The basic objective of any industrial investment project is to benefit either from the utilization of available resources or from the satisfaction of existing or potential demand for the output of the project. Market analysis is the key activity for determining the scope of an investment, the possible production programs, the technology required, and often the choice of a location.

The marketing experts should communicate and cooperate with the other members of the feasibility study team from the very beginning of the work, so as to avoid isolated marketing or engineering solutions that could prove to be financially unsound.

- **Marketing concept**

When a project strategy has been defined, a suitable marketing concept should be designed in accordance with the phases described next:

- *Strategic dimensions of marketing*

The marketing strategy to be considered involves:

1. Identification of target groups and the products to win their favor.
2. Determination of competition policies, that is, whether a low-price strategy or differentiation strategy should be pursued to defeat competitors.

- *Operative dimensions of marketing*

Main marketing tools, namely product, price, promotion, and place, should be distinguished and the combination impact on marketing concept, which is known as "marketing mix," should be specified. Table 14.3 lists the activities to be made relating the four components. Any of these components may be referred to as a sub-mix component.

Table 14.3 Marketing Mix	
Product	**Promotion**
• Scope of product mix • Department of product mix • Quality • Design • Packaging • Maintenance • Service • Warranty service • Possibility of returning a purchase	• Advertising • Public relations • Personal sale • Sales promotion • Brand polity
Price	**Place**
• Price positioning • Rebates and conditions of payment • Financing conditions	• Channels of distribution • Distribution density • Lead time • Stock • Transport

- **Marketing measures and marketing budget**

For the feasibility study, it is necessary to determine the marketing activities and to prepare a time schedule indicating the starting-point and duration of these activities.

For the development of the project strategy and marketing concept, a careful marketing research should be conducted.

The scope of marketing research required for a feasibility study should be covered to fulfill three principal aims:

- Market–project relations should be made clear for the management.
- Strategic constraints and problems should be identified.
- Strategic options for the project should be outlined.
  The work on market analysis should be organized along the following lines:
- Assessment of the target market structure.
- Customer analysis and market segmentation.
  - Analysis of the channels of distribution.
  - Analysis of the competition.
  - Analysis of the socioeconomic environment.
  - Corporate (internal) analysis.
  - Projections of marketing data.
  - Conclusions, prospects and risks.

The depth or degree of detail of the analysis should be determined according to the complexity of each problem and its importance for the project evaluation.

14.10.4 Data assessment

Two types of market information can be distinguished—"general market data" and "specific market data"—for a particular market segment (consumer group or product group). General market data should be accomplished with studies on the following:

- General economic indicators relating to product demand, such as population level and growth rate, per capita income and consumption, gross domestic product per capita and annual growth rate, and income distribution.
- Government policies, practice and legislation, to the extent directly related to consumption, productions, imports and exports of the product(s) in question, standards, restrictions, duties, taxes, as well as subsidies or incentives, credit control and foreign regulations.
- Present level of domestic production, by volume and value, including production intended for internal consumption and not placed on the market
- Present level of imports, by volume and value (CIF and local cost).
- Production and imports of substitute and near-substitutes.
- Critical imputs (see also Section IV).
- Production targets determined in national economic plans, where applicable for products in question, substitutes, and complementary products.
- Present level of exports, by volume and value (FOB).
- Behavioral patterns, such as consumer habits and responses, individual and collective, and trade practice.

The specific demand and market data for a particular market segment should be identified and their availability for feasibility study should be ascertained. Typically, the following items, as well as similar items, should be studied in this regard:

- The demand for a product may have been suppressed by market imperfections due to trade policies, high import tariffs, which would not be portable on domestic products.
- Artificially high domestic prices may be imposed on certain products whose import are severely restricted, but the pattern of demand, and consequently of product pricing, would chang2e materially once the product become available in large quantities.

It is, however, necessary to identify the specific demand and market data required and the extent to which such data are available and could be used in the feasibility study. The possibility of extending the market to other countries should be explored for most projects of any size, as export sales have to be taken into consideration in determining plant capacity.

International competition should be studied through economies of scale, for example, in production or marketing through comparative locational advantages, the development of international cooperation, access to technologies, etc. The geographical divisions of possible exports should be defined in content of a particular product.

Regarding export analysis, the feasibility study should therefore deal with the following questions:

- Will the enterprise gain strategic advantages by operating more internationally?
- What advantages will it gain in particular (example: economies of scale in production)?
- To what degree and in which fields does international competition pose a treat to the project?

- What will be the future extent of the advantage of the enterprise operating on a geographically limited field?

14.10.5 **Outline of the project strategy**

On the introduction of marketing and dimension for feasibility studies, the following main steps for defining the project strategy and corresponding marketing concept should essentially be studied.

- **Geographical area of strategy**

On the basis of the assessment of the project situation, the feasibility study should consider different strategic alternatives with regard to the geographic limitations.

- **Market share and basic strategy**

For an investment project, it is necessary to define the long-term market position or market share and corresponding profitability of the project. Since each market has its individual characteristics, the feasibility study should analyze each profitability and market share relationship very carefully.

- **Strategy of cost leadership**

The cost advantage attributable to the "learning and experience" provides protection against competition. In order to achieve cost leadership, the following assets required for this strategy should be evaluated:

- High investment capacity (i.e., access to capital)
- Process innovations and improvements
- Thorough supervisions of labor force
- Products designed for easy manufacturing
- Low-cost distribution system

- **Differentiation strategy**

Differentiation strategy aims to protect against competition, in that it binds the buyers to the brand or the firm and thus reduces price sensitivity. The following assets and similar items should be subject to differentiation strategy:

- Powerful marketing potential
- Strengths in research and development
- Customer groups with higher purchasing power
- Parts of the product line
- Tradition within the industry
- Cooperation with supply and distribution channels

- **Project strategy**

- When a project strategy is selected, the feasibility study should always consider possible alternative strategies.
- When assessing such alternatives, the following points should be considered:
  - To what extent do the strategic alternatives fulfill the original aims of the feasibility studies?

- What is the financial impact of the alternatives (profitability, return on investment)?
- What risks are linked with each alternatives (political, ecological, financial, etc.)?
- For determining project alternative strategies, the following problems should be addressed:
  - What is the geographical area in which the project will operate?
  - What basic strategy should be chosen (cost leadership, differentiation or market niche)?
  - What market position (market share) is aimed at, and how much time is required to reach the target?
  - Which product–market relation should be basis for the marketing concept?
  - What will be the product range (products, price level)?
  - Which target group of consumers will be focused?
  - Which strategy will be chosen (competition or market expansion)?
  - What core skills are required for success in respect of actual or potential competitors?
  - Will the project develop the market position exclusively by its own means, or are there possibilities for cooperation?

14.10.6 Marketing cost

The projection of marketing costs comprises all costs components resulting from the marketing activities described in this section. Depending on the scope of study and the depth of analysis, marketing costs should be projected for each product separately or for a group of products.

14.10.7 Raw materials and supplies concept

The selection of raw materials and supplies depends primarily on the technical requirements of the project and analysis of supply markets. Important determinant factors for selection of raw materials and factory supplies are environmental factors and criteria relating to project strategies, for example, minimizing supply risks and cost of material inputs.

To keep the cost of feasibility studies at a reasonable level, key aspects should be identified and analyzed in terms of requirements, availability, cost, and risk, which will be significant for feasibility of the project. The following approach should be taken to specify the requirements, check their ability, and estimate their costs:

1. Characteristics of raw materials, auxiliary materials, and utilities
2. Specification of requirements
3. Availability and supply
4. Input alternatives
5. Supply marketing and supply program
6. Cost of raw materials and supplies:
   a. The costs of materials and supplies used or kept in stock should be specified. Cost estimates for materials and inputs can be expressed either as the cost per unit produced or in terms of a certain production level.
   b. The following information should be presented:
      - Type of material and input
      - Unit of measurement (barrels, tones, cubic meters, etc.)

- Estimated costs per unit produced
- Estimated cost per input unit
- Number of input units consumed per unit produced
- Estimated cost per unit produced divided into direct (predominantly variable) and indirect (predominantly fixed) cost components
- Direct cost per unit produced divided into foreign and local currency components (although expressed in one common currency)
- Indirect cost per unit produced divided into foreign and local currency components

c. When calculating indirect costs, the amounts resulting from environmental protection, pollution control measures, etc., should be established per unit of production, or per accounting period.

14.10.8 Location, site and environment

- **Location analysis**

The feasibility study should identify locations suitable for an industrial project under consideration. The feasibility study should also indicate on what grounds alternative locations have been identified and give reasons for leaving out other locations that were suitable but not selected.

The impacts and requirements to be identified should be classified as follows:

- Natural environment, geophysical conditions, and project requirements
- Ecological impact of the project and environmental impact assessment
- Socioeconomic policies, incentives and restrictions, and government plans and policies
- Infrastructure services, such as existing industrial infrastructure, the economic and social infrastructure, the institutional framework, urbanization, and literacy

- **Natural environment**

Climatic conditions as an important locational factor should be specified in terms of air temperature, humidity, sunshine hours, winds, precipitation, hurricane risk, etc. Each of these should be specified in detail such as maximum, minimum, and average temperatures on an average day, in particular months, or over a period of 10 years.

Geodestic aspects such as soil conditions, subsoil water levels, and a number of special site hazards (earthquakes, susceptibility to flooding) should be studied.

Ecological requirements with specific regards to water supply with quality requirements, waste water disposal, pollution of area, and health risks should be studied and demonstrated in the feasibility study.

14.10.9 Environmental impact assessment

The feasibility study should include a thorough and realistic analysis of the environmental impact of industrial investment project. This impact is often of crucial importance for socioeconomic, financial, and technical feasibility of the project.

The following specific objectives of environmental impact assessment should be performed.

- To promote a comprehensive, interdisciplinary investigation of environmental consequences of the project and its alternatives for the affected natural and cultural human habit.

- To develop an understanding of the scope and magnitude of incremental environment impacts of the proposed project for each of alternative project design.
- To incorporate in the designs any existing regulatory requirements.
- To identify measures for mitigation of adverse environmental impacts and for possible enhancement of beneficial impacts.
- To identify critical environmental problems requiring further investigation.
- To assess environmental impacts qualitatively and quantitatively as required for the purpose of determining the overall environmental merit of each alternative.

Investments in export processing zones and other specified regions may be and are exempted from taxes or would benefit from other types of study. Such possibilities should be considered in the feasibility study.

14.10.10 Infrastructural conditions

The availability of a developed and diversified economic and social infrastructure is often of key importance for a project. The feasibility study should identify such key infrastructural requirements, because they are vital to the operation of a project.

Technical infrastructure for the choice of location and other characteristics such as reliability, quality, and physical aspects should be studied. Transport and communication facilities, utility requirements, labor, lands should be studied. Infrastructral services and effluent and waste disposal should also be critical factors to be studied and clearly monitored in the feasibility study report.

14.10.11 Final choice of location

The feasibility study should indicate where the project in question should be located, with specific concern to raw materials, supplies (water, power, fuel, labor, etc.), climate, available social welfare facilities (e.g., education, medical services, and recreation facilities), proximity to raw materials and downstream product-consuming markets, and so on.

As far as financial feasibility of alternative locations is concerned, the following data as well as data related to financial risks should be carefully assessed:

- Production costs (including environment protection cost)
- Marketing costs
- Investment costs (including environmental protection)
- Revenues
- Taxes, subsidies, and allowances
- Net cash flows

14.10.12 Site selection

The feasibility study should analyze and assess alternative sites on the basis of key aspects and site specific requirements. For sites available within the selected location, the following requirements and conditions should be assessed:

- Ecological conditions on site (soil, site hazards, climate, etc.)
- Environmental impact (restrictions, standards, guidelines)

- Socioeconomic conditions (restrictions, incentives, guidelines)
- Local infrastructure at site location (existing industrial, infrastructure, availability of critical project inputs such as labor and factory supplies)
- Strategic aspects (corporate strategies, regarding possible future expansion, supply, and marketing policies)
- Cost of land
- Site preparation and development, requirements, and costs

14.10.13 Cost estimates

This schedule should be used for the estimates of investment costs at the site. Examples of costs are acquisition of land, taxes, legal expenses, rights of way, site preparation, and development. Different cost items are to be identified, quantified (if relevant), estimated, and divided into components of foreign and local currency origin.

It should be carefully stated whether factory-external facilities, possibly judged as necessary (such as disposal and treatment of effluents, generation of electricity, water supply system, storage, housing, and schools etc.), should be included in the cost estimate.

14.10.14 Engineering and technology

An integral part of engineering at the feasibility stage is the selection of an appropriate technology and of the corresponding know-how. While the choice of technology defines the production processes to be used, the effective management of technology transfer requires that the technology and know-low are acquired on suitable terms and conditions and that the necessary skills are available or developed.

The required machinery and equipment should be determined in relation to the technology and processes to be used, the local conditions, the state of the art, and human capabilities and training programs at various levels.

The analysis should also outline the specific requirements of each individual technology, if selected, and specify the need for technical documentation and maintenance procedures. In particular, the analysis should include a thorough survey of spare parts and the format of necessary lists of spare parts.

14.11 Production program and plant capacity

The detailed production program should be designed in a feasibility study when the required sales program is determined.

14.11.1 Determination of production program

Full production capacity should not be considered as practical in oil and gas processing (OGP) process plants during the initial production operations. Owing to various technological, production, and commercial difficulties, most projects experience initial problems that take the form of only a gradual growth of sales and market penetration, on the one hand, and a wide range of production problems, on the other.

A production and sales target of 65% to 70% of overall capacity should be considered for the first year and 75% to 90% for the second year, with full production capacity for third year and thereafter.

14.11.2 Input requirements

Once the production program is defined, the nature and general requirements of materials should be identified, and the specific quantities needed for each stage of the production program and the costs that these entail should be determined. The costs should be assessed: basic materials such as raw materials, semiprocessed and bought-out items, major plant supplies (auxiliary materials and utilities), other supplies, and labor requirements.

14.11.3 Determination of plant capacity

A feasible normal capacity should be specified in the study under normal operating conditions, taking into account the installed equipment and technical conditions of the plant such as normal stoppages, down time, holidays, maintenance, tool changes, desired shift patterns, and management system applied.

A feasible nominal maximum capacity should also be specified, which frequently corresponds to the installed capacity as guaranteed by the supplier of the plant. This capacity should entail overtime, excessive consumption of plant supplies, utilities, spare parts, wear-and-tear parts, and deproportionate production cost increases.

14.11.4 Minimum economic size and equipment constraints

When determining the minimum economic size of a project, experience gained elsewhere should be evaluated and used. There should be full treatment of the resulting higher production costs and prices and inability to compete with a cost leadership strategy in external market.

14.12 Technology choice

The selection of appropriate technology and know-how is a critical element in any feasibility study. Such selection should be based on a detailed consideration and evaluation of technological alternatives and the selection of the most suitable alternative in relation to the project or investment strategy chosen and to socioeconomic and ecological conditions.

Technology choice should be directly related to market, resource, and environmental conditions and the corporate strategies recommended for a particular project. The form of foreign participation, national objectives and policies, industrial growth strategy, availability of local resources and skills, and several other factors that can directly impinge on technology choice should be assessed.

14.12.1 Ecological and environment impact

Ecological and environmental impacts, especially any possible hazards that may result from the use of particular technologies, should be studied and used in technology selection.

14.12.2 Assessment of technology required

The primary goals of technology assessment are to determine and evaluate the impacts of different technologies on the society and national economy (cost-benefit analysis, employment and income effects, satisfaction of human needs, etc.) impacts, on the environment, and on technoeconomic feasibility assessment from the company's point of views.

14.12.3 Technology description and project layout

A preliminary project layout should be established in the feasibility study report and should provide the overall framework of the project, which can serve as a broad basis for plant engineering, order-of-magnitude projections of civil works, machinery requirements and other investment elements and site development, plant and other buildings, transport facilities (road, railway, sidings, etc.), utility linkages including electric power substations, water connections, sewage lines, gas and telephone links, and storage and other facilities.

The preliminary project layout should include several charts and drawings (they do not need to be to scale). The functional charts and layout drawings at this stage should include the following:

- General functional layout, defining the principal physicals, or locational, features and flow relationships, civil works and constructions, and various ancillary and service facilities.
- Basic characteristics of the technology.
- Material-flow diagrams indicating the flow of materials and utilities.
- Transport layout, indicating roads, railways, and other transport facilities up to their point of connection to public networks.
- Utility lines for electric power, water, gas, telephone, sewage, and emissions (both internal and external up to the point connecting with public networks).
- Area for expansion and extension.

14.12.4 Technology evaluation and selection

Alternative techniques should be evaluated in the feasibility study to determine the most suitable technology for the plant. This evaluation should be related to plant capacity and should commence with a quantitative assessment of output, production buildup and gestation period, and a qualitative assessment of product quality and marketability.

The technology should have been proved and used in the manufacturing process, preferably in the company from which it came. While unproved, new and experimental techniques should not be considered appropriate. Obsolescent technology should be avoided.

14.12.5 Technology acquisition and transfer

When a technology has to be obtained for a specific project, the means of acquisition should be determined. This can take the form of technology licensing, outright purchase of technology, or a joint venture involving participation in ownership by the technology supplier. The implication of these methods should be analyzed.

The feasibility study should indicate the measures and action to be taken for technological absorption and adoption of the acquired technology to local conditions. The feasibility study should also identify the various categories of personnel to be recruited and the training programs they need to undergo.

14.12.6 Contract terms and conditions

The contractual terms and conditions for technology acquisition and transfer that are likely to be of particular significance to the project need to be highlighted in the feasibility study. The negotiation strategy should also be indicated, and a draft of the principal terms and conditions from the viewpoint of the prospective licensee should be prepared.

14.13 Selection of machinery and equipment

Equipment selection at feasibility study stage should broadly define the optimum group of machinery and equipment necessary for a specific production capacity.

14.13.1 Level of automation

Degree of sophistication and automation as an important issue regarding OGP production plants should be carefully assessed, and the capital cost of automation trend and competitive nature of production should be evaluated. Thus, the use of computer-aided designs for production or numerically controlled machine tools and so on should be specified in the feasibility stage of the project.

14.13.2 Categories of equipment/spare parts

The equipment should be listed and categorized. The classification should be listed under plant machinery, mechanical equipment, electrical equipment, instrumentation and controls, transport and conveying equipment, testing and research equipment, and other machinery items with estimated prices.

A list should be prepared of required spare parts and tools with their estimated prices. Spare parts needs should depend on the nature of the project, availability, and import facilities. In common practice, a stockpile of 2 years' consumption is feasible for OGP industries.

14.14 Civil engineering works

The feasibility study should provide plans and estimates for the civil works related to the project. This should cover site preparation and development, other buildings, civil engineering works relating to utilities, transport, emissions and effluent discharges, internal roods, fencing and security, and other facilities and requirements of the plant.

14.14.1 Building

Plans and estimates for buildings and structures should include the main plant buildings, buildings for ancillary production facilities, ancillary buildings for maintenance and repair, testing and research and development, storage and warehouses for stock of raw material or finished products, nonfactory or plant buildings, staff welfare facilities, residential buildings, and such other buildings.

14.15 Estimates of overall investment costs (capital cost estimates)

On the basis of estimates for technology, machinery, and equipment and civil engineering works, the feasibility study should provide an overall estimate of the capital costs of the project. Such an estimate should undergo modification in accordance with the bids and offers received from suppliers and contractors but should nevertheless provide a fairly realistic estimate of capital costs.

A physical contingency allowance should commonly be added. A typical degree of accuracy should be considered for ±10%. However, careful consideration should be given to this estimate, and in particular to the contingency allowed.

14.15.1 Cost-estimating methods

• **Process equipment cost estimating by ratio and proportion**

While there may be sophisticated software available to generate accurate cost estimates, we should never lose sight of the importance of understanding the basis for costs. Never let computer output cloud simple estimating judgment.

It is novel to be armed with simple, quick, easy-to-understand techniques to arrive at approximate equipment costs. The *Rule of Six-tenths* and the use of cost indices are two readily available and easy-to-use ratio and proportion methods to quickly estimate equipment costs.

This course provides the engineer with an understanding of the estimating technique known as *The Rule of Six-tenths* and, when appropriate, of the use of this rule in combination with cost indices. The various types of estimates are discussed as prerequisite background. Equations are provided to enable the student/engineer to escalate or otherwise adjust historical equipment cost data.

• **Cost estimate types and accuracy**

Regardless of accuracy, capital cost estimates are typically made of direct and indirect costs. Indirect costs consist of project services, such as overhead and profit, and engineering and administrative fees. Direct costs are construction items for the project and include property, equipment, and materials. This course deals with the equipment component of direct cost.

The preparation of a preliminary estimate is done by an estimator based on his or her assessment of the design, past cost estimates, in-house estimating information, and previous contracts and purchase orders. It is not normal to obtain formal quotations from equipment manufacturers in support of a preliminary estimate.

Informal telephone *budget* quotations on identified major equipment such as vessels, filters, etc., are acceptable. However, even these types of "expedient" quotations can prove to be time restrictive to obtain. Even with the advent of sophisticated estimating software, it is sometimes simply easier to manually approximate an equipment cost. That is the subject of this section.

Definitive and detailed cost estimates are full-blown exercises that are undertaken to produce a competitive bid submission or otherwise produce an accurate (±10% or higher) cost estimate for, say, a corporation's management approval for appropriation of funds. The ratio and proportion methods presented in this course would not be normally suitable for inclusion in a definitive estimate.

The equipment cost-estimating methods that will be outlined in this course are suitable for use with the first three types of estimates; definitive and detail estimates require formal, firm equipment cost quotations from equipment manufacturers and suppliers.

- **Ratio and proportion estimating**

A ratio indicates the relationship between two (or more) things in quantity, amount, or size. Proportion implies that two (or more) items are similar, differing only in magnitude. Using these well-known mathematical tools is a simple process.

When preparing preliminary estimates, two methods for estimating the cost of equipment are the *Rule of Six-tenths* and the use of cost indices to adjust historic costs to current prices. Each will be discussed and a single example will be offered to demonstrate the use of both.

- *The Rule of Six-tenths*

Approximate costs can be obtained if the cost of a similar item of different size or capacity is known. A rule of thumb developed over the years known as the Rule of Six-tenths gives very satisfactory results when only an approximate cost within $\pm 20\%$ is required. An exhaustive search in conjunction with the development of this course left this author with no indication of any individual who developed this concept. One is forced to assume that the relationship naturally evolved in the public domain after large quantities of actual cost data were analyzed retrospectively. The earliest mention of this concept was found in a reference accredited to a December 1947 *Chemical Engineering* magazine article by Roger Williams Jr. titled, "Six-tenths Factor Aids in Approximating Costs."

The following equation expresses the Rule of Six-tenths:

$$C_B = C_A \left(\frac{S_B}{S_A}\right)^{0.6} \tag{14.1}$$

where:

C_B = the approximate cost ($) of equipment having size S_B (cfm, hp, ft^2, or whatever)
C_A = is the known cost ($) of equipment having corresponding size S_A (same units as S_B)
S_B/S_A = is the ratio known as the *size factor*, dimensionless

- *The "N" exponent*

An analysis of the cost of individual pieces of equipment shows that the size factor's exponent will vary from 0.3 to unity, but the average is very near to 0.6, thus the name for the rule of thumb. If a higher degree of sophistication is sought, Table 14.1 can be used. It lists the value of a *size exponent* for various types of process equipment. The Table 14.1 values have been condensed from a vast, comprehensive tabulation of estimating cost data presented in the March 24, 1969, issue of *Chemical Engineering* magazine. This article by K. M. Guthrie is titled, "Data and Techniques for Preliminary Capital Cost Estimating." While the source for the concept and the presented exponential data is somewhat dated, i.e., 1947 and 1969, respectively, there is indication that this material is still relevant and valid.

Table 14.4 Process Equipment Size Exponent (N)

Equipment Name	Unit	Size Exponent (N)
Agitator, propeller	hp	0.50
Agitatior, turbine	hp	0.30
Air compressor, single stage	cfm	0.67
Air compressor, multiple stage	cfm	0.75
Air dryer	cfm	0.56
Boiler, industrial, all sizes	lb/h	0.50
Boiler, package	lb/h	0.72
Centrifuge, horizontal basket	dia (in)	1.72
Centrifuge, solid bowl	dia (in)	1.00
Conveyor, belt	ft	0.65
Conveyor, bucket	ft	0.77
Conveyor, screw	ft	0.76
Conveyor, vibrating	ft	0.87
Crystallizer, growth	ton/day	0.65
Crystallizer, forced circulation	ton/day	0.55
Crystallizer, batch	gallons	0.70

Using Table 14.4 size exponents transforms the previously presented formula into

$$C_B = C_A \left(\frac{S_B}{S_A}\right)^N \Bigg|$$
(14.2)

where the symbols are identical to those already described and N is the *size exponent*, dimensionless, from Tables 14.4 and 14.5

- *Cost indices*

The names and purpose of today's cost indices are too numerous to mention. Probably the most widely known cost index to the general public is the Consumer Price Index (CPI) generated by the U.S. Department of Labor, Bureau of Labor Statistics. While the CPI could probably serve our needs, more specific data are available for use in engineering and technical applications.

Cost indices are useful when basing the approximated cost on other than current prices. If the known cost of a piece of equipment is based on, for instance 1998 prices, this cost must be multiplied by the ratio of the present-day index to the 1998 base index in order to proportion the value to present-day dollars. (Incidentally, the inverse of this operation can be performed to estimate what a given piece of equipment would have cost in some earlier time.) Mathematically, this looks like,

$$C = C_O \left(\frac{I}{I_O}\right) \Bigg|$$
(14.3)

Table 14.5 Process Equipment Size Exponent (N)

Equipment Name	Unit	Size Exponent (N)
Dryer, drum and rotary	ft²	0.45
Dust collector, cyclone	cfm	0.80
Dust collector, cloth filter	cfm	0.68
Dust collector, precipitator	cfm	0.75
Evaporator, forced circulation	ft²	0.70
Evaporator, vertical and horizontal tube	ft²	0.53
Fan	hp	0.66
Filter, plate and press	ft²	0.58
Filter, pressure leaf	ft²	0.55
Heat exchanger, fixed tube	ft²	0.62
Heat exchanger, U-tube	ft²	0.53
Mill, ball and roller	ton/h	0.65
Mill, harmer	ton/h	0.85
Pump, centrifugal carbon steel	hp	0.67
Pump, centrifugal stainless steel	hp	0.70
Tanks and vessels, pressure, carbon steel	gallons	0.60
Tanks and vessels, horizontal, carbon steel	gallons	0.50
Tanks and vessels, stainless steel	gallons	0.68

where

C = current cost, dollars
C_O = base cost, dollars
I = current index, dimensionless
I_O = base index, dimensionless

Many sources exist for technical indices, but two of the more popular ones that are readily available are those published monthly in *Chemical Engineering* magazine under "Economic Indicators, Chemical Engineering Plant Cost Index (CEPCI)" and weekly in *Engineering News Record* magazine under "Market Trends." The two sources work equally well but, as with other indices, they cannot be used interchangeably.

- *About the CEPCI number*

Since we will be using the CEPCI value in an example, let's examine its makeup. The index was established in the early 1960s using the period of 1957–1959 as a base of 100. According to Couper, the value of the CEPCI index number is weighted approximately 61% toward equipment and machinery. Of that portion, fully 85% of the value comprises process equipment. These heavily weighted component values bode well for escalating chemical process equipment costs. The U.S. Department of Energy published the summary table of historical CEPCI data, shown in Table 14.6.

Table 14.6 Historical Chemical Engineering Plant Cost Index Data

Year	Annual Average
1957–1959	100
1964	103
1965	104
1970	126
1975	182
1980	261
1985	325
1990	357.6
1995	381.1
1996	381.8
1997	386.5
1998	389.5
1999	390.6
2000	394.1
2001	394.3

- *Example*

The following example illustrates a combined use of both of these ratio and proportion methods to produce an approximate cost. Please note that the costs presented here are purely hypothetical and should not be used as a basis for anything other than an illustration.

Let us assume that a rough estimate is being prepared for a project in which a 5000-gallon capacity stainless steel pressure vessel is involved. Let us further assume that our past project purchasing data shows that a 2000-gallon stainless steel pressure vessel, very similar to that currently required, was purchased in 2001 for $15,000.

We now have all of the necessary components to approximate the present day cost (C_B) of a 5000-gallon vessel. We have two dates, past and of course current; two known capacities (S_B and S_A); and one historical cost (C_O) (that of the 2001 purchased vessel).

The first step is to determine the cost index for our two dates. Referring to Table 14.6, the CECPI index for 2001 is found to be 394.3 (our base index for this example). Consulting a recent issue of *Chemical Engineering* magazine, the CECPI index for 2006 is found to be 499.6 (the current index for this example). The engineer may be interested to know that the CECPI base of $1959 = 100$ provides an astonishing indication of the amount of inflation that has taken place.

These complied data allow us to substitute,

$$C = C_O\left(\frac{I}{I_O}\right) = (\$15{,}000)\left(\frac{499.6}{3943}\right) = \$19{,}005$$

Therefore, the 2006 cost of the 2000-gallon capacity vessel is estimated to be $19,005.

Now, having determined the current estimated cost of the smaller capacity vessel, we need to adjust this amount to correspond to the larger volume (5000 gallons). Referring to Table 14.5, we find a size exponent corresponding to stainless steel vessels equal to 0.68. Substituting in the equation presented earlier results in:

$$C_B = C_A \left(\frac{S_B}{S_A}\right)^N = (\$19,005)\left(\frac{5000}{2000}\right)^{0.68} = \$35,438$$

A "rough" estimate of $35,438 is not perfect and implies a degree of accuracy that has no basis in this case; $35,000 is more sensible and just as likely to be correct in the context of a plus or minus 20% estimate. Therefore, the approximate 2006 cost of the 5000-gallon capacity vessel is $35,000.

14.15.2 Estimating based on full design

The most accurate estimate of investment costs should be based on a detailed and complete design of each project component. Since competitive quotations should be obtained for plant equipment, machineries, erection, civil works, etc., estimating based on the full design would usually not be appropriate for a feasibility study but would occur during the project implementation phase, once the project has been approved.

- **Exponential cost estimating**

This method should be used only for similar types of plants or plant items; that is, exponential factors are only valid if the technical scope of the project and process technologies are similar. If this is not the case, the margin of error of cost estimates will be greater than the required reliability of estimate.

Exponential cost factors are published but should be checked regularly against prices obtained through tendering. (Exponential factor is published periodically by some important and reliable institutions such as Institution of Chemical Engineers, Stanford Research Institution [SRI], Nelson Refinery Index.)

14.15.3 Recommended checklists and schedules

The various aspects discussed in Section VI regarding feasibility study of projects are summarized, and some additional aspects deemed necessary to be covered in the feasibility study stage are given in the form of a checklist.

The requirements under this list should adequately be weighed and assessed for various stages of project engineering and technology selection, acquisition, and management and remarked on in the feasibility study report.

14.16 Organization and overhead costs

This section is intended to describe the process organizational planning and the structure of overhead costs, which can be decisive for the financial feasibility of the project, as it is essential for feasibility of a project that a proper organizational structure should be determined in accordance with the corporate strategies and policies.

While the other sections specifically deal with direct cost for feasibility study of the projects, this section will deal with indirect or overhead costs. It should be noted that neglecting or underestimating this cost may have a significant impact on the project profitability.

14.16.1 Plant organization and management

The organizational structure of an enterprise should indicate the delegation of responsibilities to the various functional units of the project or the company and should normally be shown in a diagram.

The organizational functions should be prepared in the form of building blocks in accordance with requirement of a specific project and grouped in the following organizational units:

- General management of the project/company
- Finance, financial control, and accounting
- Personnel administration
- Marketing, sales, and distribution
- Supplies, transport, and storage
- Production:
  - Main plant
  - Service plants
  - Quality assurance
  - Maintenance and repair

14.16.2 Organizational design

A rough outline of organizational structures and related costs may be included in the prefeasibility study, of a project, an organizational setup design should be covered in the feasibility study for the following two reasons:

- First, the organization of a project should aim at the optimal coordination and control of all project input.
- Second, the organizational set-up serves to structure the investment and production costs and to determine the costs linked with the corresponding organizational unit. These costs should be treated as overhead cost unless they can be directly related to a specific product.

To facilitate cost planning and control, the project should be divided into definite cost centers.

14.16.3 Overhead costs

The overhead costs should be grouped as outlined next:

1. *Plant overhead*: Costs that occur 36 in conjunction with the transformation, fabrication, or extraction of raw materials. Typical cost items are listed:
   a. *Wages and salaries* (including benefits and social security contributions) of manpower and employees not directly involved in production.

    b. *Plant supplies*
- Utilities (water, power, gas, steam, etc.)
- Effluent disposal
- Office supplies

    c. *Maintenance*

    These cost items should be estimated by the service cost centers where they occur.

2. *Administrative overhead*: This should only be calculated separately in cases where it is of considerable importance. Otherwise, it could be included under plant overheads. Typical cost items are:

    a. Wages and salaries
    b. Office supplies
    c. Utilities
    d. Communications
    e. Engineering
    f. Rents
    g. Insurances (property)
    h. Taxes (property)

    These cost elements should be estimated for administrative cost centers such as managements, bookkeeping and account, legal services and patents, traffic management, and public relations.

3. *Marketing overhead*: Direct selling and distribution costs, such as special packaging and forwarding costs, commissioning and discounts, should be calculated separately for each product as described in Section X. Indirect marketing costs that cannot easily be linked directly with the product cost should be treated as overhead costs.

    Typical cost items are:
    a. Wages and salaries
    b. Office supplies
    c. Utilities
    d. Communications
    e. Advertising
    f. Training and others

4. *Depreciation costs*: Annual depreciation charges are frequently included under overhead costs; however, these costs could be shown separately from overhead costs (i.e., including them for calculating plant costs and for financial evaluation).

5. *Financial costs*: These costs, such as interest on term loans, should be shown as a separate item. When forecasting overhead costs, attention should be given to the problem of inflation. In view of numerous cost items in overhead costs, it will not be possible to estimate their growth individually but as a whole due to the magnitude of the overall inflation rate of overhead costs.

14.17 Human resources

Once the production program, plant capacity technological processes to be used, and plant organization have been determined, the human resources requirement at various levels and during different stages of the project should be defined as well as their availabilities and costs.

The feasibility study should identify the project needs to different categories of human resources, management, staff, and workers with sufficient skills and experiences.

14.17.1 Categories and functions

Human resources as required for implementation and operation of industrial projects should be defined by categories, such as management and supervision personnel and skilled and unskilled workers and by functions such as general management, production management, production control, machine operation etc.

Personnel requirement should be defined by categories and function for preparation of Detailed Manning Table and Estimation of Personnel Costs.

The feasibility study should define the personnel requirements by quantity on a shift-to-shift basis or at the department level and state the qualifications and experience necessary. In order to specify the minimum training and professional experience required for different posts, the kinds of professional staff, skilled labors and unskilled workers should be provided.

14.17.2 Manning tables

Labor planning should start at the department level, defining the labor and staff requirements by functions and categories.

Available working days per year considering weekends and national holidays, etc., should be specified. The manning tables should be related to a certain production level and show how requirements are expected to develop over a period of time (e.g., whether the requirements refer to the first year of operation or some future year, the production level, and the number of shifts).

The manning tables should be used to analyze availability and requirement of human resources, as well as to estimate operating costs related to these resources.

14.17.3 Availability and recruitment

The following factors should be given due consideration when availability and employment aspects are analyzed in the feasibility study:

- The general availability of relevant human resource categories in the country and project region
- The supply-and-demand situation in the project region
- Requirement policy and methods
- Training policy and program

Assessments and estimates should as far as possible be explained and justified in the feasibility study. For example, a certain technology, safety hazards, sophisticated machinery and equipment, international market orientation, and other factors may justify special skills and experience.

The study should analyze the ability of the project to attract the human resources required. The recruitment policy and methods should therefore be assessed in the feasibility study.

The feasibility study should assess the labor needs and availability of suitable domestic managerial skills and, when foreign assistance is necessary, the duration and conditions of obtaining such assistance should be prescribed.

14.17.4 Training plan

Extensive training programs should be designed and carried out as part of the implementation process of investment projects. Training programs may need considerable funds. This may well prove to be the most necessary and appropriate investment. The requirement of training for various levels of plant personnel, the duration of such training for each category and the location and arrangements for training should be defined.

14.17.5 Labor cost estimates

The manning tables prepared for each department can be used for estimating labor costs. Practically, nonproduction labor costs should be considered as fixed costs, while the production labor costs as variable costs.

The feasibility study should present the estimated labor costs for each department and function. The cost should be divided into foreign and local currency components. When estimating the total wage and salary costs, provision should be made for the following personnel overhead costs:

- Social Security, fringe benefits, and welfare cost (if any applicable within company's policy)
- Installation grants, subsistence payments, and similar cash costs that occur in connection with recruitment and employment
- Annual deposits to pension funds
- Direct and indirect costs of training
- Payroll taxes

14.18 Implementation, planning, and budgeting

The project implementation phase embraces the period from the decision to invest to the start of commercial production. A series of simultaneous and interrelated activities taking place during the implementation phase should be defined. When preparing the implementation plan for the feasibility study, it should also be borne in mind that, at a later stage, this plan will be the basis for monitoring and controlling the actual project implementation. The implementation schedule should present the costs of project implementation, as well as the schedule for the complete cash outflows, in order to allow the determination of the corresponding inflows of funds, as required for financing the investments.

14.18.1 Objectives of implementation planning

As an essential part of the feasibility study, a realistic time schedule should be drawn up for the various stages of the project implementation phase.

The objectives of project implementation planning are considered to draw the attention of the project planner to the financial implications of project scheduling and to the possibilities of the early detection of implementation delays and their financial consequences.

A comprehensive assessment should be made at the feasibility study stage on all of the implementation costs for determining the financial requirements and the accruing financial costs. The costs of this assessment should also constitute a part of the initial investment costs.

14.18.2 Implementation scheduling

Various methods of analysis and scheduling are available. The bar chart planning method can be applied to every project, without difficulty. This method gives the best overview of the main sequence of events. Particularly, in a feasibility study and in the preinvestment phase of the project, the bar chart method is considered to be a sufficient planning tool for implementation planning.

14.18.3 Projecting the implementation budget

The feasibility study should determine the cost of resources in accordance with the timing of the various stages of project implementation. The estimated implementation costs are capitalized. Pre-production costs form part of the total initial investment costs.

    The checklist and a schedule for the preparation of the implementation budget are given in this part. The cost items are based on the implementation activities and tasks determined for the projects including such cost items as housing and transportation. The estimate should include acceptable contingencies for probable price increases when the actual starting date is delayed.

14.19 Financial analysis and investment appraisal

14.19.1 Scope and objectives of financial analysis

A feasibility study, as mentioned earlier, is a tool for providing potential investors, promoters, and financiers with the information required to decide whether to undertake an investment and whether to finance such a project.

    The financial analysis and final project appraisal should involve the assessment, analysis, and evaluation of the required project inputs; the output to be produced; and the future net profits, and these should all be mentioned in financial terms.

14.19.2 Principal aspects of financial analysis and concept of investment appraisal

An important aspect to be considered when undertaking a financial analysis is that the decision makers usually give different weights to the various criteria used for investment appraisal. The financial analysis should indicate and highlight any and all of the critical impacts that would have to be considered when appraising a project. The followings are typical aspects in this regard.

- **Interest of parties involved**

The interest in future net profits is common for each party participating in a project; the financial analysis should begin with:

- Determination of the required project inputs and generated outputs, valued at market prices.
- Determination of annual and accumulated net surpluses.
- Using the methods described in this section, the net profits (yield or profitability) generated by the investment are determined in financial terms.

- **Public interest**

The public interest can be achieved only when an investment is properly integrated within the business environment (see Section III), so an industrial investment should be considered as an integrated part of a socioeconomic and ecological system within which it performs.

- **Basic criteria for investment decisions**

- Any decision on industrial investments should be based on the following criteria relating to the overall feasibility of investment project:
  - Is there any possible conflict, at present and in the long run, between the basic project (corporate) objective and the development objectives valid for the socioeconomic environment?
  - How suitable is the proposed strategy for the achievement of the project objective? Have alternative strategies been taken into consideration? And why has the proposed strategy been selected?
  - How does the project design, that is, the scope of the project, the marketing concept, the production capacity and the technology and location selected, match with the project strategy and the availability of the required resources?
  - Will the project make efficient use of economic resources, and are there better alternative uses of the main inputs required for the project?
  - Are projections of total investment costs and production and marketing costs within the acceptable confidence level?
  - Are the total investment costs within the financial limits determined by the availability of capital?
  - Does the structure of cash outflows and inflows and of the corresponding net cash returns meet with the minimum requirements and expectations of the investors and financiers?
  - Will the supply of local money and foreign exchange be sufficient to meet outstanding financial obligations at any time during the life of the project?
  - How sensitive are the accumulated discounted returns and the annual returns to the planning horizon, to errors in data assessment and project design, to inflation and relative price changes and to changes in the business environment (mainly those involving competitors, consumers, markets, supplies, and public policies)?
  - Have critical variables been identified? What risks are associated with these variables, and what strategies exist to manage or control those risks?
  - What are the financial consequences of the risks; in other words, do they entail additions to investment costs, to the funds required, to production and marketing costs, and to finance costs, or lower-than-expected production, sales volumes, and sales prices?
  - How likely is the projected scenario or business environment required as a minimum condition for the investment to be appraised by investors, by financing institutions, and/or others?

- **Accounting systems**

- Accounting systems always cover the financial status of the firm in terms of the wealth (assets) and obligations (liabilities) recorded in balance sheet, the costs accounted for over the reporting period, and the corresponding income shown in the net income statement.

- The accounting system should determine the production and marketing costs that are necessary not only for the preparation of the net income statement but also for efficient financial planning, product pricing, and cost control.
- For liquidity planning, the cash flow statement is used. It should be pointed out that depreciation allowances should not be classified among the cash outflow but should be regarded as a cost item, and not as a cash item.
- The financial costs (interest paid) should be included among the cash outflows. However, for the computation of discounted cash flow (internal rate of return (IRR) and net present value (NPV)), the financial cost should be excluded.
- Cost accounting should provide a measurement of budgeted material costs, wages and salaries, and other expenses involved in producing and marketing the goods and services generated by the project. A profit plan that defines the cost–volume–profit relationship should be constructed.
- The classification of costs is necessary in order to facilitate cost planning (budgeting) and to permit the determination of cost items that could be critical for the feasibility of a project.

- **Pricing of project inputs and outputs**

- For the purpose of the feasibility study, prices should reflect the real economic values of project inputs and outputs for the entire planning horizon of the decision makers.
- Inflation may have to be considered in financial planning, even when the relative prices remain basically unchanged, because additional equity and loan financing may be needed to deal with significant annual inflation rates, especially during the project implementation phase.
- Working capital requirements should be checked in view not only of the gradual attainment of full capacity but also of the increased inflationary pressure on the cost items to be financed from working capital. Consequently different inflation rate should be applied to local and imported materials, utilities, labors, etc., when projecting working capitals.
- Price changes should be anticipated in projecting sales price.

- **Planning horizon and project life**

- Planning consists of the anticipations and assumptions that should be made about the future need to be explicated and should be analyzed in order to find the optimal development path.
- The project planning horizon of a decision maker may be defined as the period of time over which he or she decides to control and manage his or her project-related business activities or to formulate his or her investment development plan.
- The economic life, that is, the period over which the project would generate net gains, depends basically on the technical or technological life cycle of the main plant items.
- When determining the economic life span of the project, various factors, such as the following, should be considered:
  - Duration of demand (position in the product life cycle).
  - Duration of raw materials deposits and supply.
  - Rate of technical progress.

- Life cycle of the industry.
- Duration of building and equipment.
- Opportunities for alternative investment.
- Administrative constraints.
- Considering that the accumulated not cash flows of an investment project are a function of the time period covered in the feasibility study, the planning horizon may have a considerable impact on the results of the financial analysis.
- Since the values obtained for the discounted cash flows and the various profitability and efficiency ratios vary sometimes considerably with the length of the planning period, the determination of the planning horizon of a feasibility study is often a very critical task, and should there be considered when appraising an investment project.

14.19.3 Analysis of cost estimates

It is necessary to check carefully all cost items that could have a significant impact on financial feasibility. The estimates should be grouped into local and foreign components and may be expressed in either constant or current (real or nominal terms) prices depending on the price basis used in the feasibility study, and for financial analysis, allowances for price increases (contingencies) should be provided. Since inconsistency in the use of accounting and financial terminology often causes problems for the analysis, it is recommended that the terms defined and explained later be strictly adhered to in the feasibility study.

Initial investment costs are defined as the sum of fixed assets and net working capital. *Fixed assets* constitute the resources required for constructing and equipment of an even investment project. The *net working capital* corresponds to resources needed to operate the project totally or partially.

Investment required during plant operation in order to keep a plant in operation, each item should therefore be replaced at the appropriate time and the replacement costs must be included in feasibility study.

Preproduction expenditures include but are not limited to:

1. Expenditures for preinvestment studies.
2. Opportunity, prefeasibility, feasibility, and support of functional studies.
3. Consultant fees for preparing studies, engineering, supervision of erection, and construction (if not included in fixed investment costs).
4. Other expenses for planning the project.
5. Salaries, fringe benefits, and Social Cecurity contribution of personnel engaged during the preproduction period.
6. Travel expenses.
7. Preparatory expenses, such as worker, camps, temporary offices, and stores.
8. Preproduction marketing costs, creation of sales network.
9. Training costs, including fees, travel, living expenditures, salaries, and stipends of trainees and fees payable to external institutions.
10. Know-how and patent fees.
11. Interest on loans accrued or payable during construction.
12. Insurance costs during construction.

13. Trial runs, start-up, and commissioning expenditures.
14. Important notes:
  a. Preproduction expenditures can be tabulated.
  b. In allocating preproduction expenditures, one of two practices should generally be followed:
    – All preproduction expenditures may be capitalized and amortized over a period of time that is usually shorter than the period over which equipment is depreciated.
    – A part of preproduction expenditures may be initially allocated, where attributable, to the respective fixed assets and the sum of both amortized. The rest that are not attributable are capitalized and amortized over a certain period.

14.19.4 Fixed assets

Fixed assets comprise fixed investment costs and pre-production expenditures. Fixed investments should include the following main cost items, which may be broken down further, if required:

- Land purchase, site preparation, and improvements
- Building and civil works
- Plant machinery and equipment, including auxiliary equipment
- Certain incorporated fixed assets such as industrial property rights and lump-sum payments for know-how and patents

The estimates should include the cost of supply packing and transport, duties, and installation charges. Provisions should be made for physical contingency allowances.

14.19.5 Working capital

Working capital is defined to embrace current assets (the sum of inventories, marketable securities, prepaid items, accounts receivable, and cash) minus current liabilities (accounts payable). In the analysis of investment costs, it should be carefully checked whether the initial working capital requirements as well as changes during plant operation are properly considered in the cost estimates. The amount of working capital invested should be optimal, that is, neither too large nor too small, to avoid penalties for the project. Working capital should be carefully estimated and adequately controlled and monitored.

Working capital requirements are considerably affected by the amount of capital immobilized in the form of inventories. Every attempt should be made to reduce inventories to as low a level as justifiable. If materials are locally available and in plentiful supply and can rapidly be transported, then only limited stock should be maintained. If the materials are imported and import procedures are dilatory, then inventories equivalent to as much as 6 months' consumption may be considered for working capital estimate.

Level of spare parts inventories should depend on the local availability of supplies, import procedures and maintenance facilities in the area, and the nature of the plant itself.

Finished product inventory depends on a number of factors, such as the nature of the product and trade usage. The valuation is based on the factory costs plus administrative overhead.

Calculation of working capital at the stage of feasibility study is of particular importance since it forces the project promoter, investors, and financing institutions to think about the funds needed to finance the operation of the project compared with investment funds such as preproduction expenditures and fixed investment costs.

14.19.6 Schedules for total investment costs and total assets

By summing the fixed investments, preproduction expenditures, and working capital estimates, the total initial investment costs of the project under consideration can be calculated.

It should be noted that when phasing the total investment outlay, the initial investments should be inserted in the schedule first, and then subsequent increments, until operation of full capacity is reached.

14.19.7 Production costs

The production costs should be calculated as total annual costs and preferably also as cost per unit produced. Production costs should be determined for the different levels of capacity utilization and for an operational period corresponding to the planning horizon of the investors and financing institutions interested in the project.

All cost elements required for the calculation of total production costs therefore have to be projected and scheduled in line with the production program and for the full planning period. All of the cost items entering into production costs have been described in preceding parts. These cost elements should be assembled in order to arrive at production costs.

The production cost is estimated by dividing it into following four major categories:

1. *Factory (plant) costs*: include the following cost items:
   a. Materials, predominantly variable costs, such as raw materials, factor supplies, and spare parts
   b. Labor (production personnel) (fixed or variable costs, depending on type of labor and cost elements)
   c. Factory/plant overhead costs
2. *Administrative overheads*: The composition of administrative overhead costs as well as procedures for their computation were described previously.
3. *Depreciation costs*: Depreciation costs should be used for computation of the balanced sheet and net income projections; they present investment expenditures (cash outflow during the investment phase) instead of production expenditure (cash outflow during production).
   a. Depreciation charges should therefore be added back if not cash flows are calculated from the net income statements.
   b. Depreciation costs do have an impact on net cash flows, because the higher the depreciation charges, the lower is the taxable income and the lower is the cash outflow corresponding to the tax payable on income.

14.19.8 Unit costs of production

For the purpose of cash flow analysis, it is sufficient to calculate the annual costs. At the feasibility stage, however, an attempt should be made to calculate unit costs to facilitate the comparison with sales prices per unit. Unit costs are simply calculated by dividing production costs to the number units produced (these unit costs varies with capacity utilization).

14.19.9 **Direct and indirect costs**

Production costs should be divided into direct and indirect costs. Direct costs are attributable to production materials and production labor.

Since indirect costs (plant administrative overhead costs such as management and supervision, communications, depreciation, and financial charges) cannot easily be allocated directly to a particular units of output, the direct variable and direct fixed costs should be deducted from the revenues generated by a certain product (group of products) and the remaining surplus or margin together with the margins generated by other products will be available to cover the indirect costs.

14.20 **Method of investment appraisal**

For the purpose of investment appraisal, it is necessary to assess and evaluate over a certain period (as defined herein as "the planning horizon of the decision maker") all inputs required and all outputs produced by the project. Although the information can be contained in the net income statements and projected balance sheets, they are sufficient for feasibility evaluation, and therefore the discounted cash flow concept has become the generally accepted method for investment appraisal.

14.20.1 **Definition and computation of cash flow**

Cash flow is basically either receipts of cash (cash inflow) or payments (cash outflow). For the purpose of financial planning and determination of the net cash returns of an investment, it is necessary to distinguish between the financial flow, which is related to financing of an investment, and cash flow (expenditures and revenues), representing the performance or operation of the project (operational cash flow).

Operational cash flow are shown (as discounted cash flow) in Table 14.7:

14.20.2 **Introduction to discounted cash flow analysis and financial functions**

The financial and economic analysis of investment projects is typically carried out using the technique of discounted cash flow (DCF) analysis. This module introduces concepts of discounting and DCF analysis for the derivation of project performance criteria such as NPV, IRR, and benefit-to-cost (B/C) ratios.

- **Cash flows, compounding, and discounting**

DCF analysis is the technique used to derive economic and financial performance criteria for investment projects. It is important to review some of the basic concepts of DCF analysis before proceeding to topics such as cost-benefit analysis (CBA), financial analysis (FA), linear programming, and the estimation of nonmarket benefits.

Cash flow analysis is simply the process of identifying and categories of cash flows associated with a project or proposed course of action and making estimates of their values. For example, when considering establishment of a forestry plantation, this would involve identifying and making

Table 14.7 Operational Cash Flow

Operational Cash Flow In	Operational Cash Flow Out
Revenues from selling of fixed assets	Increase in fixed assets, (investment)
Recovery of salvage valves (end of project)	Increase in net working capital
Revenues from decrease of net working capital	Operating costs (see note)[a]
	Marketing expenses
Sales revenues	Production and distribution losses
Other income due to plant operations	Corporate (income) taxes

[a]Note: It should be noted that depreciation charges (costs) and interest payments are not classified among the operational cash outflows, because inclusion of depreciation of assets would provoke a double-counting of the costs to the project, since they are already accounted for as investment costs when capitalized in the balance. However, for accounting purposes (including taxation), assets are to be depreciated over the project lifetime. This is why the depreciation of assets is a cost item in the net income statement only and must be deducted from the annual total costs of products sold (production and marketing costs) when determining the annual cash outflows. Interest and any other costs of financing are also included for the computation of the yield or return on the total capital investment, because they are part of this total yield. However, interest on loans (but not net profits distributed) is a cost item in the net income statement.

estimates of the cash outflows associated with establishing the trees (e.g., the cost of buying or leasing the land, purchasing seedlings, and planting the seedlings), maintaining the plantation (such as cost of fertilizer, labor, pruning, and thinning), and harvesting. As well, it would be necessary to estimate the cash inflows from the plantation through sales of thinnings and timber at final harvest.

DCF analysis is an extension of simple cash flow analysis and takes into account the time value of money and the risks of investing in a project. A number of criteria are used in DCF to estimate project performance, including NPV, IRR, and B/C ratios. Before discussing criteria to measure project performance, it is necessary to introduce some concepts and procedures with respect to compounding and discounting. Let us begin with the concepts of simple and compound interest. For the moment, consider the interest rate as the cost of capital for the project.

Suppose a person has to choose between receiving $1000 now or a guaranteed $1000 in 12 months' time. A rational person will naturally choose the former, because during the intervening period he or she could use the $1000 for profitable investment (e.g., earning interest in the bank) or desired consumption. If the $1000 were invested at an annual interest rate of 8%, then over the year it would earn $80 in interest. That is, a principal of $1000 invested for 1 year at an interest rate of 8% would have a future value of $1000(1.08) or $1080.

The $1000 may be invested for a second year, in which case it will earn further interest. If the interest again accrues on the principal of $1000 only, it is known as simple interest. In this case the future value after 2 years will be $1160. On the other hand, if interest in the second year accrues on the whole $1080, known as compound interest, the future value will be $1080(1.08) or $1166.40. Most investment and borrowing situations involve compound interest, although the timing of interest payments may be such that all interest is paid before further interest accrues. The future value of the $1000 after 2 years may also be derived as:

$$\$1000(1.08)^2 = \$1166.40 \tag{14.4}$$

In general, the future value of an amount $a, invested for n years at an interest rate of i is $a(1 + i)^n$, where it is to be noted that the interest rate i is expressed as a decimal (e.g., 0.08 and not "8" for an 8% rate).

The reverse of compounding—finding the present-day equivalent to a future sum—is known as discounting. Because $1000 invested for 1 year at an interest rate of 8% would have a value of $1080 in 1 year, the present value of $1080 1 year from now, when the interest rate is 8%, is $1080/1.08 = $1000. Similarly, the present value of $1000 to be received 1 year from now, when the interest rate is 8%, is

$$\$1000/1.08 = \$925.93 \tag{14.5}$$

In general, if an amount $a is to be received after n years, and the annual interest rate is i, then the present value is

$$\$a/(1 + i)^n \tag{14.6}$$

This discussion has been in terms of amounts in a single year. Investments usually incur costs and generate income in each of a number of years. Suppose the amount of $1000 is to be received at the end of each of the next 4 years. If not discounted, the sum of these amounts would be $4000. But suppose the interest rate is 8%. What is the present value of this stream of amounts? This is obtained by discounting the amount at the end of each year by the appropriate discount factor and then summing:

$$1000/1.08 + 1000/1.08^2 + \$1000/1.08^3 + \$1000/1.08^4$$
$$= \$1000/1.08 + \$1000/1.1664 + \$1000/1.2597 + \$1000/1.3605$$
$$= \$925.93 + \$857.34 + \$793.83 + \$735.03$$
$$= \$3312.13 \tag{14.7}$$

The discount factors—$1/1.08^t$ for $t = 1$–4—may be calculated for each year or read from published tables. It is to be noted that the present value of the annual amounts is progressively reduced for each year farther into the future (from $925.93 after 1 year to $735.03 after 4 years), and the sum is approximately $700 less than if no discounting (a zero discount rate) had been applied.

14.20.3 Definition of annual net cash flows

DCF analysis is applied to the evaluation of investment projects. Such a project may involve creation of a terminating asset (e.g., a forestry plantation), infrastructure (e.g., a road or plywood plant) or research (including scientific and socioeconomic research). Any project may be regarded as generating cash flows. The term *cash flow* refers to any movement of money to or away from an investor (an individual, firm, industry, or government). Projects require payments in the form of capital outlays and annual operating costs, referred to as *cash outflows*. They give rise to receipts or revenues, referred to as *project benefits* or *cash inflows*. For each year, the difference between project benefits and capital plus operating costs is known as the net cash flow for that year. The net cash flow in any year may be defined as

$$a_t = b_t - (k_t + c_t) \tag{14.8}$$

Table 14.8 Annual Cash Flows for a Hypothetical Project

Year	Project Benefits ($)	Capital Outlays ($)	Operating Costs ($)	Net Cash Flow ($)
0	0	25,000	2000	−27,000
1	15,000	0	4000	11,000
2	15,000	0	4000	11,000
3	15,000	0	2000	13,000

where:

b_t are project benefits in year t
k_t are capital outlays in year t
c_t are operating costs in year t

It is to be noted that when determining these net cash flows, expenditure items and income items are timed for the point at which the transactions takes place, rather than the time at which they are used. Thus, for example, expenditure on purchase of an item of machinery rather than annual allowances for depreciation would enter the cash flows. It is to be further noted that cash flows should not include interest payments. The discounting procedure in a sense simulates interest payments, so to include these in the operating costs would be to double-count them.

• **Example**

A project involves an immediate outlay of $25,000, with annual expenditures in each of 3 years of $4000, and generates revenue in each of 3 years of $15,000. These cash flow data may be set out, and annual net cash flow derived, as in Table 14.8.

Two points may be noted about these cash flows. First, the capital outlay is timed for Year 0. By convention this is the beginning of the first year (i.e., right now). On the other hand, only half of the first year's operating costs are scheduled for the Year 0 (the beginning of the first year).

The remaining half of the first year's operating costs plus the first half of the second year's operating costs are scheduled for the end of the first year (or, equivalently, the beginning of the second year). In this way, operating costs are spread equally between the beginning and the end of each year. (The final half of the third year's operating costs are scheduled for the end of Year 3.) In the case of project benefits, these are assumed to accrue at the end of each year, which would be consistent with lags in production or payments.

These within-year timing issues are unlikely to make a large difference to overall project profitability, but it is useful to make these timing assumptions clear. A second point to note about Table 14.1 is that net cash flows (second column less third plus fourth column) are at first negative, but then become positive and increase over time. This is a typical pattern of well-behaved cash flows, for which performance criteria can usually be derived without computational difficulties.

14.20.4 Project performance criteria

Let us now consider a number of project performance criteria, which can be obtained through discounted cash flow analysis. These criteria will be defined and then derived for the cash flow data of the above example.

- **Net present value**

The NPV is the sum of the discounted annual cash flows. For the example, taking an interest rate of 8%, this is

$$\text{NPV} = a_0 + a_1/(1+i) + a_2/(1+i)^2 + a_3/(1+i)^3 \tag{14.9}$$

A project is regarded as economically desirable if the NPV is positive. The project can then bear the cost of capital (the interest rate) and still leave a surplus, or profit. For the example,

$$
\begin{aligned}
\text{NPV} &= -27{,}000 + 11{,}000/(1.08) + 11{,}000/(1.08)^2 + 13{,}000/(1.08)^3 \\
&= -27{,}000 + 11{,}000/1.1664 + 11{,}000/1.2597 + 13{,}000/1.3605 \\
&= \$2935.73
\end{aligned}
\tag{14.10}
$$

The interpretation of this figure is that the project can support an 8% interest rate and still generate a surplus of benefits over costs, after allowing for timing differences in these, of approximately $3000.

- **Net future value**

An alternative to the net present value is the net future value (NFV), for which annual cash flows are compounded forward to their value at the end of the project's life. Once the NPV is known, the NFV may be obtained indirectly by compounding forward the NPV by the number of years of the project life. For Example 1, the net future value is

$$\text{NFV} = \text{NPV}(1.08)^3 = \$2935.73 \times 1.3605 = 3994.06 \tag{14.11}$$

- **Internal rate of return**

The IRR is the interest rate such that the discounted sum of net cash flows is zero. If the interest rate were equal to the IRR, the net present value would be exactly zero. The IRR cannot be determined by an algebraic formula, but rather has to be approximated by trial-and-error methods. For this example, we know that the IRR is somewhere above 8%. Deriving the NPV with a range of discount rates would reveal that the IRR falls between 13% and 14% but closer to the latter. In practice, a financial function can be called up to perform the trial-and-error calculations. It would be found in this case that the IRR is about 13.8%.

The IRR is the highest interest rate which the project can support and still break even. A project is judged to be worthwhile in economic terms if the IRR is greater than the cost of capital. If this is the case, the project could have supported a higher rate of interest than was actually experienced and still made a positive payoff. In this case, the project would be profitable provided the cost of capital was less than 13.8%.

The IRR as a criterion of project profitability suffers from a number of theoretical and practical limitations. On the theoretical side, it assumes that the same rate of return is appropriate when the project is in surplus and when it is in deficit. However, the cost of borrowed funds may be quite different than the earning rate of the firm. It could be more appropriate to use two rates when determining the IRR. The actual cost of capital could be used when the project is in deficit, and the earning rate (unknown, to be determined by trial-and-error) could be applied when the project is in surplus. This would give a better indication of the earning rate of the project to the firm or government.

From a practical viewpoint, the IRR may not exist or it may not be unique. This problem may be examined in terms of the NPV profile, a graph of NPV versus the rate of interest. When the IRR is well behaved, this profile takes the form as in Figure 14.1. As the interest rate increases the NPV falls, being zero where the NPV curve crosses the interest rate axis; the IRR corresponds with this discount rate.

Consider a project for which the net cash flow in each year (including Year 0) is positive. Regardless of the interest rate, the NPV will never be zero, so it will not be possible to determine an IRR. Similarly, a project with a large initial capital outlay and for which future benefits are relatively small or negative may not have a positive NPV regardless of the interest rate, so again the curve for the NPV profile may never cross the interest rate axis.

If a project generates runs of positive and negative net cash flows, the NPV profile may take the form of a rollercoaster curve, crossing the interest rate axis in several places. This indicates multiple internal rates of return, one at each interest rate where NPV is zero. It is then by no means clear which if any of the rates we should choose to call the IRR. Further, for some sections of the NPV profile (those that are upward sloping), the NPV is increasing as the interest rate increases. This implies that the greater the cost of capital the more profitable the project. Clearly, multiple internal rates of return and perverse relationships between the NPV and the discount rate are not very satisfactory.

14.20.4.4 Benefit-to-cost ratios

A number of benefit-cost ratio concepts have been developed. For simplicity, we will consider only two concepts, referred to as the gross and net B/C ratio and defined respectively as

- Gross B/C ratio = PV of benefits/(PV of capital costs + PV of operating costs)
- Net B/C ratio = (PV of benefits − PV of operating costs)/PV of capital costs

For this project, the present value of capital outlays is $25,000, since outlays are made immediately and as a single amount. The present values of project benefits and operating costs are:

$$\text{PV of benefits} = \$15,000/1.08 + \$15,000/1.08^2 + \$15,000/1.08^3 = \$38,656.45 \qquad (14.12)$$

$$\text{PV of operating costs} = \$2000 + \$4000/1.08 + \$4000/1.08^2 + \$2000/1.08^3 = \$9418.38 \quad (14.13)$$

Hence, the B/C ratios are

$$\text{gross B/C ratio} = \frac{38,656.45}{25,000 + 9418.38} = 1.12 \qquad (14.14)$$

$$\text{net B/C ratio} = \frac{38,656.45 - 9418.38}{25,000} = 1.45 \qquad (14.15)$$

A project is judged to be worthwhile in economic terms if it has a B/C ratio is greater than unity; that is, if the present value of benefits exceeds the present value of costs (in gross or net terms). If one of the above ratios is greater than unity, then the other will also be greater than unity. In this example, the ratios are greater than unity, indicating that the project is worthwhile on economic grounds. It is not clear on logical grounds which of the ratios is the most useful. Figure 14.2 shows the NPV profile for a project

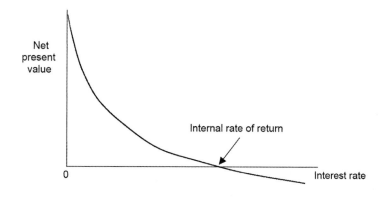

FIGURE 14.2

The net present value (NPV) profile for a project.

- **The payback period**

The payback period (PP) is the number of years for the projects to break even; that is, the number of years for which discounted annual net cash flows must be summed before the sum becomes positive (and remains positive for the remainder of the project's life). The payback period for a project with the above net cash flows can be determined as in the following table. From this table, it is apparent that the sum of discounted net cash flows does not become positive until Year 3, so the payback period is 3 years.

The PP indicates the number of years until the investment in a project is recovered. It is a useful criterion for a firm with a short planning horizon but does not take account of all the information available (i.e., the net cash flows for years beyond the PP).

- **The peak deficit**

This is a measure of the greatest amount that the project "owes" the firm or government (i.e., the farther "in the red" it goes). In Table 14.9, the largest negative value is −$27,000, so this is the peak deficit. Peak deficit is a useful measure in terms of financing a project, since it indicates the total amount of finance that will be required.

Table 14.9 Derivation of "Project Balances" and Payback Period

Year	Net Cash Flow ($)	PV of Net Cash Flow ($)	Cumulative PV of Net Cash Flow (or Project Balance) ($)
0	−27,000	−27,000	−27,000
1	11,000	10,185.19	−27,000 + 10,185.19 = −16,814.8,1
2	11,000	9430.73	−16,814.81 + 9340.73 = −7384.0.9
3	13,000	10,319.82	−7384.09 + 10,319.82 = 2935.73

Table 14.10 Summary of Definitions of Main DFC Performance Criteria

NPV	$\sum_{t=1}^{p} a_t/(1+i)^t$ where t is time, a_t is the annual net cash flow, i is the discount rate, and p is the planning horizon
IRR	The value of r such that: $\sum_{t=1}^{p} a_t/(1+r)^t = 0$
B/C ratio	Present value of project benefits/present value of project costs
PP	Number of periods until NPV becomes (and remains) positive

- **Review of DCF performance criteria**

The most commonly used discounted cash flow performance criteria—NPV, IRR, B/C ratios, and PP—may be summarized as in Table 14.10. The various criteria are closely related, but measure slightly different things. In this respect, they tend to complement one another, so that it is common to estimate and report more than one of the measures.

Perhaps the most useful measure, and the one most often reported, is the NPV. This tells the total payoff from a project. A limitation of the NPV is that it is not related to the size of the project. If one project has a slightly lower NPV than another, but the capital outlays required are much lower, then the second project will probably be the preferred one. In this sense, a rate of return measure such as the IRR is also useful.

The PP and peak deficit are useful supplementary project information for decision makers. They have greater relevance for private sector investments, for firms that cannot afford long delays in recouping expenditure, and where careful attention must be paid to the total amount of funds that will need to be committed to the project to remain solvent.

14.21 Break-even analysis

In simple words, the break-even point can be defined as a point where total costs (expenses) and total sales (revenue) are equal. Break-even point can be described as a point where there is no net profit or loss. The purpose of break-even analysis is to determine the equilibrium point at which sales revenues equal the costs of products sold. The firm just "breaks even." Any company that wants to make an abnormal profit desires to have a break-even point. Graphically, it is the point where the total cost and the total revenue curves meet.

14.21.1 Calculation (formula)

Break-even point is the number of units (N) produced that make zero profit.

$$\text{Revenue} - \text{Total costs} = 0 \tag{14.16}$$

$$\text{Total costs} = \text{Variable costs} * N + \text{Fixed costs} \tag{14.17}$$

$$\text{Revenue} = \text{Price per unit} * N \tag{14.18}$$

$$\text{Price per unit} * N - (\text{Variable costs} * N + \text{Fixed costs}) = 0 \tag{14.19}$$

So, break-even point (N) is equal

$$N = \text{Fixed costs}/(\text{Price per unit} - \text{Variable costs}) \qquad (14.20)$$

14.21.2 About the break-even point

The origins of break-even point can be found in the economic concepts of "the point of indifference." Calculating the break-even point of a company has proved to be a simple but quantitative tool for the managers. The break-even analysis, in its simplest form, facilitates an insight into the fact about revenue from a product or service incorporates the ability to cover the relevant production cost of that particular product or service or not. Moreover, the break-even point is also helpful to managers as the provided info can be used in making important decisions in business, for example preparing competitive bids, setting prices, and applying for loans.

Adding more to the point, break-even analysis is a simple tool defining the lowest quantity of sales that will include both variable and fixed costs. Moreover, such analysis facilitates the managers with a quantity that can be used to evaluate the future demand. If, in a case, the break-even point lies above the estimated demand, reflecting a loss on the product, the manager can use this information for taking various decisions. He or she might choose to discontinue the product, or improve the advertising strategies, or even re-price the product to increase demand.

Another important use of the break-even point is that it is helpful in recognizing the relevance of fixed and variable cost. The fixed cost is less with a more flexible personnel and equipment thereby resulting in a lower break-even point. The importance of a break-even point, therefore, cannot be overstated for a sound business and decision making.

However, the applicability of break-even analysis is affected by numerous assumptions. A violation of these assumptions might result in erroneous conclusions.

When sales (and the corresponding production) are below this point, the firm is making a loss, and at the point where revenues equal costs, the firm is breaking even. Break-even analysis serves to compare the planned capacity utilization with the production volume below which a firm would have losses. The break-even point can also be defined in terms of physical units produced or of the level of capacity utilization at which sales revenues and production costs are equal. The sales revenues at the break-even point represent the break-even sales value, and the unit price of a product in this situation is the break-even sales price. If the production program includes a variety of products, for any given break-even sales volume there would exist a variety of combinations of product prices but no single break-even price.

Before calculating the break-even values, the following conditions and assumptions should be satisfied:

- Production and marketing costs are a function of the production or sales volume (e.g., in the utilization of equipment).
- The volume of production equals the volume of sales.
- Fixed operating costs are the same for every volume of production.
- Variable costs vary in proportion to the volume of production, and consequently total production costs also change in proportion to the volume of production.

FIGURE 14.3

Determination of the break-even
conditions.

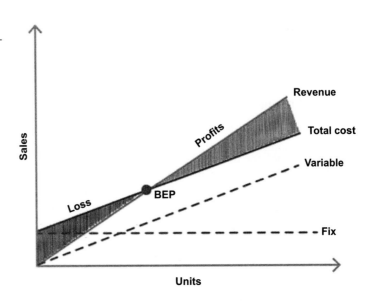

- The sales prices for a product or product mix are the same for all levels of output (sales) over time. The sales value is therefore a linear function of the sales prices and the quantity sold.
- The level of unit sales prices and variable and fixed operating costs remain constant, that is, the price elasticity of demand for inputs and outputs is zero.
- The break-even values are computed for one product; in case of a variety of products, the product mix, that is, the ratio between the quantities produced, should remain constant.

Because these assumptions may not always hold in practice, the break-even point (capacity utilization) should also be subject to sensivity analysis, assigning different fixed and variable costs as well as sales prices.

For the interpretation of the results of break-even analysis, a graphical presentation (see Figure 14.3) is very useful, because from the angle of the cost and sales curves, and the position of the equilibrium point in relation to total capacity, analysis can often identify potential weaknesses.

The break-even analysis may be carried out excluding and including costs of finance. In the latter case, the annual costs of finance need to be included in the fixed costs.

14.22 Preparation of basic engineering design data

This section covers the minimum requirements for the preparation of documents in the execution of the basic design stage of the projects applicable to the oil and gas refineries and petrochemical plants under the direction of the process engineering department.

- Preparation of basic engineering design data (BEDD)
- Data preparation of utilities (utility summary tables)

- Data preparation of effluents (preparation of data sheets in relation to gaseous and liquid effluents)
- Data preparation of catalysts and chemicals

The Basic Engineering Design Data is abbreviated to "BEDD" and should be confirmed in writing before starting the design work. The BEDD consist of a summary of basic points to be followed in the basic and detailed design that range over all specialty fields.

14.22.1 Contents of the BEDD

The contents of the BEDD can be classified as follows:

Design capacity of all process units, utility facilities, offsite, and auxiliary systems: turn down ratio may be specified if required.

System of measurements: applicable laws, codes, standards and/or design criteria to be followed and so forth

Utility conditions: conditions of utilities such as air, raw water, cooling water, steam, condensate, fuel and electric power that will be used in the plant

Flare and blow-down conditions: specifications of relieving fluid during emergency cases, depressuring to flare at emergencies, and requirements for waste disposal

Bases for equipment: interchangeability, selection basis, etc. for the standardization of equipment in the entire plant

Bases for instrumentation: basic requirements for the standardization of control systems and instruments in the entire plant

Equipment layout: for safety distances and limitations of erection and maintenance work of the equipment in the plant site

Environmental regulations: limitations on the emissions of noise, waste water, and other disposed wastes

Site conditions: weather conditions, soil conditions, sea conditions (if applicable), site location and geographical data, meteorological data and elevations

Miscellaneous: owner's requests, desires, and thoughts such as those on entire plants and plant buildings that are to be reflected in the basic design

14.22.2 Timing

Generally, all items of BEDD should be decided before starting of the process design. However, any item that is not needed to be filled at this stage should be settled with the progress of the project work. Since some detailed requirements of the detailed design cannot be covered by BEDD, detailed engineering design data (abbreviated to DEDD) are prepared in some cases to maintain the unification of equipment detailed design (if required).

14.22.3 Procedure

BEDD should be prepared by the company's or consultant's project engineer under the cooperation of specialist engineers, but many items of BEDD should be decided from the standpoints of overall plant safety and maintenance rather than from the standpoints of a single unit. Also, future/existing plants should be taken into consideration in the preparation of BEDD.

14.22.4 Explanations on individual items of BEDD

- **Design capacities**

Design capacity and/or philosophy of capacity selection for all units including process, offsite utilities, and all auxiliary facilities/systems such as air, water, fuel, product loading, flare, etc. should be indicated.

- **Standards for design and construction**

The standards/specifications to be followed by the basic designer should be clarified and a complete list of such standards/specifications should be added in BEDD.

In case the list of standards is excluded and will be provided separately, reference to the relevant document should be made.

- **Laws and codes**

Various laws, codes, and regulations are enforced by the national or local governments to secure the safety of plant facilities and around the plant and to prevent the environmental pollution (air, water, noise, etc.). In the design of plants, the legal requirements should be satisfied and the applicable laws and codes should be mentioned.

- **Design criteria**

The applicable document (if any) covering design criteria that is to be followed through the project design phase should be referred to. Design criteria normally are issued apart from BEDD and are agreed to in advance by the company and the designer.

- **Products and product specifications**

A table should be provided to demonstrate all products that are supposed to be produced during plant normal/design operations. Product specifications to be followed in the design stage should be clarified and reference to the applicable document should be made. Finished products and byproducts should be separately noted.

14.22.5 Utility conditions

In order to proceed with the design of process units, it is necessary to decide the utility conditions to equalize design bases for each process unit. The utility conditions should be decided based on the requirements on the plant design. They may be affected by the approximate consumptions, weather conditions, plot plan, waste heat recovery methods, locality conditions, etc.

Generally, as many factors remain uncertain at the stage when the utility conditions must be decided, economic studies cannot be conducted precisely at that state. Hence, when the basic plan is marked out, the utility conditions are preliminarily determined by fully studying the economics, and subsequently the utility conditions should be finally decided, so that the efficiency of each equipment can be maximized.

The following note should be added to the first sheet of the utility conditions:

"All utility information set forth in this BEDD will be confirmed during the detailed engineering stage."

- **Utility services**

The following utility services should be covered in the BEDD as applicable.

- Steam
- Water
- Condensate
- Fuel
- Air
- Nitrogen
- Electrical power
- Others

- **Steam**

Steam should include all various types of steams as foreseen in the plant design (e.g., HPS, MPS, LPS, LLPS, etc.). A table should be provided to show process and utility battery limit conditions as well as equipment mechanical design conditions for all types of steams. The process and utility battery limit conditions should cover the followings:

1. Producer battery limit (pressure and temperature)
2. Consumer battery limit (pressure and temperature)

The equipment mechanical design conditions should cover the followings:

1. Piping (design pressure and design temperature)
2. Vessels and exchangers (design pressure and design temperature)
3. Turbines (design pressure and design temperature)

Where required, the pressure and temperature mentioned under the battery limit conditions should cover minimum, normal, and maximum cases.

Design pressure specified for equipment mechanical design should not be less than system safety valve set pressure. Desuperheating conditions of any type of supplied steams in the plant should be taken into consideration and a note to be added to define the conditions where required.

Operating temperature for the reboilers for process design considerations should be noted. A separate table should be provided to present turbine inlet conditions for HPS and MPS cases in process and utility areas. The table should include pressure and temperature for the following conditions:

1. Minimum
2. Normal
3. Maximum
4. Mechanical design

- **Water operating and design conditions**

This section should include the following types of waters where applicable:

1. HP boiler feed water
2. MP boiler feed water
3. Cooling water supply
4. Cooling water return
5. Raw water
6. Plant (service) water
7. Drinking water
8. Fire water
9. Demineralized water
10. Desalinated water

A table should be provided to show process/utility battery limit conditions and equipment mechanical design conditions for each type of water.

The process and utility battery limit conditions should cover:

1. Producer battery limit (pressure and temperature)
2. Consumer battery limit (pressure and temperature)

The equipment mechanical design conditions should include design pressure and design temperature for:

1. Piping
2. Vessels and exchangers
3. Turbines
4. Compressors/pumps (if applicable)

Where required, the pressure and temperature mentioned under the battery limit conditions should cover minimum, normal, and maximum cases. Allowable pressure drop for cooler, condenser, and machinery cooling equipment to be mentioned as a note. Maximum cooling water return temperature (cooling tower design case) should be noted.

- **Water specification**

A table should be provided to cover the following characteristics for services such as circulating cooling water, cooling tower make-up, raw water/seawater, and treated boiler feed water (where applicable):

1. Source and return (if needed)
2. Availability over use (in dm^3/s)
3. Value (in dollar/1000 dm^3)
4. pH
5. Total hardness as $CaCO_3$ (in mg/kg)
6. Calcium as $CaCO_3$ (in mg/kg)
7. Magnesium as $CaCO_3$ (in mg/kg)
8. Total alkalinity as $CaCO_3$ (in mg/kg)
9. Sodium as $CaCO_3$ (in mg/kg)
10. Potassium as $CaCO_3$ (in mg/kg)

11. Sulfate as $CaCO_3$ (in mg/kg)
12. Chloride as $CaCO_3$ (in mg/kg)
13. Nitrate as $CaCO_3$ (in mg/kg)
14. Silica as SiO_2 (in mg/kg)
15. Total iron (in mg/kg)
16. Suspended solids (in mg/kg)
17. Dissolved solids (in mg/kg)
18. COD (in mg/kg)
19. Others

Cooling tower design conditions such as wet bulb temperature, type of treating system, cycles of concentration, filteration, etc., should be noted.

- **Condensate**

All various types of condensates such as HP hot condensate, LP hot condensate, cold condensate, and pump flashed condensate as foreseen in the plant design should be included. A table should be provided to show process and utility battery limit conditions as well as equipment mechanical design conditions for all types of condensates. The process and utility battery limit conditions should cover pressure and temperature for the following cases:

1. Producer battery limit
2. Consumer battery limit

The equipment mechanical design conditions should cover design pressure and design temperature for the following items:

1. Piping
2. Vessels and exchangers
3. Turbines

- **Electrical power**

- The frequency of the whole electrical system should be specified.
- The electrical system voltage levels throughout the plant should be indicated.
- Conformity of the voltages to the motors should be tabulated according to the motor size.
- Control voltage for the motor starter should be mentioned.

- **Fuel specification**

- A table should be provided to include:
  - The following types of fuels as applicable:
    - Fuel oil
    - Naphtha
    - Start-up oil
    - Blended plant fuel gas (minimum LHV conditions)
    - Blended plant fuel gas (maximum LHV conditions)
    - Natural gas

- The following characteristics for each type of the fuels.
  - API gravity for liquid fuels and relative density at 15.6 °C for all types of fuels
  - Viscosity at 100 °C for liquid fuels (in Pa s)
  - Viscosity at the burner operating temperature for liquid fuels (in Pa s)
  - Temperature at burners (in °C)
  - Lower heating value for liquid fuels (in kJ/kg)
  - Lower heating value for gas fuels (in MJ/Nm^3)
  - Availability over use (in m^3/h)
  - Vanadium/nickel (in mg/kg)
  - Sodium (in mg/kg)
  - Sulfur (in mg/kg)
  - Ash content (in mg/kg)
  - Flash point (in °C)
  - H$_2$S (in mg/kg)
  - Header pressure, in normal (bar(ga))
  - Header temperature (in °C)
- The following information should be added under the fuel specification table:
  - Maximum amount of hydrogen content for the blended plant fuel gas
  - Sources and compositions of the blended plant fuel gas
  - Source(s) of the fuel oil and start-up oil
  - Composition of the natural gas
  - Source (s) of the naphtha fuel

- **Operating and design conditions**

A table should be provided to show process and utility battery limit conditions as well as equipment mechanical design conditions for the following items and any other types of fuels as required:

- Fuel oil supply
- Fuel oil return
- Blended plant fuel gas
- Naphtha (if applicable)
- Natural gas

The process and utility battery limit conditions should cover the pressure and temperature for the following cases:

- Producer battery limit
- Consumer battery limit

The equipment mechanical design conditions should cover the design pressure and design temperature for piping, vessels, and exchangers.

- **Nitrogen gas**

- Pressure and temperature should be specified for the following requirements:
  - Producer (operating conditions) at unit battery limit
  - Consumer (operating conditions) at unit battery limit

- Mechanical equipment (design conditions)
- Nitrogen composition should be specified.
- Indication of provision for any independent and dedicated nitrogen distribution system for the catalytic units (if any)

- **Air**

A table similar to the fuel case should be provided to cover the following services:

1. Plant air
2. Instrument air
3. Catalyst regeneration air

A separate table should be provided to cover all services mentioned earlier for the following information:

1. Availability (in Nm^3/h)
2. Driver type of compressor
3. Dry air dew point
4. Oil-free air requirement

Total number of compressors and the compressors in continuous operation should be noted.

- **Flare and blow-down conditions**

Selection criteria of pressure-relieving valves for atmospheric or closed discharge blow-down should include the following requirements:

- The pressure-relieving valves, which should be discharged to the closed system.
- The pressure-relieving valves, which may be discharged to the atmosphere.
- Disposal of voluntary and involuntary liquid relief streams discharges.
- Total number of flare stacks including H_2S flare.
- Total number and service of flare KO drums.
- Status of H_2S flare stack.
- Selection criteria for pressure-relieving valves, which should be discharged into the H_2S flare (acid flare).
- Flare system design pressure and maximum allowable built-up back pressure for safety relief valve calculations.
- Number of main flare headers throughout the whole plant.
- Disposal of recovered oil and oily water from the flare KO drums and flare seal drum(s).

14.22.6 Bases for equipment

- **Vessels and columns**

The following basic design data requirements should be included in BEDD if not specified in the design criteria:

- Types of trays, packing and/or materials which are required.
- Minimum tray spacing.

- Flooding factors for hydraulic design of towers.
- Required residence time for all vessels, columns, knock out drums and all draw-offs.
- Minimum and maximum percent of normal flow rate which should be considered for design of tower hydraulic.
- Towers, vessels and vessel boots minimum diameter.
- Any known diameter, length, or mass limitation for shipping or shop fabrication of vessels (if any).
- Provision of separate steam out nozzle on all vessels.
- Vent, steam out and drain nozzles should be according to the following Table 14.11:

Vent connections must be located on the top of the vessels.

On all horizontal vessels, a blanked off ventilation nozzle should be provided on the top of the vessel near the end opposite the manway. The ventilation nozzle will be sized as follows:

- Diameter nominal (DN) 100 (4 in) nozzle for vessels up to 4450 mm tangent length;
- DN 150 (6 in) nozzle for vessels 4500–7450 mm tangent length;
- DN 200 (8 in) nozzle for vessels 7500 mm and longer tangent length.

- **Storage tanks and offsite facilities**

The following requirements should be specified on the BEDD. The number and capacity selection policy of storage tanks should be given separately for:

- Feed tanks.
- Intermediate product tanks.
- Finished product tanks.
- Maximum blending time for preparation of each finished product.
- Type of blending of the finished products.
- Basic philosophy for selection of type of the tanks.
- Height of the tanks.
- Type of firefighting facilities to be considered for various types of tanks.
- Type of product loading and maximum operating time per day of the loading facilities.
- Gas blanketing source and requirement for the storage tanks if applicable.

Table 14.11 Vent, Steam Out and Drain Nozzles Size

Vessel ID	Drain Size	Vent Size	Steam Out Nozzle
1200 mm and less	DN 40 (1½ in)	DN 40 (1½ in)	DN 25 (1 in)
1200 to 2500	DN 50 (2 in)	DN 50 (2 in)	DN 40 (1½ in)
2500 to 3500	DN 80 (3 in)	DN 80 (3 in)	DN 40 (1½ in)
3500 to 6000	DN 80 (3 in)	DN 100 (4 in)	DN 50 (2 in)
6000 and larger	DN 80 (3 in)	DN 100 (4 in)	DN 80 (3 in)

- **Air coolers**

The following notes should be specified in this section:

1. Air cooled exchangers should be used to maximum extent unless otherwise specified.
2. For air coolers, a 100 ton tower crane should be able to remove the bundle from its installed point.
3. Preferred tube length is 9114 mm (30 ft). Standard lengths are 4572 (15), 6096 (20), 7315 (25), and 9114 mm (30 ft).
4. Process fluid should be cooled to 60 °C unless otherwise noted on the process data sheet.
5. Overdesign capacity should be considered.
6. The dry bulb temperature and relative humidity for air cooler sizing should be noted.

- **Shell and tube heat exchangers**

The following requirements should be noted:

- Preferred straight tube lengths are 3048 (10), 4877 (16), and 6096 mm (20 ft). For U-tube units, the maximum nominal length (from tube ends to bend tangent) will be limited to the straight tube length.
- Preferred carbon steel and low alloy (up to and including 5 Cr-½Mo) tube size is DN 25 (1 in), 12 Birmingham Wire Gauge (BWG) and DN 20 (¾ in), 14 BWG.
- Preferred brass or admiralty tube size is DN 25 (1 in), 14 BWG and DN 20 (¾ in), 16 BWG.
- The limitation of bundle diameter is 1140 mm maximum for heat exchangers and 1524 mm for kettle type.

Positions of temperature indicators around heat exchangers should be as follow:

1. All shell and tube process/process exchangers should have a temperature indicator (TI) in the control room at the inlet and outlet of each stream.
2. For water coolers, the water side outlet should be provided with a local TI. The shell side in and out should have a TI in the control room.
3. Thermowells should be provided between each shell side and tube side of the same service.

The fouling factors of all services for air coolers and shell and tube heat exchangers should be tabulated for standardization. Provision of four-way back-flushing valves for all water cooled exchangers should be noted, and overdesign capacity should be considered.

- **Burners**

Type of the burners for all processes and utility areas should be tabulated based on the following categories:

1. Gas burners only, without provisions for the future installation of oil burners.
2. Gas burners initially, with provision for the future installation of oil burners.
3. Gas burners for on-stream operation, with oil burners for start-up and stand-by purposes.
4. Oil burners only.
5. Combination of oil and gas burners arranged to fire either or both fuels alternately or simultaneously at full load conditions.

6. Special burners designed for the process waste gas or liquid.

7. Others.

Any vertical or horizontal firing arrangement requirement for either fuel oil or fuel gas firing should be noted. The following provisions should be considered:

1. A pilot burner should be provided for each burner unless otherwise indicated.

2. When fuel oil firing is specified, the heater convection section should be bare tubes only and provision for initial installation of soot blowers in the convection section should be made.

When fuel oil firing is required, the atomizing medium and the respective pressure and temperature at the unit battery limit need to be specified.

- **Heater efficiency**

Minimum heater efficiency is to be indicated for each item. Respectively, the bases of efficiency calculations should be clarified for the following items:

- Heater throughput (e.g. normal, design, etc.)
- Low heating value of fuel
- Excess air for fuel oil and fuel gas
- Ambient temperature
- Heater maximum heat loss

As it is intended to achieve higher heater efficiency, provision of the following facilities for recovery of waste heat from flue gas for each heater should be clarified:

1. Steam generation
   a. Pressure at unit battery limit (normal and maximum)
   b. Temperature at unit battery limit (normal and maximum)
2. Air preheating
   a. Preferred type
      – Recuperative (stationary)
      – Regenerative (rotary)
      – Others
   b. Spare requirements for forced and induced draft fans
   c. Air preheater section failure would require shut down of heater. It should be indicated, if bypass of air preheat section is desired and percent of normal heater duty to be provided
3. Others

- **Stacks**

- Provision of individual or common stacks for heaters and boilers to be noted.
- Minimum stack height above grade to be specified.
- Any special heater design requirements relating to flue gas emissions such as "Low NO_x emissions" should be indicated.
- Overdesign capacity should be considered.

- **Pumps and compressors**

- Any necessary instructions relating to selection of drivers for rotating equipment should be specified.
- Spare selection philosophy for the pumps and compressors should be clarified in the BEDD.

The following information for air blower design should be specified:

1. Relative humidity
2. Dry bulb temperature

Any provision for construction of pumps and compressors building(s)/shelter(s) should be noted.

14.22.7 Basic requirements for instrumentation

The basic requirements for instrumentation should be reviewed fully and determined to meet future plant expansion and standardization policy. Further requirements such as upgradability and open system characteristics should be highly valued.

The following requirements should be clarified:

14.22.7.1 Type of control system

1. Microprocessor-based digital control system (either single loop or distributed control system-shared display). In this case, the following requirements to be specified:
   a. Maximum number of loops per controller
   b. Status of the automatic back-up controllers in case of microprocessor-based controllers
   c. Safety requirement in designing control systems such as redundancy of data highway, redundancy of consoles, etc.
   d. Extent of application if digital control system is required or mixed with analog system
   e. Any other additional requirement
2. Analog (pneumatic or electronic): extent of application in the plant if required for special cases
   a. Type of recorders
   b. Type of transmitters
   c. Type of temperature-measuring sensor required
   d. Process stream analyzers required for any specific service including environmental protection requirements
   e. Any specific requirement to be considered for location selection of control room(s)
   f. Distribution of control activities and responsibilities between control room(s) and control stations considering:
      – Number of stations per control room
      – Maximum number of loops per station
      – Number of CRT consoles per each station
   g. Extent of provision for advanced control system and optimization to be clarified

Instrument calibrations are to be specified according to the following table:

1. Pressure (in bar (ga))
2. Temperature (in °C)

3. Flow:
- **a.** Liquid (in m^3/h)
- **b.** Vapor (in Nm^3/h)
- **c.** Steam (in kg/h)
- **d.** Chemicals (in m^3/h or dm^3/s)
- **e.** Water (in m^3/h)

Any special flow metering requirements such as PD meters are to be specified.

14.22.8 Environmental regulations

Any specific environmental regulations are to be considered in the design of the plant should be noted.
A table should be provided to cover the maximum levels of the pollutants in air such as:

- H_2S (in mg/kg)
- CO (in mg/kg)
- SO_2 (in mg/kg)
- NO_x (in mg/kg)
- Hydrocarbons (in mg/kg)
- Particles (in mg/kg)

Disposal of the waste waters effluent from the plant should be clarified. The allowable limits of the following characteristics of the effluent water discharged to the public waters and/or recycled to the process should be specified:

- BOD_5 (in mg/l)
- COD (in mg/l)
- Phenol (in mg/l)
- Any toxic material (in mg/l)
- Oil (in mg/l)
- TSS in (mg/l)
- TDS (in mg/l)

- **Site conditions**

The following information should be indicated:

- Site location geographical data
- Longitude
- Latitude
- Site location with respect to the nearest city
- Site boundary (at four directions)
- Coordinates
- Accessibility (for heavy equipment and large apparatus)
- Site condition and soil report

(Reference to the site soil report and topographical survey drawings should be made.)

- **Climatic data**
- Temperature:
  - Maximum recorded
  - Minimum recorded
  - Winterizing
  - Wet bulb[a]
  - Dry bulb
- Precipitation
  - Maximum in 24 h
  - Maximum in 1 h
  - Rainy season months
- Prevailing wind direction
- Design wind velocity
- Design snow loading
- Frost line
- Water table
- Seismic conditions
- Barometric normal pressure (bar (abs))
- Humidity of air (relative humidity percent for maximum, normal, and minimum conditions)

14.22.9 Soil conditions

- Bearing value:
  - For combined dead + live load
  - For all loads + wind and seismic
- Foundation depth
- Ground water level
- Number of piles required

- **Site elevations**

- Refinery and or complex/plant site elevation above sea level.
- Designated area elevations (reference should be made to the relevant topographical drawings).
- Baseline: Baseline should be 200 mm above high point of finished grade. This figure should be used for hydraulic design calculations.
- Minimum height for finished top of foundations and high points of finished floors in building should be at baseline, unless otherwise noted.
- Units elevations.
- Elevation difference between two adjacent units.
- Sea conditions such as waves, currents, tides, etc., where applicable.

---

[a]Note: The wet bulb temperature used for cooling tower design should be based on the local conditions and effect of cooling tower vaporization.

- **Miscellaneous**
- *Buildings*

Indicate the preferred type, number, and construction of buildings for control rooms, substations, pump and compressor shelters, and other buildings as required.

- *Fireproofing*

The extent of fireproofing for process vessel skirts, supporting structural steelwork, and pipe racks should be specified.

14.23 Data preparation of utilities (utility summary tables)

The utilities such as water, steam, electrical power, etc., used in the processing plant should be specified in the "Utility Summary Tables." The summary tables should also indicate instruments and plant air, nitrogen, and inert gas, as need arises.

14.23.1 Types of utilities

Utilities, herein referred to, consist of the following items:

1. Electricity
2. Steam
3. Condensate and boiler feed water
4. Cooling water (including tempered water and cooling water for mechanical cooling)
5. Industrial water such as demineralized water
6. Fuel oil and fuel gas
7. Instrument air and plant air
8. Natural gas
9. Nitrogen (and any other inert gases)
10. Potable (drinking) water
11. Raw water

14.23.2 Operational cases

The following operation modes should be considered as required:

1. Normal operation
2. Peak operation
3. Block operation
4. Start-up operation
5. Emergency
6. Shut-down
7. Reduced operation

14.23.3 Utilities to be specified

The operational cases should be precisely defined in the design criteria that are to be used for the entire project.

The following issues should be at least specified.

- **Normal operation**

The number of operating modes as design basis according to differences in the quantity and specification of raw materials or products should be specified.

- **Peak operation**

The operation of the process units at the maximum throughput in steady state conditions and production of on specification products should be clarified.

- **Block operation**

Where the operation of part of process units is stopped extending over a long period of time, it is necessary to give definite form to such combination of process units. An example is the periodic shut-down of residue desulfurization unit for the change of catalyst with the shut-down of the hydrogen plant.

- **Start-up operation**

A Start-up sequence for each process unit should be made clear.

- Shut-down operation

Utility requirements for the normal shut-down operation should be clarified.

- **Emergency shut-down**

In most cases, power failure becomes the most severe condition for the design of utility facilities. Utility facilities, therefore, should be designed solely to cope with such conditions. However, where part of utilities is supplied by the outside facilities, it is necessary to check the conditions that such utility supply has been suspended.

- **Reduced operation**

The requirements for the operation of process units extending over long periods of time at loads lower than the design load should be made clear.

14.23.4 **Necessary information**

- **Method of preparing utility summary**

The utility summary should be prepared for all necessary utilities. Where there are several operating modes, a utility summary should be prepared for the mode that may be the most severe condition for the utility facilities. Where several operating modes become critical, a utility summary should be prepared for such modes.

- **Seasonal fluctuations**

Seasonal fluctuations in utility consumption for onsite and offsite units should be clearly prepared. Utility consumption of the following items fluctuate seasonally:

1. Heating equipment for buildings
2. Tank heaters

3. Piping traces

4. Winterizing tracing

It is necessary, therefore, to indicate steam and cooling water consumption for winter and summer seasons, respectively. Should the seasonal fluctuation of utility consumption of process units be required, due consideration should be given to such requirement and a utility summary in midwinter based on winterizing temperature should be prepared.

- **Electricity consumption**

Electricity consumption can be represented by motor rating, pump Break kilowatt Power (BkW), or motor electrical supply. Accordingly, the factor based on the electricity consumption should be clarified.

For contracted jobs, electricity consumption should be indicated in terms of supply electricity to motors. However, where electricity consumption must be calculated correctly, electricity consumption should be indicated by the value obtained by dividing BkW by the motor efficiency. Whether motor rating or pump BkW is used, the method of calculation for electricity consumption should be clearly indicated.

- **Intermittent users**

Frequency of and time of utility consumption by intermittent users and combination of those who simultaneously use the same utilities should be indicated. Intermittent users continuously using utilities for more than 8 h per day should be defined as continuous users.

The purpose of defining intermittent users is to grasp loads that must be added to the utility facilities concerned. In most cases, such an additional load can be covered by the surplus capacity of the respective utility facility. Where the frequency in use of utilities is low (several times a year), due consideration should be given to the use of spare facilities.

- **Consumption of inert gas**

The consumption of inert gases is liable to be underestimated or overestimated. In general, it is necessary to grasp the reasonable consumption not only for each piece of equipment (to be used continuously) but also for each purpose of use. Along with this, the necessity of inert gases must be fully checked. Inert gases are used in particularly large quantities to seal the shaft of rotating machinery. In most cases, the amount of inert gas used to seal the shaft of rotating machinery exceeds the design values and must be fully studied with the vendors concerned with regard to the appropriate consumption.

Users continuously using inert gases should take measures to ensure that pressures are controlled on the onsite side and that the consumption can be monitored by installing flowmeters.

- **Utilities liable to be omitted**

The following utilities are liable to be omitted from initial utility summary. This poses problems in the satisfactory execution of engineering work.

1. Atomizing steam

2. Soot blower steam

3. Steam tracing steam

4. Decoking and snuffing steam

- **Extra capacity allowance**

It is necessary for the utility side to determine the design flow rate of utility generating units by adding an allowance to the maximum necessary consumption.

14.23.5 Block operation/reduced operation

The method of preparing utility summary "Block Operation" or "Reduced Operation" may have an adverse effect solely on steam balance. Utilities other than steam show the same consumption or a tendency to decrease. Consequently, where the onsite facilities generate or consume steam in large quantities, utility summary should be prepared on the onsite side solely for steam-related items (including steam, condensate, and BFW).

- **Precautions**

In the case of block operation, it is necessary to check whether steam generation and consumption greatly fluctuate within the block concerned. If the steam generation and consumption fluctuate in large quantities, the steam balance of all facilities including the offsite facilities should be reviewed. It is desirable that due consideration be given to the amount of steam generation being balanced with steam consumption within each block as much as possible.

Special attention should be made to the existence of a large steam turbine that results in an increase in steam consumption relative to reduced operations.

During block operation, decoking or other operations are sometimes carried out for the units whose operation remains stopped.

14.23.6 Start-up operation

The utilities that may become critical during the start-up operation are steam, electricity, fuel, air, nitrogen, and inert gases. A utility summary, therefore, should be prepared for these utilities.

- **Precautions**
- *Steam*

A steam generator, where it exists within the onsite process area (if any), can be used as a steam-generating source during normal operation.

However, because it is impossible to use the steam generator existing within the onsite process area as a steam-generating source during start-up operation, it is necessary to supply steam (from the outside facility) to the user normally using steam generated by the onsite generator. Consequently, a maximum amount of steam is supplied, during the start-up operation, from the offsite steam generator.

A steam balance, therefore, should be established on the basis of the time required for start-up operation, during which a maximum amount of steam is consumed. Care should be taken in avoiding the omission of purge steam. Where start-up operation becomes the design conditions for steam boilers, the design flow rate of steam boilers should be reduced with due consideration given to the following points:

1. Stagger start-up time for each unit.
2. Slowly initiate start-up for each unit.

3. Operate the units in steady state operation at the lower limit of turn-down.

- **Inert gas/nitrogen**

Inert gases and nitrogen are used normally for purging and gas blanketing. It is possible to take sufficient time for purging prior to start-up operation. Purging prior to start-up operation, therefore, does not become critical, compared with purging during shut-down. From such standpoint, the maximum consumption of inert gas/nitrogen should be determined.

In addition, it is necessary to establish purging procedures in such a way that the simultaneous use of inert gases/nitrogen can be minimized. For N_2 purging during initial start-up, liquid N_2 can be considered to be supplied by tank trucks or N_2 cylinders from the facilities outside the plant and/or from the N_2 production unit in case of availability. Special care should be taken to provision of a dedicated N_2 source and supply header to the some of the catalytic units (e.g., continuous catalytic regeneration unit) if instructed by the licensor.

14.23.7 Shut-down operation

It is possible for the stipulations of the "Emergency Shut-down" to cover the requirements for "Shut-down Operation."

Utility consumption for the decoking of heaters should be checked.

- **Emergency shut-down**

In most cases, "emergency shut-down" plays a very vital role to establish an optimum utility facility design, particularly, the philosophy of utility facility, together with "normal operations" For instance, where there are the units or systems for which emergency shut-down must be avoided, the type of utility necessary for such units or systems and the time and amount of use should be established.

It is necessary to design utility facilities in such a manner that the onsite process units can be shut-down safely. From such a standpoint, the following cases can be cited as precautionary points.

1. Where there is the possibility of equipment being damaged due to runaway arising from exothermic reactions:
Example: Hydrocracking unit reactor and PVC polymerization reactor
2. Where it is necessary to urgently depressurize the units due to the existence of a large quantity of high-pressure flammable gases:
Example: Hydrocracking unit
3. Where solidification occurs due to cooling:
Example: Hot oil and liquid sulfur handling unit
4. Where it is necessary to urgently transfer flammable materials
5. Firefighting facilities (not onsite process units).

It is necessary to continue the supply of necessary utilities during the shut-down of plant operation.

- **Method of preparing utility summary**

The types, consumption, and time of consumption of utilities should be listed on the onsite side. In this case, care should be taken in the relevancy between each utility, in order to avoid the omission of necessary utilities. For example, should the cooling of the reaction system of the hydrocracking unit be considered, the following items of equipment must be continuously supplied with necessary utilities.

1. Recycle gas compressor
2. Cooler (AFC or water)
3. Cooling water, if the compressor driver is a condensing turbine
4. Compressor, auxiliary equipment of turbine
5. Cooling water for mechanical cooling
6. Instrument air

Items 4 through 6 are often overlooked. Care should be taken in avoiding the omission of utilities to continue the operation of the utility facilities. For instance, if steam boilers must be kept in operation, it is necessary to keep FDFs, boiler feed water pumps, etc., operating.

- **Measures**

The following measures should be taken into account according to the consumption and time of consumption of utilities:

1. Use of steam turbines or diesel engines as drivers
2. Separation of cooling system or establishment of two systems

In the case of little time (maximum of 1 h), a necessary amount of cooling water should be supplied by the use of holders, basins, etc. (to know the details, thermal calculations should be carried out). Where the time of utility consumption exceeds 1 h, the installation of emergency generators and so on should be planned.

The capacity of emergency generators (UPS or capacity of diesel engine) should be determined by adding the power consumption of the utility facilities and safety devices to the total power consumption necessary for emergency shut-down.

Large pieces of equipment and units consuming utilities in large quantities should be listed. Along with this, the equipment and units to be shut-down as a result of the shut-down of large-sized equipment and units should be detailed.

Using the steam balance sheet in normal operation, each operational case should be studied to see if the required capacity of the utility facilities can be fully covered.

- **Precautions**
- *Prevention of excessive design*

In the design of utility facilities, the supply of all utilities extending over a long period of time should be noted.

- *Inert gas/nitrogen*

For inert gas/nitrogen gas purging, it is necessary to classify units and items of equipment requiring prompt purging and those not requiring prompt purging. On the basis of such classification, the necessary consumption and time of consumption of inert and nitrogen gases should be made clear. Thus, the capacity and flow rate of the inert gas/nitrogen supply systems can be determined.

14.23.8 Other information

In addition to the requirements based on each operating mode, the following information should be prepared:

1. Classification of drivers that must be turbines
   a. Classification of drivers that must be preferably turbines
   b. Classification of drivers that may select motors or turbines
2. Selection of steam level of onsite steam generators
3. Equipment requiring boiler feed water or treated water having normal temperature (including required water quality, acceptability, and/or mixture of chemicals)

14.23.9 Utility summary

- **Preparation**

After obtaining the information listed, a utility balance sheet, which may become critical, should be prepared for each utility. Then, the capacity of each utility facility should be determined. To avoid a change in the capacity of each utility facility during the progress of job execution, the reliability of utility consumption data and particularly large-scale users should be fully checked.

In addition, after studying the steam balance of the entire plant, the necessary change of the waste heat recovery system (e.g., use of air preheaters, etc., instead of steam generators) or the necessity of change of driver specifications (change in the type of turbines from back pressure turbines to condensing turbines, etc.) should be studied.

- **Change**

It is inevitable to change the utility summary to a certain extent. However, to reduce the man-hours necessary for preparing and modifying utility balance sheets, changes should be made simultaneously where necessary.

14.24 Data preparation of effluents

Because air and water pollution controls are strictly required by legislation of the country or local regulations, the discharge amount of such pollutants should be confirmed in advance of the public application of a plant. Should such an amount exceed the specified values, appropriate treating facilities should be planned and the approval of the government authorities concerned should be duly obtained in advance. For that purpose, the effluents summary sheet is considered tp be important.

The discharge amounts of all the pollutants that will pollute the environment should be calculated prior to preparation of the effluents.

14.24.1 Units of measure

The following units should be applied:

1. Pollutants in gaseous effluents: (mg/Nm^3 or vol ppm)
2. Pollutants in liquid effluents: (g/m^3 or mg/kg; i.e., mass ppm)

14.24.2 Gaseous effluents

Regarding the gaseous effluents to be discharged to the atmosphere such as fired heater flue gas, boiler flue gas, vent gas, and etc., the discharge amounts of the pollutants described next should be calculated per source.

1. SO_x
2. NO_x
3. Solid particles
4. H_2S, NH_3, HCl, HF, etc.
5. Cl_2, F_2
6. CO
7. Hydrocarbons
8. Metal and its compounds: Hg, Cu, As, Pb, Cd, etc.

14.24.3 Liquid effluents

Regarding the liquid effluents to be discharged from processes such as process waste water, boiler blow-down water, cooling water, cooling tower blow-down water, ballast water, waste water from research laboratory, etc., the discharge amounts of the pollutants described next should be calculated per source.

1. pH
2. Oil (1)
3. COD (2), BOD
4. Total suspended solids
5. Total hardness
6. Metals: Cd, Cr, Cu, Pb, Hg, Ni, Zn, Ag, etc.
7. HCN, H_2S, HCl, NH_3, etc.
8. Phenol

14.25 Data preparation of catalysts and chemicals

Regarding all catalysts and chemicals required for the plant operation, the quantity required for the initial filling and consumption thereof should be calculated, and such catalysts and chemicals should be summarized under their type.

All requirements should be stipulated in their net value, and in the case where these catalysts and chemicals are actually purchased, orders should be placed for the same gross value, taking into account some allowances (including their filling loss and others).

All catalysts and chemicals required for the licensed units should be in accordance with the licensor's instructions.

14.25.1 Catalysts and packings

The following catalysts and packings should be specified:

1. Catalyst
2. Adsorbent, molecular sieves

3. Desiccant
4. Sand and rock salt for dehydrator
5. Ion-exchange resin
6. Ceramic balls for catalyst supporting and holding

14.25.2 Chemicals and additives

The following chemicals should be stipulated:

1. Solvents such as furfural, etc.
2. NaOH, H_2SO_4, HCl, etc.
3. Inhibitors for corrosion, fouling, polymerization, etc.
4. Antifoamer
5. Additives for lube oil, finished products, BFW, etc.
6. Amines such as MEA, DEA, DGA, etc.
7. Glycol, methanol, etc.
8. Refrigerant
9. Emulsion breaker, filter aids, etc.
10. pH control agent
11. Flocculant and coagulant

14.25.3 Others

Lube oil and seal oil required for the operation of rotary machineries or similar equipment should be summarized later by the project engineer or rotary machinery engineer, in the detail engineering stage.

Further reading

Abdul-Aziz A-R, Lee KZ. Knowledge management of foreign subsidiaries of international oil and gas contracting companies. Int J Energy Sect Manag 2007;1(1):63–83.

Anderson R, Boulanger A. How to realize LEM benefits in ultradeepwater oil and gas. Oil Gas J 2003;101(25): 36–43.

Appleton B. Good practice in controlling employees' exposure to noise and vibration in the oil and gas industries. In: Society of petroleum engineers–SPE international conference on health, safety and environment in oil and gas exploration and production 2011; 2011. pp. 551–7.

Arain FM. Critical causes of changes in oil and gas construction projects in Alberta, Canada. In: Proceedings, annual conference–Canadian society for civil engineering, vol. 3; 2011. pp. 1836–45.

Becker DL. Project management improved multiwell shallow gas development. Oil Gas J 1995;93(42):45–9.

Castillo L, Dorao CA. Decision-making in the oil and gas projects based on game theory: conceptual process design. Energy Convers Manag 2013;66:48–55.

Caulfield I, Dyer S, Gil Hilsman Y, Dufrene KJ, Garcia JF, Healy Jr JC, et al. Project management of offshore well completions. Oilfield Rev 2007;19(1):4–13.

Chen H, Jiang D. Key technologies of mainstream software sharing platform for oil & gas exploration and development. In: Proceedings-2011 IEEE international conference on control system, computing and engineering, ICCSCE 2011, art. no. 6190508; 2011. pp. 124–6.

Collins R, Durham R, Fayek R, Zeid W. Interface management. In: Society of petroleum engineers–13th Abu Dhabi international petroleum exhibition and conference, ADIPEC 2008, vol. 1; 2008. pp. 326–35.

Dejean J-P, Averbuch D, Maurel P, Gainville M, Guet S. Johnsen Ø. FAMUS I: risk-based design of offshore oil and gas production system–management of uncertainties by integrating flow assurance and reliability aspects into a stochastic Petri nets model. In: Proceedings of the European safety and reliability conference 2007, ESREL 2007-risk, reliability and societal safety, vol. 3; 2007. pp. 2443–50.

Eghbal E, Ruwanpura JY. Gorgan method: a framework to manage small oil and gas projects. In: Proceedings, annual conference–Canadian society for civil engineering; 2003. pp. 203–11.

Elsherbiny AH, Adly TA. A model for environmental risk assessment for the construction of oil/gas processing facilities in coastal areas. In: Society of petroleum engineers–9th international conference on health, safety and environment in oil and gas exploration and production 2008-"in search of sustainable excellence", vol. 2; 2008. pp. 720–5.

Finney R, Witchalls B. A project lifecycle approach to an effective and value added social, environmental and health impact management process. In: SPE/EPA/DOE exploration and production environmental conference, proceedings; 2005. pp. 299–303.

Habsi M, Ikwumonu A, Khabouri K, Rawnsley K, Ismaili I, Yazidi R, et al. The well and reservoir management strategy for the thermally assisted gas-oil gravity drainage project in Oman. In: International petroleum technology conference, IPTC 2008, vol. 4; 2008. pp. 2203–12.

Hamilton WI, Cullen L, Reeves G. Integrating people, plant, and process in the design and operation of process industry assets. In: Society of petroleum engineers–Asia Pacific health, safety, security and environment conference and exhibition 2007-"responsible performance: are we doing the best we can"; 2007. pp. 20–8.

Herbst A. Is there a role for agile management in heavy industrial projects?. In: Annual international conference of the American society for engineering management 2012, ASEM 2012-Agile management: embracing change and uncertainty in engineering management; 2012. pp. 1–8.

Herrera F. Design and construction of new projects in oil and gas industry in Mexico. In: Proceedings of the SPE international petroleum conference and exhibition of Mexico; 2002. pp. 553–8.

Jergeas G Top. 10 areas for construction productivity improvement on Alberta oil and gas construction projects. In: Construction research congress 2010: innovation for reshaping construction practice–proceedings of the 2010 construction research congress; 2010. pp. 1030–8.

Jin L, Sommerauer G, Abdul-Rahman S, Yong YC. Smart completion design with internal gas lifting proven economical for an oil development project. In: 2005 SPE Asia Pacific oil and gas conference and exhibition–proceedings, art. no. SPE 92891; 2005. pp. 103–12.

Khalsa AS, Boyd H, Slocum D. Process check: what is driving alternatives analysis in environmental-impact assessment for major capital projects?. In: SPE Americas E and P environmental and safety conference 2009; 2009. pp. 489–95.

Koottungal L. Survey shows increase in gas processing, pipeline construction. Oil Gas J 2011;109(18):28–34.

Lanquetin B, Le Marechal G. Hulls for mega FPSOS: achieving long life and performance from sound design to asset integrity management. In: International petroleum technology conference 2007, IPTC 2007, vol. 2; 2007. pp. 979–93.

Liang L, Wang N, Ma X, Liu Z. Design of on-line monitoring system for transformer oil chromatogram and its data analysis. Adv Mater Res 2012;462:281–6.

Mann J, Dick B, Pearce R. Nansen/boomvang field development-oil & gas export systems. In: Proceedings of the annual offshore technology conference; 2002. pp. 901–23.

McBride W, Bockle HA, Kuczka JH. First zone classified oil and gas facility in North America. In: Record of conference papers–annual petroleum and chemical industry conference; 1999. pp. 1–7.

McHugh S, Manning A, Heagney K. Integration of HES into the management of a major capital project. In: Society of petroleum engineers–SPE/APPEA international conference on health, safety and environment in oil and gas exploration and production 2012: protecting people and the environment–evolving challenges, vol. 2; 2012. pp. 1375–9.

Mckellar JM, Bergerson JA, Kettunen J, Maclean HL. Predicting project environmental performance under market uncertainties: case study of oil sands coke. Environ Sci Technol 2013;47(11):5979–87.

McPhail K. The revenue dimension of oil, gas and mining projects: issues and practices. In: International conference on health, safety and environment in oil and gas exploration and production; 2002. pp. 938–46.

New ME, Smith JE. The Belanak FPSO project–a description of the challenges in designing and building one of the most complex FPSOS to date and the first to incorporate LPG processing and export. In: 2005 international petroleum technology conference proceedings; 2005. p. 1883.

O'Dell PM, Lindsey KC. Uncertainty management in a major CO2 EOR project. JPT, J Pet Technol 2011;63(7): 112–3.

Paik JK, Czujko J. Engineering and design disciplines associated with management of hydrocarbon explosion and fire risks in offshore oil and gas facilities. Transactions–Soc Nav Archit Mar Eng 2013;120:167–97.

Petrone A, Scataglini L, Fabio F. A structured approach to process safety management. In: Society of petroleum engineers–SPE international conference on health, safety and environment in oil and gas exploration and production, vol. 1; 2010. pp. 496–511.

Petrone A, Scataglini L, Fabio F. A structured approach to process safety management. In: Proceedings–SPE annual technical conference and exhibition, vol. 1; 2010. pp. 9–24.

Platt JD. Minimizing environmental impacts in the Arctic: 30 years of oil development on the North slope of Alaska. In: Society of petroleum engineers–9th international conference on health, safety and environment in oil and gas exploration and production 2008-"in search of sustainable excellence", vol. 4; 2008. pp. 1987–91.

Pushkarev MA, Tyutyaev EA. Experience of designing of the automated control systems and safety of objects of oil-and-gas fields construction. Neft Khoz–Oil Ind 2008;10:110–1.

Rawnsley K, Ikwumonu A, Bettembourg S, Putra P, Al-Ismaili I. A reservoir management system for the world's first thermal gas oil gravity drainage project. In: Society of petroleum engineers–SPE EOR conference at oil and gas West Asia 2012, OGWA–EOR: building towards sustainable growth, vol. 1; 2012. pp. 185–200.

Reymert D. Malampaya deep water gas-to-power project: onshore gas plant value engineering and flawless start up. In: Proceedings of the annual offshore technology conference; 2002. pp. 465–77.

Saadawi H. Facilities engineering for a full scale gas injection gas project. In: Proceedings of the middle east oil show, vol. 2; 1995. pp. 45–57.

Sahnoun S, editor. salah gas, a fully integrated project. In: 2005 international petroleum technology conference proceedings; 2005. p. 1871.

Sayda AF, Taylor JH. A multi-agent system for integrated control and asset management of petroleum production facilities–part 1: prototype design and development. In: IEEE international symposium on intelligent control–proceedings, art. no. 4635950; 2008. pp. 162–8.

Sayda AF, Taylor JH. A multi-agent system for integrated control and asset management of petroleum production facilities–part 2: prototype design verification. In: IEEE international symposium on intelligent control–proceedings, art. no. 4635951; 2008. pp. 169–75.

Scataglini L, Carnevale P, Morganti M. HSE minimum design requirements in oil and gas upstream: one company's experience. In: Society of petroleum engineers–SPE international conference on health, safety and environment in oil and gas exploration and production, vol. 1; 2010. pp. 267–75.

Srivastava SK, Takeidinne O. Feed endorsement–a challenge to EPC contractor's engineering team. In: Society of petroleum engineers–Abu Dhabi international petroleum exhibition and conference 2012, ADIPEC 2012-sustainable energy growth: people, responsibility, and innovation, vol. 1; 2012. pp. 301–11.

Tam V, Moros T, Webb S, Allinson J, Lee R, Bilimoria E. Application of ALARP to the design of the BP Andrew platform against smoke and gas ingress and gas explosion. J Loss Prev Process Ind 1996;9(5):317–22.

Taylor JH, Sayda AF. Prototype design of a multi-agent system for integrated control and asset management of petroleum production facilities. In: Proceedings of the American control conference, art. no. 45871, 79; 2008. pp. 4350–7.

Urra S. The formal adoption of a process safety management methodology within an international oil & gas pipeline company. In: Proceedings of the biennial international pipeline conference, IPC, vol. 1; 2012. pp. 395–402.

Van Thuyet N, Ogunlana SO, Kumar Dey P. Risk management in oil and gas construction projects in Vietnam. Int J Energy Sect Manag 2007;1(2):175–94.

Vindasius J. The integrated collaboration environment as a platform for new ways of working: lesson learned from recent projects. In: Society of petroleum engineers–intelligent energy conference and exhibition: intelligent energy, vol. 2; 2008. pp. 893–902.

Yan Q, Zhou X, Sun X, Zhang Y. Design and implementation of oil and gas pool-forming simulation project database based on oilfield database. Int J Oil, Gas Coal Technol 2013;6(1–2):31–9.

Yang J. Approaches to delicacy management of engineering cost in oil and gas fields: a case study of PetroChina Southwest Oil & Gasfield Company. Nat Gas Ind 2012;32(1):108–12.

Detailed Engineering and Design for Natural Gas Processing Projects

15

This process engineering specification, which should be regarded as a recommended practice, specifies the minimum requirements for handling of a project in the detail design and procurement stages. However, depending on the nature and extent of the contract between the Company and Contractor, some parts/sections may be added, modified, or deleted as required. The main activities for implementation of the detailed engineering, procurement services, and supply of equipment and materials are covered in this chapter.

This specification does not deal with the construction activities and/or efforts that normally should be made after or in parallel with the engineering phase for completion of the project in the site.

This specification includes all activities pertaining to the production of drawings, data sheets, specifications, etc., covering all technical aspects of the job, including the execution of the studies, analysis, and detailed designs that are necessary to allow the designer to place purchase orders for the supply of equipment and materials, and to award such subcontracts as are planned for fabrication, installation, construction, and pre-commissioning of the facilities. The basis of the Works to be executed during the detailed design phase shall be the basic engineering design packages and specifications.

15.1 Detailed implementation plan

- The detailed design and engineering and procurement services (hereinafter referred to as Detailed Services) that shall be performed by the Contractor for the realization of the Unit shall be based on Basic Engineering Design Specifications
- In addition to the services stipulated in other parts of the Contract, the following services shall be performed as defined in the Scope of Work of Contract. The activities specified herein below are minimum requirements and the Contractor should make all his efforts and engineering capabilities to the maximum extent possible in order to complete the Works.
- The main activities of project implementation are covered in the following sections:
  - Project Management.
  - Quality Assurance and Control.
  - Project Controls.
  - Detailed Design and Engineering.
  - Procurement.
- In general the Contractor usually provides the following main services as minimum requirement:
  - Review project scope and objectives.
  - Provide overall management of the Detailed Services.
  - Perform Detailed Engineering Works.

- Provide drawings, data sheets, specifications, material requisitions, manuals, and other documents as described under project management for approval and record purposes according to the project schedule to enable the Company to review, check, and approve the Detailed Services and all materials, apparatus and equipment, and other goods, including all spare parts required for commissioning and for 2 years of operation and chemical and catalysts for initial loading and for 2 years of operation (herein individual or together referred to as Materials), which enter into the realization of the Unit with the conditions as defined in the Contract.
- Review, check, and approve any and all Specifications for compliance with the Contract requirements.
- Review Vendor's drawings and other technical documents of Materials for compliance with purchase order requirements.
- Ensure that all drawings, data, specifications, and other information are specifying applicable codes and standards, which will form the basis for purchase order and construction activities.
- Prepare computerized construction planning schedule, and construction work content, derivation including estimate of manpower required and bills of materials value, and progress schedules.
- Provide supporting procedures and Standard Documentation to illustrate day-to-day running of the Project.
- Furnish Company hard copies and electronic files of the Operating and Maintenance Instructions Manuals together with drawings for the Materials in sufficient details to enable the Company to operate and maintain the equipment, and where applicable, dismantle, reassemble, and adjust all parts of the equipment.
- Furnish Company with the revised sheets/drawings of any and all documents and Manuals as may be requested by Company and/or due to changes made by the Contractor.
- Furnish the Basic Designers and/or Licensors with those portions of Detailed Design that are required to be reviewed and or approved by them, as specified in the relevant Basic Design documents.
- Furnish Company as-built drawings in hard copies and electronic files not later than 3 months after Completion and in any case before the Provisional Acceptance. However, in case any change(s) occur after issuance of the Provisional Acceptance Certificate, the as-built drawings shall be revised accordingly. Contractor shall furnish Company with the said revised as-built drawings in the same number as mentioned above, not later than the date of Final Acceptance Certificate.
- Provide replies to any questions and queries that may arise from respective authorities, Basic Designers, Licensors, and Contractors with respect to the services included in the Contract with prior coordination and finalization with Company.
- Provide technical assistance services during construction/erection, precommissioning, commissioning, and start-up.

15.2 Project schedule and control services

Normally, planning, scheduling, controlling, and coordination of the Project include but are not limited to the following activities:

- Establishment of overall communication and coordination procedures with approval of Company.
- Establishment of overall Project execution policies, Project schedules, and procedures with approval of Company.

- Integration of activities to ensure overall uniformity of job philosophy and execution.
- Being sure that all Work is executed in accordance with the Project Specifications and within time schedule of the project.
- Establishment of risk management process and mitigation throughout all Project development and implementation phases.
- Submitting original electronic files of all information, i.e., progress calculation sheets and reports, in addition to hard copies.
- Filling the blank information formats sent by Company with respect to planning, scheduling, and controlling.
- Documentation of all project phases by using the above said software. It shall be delivered to Company as a package at the end of the Work/Project.

15.2.1 Planning

Upon receipt of the Contract, a project planning task force shall be established, which should include Project Director, Project Manager, and the Directorate Coordinators of Planning, Quality Assurance, Engineering, and Procurement in order to:

- Confirm the Master Project Schedule.
- Set-up the Quality Assurance Plan for the Project.
- Identify long-lead equipment, materials, and activities.
- Establish fabrication and contracting plans and sequences so that engineering, procurement, and all other project activities can be planned accordingly.

15.2.2 Work plan

The work plan covers at least:

- Ratify the basis, on which project activity is to commence,
- Detail project scope and measurable objectives,
- Prepare work plans, schedules, budgets, and project procedures and allocate resources.
- Arrange necessary meetings with all parties involved in the Project to ensure that the Project objectives, schedules, priorities, and all other criteria required making the Project a success to be defined clearly.

15.2.3 Definitive project execution plan

For execution of the work, these items at a minimum have to be covered:

- Set up and maintain throughout the Project the appropriate communications and transfer of information with the Company.
- Organize and staff the Project teams. Place major emphasis on the selection of a balanced team with expert knowledge of the Project's requirements.
- Administer the Contract to fulfill its terms and conditions.
- Ensure that the requirements of the Company, governmental regulatory agencies, certifying authorities, insurance underwriters, and others are complied with.
- Control the scope, cost, schedule, and quality of the Project Works.

15.2.4 Procedures

It is necessary to prepare, develop and implement complete project control procedures for Planning and Scheduling, including WBS (Work Breakdown Structure) and OBS (Organization Breakdown Structure) procedures. WBS procedures including weight factors up to agreed level by Company (approximately 100 activities) shall be detailed. The procedures shall also specify clearly the methods that are used for planning, scheduling, monitoring, controlling, and reporting of progress to comply with the plan.

Procedures shall include samples of all standard formats intended to be used by Contractor during execution of the Project. Contractor shall prepare the following procedures as minimum:

- Progress measurement procedure (measuring the physical progress)
- Schedule control procedure
- WBS procedure including weight factor calculation procedure.
- OBS procedure
- Document control procedure
- Requisition control procedure
- Material control procedure
- Cost control procedure
- Reporting procedure (including blank formats)
- Coordination procedure (including primary duties and responsibilities of their staffs vis-à-vis OBS and etc.)
- Schedule revising procedure
- Inventory control procedure
- Value engineering procedure
- Technical Query (TQ) procedure (including blank format)

15.2.5 Scheduling

Contractor shall develop and implement project execution policies, overall plans, and schedules for the Work based on Project Master Schedule, Milestones, and target dates specified in the Contract.

1. Master Schedule

The Master Schedule is the schedule showing the commencement and completion dates of each major phase of the Work and dates of Milestone events/target dates as agreed between the parties in the Contract.

2. Initial Schedule

Within one week from the effective date, the Contractor shall issue to Company a schedule in bar-line form covering the initial 14 weeks or agreed duration of project work. This schedule shall include steps for initial resource mobilization, Basic and Detail engineering, procurement services, supply of the equipment for preparation of the Overall Project Schedule, and Detailed Schedule. This schedule shall be updated every two weeks until issue of Detailed Schedule.

3. Overall Project Schedule

Contractor shall issue an Overall Project Schedule for implementation of the Work, which shall be agreed upon by the parties in Contract. This schedule shall be detailed up to agreed level of the WBS.

This schedule shall cover Contractor's entire activities in respect of Basic engineering, Detail engineering, procurement activities, and estimated manufacturing time. This schedule shall also include all key milestone activities as defined in the Contract. The schedule would form the basis for developing the detailed schedule and shall indicate delivery time for engineering documents (discipline wise), site preparation, and other activities.

4. Detailed Schedule

a. Preliminary Detailed Schedule

Contractor shall issue a preliminary detailed schedule using labor and machine allocation, activity/event-oriented network analysis method on a time scale, covering all engineering and procurement activities.

- The schedule shall identify probable critical items and shall have all key milestones during implementation of the Work agreed by Parties.
- Contractor shall issue a graphical S curve for the total Project with preliminary detailed schedule. This S curve shall not be changed after the final detailed schedule is approved by Company.

b. Final Detailed Schedule

Contractor shall issue a detailed schedule for implementation of the whole Work. This schedule shall cover Contractor's entire activities in respect of basic design, detail engineering activities, manufacturing time, time of delivery of equipment including custom clearance and transportation to the site. Detailed schedule shall cover all key milestones activities, including subcontractors' activities, and shall also indicate the critical path of the Project.

- The schedule shall be prepared by considering the available resources, i.e., workforce, budget during execution of the project.
- Contractor shall submit to Company the Critical Path Method (CPM) network diagram including critical activities.
- Final detailed schedule shall be established according to planning logic and Critical Path.
- Every month, for each activity, the actual monthly progress shall be entered/incorporated in final detailed schedule and update it. Obviously the baseline and S curve of schedule shall not be changed.

5. Preliminary construction schedule

Based on the project master schedule, Contractor shall prepare preliminary construction/erection, precommissioning, and commissioning schedules, which will be developed in next phase by construction Contractor.

15.2.6 Monitoring and controlling

Project control is that element of a project that keeps it on-track, on-time, and within budget. Project control begins early in the project with planning and ends late in the project. The Contractor is responsible for using proven established project control techniques and procedures to monitor and control execution of the Work. These techniques and procedures should include cost estimating, cost control, and planning and scheduling control, and encompass the engineering, procurement, and construction (if required) phases of the Project. The principal methods used to plan and control the progress of the job shall include but not be limited to the following activities:

- Develop a control system based on Project control techniques and procedures.
- Measure progress performance for cost, time, quality, and human resources and compare them to planned progress on a timely immediately.
- Determine the effects of actual schedule performance on the Project schedule and update it.
- Incorporate project changes into the schedule.
- Control and analyze the Project schedule.
- Recalculate the schedule and allocate the resources.
- Develop detailed sub-networks where required to meet the objectives shown as critical on the master schedule. These detailed sub-networks should be produced for all activities required controlling the satisfactory progress of the Project and should be analyzed by the most appropriate computer systems.
- Prepare weekly progress reports for the weekly meetings. The reports should show achievements, activities to be achieved during the following week, and critical activities. The report should be used as an agenda for the weekly meeting.
- Issue a monthly progress report, which should incorporate progress made by all groups during the period. Particular attention should be paid to any critical activities and problem areas together with recommended action would be included.

15.2.7 Reporting

The Contractor shall submit a monthly consolidated report, in an agreed format, prepared as per WBS and schedule control procedures, including:

1. Technical review.
2. Key milestone table as plan and actual in agreed format.
3. Goals achieved last month.
4. Next month's goals and look-ahead schedule.
5. Progress, actual versus scheduled, with reasons for shortfall and actions to be taken by the Contractor to overcome any shortfall or delay at earliest.
6. The reasons for delay if it occurred and percentage of each reason for delay.
7. The Contractor shall give reasonable resolutions to overcome any problems raised in the areas of concern.
8. Statement of any reimbursable costs incurred.
9. Document Control Index (DCI) and summary status table of the issued documents.
10. Material Control Index (MCI).
11. Requisition Control Index (RCI) and summary status table of the issued requisitions. Material requisition for quotations (MRQs) shall be categorized corresponding to the equipment categories as listed in the Contract.
12. Financial status including Change Orders, if any.
13. Summary of any Contract changes, pending and agreed issues, including cash requirement.
14. Correspondence status table exchanged by the Contractor and the Company.

The report shall include Project progress (scheduled and actual) broken down into engineering (up to each document) and procurement (up to each equipment and bulk material category).

Progress shall be measured physically according to "progress measurement procedure" for reporting purpose. Reports shall include curves showing planned and actual progress percentages, together with a statement of work that has not been achieved to schedule and the actions being taken by the Contractor to remedy delays.

Actual physical progress in the field shall be measured based on physical progress measurement procedure prepared by the Contractor and approved by the Company. Actual physical progress shall be calculated for each activity from lowest level (job phase) up to highest level of each category of the Work.

Procedures to measure the physical progress shall be prepared by the Contractor and be approved by the Company. The said procedure shall include the weight of each activity. These weights shall be approved by the Company based on the information submitted by the Contractor. After approving of weight factor calculation and WBS procedures, the Contractor shall make the utmost endeavor to apply any changes in weights of activities, if so requested by the Company.

Monthly consolidated reports shall include any further detailed analysis of any important aspect(s) of the Project as may be requested by the Company.

The sample of progress reports will be reviewed during the Kick-off Meeting.

The Contractor shall send monthly Vendors status report to the Company as per agreed format.

The report shall include Risk Analysis comprising Risk Events and Recommended Response Plan for each Risk Event (if any).

15.2.8 Status review meeting

Periodic Project status review meetings shall be held (monthly/biweekly/weekly/daily or at any time requested by the Company) as desired by Company. It may also be necessary to hold review meetings at regular intervals at management levels as deemed necessary by the Company. Such meetings shall generally be arranged at the place of activity concerned.

- **Additional Efforts**

The Contractor shall provide additional efforts whenever the CPM diagram indicates a possible delay in the target dates. Such additional efforts may require supplementing of equipment, personnel, work in excess of the normal work per day/week, or other resources. All extra costs incurred by the Contractor for such additional efforts in order to prevent an actual and/or possible delay in the target dated shall be borne by the Contractor.

- **Risk Analysis and Contingency**

A mathematical model of a project and its various cost components shall be used by estimators. Each component will be assigned to variable probabilities, and these are used to simulate the most probable outcome of the project cost. The program analyzes the distribution showing the percent estimate costs (or forecast totals) with overrun/underrun probabilities for each component.

- **Access to Documents**

The Company shall have on-line access through an allocated password to all necessary documents, original electronic files, and work centers of a non-confidential nature of the Contractor related to this Project necessary for execution of the Work and for assessment of the progress and monitoring.

The Company shall have access to the work progress files (original files) whenever it wishes during the course of the project.

The Contractor shall establish a database program with suitable and secure software. This program shall have abilities for save, maintain, search, and access to all documents of Project.

- **Change proposals**

The Contractor shall provide to the Company a procedure for handling changes in the Scope of Work. The objective of this procedure shall be to permit the timely evaluation of cost and schedule impacts associated with proposed scope changes prior to their implementation.

15.3 Quality assurance and control

The Contractor shall be responsible for all quality activities associated with the Project. These activities shall have two prime objectives summarized as below:

- The establishment of a Quality System for their part of the Project in compliance with the Contract requirements.
- The verification of compliance (or otherwise) with the Quality System by all personnel assigned to the project.

The main parameters necessary to secure the prime quality objectives of the Project by the Contractor shall be as follows:

- To establish a Quality Control Organization whose sole duty shall be to ensure conformance to the Contract of all contractual activities.
- To establish a Quality Control System to perform sufficient inspection and tests of all items of work.
- To specify the components of the Quality System by the production of the Project Quality Plan.
- To establish a Schedule of Quality System Audits.
- To make sure that the documentary evidence of the Quality System (the Quality Program) is established and complete.
- To monitor compliance by project personnel (both Home Office and Construction Site if required) with the project Quality System by preparing, conducting, and closing out audits of specified activities in accordance with the Audit Schedule.
- To see that Project personnel for their part of the project are fully aware of the Quality System and understand all quality requirements applicable to them.
- To respond the reviews of the Project Quality System by the Contractor's Quality Assurance Department via the Project Directorate.
- To advise the Contractor's Quality Assurance Managers and their respective Task Force Managers of Project progress/status by issuance of regular departmental reports.
- To liaise directly with the Construction Site Quality Control Groups (if required by the Company) which will be under the jurisdiction of the Field Engineering Manager(s) and report all pertinent matters regarding quality to the Construction Manager on a regular basis.
- To liaise with the Company management representatives on quality-related matters on a regular basis.

15.3.1 **Quality control system**

The Quality Control System consists of the Quality Controller and Quality Control Engineer as required to meet the specifications and to ensure qualified inspection of work.

Quality Control shall perform or coordinate and supervise the performance of all required inspections, testing, and document checking and approval. In addition Quality Control will keep complete, updated records on submittals of documents. As a general procedure, Quality Control shall:

1. Review the Contract requirements.
2. Check to ensure that the required submittals have been prepared and approved.
3. Make sure that the required materials and equipment are on the site.
4. Check to ensure that the required off-site inspections and tests have been accomplished and approved.
5. Coordinate and arrange for the required on-site inspection and tests (if required by the Contract).
6. Determine that all preliminary work has been completed.
7. Recheck materials and equipment for compliance.
8. Prepare the schedule of inspection.

Quality Control shall continue inspecting the work daily, or as required, to ensure continuing compliance with the plans and specifications until the Work is completed. Upon completion of an item of work, required operational or performance testing shall be supervised by Quality Control and required certification and/or approval submitted.

When materials being used do not comply with the specifications, or workmanship is not satisfactory, Quality Control shall stop the work immediately and ensure the corrective actions.

As soon as a representative segment of an item of work is accomplished, Quality Control shall inspect workmanship, dimensional accuracy, and ensure use of approved materials. In addition, Quality Control shall review the testing and inspection operations to ensure compliance with the specification.

The Quality Control Program shall be prepared by the Contractor and shall be submitted to the Company for review and approval.

15.4 **Detailed design and engineering**
15.4.1 **Work sequence and procedures**

The Contractor shall develop Project procedures to cover all aspects of the design and procurement phases of the project. These procedures shall be based on the Company's Standard procedures, modified as necessary to suit the project requirements. In case of lack of the Company's Standard procedures, the Contractor can utilize either his own or other international procedures upon approval of the Company. These procedures should include but not be limited to the following:

- Filing System.
- Document Distribution.
- Standards and Codes (Data Base).
- Engineering Symbols, Scales, and Units.
- Numbering Procedures.

- Drafting Procedures.
- Specification for Handling of the Technical Documents.
- Specification Preparation.
- Progress Measurement Procedure for Engineering and Procurement Services.
- Design Interface Control.
- Safety and Operability Review.
- Control of Engineering Budget and Schedule.
- Document Control Center.
- Engineering Document Checks and Reviews.
- Requisitions.
- Testing and inspection.
- Quality assurance plan.

15.4.2 Work methods

The Contractor shall:

- Monitor the progress in all areas against the Project schedule to detect early deviations to schedule and to arrange for corrective action, e.g. additional staff, computing facilities or other measures. A biweekly progress meeting may be held to outline the progress achieved, problems encountered, and solutions intended.
- Monitor, identify, and resolve any non-conformity with the Contract requirements and potential problems.
- Be responsible for any and all Specifications prepared by Vendors/Subcontractors. In this connection the Contractor shall review and check and approve the said Specifications compliance with the Contract requirements.
- Be responsible for the quality and completeness of Work and shall review and sign all drawings, data sheets, Specifications, and acquisitions.
- Establish format of all data sheet forms.
- Confirm that Works are all in accordance with Contract requirements and design guides, and shall act to identify and resolve problems. The Contractor shall also monitor any possible trends involving design changes and shall alert the Company of these potential changes.

15.4.3 Process, utilities, and safety engineering

- **Process and utilities engineering**

Based upon the Basic Design Process Piping and Instrument Diagrams (P&ID), Contractor shall develop and prepare detail design P&IDs to be approving for design and then follow through to approve for construction incorporating vendors' information.

The P&IDs shall show the interfaces with other drawings included but not limited to those supplied parts by the Vendor.

In case any P&ID are prepared by Vendors, the said P&IDs should comply with the above mentioned requirements.

The Contractor shall perform the following main activities as minimum requirement relevant to Process and Utilities:

- Develop complete (inclusive auxiliary system) P&IDs for each Unit to be approved for design based upon the Basic Engineering P&IDs and then follow through and complete the said P&IDs to be approved for construction incorporating Vendors' information.
- Develop definitive Plant General Plot Plan considering plant safety aspects, operability, and maintenance.
- Develop definitive detailed plot plan drawings for the Units taking into account safety, ease of operation, and maintenance of individual equipment and accessories and parts included in the plant.
- Complete where necessary as a basis for detailed design and issue process data sheets.
- Develop process engineering specifications and drawings for each individual equipment.
- Review and develop basic requirements for plant drainage and disposal systems.
- Provide equipment list/index and schedule for all equipment including driver where applicable as per requirements outlined in relevant standards.
- Prepare utility data including effluent data and utility balances diagrams.
- Develop safety data related to P&ID and review process design safety and conduct P&ID safety review.
- Review equipment arrangement drawings.
- Prepare and develop procedures for preservation of equipment during short/long time of nonoperation.
- Review flares and relieving philosophy (in conjunction with different emergency cases) and finalize size of the flares headers and approve flares load data.
- Check and verify all tower capabilities in design, normal, and turn-down throughputs based on the tower load calculations performed by the tray or packing supplier.
- Prepare line schedules for all piping, including line numbers, unit number, fluid symbol, origin and termination, size, material specification, operating and design conditions field test pressure, insulation type and thickness, special requirements (e.g. stress relieving), and tracing design conditions.
- Prepare piping lists for hydraulic review of piping engineering.
- Prepare hazardous area drawings.
- Prepare piping classification data.
- Develop instrument control system and develop safety safeguarding system basic requirements.
- Complete utility summary tables. The summaries shall be provided all required utilities such as but not limited to for the following utilities:
  - Electrical Load.
  - Steam (all types).
  - Condensate (all types).
  - Boiler Feed Water (BFW).
  - Cooling Water (all types), Demineralized Water, Fire Water, Desalinated Water, Plant Water, and Potable Water.
  - Instrument and Plant Air.
  - Nitrogen.
  - Fuels (Gas and Oil).

- Prepare Utility Distribution P&I Diagrams for each Unit showing distribution of the all utility services as mentioned above. All headers, branches to the users, and all miscellaneous items such as utility stations, safety showers, and eye washes, etc., with full details shall also be shown.
- Provide system hydraulic design calculations. The Contractor shall perform a complete hydraulic design at rated (design) capacity and at the defined turndowns (i.e. Lower Operating Levels) for each part of the Units within the Units Battery Limits. Hydraulic Design shall be based on the procedure established by the Contractor and approved by the Company and shall include, but not be limited to, the following:
  - Calculation of line sizes.
  - Control valve process design specifications (e.g. differential pressure across the control valve, etc.).
  - Pump suction and discharge pressure and net positive suction head (NPSH).
  - Equipment elevations.
  - Compressor inlet/discharge pressure.
  - Equipment and piping design pressures.
  - Liquid flows in towers and vessels to ensure satisfactory hydraulic flows.
  - Relief systems including relief valve specifications.
  - Equipment to be purchased, to ensure that such equipment will perform satisfactorily within the system for which it is specified.
- Prepare, for the Company's review, the pressure profiles for all systems comprising the Unit based on the hydraulic design calculations.
- Develop emergency shutdown philosophy and review P&I Diagrams for Advanced Process Optimization start-up, shut-down, and emergency operations of each Unit and catalyst regeneration (where applicable) to ensure that all necessary processing, utility, and blow-down piping and facilities are included for safe operation.
- Review alternative operations of the Units when associated Units may be shut down to ensure continuous operation of each Unit.
- Prepare each Unit battery limit conditions (operating and design) for any and all lines inclusive of operating and design flow rates, temperature, pressure, and destination/sources.
- Complete process information on all equipment data sheets, including instruments, vessels, heat exchangers, heaters, electrical motors, fans, and blowers and all other miscellaneous equipment.
- Define the philosophy and the functionality for Advanced Process Control system for fired heaters and multi products column.
- Develop process duty specifications for the packaged Units.
- Prepare catalyst and chemicals summary.
- Prepare effluent summary for each Unit separately.
- Prepare chemical hazard report.
- Prepare start-up, shutdown, catalyst regeneration (if applicable), and normal operation procedures.
- Prepare normal and emergency shutdown procedures.
- Supply all other services required to do process and utilities works.
- Supply all other services as may be required to complete the above.

- **Safety engineering**
- Make sure that applicable safety and loss prevention codes as well as the Company's special requirements as expressed in the Safety Rules as mentioned in the Contract are applied in a systematic and effective manner by safety audits during the engineering design phase.
- Provide necessary documentation to support safety case and certification submissions as required by the applicable legislation.
- Prepare and/or complete the overall safety philosophy and based on this philosophy, the Contractor shall prepare separate detailed safety documents for each section of the Project.

The said documents among other necessary information and Specifications shall includes hazards and loss prevention data, including plant layouts and arrangements, hazard sources and evaluation, area classifications, detection and alarm systems for specific events, e.g. fire, gas release, shutdown, ESD (Emergency Shut Down) systems, toxic gas release, fire protection systems both active and passive, firefighting equipment, means of escape, lifesaving appliances, drainage systems, ventilation, communication systems, navigational aids, regulations for effluent discharge, emergency power supply, sick bay. and first aid requirements.

- Foresee the following main sequences of safety and loss prevention work:
  - Preparation of logic diagram, Cause and Effect Charts; preparation of safety documentation.
  - Preparation of layouts of fire and gas detection systems as well as fixed firefighting equipment; collection of up-to-date Vendor information; preparation of inquiry packages for loss prevention systems; review and approval of Vendor drawings and documentation.
  - Provide all other services as may be required to complete the above.

- **Hazard and operability analysis (HAZOP) Study**

In support of safety and safe operation obligations, full HAZOP studies that allow a systematic approach to identifying hazards and potential operating problems are required to be conducted at the detail engineering design early stage.

HAZOP will be undertaken at the start of detail design and another following the approved for construction issue of engineering documents. Actions arising from HAZOP studies shall be recorded and implemented in according with HAZOP procedure.

The Contractor is responsible for organizing the HAZOP and following up all actions outstanding as well as provision of any and all facilities for the Works, to ensure that all the process, utility, offsite, miscellaneous, and other units and equipments to be operated and maintained safely without endangering personnel and equipment at all times.

- **HAZOP Outputs**

The information provided by Contractor shall include but be not limited to:

1. Hazard identification report: Defining objectives, methods, and scope of hazards and operability.
2. Hazard Assessment: The hazard identification studies shall identify areas for hazard assessment and appropriate actions for elimination of the hazard.

The following major hazards are to be considered as minimum requirements:
a. Fire,
b. Explosions,
c. Hazardous substances release and their environmental impact,
d. Events causing escalation of incident, including process emergency systems,
e. Power supply failure,
f. Human error,
g. Others, specific to a Unit or equipment.
3. Supply all other services required to do HAZOP study and review.

15.4.4 Civil and structure engineering

- Review the Site survey provided by the Company and perform Site survey if additional data are required per Contract requirements. In case no Site survey data are provided by the Company, the Contractor is responsible for performing the Site survey to complete the required data for implementation of the Project.
- Prepare Site preparation information, including drawings and data that are necessary for construction activities.
- Establish engineering and construction specific job specifications.
- Establish specific job requirement for civil work, including structure and fireproofing.
- Design and prepare detailed drawings for all foundations, elevated concrete, floors, roads, sewer system, basins, sumps, cable trenches, underground piping, etc., including arrangement and detailing of reinforcing and piling (if required), complete with relevant specifications and rebar bending schedule for equipment foundations.
- Design and prepare general arrangement drawings, specific details, and design computations for all reinforced concrete piperacks, steel structures such as equipment structures and platforms, steel buildings, etc. in sufficient detail.
- Develop and design special equipment, which may be necessary for handling of Materials.
- Design and prepare general arrangement plans, elevations, and specific details for concrete structures, including concrete buildings.
- Perform checks for drawings of equipment and structures that are bolted to foundations.
- Prepare key plan showing location and orientation of the Units, buildings, shelters, structures, etc.
- Design and prepare arrangement of reinforcing for concrete structures and the necessary details.
- Prepare foundation location plan.
- Design special pipe supports and assist in preparation of pipe support drawings.
- Coordinate foundation and structural steel drawings.
- Prepare bills of material.
- Prepare drawings for ladder and platform of vessels.
- Prepare drawings covering specific details for fireproofing.
- Provide fire alarm and firefighting systems for buildings.
- Provide loading diagram and calculations results for structures.
- Provide loading diagram and calculations results for foundations.
- Design and prepare detail drawings for industrial buildings as per Project requirements.
- Coordinate with job Site.

- Incorporate Vendor information on drawings.
- Review, comment, and approve Vendor documents.
- Design buildings and all aspects of the buildings, including:
  - Architectural layouts,
  - HVAC,
  - Building services,
  - Structural engineering.
- Design and prepare drawings and specific details for boundary and fencing, retaining walls, lift stations, evaporation ponds, etc.
- Carry out surveying, engineering, and design of temporary access roads, diversion channel (if required), and other facilities to the plant, required for Materials handling as well as construction activities.
- Design and prepare detail drawings for area paving, sumps, and drainage drawing complete with bills of Material and Specifications with due consideration to Contract requirements.
- Prepare any other drawing and detailed Specifications as required.
- All civil and structure drawings should be Approved for Construction (AFC).
- Supply all other services required to do civil and structure works.
- Supply all other services as may be required to complete the above.

15.4.5 Vessels, towers, reactors, and storage tanks

- Establish specific job Specifications for towers, pressure vessels, and storage tanks.
- Design each vessel (reactors, towers, storage tanks, etc.) and prepare detail drawings showing wall thickness, heads, shells, nozzles, supports, internals including number and locations of caps/valves, risers, baffles, weir supports, down-flow section, platform clips, insulation clips and angles, etc. in sufficient details to permit vendors to prepare shop details.
- Check Vendor's drawings for conformance with Specifications.
- Compile Vendor information on the drawings, data sheets, and Specifications.
- Finalize vessel drawings with orientation and lugs.
- Check all drawings including vendor's drawings to be virtually complete and issue for AFC.
- Perform checks for:
  - Vessel foundation drawings.
  - Drawings for steel work and platform supporting vessels.
  - Nozzle sizes and location/orientation.
- Supply all other services required for vessel, tower, reactor, and storage tank works.
- Supply all other services as may be required to complete the above.

15.4.6 Heat transfer equipment (including heaters, heat exchangers, water and air coolers, condensers, reboilers, coils, etc.)

- Establish specific job Specifications for heat transfer equipment.
- Prepare and complete data sheets.
- Perform thermal and mechanical optimization.
- Supply thermal and mechanical design.

- Supply bills of Materials.
- Supply setting plans.
- Prepare detailed drawings to enable Vendors to prepare shop detail drawings.
- Review Vendors' drawings, data sheet, and setting plants for conformance with Specifications, orientation of nozzles, and location of supports.
- Compile Vendor information on the drawings, Specifications, data sheets, and other Project documents.
- Assist in preparation of plant technical and equipment manuals.
- Supply all other services required to do heat transfer equipment works.
- Supply all other services as may be required to complete the above.

15.4.7 Machineries (pumps, compressors, blowers, etc. and their drivers)

- Establish specific job Specifications for machineries.
- Provide and complete data sheets and make NPSH and machinery discharge systems hydraulic calculations.
- Review Vendors' data sheets and drawings for conformance with Specifications and Project requirements.
- Compile Vendor information on the drawings, Specifications, and other Project documents.
- Perform checks for:
  - Machinery foundation drawings.
  - Nozzle sizes and location/orientation.
- Supply all other services required to do machinery works.
- Supply all other services as may be required to complete the above.

15.4.8 Piping

- Establish specific job Specifications for piping.
- Prepare general and Unit plot plans.
- Prepare piping layout and general arrangement drawings.
- Establish mechanical and material Specification for each section of piping, including Specifications and data sheets for expansion joints, spring support, shock arrestors, and other special items.
- Prepare line numbering schedule.
- Prepare and complete line list.
- Review and check technically all packaged Units' inquiries and purchase order requisitions where piping is to be furnished by a Vendor as part of the packaged Unit with conformity with Contract requirements.
- Check Vendor drawings and specifications for piping and piping components for compliance with Contract requirements.
- Design all piping system including special piping items (steam jacketing included) and prepare all necessary arrangement and detail drawings including tie-in points.
- Where steam tracing is required, design the steam tracing system and provide details and Specifications of steam tracing and traps materials and details plus isometric drawing.

- Design underground piping systems and prepare all necessary arrangement and detail drawings.
- Design utility piping and prepare drawings showing arrangement of utilities distribution system.
- Prepare isometric drawings and spool drawings inclusive of complete bill of materials suitable for fabrication of small and large bore piping, except for underground pressurized lines of below 2 inches and for skids to the extent they are shop assembled.
- Prepare P&ID for pressure testing giving required information for testing.
- Prepare bill of material sheets for each isometric in the same drawing.
- Prepare stress analysis calculations and pipe support details.
- Finalize layout arrangement drawings.
- Check and coordinate equipment nozzle orientation.
- Prepare plant three-dimensional computer models for the new Units. For the Units that are duplicated, one computer model may be prepared and in this case the interconnecting piping between the identical Units shall also be shown.
- Perform checks for:
  - Drawings of equipment and terminal point of package Units to which piping is connected.
  - Layout drawings of foundations.
  - Layout and elevations on structural steel drawing.
- Assist in preparation of plant technical and equipment manuals.
- Supply all other services required to do piping works
- Supply all other services as may be required to complete the above.

Stress relief and branch reinforcement requirements shall be included in piping Specifications.

15.4.9 Instrumentation and control system

- Establish specific job specifications for field instruments and analytical instruments.
- Prepare detail specifications for Process Control System (PCS), Emergency Shut-down Systems (ESD), Safe Guarding System (SGS), and Fire and Gas (F&G) Detection systems.
- Prepare detail Specification for advanced Process Control System.
- Prepare detail specifications for Maintenance Management System.
- Prepare detail specification for Management Information System.
- Prepare detail specifications for Asset Management System.
- Provide detail specification for tank gauging system.
- Provide computer system and application software for the above-mentioned systems.
- Develop process P & IDs and utility distribution diagrams including, instruments numbering symbols and identification.
- Provide computerized data base inclusive of data and information for any and all equipment and other items to be utilized for the said application software.
- Minimum two dedicated consoles shall be provided for start-up and shutdown in main control system. Allocation of the various Units to each console shall be proposed by the Contractor and approved by the Company.
- Prepare instrument lists, comprising all loop components, instrument tag numbers, and relevant drawing cross-references.
- Prepare data sheets for all instrumentation components.

- Size control valves.
- Size safety and relief valves.
- Size orifice plates and other flow elements.
- Size uninterruptible power supplies (UPS), battery, and chargers.
- Size pressure differential (PD) meters, turbine meters.
- Prepare cable schedule.
- Prepare layouts for instrument panels, etc.
- Review and check technically all packaged Units' inquiries and purchase order requisitions where instrumentation is to be furnished by a Vendor as part of the packaged Unit with conformity with Contract requirement.
- Prepare drawings showing location of instruments, cable route, and utilities distribution systems.
- Layout and develop electrical systems, for instrumentation.
- Check instrumentation Vendor prints/drawings and Specifications for compliance with Contract requirement.
- Perform checks on vessel drawings for instrumentation.
- Prepare panel layout drawings to scale with overall dimensions and show the locations of instruments, push buttons, lights, annunciators, alarms, etc.
- Prepare instrument location drawings, using piping drawings as background.
- Prepare logic diagrams for interlock and alarm systems.
- Prepare logic diagrams for sequence and program control.
- Prepare cause and effect tables showing all causes with their consequences.
- Establish Safety Integrity Level (SIL) requirements.
- Prepare SIL assessment report.
- Prepare all other data sheets, drawings, and diagrams required for installation maintenance and operation of instrument and control items.
- Prepare instrument cable/tubing schedules.
- Prepare junction box locations.
- Prepare instrument hook-up drawings with bill of material.
- Prepare instrument transmission loop details.
- Prepare bulk items Specifications.
- Prepare all analyzer Specifications.
- Prepare detail test procedure of the interlocks and sequential loops.
- Prepare basic logic scheme and function description for start-up, shutdown, emergency shutdown procedure, anti-surge control, etc.
- Prepare comprehensive drawings and Specifications index sorted by document number as well as document title.
- Prepare list of all set-point values for alarm and shutdown system.
- Prepare full instrument instruction manuals including operation, installation, calibration, troubleshooting, and maintenance.
- Prepare instrument inspection report.
- Prepare instrument summaries for checking, cross-checking, and reference.
- Provide comprehensive calculation sheets and selection philosophy, bounded separately for each category of control valves, flow elements, safety valves, pressure control valves, PD meters, turbine meters, UPS, and batteries and charges, including method of calculations.

- Prepare schematic wiring diagram of alarms and inter-locks showing functional sequence of start and stop buttons, relays, alarms, solenoid, and shutdown switches.
- Prepare initial and final material take-off for all instruments and instrument material.
- Supply all other services required to do instrument and control system works.
- Supply all other services as may be required to complete the above.

15.4.10 Electrical

- Establish specific job Specifications for all electrical systems (inclusive of electrical tracing system and cathodic protection system).
- Prepare single line diagrams for the whole electrical generation and distribution systems, and also for each area substation.
- Design electrical distribution system.
- Design electrical tracing where required.
- Prepare lighting systems and prepare drawing showing arrangement of lighting panels, lighting requirement at grade, on platforms and structures, and electrical trays with specific details as required.
- Prepare data sheets for electrical equipment (including motors).
- Prepare layout drawings of power cables and specific requirements for switchgear and motor control center.
- Prepare grounding drawings and details.
- Layout and prepare electrical power supply systems for instrumentation.
- Prepare area classification drawings.
- Perform checks for underground drawings of piping and civil where underground electrical cables are to be laid.
- Prepare electrical load list including motors and other consumers.
- Prepare relay setting schedule.
- Prepare electrical system design report, including voltage profile, reacceleration, and fault studies.
- Prepare electrical system study and short circuit calculations.
- Prepare electrical cable schedules and routing.
- Prepare schematic wiring diagrams for all circuit breakers and electrical items having internal wiring or relays.
- Prepare layout of switch rooms showing the location of major equipment, battery charger room, and classification of hazardous locations.
- Prepare list of all starters and switchgears with capacity requirements and specifications for each.
- Prepare all earthing, control station, and other miscellaneous fixing and mounting details.
- Prepare material take-off for all electrical material.
- Prepare power control house building (substation) layout.
- Prepare substation and switchgear drawings.
- Prepare system shutdown connection diagrams.
- Prepare electrical instrument drawings.
- Prepare electrical heat tracing drawings.
- Prepare cathodic protection system with detailed Specifications and drawings.
- Supply load flow calculations in start-up and steady-state operation of the electrical system.
- Design emergency supply, including uninterrupted power supply system.

- Prepare initial and final material take-off for all electrical equipment, accessories, and materials.
- Prepare material requisitions for all electrical accessories, equipment, and materials, including heat-tracing material, if any.
- Prepare block diagrams, connection diagrams, design philosophy, and instruction manuals for interlocking systems, alarm system, and other complicated power and control systems.
- Prepare physical location of electrical equipment and wiring installed and installation details.
- Prepare physical location of grounding electrodes, equipment to be grounded, and wiring layouts as well as their installation details.
- Prepare engineering, manufacturing, inspection requirements, construction/erection, precommissioning, and commissioning Specifications and procedures for all electrical components, equipment, accessories, and materials.
- Prepare cable cutting schedule.
- Prepare cable orientation on trays and/or trenches.
- Prepare cable room tray orientation.
- Prepare power distribution control system (PDCS) system installation.
- Prepare PDCS I/O list (including serial links).
- Prepare logic sequence diagram for PDCS system.
- Prepare PDCS software details.
- Prepare operation and maintenance manual for PDCS complete with illustrated spare list.
- Check Vendor's drawings and data for conformance with Contract requirements.
- Supply all other services to do the electrical works.
- Supply all other services as may be required to complete the above.

15.4.11 Telecommunication

- Establish specific job Specifications for the telecommunication systems.
- Prepare detailed drawings and Specifications for the telecommunication systems.
- Perform checks for underground drawings of piping and civil where underground telecommunication cables are to be laid.
- Check Vendor's drawings and data for conformance with Contract requirement.
- Supply all other services required to do telecommunication works.
- Supply all other services as may be required to complete the above.

15.4.12 Miscellaneous and/or special equipment

- Establish specific job Specifications for miscellaneous and/or special equipment.
- Prepare detailed Specifications, data sheets, duty Specifications (where applicable) for each item.
- Review Vendor's drawings and technical documents for conformance with Contract requirements.
- Supply all other services to do miscellaneous/special equipment.
- Supply all other services as may be required to complete the above.

15.4.13 **Insulation and painting**

Establish specific job Specifications for all insulation and painting items.

Prepare insulation schedules for equipment and piping showing operating temperature, insulation, service, type and thickness of insulation, and reference to the applicable Specifications.

- Prepare painting schedule and paint/painting application Specifications
- Prepare insulation and painting bill of material.
- Prepare methods and procedures of surface preparation in detail.
- Prepare methods and procedures of painting of equipment and material in the manufacturer workshop and at Site in detail.
- Supply all other services required to do insulation and painting works.
- Supply all other services as may be required to complete the above.

15.4.14 **Fireproofing**

- Prepare fireproofing specifications for steel structure and vessel skirt or supports.
- Establish specific job Specifications for fireproofing.
- Prepare drawings covering specific details for fireproofing.
- Prepare fireproofing material specification and bill of quantities.
- Supply all other services required to do fireproofing work.
- Supply all other services as may be required to complete the above.

15.4.15 **Firefighting system**

- Establish specific job Specifications for firefighting system.
- Prepare detail Specification, data sheet, and detail drawings for firefighting system.
- Perform checks for underground drawings of piping and civil where underground firefighting system is to be laid.
- Check Vendor's drawings and data for conformance with Specifications.
- Prepare overall design basis and Specification of fixed fire and gas systems and firefighting equipment.
- Supply all other services required to do firefighting system works.
- Supply all other services as may be required to complete the above.

15.5 **Procurements services**

- The procurement of Materials shall be carried out based on the engineering works inclusive but not limited to data sheets, Specifications, and drawings. The issued drawings and documents to Vendors for requisition and purchasing shall be approved by the Company.
  - Perform all procurement services for the purchase of all equipment and materials.
  - Procure all of the Equipment and Materials, including bulk Materials, of the project based on the engineering works approved by the Company. The spare parts shall be in congruent with the Material codes indicated in the Company's Material and Equipment Standards and Code (MESC) Books, unless otherwise approved by the Company.

- Utilize his own resources to carry out any expediting and inspection that may be necessary to ensure that the Materials are delivered in accordance with the Project schedule and the relevant Specifications.
- Prepare Material Selection Guides based upon final process flow diagrams and submit to the Company for approval.
- Prepare Material Selection Guides based upon project Specification, Materials requirements, and data given in the Basic Design Package and published corrosion data by the acceptable international committee (e.g. NACE) approved by the Company. In general, the Materials shall be selected to reflect up-to-date, proven, and established technology if not specified in the Basic Design Packages.
- Arrange third party inspection as foreseen in the Contract.
- The procurement services main activities shall include but not be limited to:
  - Establishing specific job Specifications.
  - Compiling all relevant Specifications and data sheets.
  - Preparation of the requisitions for purchase of Materials
  - Preparation of technical bid evaluation.
    - Technical and commercial negotiations with Vendors, including clarification of technical questions that may arise from Vendors.
  - Preparation and issue of purchase order after Company approval.
  - Checking of orders confirmation.
  - Preparation and issue of variations to purchase order if necessary.
  - Checking Vendors' documents and drawings for conformance with Contract/purchase order requirements.
  - Control, tracing, and monitoring of Materials.
  - Desk and outside expediting.
  - Expediting of documents, drawings, and spare part lists with Vendors.
  - Proper inspection of the Materials.
  - Checking invoices.
  - Provision and arrangements for all risk insurance up to Site where the Materials are to be installed.
  - Cargo inspection.
  - Arrangement for packaging.
  - Arrangement for shipment with shipping agency.
  - Preparation of necessary documentation for shipment/custom clearance/exportation/importation.
  - Planning and coordination of all the procurement activities.
  - Reporting on progress.
  - Expediting delivery in such manner as to allow completion of the Works on time.
  - Supply all other required services.
  - Supply all other services as may be required to complete the above.

15.5.1 Enquiries/request for quotation

The Contractor shall prepare enquiries that shall be in accordance with the Contract requirements and shall state that Materials shall be required for the Company and for the Project.

Enquiries for Materials shall include the provision of approved job specifications, process duty specifications, data sheets, all other necessary documents and drawings as required, and whenever applicable spare for construction, commissioning, and one year of operation.

The Vendor assistance required is to be defined. In general, Vendor assistance during commissioning is to be requested for major equipment such as major rotating machinery, boilers, furnaces, internal of towers, package Units, high and medium voltage panels, etc., that shall be agreed upon by the Company before issuance of purchase. Enquiries shall also state requirements for items such as drawings and documentation.

Enquiries for major Equipment shall be issued to the Company for review and approval.

The Contractor shall solicit competitive quotations from a sufficient number of approved Vendors (listed in the Contract's attachments) for the respective Materials to ensure the receipt of a minimum of three acceptable bids. Exceptions to this would be subjected to approval of the Company.

All Requests For Quotation are to contain the following note:

Review all attachments with care. Strict compliance as requested is mandatory for bid evaluation. Any exceptions/deviation to Specifications and/or "Terms and Conditions of Purchase," must be clearly stated in a separate section titled "Exception/Deviation in the quotation."

All engineering and design data including technical information of all quotations shall be in English with consistent use of metric units of measurements. All drawings, instructions for installation, operating manuals, maintenance manuals, and any other printed matter pertaining to the equipment furnished by the Contractor/Vendor/and Sub-vendor shall be in the English language"

Requests for Quotation shall state requirements for and shall request bidder to furnish of the following as well:

- Shipping promise.
- Shipping point.
- Method of shipment.
- Delivery Point.
- Spare parts (for two years operating the spare parts list and interchangeability record (SPIR) list shall be furnished by the Vendors in the quotation).
- Free access to expediters and inspectors, including the Company's representatives.
- Drawings and data sheet.
- Installation, operating and maintenance instructions.

15.5.2 Bid analysis

The Contractor shall prepare technical bid analysis and evaluation reports for the Materials inclusive of necessary tabulation from the three Vendors for each case and submit the same to Company for review and approval. The said bid evaluation reports shall also include:

- Contractor-proposed selected Vendors with indication of reason for selection.
- In case where a Vendor's bid is not acceptable, a statement for the reason for its rejection.
- Un-priced Vendor quotation.
- List of rejected Vendors because of nonconformity with the Contract requirements with nonconformance items in detail.

- Differences in operating costs and maintenance costs, including the Contractor's procedures and methods to calculate all of the said parameters if requested by the Company.
- Special tools and tackles required for erection, precommissioning, commissioning, and operation and maintenance of the equipment wherever necessary.

Consideration of standardization of equipment in the interest of simplifying and minimizing the stock of spare parts of particular interest to the Company will be the Contractor's efforts in dealing realistically with the following:

1. Grouping together the various types and models of related functional equipment to encourage selection of a single Vendor.
2. Maximizing interchangeability in each unit.

The Contractor shall submit one copy of all technical communications with Vendors to the Company.

15.5.3 The company review and approval

The Company will review and check the submitted technical analysis and evaluation to ascertain the compliance of the quotations with the Contract requirements and approve Contractor Vendor for respective Materials or otherwise will indicate the Company-preferred Vendor among the above-mentioned three bidders provided that any cost consequent cost-effective (if any) to be compensated as per the Contract. For any deviation from the Contract requirements the Contractor shall obtain the approval of the Company before the issuance of technical analysis and evaluation report. Such approval or ascertainment shall not relieve the Contractor from his responsibility and obligation under the Contract.

15.5.4 Purchase orders

Purchase orders shall give a complete description of items required. Purchase orders shall include:

1. Guarantees of performance
2. Required drawings and data sheets.
3. Schedule for drawings and data submitting.
4. Expediting, inspection and shipping requirements (including test certificate required).
5. Spares parts information as per SPIR form in accordance with the original manufacturer's documents.
6. Storage recommendations.
7. Operating and maintenance manuals.
8. Any other required mechanical catalog information.
   The Contractor shall indicate the following notes on all purchase orders:
   a. Progress Report:

   Unless otherwise the delivery schedule is included in this purchase order, within 30 days from date of order, the vendor shall submit the detail program for the engineering, Materials procurement, fabrication, and delivery for this purchase order followed by a progress report issued each month until final delivery.
   Items to be purchased are subject to inspection and approval by the Company and any items not conforming to approved Specifications shall be rejected.

b. Complete Order:

This order will not be considered complete until the Contractor is in receipt, in proper form, of all the engineering data requirements, drawings, spare parts lists, and instruction manuals. Payment, or in the case of progressive payment, final payment, will be withheld pending receipt of any or all of the above data.

The following statement should be included on each purchase order as the last paragraph:

c. Notice of Shipment:

Supplier shall notify the Contractor and the shipping agent by facsimile and/or telex on the day of each shipment goes forward. Such notification must be identified by purchase order number.

All applicable specifications, standards, drawings and standard notes shall be listed on the purchase order and shall be attached to it unless it has been verified and confirmed jointly by the Contractor's respective vendors in writing that current issues are already in the vendor's possession.

Purchase orders shall state the delivery date as a specific date and not as a period of time from the order date.

Purchase orders shall include following requirements:

- Place of delivery.
- Required spares for precommissioning, commissioning, two years of operation, and capital spare parts (if applicable).
- Special tools and tackles required for erection, precommissioning, commissioning, operation and maintenance of the equipment wherever necessary.
- Any other required mechanical catalogs/information to enable the Contractor to fulfill his contractual obligations stipulated in Contract documents.

The Contractor shall ensure that the vendor provides all drawings and data requirements as stated on the purchase order. An order shall not be considered as completed until every such document has been received and accepted as satisfactory by the Contractor and the Company.

15.5.5 Inspection

- Inspection shall be made in accordance with the Contract requirements.
- The Contractor shall not accept from vendors any Materials that are not of acceptable quality of workmanship, or fail to comply with the Specifications, and/or that are required to have the inspection as above but which have not performed satisfactory in such inspection.
- A shop inspection and testing report shall be delivered to the Company not later than 30 days after performance of the inspection. However, it must be delivered before issue of the Contractor's invoice.
- The Company reserves the right for his representative and/or agent to visit vendor's shop, and those of their sub-suppliers, at any time, for the purpose of inspecting Materials. Arrangements for these visits shall in all cases be made through the Contractor. Such visits will not relieve the Contractor or any vendor or sub-contractor of his responsibility for inspection as detailed above. If the Contractor intends to subcontract inspection at vendor's shop partly or in whole, then the inspection agency to be used shall have the prior approval of the Company. The Contractor shall,

in conjunction with the Company, make any necessary arrangements to obtain the approvals of statutory authorities where required in connection with inspection of Materials.

- The Contractor shall give to the Representative Engineer advance 60 calendar days' notice of any test per coordination procedure to be made by the vendor and/or the Contractor in order that Company representative may witness any such test.
- Third party inspection shall be arranged by the Contractor as foreseen in the Contract.
- The Contractor shall supply all other services as may be required to complete the above.

15.5.6 Shipment

- The Contractor shall establish maximum allowable shipping dimensions and masses (weights and volumes), and any other specific requirements that may be needed for shipment of Materials.
- The Contractor shall prepare a detailed Shipping Procedure for review and approval of the Company.
- The Contractor shall provide for shipment all risk insurance for Materials shipment to the Site where the respective Materials are to be installed.
- Purchase orders shall be accompanied by copies of shipping instructions (i.e. Packing, Marking and Documentation attachment).
- The Contractor shall ensure that all necessary customs clearance documentations are available at the port of entry prior to arrival to customs to avoid delay in customs clearance. Materials going by air or over land shall be accompanied by the necessary customs clearance documentation.

15.5.7 Progress control and expediting

- The Contractor shall expedite delivery in such manner as to satisfy the time for completion and to maintain the integrity of the Works.
- Any threatened delays that may affect the Project schedule shall immediately be brought to the attention of the Company, together with actions to be taken to improve the situation.
- Materials shall be inspected regularly and as a minimum monthly inspection shall be made to all vendors and sub-suppliers of major equipment and other Materials likely to be critical for the completion of the Project.
- The Company reserves the right for his representatives and/or agent to visit Vendor's and sub-suppliers' works, for the purpose of expediting Materials and Equipment. Arrangements for these visits shall in all cases be made through the Contractor. Such visits shall in no way be construed as relieving Contractor of his responsibility for expediting as detailed above.
- The Contractor shall maintain a Materials Progress Report to record the status of all items procured by the Contractor. These reports shall be issued to the Company at least monthly throughout the Project execution period until all Materials have been received at the Site where the Materials shall be installed.
- The Contractor shall establish the format of the Materials Progress Report.
- The Materials Progress Report shall include, as applicable, at least the following:
- Requisition number, revision number, and revision issue date.
- Equipment item number.

- Brief description of Materials.
- Date of issue of enquiry.
- Due dates for bids.
- Planned and actual order placement date.
- Purchase order number.
- Name of Vendor, and country of origin.
- Delivery schedule viz. Contractual promise, latest promise. Total slippage, slippage at last check.
- Expediting activity details (viz. date of last check and nature. Of check-by telephone, cable, letter or personal visit).
- Inspection date (or a note if inspection has been waived).
- Date of issue of shipping instructions to the Vendor.
- Ex-works date (actual).
- Site (where the Materials shall be installed) and receipt date (estimated and actual).
- Any problem (anticipated or existing) at the date of report and Contractor solution and arrangement for solving problem.
- The promised delivery date must take into account time allocated for shop inspection, preparation for dispatch, and any intervening holidays.
- Separate records shall be maintained for spare parts.
- Separate records shall be maintained for Overage, Shortage, and Damage (OSD) of delivered items.

15.5.8 Vendor specialists

The Contractor shall provide the services of vendor specialists as required for supervision and assistance of erection, pre-commissioning, and commissioning of the Materials, e.g. Compressors, Rotating Machinery, Tower Internals, Boilers, Fired Heaters, high and medium voltage Panels, Packaged Units, and complex instrumentation such as PCS/SGS/F&G/PLC, and Telecommunication equipment.

15.5.9 Guarantee

- Prior to order placement, the Contractor shall obtain guarantee from vendors, and shall ensure that these guarantees are fully transferable to the Company. The guarantee shall not be less than the requirement stipulated in the conditions of the Contract.
- In the event that Materials require repair/replacement but are not covered by vendor's guarantee, e.g. damage during transport or construction,
- The Contractor shall ensure that necessary repair/replacement is carried out at no cost to the Company.

15.6 Supply of materials

- The Contractor shall supply on the basis of delivery terms specified in the Contract.
- Packaging, loading, unloading, transportation, insurance, customs clearance of Materials shall be carried out by the Contractor.
- All Materials shall be new.

- The Contractor shall not submit to the Company shipping documents and shall not accept and supply any Materials subject to the conditions of the Contract.
- The Contractor shall prepare and submit to the Company for review and the approval procedures defining controls for the receipt, storage, custody, and distribution of all Materials. Such procedures shall include as a minimum the followings:
  - Receipt of Materials and documentations,
  - Inspection at point of receipt and identification of overage/shortage and damaged (OS & D) items,
  - Stock control,
  - Storage and custody of Materials, external and internal, including environmentally controlled areas,
  - Materials handling,
  - Checking for conformity with Contract requirements.
  - Quality control of vendor/supplier furnished item,
  - Return or replacement of uncertified Materials.
- The Contractor shall submit Materials certificates for Company approval. Materials survey reports shall also be included as part of test and inspection dossiers.

15.7 Detail design & engineering documents
15.7.1 Documents mutually explanatory

The several documents forming the Contract are to be taken as mutually explanatory. In case of discrepancies between the several documents forming the Contract, precedence shall take place in the order designated in the Contract. Any ambiguities in the documents and/or drawings shall be referred to the Engineer for correct interpretation.

- Produce the required copies of maintenance schedule for each unit of the plant. This schedule shall state routine inspection and maintenance requirements during operation, routine shutdowns, and general overhauls. Schedules shall be in English and in a format agreed with the Company. The schedule shall give the timing and outline the action for inspection and maintenance for items such as:
  - Lubrication of machinery and moving parts,
  - Topping up or refilling of filled systems,
  - Checking of flow in Equipment and piping in services liable to fouling,
  - Checking of wear in moving parts and electrical contacts,
  - Checking of calibration of instrumentation,
  - Checking of relay settings and operations,
  - Checking of wall thickness of Equipment and piping in erosive or corrosive services,
  - Electrical insulation tests,
  - Checking of vibration in rotating Equipment,
  - Scale formation in boilers,
  - Correct functioning of steam traps,
  - Checking of relief valve settings.
  - Coordinate with Basic Designer/Licensor of unit in case of discrepancies in Technical documents provided by them for rectification the same.

15.7.2 **Project control and scheduling methods and procedures (typical)**

Level 1: Milestone Level Control
1. Project Master Schedule.
2. Progress Curves.

Level 2: Summary Level Controls
1. 90-day Kick-Off Schedule.

Level 3: Detail Level Controls
1. Critical Activity Listing.
2. Manpower Histograms.
3. Schedule Trend Meetings.
4. 90-day Bar Charts.
5. Subcontract Preparation Schedule (if required for the construction phase).

Level 4: Working Level Controls
1. Physical Progress Measurement.
2. Weekly Work Schedules.
3. Deviation List.

- **Project master schedule**

  The project master schedule shall highlight all major project milestones as agreed with the Company and provide sufficient information to allow management to evaluate the overall status of the Project at a glance. It shall be updated to show progress achieved and issued as part of the Monthly Progress Report. It shall clearly identify:
  - Start and complete dates for Engineering by discipline and for Procurement.
  - Material delivery and shipping periods.
  - Site mobilization and construction activities (if required).
  - Mechanical Completion and ready for startup dates (if required).

- **Progress curves**

  As an aid in planning and control, overall progress and manpower curves shall be developed to monitor Engineering and Procurement progress.
  The development of these curves shall be based on the approved Master Schedule and shall include breakdowns for all disciplines and sub-contracts to cover:
  - Engineering document release.
  - Equipment and material purchase, manufacture, and delivery.
  - Fabrication/construction progress (if required).

The curves shall be updated monthly to display actual progress throughout the Project duration and issued with the Monthly Progress Report.

- **90-day kick-off schedule**

At the start of the project, a 90-day Kick-Off schedule shall be issued, which shall illustrate in detail all work anticipated within that period. This schedule shall be updated weekly and used for Project control until the detailed planning effort has been implemented.

- **Critical activity listing**

The computer-produced Critical Path printout shall include a listing of activities at negative and zero float or up to a predetermined level of positive float that enables detailed analysis of the activities, and the formulation of action plans to reduce criticality and ensure completion of the Project on schedule.

- **Manpower histograms**

Actual manpower should be plotted against current forecast requirements, as generated by the CPM network, so that sufficient resources are being mobilized to accomplish the work planned. The histograms indicate manpower for Engineering, Procurement and Construction (if required).

- **Schedule trend meetings**

  Weekly meetings should hold between all key members of the Project team to:
  - Review possible scope changes.
  - Review the critical activity list.
  - Review progress achieved.
  - Identify current and potential future problem areas.
  - Formulate action plans to resolve problem areas.
  - Follow-up on previous action plans.

- **90-day bar charts**

The action plans identified at the schedule trend meetings to eliminate negative float such that activities achieved to schedule and any scope changes shall be entered into the computer system. The computer shall then be used to generate a bar chart schedule, sorted by discipline, showing the planned activities for the next 90-day period. The bar charts shall be used so that all personnel on the project implement the agreed activities for completion of the overall project on schedule. The 90-day bar charts shall be reissued on a monthly basis to the Engineering and Procurement task force teams.

- **Subcontract preparation schedule (if required)**

A subcontract preparation schedule shall be printed in bar chart form and shall show the activities required by all Project groups to formulate subcontracts, review bids received, and place the subcontracts according to the agreed milestone dates for the commencement of construction.

- **Physical progress measurement**

All disciplines on the project shall prepare detailed control documents for major activities. The following areas of effort shall be measured and compared with the detailed schedules:

- **Engineering**
  - Drawings to be issued for approval.
  - Drawings to be issued for construction.
  - Requisitions to be issued for quotation.
  - Requisitions to be issued for purchase.
  - Model progress.
  - Piping isometrics issued.

- Physical percentage complete by discipline.
- Overall percent complete.
- **Procurement**
  - Material requisitions enquiries issued.
  - Purchase orders issued.
  - Pieces of equipment and bulk materials delivered to site.
  - Subcontracts placed.

15.8 Supply of spare parts, miscellaneous equipment and materials, chemicals and catalysts

Supply of spare parts and other special items may be included in the Contract in the lump sum price and/or as reimbursable items. However, in any case, an expert team of the Company consisting of the necessary specialists in the various fields is required to review the Contractor and/or Vendor(s) recommendations and issue final Company approval for supply of spare parts.

For the supply of materials and equipment hereinafter provided, the Contractor shall act as purchasing agent of the Company, providing also for the relevant payments to the supplier, forwarding agent, shipper, and insurer. The Contractor shall submit to the Company a program of his services under this section and, when this is agreed by the Company, it shall be included in the Contract.

The Contractor shall not be responsible for:

- The supply of any guarantee other than those of the manufacturer. These guarantees to be transferred directly to the Company.
- The time of delivery.

The Contractor shall report to the Company monthly the value of materials and equipment ordered.

The Contractor shall invoice monthly to the Company in the currency requested the amount of the fee due to the Contract for the services rendered based upon the purchase orders issued during that month.

The Contractor shall arrange for marine insurance in accordance with the provisions of the Contract.

Spare parts, miscellaneous equipment and materials, and catalysts and chemicals may be purchased on reimbursable basis if required by the Contract. In this case, those items to be supplied as lump sum price, detail specification, and conditions of purchasing are to be defined in the Contract.

15.8.1 Supply of spare parts

The Contractor shall arrange as a condition of each purchase order for suppliers of equipment and from each manufacturer of parts bought by suppliers, complete recommendation and interchange ability record forms according to the SPIR forms. Completion of these forms shall be based on the manufacturing drawings for the order and the completed forms shall be accompanied by copies of such drawings and the parts lists certified as being applicable to the order.

The Contractor shall include his own recommendation in the space provided on the form referred to above after reviewing the suppliers' proposals and quotation and interchange ability between all orders placed on the same supplier except as provided for herein below.

The Contractor shall prepare interchange ability charts for those items that occur in significant numbers between different suppliers and between different categories of equipment and are of significant cost such as seals, furnace and exchanger tubes, safety valves, and instruments. The Contractor shall compile a consolidated summary of spare parts for all equipment to determine maximum interchangeability and minimize spare parts inventory.

The Contractor shall prepare his own recommendations for spares for general materials such as pipes, valves, pipe fittings, bolts, gaskets packing, electrical material, etc.

The Company shall complete the form with purchasing instructions after consultation with his coding adviser.

The Company shall provide the Contractor with an identification code and the Contractor shall arrange with the supplier for each item to be tagged according to this identification code and for inclusion of the same identification in packing lists and invoices.

When submitting bid analyses for engineered equipment, the Contractor shall also submit priced spare parts list(s) including lubricants, chemicals, etc. and recommended spare parts list for start-up and two years operation.

The Company may withhold consideration of any bid tabulations unless they are recommended by the required spare parts information, and will not be responsible for any delays that result due to this requirement not being met.

Vendor spare parts recommendations must be complete with prices, complete parts list, and sectional drawings showing interchangeability of parts. Standard items, (e.g. bearings, oil seals, gaskets, packing, valves, fittings, etc.) shall be identified with manufacturers' size, catalog or part number, and description.

The Contractor shall expedite the delivery, inspect the major spare parts, inspect all tagging of the spare parts, check invoices, and forward one copy of each packing list to the Company as soon as possible after shipment. One copy of the packing list shall be packed with each shipment.

15.8.2 Supply of miscellaneous equipment and materials

The Contractor shall prepare and issue enquiries and purchase orders for the items as the Company will require and for their spares. The Contractor shall follow the same procedure for spare pasts.

The Contractor shall expedite the delivery, inspect, accept the equipment and spares, check invoices, arrange shipment and insurance, and forward one copy of each invoice and of each packing list to the Company as soon as possible after shipment.

The Contractor shall provide shipping documents made out in the name of the Company.

15.8.3 Supply of chemicals and catalysts

The Company shall provide the Contractor with a detailed list stating quantities and types of catalysts and quantities and specification of chemicals.

The Contractor shall prepare and issue inquiries and purchase orders for the items according to the above list.

The Contractor shall submit the quotations to the Company who shall issue purchasing instructions to the Contractor.

The Contractor shall expedite the delivery and accept the catalysts, check invoices, arrange shipping and insurance, and forward one copy of each invoice and of each packing list to the Company as soon as possible after shipment.

The insurance shall include the platinum content of platformer (Catalytic Reformer) catalyst and/or any high value material content of other catalysts as required by the Company. The Contractor shall also arrange for guarding as may be required during loading.

The Contractor shall provide shipping documents made out in the name of the Company.

15.8.4 Spare parts-MESC numbering and SPIR forms

Contractor shall be responsible for preparing a spare parts interchangeability list, as follows:

- **Contractor's responsibility**

The Contractor shall give procurement services, and assist in shipping FOB (1) all construction/commissioning spares and (2) a recommended two years operating supply.

- **Delivery**

The Contractor shall assist the Company for the timely delivery FOB of all spares in order that by means of regular and normal transportation they will be on hand at the Site in advance of the time for the start of precommissioning.

- **Recommendations**

The Contractor shall assist in obtaining Vendors' recommendation of the amount of all spares required. He shall pass on to the Company his own independent recommendation.

- **SPIR form**

All spare parts recommendations shall be made on a SPIR form; this applies to construction/commissioning spares as well as operating spares.

The Contractor shall assist in having the SPIR form filled out by the manufacturer/supplier of the equipment.

The Contractor shall verify the completed form from the manufacturer/supplier for accuracy and completeness, fill in his spare parts recommendation, and forward it to the Company.

The completed SPIR forms shall be supported by the following documents:
1. Complete manufacturer's parts list.
2. Relevant drawings, catalogs, pamphlets, and bulletins of the primary equipment.
3. Relevant parts lists, drawings, sketches, and pamphlets of subsuppliers of auxiliary equipment.

15.9 Reimbursable items

If required by the Contract, the following items purchased in accordance with the provisions of this section will be payable on reimbursable basis. Detail specifications and conditions of purchasing in each case shall be foreseen in the Contract if these items to be supplied based on lump sum price.

- One or two years spare parts for permanent plant equipment and material.
  - Laboratory equipment.
  - Laboratory chemical.

- Laboratory books, manuals, etc.
- Laboratory furnishings, consisting of permanently installed counters, sinks, cabinets, and vent hoods.
- Fire trucks, trailers, and accessories.
- Foam liquid
- Portable monitors
- Firefighting tools and spares.
- Fire extinguishers.
- Fire hose reels for buildings.
- Firefighting supplies and spares.
- Main workshop equipment and spares.
- Mobile maintenance equipment and spares.
- Safety equipment and spares.
- Inspection tools and spares.
- Instrument shop fixed and portable equipment, tools and spares.
- Electrical shop equipment, tools and spares.
- Miscellaneous maintenance tools and spares.
- Catalysts for Licensed Units as may be requested by the Company.
- Portable maintenance tools that may be required for specific equipment, such as:
- Diaphragm seal cutting tool for high-pressure heat exchangers in accordance with the Company's specifications.
  - Portable sky climbers and cables for stacks.
  - Portable equipment and materials for loading and unloading catalysts, etc.
  - Bolt tensioners, with accessories and spares.
  - Test pumps.
  - Bins and racks for workshop and stores building.
- Clinic equipment.
- Cafeteria and kitchen equipment (fixed and movable).
- Transport and Mobile Plant equipment (fixed and movable).

The followings are excluded from the reimbursable items and may be included in the Lump-Sum Prices as required by the Company:

- Cost of initial charge of chemicals and lubricating oils for the first year of operation.
- Cost of catalysts involving precious materials such as molecular sieves/catalysts for Pressure Swing Adsorption Unit, etc.
- Cost of spare parts for construction period and refinery and/or plant commissioning.

15.10 Process flow diagram (PFD) and piping and instrumentation diagrams

This section is intended for convenience of use and pattern of follow-up and also guidance. Also, it indicates the check points to be considered by the process engineers for assurance of fulfilment of prerequisitions at any stage in the implementation of process plant projects.

15.10.1 Definition of PFD

A process flow diagram (Figure 15.1) mainly defines:

1. A schematic representation of the sequence of all relevant operations taking place during a process and includes information considered desirable for analysis.
2. The process presenting events that occur to the material(s) to convert the feedstock(s) to the specified products.
3. An operation occurring when an object (or material) is intentionally changed in any of its physical or chemical characteristics, is assembled or disassembled from another object, or is arranged or prepared for another operation, transportation, inspection, or storage.

15.10.2 Purpose of PFD

The purpose of a PFD is generally as follows:

1. Plant design basis

The PFD shows the plant design basis indicating feedstock, product, and mainstream flow rates and operating conditions.

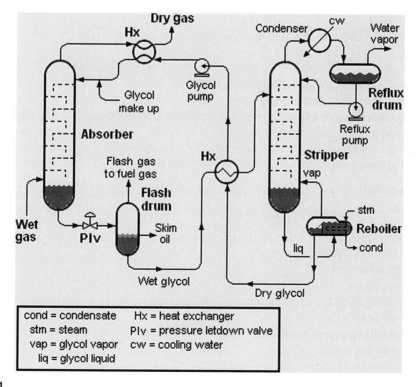

FIGURE 15.1

A simplified process flow diagram for a natural gas dehydration system (without material balance table).

2. Scope of process

The PFD serves to identify the scope of the process.

3. Equipment configuration

The PFD shows graphically the arrangement of major equipment, process lines, and main control loops.

4. Required utilities

The PFD shows utilities that are used continuously in the process.

15.10.3 Contents of PFD

- **Inclusive**

The PFD shall comprise but not be limited to the following items:

1. All process lines, utilities, and operating conditions essential for material balance and heat and material balance.
2. Type and utility flow lines that are used continuously within the battery limits.
3. Equipment diagrams to be arranged according to process flow, designation, and equipment number.
4. Simplified control instrumentation pertaining to control valves and the like to be involved in process flows.
5. Major process analyzers.
6. Operating conditions around major equipment.
7. Heat duty for all heat transfer equipment.
8. Changing process conditions along individual process flow lines, such as flow rates, operating pressure and temperature, etc.
9. All alternate operating conditions.
10. Material balance table.

- **Exclusive on PDF**

The following items are generally not shown on the PFD, except in special cases:

1. Minor process lines that are not usually used in normal operation and minor equipment, such as block valves, safety/relief valves, etc.
2. Elevation of equipment.
3. All spare equipment.
4. Heat transfer equipment, pumps, compressor, etc., to be operated in parallel or in series shall be shown as one unit.
5. Piping information such as size, orifice plates, strainers, and classification into hot or cold insulated of jacket piping.
6. Instrumentation not related to automatic control.
7. Instrumentation of trip system, (because it cannot be decided at the PFD preparation stage).

8. Drivers of rotating machinery except where they are important for control line of the process conditions.
9. Any dimensional information on equipment, such as internal diameter, height, length, and volume. Internals of equipment shall be shown only if required for a clear understanding of the working of the equipment.

15.10.4 General drafting instructions

• **Scale**

PFDs should not be drafted to scale. However, their size should be compatible with that of equipment drawings.

• **Flow Direction**

As a rule, PFDs should be drawn from the left to the right in accordance with process flows (See Figure 15.2).

• **Process and Utility Lines in General**

The main process flow shall be accentuated by heavy lines.
  Process utility lines shall be shown only where they enter or leave the main equipment.
  Pipelines shall not be identified by numbers.
  Valves, vents, drains, bypasses, sample connections, automatic or manual control systems, instrumentation, electrical systems, etc. shall be omitted from the schemes.
  The direction of the flow shall be indicated for each line.
  Where a PFD consists of two or more divided sheets, drawing numbers should be indicated.

15.10.5 Direction of flow

The direction of flow should be indicated by arrows. In principle all flow lines should be denoted by arrows located at the inlet of equipment, at merging points, and at the corners of the lines. Where a process line is long, however, the process flow may be denoted by arrows located at intermediate points.
  The number of arrows used to denote one process flow line is not restricted. However, care should be taken not to clutter the drawing with excessive arrows.
  Arrows at corners may be suitably omitted.
  Where a PFD must be divided into two or more sheets, it should be divided at portions where division is easiest from the process standpoint and each divided section should be drawn on a separate sheet.
  Where there are two or more identical trains of process flows, one representative train may be given in the PFD and the others omitted. However, notations pointing out such omissions must be clearly indicated in the titles of all relevant PFDs to avoid confusion.
  As a rule, base lines should not be drawn. Similar items of equipment, however, should be aligned at the same level as far as possible.

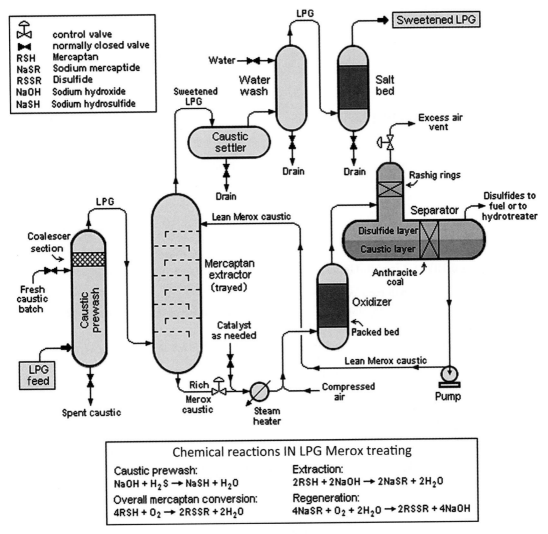

FIGURE 15.2

A simplified process flow diagram for sweetening of LPG (without material balance table).

15.11 Identification and numbering of equipment

15.11.1 Process equipment

- **Letter of group**

Each item of equipment shall be identified by an identifying or a tag number composed of letter.

- **Equipment number and name**

The equipment number and name should be given in the PFD, as a rule, at the upper or the lower part of the sheet, preferably in a space close to the center line of the equipment that is to be denoted. However, depending on the space, either the number or the name can be omitted.

- **Installed spare equipment**

Installed spare equipment, such as pumps, shall be indicated by a suffix letter like "A" or "B".

- **Equipment drivers**

Equipment drivers shall carry the same designation as the driven equipment.

- **Instrumentation**

It is not necessary to assign an identifying number in the PFD.

15.12 Description of equipment

15.12.1 Symbols of equipment and operating conditions

1. As a rule, piping and equipment symbols that are common to individual processes should be unified (see Figures 15.3–15.6).
   Symbols to denote other equipment not specified in this section shall be decided during Project execution upon the Company's approval.
2. Decimal numbers should be used inside the symbols to denote operating conditions.
3. The position of the operating condition denotation should be as close as possible to the point requiring indication. Where it is difficult to find space for such denotation, however, an auxiliary line should be used to indicate it.

15.12.2 Minimum information requirements for equipment

- **Designated streams**
  1. Stream numbers should be serially denoted by decimal numbers.
  2. Fluid name.
  3. Total flow rate.
  4. Density and/or molecular mass (weight) if required.
  5. Operating pressure and temperature if required.
- **Heat exchangers**
  1. Identification number and service name.
  2. Operating heat duty.
  3. Inlet and outlet temperatures on both shell and tube sides.
- **Furnaces**
  1. Identification number and service name.
  2. Operating absorbed heat duty.
  3. Inlet and outlet operating temperatures on tube side.

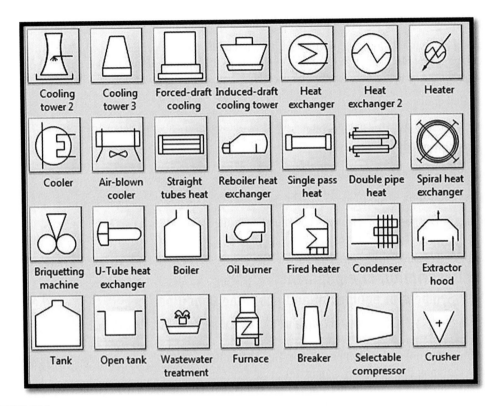

FIGURE 15.3

Typical symbols in process flow diagrams (part 1).

- **Reactors**
  1. Identification number and service name.
  2. Inlet and outlet operation temperature.
  3. Inlet and/or outlet pressure.
- **Columns**
  1. Identification number and service name.
  2. Tray numbers, operating temperature and pressure for top and bottom trays and also for special trays such as feed and draw-off, etc.
  3. Trays shall be numbered from bottom to top.
- **Drums**
  1. Identification number and service name.
  2. Operating temperature.
  3. Operating pressure.
- **Pumps**
  1. Identification number and service name.
  2. Normal operating capacity and differential pressure.

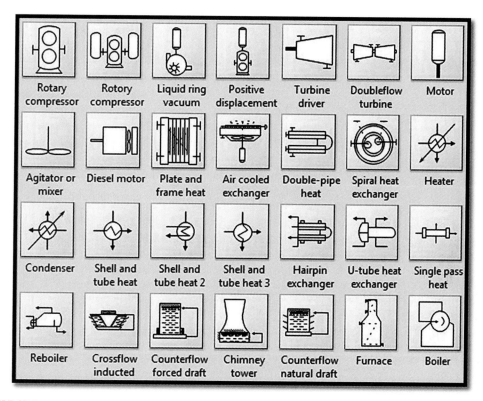

FIGURE 15.4

Typical symbols in process flow diagrams (part 2).

- **Compressors and blowers**
  1. Identification number and service name.
  2. Normal operating capacity and differential pressure.
- **Ejectors**
  1. Identification number and service name.
  2. Inlet and outlet operating pressure for ejector system.
- **Tanks**
  1. Identification number and service name.
  2. Operating temperature.
  3. Operating pressure.

15.13 Description of instrumentation

Instrumentation to be denoted are instruments, measuring devices, and control valves.

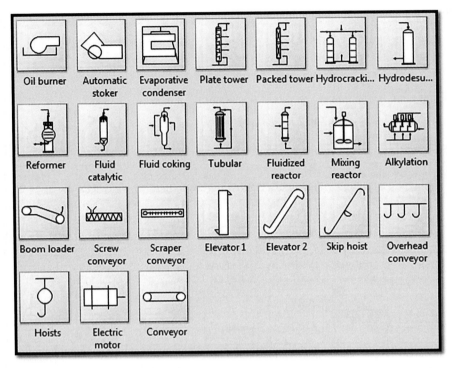

FIGURE 15.5

Typical symbols in process flow diagrams (part 3).

15.13.1 Instruments

- **Symbols for instrument**
  1. The symbol for an instrument (Figure 15.7) is a circle that shall be connected to the line nearest the point of measurement.
  2. Where the instrument is a controller, a dotted line representing the control impulse shall connect the instrument circle to the controller valve.
  3. The denotation of such functional symbols as "R" for recorder, "I" for indicator, and "A" for alarm, etc. should be omitted except for the functional symbol "C" for control.

There should be no distinction as to whether instruments should be locally installed or mounted on the main instrument panel.

15.13.2 Functional symbols for control

The following symbols are shown inside the circle representing the instrument.

Flow Controlling FC
Flow Ratio Controlling FRC
Level Controlling LC

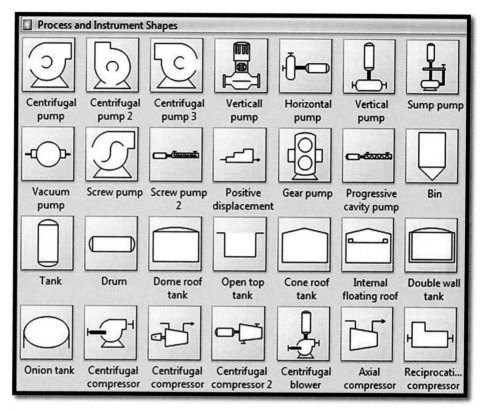

FIGURE 15.6

Typical symbols in process flow diagrams (part 4).

Pressure Controlling PC
Pressure Differential Controlling PDC
Temperature Controlling TC
Temperature Differential Controlling TDC
Speed Controlling SC
Mass (Weight) Controlling MC

- **Cascade control**

Where one controller alters the desired value of one or more other controllers, the instruments circles shall be connected by a dotted line.

- **Compound control**

Where the control actions of two or more controllers combine to operate one or more control valves, the instrument circles representing the controllers shall be joined by dotted lines to the instrument circle representing the combining device.

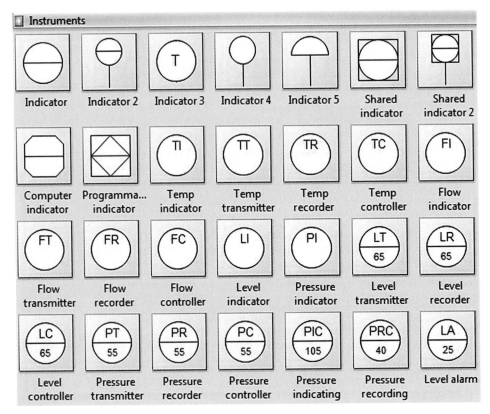

FIGURE 15.7

Symbols for instrument.

15.13.3 Measuring devices

The connecting line between the circle representing the instrument and the stream line represents the measuring device such as for temperature measurement, pressure measurement, flow rate measurement, etc.

- **Flow rate measuring**
  1. Regarding flow rate measurement, definitions of apparatus type such as rota meter, pitot tube, turbine meter, should be shown.
  2. Valves associated with the device need not be shown.
- **Level measuring**
  1. Definitions of apparatus type such as ball float, displacement, difference pressure, etc., need not be shown.
  2. A distinction should not be made as to whether the apparatus is an internal or external type.
  3. Valves associated with the device need not be shown.

- **Measurement of pressure, temperature, etc.**
  1. No distinction should be made regarding measuring type.
  2. Valves associated with measuring devices need not be shown.

15.13.4 Control valves

- **Actuator**
  1. The symbol for an actuator is a half circle, which shall be connected with a dotted line representing the control impulse.
  2. There should be no distinction made as to whether the actuator is a diaphragm type, electric motor type, or oil cylinder type, etc.

15.14 Material balance table
15.14.1 Contents of material balance table

A typical material balance table as shown in Table 15.1 shall consist of at least the following information:

- **Stream information**

Stream number, name of stream, flow rate, composition.

- **Operating conditions**

Operating temperature and pressure.

- **Basic physical properties**

Molecular mass (weight), API gravity, relative mass density (specific gravity or Sp.Gr.), etc.

- **Data concerning hydraulic calculation**

Density, viscosity, etc. if required.

15.14.2 Denotation of material balance table

In preparation of material balance table, care should be given to the following points:

- **Position of denotation**

As a rule, the material balance table should be inserted at the lower part of the PFD.

- **Number of digits of numerals and denotation of small quantities**

As a rule, percentages should be expressed down to 0.01%. Where traces of components are concerned, special units, such as ppm, should be used.

Table 15.1 Typical Material Balance Table

Stream Description		Treated Flue Gas to Power Plant Stack		Make-up Wash Water to Absorber		Low Pressure CO₂ to Compression		Compressed CO₂ Product		Reclaimer Efficient (Note 3)	
Stream number		202		203		303		310		404	
Temperature, °C		34.9		15.0		40.0		60.0		139.7	
Pressure, bar(a)		1.00		1.00		1.65		129.00		1.00	
Component Flows	MW	kgmol/h	mol %	kgmol/h	mol%	kgmol/h	mol %	kgmol/h	mol %	kgmol/h	mol %
H_2O	18.02	1456	5.7%	409	100.0%	183	4.60%	0	<30 ppmv	--	--
CO_2	44.01	427	1.7%	0	0.0%	3841	95.4%	3841	99.9%	--	--
EFG + solvent	61.08	0	1 ppmv	0	0.0%	0	0.0%	0	0.0%	--	--
N_2	28.02	22,085	87.4%	0	0.0%	1	261 ppmv	1	0.0%	--	--
Ar	39.95	249	1.0%	0	0.0%	0	6 ppmv	0	0.0%	--	--
O_2	32.00	1059	4.2%	0	0.0%	0	24 ppmv	0	0.0%	--	--
SO_3	80.06	0.07	3 ppmv	0	0.0%	0	0.0%	0	0.0%	--	--
SO_2	64.06	0	0.0 ppmv	0	0.0%	0	0.0%	0	0.0%	--	--
NO_2	46.01	0	0.1 ppmv	0	0.0%	0	0.0%	0	0.0%	--	--
NO	30.01	1.46	58 ppmv	0	0.0%	0	0.0%	0	0.0%	--	--
HSS	—	0	0.0%	0	0.0%	0	0.0%	0	0.0%	--	--
Total molar flow, kgmol/h		25,276		409		4025		3842		1.2	
Total mass flow. kg/h		707,560		7,370		172,370		169,040		50	
Molecular weight		28.0		18.0		42.8		44.0		39.0	
Density, kg/m³		1.097		999.3		2.732		433.8		1200.0	
Liquid flow, m³/h				7.4				389.6		0.04	
Vapor flow, m³$_{LN}$/h (dry)		533,903				86,115					
Vapor flow, m³$_{LN}$/h (wet)		566,538				90,216					
Vapor flow, m³/h (actual)		644,760				63,099					
pH				7.2						8.4	

15.14.3 **Examples**

A typical material balance table is shown in Table 15.1. The following comments should be noted prior to preparation of the sheet:

1. The total flow rate is described as m^3/h and/or bbl/sd (stream day) for liquid flow and as Nm^3/h for gas flow as mentioned in Table 15.1.
2. The molecular mass (weight) and/or mole percent and/or melting point of each component are often inserted near each component

When preparing the heat and material balance sheets in addition to material balance tables, necessary reference should be made to PFD stream numbers.

15.15 **Piping and equipment symbols**

- Pipeline Symbols
  - Piping-Pipelines symbols
  - Piping-Trap functions symbols
  - Piping-General equipment symbols
- Shell and Tube Type Heat Exchanger Symbols
- Double-Pipe Heat Exchanger Symbol
- Reboiler Symbols
- Air Fin Cooler Symbols
- Box Cooler Symbols
- Furnace Symbols
- Pump Symbols
- Compressor, Expander, and Blower Symbols
- Column Symbols
- Drum Symbols
- Reactor Symbol
- Ejector Symbol
- Tank Symbols
- Other Equipment Symbols
- Atmospheric storage tanks symbols
- Vessels, including pressure storage vessels (Figure 15.8)

Figure 15.8 shows typical symbols, lines, and connection for PFDs.

15.16 **Piping & instrumentation diagrams (P&IDs)**

The P&ID, based on the PFD, represents the technical realization of a process by means of graphical symbols for equipment and piping together with graphical symbols for process measurement and control functions.

The Utility Flow Diagram (UFD) is a special type of a P&ID that represents the utility systems within a process plant, showing all lines and other means required for the transport, distribution, and

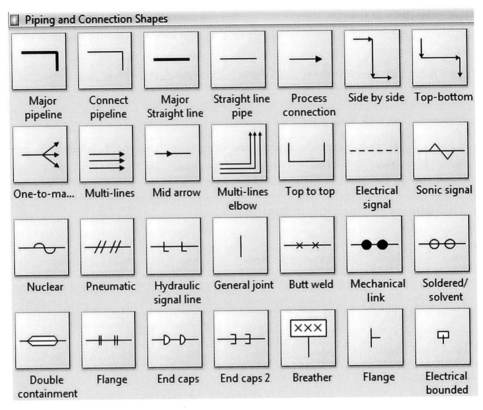

FIGURE 15.8

Typical symbols, lines, and connection for PFDs.

http://www.edrawsoft.com/pidsymbols.php.

collection of utilities. The process equipment in the UFD can be represented as a box with inscription (e.g. identification number) and with utility connections.

15.16.1 Representation

The representation and designation of all the equipment, instrumentation, and piping should comply with the requirements of Standards. Auxiliary systems may be represented by rectangular boxes with reference to the separate diagrams.

Dimensions of the graphical symbols for equipment and machinery (except pumps, drivers, valves, and fittings) should reflect the actual dimensions relative to one another as to scale and elevation. The graphical symbols for process measurement and control functions for equipment, machinery and piping, as well as piping and valves themselves, shall be shown in the logical position with respect to their functions.

All equipment shall be represented such that the consistency in their dimensions is considered if not in contrast to the good representation of the equipment.

15.16.2 Drafting

- **General rules**

Drafting shall be in accordance with the requirements outlined in standard. The drafting must be of sufficiently high quality to maintain legibility when the drawing is reduced to an A3 size sheet.

- **Drawings sheet sizes**

Diagrams shall be shown on A0 size (841 × 1189 mm) tracing paper. A1 size (591 × 841 mm) may be used for simple P&IDs and UFDs per Company's approval (see Article 6.3.3.2 for drawing dimensions and title block sizes).

- **Drawing title block**

The following requirements shall be shown on the title block of each drawing (Figure 15.9):

- Name of Company Relevant Organization, (if any), (e.g. Refineries Engineering and Construction);

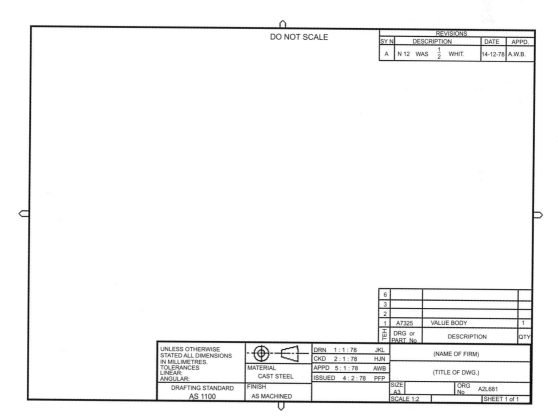

FIGURE 15.9

Drawing title block.

- name of refinery or plant
- Company's emblem;
- Contractor's name;
- drawing title;
- Company's project No.;
- Contractor's job No. (optional);
- Contractor's drawing No. (optional);
- Company's drawing No.

- **Line spacing**

The space between parallel lines shall not be less than twice the width of the heaviest of these lines with a minimum value of 1 mm. A spacing of 10 mm and more is desirable between flow lines.

- **Direction of flow**

In general, the main direction of flow proceeds from left to right and from top to bottom. Inlet and outlet arrows are used for indicating the inlet and outlet of flows into or out of the diagram.

Arrows are incorporated in the line for indicating the direction of the flows within the flow diagram. If necessary for proper understanding, arrows may be used at the inlets to equipment and machinery (except for pumps) and upstream of pipe branches. If a diagram consists of several sheets, the incoming and outgoing flow lines or piping on a sheet may be drawn in such a manner that the lines continue at the same level when the individual sheets are horizontally aligned.

- **Type of lettering**

Lettering in accordance with ISO 3098 Part 1, Type B vertical, is to be used.

- **Height of lettering**

The height of letters should be:

- 7 mm for drawing number;
- 5 mm for drawing title and identification numbers of major equipment;
- 3 mm for other inscriptions.
- **Arrangement of inscription**
  1. **Equipment**
     Identification numbers for equipment should be located close to the relevant graphical symbol, and should not be written into it. Further details (e.g. designation, design capacity, design pressure, etc.) may also be placed under the identification numbers.
  2. **Flow lines or piping**
     Designation of flow lines or piping shall be written parallel to and above horizontal lines and at the left of and parallel to vertical lines. If the beginning and end of flow lines or piping are not immediately recognizable, identical ones should be indicated by corresponding letters.
  3. **Valves and fittings**
     Designation of valves and fittings shall be written next to the graphical symbol and parallel to the direction of flow.

4. Process measurement and control functions

The representation should be in accordance with the requirements stipulated in ISA-S5.1 and ISO 3511, Parts 1 and 4, latest revisions unless otherwise specified in Standards.

• **Equipment Location Index**

P&IDs (Figure 15.10) shall be divided into equivalent intervals (each in 50 mm) either in length or width. The intervals shall be designated with numbers in length and alphabets in width. Equipment location on each diagram shall be addressed by the relevant coordinates where required. In the upper right-hand area of the flow diagram under title of Item Index all main equipment shall be listed by equipment number, alphabetically and numerically, and equipment location coordinates. In a separate sheet apart from P&IDs, an Item Index shall be prepared to summarize all equipment of the Unit/Plant with reference P&IDs and equipment location.

• **Arrangement**

The preferred arrangement is such that towers, vessels, and fired heaters be shown in the upper half of the diagram, heat exchange equipment in the upper three-quarters as practical, and machinery equipment in the lower quarter. The spacing of equipment and flow lines shall permit identification and tracing of the lines easily.

The area above the title block on each sheet shall be completely left open for notes.

The general flow scheme shall be from left to right. Unnecessary line crossing should be avoided.

Process lines entering and leaving the diagram from/to other drawings in the Unit shall be terminated at the left-hand or right-hand side of the drawing. Lines from/to higher number drawings shall enter and leave the drawing on the right-hand end and vice versa.

Each process line entering or leaving the side of the drawing should indicate the following requirements in an identification box:

• The service
• The origin or destination equipment item number
• Continuation drawing number with the relevant coordinates.

Process lines to/from other Units shall be terminated at the bottom of the drawing at a box indicating

• The service
• Source or destination Unit name and number
• The drawing number of the connecting flow diagrams with the relevant coordinates.

All utility lines entering or leaving the diagram shall be terminated at any convenient location at a box indicating the relevant utility service abbreviation (e.g. CWS, CWR, ISA). A Utilities Identification Table showing utility services with the reference drawings should be provided at top or left-hand side of each drawing title block.

Instrument, control system, and software linkage signals from sheet to sheet shall be terminated preferably at the side of sheet or in an appropriate location at a box indicating the continuation instrument number, location, and drawing number.

Equipment descriptions of towers, vessels, tanks, furnaces, exchangers, mixers, and other equipment except machinery shall be located along the top of the flow diagram. Machinery descriptions shall be along the bottom.

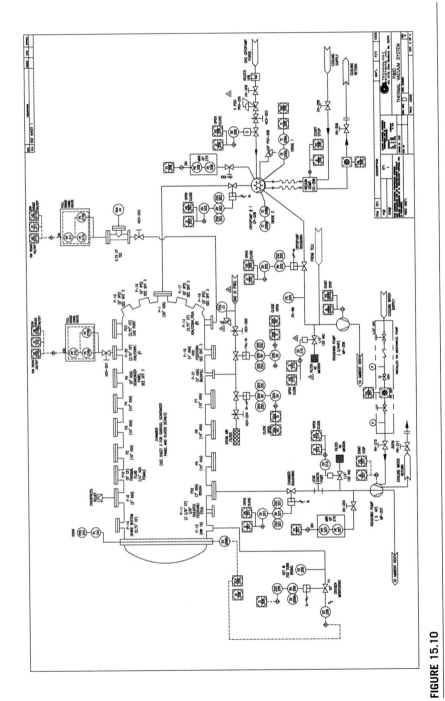

FIGURE 15.10

A typical P&ID diagram.

15.17 Minimum information to be shown on P&IDs

Each P&ID shall present all information as required herein below during implementation of a Project in detailed design phase. Extent of information shown on each P&ID in the basic design stage shall be agreed by Company in advance.

Vendor-supplied packages with an outline of the main components shall be shown in a dashed/dotted box. The letter P referring to package shall be indicated adjacent to each equipment and instrument of the package.

Equipment, instruments, or piping that are traced or jacketed shall be shown.

The identification number and service presentation shall be shown for each piece of equipment. This information shall be indicated in or adjacent to towers, drums, heaters, tanks, and heat exchangers, etc.

15.18 Equipment indication
15.18.1 Vessels, towers, drums

The following requirements shall be shown:

- changes of shell diameter (if any);
- top and bottom trays, and those trays that are necessary to locate feed, reflux, and product lines;
- all draw-off trays with tray number and diagrammatic representation of the downcomer position (e.g. side or center);
- all nozzles, manholes, instrument connections, drains, vents, pump-out and steam-out connections, blank-off ventilations, vortex breakers, safety/relief valve connections, sample connections, and handholds;
- skirt or legs, top and bottom tangent lines;
- elevations above base line to bottom tangent line of column or to bottom of horizontal drum;
- the position of high high liquid level (HHLL), high liquid level (HLL), normal liquid level (NLL), low liquid level (LLL), and low low liquid level (LLLL);
  Notes:
- For draw-offs, only NLL shall be shown. The other liquid positions will be shown as required.
- Indication of HHLL and LLLL shall be made only when they are actuating start/stop of an equipment or machinery through a switch.
- HLL, NLL, and LLL shall be shown for all cases except as specified under item 1 above.
- all flanged connections; [all connections whose purpose is not readily evident shall indicate the purpose (e.g. spare inlet, catalyst draw-off)];
- catalyst beds, packings, demisters, chimney trays, distributors, grids, baffles, rotating discs, mixers, cyclones, tangential inlet, and all other internals;
- water drop-out boots;
- maintenance blinds for the vessel nozzles.
- **Important notes:**
- All nozzles and connections indicated on the equipment data sheet shall be shown in their correct positions.

- All indications shall be such that the consistency in the dimensions is considered, although not necessarily to scale.
- Numbering of the trays shall be from bottom to top.
- Height of the vessel bottom tangent line shall be indicated.
- A valved drain for all columns and vessels shall be indicated. Generally, this valve is to be located on the bottom line outside the skirt and between the vessel and the first pipe line shut-off location (valve or blinding flange). The drain valve shall be located on the bottom of the vessel when:
  - No bottom line is present.
    or
  - The bottom line is not flush with the lowest point of the vessel.
- The valved vent with blind flange for all columns and vessels provided on the top of the vessel should be indicated.
- Relief valves generally located on the top outlet line downstream of the vessel blinding location or directly connected to the vessel should be indicated.
- Utility connections on all vessel/columns shall be shown.
- One local pressure indicator (PI) shall be indicated on top of vessel/column.
- One local temperature indicator (TI) shall be indicated on the top outlet line of vessel/column.
- Nozzles identifications on vessels, reactors, and towers shall be according to Standards.

15.18.2 Equipment description

The following requirements shall be described under equipment description:

1. vessel item number (this number will also appear adjacent to the vessel);
2. service;
3. size [inside diameter(s) and tangent to tangent length];
4. design pressure (internal/external) and design temperature;
5. indication of insulation;
6. line number of vessel trim (this applies to level gauge (LG) & level controller (LC) connections, vents, sample connections, etc.);
7. indication of cladding and lining (if any).

15.18.3 Tanks

The following requirements shall be shown:

1. all nozzles, manways, instrument connections, drains, vents, vortex breakers, and safety/relief valve connections;
2. all internals such as steam coils, air spargers, tank heaters, etc.

15.18.4 Equipment description

1. equipment item number (this number also appears adjacent to the tank);
2. service;
3. inside diameter and height;
4. nominal capacity, in m^3;
5. design pressure and temperature;
6. indication of insulation.

15.18.5 Fired heaters, boilers, incinerators

The following requirements shall be shown:

1. all nozzles, instrument connections, drains, vents, and dampers;
2. ducting arrangement including damper actuators where required;
3. detail of draft gauges piping and arrangement;
4. waste heat recovery system (if present), such as economizer, air preheater, forced draft fan, induced draft fan, etc.;
5. decoking connections;
6. detail of one complete set of burners for each cell and total burner number required for each type of burner;
7. tube coils schematically in correct relative positions and all skinpoint thermocouples;
8. logic diagram of shutdown system (heat off sequence);
9. number of passes and control arrangement;
10. snuffing steam nozzles and piping arrangement;
11. blow-down and steam-out connections;
12. testing facilities;
13. convection section (where applicable).

15.18.6 Equipment description

1. item number (this number will also appear adjacent to the equipment);
2. service;
3. duty (kJ/s);
4. design pressure and temperature of coils;

15.18.7 Heat exchangers, coolers, reboilers

The following requirements shall be shown:

1. all nozzles, instrument connections, drains and vents, chemical cleaning connections, and safety/relief valves as indicated on the equipment data sheet;
2. spectacle blinds for the isolation;
3. elevations required for process reason (e.g. reboilers, condensers);
4. the connections that allow pressure and temperature survey of heat exchanger facilities;
5. the position of HLL, NLL, and LLL for kettle type reboilers;
6. direction of flow in each side of exchanger.

- **Important notes**

Due considerations should be made for proper indication in the following requirements:

1. Generally, direction of flow shall be downflow for cooled media and upflow for heated media.
2. Isolation valves shall be provided on inlet and outlet lines where maintenance can be performed on the exchanger with the Unit operating. Provision of bypassing is required for this case.
3. Shell and channel piping shall be provided with a valved vent connection and a drain connection unless venting and draining can be done via other equipment.

4. At exchangers with circulating heat transfer media, the outlet valve shall be of a throttling type for control of heat duty.
5. An inlet and outlet TI shall be provided on each exchanger (on either shell or tube side) to facilitate checking of heat balance around exchanger.

   Type of TI shall be as follows:

   a. A board-mounted TI shall be provided at the inlet and outlet of all shell and tube process/process exchangers.
   b. For water coolers, the water side outlet shall be provided with a local TI only. The shell side in and out shall be provided with board-mounted TIs.
   c. Thermowells (TWs) shall be provided between each shell side and tube side of the same services when the exchangers are in series.
   d. Local indicator type shall be provided for the requirement of local temperature control, such as manual bypass control.

15.18.8 Equipment description

1. Equipment item number (this number also appears adjacent to the equipment);
2. service;
3. duty (kJ/s);
4. shell side design pressure and temperature;
5. tube side design pressure and temperature;
6. indication of insulation.

Sequence of numbering for stacked exchangers/coolers shall be from top to bottom.

15.18.9 Air fin coolers

The following requirements shall be shown:

1. all nozzles and instrument connections;
2. blinds for the isolation;
3. any automatic control (fan pitch control or louver control) and any alarm (vibration alarm, etc.);
4. configuration of inlet and outlet headers and the branches. Only one bundle and fan shall be shown; total number of fans and bundles shall be indicated. When multiple bundles are required, header's arrangement as separate detailed sketch shall be indicated;
5. steam coil and condensate recovery system (if required);
6. isolation valves (if required); isolation valves shall be provided in corrosive and fouling services where individual bundles can be repaired and maintained with the Unit operating;
7. valved vent and valved drain connection for each header, vent header should be connected to closed system for volatile services;
8. a board-mounted TI at inlet and outlet, (the TI will monitor the process side of each air fin service). If multiple bundles to be used for fouled services, provide TWs on the outlet of each bundle.

15.18.10 Equipment description

1. equipment item number (this number will also appear adjacent to the equipment);
2. service;
3. duty (kJ/s);
4. tube side design pressure (internal and external) and design temperature.

15.18.11 Rotary machineries

The following requirements shall be shown:

- **Pumps**

    1. all nozzles including instrument connections;
    2. pump suction valve and strainer, and discharge valve and check valve. Provision of wafer type check valve should be avoided unless otherwise specified;
    3. pump drains and vents piping and destination.
    4. the type of pump;
    5. pump auxiliary system connections such as cooling water, seal oil and lube oil, steam;
    6. detail of lube and seal oil/sealing systems, cooling water piping arrangement, and minimum flow bypass line requirement for pumps;
    7. winterization and/or heat conservation (steam or electrical) where required;
    8. warm-up and flushing oil lines detail; a DN20 (¾ inch) bypass/drain from the check valve to the pump discharge line shall be provided as warm-up line for the cases.
    9. pressure gauge located on the discharge of each pump; the gauge shall be installed between the pump discharge nozzle and the check valve;
    10. pressure relief safety valves (if any);
    11. automatic start-up of standby Unit (if required);
    12. balanced or equalized line for vacuum service.

- **Compressors and blowers**

    1. type of compressor or blower;
    2. start-up facilities (i.e. inert gas purge system);
    3. safety/relief valves;
    4. suction and discharge valves;
    5. suction strainer (filter) and discharge check valve;
    6. suction and discharge pulsation dampener where required;
    7. valved vents and casing drains;
    8. winterization (steam or electrical tracing on suction piping) where required;
    9. lube and seal oil/sealing system and cooling water systems detail arrangement;
    10. interstage coolers where required;
    11. surge protection (where required);
    12. inlet and outlet nozzles;
    13. all instrument connections.

- **Steam and gas turbine drivers**

1. all nozzles and connections;
2. detail of all auxiliary systems for steam turbine drivers such as steam supply, condensate return, surface condenser;
3. detail of lube oil, cooling water, etc.;
4. all instrumentations such as PI, TI;
5. safety/relief valves; relief valves shall be located between the discharge nozzle and the outlet isolation valve; weep hole at exhaust of the relief valve that opens to the atmosphere shall be provided to draw off the condensate drain.
6. warming bypass around inlet isolation valve for steam turbines; the valve on warm-up line shall be DN25 (1 inch) globe type;
7. steam traps and condensate recovery system for the steam turbine casing drain and upstream of isolation valve at inlet of the turbine;
8. vent line to atmosphere at turbine exhaust; the vent is required for the start-up/test operation of the turbine.
9. detail of all firing and control systems for gas turbine drivers.

15.18.12 Equipment description

- **Pumps**

1. pump item number (this number also appears below the pump);
2. service;
3. capacity, (m^3/h, dm^3/h for injection pumps);
4. differential pressure, (kPa);
5. relative density (specific gravity) of pumped fluid at pumping temperature;
6. indication of insulation and tracing;
7. miscellaneous auxiliary piping (CW, flushing oil, seal oil, etc.).

- **Compressors and blowers**

1. equipment item number and stage (this number also appears below the compressor);
2. service:
3. capacity, (Nm^3/h);
4. suction pressure and temperature, [kPa (g)], (°C);
5. discharge pressure and temperature, [kPa (g)], (°C);
6. miscellaneous auxiliary piping (CW, lube, oil, seal oil/sealing system, etc.);
7. gas horse power, (kW).

- **Other requirements**

1. When a pump or compressor is spared, the data are listed once commonly for both pumps at the bottom of the flow diagram. The spare is identified by the word "Spare" below the pump or compressor. The operating equipment and the spare have the same number but with suffixes "A" and "B".

2. Stage numbers are shown only for multistage compressors. All compressor data for the first stage shall be indicated. For subsequent stages only Nm^3/h may be omitted.

15.18.13 Miscellaneous equipment

Depending on the type of equipment (silensor, flame arrestor, filter, etc.) the following information shall be presented:

1. all nozzles, instrument connections, vents, drains, etc.;
2. equipment description at top of the flow diagram and including:
   a. equipment item number;
   b. service;
   c. tracing/insulation requirements;
   d. design pressure and temperature;
   e. capacity.

15.19 Instrumentation

The following requirements shall be shown:

- all instrumentation including test points;
- isolation valves connecting to instruments (primary connection valve);
- control valve sizes and air failure action (fully closed (FC), fully open (FO));
- block and bypass valve sizes at control valve stations;
- level gauges connection type and range, and level controllers connection type, range and center of float (where NLL is not shown). Type, material, and tracing requirement of level gauges shall be shown;
- sequence of opening and closing for the split range control valves;
- solenoid shutdown devices at control valves/shutoff valves;
- tight shutoff valves requirements (where required);
- handwheels when provided on control valves;
- limit switches on control valves when required;
- mechanical stopper and/or signal stopper on control valves when required;
- push buttons and switches associated with shutdown systems;
- the instrument tag number for each instrument;
- analyzer loop details and special notes as required;
- winterization of instruments;
- compressor local board-mounted instrumentation;
- software linkage and alarm and shutdown logic system. Complex shutdown systems shall be shown as a "black box" with reference made to the logic diagram shown on a separate sheet. All actuating and actuated devices shall be connected to the black box;
- all elements of advance control and optimization systems;
- indication of "Readable From" for all local indicators and/or gauges that shall be readable from a designated valve.

15.20 Piping

- All piping shall be shown on P&IDs, including:
  - process lines;
  - utility/common facility branch lines (e.g. sealing and flushing lines, cooling water lines, steam-out lines and connection, nitrogen lines);
  - flare lines, including safety/relief valves discharge lines;
  - start-up and shutdown lines;
  - pump-out lines;
  - drain and vent lines and connections;
  - purge and steam-out facilities;
  - catalyst regeneration lines;
  - catalyst sulfiding lines;
  - catalyst reduction lines;
  - equipment and control valve bypasses;
  - detail of spool pieces, equipment internals, etc., when required;
  - steam tracing and steam jacketing.
- The direction of normal flow shall be shown for all lines.
- The points or breaks at which line sizes or line specifications change shall be clearly indicated.
- All blinds shall be indicated on the drawings, and the symbols used shall distinguish between tab blinds and spectacle blinds.
- All vent and drain connections shall be identified whether screw caped or blind flanged, if required.
- Steam-traced lines and steam-jacketed lines shall be so indicated.
- All equipment flanges, all reducers, and non-standard fittings, such as expansion bellows, flexible tubes, shall be shown.
- All valves shall be shown by a symbol representing the type of valve. Any special orientation or location required for process reason and/or operability shall be shown. It is not necessary to show flanges at flanged valves except for those cases where the flanges deviate from the piping specification for the line in question, in which case flange and rating shall be shown. Any isolating valve shall be shown locked, normally open, or closed.
- Control valve sizes shall be shown.
- All valves shown on the flow diagram shall have their size indicated by the valve, if different from line size.
- Valve boxes/valve pits shall be shown by two embraced squares or rectangles with indication of Valve Box or Valve Pit.
- Safety relief valves type, inlet and outlet size, and rating and set pressure should be shown.

15.21 Special requirements

- High-point vents and low-point drains are shown only when they are connected to a closed system or are required for process reasons.

- Utility lines originating and terminating adjacent to the equipment involved shall be shown. Only the length of line necessary for valving, instrumentation, and line numbering is shown. Utility line origin and terminus are indicated by reference symbol or abbreviation only. Main utility headers are not shown on the P&IDs; they are shown on the utility system flow diagrams.
- Pertinent information regarding a line such as "do not pocket" or "slope" shall be noted adjacent to the line.
- Typical air cooler manifold piping arrangement should be shown.
- Connections on process lines that require to be blanked or deblanked for flow direction under special circumstances to be shown on P&ID.
- Reduction and enlargement in line size are indicated by line size designation, and reducer and expander symbols.
- Calculated wall thicknesses and/or schedules not already prespecified in the individual line classes shall be shown on the flow diagrams.
- Corrosion allowances other than the nominal allowances indicated in the individual line classes shall be shown on the diagrams.
- All operating drains shall be noted and sized on the flow diagrams and shall be routed to a drain funnel. Destination of the drains shall be according to the relevant specifications. All drains carrying light hydrocarbons (Reid vapor pressure 34.5 kPa absolute or greater) shall be segregated from the oily sewer system, and shall be connected to the flare system.
- Sample and test connections shall be shown on the diagrams where required. Samples that require cooling and connections to the flare shall be shown with the cooling and flare lines connections.
- Emergency showers, eye wash fountains, and utility stations shall be shown on the Utility Distribution Flow Diagrams.
- Any locations where slopes, straight runs, minimum mixing runs, etc., are required for process reasons must be indicated.
- The necessary instrumentation and piping for start-up, control and shutdown, etc., should be shown for any equipment on P&IDs wherever applicable.
- Break points between underground and aboveground piping with insulating flanges (if required) shall be shown.
- Minimum distance requirement for in-line blending is to be indicated.
- Weep hole requirement is to be shown.

15.21.1 Piping specialty items

- Piping components not identified by instrument or mechanical equipment numbers, etc., and not covered by the piping material specification, shall be identified by assigning a Specialty Item Number or an Item Code Number for identification symbol and shall be shown on the diagrams.
- Symbol "M" standing for "monel trim" should be mentioned on the valves on the P&IDs
- Services where there is a possibility of condensed water and H_2S being present except for the line classes that provide monel trim valves and other features. Where it is intended that the whole line should have monel trim valves should also be indicated on the line list.
- ASME and non-ASME Code change should be indicated for connection wherever applicable.

15.21.2 Steam traps and winterizing system

The following requirements shall be followed:

- Steam traps pertaining to the winterizing systems (steam tracing) are not shown on the P&IDs except for the following cases:
  - at dead ends/pockets on steam lines;
  - at upstream of the Unit battery limit main block valves on steam lines;
  - at all points where there is possibility of condensation;
  - at upstream of the first block valve of steam line going to the steam turbine drivers, steam coils, or steam reboilers.
- Steam trap and the relevant steam and condensate lines are to be shown for all steam reboilers, heaters, coils, etc.
- Steam/electrical tracing requirement shall be noted on P&IDs by a dashed line parallel to the line to be traced.

15.22 General notes

General notes to be put on the front sheet of P&I Diagrams of each Unit under title of General Notes. Reference should be made to the front sheet drawing number showing General Notes on each P&ID.

The following general notes shall be specified as minimum requirement:

- All dimensions are in millimeters except as noted.
- Elevations shown are above the highest point of paving.
- All valves are line size unless otherwise shown.
- This flow diagram is diagrammatic only. Design of pipe lines must be investigated for venting of gas and vapor pockets in piping and equipment, low points in piping, pumps and equipment for freezing and draining and accessibility of all valves, flanges, and instruments including thermocouples, etc.
- All electronic instrumentation shall be installed away from steam lines and high-temperature heat sources.
- For level transmitter, center of float is NLL. The range shall cover the difference between LLL and HLL.
- Sample tappings for gas samples shall be from the top of the main line. Liquid samples tapping shall be done from the side.
- Except for process reasons, low-point drains and high-point vents are not shown.
- All items marked P can be supplied as part of package Units.
- Temperature instruments shown with M are provided with monel well.

15.22.1 Piping drains and vents

Low-point drains and high-point vents of piping shall be provided in accordance with the following:

1. **Drains for all sizes**
   a. alloy piping: DN 20 (¾ inch) gate valve with blind flange.
   b. carbon steel piping: DN 20 (¾ inch) gate valve with threaded plug.

2. **Vents for DN 50 (2 inch) and larger**
   High-point vent shall be provided for the piping of DN 50 (2 inch) and larger. Size and type are based on the following:
   a. alloy piping: DN 20 (¾ inch) gate valve with blind flange;
   b. carbon steel piping: DN 20 (¾ inch) gate valve with threaded plug;
   c. the vent provided for hydrostatic testing shall be DN 20 (¾ inch) boss with threaded plug.

15.22.2 Block valves on orifice tap

1. DN 15 (½ inch) single gate valve shall be provided for the all orifices of the piping class of PN 100 (600 #) and less.
2. DN 20 (¾ inch) single gate valve shall be provided for the all orifices of the piping class of PN 150 (900#) and over.

Drain valve of level gauges and instruments:

1. Drain valves [DN 20 (¾ inch) gate valve] shall be provided.
2. The provisions should be made for routing the drain of liquids with RVP of greater than 34.5 kPa (abs) to flare.

Figure 15.11 provides symbols for various valves in P&IDs.

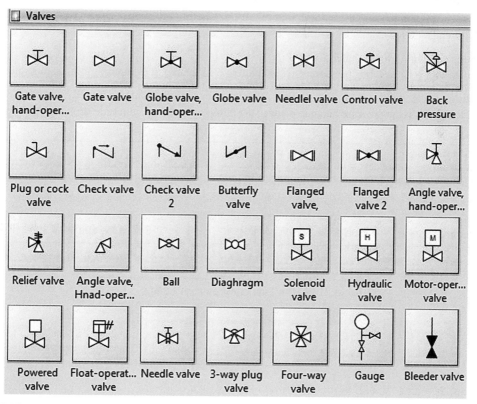

FIGURE 15.11

Symbols for various valves in P&IDs.

15.23 Design criteria for preparation of P&IDs

The following design criteria shall be applied for preparation of P&IDs unless otherwise specified in the relevant piping and/or equipment Specifications of the Company. In case of any conflict, the specific piping and/or equipment Specifications will govern.

15.23.1 Assembly piping of pumps

- **Valve size selection basis for pumps**

Generally, the size is likely different between pump suction line and pump suction nozzle, or pump discharge line and pump discharge nozzle.

If the pump nozzle is one or more sizes smaller than the line size, the size of block valve shall be in accordance with Table 15.2:

- **Pump strainer**

The suction strainer of pumps shall be selected in accordance with the criteria in Table 15.3:

    2 Strainers DN150 (6 inch) and larger shall have DN 25 (1 inch) drain valve.

- **Pump vents and drains**

Vent gas from pump casing drains and vents shall be routed to the closed system such as the flare for the following services:

1. fluids containing toxic material;
2. fluids with a Reid vapor pressure greater than 34.5 kPa (abs) at pump operating temperature.

In addition to the above, the vent of casing for the vacuum service should be routed back to the suction vessel to make out the pressure balance prior to the pump operation. Drain of hydrocarbon pumps shall

Table 15.2 The Size of Block Valves for Pump Nozzles		
	Nozzle	**Block Valve**
At pump suction	1—One size smaller than line 2—Two or more sizes smaller than line	1—Same as suction line size 2—Select one size smaller than line
At pump discharge	Smaller than discharge line	Select one size smaller than line

Table 15.3 The Suction Strainer of Pumps	
Line Size	**Strainer Type**
DN 80 (3 inch) and larger	T
DN 50 (2 inch) and smaller	Y

also have disposal to the oily water sewer in all cases in addition to the above requirements unless otherwise specified.

- **Warming-Up line**

The provisions for warming up of a pump is required for the pump operated at 170 °C and higher or when the process fluid solidifies at ambient conditions or the fluids are corrosive or toxic.

- **Auxiliary piping of pump**

Details of auxiliary piping such as cooling water, plant water, steam and condensate, and mechanical seal flush fluid that are required as per the pump data sheet shall be shown on a separate drawing. Reference to the auxiliary piping drawing shall be noted under the pump description.

15.23.2 Steam-out, drain, and vent for vessels

Vent connections must be located on top of the vertical and horizontal vessels.
   The drain valve will be provided as follows:

- For low-pressure services up to design pressure of 3800 kPa, provide single block valve with blind plate.
- For high-pressure services over design pressure of 3800 kPa, or where the nature of liquid requires it, provide double block valves with blind plate.

15.24 Bypass for safety/relief valve

The bypass shall be provided for venting the hydrocarbon gas or toxic gas to flare system while plant shutdown or start-up. Provision of bypass shall be as per following criteria:

15.24.1 Vessels

Bypass shall be provided unless otherwise specified in the relevant Company's specifications.

15.24.2 Piping/equipment

- **Gas service**
   1. If there is other purge line to flare on same stream line, bypass is not required for safety/relief valve.
   2. In case of no purge line to flare for toxic or flammable hydrocarbon, bypass valve shall be provided. The size of bypass valve and line shall be same as the vent size of piping/equipment.

- **Liquid service**

Bypass valves are generally not provided for liquid service unless otherwise specified.

- **Line Numbering**
  Line numbers shall be assigned to all lines with the following origins and destinations:
  - from individual equipment item to individual equipment item;
  - from line to individual equipment item and vice versa. Another number is required for the line located the downstream of equipment;
  - from line to line (exceptions: control valve bypass, block valve warm-up and equalizing bypasses, and safety/relief valve bypass);
  - from unique equipment to the same unique equipment item (except level standpipes);
  - from line or equipment to atmosphere, funnel, or closed drainage system (exception: continuous process vent stacks and process drains).

Pipe line numbers shall be prefixed from source to Unit battery limit with the Unit number of the Unit of origin.

A new line number is required when the pipe design condition can vary (e.g. downstream of the control valve assembly) or when a new piping class is to be specified.

Line number shall be held up to the point where the line ends to the header or Unit battery limit block valve. All branches to and from header shall have an individual line number.

All utility headers (systems) including all steam, water, and sewer lines shall be numbered with their respective Units. All branches serving a specific Unit will be numbered with that Unit.

Line numbers shall be selected so that consecutive line numbers are grouped first by common service. Spare line numbers may be left between the groupings.

All process lines routed from Unit to Unit shall be assigned on interconnecting line number. Within the process Unit(s), Unit line numbers are to be assigned. The interconnecting Unit P&ID is to show every interconnecting process line and indicate the line numbers inside the process Units at the Units battery limits.

15.24.3 Utility connections

Utility connections to process line and equipment for steam and nitrogen shall be as follows:

- **Notes on utility tie-in:**
  1. The isolation valve may be omitted if the process line is open to atmosphere.
  2. Provide a drain downstream of the check valve to check the leakage.
  3. Provide a spectacle blind and block valve for N_2 service.
  4. Main block valve for steam service shall be at the branch point from steam header.
  5. This configuration shall be used for low-pressure steam (all sizes). For medium- and high-pressure steam double block valves with bleeder between the valves is required.

15.24.4 Unit battery limit installation

- **Notes on Unit battery limit installation requirements:**
  1. Provide for hydrogen, nitrogen, toxic gases, and all high-pressure fluids ($P > 3800$ kPa), double block valves, spectacle blind, and drain.

2. Provide for each process line (not included in item 1 above) an isolation valve, spectacle blind, and drain.
3. Provide a flow indicator and recorder shown on board for each process stream entering and leaving each Unit. Do not duplicate measuring elements in the same stream within one block area.
4. Provide a board-mounted TI on each process stream entering and leaving the Unit where a flow integrator is provided. Do not duplicate with TIs required for other purposes. Generally, the TI is to be located at downstream of the flow element.
5. Provide a sample station for all products leaving and/or entering the Unit.
6. Product streams leaving Units shall be piped at the Unit limits to the relevant slops header (light or heavy slops) as well as for the start-up (off-spec.) operation.
7. Provide a local PI on each process stream entering and/or leaving the Unit. Do not duplicate with PIs required on the same streams. PIs may be board mounted as required.
8. Special attention should be made to the possibility of avoiding duplication of some or all of the above- mentioned hardware on the adjacent Units.

15.25 Criteria for utility flow diagrams

The UFDs shall be prepared as separate drawing titled as Utilities Distribution Flow Diagram. The distribution of utilities for plant operation shall be shown on the drawing. The utilities for plant operation are generally classified as follows where applicable:

- several grades of steam;
- several grades of condensate;
- boiler feed water;
- cooling water and sea water;
- raw (fresh) water;
- plant and potable water;
- fuel oil and fuel gas;
- instrument and plant air;
- nitrogen;
- inert gas;
- seal oil/flushing oil;
- closed circuit hot oil system;
- flare and blow-down;
- chemical system such as caustic and ammonia.

The above utilities are classified into several groups and shown on diagram(s) in accordance with the next articles. A dedicated drawing shall be prepared for Flare and Blow-down.

Utility Flow Diagrams shall be presented in accordance with the requirements stipulated in this section for P&IDs where applicable.

Utility Flow Diagrams shall show main distribution/collection headers and finger headers with their isolating facilities and instrumentation. The branch line and subheader arrangement shall be shown as practical as possible.

Indication criteria of connection between P&IDs and UFDs is according to the following general philosophy:

1. The indication of isolation valve shall not be duplicated on the P&ID and UFD.
2. Valve and instrument that will be used for the normal operation shall be indicated on the P&ID, such as:
   a. block valves for water cooler inlet and outlet;
   b. block valves for snuffing steam of heater;
   c. globe valve for steam injection control;
   d. control valves for fuel control.
3. Valves that will be used only for start-up and shutdown shall be indicated on the UFD such as:
   a. header isolation valve for steam purge connection;
   b. isolation valve for fuel gas or fuel oil.

Utility/common facility branch line header valves at the process Unit battery limit shall be shown. The UFD shall also indicate any valve in utility/common facility individual branch lines required for process and maintenance operations even if these valves may be physically located in the pipe rack or the sequence of branches may allow in the future for a single valve to serve several branch lines.

Isolation facilities shall be indicated for:

- finger areas;
- process Unit block areas;
- at position of change from pipe rack to pipe rack.

The finger area is defined as being the area that serves a particular process area that may consist of one or more process Units. In addition to the equipment that is located alongside the finger pipe rack, the finger area also includes the equipment located alongside the main pipe rack.

The UFDs shall be arranged to cover the whole refinery/plant area, and these are divided into separate sheets each with corresponding match lines. Depending on the complexity and extent of the particular utility/common facility, sheets may be combined, extended, or omitted as required.

All equipment that is supplying a particular utility common facility either from the system (e.g. steam boilers) or from a process Unit (e.g. waste heat boilers) shall be shown in a "box" in a geographical location. This box shall give relevant equipment number(s), Unit number, and sheet number of the drawing in which the equipment is detailed.

Graphical symbols shall be used throughout the oil, gas and petrochemical project to establish uniform symbols for equipment, piping and instrumentation on P&IDs and UFDs. This include also Vendor drawings with the same purpose.

The graphical symbols shown for equipment may be turned or mirrored, if their meaning does not depend on the orientation. The representation of some graphical symbols (i.e. columns, vessels, etc.) can be adjusted to the actual scale with respect to the process plant.

The instrumentation symbol size may vary accordingly as required and as per type of document. However, consistency should be followed in all similar documents.

15.26 Preparation of P&IDs

Since the P&ID contains a large amount of plant design information, its revision will have a great effect on the subsequent engineering works.

Accordingly, for the purpose of minimizing the revisions and avoiding unnecessary works, the steps for preparing the P&IDs shall be established. The following steps should be realized in preparing the P&IDs. Concerning the information that can be prepared as engineering work proceeds, steps 2, 3, and 4 may be combined or extended to more steps as required.

- **Step 1:** Preparatory Step for Preparation of the P&IDs
- **Step 2:** P&IDs for Engineering Start
- **Step 3:** P&IDs for Piping Layout
- **Step 4:** P&IDs for Piping Drawings
- **Step 5:** P&IDs for Construction
- **Step 6:** P&IDs As-built

In the case where the P&IDs are prepared by the Licensor, only a part of the above-mentioned steps is applied and the main Contractor shall be responsible to complete the P&IDs preparation steps. The extent of the Licensor's and Contractor's scope of work will be according to the relevant contracts.

15.26.1 Establishment of P&ID preparation steps

- **Step 1: preparatory step for preparation of P&IDs**

Through step 1, the basic design philosophy concerning those basic items for the preparation of P&IDs such as mode of indication, applicable standards, numbering system, valve arrangement, and those other basic items on which agreements shall be made by the Company prior to the preparation of the P&ID should be clarified.

The basic items (Figure 15.12) that should be taken into consideration in step 1 are listed below, but should not be limited to the following items:

1. **Vellum and drafting**
   a. size and vellum of drawing;
   b. title;
   c. drafting;
   d. arrangement;
   e. equipment description;
   f. interconnection.
2. **Numbering System**
   a. drawing number;
   b. equipment number;
   c. instrument tag number;
   d. line number.
3. **Symbol**
   a. equipment;
   b. piping components;
   c. instrument symbol;
   d. process stream symbol;
   e. utility symbol.

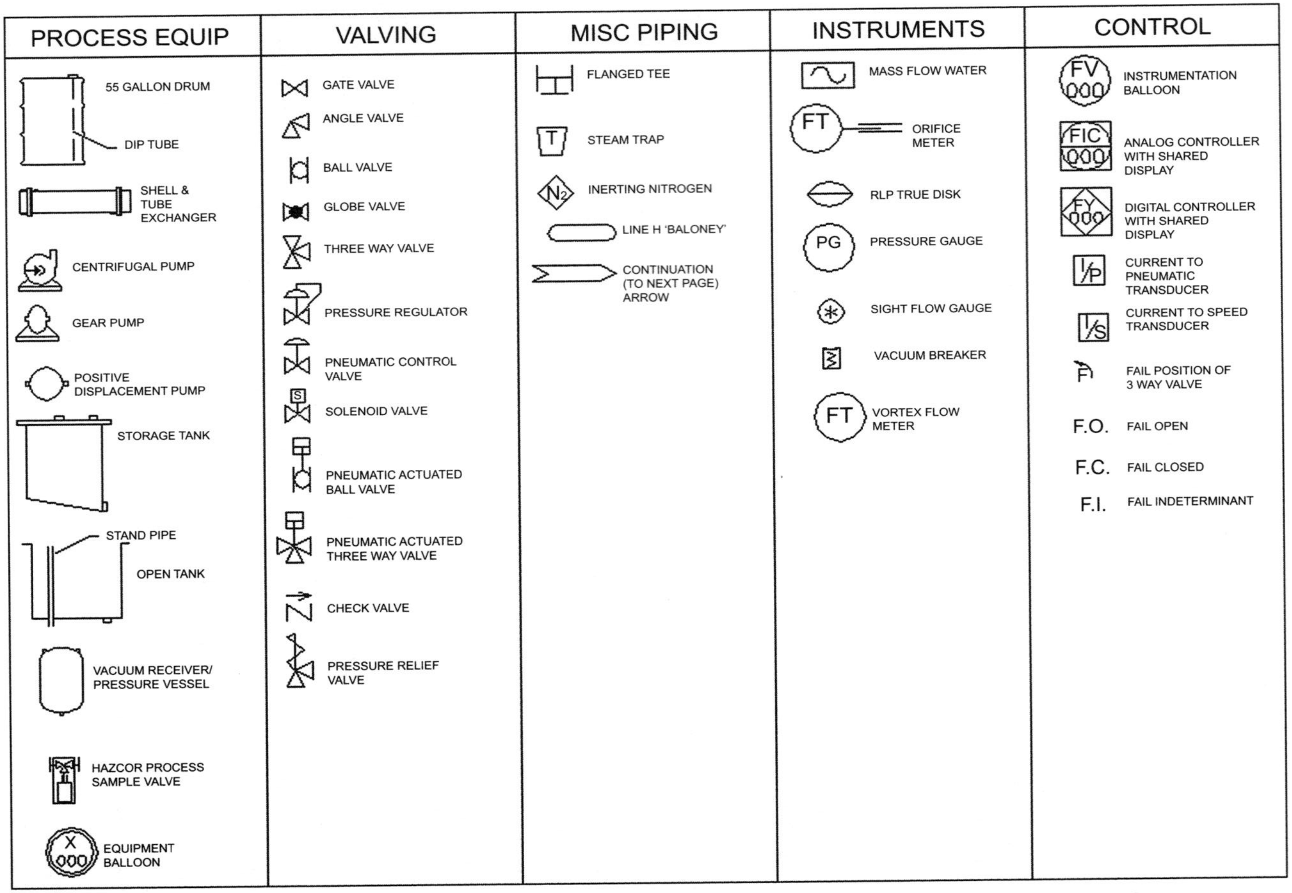

FIGURE 15.12

Typical symbols for P&IDs

4. **Valve arrangement around equipment**
   a. valve arrangement for drain, vent, and purge;
   b. valve arrangement for steam-out;
   c. sizes of the nozzles for installing the instruments;
   d. valve arrangement around the heater and exchanger;
   e. valve arrangement around the pump and compressor;
   f. valve arrangement around the steam turbine.
5. **Piping**
   a. piping classification standards;
   b. valve arrangement at the battery limit;
   c. valve arrangement for drain, vent, purge, and steam-out on piping;
   d. valve arrangement around the steam trap;
   e. valve arrangement around the sample point/sample connections;
   f. blow-down;
   g. valve type selection criteria/standards;
   h. strainer type selection standards;
   i. pipeline sizing criteria.
6. **Instrumentation**
   a. valve arrangement around the control valve;
   b. valve arrangement around the safety/relief valve;
   c. valve arrangement around other instruments;
   d. instrument type selection standards;
   e. mode of indication concerning computer control;
   f. software linkage and DCS presentation.
7. **Miscellaneous**
   a. winterizing and heat conservation;
   b. recovery of steam condensate;
   c. disposal of drains and waste water effluent.

For the purpose of obtaining a unified design philosophy and appropriate design relations among the Units, the illustrated process considerations concerning operation (start-up, normal, shutdown), safety and other features of the Unit shall be achieved by indicating them on the PFD. Where it does not suffice to give more illustrations, additional brief written explanations shall be provided. The items that should be covered to complete the design and operation philosophy and shown on P&ID (as required) shall include but not be limited to the following requirements (where applicable):

1. **Precommissioning and start-up operations:**
   a. flushing;
   b. purging;
   c. soda washing(where required);
   d. chemical cleaning;
   e. steaming-out;
   f. evacuation;
   g. drying;
   h. water operation;

  i. cold circulation;
  j. hot circulation;
  k. catalyst pretreating such as sulfiding, reduction, etc.;
  l. feed cut-in;
  m. off-spec. product handling.
2. **Normal operation:**
  a. recorder and indicator points;
  b. stream analyzer point;
  c. sampling point and type;
  d. control valve block and bypass;
  e. driver type;
  f. chemical injection point and types of chemical;
  g. batch operation;
  h. local start;
  i. instrumentation and control system needed for optimization and/or process control.
3. **Shutdown operation**
  a. depressuring;
  b. feed cut-out;
  c. cooling;
  d. purging;
  e. steaming-out and flushing;
  f. decoking;
  g. catalyst regeneration.
4. **Safety operation**
  a. location of safety/relief valves;
  b. failure action of control valves;
  c. prealarm system;
  d. emergency shutdown system;
  e. auto start of equipment/system.
  f. Results of hazard analysis and operability (HAZOP) study (if any)

The Contractor shall perform the HAZOP study (if required by the owner) using the PFD, P&ID, and plot plan together with equipment data sheets and related safety equipment checklist. The Contractor shall provide information about the reported accidents in similar process units in the world during HAZOP meetings.

  The Contractor should prepare both the draft of the basic items for preparation of the P&IDs and all necessary operation and safety features as mentioned above to the Company's review and approval before issuance of official revision of the P&ID for engineering start.

- **Step 2: P&ID for engineering start**

The following information as minimum requirement shall be reviewed and completed at this stage:

1. **Equipment**
  a. number of equipment;

    b. type of equipment;

    c. equipment number and name.

2. **Piping**

    a. size of main piping;

    b. winterizing/heat conservation requirement;

    c. valve type;

    d. provision of drain and vent;

    e. provision of purge, steam-out, chemical injection and water injection connections and valving;

    f. line number;

    g. utility services connected to each equipment, piping, and packaged Units.

3. **Instrumentation**

    a. type of instrument and location of the primary element;

    b. location and discharge destination of the safety/relief valve;

    c. location, type, and valve functioning (failure action) of the control valve;

    d. measurement and control method;

    e. instrument tag number.

The draft of the P&ID for engineering start shall be sent to the Company's review. After the joint meeting between the Company and Contractor, the P&ID for engineering start can be officially issued based on the established Company's comments as per the agreed items mentioned in the relevant minutes of meeting.

- **Step 3: P&ID for piping layout**

The purpose of issuing the P&ID for piping layout is the Company's approval on the basis of detailed design for piping layout.

The minimum information which should be added on the P&ID at this stage shall be as follows:

1. **Equipment**

    a. elevation of equipment;

    b. size of equipment;

    c. internal of equipment.

2. **Piping**

    a. line class;

    b. miscellaneous piping size (except around the safety/relief valve and control valve);

    c. thermal and cold insulation;

    d. precautions concerning piping layout;

    e. correct orientation of piping around equipment.

3. **Instrumentation**

    a. size of main control valves;

    b. additions and revisions on the basis of detailed design.

4. **Vendors' packaged Units**

    a. The details of some available information concerning the Vendors shall be indicated.

- **Step 4: P&ID for piping drawings**

The following information shall be added on the P&IDs at this stage:

1. **Piping**
   a. piping around the safety/relief valve and control valve;
   b. size of all valves;
   c. additional review of the pipe size and branch by the checking of the piping layout;
   d. hydraulic of system (checking and implementation of the necessary notes).
2. **Instrumentation**
   a. sizes of the safety/relief valves;
   b. sizes of the control valves;
   c. details concerning level transmitters and level gauges;
   d. logic diagram for heaters, incinerators, compressors, and all other main equipment (where applicable).
   e. Consequences of details of cause and effect tables.
3. **Vendor's information**
   a. The necessary information concerning the Vendors' equipment shall be indicated.
- **Step 5: P&ID for construction**

At a stage where detailed design has been nearly completed, upon approval of the Company, the P&ID shall be frozen for the purpose of smooth execution of the construction work.

The P&ID shall be issued for construction after completion of the following activities:
   a. piping material table;
   b. piping class and all relevant job specifications;
   c. all job Specifications and standard drawings in relation to the preparation of P&IDs;
   d. logic diagram of the main equipment;
   e. hydraulic of system;
   f. size of all piping, valves, and instrumentation components;
   g. Vendors' information.

Required revisions after freezing of the P&IDs shall be made only by conducting design activities using the relevant field sketches and executing the required modifications approved by the Company. This is applicable to revisions called for by the design department. The frozen P&ID shall not be revised.

- **Step 6: P&ID as-built**

The P&ID as-built shall be prepared upon completion of the project for filing and submission to the Company. Since the P&ID is intended for use in conducting operation control, maintenance, or revamping, the prepared drawing shall be entirely in conformity with the completed facilities.

The P&ID as-built shall be prepared in accordance with the results of line checking and the final edition of the field sketches.

The specified piping and instrument take-off and branch points shall be observed as strictly as possible and shall be implemented on the P&ID. Although bearing no relationship to piping layout, none of the flange, cap, drain pot, spectacle blind, and other miscellaneous piping designed for installation at the ends of the drain and vent required for operational purposes shall be omitted.

15.27 Handling of licensed process

Where a licensed process or basic design should be prepared by a Licenser, the Contractor's scope of work concerning the completion of the P&ID will be dependent on the type of contract with the Licenser and Contractor.

15.27.1 Licensing contract via the contractor

In this case, the Contractor and the Licenser jointly and severally shall give a process performance guarantee to the Company.

- **Case 1: licenser prepares P&ID**
  The P&ID supplied by the Licenser shall be equivalent to the "P&ID for piping layout" given in this section, and shall contain all design philosophies concerning process. The Contractor shall carry out mainly the following activities:
  - prior to the Licenser's commencing the preparation of the P&ID, the Contractor shall establish the basic items for the preparation of the P&ID and shall submit them to the Company for approval. The Licenser should prepare the P&ID based on the above-mentioned items;
  - based on the P&ID and operational guides prepared by the Licenser, review shall be made with regard to operability, safety, conformity to design of the Unit, etc.;
  - checking of the above-mentioned P&ID against the basic items for preparing the P&IDs and relevant design data;
  - general review and checking of the drawings against the project requirements;
  - establishing the result of above-mentioned checking and reviews in a joint meeting with the Licenser. The Licenser should implement all necessary Contractor's engineering comments and issue the revised P&ID;
  - the following items of review shall be made by the Contractor on the revised P&ID by the Licenser:
    - review in accordance with the results of detailed design hydraulic review;
    - review in accordance with the results of detailed design;
    - review on the basis of information concerning vendors.

The Licenser's approval should be obtained on any revision which should be made during the execution of the above-mentioned reviews by the Contractor, if it is expected to have an effect on the process performance.

- **Case 2: the contractor prepares P&ID**
  The P&ID shall be prepared by the Contractor in accordance with the following requirements:
  - required sufficient information for the preparation of the P&ID shall be obtained from the Licenser;
  - the prepared P&ID shall be subject to the Licenser's review and approval.

15.27.2 Direct contract between company and licenser

In this case, the Licenser shall give a process performance guarantee to the Company. The Contractor will be responsible for hydraulic of system and mechanical guarantee.

- **Case 1: the Contractor's verification is required**

Usually, the verification is limited to mechanical and hydraulic matters. However, extent of the Contractor's verification should be established in detail by the Company.

> For the purpose of conducting verification, the Contractor shall carry out mainly the following basic items:
> - the basic items for preparation of the P&ID shall be prepared and finalized with the Company;
> - the required activities shall be performed to complete all design philosophies in relation to the process, operation, safety and other features based on the operational guides and/or the P&ID prepared by the Licenser;
> - The P&ID prepared by the Licenser shall be checked against the above-mentioned finalized basic items and design philosophies;
> - checking of the P&ID should be performed against the hydraulic of system and detailed design data;
> - the results of the above-mentioned activities shall be finalized with the Company and shown on the P&ID as required;
> - upon the completion of the above-mentioned items, the required steps for preparation of the P&ID shall be followed to complete detailed design activities.
- **Case 2: verification is not required by contractor**
  In this case, the following activities shall be conducted by the Contractor:
  - review for the detailed hydraulic of system;
  - review for implementation of results of the detailed design;
  - review for information concerning vendors;
  - completion of the P&ID preparation steps.

15.28 Revisions of P&ID

Generally the P&ID can be revised in the following conditions if complied with the requirements as outlined in Standards:

- for correction of typographical and/or engineering errors;
- as per the Company's instructions;
- implementation of pertinent information in the course of execution of the relevant engineering work on the P&ID;
- addition of information concerning vendors.

Upon agreement with the Company, revisions made after the issuance of the P&ID for piping layout may not be needed by directly revising the P&ID but by issuing the NPIC or Notification of P&ID Change.

15.28.1 Revisions

The P&ID shall be revised depending on necessity at each step in addition to the required edition(s) that shall be issued per each step. Accordingly, it does not follow that the step number and revision number coincide with each other. At the time of revising the P&ID, the NPIC issued up to that time and information concerning vendors obtained up thereto shall be incorporated on the P&ID.

15.28.2 **NPIC**

Issuance of NPIC and manner of presentation shall be agreed in advance with the Company. The NPIC shall be issued in an NPIC form finalized with the Company. In general, issuance of the NPIC should consider:

- minimization of P&ID revisions;
- not accumulation of a large amount of additions/changes that should be incorporated on the new revisions of P&ID.

15.28.3 **Approval of P&ID**

1. The Company's approval of the basic items for preparation of P&ID (see Article 11.2.1 above) shall be obtained prior to commencement of the P&ID preparation work.
2. The Company's approval at step 2, P&ID for engineering start, shall be obtained regardless of the cases where the P&ID is prepared by the Contractor or Licenser or both.
3. Where the P&ID prepared by Licenser has been reviewed or verified in step 3. P&ID for piping layout, by the Contractor, the Company's approval is needed before any official revision.
4. In general, the Company's approval is required for any change, deletion, and/or addition on the P&ID through all steps of preparation of P&ID.

15.29 **Block and bypass valves for control valve**
15.29.1 **Without block and bypass valves**

Block and bypass valve system may not be necessary where the process can be shut down to repair the control valve without significant economic loss or where the process cannot be feasibly operated through the bypass. However, the consequences of shutting down a process Unit to perform a simple task (such as replacing control valve packing) should always be considered. In cases where the block and bypass valves are not used, the control valve should be equipped with a handwheel or other operating devices.

Block and bypass valves are not always necessary in the following cases:

1. In instances where it is desirable to reduce the sources of leakage of hazardous fluids, such as hydrogen, phenol, or hydrofluoric acid;
2. In clean service where the operating conditions are mild, and mission of valves will not jeopardize the safety or operability of the Unit;
3. In temporary services such as start-up or shutdown, and where the other operation modes are possible while the repairing of control valve, such as blending system of oil;
4. Pressure self regulating valves;
5. Shut-off valves

15.29.2 With block and bypass valves

The following services should be provided with block and bypass valves:

1. Services where omission of valves will jeopardize the safety or operability of the Unit;
2. Services containing abrasive solids or corrosive fluids result in damage of trim of control valve, and require the repair;
3. In lethal services;
4. In product rundown and feed supplying services;
5. In fuel supply system;
6. In cooling-medium supply service;
7. Control valves less than DN 50 (2 inch) size. The block and bypass valves are required due to small diameter of trim, and may have a possibility of plugging of sludge or foreign matters;
8. In services that are flashing or at high differential pressure.

15.29.3 Additional requirements for control valves

The following notes should also be considered:

1. Provide an upstream isolation valve for all control valves unless the upstream system is to be shut down on control valve failure.
2. Provide a downstream isolation valve whenever the downstream side of the control valve cannot be isolated from other continuously operating pressure sources.
3. Provide a drain valve upstream of all control valves.
4. Provide a drain valve downstream of the control valve only when the process fluid is toxic or corrosive and for tight shut-off services.

15.30 Philosophy of instrumentation installation

15.30.1 Flow and quantity

Sufficient flow metering, temperature, and pressure indications shall be installed in feed, rundown, and utility streams to provide information for the operation and the calculation of heat, pressure, and material balances for each individual Unit.

15.30.2 Alarm and safeguarding system

If failure of any piece of plant equipment or its associated instrumentation may give rise to hazards for personnel, to consequences with considerable economic loss, or to undue environmental pollution, alarm and/or safeguarding instruments shall be installed. Where appropriate, safeguarding equipment shall be automatically bring the relevant plant or part of the plant to a safe condition when a desired measurement reaches an unacceptable value.

15.30.3 Separate instrument connections

Depending on potential hazards, operational importance, instrument reliability, plugging of connections, etc., the need for separate connections from those for normal operation shall be decided upon in the design stage and indicated on P&IDs.

Separate connections are especially required for instrument of shutdown systems, such as:

- High- or low-pressure point that actuates shutdown system.
- High- or low-temperature connection that actuates shutdown system.
- HHL or LLLL connection that actuates shutdown system.

Further reading

Abraham KS. World of oil: front-end projects awarded for Angolan LNG plant. World Oil 2005;226(6):14.

Asmoro TH, Putri AM. Balancing project schedule & cost on sour gas development project case study. In: SPE Production and Operations Symposium. Proceedings 2012;1:54–67.

Bhatia S, Fiaz M. Debottlenecking and expansion of upper zakum GTP plant at Zadco. In: 12th Abu Dhabi International Petroleum Exhibition and Conference, ADIPEC 2006: Meeting the Increasing Oil and Gas Demand Through Innovation, 2; 2006. pp. 522–31.

Brennan MF. A strategic approach to high-reliability pipeline power applications. Pipeline Gas J 2004;231(9): 67–8.

BSI (British Standards Institute) BS EN ISO 9001. Quality Systems Models for Quality Assurance in Design, Development, Production, Installation and Servicing; 2000.

Carr H. Once-booming petroleum market falters under economic uncertainty. ENR Eng News-Record 2008; 261(19):60–2.

Chandran A. Why quiet? Legislative, commercial and ethical drivers behind the design of quieter offshore facilities in Australia. In: Society of Petroleum Engineers—SPE/APPEA Int. Conference on Health, Safety and Environment in Oil and Gas Exploration and Production 2012: Protecting People and the Environment—Evolving Challenges 1; 2012. pp. 818–26.

Chia S, Walshe K, Corpuz E. Application of inherent safety challenge to an offshore platform design for a new gas field development—Approaches and experiences. In: Institution of Chemical Engineers Symposium Series, 149; 2003. pp. 257–65.

Eweje J, Turner R, Müller R. Maximizing strategic value from megaprojects: the influence of information-feed on decision-making by the project manager. Int J Proj Manag 2012;30(6):639–51.

Gopikrishnan M, Awda H, Sherif A, Srinivasa R, Kotrappa M, Gupta M, Ismail S, Telepurath S. Modification of existing offshore process facilities (engineering, projects & contracting—execution of brown field projects). In: Society of Petroleum Engineers—14th Abu Dhabi International Petroleum Exhibition and Conference 2010, ADIPEC 2010, 3; 2010. pp. 2188–202.

Hwang J, Ahn Y, Min J, Jeong H, Lee G, Kim M, Kim H, (…), Lee K-Y. Application of an integrated FEED process engineering solution to generic LNG FPSO topsides. In: Proceedings of the International Offshore and Polar Engineering Conference; 2009. pp. 136–43.

Hwang J, Lee K, Roh M, Cha J, Ham S, Kim B. Establishment of offshore process FEED (Front End Engineering Design) method for oil FPSO topsides systems. In: Proceedings of the International Offshore and Polar Engineering Conference; 2009. pp. 144–50.

Lanquetin B, Le Marechal G. Hulls for mega FPSOS: Achieving long life and performance from sound design to asset integrity management. In: International Petroleum Technology Conference 2007, IPTC 2007, 2; 2007. pp. 979–93.

McHugh S, Manning A, Heagney K. Integration of HES into the management of a major capital project. In: Society of Petroleum Engineers—SPE/APPEA Int. Conference on Health, Safety and Environment in Oil and Gas Exploration and Production 2012: Protecting People and the Environment - Evolving Challenges, 2; 2012. pp. 1375–9.

McNeilly CC, Will SA. Engineering the Benguela-Belize compliant piled tower. World Oil 2006;227(9):73–5.

Monahan M, Blackwell V, Hoffmann R, Julliand M, Iannuzzi M, Conrad D, Craig A. Tools for standardizing environmental designs for capital projects. In: Society of Petroleum Engineers—SPE/APPEA Int. Conference on Health, Safety and Environment in Oil and Gas Exploration and Production 2012: Protecting People and the Environment—Evolving Challenges 2; 2012. pp. 1380–3.

Osaghae EO. Economic model for gas-to-power project capitalization in developing economy. In: SPE Hydrocarbon Economics and Evaluation Symposium; 2003. pp. 173–81.

Saadawi H. HAZOP studies for grass-roots field development project. In: SPE Middle East Oil and Gas Show and Conference, MEOS, Proceedings; 2005. pp. 1367–73. Art. no. SPE 93723.

Sharma S, Cook P, Berly T, Lees M. The CO2CRC Otway Project: Overcoming challenges from planning to execution of Australia's first CCS project. Energy Procedia 2009;1(1):1965–72.

Slangen F, Bal W, Riemers M. An innovative self installing platform (SIP) concept. In: Proceedings of the International Conference on Offshore Mechanics and Arctic Engineering—OMAE, 1; 2011. pp. 849–54.

Srivastava SK, Takeidinne O. Feed endorsement—a challenge to Epc contractor's engineering team. In: Society of Petroleum Engineers—Abu Dhabi International Petroleum Exhibition and Conference 2012, ADIPEC 2012-Sustainable Energy Growth: People, Responsibility, and Innovation, 1; 2012. pp. 301–11.

Stevenson K. Care and concern for people we'll never meet—process safety in the design of the Wheatstone platform. In: Society of Petroleum Engineers—SPE/APPEA Int. Conference on Health, Safety and Environment in Oil and Gas Exploration and Production 2012: Protecting People and the Environment—Evolving Challenges, 2; 2012. pp. 942–7.

Stopek DJ, Smith R, McHone S, Armpriester A. CCUS demonstration project at WA parish station—results of FEED study. In: Air and Waste Management Association—Power Plant Air Pollutant Control "MEGA" Symposium, 2; 2012. pp. 767–78.

Swatton MJR, Van Soest-Vercammen E, Nagelvoort RK. Innovation and integration in LNG technology solutions. In: Society of Petroleum Engineers—International Petroleum Technology Conference 2009, 2; 2009. pp. 1324–36. IPTC 2009.

Thiabaud P, Gros F, Tourdjman J, Guaraldo M. Benefits of life cycle dynamic simulation for FPSO projects & offshore developments. In: Society of Petroleum Engineers—Brazil Offshore Conference, 2; 2011. pp. 590–602.

Van Wijngaarden W, Meek HJ, Schier M. The generic LNG FPSO—a quick & cost-effective way to monetize stranded gas fields. In: SPE Asia Pacific Oil and Gas Conference and Exhibition 2008—"Gas Now: Delivering on Expectations", 2; 2008. pp. 655–66.

Zhao Y, Wang G. Design and performance tests of a 1.5 MM TPY hydrocracking unit. Pet Refin Eng 2006;36(9): 36–41.

Start-up Sequence and Commissioning Procedures

This chapter specification covers minimum process requirements for plant start-up sequences and general commissioning procedures for non-licensed units or facilities. For licensed units, instructions given by the licenser shall be followed.

Although, the start-up sequences and commissioning procedures differ to some extent from process to process, the basic philosophy and general aspects shall conform to the concepts of this chapter.

However, considering the general commissioning and all activities to be performed prior to initial start-up, as outlined in this chapter, feed introduction to the unit shall be according to the stepwise start-up procedure provided by the contractor in the unit operating manual.

The chapter is intended for convenience of use and pattern of follow-up and also as guidance. This also indicates the check points to be considered by the process engineers for assurance of fulfilment of prerequisites at any stage in the implementation of process plant projects.

Commissioning shall start from the point at which steps are taken to bring the unit/facility up to operating pressure and temperature and to cut in the feed. It shall be complete when the unit/facility is operating at design capacity and producing products to specification.

For a project involving a number of process units and offsite facilities, it shall be agreed between the contractor and the company in the earliest stages of a project the sequence of the commissioning of the units. It shall be necessary to commission utilities and some of the units in advance of others, because of their interdependence from a process point of view.

The responsibilities of the contractor and company during the commissioning stage should be clarified for the provision of labor, operators, specialists, and service engineers, and also for the correction of faulty equipment, etc.

The general requirements for testing of equipment/lines shall be followed. The detail procedures for testing of equipment and lines and other pre-commissioning steps are not included and shall be prepared in accordance with the company's Engineering Standards by the contractor and submitted to the company for approval. However, on completion of testing, vessels, equipment, and piping should be vented and drained, and, where necessary, cleaned and dried to the satisfaction of the company.

Spades, blanks, and other equipment installed for testing shall be removed on completion of testing. Wherever a flange joint is broken after testing (e.g., on heat exchangers, pipework, fired heaters, and at machinery), then the joint rings or gaskets must be renewed. Particular attention must be given to heat exchangers using solid metal or filled gaskets, and great care should be taken to ensure that all gaskets are renewed after testing of heat exchangers and are correctly fitted before tightening flanges.

Where required, valves are to be repacked with the appropriate grade of material.

Any temporary bolting that has been used shall be replaced and any temporary fittings that may have been installed to limit travel (e.g., in expansion joints and pipe hangers) shall be removed.

Prior to commissioning, each item of equipment should have its name, flow sheet number, and identification number painted and/or stamped on it according to the company's specifications.

A typical implementation start-up path is shown in Figures 16.1 and 16.2.

The start-up steps are as follow in Table 16.1:

16.1 Preparation prior to initial start-up

The procedures described in this section shall be carried out at the completion of construction and before initial operation of the unit. Appropriate phases should be repeated after any major repair, alteration, or replacement during subsequent shutdowns. The phases of preparation for initial start-up shall be according to the following steps:

- Operational checkout list.
- Hydrostatic testing.
- Final inspection of vessels.
- Flushing of lines.
- Instruments.
- Acid cleaning of compressor lines.
- Breaking in pumps.
- Breaking in compressors.
- Dry-out and boil-out.
- Catalyst loading.
- Tightness test.

16.1.1 Operational check-out

1. Check line by line against flow sheet and locate all items.
2. Identify the location of instruments.
3. Indicate the location of all critical valves including valves at critical vent and drain locations.
4. Check control valves, valves, and globe valves to see that they are installed properly with respect to flow through their respective lines. Special attention must be given to check valves regarding their direction of flow.
5. Review all piping and instrument connections for steam tracing.
6. Check that the following facilities have been installed so that the plant can be commissioned and put on stream:
   a. Start-up bypass lines.
   b. Purge connections.
   c. Steam-out connections.
   d. Drains.
   e. Temporary jump overs.
   f. Blinds.
   g. Check valves.
   h. Filters and strainers.
   i. Bleeders.
   j. Etc.

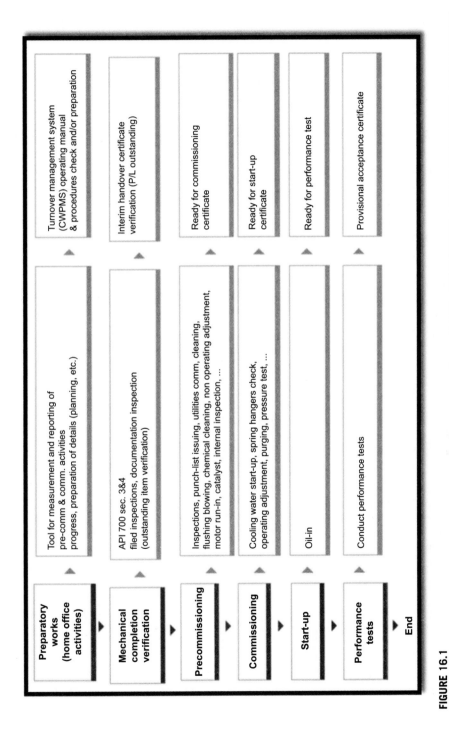

FIGURE 16.1

A typical implementation start-up path.

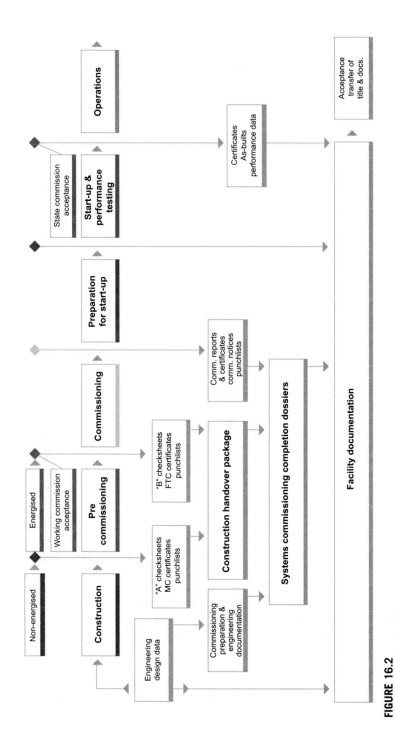

FIGURE 16.2

A project from start-up to completion.

Table 16.1 The Start-up Steps

Step	Remarks
Pre-commissioning	Pre-commissioning activities start when the plant or system achieves mechanical completion. Frequently, however, pre-commissioning activities overlap mechanical completion activities and, for this reason, the plant should be separated into easily manageable system packages, each system package will be pre-commissioned as a whole and isolations at the boundaries of the system package will be maintained until the completion of pre-commissioning activities. Pre-commissioning activities include: checking for design conformity, checking the status of electrical, mechanical, and instrument installations, running-in of equipment, flushing and cleaning activities, and drying, etc.
Unit ready for commissioning	This is the point in a project where all utilities are commissioned and operational and the unit is ready to accept the introduction of hydrocarbons.
Commissioning	Commissioning is the phase in a project when design process fluids are introduced to the process. Note that for hydrocarbon systems, the commissioning activities commence with inerting operations. Commissioning activities normally consist of activities associated with running or operating the plant and include operating adjustments necessary for satisfactory operation of the plant or part thereof. Also included are "Functional checks" which are methods used to prove that an item of mechanical equipment or control system functions correctly.
Unit ready for start-up	This is the point in a project where the unit is ready to establish process conditions with the intent of making products.
Start-up	Start-up is the point in a project where process fluids and conditions are established with the intent of making products.

7. Check pumps and compressors for start-up.
8. Check sewer system for operability.
9. Check blow down systems.
10. Check heater for burner installation, refractories, stack damper controls, burner refractories, etc.

16.1.2 Hydrostatic testing

Hydrostatic pressure testing of the unit shall be performed to prove strength of the materials and weld integrity after completion of the construction. The tests shall be made on new or repaired equipment and piping. The initial testing is ordinarily done by the contractor in the course of erection.

Detail procedure for testing the equipment and lines shall be prepared by the contractor and submitted to the company for approval.

Fresh water containing a corrosion inhibitor which meets the company's approval shall be used for hydrostatic test purposes unless otherwise specified in Standards. In systems where residual moisture cannot be tolerated (e.g., in SO_2, acid, ammonia, and LPG service), and where certain catalysts are used, oil is the preferred test medium. If water has to be used, the system should afterward be dried out with hot air. Special attention should be given to the points where water may be trapped, such as in valve bodies or low points.

If for any reason it is not practical to carry out a hydraulic test, a pneumatic or partially pneumatic test may be substituted subject to prior agreement with the company. Full details, including proposed safety precautions, will be required. The following are usually excluded from hydrostatic testing, and are usually tested with compressed air and soap suds:

1. Instrument air lines (test with dry air only, if possible).
2. Air lines to air-operated valves (test with dry air only).
3. Very large (usually over DN 600 or 24 in) gas or steam overhead lines.
4. Pressure parts of instruments in gas or vapor service.

When austenitic or austenitic acid clad or lined equipment and piping are tested, the test fluid chloride ion content shall meet the following conditions:

1. If the piping and equipment metal temperature never exceeds 50 °C during commissioning, operation or nonoperation, water containing up to 30 ppm (by mass) chlorides ion shall be used. The chlorides ion content might be increased up to 150 ppm (by mass) if the equipment or piping can be thoroughly washed out using water containing less than 30 ppm (by mass) chlorides ion as soon as testing is complete, if allowed by the company. In any case, the water must be drained and the equipment thoroughly dried immediately thereafter.
2. If the piping and equipment metal temperature exceeds 50 °C during commissioning, operation or nonoperation, the piping shall be tested using condensate water, demineralized water or oil with minimum flash point of 50 °C.

The testing medium should not adversely affect the material of the equipment or any process fluid for which the system has been designed. Reference should be made to the applicable codes in the case of pressure vessels to determine the minimum ambient and fluid temperatures at which testing may be carried out. If it is desired to test vessels, tanks, or piping at temperatures below 16 °C, attention shall be made to the danger of brittle fracture occurring in carbon steels and ferritic alloy steels unless the materials have adequate notch ductility properties. For any equipment or piping, water should not be used for testing when either the water temperature or the ambient temperature is below 5 °C. Hydrostatic testing at temperatures below this value may be carried out using gas oil, kerosene, or antifreeze solution at the appropriate strength, provided that the fluid used is agreed upon with the company. When flammable liquids, including gas oil or kerosene, are to be used, appropriate safety measures must be observed and a work permit shall be required.

Sea water shall not be used for the testing of equipment and lines on process units and steam generating plants. Proposals for the use of sea water for the testing of storage tanks and offsite lines shall receive the company's approval.

During the hydrostatic test pressure with water, the system loss should not exceed 2% of the test pressure per hour, unless otherwise specified. Evidence of water at valves, flanges, etc., will indicate the leaking areas to be repaired if the system fails the test. All welds and piping must be inspected for defects by looking for wet spots, therefore, they should be tested before they are insulated.

If piping is tested pneumatically, the test pressure shall be 110% of the design pressure. Any pneumatic test shall be increased gradually in steps, allowing sufficient time for the piping to equalize strains during the test. All joints welded, flanged, or screwed shall be swabbed with soap solution during these tests for detection of leakage.

Vents or other connections shall be opened to eliminate air from lines that are to receive hydrostatic test. Lines shall be thoroughly purged of air before hydrostatic test pressure is applied. Vents shall be open when systems are drained so as not to create buckling from a vacuum effect.

Relief valves must be removed or blinded prior to hydrostatic testing.

After completion of hydrostatic testing, all temporary blanks and blinds shall be removed and all lines completely drained. Valves, orifice plates, expansion joints, and short pieces of piping that have been removed shall be reinstalled with proper and undamaged gaskets in place. Valves that were closed solely for hydrostatic testing shall be opened. After lines have been drained with the vents open, temporary piping supports shall be removed so that insulation and painting may be completed.

Extreme care shall be taken in field testing of heater and furnace tubes, which are normally field fabricated and must be hydrostatically tested. Where possible, piping and heater tubes shall be tested together.

Heater tubes shall be tested according to the manufacturer's recommended test pressure. The test shall be coordinated with the heater erection by the contractor.

Vessels constructed in accordance with American Society of Mechanical Engineers Code Section VIII will not require individual pressure tests at the site except in the following cases:

1. Vessels whose condition resulting from transport, storage, handling, or for any other cause, is suspect in the opinion of the company.
2. Vessels that have had any site modification, which, in the opinion of the company, necessitate a site pressure test.

With the exception of the relief valves, all valves and fittings should be installed on the vessels and included in the test. Relief valves must be removed or otherwise isolated.

On satisfactory completion of a pressure test, the vessel should be drained completely, any blinds inserted for test purposes removed, and these joints remade to the satisfaction of the company.

At the conclusion of the test, the system must be drained. If pumps have been included, they must each be drained and refilled with oil to prevent rust forming in the seals. If fractionating columns are included, the water must be displaced with sweet gas, nitrogen, or an inert atmosphere rather than air to avoid corrosion and sticking of the valves on the fractionating trays.

Provided that the manufacturer's test certificates are available, only the following pressure tests should be carried out on heat exchangers.

1. On floating head shell and tube type exchangers, on tube side with bolted bonnets removed.
2. On tube-in-tube types, on both sides, in conjunction with associated pipework.
3. On air-cooled types, the bundles are to be isolated and tested separately from associated pipework.

In the case of tubular exchangers whose condition, resulting from transport or other causes, is suspect in the opinion of the company, a pressure test on the shell side at least equal to the maximum allowable pressure should be carried out. Tests may be carried out individually or on groups of exchangers having similar operating conditions. Test pressure for shell and tube and air-cooled types should be not less than the maximum operating pressure in the tubes. For tube-in-tube types, the test pressure should be as for the associated pipework.

After completion of the test, all equipment must be thoroughly drained and, if necessary, dried out to prevent scaling of tubes before commissioning.

After completion of the test, all bonnets and covers should be replaced, temporary blanks removed, and all joints remade to the satisfaction of the company. All compressed asbestos fiber joints should be renewed.

All tanks shall have bottoms, shells, and roofs tested in accordance with API 650, latest edition. Shells shall be tested by filling tanks with water. Vacuum testing shall be used for all tank bottoms and roofs.

The blow off (set) and blow down (re-set) pressure of all relief valves should be set to the satisfaction of the company. After approval of the company, a seal is to be fixed to each valve. The company's inspector shall check each valve after it has been reinstalled to ensure that its seal is intact.

Inlet lines to relief valves should be cleared before the valves are finally installed.

In cases where the valve exhausts to a pressure system, the downstream side should be tested to a pressure equal to the test pressure rating of the outlet system.

Flame arrestors and other miscellaneous equipment that does not have test pressure indicated shall be isolated from the test.

Certain types of instruments with their connecting process lead pipelines shall be tested at the same pressure as the main pipelines or the equipment to which they are connected. Such instruments normally include the following types:

1. Displacer type level instruments.
2. Gage glasses.
3. Rotameters.
4. Control valves.
5. Flow meter pots.

Other types of instruments shall not be tested at line pressure, but shall have process lead lines tested to the first block valve or valves nearest the instrument. Care shall be taken that this equipment is protected by removal, or by blocking the instrument lead line and disconnecting or venting the instruments. These types will normally include the following:

1. Analyzers.
2. Diaphragm type level instruments.
3. Differential pressure type flow instruments.
4. In-line type flow switches.
5. Direct connected regulators.
6. Open-float type level indicators and alarm switches.
7. Positive displacement type flow meters.
8. Pressure indicators, recorders, and transmitters.
9. Pressure switches.
10. Pressure-balanced control valves.
11. Pressure gages.
12. Turbine-type flow sensors.

Special precautions shall be taken to insure that instruments and instrument lead lines to be tested are vented and completely filled before testing and are thoroughly drained after testing.

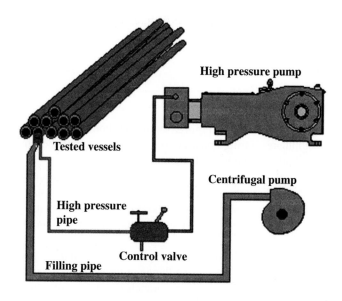

FIGURE 16.3

A schematic for hydrostatic test.

Hydrostatic test is done by using the high pressure pump to pressurize the pressure vessel that requires a pressure test (Figure 16.3). The system utilizes a special valve that allows the flow of high-pressure liquid to move only to one direction, to the test vessel until the vessel reaches the desired pressure. When big vessels are used, a large amount of liquid is required to fill them. The centrifugal pump that provides high flow of liquid at a relatively low pressure fills the vessel. When the vessel is filled, the high pressure pump generates a low flow under high pressure to pressurize the vessel. The control valve is either automatically operated or manually. When an automatic control valve is operated, a pressure sensor will detect the pre-set pressure and then hold the pressure to a pre-programmed time, then release the pressure and drain the liquid.

16.2 Final inspection of vessels

Before the final bolting of cover plates and manholes, vessel interiors should be inspected for cleanliness, completeness, and proper installation of internal equipment. The inspection list shall include at least the following items, where applicable:

1. Tray installation; satisfactory testing (where required).
2. Internal draw off piping.
3. Catalyst supports and screens.
4. Internal distributors.
5. Liquid entrainment separators.
6. Internal risers and vortex breakers.
7. Thermo well location and length.

8. Level instruments, location and range, internal float, external displacement, and differential pressure type.
9. Internal cement lining.

16.3 Flushing of lines

All fluid handling equipment, particularly piping, should be thoroughly cleaned of scale and the internal debris, which accumulates during construction. This is accomplished by blowing or washing with air, steam, water, and other suitable medium.

Some utility systems, such as water and high-pressure steam, may be satisfactorily cleaned with their normal media, introduced through normal channels. Other systems must, or preferably should, be flushed with foreign media, admitted via temporary hose or pipe connections. Thus, flushing may be accomplished on exhaust steam lines with high-pressure steam or air, and on fuel oil lines with steam followed by air.

The best way to clean a typical utility or auxiliary system is to flush the supply main from source to end at first and making an end outlet by breaking a flange or fitting, if necessary, then flush each lateral header in the same manner, and, finally, flush the branch that takes off from the headers. In some instances, if required, the weld-cap on the end of each header could be cut off and re-welded after the line has been flushed.

16.3.1 Steam piping

- Before blowing a steam line for cleaning, an open end for a free passage of the steam and debris shall be provided. If the open end is a temporary provision, be certain that the end is tied down to prevent possible whipping caused by high velocity flow. All drains must be open and the line free of water. Valving to steam traps shall be closed and the traps shall remain out of service until the line cleaning is completed. Prior to starting stream flow, all safety precautions shall be taken. The area should be cleaned and guarded to prevent injury to personnel. Steam shall slowly be introduced from the source and the line heated gradually.
- The steam flow rate shall be limited to the condensate drainage flow rate. Condensate must not be permitted to accumulate in order to avoid water hammer. As the piping heats and the condensate diminishes, the steam rate shall be increased. As the line heats, observe its expansion and determine that there is no binding or distortion. After the piping is thoroughly heated, the steam rate shall be increased to provide a hard blow. After blowing, the line shall be permitted to cool and contract, then the preceding blowing is to be repeated. The cleaning procedure should be repeated three times and unobstructed.

16.3.2 Process and utility lines

1. Process, utility, and auxiliary systems shall mainly be washed with water. Any line not accessible to water, or lines which would trap water in such a way that it could not be drained, may be blown out with air. Most of the process and auxiliary lines may be flushed through established circuits from vessels filled with water for the purpose. Machinery auxiliary lines should be flushed with oil.

To the maximum extent, the lines connected may be flushed with water contained within vessels after hydrostatic testing. A single filling of a vessel may not provide sufficient water to flush all lines for which it is the reservoir, in which case, a continuous or intermittent flow of water into the vessel should be maintained.

2. When washing exit lines of a vessel, it should be noted that the vessel is adequately vented to prevent a vacuum condition. Inadequate venting of vessels not designed for vacuum service could rupture them.

3. In any system, to the maximum extent possible, flushing should be made downward or horizontally and out at low points. The low point discharge opening may be temporary openings made by disconnecting flanges or fittings. Normal drains may be used for flushing outlets, provided they are equal to line size or nearly so. For the best results, there should be no restriction of the outlet or at any other point in a line undergoing cleaning.

4. Too many circuits or openings shall not be flushed simultaneously.

5. Some lines, such as buried pump suction lines, must be flushed through a temporary spool-piece and line-sized valve flanged to one end, and pointed to a safe location (not to the pump suction). Compressed air is introduced into the other end and pressure is allowed to build. When the valve is suddenly opened, the sudden release of pressure will effectively clean the line. Flushing shall be done through all vents, drains, and other side connections, bypasses, and their main channels alternatively.

6. Flushing line debris into equipment where it may become trapped or lodged shall be avoided.

7. All control valves should be blocked off and bypassed until the major part of the foreign matter has been removed from their systems. Then, remove the bottom plate of the control valve or, if the valve is closed, remove the valve itself from the line, and flush through the opening thus created. Finally, replace the plate or valve and flush through the valve in normal alignment (if the valve is opened or can be opened).

8. Flow meter and restriction orifices should not be installed until lines are clean. Any orifices installed before cleaning should be removed.

9. All connections at pumps and drivers must be closed off or disconnected while the lines running there are thoroughly flushed. This applies to the main pump suction and discharge lines, drive and exhaust steam, jacket cooling water, gland oil, and other auxiliaries.

The flushing outlet point should be as near the pump or drivers as conveniently possible. Generally, this will require disconnection of a flange or fitting. Where this is done on the pump (or driver) side of the block valve, cover the open pump connection, if necessary, to prevent entry of the flushing flow. In the main discharge lines containing a check valve immediately adjacent to the pump (generally the case with centrifugal pumps), a flush outlet may be made by removing the check valve cover plate, provided the flapper or disc remains in place to seal off the pump itself.

Where it is desirable to flush through a check valve, the flapper must be removed and the cover replaced.

10. All connections to instruments should be closed off during flushing, or disconnected at the instrument and flushed. Instrument air lines must be blown with special thoroughness using clean dry air.

Steam traps, until the lines are fairly clean, should be bypassed. Check the operation of traps after they have been opened to the flow, and remove for inspection and cleaning those that are not working properly.

11. Disconnect heater burners until the fuel gas line to them is clean; then reconnect them and blow through the burner and pilot.
12. At the conclusion of flushing any system, it shall be checked carefully to see that normal alignments are restored, temporary connections are broken, and temporary breaks reconnected, check valve flappers and/or cover plates replaced, and orifices installed, etc. In the case of lines that will receive further cleaning during subsequent pump break-in, this instruction may be qualified in part.

When flushing of process lines is finished, drain water from the system as completely as possible. Provide ample top venting during the draining operation, or whenever the level is being lowered in a vessel, to avoid pulling a vacuum on the equipment. Blow lines with air to effect further water removal.

The basic utility systems, such as steam, water, and air, should be put in normal working order after they have been cleaned so that supplies will be available for further operations.

16.4 Instruments

In general, the instruments should be checked against the design data as well as to be checked for installation, calibration, and operation, as per the following list:

1. Installation check
   a. The instrument is properly installed and accessible for operation and maintenance.
   b. All wiring is checked out.
   c. All loop checks are completed. The loop check procedure shall be prepared by the contractor/vendor and shall be submitted to the company for approval.
2. Calibration check
   a. The instrument is calibrated for operation.
   b. Orifice plates are installed after hydrotest and line flushing.
   c. The proper charts are installed on all recorders.
3. Operation check
   a. Power is supplied to all instruments.
   b. All alarms are test actuated and interlock systems checked.
   c. Instruments, connecting piping, and pneumatic tubing are checked for leaks.

16.5 Acid cleaning of compressor lines

Mill scale, dirt, heavy greases, and other foreign materials that could enter the compressor and result in operating and maintenance problems must be removed from the compressor system as required in the Piping & Instrumentation Diagrams (P & IDs). The following items must be acid cleaned:

- All make-up gas piping including spill-back lines.
- Make-up compressor suction lines and drums. Suction drums may be acid cleaned where practical.

- Make-up gas coolers and intercoolers.
- Fuel gas lines to gas turbine burners.

The procedure to be followed should be supplied by the cleaning contractor who must accept the responsibility of proposing and carrying out an acceptable and proven procedure for the entire cleaning operation.

16.6 Breaking-in pumps

New pumps should be given a preliminary run with strainers in their suction lines in order to test mechanical performance and reveal any defects before attempting to start up the unit. This preliminary circulation also serves as a supplemental cleaning operation, instrument and control valve check, a leak check, and allows a general performance test of the lines and equipment involved in the flow. Flow meter orifices should be installed.

Pumps not equipped with permanent strainers or strainer holders should be furnished with temporary basket-type screens inserted in the nearest accessible suction flange. This pump "run-in" is usually done by circulating water through the unit, which requires temporary piping. Before starting the pumps, go through the following check-list:

- Check driver rotation.
- If turbine drivers, check the turbine over speed trip and governor for operation with the turbine disconnected from the pump.
- Be sure there is a bleeder on the pump so that the case can be filled with liquid before starting. Being asphalt plant pump or rotary type, their priming shall be done according to the manufacturer's instructions.
- Check lube oil in bearing housing of both pump and driver.
- Check out pump cooling water and seal oil systems, where applicable.
- If pumps are run on a fluid other than that for which the pump was specified, it may be necessary to throttle the pump discharge to keep from overloading the driver.
- Be sure mechanical seals or packing is installed.

For reciprocating pumps, the following should be checked:

- On power pumps, check driver rotation and lube oil in the gearbox.
- Check oil level and operation of piston rod lubricators.
- On direct-acting steam pumps, check out steam piping and cylinder drains.

Warm up steam end before starting pump.

Before starting a centrifugal pump, rotate it by hand to make sure that it turns freely. This is a good practice always, but it is especially important during the first few starts. If the pump is at all tight, it should be dismantled and inspected for grit in the packing box, or inadequate clearance in the rotors or bearings.

If the pump is free, open its suction valve wide and vent the casing to ensure that it fills with liquid. Leave the discharge valve closed or open it very slightly.

The pump is now ready for starting, assuming that its suction and discharge circuits are properly aligned and that its driver is also in readiness.

Once it has been put in rotation, a centrifugal pump must be quickly brought up to normal speed—or at least to a speed that develops substantial discharge pressure, as read on the local gage—in order to provide internal lubrication of the rotor.

The pump discharge valve should remain in its initial closed or throttled position until the pump comes up to speed and establishes full discharge pressure. The valve may then be opened gradually until the desired liquid flow is obtained, or until an automatic valve assumes control; in the latter instance, the manual valve is then opened wide.

If the pump does not quickly develop a discharge pressure close to its normal value, try venting the pump while it is running. If this does not help, stop the pump and vent it freely before starting again.

In case of persistent loss of suction (low or fluctuating discharge pressure), look for closed or throttled valves, plugged strainer, or other restrictions in the suction line. If a difficulty of this sort is not located, the system must be examined for sufficient net suction head.

A reciprocating pump is started with its suction and discharge valves wide open.

Rotary pumps installed in the plant must be started strictly according to the manufacturer's instructions.

All pumps should be watched closely during preliminary circulation, and particularly when first started. If bearings or packing boxes begin to overheat (usually this means hot to the touch) or give other evidence of trouble, such as noisy bearings, shut the pumps down for inspection.

To ensure against overloading, electricians should measure current with "tong"-type ammeters.

Pumping rates should be kept as high as possible for thoroughness of line flushing. The collection of solids on the strainer screen restricts suction line flow, and may eventually lead to loss of suction.

A pump should be stopped after running for several hours or whenever it shows signs of losing suction, for removal, inspection, and cleaning of the strainer. The strainer is then replaced, the pump started, and the process repeated until the strainer shows itself clean. Standby pumps should be run alternately with their counterparts. It will be found helpful to keep a record for each pump of running time, screen inspections and condition, and final disposition of strainer.

Following the run-in period, the pump should be isolated, drained, and filled with oil to protect the mechanical seal against rusting.

16.7 Breaking-in compressors

Compressors and drivers of all types must be properly installed and operated for "run-in," similarly to pumps, to assure their satisfactory service.

The detailed instructions issued by the manufacturer for the installation and operation of the machine must be closely followed. The following checklist should be observed:

- Check lube and seal oil system on compressor and driver. Circulate "run-in" oil in lube and seal oil system to clean all lines. Heat oil as necessary. "Run-in" oil should be changed. See manufacturer's instructions.
- On turbine driven machine, check over speed trip and governor with turbine disconnected.
- On motor driven centrifugal and turbine driven reciprocating compressor, check gearbox lubrication.

- Check out suction and discharge piping vent and drain connections, block valves, bypasses, and relief valves.
- On gas engine driven compressors, check out fuel system and starting air system.
- On turbine driven equipment, check out all the turbine and condensing equipment.
- Check operation of valve unloaders and clearance pockets on reciprocating machines.
- Check out intercooler and interstage knock-out drums.
- Be sure there is no liquid in suction piping and bottles on reciprocating machines before starting compressor.
- Check out control system and emergency shutdown devices.
- Both centrifugal and reciprocating compressors may be run in on gas different from that specified for the design of the machine. However, if there is an appreciable difference in gas properties, the manufacturer should be consulted. This is especially true for centrifugal machines.
- Check both reciprocating and centrifugal machines for adequate instrumentation.
- Cooling water to oil coolers, intercoolers, and aftercoolers must be made available.
- If permanent strainers are not installed, temporary strainers at each suction flange must be installed.

Reciprocating compressors are initially operated at no load, with valves removed to check lubrication, piston clearances, and operability of moving parts without undue noise or rubbing.

During the run-in, all controls of the lubricating system must be made to function correctly and satisfactorily. All alarms and safety devices must be verified for proper setting and correct functioning. If the compressor has carbon parts, the run-in period is used to polish and seat the carbon parts before pressure is applied to them.

16.8 Dry-out and boil-out

16.8.1 Heater dry-out

Before a heater is put into service for the first time, it will be necessary to slowly expel the excess moisture from the insulating concrete (setting) by gradually raising its temperature before any appreciable load is put on the heater.

Heater dry-out procedure shall be in accordance with the manufacturer's drying procedure. The heaters served in the reactor section of the licensed units shall be dried-out following the licensers instructions.

16.8.2 Chemical boil-out of steam generation facilities

Chemical boil-out is desirable as a means of internally cleaning the system parts with a mixture of chemicals and hot water to remove oil and other deposits that may have accumulated during fabrication and erection of the components. It is most important to remove any oil, grease, or oily scale from the inside of the pressure parts in order to avoid foaming and priming during subsequent operation.

The boil-out is performed as the final step prior to bringing the unit on stream. It should be done after all physical inspections and check-out procedures have been performed. Several combinations of chemicals may be used to achieve a satisfactory job of cleaning. Two satisfactory mixtures for chemical boil-out are the following; use of any other types shall receive approval of the manufacturer in advance:

1. A mixture of soda ash (Na_2CO_3) and caustic soda ($NaOH$) in equal proportion for a total of 6 kg of chemicals per cubic meter of boiler water.
2. An equal mixture of tri-sodium phosphate and soda ash for a total of 8 kg of chemicals per cubic meter of boiler water.

These chemicals should be well mixed and thoroughly dissolved in hot water and should be introduced into the steam drum through the chemical feed connection. If a portable pumping system is not available, the chemical solution may be fed directly into the drum through the manway prior to boil-out.

If possible, it is desirable to introduce part of the chemical solution into the feed line leading to the boiler feed water preheat coil in order to obtain some cleaning action in the preheat coil itself.

With the system in readiness and the chemical solutions prepared, the actual boil-out can begin. A typical chemical boil-out sequence can be used.

During each blow down period, the superheater drains should also be opened wide to assure that no condensate has accumulated in the coil or outlet header.

The duration of boil-out will normally vary between 24 and 72 h, depending on the type and initial cleanliness of the unit, as well as the chemical concentration and temperature maintained during boil-out. It is to be expected that 24–36 h should suffice to prepare the system for satisfactory operation.

Blow down water condition is one indication of whether the boil-out has achieved satisfactory results. The only conclusive determination of boil-out effectiveness, however, is by a visual internal inspection of the steam drum.

Upon completion of boil-out, the fires should be extinguished and the unit permitted to cool. After the drum pressure has reached zero, open the vent, and completely drain the unit of chemical solution.

After draining, the unit should be refilled with clean, fresh water to high drum level and flushed through the drum blow downs, bottom drains, and the steam generation coils.

After flushing, open the steam drum for internal inspection to check on the effect of the boil-out. The flushing procedure should remove practically all residual chemicals and any accumulation of sludge.

If the interior of the drum is adequately clean, the boil-out may be considered complete. If cleaning is not adequate, the boil-out should be repeated. It must be emphasized that the satisfactory operation of the unit depends, to a considerable extent, on a complete and thorough job of chemical cleaning.

After chemical cleaning is complete, the following steps should be taken:

1. The gage glasses should be removed, cleaned and reassembled. Care should be taken to be sure that all connections between the drum, gage glasses, and water columns are unobstructed.

Note: If desired, spare gage glasses may be used during boil-out to be replaced by new glasses prior the normal operation.

2. All manhole and other gaskets which were exposed during inspection, blinding, etc., should be replaced with new operational gaskets.

3. All connections and closures, which were opened after hydrotesting, should be checked and tightened securely. All such connections should be checked carefully, as the boiler is brought on the line for normal operation.

16.8.3 Reactor section dry-out

Before loading the catalyst into the reactors in the licensed units, it is necessary that lines and equipment be dried to remove any traces of water that might have remained from the construction and flushing.

Under most circumstances, free water can be adequately removed by allowing it to drain from the lines, heaters, reactors, exchangers, etc., while blowing through the lines with air. However, the dry-out of the refractory in the charge heaters might be convenient to be performed with the run-in of the recycle gas compressor.

All components existed in the reaction section shall be connected together for their dry-out. Before beginning the dry-out, a tightness test may be required.

Following the tightness test, a test under vacuum is to be performed. Then, a pressurization by nitrogen is to be done. The vacuum test is to be repeated and the system should be pressurized again with nitrogen. In any case, the detailed dry-out procedure specified by the licenser shall be followed incorporating the heater refractory dry-out procedure provided by the heater's manufacturer. The use of air for dry-out is not allowed because of the possibility of an oxygen/seal oil explosive mixture in the compressor at the elevated temperature.

Nitrogen, inert gas, fuel gas, or a mixture of hydrogen and fuel gas can be used as a dry-out purpose upon approval of the licenser. If fuel gas is used, it shall be sweet, light, and contain no contaminants, which might cause equipment coking or catalyst poisoning.

Being that the compressor is designed for light H_2-rich recycle gas, the compressor discharge operating conditions will be higher than normal operating ones when using nitrogen or other heavier gases. It may be necessary to reduce the driver (steam turbine) speed to maintain said conditions within normal. Special attention shall be paid to the heaters equipped with a waste heat recovery steam generation section that shall be protected during the dry-out procedure. All of the reactor internals, including the catalyst flow pipes between the reactors, should be installed prior to the dry-out operation.

16.8.4 General notes for dry-out and boil-out

1. The heater refractory dry-out, plant dry-out, and chemical boil-out procedures can be combined in the interests of saving time. It is likely, however, that the heat input during either the heater refractory or the unit dry-out procedures may be excessive for controlled chemical boil-out. Therefore, combining the two or three operations may require a temporary interruption of the drying-out procedure while the flushing operation is completed and the steam generation facilities are made completely operational.
2. The drying-out and boil-out periods present an opportune time to check out the operation of auxiliary equipment and instrumentation.

The circulating pumps should be operated in rotation to make certain that there will be no difficulties during final start-up.

16.9 Catalyst loading

Initial catalyst loading activities shall be performed according to the licenser's and/or catalyst manufacturer's procedures under supervision of the licenser's representatives. Full attendance and cooperation of the licenser's responsible authorities is required.

16.10 Tightness test

- During equipment cleaning, lines have been disconnected, and orifice plates and blinds removed and reinstalled. As a consequence, tightness tests shall be required to eliminate leakages due to gaskets that have been damaged, and flanges or drains that have been left open.
- Tightness test is to be carried out after final installation by air, steam, nitrogen, or the proper process fluids.
- All joints, flanges, packing glands, etc., must be checked for leaks by means of a sensitive detector, for example, by using a soapy solution.
- Thermal insulation of flanges shall be done only after the final pressure test of the section is completed.
- Air or nitrogen is commonly used for tightness test purpose. Tightness test pressure will depend on the operating pressure of the system under test. Normally, it shall be 1.2 times the normal operating pressure provided that this does not exceed the set of the PSV existing in the section. In any event, the maximum test pressure is normally limited by the available plant or instrument air pressure. When the operating pressure is considerably higher than that of said air, a preliminary tightness test is done with air, and the final tightness test is done with process fluid during start-up.
- The acceptability of the test is given by the pressure drop that can occur in a limited period of time. It is normally accepted that the tightness of a section is reasonably good when said pressure drop is less than 1% in 2 h time.
- When the unit, or section, is subject to special conditions of pressure, temperature, and type of process fluid, a more tight condition of pressure drop measurement will be given.

16.10.1 Tightness test of sections operating under vacuum

1. When sections of units are normally operated under vacuum conditions, it is necessary to test for proper tightness.

Being almost impossible to find leaks while the section is under vacuum, the most common way to do the test is to check all flanges, joints, valves packing, pumps seal, and any other possible point of leaks by pressurizing the system up to 80% of the lower design pressure of equipment included in the system to be tested or 80% of pressure relieve valves installed. Whichever is the lower will govern the pressure test.

2. Once the system or section is at the selected pressure conditions, all joints shall be carefully checked with soapy solution, or equivalent, and even minimum leaks shall be repaired. Then, the vacuum shall be pulled at the selected value. Stop vacuum pulling devices and check for pressure drop in the selected period of time (normally 1 h).

If the pressure drop is higher than specified, the system shall be pressurized again and rechecked to find the leaks.

Repair possible leaks and pull the vacuum again. The operation shall be repeated until the pressure drop is within given limits.

3. When the system to be vacuum-tested is very large, it may be convenient to split it in two or more sections by closing isolating valves. Installation of blinds is not recommended.
4. Only when the system does not show any leak from the joints, possible internal leak due to bad-seal of valves may be suspected. In this case, it is convenient to blind-off suspect valves.

When suspect bad-sealing valves are the welded type, it will be necessary to open their cover to inspect the internal disc, and/or seat for proper smoothness and/or full travel of the wedge into the ring seats.

16.11 **Normal start-up procedures**

Placing the new unit in operation can be made through several methods depending on the experiences of the operational crew.

Due to the variety in the types of the process and facilities included in the unit, a standard procedure for start-up cannot be established. The procedure and important points outlined in Standard Specifications are general start-up main items that shall be followed by the start-up crew regardless of the type of the process. The detailed procedure for placing the unit in operation shall be provided by the contractor and/or licenser, where required.

The procedure should be studied carefully before start-up, because several operations described must be done simultaneously and the crew should be prepared to act upon contingencies.

The normal start-up activities can be proceded when the following main commissioning and pre-commissioning steps are completed:

- Heater refractory dry-out.
- Line and equipment flushing.
- Rotating equipment run-in.
- Other activities.

16.11.1 **Prestart-up check list**

The procedure describes, in general terms, the steps to be followed for placing the unit on stream. The exact sequence of events depends on the flow scheme of the particular unit; however, the following steps must be completed before charging feed to the unit.

- All unnecessary blinds are removed.
- All relief valves are tested and installed.
- All temporary lines have been removed.
- The flare header is purged and in service.
- The sewers are in service.
- The heaters are steamed out.
- The fuel oil and fuel gas lines are in service.

- The pilots are lit in all heaters.
- The orifices are installed and are correct in direction of flow.
- All instruments are ready for service.
- All utilities are in service.
- All drains and vents are closed.
- Control valves and bypasses are blocked in.
- All compressors are blocked in.
- The chemical systems are ready for operation.
- The catalysts have been filled or regenerated.
- The unnecessary connections, such as pump out, etc., are closed. All fire fighting facilities are ready for operation.

16.11.2 Make area safe

Remove all welding gear from the area and all maintenance tools not of a non-sparking nature, and clear away all planks and scaffolding. No further work is to be done except with work permits and, if necessary, specific welding permits. "No Smoking" signs must be re-installed, if required.

16.11.3 Utilities commissioning

All utilities, such as various types of steam, condensate, boiler feed water, fuels and plant air, instrument air, nitrogen, and plant water shall be commissioned. Do not put cooling water into exchangers, which are to be left with water side drains and vents open and drained until after steam-out.

16.11.4 Air purging and gas blanketing of non-catalytic sections

Special attention should be paid that oil or flammable gas never be charged into process lines or vessels indiscriminately. The unit must be purged before admitting hydrocarbons. Air purging can be performed by nitrogen or inert gas, displacement of air by liquid filling followed by gas blanketing, or steaming out followed by gas blanketing. Air purging by nitrogen normally is dictated to be followed for the reactor section in the catalytic units by the licenser. For the remainder of the unit, other than the reactor section, steam purging followed by fuel gas blanketing can be used to air free the unit. The following steps and potential problems or hazards should be taken into consideration during the steam purge:

1. Collapse due to vacuum

Some of the vessels are not designed for a vacuum and, therefore, should not be allowed to stand blocked in with steam due to developing a vacuum by condensation of the steam. Thus, the vessel must be vented during steaming and immediately followed up with fuel gas purge at the conclusion of the steam-out.

2. Flange and gasket leaks

Thermal expansion and stress during warm-up of equipment along with dirty flange faces can cause small leaks at flanges and gasket joints, which shall be corrected accordingly.

3. Water hammering

Care must be taken to prevent "water hammering" when steam is purging the unit. Severe equipment damage can be resulted from water hammering.

4. Block in the cooling water to all coolers and condensers

Shut down fans on fin-fan coolers and condensers. Open high point vents and low point drains on the vessels to be steam purged.

5. Start steam-out operation

Start introducing steam into the bottom of the columns, towers, and at low points of the various vessels. It may be necessary to make up additional steam connections to properly purge some piping that may be dead-ended. Thoroughly purge all equipment and associated piping of air. Be sure to open sufficient drains to drain condensate that will accumulate in low spots and receivers. Pumps and instruments shall not be steamed out. Steam-out duration shall not be less than 12 h.

6. Start fuel gas injection

When purging is completed, close all vents and drains and start introducing fuel gas into all vessels and cut back the steam flow until it is stopped completely when the systems are pressured. Regulate the fuel gas flow and the reduction of steam so that a vacuum due to condensing steam is not created in any vessel and/or that the refinery/plant fuel gas system pressure is not appreciably reduced. A minimum pressure of 80 kPa [0.8 bar (ga)] in the vessels should be maintained throughout the steaming-out and gas purge procedures.

16.12 Catalytic units reactor section air purging and gas blanketing

The following procedure is normally applied for the reactor section unless otherwise instructed by the licenser.

Evacuate the reactor circuit to the lowest attainable pressure.

Hold the vacuum for 1–2 h to check for tightness of the unit.

If a good vacuum cannot be held, introduce nitrogen to the system at the discharge of recycle gas compressor and pressure the reactor circuit to a positive pressure and repair all leaks that are discovered. Afterward, repeat the vacuum test until a good vacuum can be held.

Introduce nitrogen at the discharge of the recycle gas compressor and pressure the reactor circuit to a positive pressure of 10–35 kPa [0.1–0.35 bar (ga)]. The purge nitrogen should contain less than 100 mol ppm oxygen.

Pull the vacuum on the system twice more, pressuring each time with nitrogen.

After pulling the last vacuum, pressure the reactor circuit with nitrogen to at least the minimum pressure required for recycle gas compressor operation, but not higher than the allowable percentage of the reactor shop test pressure.

At the same time that the reactor circuit is evacuated, the compressor suction drums should also be evacuated and purged with nitrogen. Do not pull a vacuum on the compressors, but leave them blocked in. The compressors must be thoroughly purged with nitrogen and maintained under nitrogen pressure before admitting hydrogen into the unit.

16.13 Heat exchanger activation

Water should now be put through the water-cooled exchangers. Vents at high points are to be opened for releasing of air. Start one fan on each section of the air-cooled exchangers.

16.14 Vacuum test

A vacuum test is to be performed for the equipment designed for vacuum and to be operated under vacuum conditions, such as the vacuum column in the vacuum distillation unit. All sections and facilities not designed for vacuum should be blocked off and segregated from the section to be tested for vacuum. Check the system for leakage after the vacuum approaches minimum and perform the following activities.

Block off the inlet to the ejector facilities, which are producing a vacuum, and hold for 1 h.

If leakage does not exceed 50 mm/h, it will be acceptable, otherwise the system should be investigated for the leakage sources.

16.15 Establish flow in the unit

The unit should be lined up and a final check shall be carried out to introduce the feed to the unit. All safety precautions should be taken into account. The feed is to be charged at the turn down ratio of the unit design throughput, according to the stepwise start-up procedure developed by the contractor. Hydrocarbon circulation in the coils of reboiler heaters shall be maintained at all times at a flow rate close to the design values in order to avoid coke deposition inside the coils.

16.16 Inhibitor/chemical injections

Corrosion inhibitor and/or chemical injection shall be started where required when the temperature and pressure of the system are rising to the values close to operating conditions. The inhibitor and chemical injection rate shall be as indicated by the contractor/licenser.

On the basis of laboratory tests, operating conditions must be adjusted to meet specifications on the products as well as product yields.

16.17 Typical acid cleaning procedure for compressor lines

16.17.1 The preparation

A list of metals, alloys, and nonmetallic materials in the sections to be cleaned, including block valve trims, gaskets, valve packing, nuts, and exchanger tubing, as well as major equipment and piping must be made.

Assurance must be obtained from the cleaning contractor that the chemicals and chemical solutions used in the operation will not be injurious to these materials.

A list must be made of the safe operating pressures of all components in the sections to be cleaned.

Assurance must be obtained by the cleaning contractor that these pressures will not be exceeded (especially if the safety valves in these sections are going to be blinded off; in this case, the cleaning contractor should provide safety valves with his equipment).

Spool pieces must be made and substituted for turbine meters and for any valves that must be protected from any chemical solutions. Valves that are removed should be cleaned separately and their openings sealed off.

Orifice plates must be removed from the lines.

All instrument taps in the system must be disconnected or blocked off. Drain points must be provided in the taps to drain off solution, and all instrument drain valves should be opened.

All externally mounted liquid level instruments, such as displacement type level transmitters and gage glasses, should have all block valves adjacent to the vessel closed and all drain valves opened.

Pressure gages and thermowells should not be in place and their connections should be blocked off.

All piping strainer screens must be removed.

All high points must be provided with vent valves. These vent valves should be opened periodically during the cleaning operations.

Major items of equipment, such as compressors, pulsation dampeners, etc., must be blinded off.

Cleaning circulation circuits must be determined. For a three stage make-up compressor system, this might mean four separate circulation circuits, one for each stage and one for the incoming fresh hydrogen line.

16.18 The acid-cleaning operation

The acid-cleaning operations can be generally divided into the following steps:

16.18.1 Flushing

All sections should be water flushed to remove all loose dirt, debris, and other foreign material in the lines. It should be noted that process pumps must not be used to circulate any of the flushing, rinsing, or chemical solutions. All transfer and circulating pumps for handling these solutions must be furnished by the chemical cleaning contractor.

16.18.2 Degreasing

All sections should be flushed with a degreasing solution (generally an alkaline solution, such as a soda ash solution) to remove all grease or oil that may have been applied to the lines and vessels as a rust preventative measure. The cleaning contractor should specify the type and concentration of the solution to be used.

During this and other phases of the operation, the contractor may want to heat the circulating solutions. In doing so, reboilers or exchangers must not be used as a means of heating them. All heating is to be external to the systems being cleaned and by equipment furnished by the chemical cleaning contractor.

After this step, all sections should be rinsed with water.

16.18.3 **Chemical cleaning**

All sections must be treated with an acid solution to remove all rust and scale from the metal surface. There are several types of cleaning solutions that can be used to do this step (such as inhibited hydrochloric acid or inhibited phosphoric acid); it is the responsibility of the cleaning contractor to select one that has been proven by experience. A suitable inhibitor must also be chosen to reduce the attack on metal. The contractor should specify the concentration to be used and the percentage of metal components (such as iron) to be allowed in solution. Afterward, the acid circulation should be followed by a water rinse.

16.18.4 **Neutralizing**

All sections must be flushed with a neutralizing solution (perhaps a soda ash solution) to neutralize all traces of acid left in the system. The cleaning contractor should specify the type and concentration of the solution to be used. After this step, all sections should be rinsed with water.

16.18.5 **Passivating**

In order to form an anti-rust skin, a solution with a passivating agent must be circulated through each section. Afterward, each system is allowed to dry. Note that any passivating agent used must meet with the licenser's approval and must be flushed from the system prior to start-up.

After completing the cleaning operation, the vessels and lines should be inspected to determine the quality of the cleaning. Treated surfaces should be clean, rust-free, and dull gray in color. In-line turbine meters, valves, strainers, and all other equipment that was removed must be installed. Afterward, the make-up system must be nitrogen purged and left under nitrogen pressure until the start-up.

16.19 **Typical heater dry-out procedure**

During the initial heater refractory drying out period, it is preferable that no material be flowing through the tubes.

Make a temporary installation of thermocouples through the pipe sleeves in the hip section of the heater. The tips of these thermocouples should extend 150 mm (6 inches) beyond the inside of the insulating concrete, but should not contact the tubes.

It is preferable to use gaseous fuel (refinery gas or LPG) for drying out the setting. If no gas is available, liquid fuel may be used, but it should be free of sediment and heated as required to give the proper viscosity about 43 mm^2/s or 43 cSt (about 200 SSU) for good atomization and clean combustion.

Light one or more burners, as required, in each section of the heater and fire slowly, so that the temperature, as indicated by the hip thermocouples, is increased at a rate of about 14 °C (25 °F) per hour until it reaches 482 °C (900 °F). Hold this temperature for 10 h, or 2 h per 25.4 mm (per inch) of refractory thickness, whichever applies.

While increasing the temperature, the burner operation should be rotated frequently in order to distribute the heat as evenly as possible over the entire length of the setting.

After the 10-hour holding period, all burners should be shut off and the heater setting allowed to cool slowly by keeping the air inlet doors and stack damper(s) fully closed.

After drying has been accomplished, the temporary hip thermocouples should be removed and the plugs replaced in the pipe sleeves. If the setting has been dried as outlined above, the temperature may be subsequently raised or lowered at any desired rate within the design limits of the heater.

16.19.1 For gas-fired heaters

When the unit is shut down, always blind off the fuel gas supply line because gas may leak through the block valves at the heaters and fill a furnace.

Before starting to light any pilot burner, see that all individual burner block valves are closed and steam-out the firebox to remove any gas accumulation. Make sure the damper is opened. Steam-out the box until a steady plume of steam can be seen rising out of the stack. Stop steaming and pinch in the damper.

When all pilot burners are lit, light each burner individually by opening the gas valve to each burner after the torch is inserted in front of the burner. After a few burners are lit, it will be necessary to open the damper to provide enough draft to light the remainder of the burners.

Burners should be fired to produce a blue flame with a yellow tip obtained by regulating the primary and secondary air supply. The heaters should be checked frequently for dirty burners, which might give either too long, too short, or a misdirected flame. There must be some excess of air to the burners so that an increase in fuel gas flow will have sufficient air to produce complete combustion. If for any reason the fires in a heater go out:

1. Shut off the gas supply immediately by closing the block valves at the fuel gas control valves. Bypass and pilot lines that might be open around the control valves must also be closed.
2. Put snuffing steam in the firebox.
3. Close all individual burner valves.
4. As in all heaters, care should be taken that no flame impingement on the tubes is permitted.

16.19.2 For oil-fired heaters

When the unit is shut down and before entering heaters, always double block the oil supply line on both the supply and return headers and pull the oil guns from the burners as oil may leak through the block valves at the heaters and fill a furnace.

Before starting to light any pilot burners, see that all individual oil guns are removed from the burners, and steam-out the firebox and header to remove any gas accumulation. Make sure that the dampers are opened slightly.

Oil burners without gas pilots should be lit from a regulation torch. When there is a gas pilot, light it first and then light the oil from the pilot. Have fuel oil circulating through the fuel oil return at normal operating temperatures.

Burners should be fired to produce a yellow flame with good pattern obtained by regulating the primary and secondary air supply. The furnaces should be checked frequently for dirty burners that might give either too long, too short, or a misdirected flame. There should be some excess air to the burners so that an increase in fuel flow will have sufficient air to produce complete combustion. If for any reason the fires in the furnace go out, then:

1. Shut off the fuel supply immediately. Do this by closing the main block valve in the fuel supply to the furnace. This will take care of any bypass lines that might be open around the control valves. Be sure the check valve on the fuel oil return does not leak allowing fuel to back into the firebox.
2. Put snuffing steam in the firebox.
3. Block in the pilot gas line. Close individual burner valves.

As in all heaters, care should be taken that no flame impingement on the tubes is permitted.

16.19.3 Safe procedure for lighting oil burners

Push the oil gun forward, and then turn on steam by fully opening the steam block valve and the steam control valve. Close off when the steam is dry.

Make sure the oil block valve is closed, and then open the steam bypass valve to clean and warm the burner.

When condensate has been removed and the steam is dry (dry steam is invisible), close the bypass steam valve.

Adjust atomizing steam valve for a small flow of steam.

Open oil block valve gradually until the oil starts burning. The oil will ignite from the pilot gas flame or an oil torch. Take care to see that unburned oil is not put into the firebox. Accumulated oil will become hazardous as the firebox heats up.

Adjust the atomizing steam valve and oil valve to obtain correct flame pattern. Never let the flame touch the tubes.

16.20 Typical chemical boil-out sequence

Fill the system to normal drum level, using the feed-water pump and the regular feed connections. Introduction of some chemical compound into the feed water, as outlined above, would be desirable. The feed water temperature should be limited to approximately 80–90 °C.

Make certain that the superheated steam line is closed to the refinery/plant steam system. Open the superheated steam coil outlet vent and drain lines that are open to the atmosphere and leave them partially open during boil-out. This will assure flow through the superheater tubes and avoid any accumulation of condensate.

Open the vent valve on top of the steam drum.

Start one of the water circulation pumps to provide flow through the steam-generating coils.

Light the fire in the heater. The boil-out procedures can be combined with the heater refractory dry-out or the unit reactor section dry-out. Bring the heater firebox temperature up at 50 °C/h until the steam drum pressure is established at approximately 3.5–7 bars (3.5–7 kg/cm^2). Care must be taken to maintain a level in the steam drum at this time.

Flow must be maintained through the boiler feed water preheat coil. Because steam will be venting into the atmosphere, this will be accomplished by charging make-up water to maintain the level in the steam drum. If possible, a temporary line should be connected from the discharge of the water circulating pumps to the boiler feed water preheat coil inlet line to provide chemicals to clean this section also.

Approximately once each shift, the unit should be blown down using the blow down connections. The water should be dropped to the bottom of the gage glass and then fresh water added to bring the water back to the normal operating level. Chemical solution should be added through the chemical feed connection to maintain the concentration in the boiler water as close to the recommended levels as possible.

16.21 **Basic considerations in preparing operating manuals**

This section covers the minimum requirements of the format in preparing process and/or utility units operating manuals, including essential instructions and noteworthy points.

The purpose of this section is to standardize the content and format of operating manuals, which shall be prepared by the contractor. Although operating manuals differ to some extent from process to process, the basic philosophy and general aspects shall conform to the concepts of this section.

A general index for the plant operating manual is shown below in Table 16.2.

Table 16.2 A General Index for Plant Operating Manual

Typical Operating Manual General Index

1. Introduction
   1.1 General
2. Basis of design
   2.1 Duty of plant
   2.2 Environmental conditions
   2.3 Feedstock and product specifications
   2.4 Battery limit conditions
   2.5 Specifications and consumptions of utilities, chemicals and catalysts
3. Process description
   3.1 Process theory
   3.2 Description of flow
   3.3 Process variables
4. Preparation for initial startup
   4.1 General
   4.2 Plant inspection
   4.3 Cleaning of piping and equipment
   4.4 Specific pre-startup operations
5. Startup procedure (for each configuration)
6. Normal operation
7. Normal shutdown procedure
8. Emergency shutdown procedure
9. Analytical requirements
10. Attachments

16.22 Safety manual/quality manual

A recommended general index for plant safety manuals is shown below. The index consists of section titles (one digit numbering) and paragraph titles (two digits numbering). A detailed section index listing all subparagraphs titles is given (Table 16.3) before each section of the manual.

Table 16.3 General Index for Plant Safety Manuals

Typical Safety Manual/Quality Manual Contents:

1. Introduction
   1.1 General
2. Pre-commissioning and commissioning emergency procedures
   2.1 Emergency contact numbers
3. General site hazard
   3.1 Pressure testing and air freeing of process plant on site
   3.2 Safety in plant commissioning
   3.3 General fire protection and prevention
4. Protective clothing and equipment
   4.1 Respiratory protective equipment
   4.2 Compressed air breathing apparatus
   4.3 Escape filter
5. Pre-commissioning and commissioning safety training
   5.1 Safety training concept
   5.2 Training modules common to pre-commissioning and commissioning activities
   5.3 Safety training modules content
6. Permit to work procedure
   6.1 Lockout and tagging
   6.2 Working inside energized buildings
   6.3 Vessels/confined space entry
7. Project safety forms
   7.1 Permits to work
   7.2 Lockout/tagging
   7.3 General
8. Safety concepts
   8.1 General
   8.2 Codes and standards
   8.3 Climatic condition
   8.4 Hazardous area classification
   8.5 Fire protection
   8.6 Personal protection
   8.7 Fire and safety point shelters
   8.8 Fire & gas detection
   8.9 Fire proofing
9. General information
   9.1 Fire water network—overview
   9.2 Deluge system—overview
   9.3 Inert gas system—overview
   9.4 Fire equipment—overview
10. Material safety data sheets

16.23 **Non-licensed processes**

The purpose of an operating manual is not only to help the operation engineers and staff at the customer side to operate the plant safely, but also to present all detailed procedures for the plant start-up and shut-down in the various operation cases.

- Extent of description

Operating manuals should contain all operating procedures, guidances, hints, cautions, and trouble-shooting guides necessary for safe and correct plant operation.

The detailed operation procedures should also include the sequence of valve operation, time schedule, etc.

The contractor should carefully study the history of troubles experienced and countermeasures used in similar processes and provide the latest instructions.

Abnormal levels of operating variables (temperature, pressure, flow rate, fluid level in vessel, etc.) together with appropriate countermeasures should be listed in the operating manual as far as possible to avoid similar troubles.

- Final check of basic design

The operating manual should be checked carefully against the basic design's latest revision.

The operation philosophy, which was prepared at the early stages of the basic design, shall be reviewed.

The necessary facilities, equipment, instruments, lines, etc., for each operation mode (namely start-up, normal shut-down, emergency, regeneration, maintenance, etc.) must be taken into consideration.

16.23.1 **Licensed processes**

The detail designer shall prepare an operating manual for the licensed unit in accordance with the conditions stipulated in standards for non-licensed units and based on the contents of the operating manual furnished by the licensor. All modifications/changes resulting from the detailed engineering activities should be reflected in the final revision of the operating manuals.

Before making any unavoidable modifications/changes by the contractor, it is necessary to obtain the approval of the process licensor.

16.24 **Noteworthy points**

Generally, the necessary documents providing a guide to operation comprise:

1. Operating manual.
2. Analytical manual.
3. Vendor's instruction manuals.
4. Safety manual.

The vendor's instruction manuals have priority over the operating manual prepared by the contractor in order to place the responsibility for maloperation on the vendors.

The following figures and tables are to be inserted in the operating manual for the operator's convenience and easy comprehension.

1. Figures
   a. Trip sequence (flow) diagram.
   b. Simplified flow scheme of plant heat-up.
   c. Simplified flow scheme of catalyst reduction, activation, oxidation, and regeneration (generally for licensed units).
   d. Simplified flow scheme of feed cut-in, shut-down, and other operation modes.
   e. Furnace drying curve.
   f. Other charts as needed.
2. Tables
   a. Pressure relief valves load summary tables.
   b. Setting point list for instruments (especially alarm and trip elements).
   c. Analytical schedule.
   d. Utility summary tables.
   e. Heat and material balance tables.
   f. Major equipment specification summary.

16.25 Design basis

- Sections A, B, and C

Type and source of feed and unit different operating modes are to be specified.
  The characteristics of feed, products, and byproducts (if necessary) shall be specified.

- Section D

Utility conditions shall cover operating pressure and temperature as well as application of each type for all utilities concerned in the plant.

- Section E

Heat and material balance tables, including the following characteristics of each stream, as marked on the relevant process flow diagram to be covered in this section.
  Enthalpy basis (datum level) for all fluids are to be identified. Computer program used for preparation of heat and material balance tables shall also be clarified.

- Section F

Utility summary tables shall cover the following requirements (where applicable) as shown below typical tables.

- Item number.
- Service.
- Load BkW (BhP), kW.

- Electrical power, kW.
- Steam, 1000 kg/h.
- HP steam, pressure in, bar (ga).
- MP steam, pressure in, bar (ga).
- LP steam, pressure in, bar (ga).
- LLP steam, pressure in, bar (ga).
- Condensate, 1000 kg/h.
- Cold cond., pressure in, bar (ga).
- HP hot cond., pressure in, bar (ga).
- LP hot cond., pressure in, bar (ga).
- LLP hot cond., pressure in, bar (ga).
- BFW, 1000 kg/h
- Pure demineralized water, 1000 kg/h.
- Loss (steam, condensate, BFW, etc.), 1000 kg/h.
- Cooling water.
- Tempered water, m^3/h.
- Fresh water, m^3/h.
- Temperature rise, °C.
- Fuel (LHV).
- Oil, 1000 kJ/s.
- Gas, 1000 kJ/s.
- Nitrogen, Nm^3/h.
- Air, Nm^3/h.
- Instrument.
- Plant.
- Plant water, m^3/h.
- Potable water, m^3/h.
- Inert gas, Nm^3/h.
- Natural gas, Nm^3/h.

Utility summary tables shall be provided separately for summer and winter operating cases when the unit is operated under design flow rate. Additional cases may be included upon the company's request.

- Section G

Effluent summary shall cover all unit effluents except those streams considered as the unit products/byproducts as presented in section C. The effluent summary shall include the following streams, where applicable.

- Sour water.
- Oily water.
- Spent caustic solution.
- Chemical sewer.
- All other disposed liquid and solid wastes.

The following characteristics for each effluent shall be specified.

- Quantity, kg/h and/or m^3/h.
- Impurities, such as H_2S, NH_3, oil, Cl^-, Na^+, etc. in mass ppm (wt), (mg/kg).
- Sources, including all equipment involved.
- Destinations, such as oily water sewer, non-oily water sewer, chemical sewer, etc.

16.25.1 Process description

- Section A

The following requirements shall be included under the "nature of process."

- Introduction.
- Chemistry of the process.
- Typical reactions.
- Reaction rates and heats of reaction.
- Section B

Detailed line up of the process flow separately for each section of the unit to be provided.

16.25.2 Operating variables and controls

- Section A

This section embodies the main process features and is prepared to help the plant operators overcome troubles not mentioned in the next chapters (start-up and shut-down procedures). Any operating variable, such as pressure, temperature, chemical additions, feedstock properties, hydrogen to hydrocarbon ratio (if any), etc., which has a significant effect on the unit operation and main product specifications shall be mentioned.

- Section B

Any significant differences in product quality and/or unit operation resulting from any changes of operating variables shall be elaborated in this section.

- Section C

Special attention should be paid to specify all the possible troubleshooting that the unit operators may face during the operation. The causes of and preventative actions for any troubleshooting shall be clearly demonstrated.

16.25.3 Auxiliary systems

The following sections shall be included with full operation description and useful operating guidelines, where applicable.

1. Tempered water system.
2. The ram pump.
3. The flushing oil circuit.

4. Soot blowers.
5. Fuel oil and fuel gas systems.
6. Chemical injection systems.
7. Chloride and/or condensate injection to the reactor system.
8. Any other auxiliary system as applicable.

16.25.4 Equipment operation

The detailed start-up, operation, and inspection prior to operation for all main equipment, as well as packaged units to be outlined in this section. Reference to the operating and maintenance instructions prepared by the equipment manufacturer for each item shall be given. Main operating points, all useful operation guidelines, and description of all equipment accessories shall also be pointed out. Compressor auxiliary systems, such as lube oil, seal oil, tempered water, etc., and fired heater burners, forced draft, and induced draft fans operations should be explained in detail and all troubleshooting that may occur during operation of such systems is to be described.

16.25.5 Instrumentation and control

Simplified logic diagrams for the major equipment with step-wise operation guidelines are to be given. Main features of the advanced control systems and optimization to be specified, where applicable. Reference to the relevant specifications shall be made. Set points of all alarms and shut-down switches shall be listed.

16.25.6 Start-up and shut-down

Several operating activities shall be conducted at the same time during the start-up and shut-down period. Therefore, all such activities that shall be performed in parallel for a safe and reliable start-up and shut-down operation should be described. Special attention shall be made to the vendor's or licensor's instruction manuals and the operating manual shall be reviewed carefully to be in congruent with the vendor's or licensor's instructions. All start-up and shut-down procedures shall be prepared in detailed step-wise activities, which will be performed by the operators with simplified start-up/shut-down sketches.

16.25.7 Emergency shut-down procedure

Safeguarding systems and equipment provided to protect the plant during emergency cases, such as emergency shut-down plant depressurizing, shall be elaborated. Because measures to be considered in an emergency vary according to the type, degree, and duration of the emergency encountered, the operation supervisor (or chief operation engineers) dispatched by the contractor shall be responsible for determining measures to be taken into account during the commissioning and performance test period. This should be emphasized in the operating manual. Steps to be taken in each emergency case shall be outlined in full description such that to help operators to recognize and act upon immediately. Because hard and fast rules cannot be made to cover all situations that may arise, examples of some typical emergencies are taken up in this chapter. Generally, a double failure or multiple failure cannot be discussed because of its complexity, although it often occurs in an actual emergency. However, the contractor shall take into account the necessary precautions and prepare the required procedures for such emergency cases.

Miscellaneous procedures called by the nature of the unit process and/or the special equipment operation not included in the other chapters shall be covered as required. Safety instructions issued by the competent local authorities shall take precedence over this chapter.

16.25.8 Analytical tests

The analytical plan shall comprise the following requirements.

- Stream name.
- Test name.
- Test number (analytical methods).
- Sampling point.
- Normal sampling frequency.
- Start-up sampling frequency.

Detailed analytical procedures shall be prepared separately in the form of an analytical manual.

16.25.9 Catalysts, chemicals, and packings

A summary of catalysts, chemicals, and packings requirements comprising of the following information shall be tabulated.

- Description of the catalyst/chemical/packing.
- Manufacturer name and type.
- Quantity required for initial charge, m^3 or kg.
- Vessel number/where used.
- Vessel name.
- Estimated consumption rate (daily (day) and/or yearly (year)), m^3 or kg.

16.25.10 Drawings

An equipment item index shall be provided to show at least the following requirements in section A for the operator's easy reference.

- Equipment category (e.g., tower, vessel, heat exchanger, pumps, compressors, etc.).
- Equipment number.
- Equipment service name.
- Quantity.
- Referenced P & IDs.

A project general legend diagram (if any) shall be inserted in this section.

16.26 **Plant technical and equipment manuals**

Plant technical manuals shall be prepared for each process and/or utility unit. All subjects common to the refinery/gas process and/or utility units shall be presented in separate plant technical manuals. Plant equipment manuals, including vendor dossiers, shall be prepared in accordance with the equipment categories and vendor information throughout the refinery/gas processing plant.

16.26.1 **Contents of the manuals**

In order to standardize the quality of the manuals, the contents shall include but not be limited to the following information.

- Utility summary tables based on operation and design conditions for the alternate operations and utility balance drawings.
- Basic engineering design data (BEDD).
- Plant/complex block flow diagram.
- Unit process flow diagrams.
- Piping and instrumentation diagrams.
- Utilities distribution diagrams including: steam, condensate, water, power, fuel, air, nitrogen, etc.
- Plot plans and general arrangement diagrams showing location of all equipment and extent and type of structure, where required.
- Earth work, such as rough-grading, surface preparation, landscaping, etc.
- Unit process description.
- Flare load summary tables.
- Power system single line diagrams.
- Electrical power load diagrams.
- Equipment specifications, data sheets, and curves (where required) containing design information for vessels and tanks, compressors, pumps, heaters, heat exchangers (including water and air coolers, condensers, reboilers, chillers, tank heaters, etc.), motors, turbines, generators, piping, instrumentation, packaged units, and other miscellaneous equipment and/or devices.
- Job specifications, including all project technical specifications for excavation and grading, concrete, steel structure, buildings, tanks, vessels, fired heaters, boilers, heat exchanging equipment, compressors and generators, pumps and turbines, motors, mixers, miscellaneous equipment (such as filters, silencers, desuperheaters, etc.), electrical, instrumentation and control system, piping, insulation, painting, welding, inspection, etc.
- Analyzers specification and data sheets.
- Electrical hazardous area classification.
- Pipeline lists and line schedules for all piping, including piping specialty items.
- Fire fighting facilities drawings and data sheets.
- Pump and compressor performance curves stamped by the manufacturer including head, capacity, efficiency, net positive suction head, and brake horse power in kilowatts (kW). The curves shall include the information for the diameter of the impeller furnished and the maximum size impeller that can be used.
- Index of contractor's drawings.

- Index of manufacturer's drawings.
- Lubricating oil schedule and specification.

All general data/specifications applicable to all units throughout the refinery/complex shall be presented as common technical information in the separate volumes apart from the individual items for each unit. The common subjects shall not be repeated in the manuals prepared for each individual process, offsite, or utility unit. Some of the items may be modified and/or changed depending upon the particular process or utility unit.

16.26.2 Plant equipment manuals

Plant equipment manuals shall contain all specifications, data sheets, drawings and equipment operation, maintenance and safety instructions, and all other engineering documents necessary for safe operation of the equipment produced by the equipment manufacturer.

The manuals shall contain all engineering documents prepared by the manufacturers/vendors in accordance with the equipment purchased order numbers.

A general index should be provided for all plant equipment manuals containing purchase order numbers, description, vendor's or manufacturer's name, and volume number.

Contents of the manuals shall include the subjects according to the following categories in order of precedence:

- Civil including structure, concrete, and buildings.
- Tanks.
- Vessels, towers, and reactors (including all vessel internals, such as trays, packing, etc.).
- Fired heaters and boilers.
- Heat exchangers (including water and air coolers, chillers, reboilers, condensers, tank heaters, coils, etc.).
- Compressors and power generators.
- Pumps drivers and turbines.
- Filtration units.
- Mixers.
- Dryers.
- Refrigeration units.
- Package equipment.
- Other miscellaneous equipment (if any).
- Instrumentation.
- Control system.
- Electrical.
- Piping and miscellaneous piping items.
- Insulation.

16.26.3 Timing

Plant equipment and plant technical manuals shall be furnished by the contractor to the employer/company before the works are taken over. The manuals should be provided together with drawings (other than shop drawings) of the project as completed, in sufficient detail to enable the company to operate, maintain, dismantle, reassemble, and adjust all parts of the works. The project shall not be considered as completed for the purpose of taking over under the conditions as required in the contract for test and acceptance until such requirements are met as follows.

1. At least 90 days before commencement of the performance test, the contractor shall prepare in draft and deliver to the company for approval two copies of complete plant technical manuals and two copies of plant equipment manuals.
2. The contractor shall carry out all corrections; amendments, and additions to such manuals as may be instructed by the company. Within 15 days after the company's approval, the contractor shall deliver to the company the required numbers of copies of approved plant technical and plant equipment manuals.

Further reading

Aguilar R. Successful implementation of advanced process control in cryogenic plants. In: 12AIChE – 2012 AIChE spring meeting and 8th global congress on process safety, conference proceedings; 2012.

Al-Bidaiwi M, Beg MS, Sivakumar KV. Deal with start-up and commissioning threats and challenges at an early stage of the project for a successful handover and project completion. In: SPE production and operations symposium, proceedings, vol. 1; 2012. pp. 425–34.

Al-Khaledi S, Waheed F. Execution of brownfield projects. In: Society of Petroleum Engineers – 14th Abu Dhabi international petroleum exhibition and conference 2010, ADIPEC 2010, vol. 2; 2010. pp. 1327–33.

API (American Petroleum Institute) API Std. 650, Welded steel tanks for oil storage, English edition latest edition.

ASME (American Society of Mechanical Engineers) ASME Code, Section VII.

Bahmannia G. What are the major issues in a new built gas processing plant: lesson learned in a large scale gas plant in Iran. In: International gas research conference proceedings, vol. 4; 2011. pp. 3030–7.

Chen Y, Gong J, Li X, Zhang N, He S, Liu J, et al. Simulation of liquid-gas replacement in commissioning process for large-slope crude oil pipeline. In: Proceedings of the biennial international pipeline conference, IPC, vol. 1; 2012. pp. 175–81.

Cicerone D. Shutdown systems and their effect on commissioning. In: 11AICHE – 2011 AICHE spring meeting and 7th global congress on process safety, conference proceedings; 2011.

Cramer R, Griffiths N, Kinghorn P, Schotanus D, Brutz J, Mueller K. Virtual measurement value during start- up of major offshore projects. In: Society of Petroleum Engineers – international petroleum technology conference 2012, IPTC 2012, vol. 2; 2012. pp. 996–1005.

El-Reedy MA. New project management approach for offshore facilities rehabilitation projects. In: Society of Petroleum Engineers – Abu Dhabi international petroleum exhibition and conference 2012, ADIPEC 2012 – sustainable energy growth: people, responsibility, and innovation, vol. 1; 2012. pp. 558–69.

Feitosa SR, Estites C, Everard J. Pre-commissioning of first interfield pipeline of Brazilian pre-salt projects. In: Proceedings of the annual offshore technology conference, vol. 2; 2011. pp. 1353–60.

Fournié F. AKPO project start-up and operation. In: Proceedings of the annual offshore technology conference, vol. 4; 2010. pp. 3169–79.

Jakobsen P, Peck G, Snow E. Electrical power system challenges during the expansion of offshore oil & gas facilities. In: Petroleum and chemical industry conference Europe conference proceedings, PCIC EUROPE, art. no. 6243263; 2012.

Liu J, Zhang L, Yan X, Xu C. Adaptation and optimization of oil pipeline nets in Chengdao Oilfield. In: ICPTT 2011: sustainable solutions for water, sewer, gas, and oil pipelines – proceedings of the international conference on pipelines and trenchless technology 2011; 2011. pp. 218–28.

Ma A. Engineer's role in resolving disputes in offshore projects. Proc Inst Civ Eng Manage Procure Law 2009; 162(4):191–6.

Maddirala K, Ghael A, Kanithai S. Best practices systems processes implementation KGD6 deep water fields India. In: Society of Petroleum Engineers – SPE oil gas India conf exhib 2012, OGIC – furth deep tougher: quest continues; 2012. pp. 258–83.

Marchini R, Calvarano MP, Cocchi G, Botta M. A new approach for gas fields development within uncertain business context. In: Society of Petroleum Engineers – international petroleum technology conference 2012, IPTC 2012, vol. 3; 2012. pp. 2541–9.

McCulloch B. Transitioning safely from greenfield to operations: case study of a major capital project. In: Society of Petroleum Engineers – SPE international conference on health, safety and environment in oil and gas exploration and production 2010, vol. 3; 2010. pp. 2129–34.

McLean D, Stott C, Foster S, Hesse E. Commissioning and operational history of Nkossa II LPG FSO. SPE Proj Facil Constr 2011;6(3):138–44.

Paisie JE. Three-point system compares US LNG export projects. Oil Gas J 2012;110(12):116–23.

Pallanich J. Optimizing opportunities in Oz. Offshore Eng 2010;35(5):40–2.

Peng K-YB. The production enhancement of NGL and reduction of pipeline liquid condensate flaring with successful implementation of triethylene glycol dehydration unit. In: Society of Petroleum Engineers – international petroleum technology conference 2013, IPTC 2013: challenging technology and economic limits to meet the global energy demand, vol. 3; 2013. pp. 1987–96.

Ramachandran S, Engel D. Reliable & efficient feed gas preparation – a key enabler to pearl GTL. In: SPE production and operations symposium, proceedings, vol. 2; 2012. pp. 1187–95.

Saadawi H. Commissioning of Northeast Bab (NEB) development project. In: 12th Abu Dhabi international petroleum exhibition and conference, ADIPEC 2006: meeting the increasing oil and gas demand through innovation, vol. 2; 2006. pp. 898–904.

Salman Y, Wittfeld C, Lee A, Yick C, Derkinderen W. Use of dynamic simulation to assist commissioning and operating a 65-km-Subsea-tieback gas lift system. SPE Prod Oper 2009;24(4):611–8.

Seethepalli C. Safety lifecycle for a separator/desalter plant – lessons learnt. In: 11AIChE – 2011 AIChE spring meeting and 7th global congress on process safety, conference proceedings; 2011. pp. 1–12.

Thiabaud P, Gros F, Tourdjman J, Guaraldo M. Benefits of life cycle dynamic simulation for FPSO projects & offshore developments. In: Society of Petroleum Engineers – Braz offshore conf 2011, vol. 2; 2011. pp. 590–602.

Unnikrishnan G. Managing process safety in facilities design. In: Society of Petroleum Engineers – international petroleum technology conference 2013, IPTC 2013: challenging technology and economic limits to meet the global energy demand, vol. 1; 2013. pp. 372–84.

Glossary of Terms

Absorbed Gas Absorbed gas is natural gas that has been dissolved into the rock and requires hydraulic fracturing to be released.

Absorption Absorption is a process in which the liquid (absorbent) flows countercurrent to a gas stream for the purpose of removing one or more constituents (absorbate) from that gas.

Accumulation Pressure increase over the maximum allowable working pressure of the vessel during discharge through the pressure relief valve (expressed as a percent of that pressure) is called accumulation.

Acid Gas Loading Acid gas loading is the amount of acid gas, on a molar or volumetric basis, that will be picked up by a solvent.

Acid Gases Acid gases are impurities in a gas stream usually consisting of CO_2, H_2S, COS, RSH, and SO_2. Most common in natural gas are CO_2, H_2S, and COS.

Actual Cubic Meters per Hour (Am^3/h) Am^3/h refers to the flow rate at flowing conditions of temperature and pressure at any given location. Because this term describes flow at a number of locations, it should not be used inter-changeably with inlet m^3/h.

Adsorption Adsorption is a separation process involving the removal of a substance from a gas stream by physical binding on the surface of a solid material.

Antifoam Antifoam is a substance, usually a silicone or long-chain alcohol, added to the treating system to reduce the tendency to foam.

Aquifer Storage Field An aquifer storage field is a subsurface facility for storing natural gas consisting of water-bearing sands topped by an impermeable caprock.

Atmospheric Discharge Discharge is the release of vapors and gases from pressure-relieving and depressuring devices to the atmosphere.

Autorefrigeration Autorefrigeration is the reduction in temperature as a result of pressure drop and subsequent flashing of light hydrocarbon liquids.

Back Pressure Pressure on the discharge side of safety-relief valves is back pressure. Back pressure is the pressure that exists at the outlet of a pressure relief device as a result of pressure in the discharge system.

Balancing Item A balancing item represents differences between the sum of the components of natural gas supply and the sum of the components of natural gas disposition. These differences may be due to quantities lost or to the effects of data reporting problems. Reporting problems include differences due to the net result of conversions of low data metered at varying temperature and pressure bases and converted to a standard temperature and pressure base; the effect of variations in company accounting and billing practices; differences between billing cycle and calendar period time frames; and imbalances resulting from the merger of data reporting systems that vary in scope, format, definitions, and type of respondents.

Base Gas The quantity of gas needed to maintain adequate reservoir pressures and deliverability rates throughout the withdrawal season is base gas. Base gas usually is not withdrawn and remains in the reservoir. All gas native to a depleted reservoir is included in the base gas volume.

Basic Engineering Package A basic engineering package is the basic engineering specifications and preliminary operating and laboratory manuals for a project that shall be provided by the basic engineering designer.

Biomass Biomass is organic non-fossil material of biological origin constituting a renewable energy source.

Black Shale Black shale is thinly bedded shale that is rich in carbon, sulfide, and organic material, formed by anaerobic (lacking oxygen) decay of organic matter. Black shales occur in thin beds in many areas at various depths and are of interest both historically and economically.

Blowdown The difference between set pressure and reseating pressure of a safety valve expressed in percent of the set pressure or in bar or kPa is called blowdown.

Boundary Boundary of the equipment is the term used in a processing facility, for an imaginary line that completely encompassed the defined site. The term distinguishes areas of responsibility and defines the processing facility for the required scope of work.

British Thermal Unit A Btu is the quantity of heat required to raise the temperature of 1 pound of liquid water by 1 °F at the temperature at which water has its greatest density (approximately 39 °F).

Built-Up Back Pressure Built-up back pressure is the pressure in the discharge header that develops as a result of flow after that the safety relief valve opens.

Carbon Capture and Storage CCS is a combination of a number of existing technologies, with the potential to play a major role in the management and reduction of global carbon dioxide (CO_2) levels. The CCS process allows for CO_2 emissions released during energy production to be captured and stored underground.

City Gate A point or measuring station at which a distributing gas utility receives gas from a natural gas pipeline company or transmission system is a city gate.

Closed Disposal System A closed disposal system is capable of containing pressures different from atmospheric pressure without leakage.

Coalescer A mechanical process vessel called a coalescer has a wettable, high-surface area packing on which liquid droplets consolidate for gravity separation from a second phase (for example, gas or immiscible liquid).

Coke Oven Gas Coke oven gas is the mixture of permanent gases produced by the carbonization of coal in a coke oven at temperatures in excess of 1000 °C.

Company or Employer/Owner Company or employer/owner refers to one of the related and/or affiliated companies of the oil and gas industries in the world.

Completion Certificate The completion certificate is issued by the engineer stating that part of the permanent works (as defined in the contract) specified in the certificate has been completed.

Compressed Natural Gas *CNG is Natural gas* compressed to a pressure at or above 200–248 bar (i.e. 2900–3600 pounds per square inch) and stored in high-pressure containers. It is used as a fuel for natural gas-powered vehicles.

Contractor Contractor refers to the persons, firm, or company whose tender has been accepted by the employer, and includes the contractor's personnel representative, successors, and permitted assigns.

Control Volume Control volume is a certain liquid volume necessary for control purposes and for maintaining the velocity limit requirement for degassing and to counter foam in separators.

Conventional Gas Conventional gas is natural gas that is extracted from underground reservoirs using traditional exploration and production methods.

Conventional Gas-Liquid Separator Conventional gas-liquid separators are vertical or horizontal separators in which gas and liquid are separated by means of gravity settling with or without a mist eliminating device.

Conventional Safety-Relief Valve A conventional safety-relief valve is a closed-bonnet pressure relief valve whose bonnet is vented to the discharge side of the valve. The valves performance characteristics, opening pressure, closing pressure, lift, and relieving capacity are directly affected by changes of the back pressure on the valve.

Conversion Efficiency The ratio of useful energy output versus the energy required to perform the process is conversion efficiency.

Critical Diameter Critical diameter or cut point is the diameter of particles beyond which larger particles will be eliminated in a sedimentation centrifuge.

Defects Defects are all items that require replacement or repair but could not have been replaced or repaired before takeover and in no way hinder or affect the requirements for substantial completion.

Degradation Products Degradation products are impurities in a treating solution that are formed in both reversible and irreversible side reactions.

Delivered (Gas) Delivered describes the physical transfer of natural, synthetic, and/or supplemental gas from facilities operated by the responding company to facilities operated by others or to consumers.

Depleted Storage Field Depleted storage field is a sub-surface natural geological reservoir, usually a depleted gas or oil field, used for storing natural gas.

Design Pressure Design pressure is used in design of equipment, a vessel, or tank for the purpose of determining the minimum permissible thickness or physical characteristics of its different parts. When applicable, static head shall be included in the design pressure to determine the thickness of any specific part.

Dike A dike is an earth or concrete wall providing a specified liquid retention capacity.

Direct Vaporizer In a direct vaporizer, heat is furnished by a flame directly applied to some form of heat exchange surface in contact with the liquid LP-gas to be vaporized.

Disengaging Height Disengaging height is the height provided between under the wire mesh pad and liquid level of a vapor–liquid separator.

Diversion Wall A diversion wall is an earth or concrete wall that directs spills to a safe disposal area.

Dossier A dossier combines all inspections and test certificates and all other documents that record the system and/or unit completion status in accordance with terms of the contract. The dossier will be prepared individually for each system and/or unit.

Dry Natural Gas Dry natural gas is what remains after (1) the liquefiable hydrocarbon portion has been removed from the gas stream (i.e. gas after lease, field, and/or plant separation); and (2) any volumes of nonhydrocarbon gases have been removed where they occur in sufficient quantity to render the gas unmarketable. Note: Dry natural gas is also known as consumer-grade natural gas. The parameters for measurement are cubic feet at 60 °F and 14.73 pounds per square inch absolute.

Dry Natural Gas Production Dy natural gas production is the process of producing consumer-grade natural gas. Natural gas withdrawn from reservoirs is reduced by volumes used at the production (lease) site and by processing losses. Volumes used at the production site include (1) the volume returned to reservoirs in cycling, repressuring of oil reservoirs, and conservation operations; and (2) gas vented and flared. Processing losses include (1) nonhydrocarbon gases (e.g., water vapor, carbon dioxide, helium, hydrogen sulfide, and nitrogen) removed from the gas stream; and (2) gas converted to liquid form, such as lease condensate and plant liquids. Volumes of dry gas withdrawn from gas storage reservoirs are not considered part of production. Dry natural gas production equals marketed production less extraction loss.

Effective Date of the Contract The effective date is when all the necessary formalities mutually agreed upon including signing of all the agreement between the company and the contractor take place in accordance with the contract.

Energy Security Energy security is the term used to refer to the reliability of future energy supply. This depends on a number of different factors, including the availability of energy supplies, their affordability, and the capacity to extract them in an environmentally sustainable manner.

Evaporation Ponds These artificial ponds have very large surface areas and are designed to allow the efficient evaporation of water through exposure to sunlight and ambient surface temperatures.

Extraction Loss Extraction loss is the reduction in volume of natural gas due to the removal of natural gas liquid constituents such as ethane, propane, and butane at a natural gas processing plant.

Extraction Process Extraction process is the method used for drawing out resources from the ground. For unconventional gas, this is mainly hydraulic fracturing.

Fabric Filter Commonly termed bag filters or baghouses, fabric filters are collectors in which dust is removed from the gas stream by passing the dust-laden gas through a fabric of some type.

Final Acceptance The final acceptance certificate is to be issued by the engineer stating that all of the contractor's guarantees under the contract have been satisfactorily met or discharged subject to contractor's obligations and after completion of works, tests on completion, taking over, and the remedy of defects period.

Filter Medium The filter medium or septum is the barrier that lets the flow pass while retaining most of the solids; it may be a screen, cloth, paper, or bed of solids.

Filtrate The liquid that passes through the filter medium is called the filtrate.

Fire Resistive Fire resistive describes the fire resistance rating, as the time in minutes or hours, that materials or assemblies have to withstand a fire exposure as established in accordance with the test of NFPA 251.

Flare Flare is a means of safe disposal of waste gases by combustion. With an elevated flare, the combustion is carried out at the top of a pipe or stack where the burner and igniter are located. A ground flare is similarly equipped, except combustion is carried out at or near ground level. A burn pit differs from a flare in that it is primarily designed to handle liquids. A tall stack equipped with burners used as a safety device at wellheads, refining facilities, gas processing plants, and chemical plants. Flares are used for the combustion and disposal of combustible gases. The gases are piped to a remote, usually elevated, location and burned in an open flame in the open air using a specially designed burner tip, auxiliary fuel, and steam or air. Combustible gases are flared most often due to emergency relief, overpressure, process upsets, startups, shutdowns, and other operational safety reasons. Natural gas that is uneconomical for sale is also flared. Often natural gas is flared as a result of the unavailability of a method for transporting such gas to markets.

Flare Blowoff/Flame Lift-Up Blowoff or left-up is the lifting of flame front from the flare tip.

Flare Blowout Blowout is the extinguishing of flare flame.

Flash Tank A flash tank is a vessel used to separate the gas evolved from liquid flashed from a higher pressure to a lower pressure.

Flow Rate Flow rate expresses the volume of fluid or gas passing through a given surface per unit of time (e.g., cubic feet per minute). Usually represented by the symbol Q.

Free Gas Free gas is natural gas that is trapped within spaces in the rock, making it more accessible.

Gas Condensate Well Gas Natural gas remaining after the removal of the lease condensate is called gas condensate well gas.

Gas Well A gas well is completed for production of natural gas from one or more gas zones or reservoirs. Such wells contain no completions for the production of crude oil.

Gross Withdrawals Gross withdrawal described the full well stream volume from both oil and gas wells, including all natural gas plant liquids and nonhydrocarbon gases after oil, lease condensate, and water have been removed. Also includes production delivered as royalty payments and production used as fuel on the lease.

Heating Value Heating value is the average number of British thermal units per cubic foot of natural gas as determined from tests of fuel samples.

High Flash Stock High flash stock are those having a closed up flash point of 55 °C or over (such as heavy fuel oil, lubricating oils, transformer oils). This category does not include any stock that may be stored at temperatures above or within 8 °C of its flash point.

Hold-Up Time The time period during which the amount of liquid separated in a gas–liquid separator is actually in the vessel for the purpose of control or vapor separation is called the hold-up time.

Horizontal Drilling Horizontal drilling is a hydrocarbon well drilling technique that allows for multiple underground wellbores to be deviated from one surface well site.

Hydraulic Fracturing In hydraulic fracking, small fractures are made in impermeable rock by a pressurized combination of water, sand, and chemical additives. The small fractures are held open by grains of sand, allowing the natural gas to flow out of the rock and into the wellbore.

Indirect Vaporizer An indirect vaporizer has heat furnished by steam, hot water, the ground, surrounding air, or other heating medium applied to a vaporizing chamber or to tubing, pipe coils, or other heat exchange surface containing the liquid LP-gas to be vaporized; the heating of the medium used is at a point remote from the vaporizer.

Inlet Cubic Meters per Hour Im^3/h refers to flow rate determined at the conditions of pressure, temperature, compressibility, and gas composition, including moisture, at the compressor inlet flange (substitution to API Std. 617, 1.5.14).

International Energy Agency The IEA is an autonomous organization linked with the Organisation for Economic Co-operation and Development. Based in Paris, it works to ensure reliable, affordable, and clean energy for its 28 member countries and beyond. Founded in response to the 1973–74 *oil* crisis, the IEA's initial role was to help countries coordinate a collective response to major disruptions in oil supply through the release of emergency oil stocks to the markets. Today, the IEA's four main areas of focus are maintaining energy security, achieving economic development, raising environmental awareness, and engaging non-member countries to find solutions to shared energy and environmental concerns.

Knock-Out A knock-out is a separator used for a bulk separation of gas and liquid, particularly when the liquid volume fraction is high.

Lean Oil (Absorbent) Those hydrocarbons that have a molecular mass of about 100–180 and a maximum final boiling point of 160 °C (e.g., naphtha) are called lean oils.

Licenser a licenser refers to a company duly organized and existing under the laws of the said company's country and as referred to in the preamble to the contract.

Lease and Plant Fuel Lease and plant fuel is natural gas used in well, field, and lease operations (such as gas used in drilling operations, heaters, dehydrators, and field compressors) and as fuel in natural gas processing plants.

Lease Fuel Lease gas is natural gas used in well, field, and lease operations, such as gas used in drilling operations, heaters, dehydrators, and field compressors.

Lease Separator A lease separator is a facility installed at the surface for the purpose of separating the full well stream volume into two or three parts at the temperature and pressure conditions set by the separator. For oil wells, these parts include produced crude oil, natural gas, and water. For gas wells, these parts include produced natural gas, lease condensate, and water.

Lift The rise of the disk in a pressure-relief valve is called lift.

Line Drip A line drip is a device typically used in pipelines with very high gas-to-liquid ratios to remove only free liquid from a gas stream, and not necessarily all the liquid.

Liquefied Natural Gas LNG is natural gas (primarily methane) that has been liquefied by reducing its temperature to −260 °F at atmospheric pressure.

Liquefied Petroleum Gas LP gas or LPG is any material having a vapor pressure not exceeding that allowed for commercial propane composed predominantly of the following hydrocarbons, either by themselves or as a mixtures: propane, propylene, butane (normal butane or isobutane) and butylene, as a byproduct in petroleum refining or natural gasoline manufacture.

LNG Liquefied natural gas is natural gas that has been cooled to −162 °C to convert the gas temporarily to a liquid state for storage or transportation purposes. As a liquid, natural gas occupies about 600 times less space than when in its gaseous form. LNG is principally used for transporting natural gas to distant markets, where it is regasified and distributed via pipelines to consumers. A fluid in the liquid state composed predominantly of methane may contain minor quantities of ethane, propane, nitrogen, or other components normally found in natural gas.

Low-Flash Stocks Low flash stocks are those having a closed up flash point under 55 °C, such as gasoline, kerosene, jet fuels, some heating oils, diesel fuels, and any other stock that may be stored at temperatures above or within 8 °C of its flash point.

Mach Number The ratio of vapor velocity to sonic velocity in that vapor at flowing conditions is called the Mach number.

Manufactured Gas Manufactured gas is gas obtained by destructive distillation of coal or by the thermal decomposition of oil, or by the reaction of steam passing through a bed of heated coal or coke. Examples are coal gases, coke oven gases, producer gas, blast furnace gas, blue (water) gas, and carbureted water gas. Btu content varies widely.

Marketed Production Marketed production is gross withdrawals less gas used for repressuring, quantities vented and flared, and nonhydrocarbon gases removed in treating or processing operations. It includes all quantities of gas used in field and processing plant operations.

Maximum Allowable Working Pressure As defined in the construction codes for pressure vessels, the maximum allowable working pressure depends on the type of material, its thickness, and the service conditions set as the basis for design. The vessel may not be operated above this pressure or its equivalent at any metal temperature other than that used in its design. Consequently, for that metal temperature, it is the highest pressure at which the primary pressure-relief valve is set to open.

Mercaptan   Mercaptan is a hydrocarbon group (usually a methane, ethane, or propane) with a sulfur group (-SH) substituted on a terminal carbon atom.

Mesh   The "mesh count" (usually called "mesh") is effectively the number of openings of a woven wire filter per 25 mm, measured from the center of one wire to another 25 mm from it.

Millidarcies   A darcy (or darcy unit) and millidarcies (mD) are units of permeability.

Mist Extractor   A mist extractor is a device installed in the top of scrubbers, separators, tray or packed vessels, etc. to remove liquid droplets entrained in a flowing gas stream.

Modern Cycle Gas Turbine   A gas turbine, also called a combustion turbine, is a rotary engine that extracts energy from a flow of combustion gas. It has an upstream compressor coupled to a downstream turbine, and a combustion chamber in-between. It is often used for power generation. Gas turbine may also refer to just the turbine element.

Native Gas   Native gas is gas in place at the time that a reservoir was converted to use as an underground storage reservoir in contrast to injected gas volumes.

Natural Gas   A gaseous mixture of hydrocarbon compounds, the primary one being *methane*, is called natural gas.

Natural Gas Field Facility   A field facility is designed to process natural gas produced from more than one lease for the purpose of recovering condensate from a stream of natural gas; however, some field facilities are designed to recover propane, normal butane, pentanes plus, etc., and to control the quality of natural gas to be marketed.

Natural Gas Gross Withdrawals   The full well-stream volume of produced natural gas, excluding condensate separated at the lease, is called gross withdrawal.

Natural Gas Hydrates   Natural gas hydrates are solid, crystalline, wax-like substances composed of water, methane, and usually a small amount of other gases, with the gases being trapped in the interstices of a water-ice lattice. They form beneath permafrost and on the ocean floor under conditions of moderately high pressure and at temperatures near the freezing point of water.

Natural Gas Lease Production   Natural gas lease production involves gross withdrawals of natural gas minus gas production injected on the lease into producing reservoirs, vented, flared, used as fuel on the lease, and nonhydrocarbon gases removed in treating or processing operations on the lease.

Natural Gas Liquids   Those hydrocarbons in natural gas that are separated from the gas as liquids through the process of absorption, condensation, adsorption, cooling in gas separators, gas processing, or gas cycling plants are called NGL. Generally, natural gas liquids include *natural gas plant liquids* and *lease condensate*. NGL is a mixture of liquefied hydrocarbons extracted from natural gas by various methods and stabilized to obtain a liquid product.

Natural Gas Liquids Production   The volume of natural gas liquids removed from natural gas in lease separators, field facilities, gas processing plants, or cycling plants during the report year is called NGL production.

Natural Gas Marketed Production   Marketed production is gross withdrawals of natural gas from production reservoirs, less gas used for reservoir pressuring, nonhydrocarbon gases removed in treating and processing operations, and quantities vented and flared.

Natural Gas Marketer   A company that arranges purchases and sales of natural gas is a natural gas marketer. Unlike pipeline companies or local distribution companies, a marketer does not own physical assets commonly used in the supply of natural gas, such as pipelines or storage fields. A marketer may be an affiliate of another company, such as a local distribution company, natural gas pipeline, or producer, but it operates independently of other segments of the company. In states with residential choice programs, marketers serve as alternative suppliers to residential users of natural gas, which is delivered by a local distribution company.

Natural Gas Plant Liquids Those hydrocarbons in natural gas that are separated as liquids at natural gas processing plants, fractionating and cycling plants, and in some instances, field facilities are called natural gas plant liquids. Lease condensate is excluded. Products obtained include liquefied petroleum gases (ethane, propane, and butanes), pentanes plus, and isopentane. Component products may be fractionated or mixed.

Natural Gas Policy Act of 1978 Signed into law on November 9, 1978, the NGPA is a framework for the regulation of most facets of the natural gas industry.

Natural Gas Processing Plant Facilities designed to recover natural gas liquids from a stream of natural gas may or may not have passed through lease separators and/or field separation facilities. These facilities control the quality of the natural gas to be marketed. Cycling plants are classified as gas processing plants.

Natural Gas Used for Injection Natural gas is used to pressurize crude oil reservoirs in an attempt to increase oil recovery or in instances where there is no market for the natural gas. Natural gas used for injection is sometimes referred to as repressuring.

Natural Gas Utility Demand-Side Management Program Sponsor A DSM (program sponsored by a natural gas utility suggests ways to increase the energy efficiency of buildings, to reduce energy costs, to change the usage patterns, or to promote the use of a different energy source.

Natural Gasoline Natural gasoline is a term used in the gas processing industry to refer to a mixture of liquid hydrocarbons (mostly pentanes and heavier hydrocarbons) extracted from natural gas. It includes isopentane.

Natural Gasoline and Isopentane A mixture of hydrocarbons, mostly pentanes and heavier, is extracted from natural gas; it meets vapor pressure, end-point, and other specifications for natural gasoline set by the Gas Processors Association. It includes isopentane, which is a saturated branch-chain hydrocarbon (C_5H_{12}), obtained by fractionation of natural gasoline or isomerization of normal pentane.

Non-Combustible Material incapable of igniting or supporting combustion is considered noncombustible.

Nonhydrocarbon Gases Typical nonhydrocarbon gases may be present in reservoir natural gas, such as carbon dioxide, helium, hydrogen sulfide, and nitrogen.

Normal Cubic Meters Nm^3 refers to capacity at normal conditions (101.325 kPa and 0 °C) and relative humidity of 0%.

Normal Cubic Meters per Hour Nm^3/h refers to a flow rate at any location corrected to the normal atmospheric pressure and a temperature of 0 °C with a compressibility factor of 1.0 and in dry conditions.

Organic Richness Organic richness refers to the density of organic material within the rock.

Overflow The stream being discharged out of the top of a hydrocyclone, through a protruding pipe, is called overflow. This stream consists of bulk of feed liquid together with the very fine solids.

Overpressure Pressure increase over the set pressure of the relieving device is overpressure. It is the same as accumulation when the relieving device is set at the maximum allowable working pressure of the vessel and may be greater than the allowable accumulation if the valve is set lower than the vessel MAWP.

Performance Test The performance test is conducted to demonstrate and ratify performance of unit or units meeting all process and utilities guarantees as requested and defined in the contract.

Permanent Works Permanent works includes all works that will be incorporated in and form part of the project to be handed over to the company by the contractor.

Permeability Permeability is the measure of how fluids or natural gas move through the rock (typically measured in millidarcies).

Pipeline (natural gas) A pipeline is a continuous pipe conduit, complete with such equipment as valves, compressor stations, communications systems, and meters for transporting natural and/or supplemental gas from one point to another, usually from a point in or beyond the producing field or processing plant to another pipeline or to points of utilization. Also refers to a company operating such facilities.

Pipeline Fuel Gas consumed in the operation of pipelines, primarily in compressors, is pipeline fuel.

Pressure Relief Valve Pressure relief valve is a generic term applied to relief valves, safety valves, and safety-relief valves. A pressure-relief valve is designed to automatically recluse and prevent the flow of fluid.

Progress Report Progress reports are provided in writing by the contractor to the company's authorized representative specifying the amount of progress of the services and works, respective values and project area of concerns.

Pipe Rack The pipe rack is the elevated supporting structure used to convey piping between equipment. This structure is also utilized for cable trays associated with electric-power distribution and for an instrument tray.

Plot Plan The plot plan is the scaled plan drawing of the processing facility.

Porosity Porosity measures the open space (pore space) between grains. The lower the porosity, the more important hydraulic fracturing is to extracting natural gas.

Process Flow Diagram Process flow diagram mainly defines: (a) A schematic representation of the sequence of all relevant operations occurring during a process and includes information considered desirable for analysis. (b) The process presenting events that occur to the material(s) to convert the feedstock(s) to the specified products. (c) An operation occurring when an object (or material) is intentionally changed in any of its physical or chemical characteristics, is assembled or disassembled from another object, or is arranged or prepared for another operation, transportation, inspection, or storage.

Project Project refers to the equipment, machinery, and materials to be procured by the contractor and the works and/or all activities to be performed and rendered by the contractor in accordance with the terms and conditions of the contract documents.

Provisional Acceptance Provisional acceptance means that an operability test has been satisfactorily completed with the system operating at the capacity as defined in the contract for a continuous period as defined. Substantial completion shall be evidenced by issuance of a Provisional Acceptance Certificate per contract.

Quenching Quenching is the cooling of a hot vapor by mixing it with another fluid or by partially vaporizing another liquid.

Recovery Rate The rate at which natural gas can be removed from a reservoir is the recovery rare.

Refinery Gas Refinery gas is noncondensate gas collected in petroleum refineries.

Reid Vapor Pressure RVP is the pressure of the vapor in equilibrium with liquid at 37.8 °C (100 °F).

Relief Valve A relief valve is an automatic pressure-relieving device actuated by the static pressure upstream of the valve. The valve opens in proportion to the increase in pressure over the opening pressure. It is used primarily for liquid service.

Reseating Pressure of a Safety Valve Reseating pressure is the value of inlet static pressure at which the disc reestablishes contact with the seat or at which lift becomes zero.

Residence Time Residence time is the time period in which a fluid will be contained within a certain volume.

Rupture Disk A rupture disk consists of a thin metal diaphragm held between flanges and bursts when a predetermined pressure is reached below the disk, so preventing a predetermined safe pressure being exceeded in the vessel to be protected.

Safety-Relief Valve A safety-relief valve is an automatic pressure-relieving device suitable for use as either a safety or relief valve, depending on application. It is used in either gas and vapor or liquid services.

Safety-Relief Valve (Direct Loaded) A safety-relief valve in which the loading due to the fluid pressure underneath the valve disk is opposed only by direct mechanical loading such as a mass, a lever and mass, or a spring.

Safety-Relief Valve (Indirect Loaded) A safety-relief valve in which the operation is initiated and controlled by the fluid discharged from a pilot valve, which is itself a direct loaded safety-relief valve subject to the requirements.

Safety-Relief Valve (Balanced Bellows) A safety-relief valve that incorporates a bellows that has an effective area equal to that of the valve seat to eliminate the effect of back pressure on the set pressure of the valve and which effectively prevents the discharging fluid entering the bonnet space.

Safety-Relief Valve (Conventional) A safety-relief valve of the direct loaded type, the set pressure of which will be affected by changes in the superimposed back pressure.

Safety Valve A safety valve is an automatic pressure-relieving device actuated by the static pressure upstream of the valve and characterized by rapid full opening or pop action. It is used for gas or vapor services.

Scrubber A scrubber is a type of separator that has been designed to handle flow streams with unusually high gas-to-liquid ratios.

Selective Treating Selective treating is preferential removal of one acid gas component, leaving other acid components in the treated gas stream.

Set Pressure Set pressure is the inlet pressure to which a safety valve is adjusted, in a test stand or other source of pressure, to open with an atmospheric discharge or atmospheric back pressure.

Shale Basins Shale basins are the underground rock formations that serve as both natural gas generators and reservoirs.

Shale Gas Shale gas is natural gas extracted from fine-grain rock composed mainly of clay flakes and tiny fragments of other minerals. Natural gas can be produced from wells that are open to shale formations. Shale is a fine-grained, sedimentary rock composed of mud from flakes of clay minerals and tiny fragments (silt-sized particles) of other materials. The shale acts as both the source and the reservoir for the natural gas.

Site The lands and other places on, under, or through which the works are to be executed or carried out, and any other lands or places provided by the company for the purposes of the contract together with such other place as may be specially designated in the contract as forming part of the site.

Sleepers The sleepers comprise the grade-level supporting structure for piping between equipment for facilities, e.g., tank farm or other remote areas.

Slickwater Slickwater fracturing is a method or system of hydraulic fracturing that involves adding chemicals to water to reduce friction and increase the fluid flow. Slickwater increases the speed at which the pressurized fluid can be pumped into the wellbore.

Slug Catcher A particular separator design, a slug catcher is able to absorb sustained in-flow of large liquid volumes at irregular intervals.

Sour Gas Sour gas is any gas stream that contains acid gas components.

Specific Volume Specific volume is the volume per unit mass or volume per mole of material.

Specifications Specifications include drawings, specifications, bills of materials, and any other technical documents, whatever they may be, issued with the contract documents, including any revisions or additions from time to time to the drawings, specifications, bills of material, and any other technical documents.

Standard Condition The standard condition is a temperature of 15 °C and a pressure of 1 atm (101.325 kPa), which also is known as standard temperature and pressure (STP).

Standard Cubic Meter per Hour Sm^3/h refers to the flow rate at any location corrected to a pressure of 101.325 kPa and at a temperature of 15 °C with a compressibility factor of 1.0 and in a dry condition.

Steric Hindrance Steric hindrance is chemically attaching a bulk molecule (such as benzene) to the hydrocarbon chain of an amine to inhibit CO_2 reacting to form a carbonate.

Subcontractor A subcontractor is any person, firm, or company (other than contractor) named in the contract for any part of the works or any person to whom any part of the contract has been sub-let with the consent in writing of the engineer and the legal personal representatives, successors, and assigns of such person.

Superimposed Back Pressure Superimposed back pressure is the pressure in the discharge header before the safety-relief valve opens.

Supplemental Gaseous Fuel Supplies Synthetic natural gas, propane-air, coke oven gas, refinery gas, biomass gas, air injected for Btu stabilization, and manufactured gas commingled and distributed with natural gas are supplemental gaseous fuel supplies.

Sweet Gas Sweet gas is a gas stream that has acid gas components removed to an acceptable level.

Synthetic Natural Gas (SNG) SNG is also referred to as substitute natural gas. It is a manufactured product, chemically similar in most respects to natural gas, resulting from the conversion or reforming of hydrocarbons that may easily be substituted for or interchanged with pipeline-quality natural gas.

Tank Diameter Where tank spacing is expressed in terms of tank diameter, the following criteria governs: (a) If tanks are in different services, or different types of tanks are used, the diameter of the tank which requires the greater spacing is used. (b) If tanks are in similar services, the diameter of the largest tank is used.

Tank Spacing Tank spacing is the unobstructed distance between tank shells, or between tank shells and the nearest edge of adjacent equipment, property lines, or buildings.

Target Efficiency The fraction of particles or droplets in the entraining fluid of a separator, moving past an object in the fluid, which impinge on the object, is called target efficiency.

Temporary Works All temporary works of every kind are required in or about the execution or remedy of defect of the works but does not include contractor's equipment.

Terminal Velocity or Drop-Out Velocity Terminal velocity is the velocity at which a particle or droplet will fall under the action of gravity, when drag force just balances gravitational force and the particle (or droplet) continues to fall at constant velocity.

Tests on Completion Completion tests are made by the contractor before the works' are taken over by the company as provided for in the contract and such other tests as may be agreed to by the company and the contractor.

Thermal Maturity Thermal maturity is the amount of heat, in relative terms, to which a rock has been subjected. A thermally immature rock has not been subjected to enough heat to begin the process of converting organic material into oil and/or natural gas. A thermally over-mature rock has been subjected to enough heat to convert organic material to graphite. However, these are the two extremes, and there are many intermediate stages of thermal maturity.

Thermogenic Thermogenic means generated or formed by heat, especially via physiological processes.

Threshold Limit Value Threshold limit value is the amount of a contaminant to which a person can have repeated exposure for an 8 h day without adverse effects.

Tight Gas Tight gas is natural gas found trapped in impermeable rock and non-porous sandstone or limestone formations, typically at depths greater than 10,000 feet below the surface.

Toe Wall A toe wall is a low earth, concrete, or masonry unit curb without capacity requirements for the retention of small leaks or spills.

Total Organic Carbon TOC is the concentration of material derived from decaying vegetation, bacterial growth, and metabolic activities of living organisms or chemicals in the source rocks.

Total Storage Field Capacity (Design Capacity) Total storage field capacity is the maximum quantity of natural gas (including both base and working gas) that can be stored in an underground storage facility in accordance with its design specifications, the physical characteristics of the reservoir, installed compression equipment, and operating procedures particular to the site. Reported storage field capacity data are reported in 1000 cubic feet at standard temperature and pressure.

Unaccounted for (Natural Gas) The differences between the sum of the components of natural gas supply and the sum of components of natural gas disposition are unaccounted for. These differences may be due to quantities lost or to the effects of data reporting problems. Reporting problems include differences due to the net result of conversions of flow data metered at varying temperatures and pressure bases and converted to a standard temperature and pressure base; the effect of variations in company accounting and billing practices; differences between billing cycle and calendar-period time frames; and imbalances resulting from the merger of data reporting systems that vary in scope, format, definitions, and type of respondents.

Underflow The stream containing the remaining liquid and the coarser solids that is discharged through a circular opening at the apex of the core of a hydrocyclone is referred to as underflow.

Underground Gas Storage The use of subsurface facilities for storing natural gas for use at a later time is termed underground storage. The facilities are usually hollowed-out salt domes, geological reservoirs (depleted oil or gas fields), or water-bearing sands (called aquifers) topped by an impermeable cap rock.

Underground Storage Injections Gas can be put (injected) into underground storage reservoirs.

Underground Storage Withdrawals Gas can be removed from underground storage reservoirs.

Unit(s) The term unit refers to one or all process, offsite, and/or utility units and facilities as applicable to form a complete operable oil, gas, and/or petrochemical plant.

Unit Value, Consumption The consumption unit value is the total price per specified unit, including all taxes, at the point of consumption.

Unit Value, Wellhead The wellhead sales price includes charges for natural gas plant liquids subsequently removed from the gas; gathering and compression charges; and state production, severance, and/or similar charges.

Vacuum Relief Valves Vacuum relief valves are usually installed on storage tanks and shall normally be of the mass loaded or pilot operated type. For full description and determination of size of vacuum relief valves, reference shall be made to API 2000 "Venting Atmospheric and Low Pressure Storage Tanks (Non-Refrigerated and Refrigerated)."

Vaporizer A vaporizer is a device other than a container that receives LP-gas in liquid form and adds sufficient heat to convert the liquid to a gaseous state.

Vent Stack A vent stack is the elevated vertical termination of a disposal system that discharges vapors into the atmosphere without combustion or conversion of the relieved fluid.

Vented Natural gas that is disposed of by releasing to the atmosphere is called vented.

Vessel Diameter Where vessel spacing is expressed in terms of vessel diameter, the diameter of the largest vessel is used. For spheroids, the diameter at the maximum equator is used.

Vessel Spacing Vessel spacing is the unobstructed distance between vessel shells or between vessel shells and the nearest edge of adjacent equipment, property lines, or buildings.

Vapor Space Vapor space is the volume of a vapor liquid separator above the liquid level.

Wellbore A wellbore is the channel created by the drill bit.

Wellhead The wellhead is the point at which the crude (and/or natural gas) exits the ground. Following historical precedent, the volume and price for crude oil production are labeled as wellhead, even though the cost and volume are now generally measured at the lease boundary. In the context of domestic crude price data, the term wellhead is the generic term used to reference the production site or lease property.

Working Gas The quantity of gas in the reservoir that is in addition to the cushion or base gas is the working gas. It may or may not be completely withdrawn during any particular withdrawal season. Conditions permitting, the total working capacity could be used more than once during any season. Volumes of working gas are reported in 1000 cubic feet at standard temperature and pressure.

Index

Note: Page Numbers followed by *"f"* indicate figures; *"t"*, tables; *"b"*, boxes.

665.73 B151   CEN
Bahadori, Alireza,
Natural gas processing :technology
 and engineering design /
CENTRAL LIBRARY
04/15